FARIBORZ M. FARHAN
GA. TECH BOX 36030
ATL, GA 30332

فریبرز محمدزاده فرحان

# ELEMENTARY SOLID STATE PHYSICS: Principles and Applications

**M. A. OMAR**
Lowell Technological Institute

# ELEMENTARY SOLID STATE PHYSICS: Principles and Applications

 **ADDISON-WESLEY PUBLISHING COMPANY**
Reading, Massachusetts · Menlo Park, California
London · Amsterdam · Don Mills, Ontario · Sydney

This book is in the
ADDISON-WESLEY SERIES IN SOLID STATE SCIENCES

Consulting Editor
David Lazarus

Cover: The conduction band of germanium

*Second printing, July 1978*

 Printed in the United States of America. Published simultaneously in Canada. Library of Congress Catalog Card No. 73-10593.

ISBN 0-201-05482-5
ABCDEFGHIJK-MA-798

بسم الله الرحمن الرحيم

*To my son, Riyad*

# PREFACE

This volume is intended to serve as a general text in solid state physics for undergraduates in physics, applied physics, engineering, and other related scientific disciplines. I also hope that it will serve as a useful reference tool for the many workers engaged in one type of solid state research activity or another, who may be without formal training in the subject.

Since there are now many books on solid state physics available, some justification is needed for the introduction of yet another at this time. This I can perhaps do best by stating the goals I strove to achieve in the writing of it, and let the reader judge for himself how successful the effort may have been.

First, I have attempted to cover a wide range of topics, which is consistent with my purpose in writing a general and complete text which may also serve as an effective general reference work. The wide coverage also reflects the immensely wide scope of current research in solid state physics. But despite this, I have made a determined effort to underline the close interrelationships between the disparate parts, and bring the unity and coherence of the whole subject into perspective.

Second, I have tried to present as many practical applications as possible within the limits of this single volume. In this not only have I taken into consideration those readers whose primary interest lies in the applications rather than in physics *per se*, but I have also encouraged prospective physics majors to think in terms of the practical implications of the physical results; this is particularly vital at the present time, when great emphasis is placed on the contribution of science and technology to the solution of social and economic problems.

Third, this book adheres to an interdisciplinary philosophy; thus, in addition to the areas covered in traditional solid state texts in the first ten chapters, the last three chapters introduce additional material to which solid state physicists have made many significant contributions. The subjects include metallurgy, defects in solids, new materials, and biophysics and are of great contemporary importance and practical interest.

Fourth, I have made every effort to produce a modern, up-to-date text. Solid state physics has progressed very rapidly in the past two or three decades, and yet many advances have thus far failed to make their way into elementary texts, and remain scattered haphazardly throughout many different sources in the literature. Yet it is clear that early and thorough assimilation of the concepts underlying these advances, particularly by the young student, is essential to the growth and development in this field which await us in the future.

Fifth, and of greatest importance, this book is elementary in nature, and I have made every effort to ensure that it is thoroughly understandable to the well-prepared undergraduate student. I have attempted to introduce new concepts gradually, and to supply the necessary mathematical details for the various steps along the way. I have then discussed the final results in terms of their physical meaning, and their relation to other more familiar situations whenever this seems helpful. The book is liberally illustrated with figures, and a fairly complete list of references is supplied for those readers interested in further pursuit of the subjects discussed here.

Chapter 1 covers the crystal structures of solids, and the interatomic forces responsible for these structures. Chapter 2 includes the various experimental techniques, such as x-ray diffraction, employed in structure analysis. Except at very low temperatures, however, the atoms in a solid are not at rest, but rather oscillate around their equilibrium positions; therefore, Chapter 3 covers the subject of lattice vibrations, together with their effects on thermal, acoustic, and optical properties. This is followed in Chapter 4 by a discussion of the free-electron model in metals, whereby the valence electrons are assumed to be free particles. A more realistic treatment of these electrons is given in Chapter 5, on energy bands in solids. Before beginning Chapter 5, the student should refresh his understanding of quantum mechanics by reference to the appendix. The brief treatment of this complex subject there is not intended to be a short course for the uninitiated, but rather a summary of its salient points to be employed in Chapter 5, on the energy bands in solids. This is, in fact, the central chapter of the book, and it is hoped that, despite its somewhat demanding nature, the reader will find it rewarding in terms of a deeper understanding of the electronic properties of crystalline solids.

Semiconductors are discussed in Chapter 6. The detailed coverage accorded these substances is warranted not only by their highly interesting and wide-ranging properties but also by the crucial role played by semiconductor devices in today's technology. These devices are discussed at length in Chapter 7. When an electric field, static or alternating, penetrates a solid, the field polarizes the positive and negative charges in the medium; the effects of polarization on the dielectric and optical properties of solids are the subject of Chapter 8. The magnetic properties of matter, including recent developments in magnetic resonances, are taken up in Chapter 9, and the fascinating phenomenon of superconductivity in Chapter 10.

Chapter 11 is devoted to some important topics in metallurgy and defects in solids, and Chapter 12 features some interesting and new substances such as amorphous semiconductors and liquid crystals, which are of great current interest; this chapter includes also applications of solid state techniques to chemical problems. Chapter 13 is an introduction to the field of molecular biology, presented in terms of the concepts and techniques familiar in solid state physics. This is a rapidly expanding and challenging field today, and one in which solid state physicists are making most useful contributions.

Each chapter concludes with a number of exercises. These consist of two types: *Questions*, which are rather short, and intended primarily to test conceptual understanding, and *Problems*, which are of medium difficulty and cover the entire chapter. Virtually all the problems are solvable on the basis of material presented in the chapter, and require no appeal to more advanced references. The exercises are an integral part of the text and the reader, particularly the student taking a solid state course for the first time, is urged to attempt most of them.

**ACKNOWLEDGEMENTS**
Several persons helped me directly or indirectly in this work. Professor Herbert Kroemer of the University of Colorado has given me the benefit of his insight and incisive opinion during the early stages of writing. Professor Masataka Mizushima, also of the University of Colorado, gave me unfailing encouragement and support over a number of years. Chapter 5 on band theory profited from lectures by Professor Henry Ehrenreich of Harvard University, and Chapter 9 on magnetism reflects helpful discussions with Professor Marcel W. Muller of Washington University. Professor J. H. Tripp of the University of Connecticut made several useful comments and pointed out some editorial errors in the manuscript.

Professor David Lazarus of the University of Illinois-Urbana read the entire manuscript with considerable care. His comments and suggestions, based on his wide experience in teaching and research, resulted in substantial improvement in the work and its usefulness as a textbook in solid state physics.

To these distinguished scholars my sincerest thanks. The responsibility for any remaining errors or shortcomings is, of course, mine.

Joyce Rey not only typed and edited the manuscript with admirable competence, but always went through the innumerable revisions with patience, care, and understanding. For her determined efforts to anglicize the style of the author (at times, perhaps, a little over-enthusiastically), and for keeping constant track of the activities of the main characters, those "beady-eyed" electrons, despite her frequent mystification with the "plot," I am most grateful to "Joycie."

In closing, the following quotation from Reif (*Fundamentals of Statistical and Thermal Physics*) seems most appropriate: "It has been said that 'an author never finishes a book, he merely abandons it.' I have come to appreciate vividly the truth of this statement and dread to see the day when, looking at the manuscript in print, I am sure to realize that many things could have been done better and explained more clearly. If I abandon the book nevertheless, it is in the modest hope that it may be useful to others despite its shortcomings."

*Lowell, Massachusetts* M. A. O.
*July 1974*

# CONTENTS

# CHAPTER 1 CRYSTAL STRUCTURES AND INTERATOMIC FORCES



---

*Good order is the foundation of all good things.*

Edmund Burke

## 1.1 INTRODUCTION

To the naked eye, a solid appears as a *continuous* rigid body. Experiments have proved, however, that all solids are composed of *discrete* basic units—atoms. These atoms are not distributed randomly, but are arranged in a highly ordered manner relative to each other. Such a group of ordered atoms is referred to as a *crystal*. There are several types of crystalline structure, depending on the geometry of the atomic arrangement; a knowledge of these is important in solid-state physics because these structures usually influence the physical properties of solids. This statement will be amply illustrated in the following chapters.

In the first part of this chapter, we shall expand on the meaning of the crystalline structure, and introduce some of the basic mathematical definitions employed in describing it. We shall then enumerate the various structures possible, and introduce the concept of Miller indices. We shall also present a few examples.

The atoms in some solids appear to be randomly arranged, i.e., the crystalline structure is absent. Such noncrystalline—or *amorphous*—solids will also be described briefly.

The chapter closes with an account of the interatomic forces that cause bonding in crystals.

Chapter 2 will discuss the experimental determination of crystal structure by x-rays.

## 1.2 THE CRYSTALLINE STATE

A solid is said to be a *crystal* if the atoms are arranged in such a way that their positions are *exactly periodic*. Figure 1.1 illustrates the concept. The distance between any two nearest neighbors along the $x$ direction is $a$, and along the $y$ direction is $b$ (the $x$ and $y$ axes are not necessarily orthogonal). A *perfect* crystal maintains this periodicity (or repetitivity) in both the $x$ and $y$ directions from $-\infty$ to $\infty$. It follows from the periodicity that the atoms $A$, $B$, $C$, etc., are *equivalent*. In other words, to an observer located at any of these atomic sites, the crystal appears exactly the same.

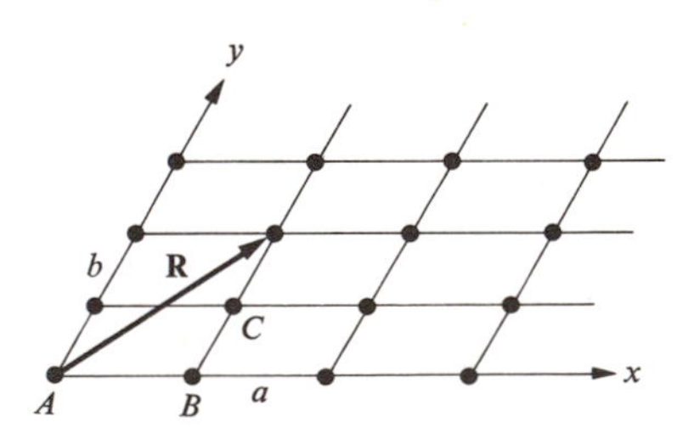


**Fig. 1.1** A crystalline solid. All the atoms are arranged periodically.

The same idea is often expressed by saying that a crystal possesses a *translational symmetry*, meaning that if the crystal is translated by any vector joining two atoms, say **R** in Fig. 1.1, the crystal appears exactly the same as it did before the translation. In other words, the crystal remains *invariant* under any such translation. The consequences of this translational symmetry or invariance are many, and a great portion of this book will be concerned with them.

Strictly speaking, one cannot prepare a perfect crystal. For example, even the surface of a crystal is a kind of *imperfection* because the periodicity is interrupted there. The atoms near the surface see an environment different from the environment seen by atoms deep within the crystal, and as a result behave differently. Another example concerns the thermal vibrations of the atoms around their equilibrium positions for any temperature $T > 0°K$. Because of these vibrations, the crystal is always distorted, to a lesser or greater degree, depending on $T$. As a third example, note that an actual crystal always contains some foreign atoms, i.e., impurities. Even with the best crystal-growing techniques, some impurities ($\simeq 10^{12}\ cm^{-3}$) remain, which spoils the perfect crystal structure.

Notwithstanding these difficulties, one can prepare crystals such that the effects of imperfections on the phenomena being studied are extremely minor. For example, one can isolate a sodium crystal so large ($\simeq 1\ cm^3$) that the ratio of surface atoms to all atoms is small, and the crystal is pure enough so that impurities are negligible. At temperatures that are low enough, lattice vibrations are weak, so weak that the effects of all these imperfections on, say, the optical properties of the sodium sample are negligible. It is in this spirit that we speak of a "perfect" crystal.

Imperfections themselves are often the main object of interest. Thus thermal vibrations of the atoms are the main source of electrical resistivity in metals. When this is the case, one does not abandon the crystal concept entirely, but treats the imperfection(s) of interest as a small perturbation in the crystalline structure.

Many of the most interesting phenomena in solids are associated with imperfections. That is why we shall discuss them at some length in various sections of this book.

## 1.3 BASIC DEFINITIONS

In order to talk precisely about crystal structures, we must introduce here a few of the basic definitions which serve as a kind of crystallographic language. These definitions are such that they apply to one-, two-, or three-dimensional crystals. Although most of our illustrative examples will be two-dimensional, the results will be restated later for the 3-D case.

### The crystal lattice

In crystallography, only the geometrical properties of the crystal are of interest, rather than those arising from the particular atoms constituting the crystal. Therefore one replaces each atom by a geometrical point located at the equilibrium position of that atom. The result is a pattern of points having the same geometrical properties as the crystal, but which is devoid of any physical contents. This geometrical pattern is the *crystal lattice*, or simply the *lattice*; all the atomic sites have been replaced by lattice sites.

There are two classes of lattices: the *Bravais* and the *non-Bravais*. In a Bravais lattice, all lattice points are equivalent, and hence by necessity all atoms in the crystal are of the same kind. On the other hand, in a non-Bravais lattice, some of the lattice points are nonequivalent. Figure 1.2 shows this clearly. Here the lattice sites $A$, $B$, $C$ are equivalent to each other, and so are the sites $A'$, $B'$, $C'$ among themselves, but the two sites $A$ and $A'$ are not equivalent to each other, as can be seen by the fact that the lattice is not invariant under a translation by $AA'$. This is so whether the atoms $A$ and $A'$ are of the same kind (for example, two H atoms) or of different kinds (for example, H and Cl atoms). A non-Bravais lattice is sometimes referred to as a *lattice with a basis*, the basis referring to the set of atoms stationed near each site of a Bravais lattice. Thus, in Fig. 1.2, the basis is the two atoms $A$ and $A'$, or any other equivalent set.

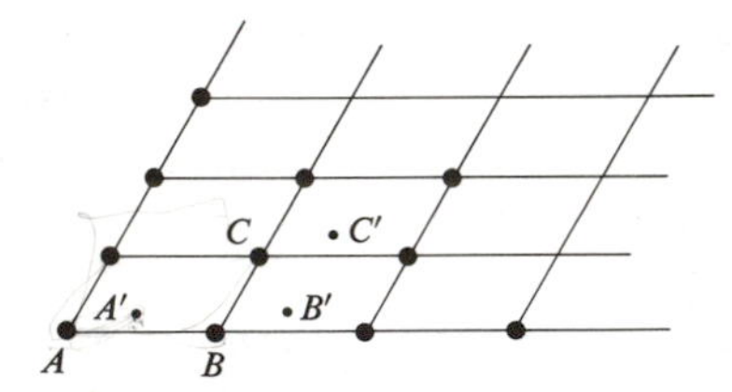


**Fig. 1.2** A non-Bravais lattice.

The non-Bravais lattice may be regarded as a combination of two or more interpenetrating Bravais lattices with fixed orientations relative to each other. Thus the points $A$, $B$, $C$, etc., form one Bravais lattice, while the points $A'$, $B'$, $C'$, etc., form another.

**Basis vectors**

Consider the lattice shown in Fig. 1.3. Let us choose the origin of coordinates at a certain lattice point, say $A$. Now the position vector of any lattice point can be written as

$$\mathbf{R}_n = n_1\mathbf{a} + n_2\mathbf{b}, \tag{1.1}$$

where $\mathbf{a}$, $\mathbf{b}$ are the two vectors shown, and $(n_1, n_2)$ is a pair of integers whose values depend on the lattice point. Thus for the point $D$, $(n_1, n_2) = (0, 2)$; for $B$, $(n_1, n_2) = (1, 0)$, and for $F$, $(n_1, n_2) = (0, -1)$.

The two vectors $\mathbf{a}$ and $\mathbf{b}$ (which must be noncolinear) form a set of *basis vectors* for the lattice, in terms of which the positions of all lattice points can be conveniently expressed by the use of (1.1). The set of all vectors expressed by this equation is called the *lattice vectors*. We may also say that the lattice is invariant under the group of all the translations expressed by (1.1). This is often rephrased by saying that the lattice has a *translational symmetry* under all displacements specified by the lattice vectors $\mathbf{R}_n$.

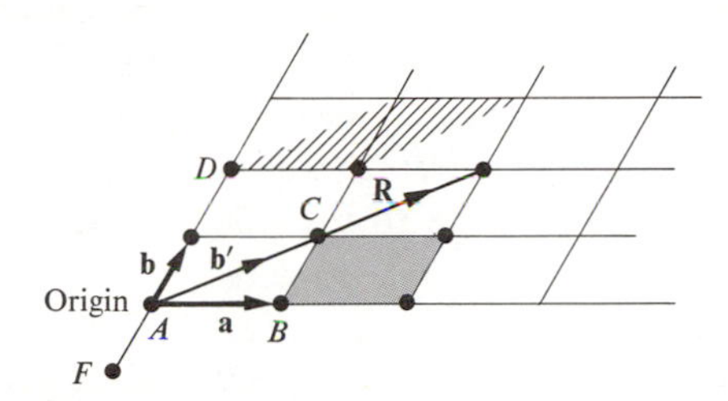


**Fig. 1.3** Vectors **a** and **b** are basis vectors of the lattice. Vectors **a** and **b**′ form another set of basis vectors. Shaded and hatched areas are unit cells corresponding to first and second set of basis vectors, respectively.

The choice of basis vectors is not unique. Thus one could equally well take the vectors **a** and **b**′ ($= \mathbf{a} + \mathbf{b}$) as a basis (Fig. 1.3). Other possibilities are also evident. The choice is usually dictated by convenience, but for all the lattices we shall meet in this text, such a choice has already been made, and is now a matter of convention.

**The unit cell**

The area of the parallelogram whose sides are the basis vectors **a** and **b** is called a *unit cell* of the lattice (Fig. 1.3), in that, if such a cell is translated by all the lattice vectors of (1.1), the area of the *whole* lattice is covered once and only once. The unit cell is usually the *smallest* area which produces this coverage. Therefore the lattice may be viewed as composed of a large number of equivalent unit cells placed side by side, like a mosaic pattern.

The choice of a unit cell for one and the same lattice is not unique, for the same reason that the choice of basis vectors is not unique. Thus the parallelogram formed by **a** and **b**′ in Fig. 1.3 is also an acceptable unit cell; once again the choice is dictated by convenience.

The following remarks may be helpful.

i) All unit cells have the same area. Thus the cell formed by **a**, **b** has the area $S = |\mathbf{a} \times \mathbf{b}|$, while that formed by **a**, **b**′ has the area $S' = |\mathbf{a} \times \mathbf{b}'| = |\mathbf{a} \times (\mathbf{a} + \mathbf{b})| = |\mathbf{a} \times \mathbf{b}| = S$, where we used the result $\mathbf{a} \times \mathbf{a} = 0$. Therefore the area of the unit cell is unique, even though the particular shape is not.

ii) If you are interested in how many lattice points belong to a unit cell, refer to Fig. 1.3. The unit cell formed by $\mathbf{a} \times \mathbf{b}$ has four points at its corners, but each of these points is shared by four adjacent cells. Hence each unit cell has only one lattice point.

### Primitive versus nonprimitive cells

The unit cell discussed above is called a *primitive cell.* It is sometimes more convenient, however, to deal with a unit cell which is larger, and which exhibits the symmetry of the lattice more clearly. The idea is illustrated by the Bravais lattice in Fig. 1.4. Clearly, the vectors $\mathbf{a}_1$, $\mathbf{a}_2$ can be chosen as a basis set, in which case the unit cell is the parallelogram $S_1$. However, the lattice may also be regarded as a set of adjacent rectangles, where we take the vectors **a** and **b** as basis vectors. The unit cell is then the area $S_2$ formed by these vectors. It has one lattice point at its center, in addition to the points at the corner. This cell is a *nonprimitive* unit cell.

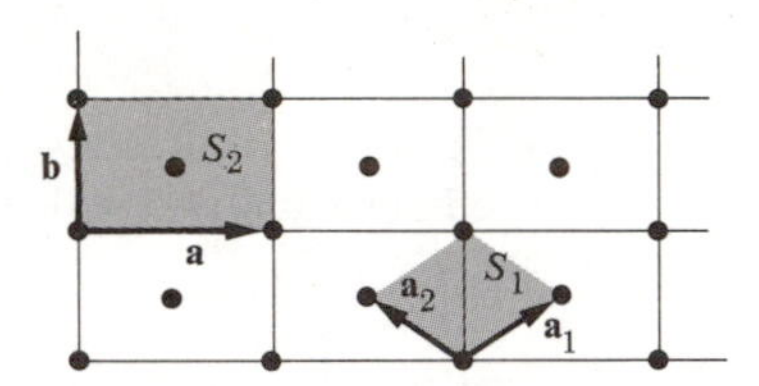


**Fig. 1.4** Area $S_1$ is a primitive unit cell; area $S_2$ is a nonprimitive unit cell.

The reason for the choice of the nonprimitive cell $S_2$ is that it shows the rectangular symmetry most clearly. Although this symmetry is also present in the primitive cell $S_1$ (as it must be, since both refer to the same lattice), the choice of the cell somehow obscures this fact.

Note the following points.

i) The area of the nonprimitive cell is an integral multiple of the primitive cell. In Fig. 1.4, the multiplication factor is two.

ii) No connection should be drawn between nonprimitive cells and non-Bravais lattices. The former refers to the particular (and somewhat arbitrary) choice of basis vectors in a Bravais lattice, while the latter refers to the physical fact of nonequivalent sites.

### Three dimensions

All the previous statements can be extended to three dimensions in a straightforward manner. When we do so, the lattice vectors become three-dimensional, and are expressed by

$$\mathbf{R}_n = n_1\mathbf{a} + n_2\mathbf{b} + n_3\mathbf{c}, \tag{1.2}$$

where **a**, **b**, and **c** are three *noncoplanar* vectors joining the lattice point at the origin to its near neighbors (Fig. 1.5); and $n_1$, $n_2$, $n_3$ are a triplet of integers 0, $\pm 1$, $\pm 2$, etc., whose values depend on the particular lattice point.

The vector triplet **a**, **b**, and **c** is the basis vector, and the *parallelepiped* whose sides are these vectors is a unit cell. Here again the choice of primitive cell is not unique, although all primitive cells have equal volumes. Also, it is sometimes convenient to deal with nonprimitive cells, ones which have additional points either inside the cell or on its surface. Finally, non-Bravais lattices in three dimensions are possible, and are made up of two or more interpenetrating Bravais lattices.

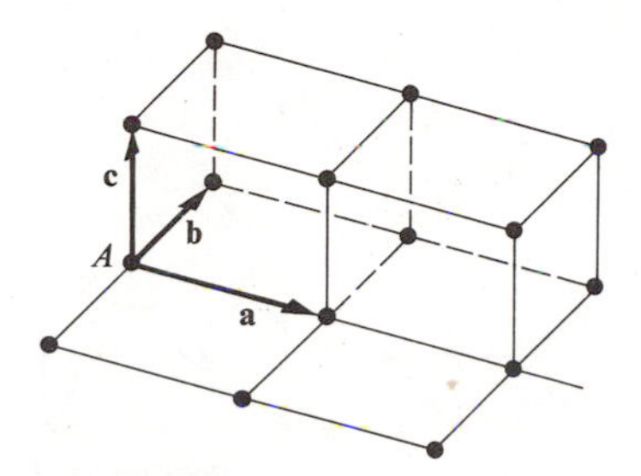


**Fig. 1.5** A three-dimensional lattice. Vectors **a**, **b**, **c** are basis vectors.

## 1.4 THE FOURTEEN BRAVAIS LATTICES AND THE SEVEN CRYSTAL SYSTEMS

There are only 14 different Bravais lattices. This reduction to what is a relatively small number is a consequence of the translational-symmetry condition demanded of a lattice. To appreciate how this comes about, consider the two-dimensional case, in which the reader can readily convince himself, for example, that it is not possible to construct a lattice whose unit cell is a regular pentagon. A regular pentagon can be drawn as an isolated figure, but one cannot place many such pentagons side by side so that they fit tightly and cover the whole area. In fact, it can be demonstrated that the requirement of translational symmetry in two dimensions restricts the number of possible lattices to only five (see the problem section at the end of this chapter).

In three dimensions, as we said before, the number of Bravais lattices is 14. The number of non-Bravais lattices is much larger (230), but it also is finite.

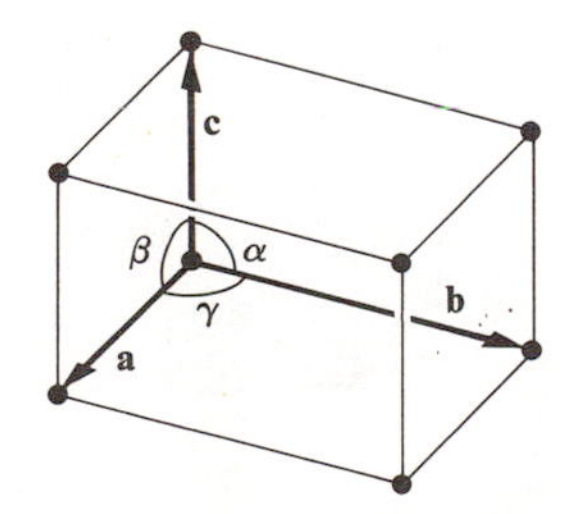


**Fig. 1.6** Unit cell specified by the lengths of basis vectors **a**, **b**, and **c**; also by the angles between the vectors.

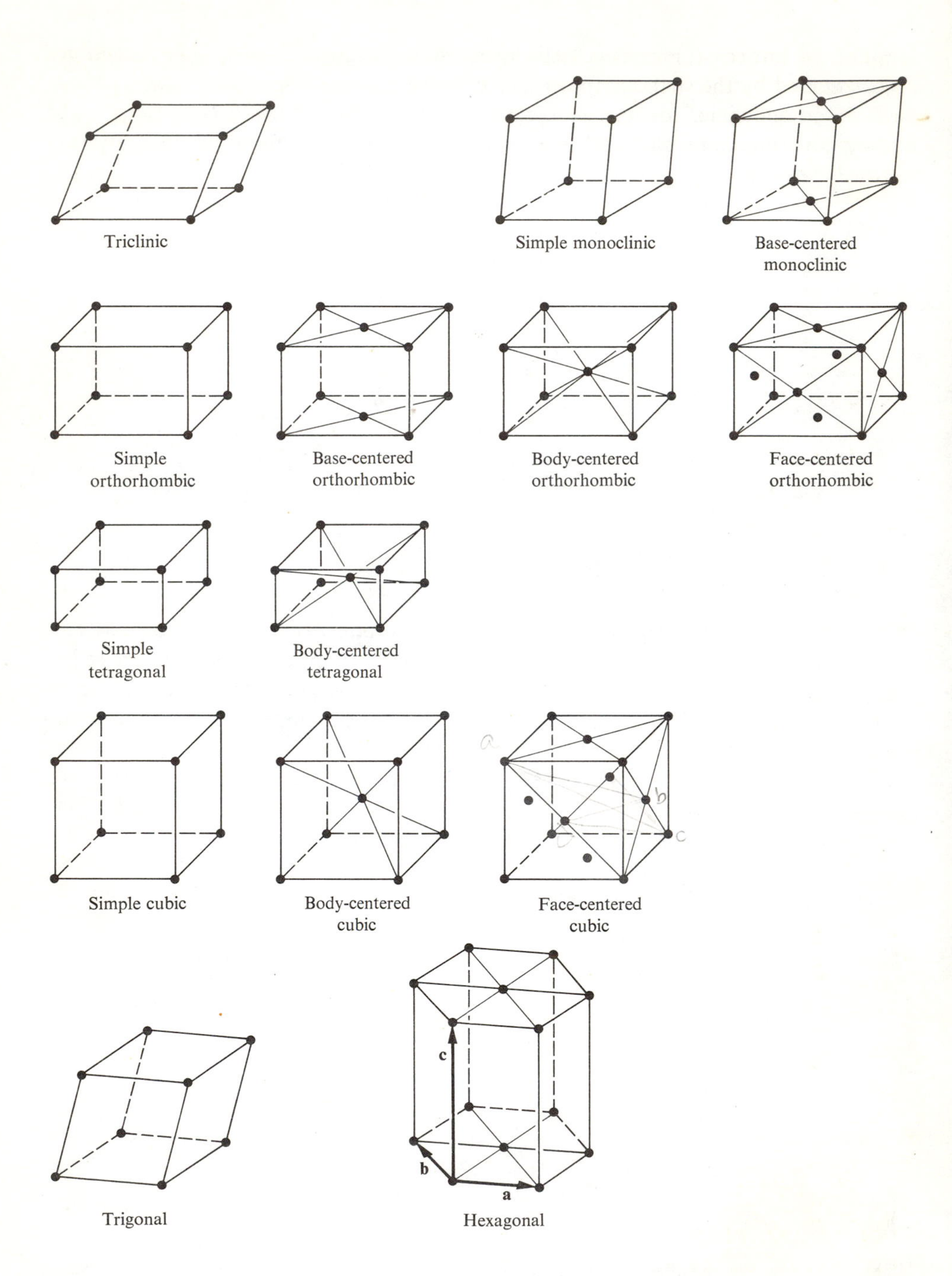


**Fig. 1.7** The 14 Bravais lattices grouped into the 7 crystal systems.

The 14 lattices (or crystal classes) are grouped into seven crystal systems, each specified by the shape and symmetry of the unit cell. These systems are the triclinic, monoclinic, orthorhombic, tetragonal, cubic, hexagonal, and the trigonal (or rhombohedral). In every case the cell is a parallelepiped whose sides are the bases **a**, **b**, **c**. The opposite angles are called $\alpha$, $\beta$, and $\gamma$, as shown in Fig. 1.6. Figure 1.7 shows the 14 lattices, and Table 1.1 enumerates the systems, lattices, and the appropriate values for **a**, **b**, **c**, and $\alpha$, $\beta$, and $\gamma$. Both Fig. 1.7 and Table 1.1 should be studied carefully, and their contents mastered. The column referring to symmetry elements in the table will be discussed shortly.

**Table 1.1**

The Seven Crystal Systems Divided into Fourteen Bravais Lattices

| System | Bravais lattice | Unit cell characteristics | Characteristic symmetry elements |
|---|---|---|---|
| Triclinic | Simple | $a \neq b \neq c$<br>$\alpha \neq \beta \neq \gamma \neq 90^\circ$ | None |
| Monoclinic | Simple<br>Base-centered | $a \neq b \neq c$<br>$\alpha = \beta = 90^\circ \neq \gamma$ | One 2-fold rotation axis |
| Orthorhombic | Simple<br>Base-centered<br>Body-centered<br>Face-centered | $a \neq b \neq c$<br>$\alpha = \beta = \gamma = 90^\circ$ | Three mutually orthogonal 2-fold rotation axes |
| Tetragonal | Simple<br>Body-centered | $a = b \neq c$<br>$\alpha = \beta = \gamma = 90^\circ$ | One 4-fold rotation axis |
| Cubic | Simple<br>Body-centered<br>Face-centered | $a = b = c$<br>$\alpha = \beta = \gamma = 90^\circ$ | Four 3-fold rotation axes (along cube diagonal) |
| Trigonal (rhombohedral) | Simple | $a = b = c$<br>$\alpha = \beta = \gamma \neq 90^\circ$ | One 3-fold rotation axis |
| Hexagonal | Simple | $a = b \neq c$<br>$\alpha = \beta = 90^\circ$<br>$\gamma = 120^\circ$ | One 3-fold rotation axis |

Note that a *simple* lattice has points only at the corners, a *body-centered* lattice has one additional point at the center of the cell, and a *face-centered* lattice has six additional points, one on each face. Let us again point out that in all the *nonsimple* lattices the unit cells are nonprimitive.

The 14 lattices enumerated in Table 1.1 exhaust all possible Bravais lattices, although a complete mathematical proof of this statement is quite lengthy. It may be thought, for example, that a base-centered tetragonal should also be included in the table, but it can readily be seen that such a lattice reduces to the simple tetragonal by a new choice of a unit cell (Fig. 1.8). Other cases can be treated similarly.

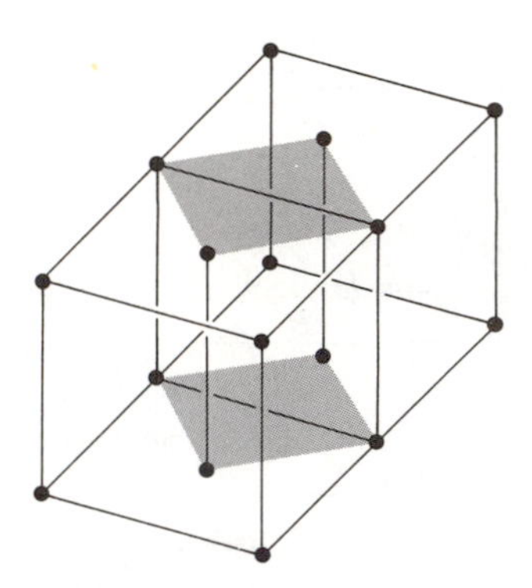

**Fig. 1.8** A base-centered tetragonal is identical to a simple tetragonal of a different unit cell. Shaded areas are the basis of the simple tetragonal cell.

The system we shall encounter most frequently in this text is the cubic one, particularly the *face-centered cubic* (fcc) and the *body-centered cubic* (bcc). The hexagonal system will also appear from time to time.

### 1.5 ELEMENTS OF SYMMETRY

Each of the unit cells of the 14 Bravais lattices has one or more types of symmetry properties, such as inversion, reflection, or rotation. Let us consider the meanings of these terms.

*Inversion center*. A cell has an inversion center if there is a point at which the cell remains invariant when the mathematical transformation $\mathbf{r} \rightarrow -\mathbf{r}$ is performed on it. All Bravais lattices are inversion symmetric, a fact which can be seen either by referring to Fig. 1.7 or by noting that, with every lattice vector $\mathbf{R}_n = n_1\mathbf{a} + n_2\mathbf{b} + n_3\mathbf{c}$, there is associated an inverse lattice vector $\mathbf{R}_n \equiv -\mathbf{R}_n = -n_1\mathbf{a} - n_2\mathbf{b} - n_3\mathbf{c}$. A non-Bravais lattice may or may not have an inversion center, depending on the symmetry of the basis.

*Reflection plane*. A plane in a cell such that, when a mirror reflection in this plane is performed, the cell remains invariant. Referring to Fig. 1.7, we see that the triclinic has no reflection plane, the monoclinic has one plane midway between and parallel to the bases, and so forth. The cubic cell has nine reflection planes: three parallel to the faces, and six others, each of which passes through two opposite edges.

*Rotation axis.* This is an axis such that, if the cell is rotated around it through some angle, the cell remains invariant. The axis is called $n$-fold if the angle of rotation is $2\pi/n$. When we look at Fig. 1.7 again, we see that the triclinic has no axis of rotation (save the trivial 1-fold axis), and the monoclinic has a 2-fold axis ($\theta = 2\pi/2 = \pi$) normal to the base. The cubic unit cell has three 4-fold axes normal to the faces, and four 3-fold axes, each passing through two opposite corners.

We have discussed the simplest symmetry elements, the ones which we shall encounter most frequently. More complicated elements also exist, such as rotation–reflection axes, glide planes, etc., but we shall not pursue these at this stage, as they will not be needed in this text.

You may have noticed that the symmetry elements may not all be independent. As a simple example, one can show that an inversion center plus a reflection plane imply the existence of a 2-fold axis passing through the center and normal to the plane. Many similar interesting theorems can be proved, but we shall not do so here.

**Point groups, space groups, and non-Bravais lattices**

A non-Bravais lattice is one in which, with each lattice site, there is associated a cluster of atoms called the *basis*. Therefore one describes the symmetry of such a lattice by specifying the symmetry of the basis in addition to the symmetry of the Bravais lattice on which this basis is superimposed.

The symmetry of the basis, called *point-group symmetry*, refers to all possible rotations (including inversion and reflection) which leave the basis invariant, keeping in mind that in all these operations one point in the basis must remain fixed (which is the reason for referring to this as point-group symmetry). A close examination of the problem reveals that only 32 different point groups can exist which are consistent with the requirements of translational symmetry for the lattice as a whole. One can appreciate the limitation on the number of point groups by the following physical argument: The shape or structure of the basis cannot be arbitrarily complex, e.g., like the shape of a potato. This would be incompatible with the symmetry of the interatomic forces operating between the basis and other bases on nearby lattice sites. After all, it is these forces which determine the crystal structure in the first place. Thus the rotation symmetries possible for the basis must be essentially the same as the rotational symmetries of the unit cells of the 14 Bravais lattices which were enumerated in Section 1.4.

When we combine the rotation symmetries of the point groups with the translational symmetries, we obtain a *space-group symmetry*. In this manner one generates a large number of space groups, 72 to be exact. It appears that there are also in addition some space groups which cannot be composed of simple point groups plus translation groups; such groups involve symmetry elements such as screw axes, glide planes, etc. When one adds these to the 72 space groups, one obtains 230 different space groups in all (Buerger, 1963). Figure 1.9 shows a tetragonal $D_{2\alpha}$ space group. However, further discussion of these groups lies outside the scope of this book.

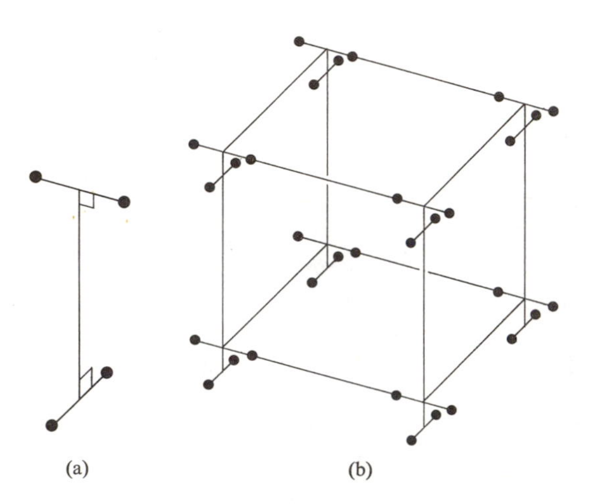


**Fig. 1.9** (a) A basis which has a $D_{2d}$ point group symmetry (two horizontal 2-fold axes plus two vertical reflection planes). (b) A simple tetragonal lattice with a basis having the $D_{2d}$ point group.

## 1.6 NOMENCLATURE OF CRYSTAL DIRECTIONS AND CRYSTAL PLANES; MILLER INDICES

In describing physical phenomena in crystals, we must often specify certain directions or crystal planes, because a crystal is usually anisotropic. Certain standard rules have evolved which are used in these specifications.

### Crystal directions

Consider the straight line passing through the lattice points $A, B, C$, etc., in Fig. 1.10. To specify its direction, we proceed as follows: We choose one lattice point on the line as an origin, say the point $A$. Then we choose the lattice vector joining $A$ to any point on the line, say point $B$. This vector can be written as

$$\mathbf{R} = n_1\mathbf{a} + n_2\mathbf{b} + n_3\mathbf{c}.$$

The direction is now specified by the integral triplet $[n_1n_2n_3]$. If the numbers $n_1, n_2, n_3$ have a common factor, this factor is removed, i.e., the triplet $[n_1n_2n_3]$ is the smallest integer of the same relative ratios. Thus in Fig. 1.10 the direction shown is the [111] direction.

Note that, when we speak of a direction, we do not mean one particular straight line, but a whole set of parallel straight lines (Fig. 1.10) which are completely equivalent by virtue of the translational symmetry.

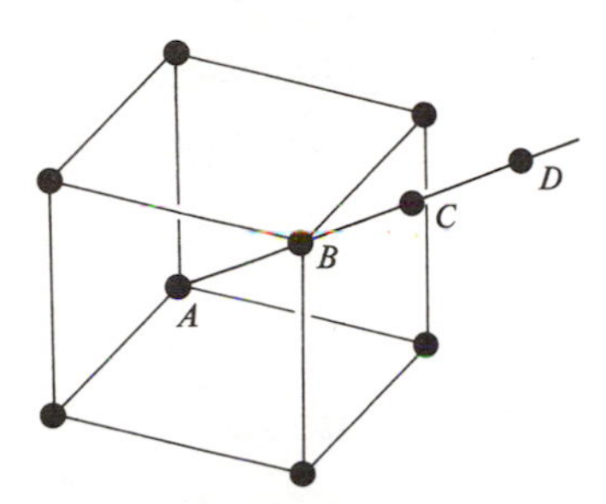


**Fig. 1.10** The [111] direction in a cubic lattice.

When the unit cell has some rotational symmetry, then there may exist several nonparallel directions which are equivalent by virtue of this symmetry. Thus in a cubic crystal the directions [100], [010], and [001] are equivalent. When this is the case, one may indicate collectively all the directions equivalent to the $[n_1n_2n_3]$ direction by $\langle n_1n_2n_3 \rangle$, using angular brackets. Thus in a cubic system the symbol $\langle 100 \rangle$ indicates all six directions: [100], [010], [001], $[\bar{1}00]$, $[0\bar{1}0]$, and $[00\bar{1}]$. The negative sign over a number indicates a negative value. Similarly the symbol $\langle 111 \rangle$ refers to all the body diagonals of the cube. Of course the directions $\langle 100 \rangle$ and $\langle 111 \rangle$ are not equivalent.

Note that a direction with large indices, e.g., [157], has fewer atoms per unit length than one with a smaller set of indices, such as [111].

### Crystal planes and Miller indices

The orientation of a plane in a lattice is specified by giving its *Miller indices*, which are defined as follows: To determine the indices for the plane $P$ in Fig. 1.11(a), we find its intercepts with the axes along the basis vectors **a**, **b**, and **c**. Let these intercepts be $x, y$, and $z$. Usually $x$ is a fractional multiple of $a$, $y$ a fractional multiple of $b$, and so forth. We form the fractional triplet

$$\left(\frac{x}{a}, \frac{y}{b}, \frac{z}{c}\right),$$

invert it to obtain the triplet

$$\left(\frac{a}{x}, \frac{b}{y}, \frac{c}{z}\right),$$

and then reduce this set to a similar one having the smallest integers by multiplying by a common factor. This last set is called the *Miller indices* of the plane and is indicated by $(hkl)$. Let us take an example: Suppose that the intercepts are $x = 2a$, $y = \frac{3}{2}b$, and $z = 1c$. We first form the set

$$\left[\frac{x}{a}, \frac{y}{b}, \frac{z}{c}\right] = (2, \tfrac{3}{2}, 1),$$

then invert it $(\frac{1}{2}, \frac{2}{3}, 1)$, and finally multiply by the common denominator, which is 6, to obtain the Miller indices (346) (pronouced as "three four six").

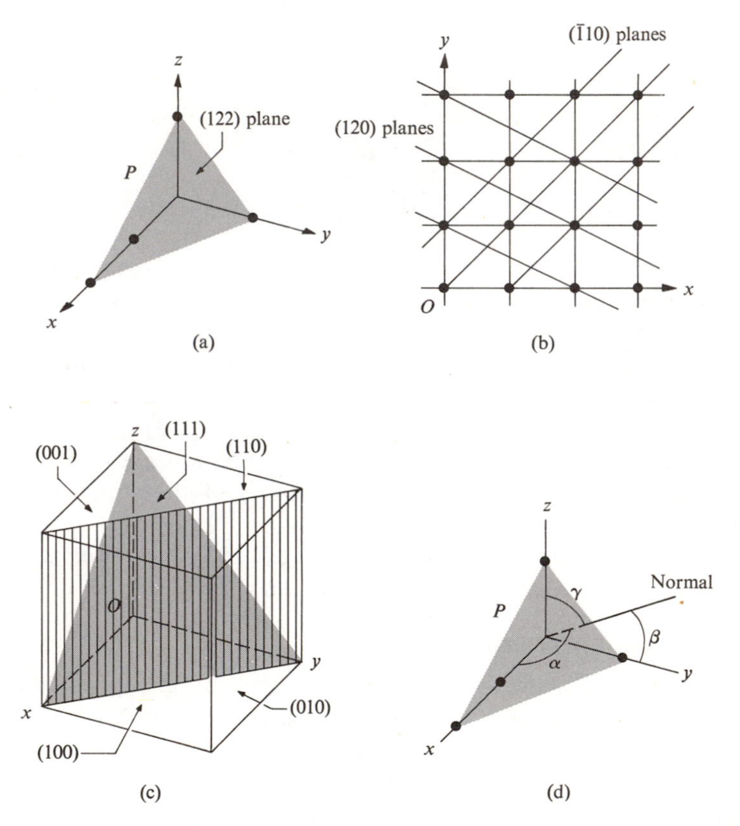


**Fig. 1.11** (a) The (122) plane. (b) Some equivalent, parallel planes represented by the Miller indices. (c) Some of the planes in a cubic crystal. (d) Finding the interplanar spacing.

We note that the Miller indices are so defined that all equivalent, parallel planes are represented by the same set of indices. Thus the planes whose intercepts are $x, y, z; 2x, 2y, 2z; -3x, -3y, -3z$, etc., are all represented by the same set of Miller indices. We can prove this by following the above procedure for determining the indices. Therefore a set of Miller indices specifies not just one plane, but an infinite set of equivalent planes, as indicated in Fig. 1.11(b). There is a good reason for using such notation, as we shall see when we study x-ray diffraction from crystal lattices. A diffracted beam is the result of scattering from large numbers of equivalent parallel planes, which act collectively to diffract the beam. Figure 1.11(c) shows several important planes in a cubic crystal.

[The reason for inverting the intercepts in defining the Miller indices is more subtle, and has to do with the fact that the most concise, and mathematically convenient, method of representing lattice planes is by using the so-called reciprocal lattice. We shall discuss this in Chapter 2, where we shall clarify the connection.]

Sometimes, when the unit cell has rotational symmetry, several nonparallel planes may be equivalent by virtue of this symmetry, in which case it is convenient to lump all these planes in the same Miller indices, but with curly brackets. Thus the indices $\{hkl\}$ represent all the planes equivalent to the plane $(hkl)$ through rotational symmetry. As an example, in the cubic system the indices {100} refer to the six planes (100), (010), (001), $(\bar{1}00)$, $(0\bar{1}0)$, and $(00\bar{1})$.

**Spacing between planes of the same Miller indices**

In connection with x-ray diffraction from a crystal (see Chapter 2), one needs to know the interplanar distance between planes labeled by the same Miller indices, say $(hkl)$. Let us call this distance $d_{hkl}$. The actual formula depends on the crystal structure, and we confine ourselves to the case in which the axes are orthogonal. We can calculate this by referring to Fig. 1.11(d), visualizing another plane parallel to the one shown and passing through the origin. The distance between these planes, $d_{hkl}$, is simply the length of the normal line drawn from the origin to the plane shown. Suppose that the angles which the normal line makes with the axes are $\alpha$, $\beta$, and $\gamma$, and that the intercepts of the plane $(hkl)$ with the axes are $x$, $y$, and $z$. Then it is evident from the figure that

$$d_{hkl} = x\cos\alpha = y\cos\beta = z\cos\gamma.$$

But there is a relation between the directional cosines $\cos\alpha$, $\cos\beta$, and $\cos\gamma$. That is, $\cos^2\alpha + \cos^2\beta + \cos^2\gamma = 1$. If we solve for $\cos\alpha$, $\cos\beta$, and $\cos\gamma$ from the previous equation, substitute into the one immediately above, and solve for $d_{hkl}$ in terms of $x$, $y$, and $z$, we find that

$$d_{hkl} = \frac{1}{\left(\dfrac{1}{x^2} + \dfrac{1}{y^2} + \dfrac{1}{z^2}\right)^{1/2}}. \tag{1.3}$$

Now $x$, $y$, and $z$ are related to the Miller indices $h$, $k$, and $l$. If one reviews the process of defining these indices, one readily obtains the relations

$$h = n\frac{a}{x}, \qquad k = n\frac{b}{y}, \qquad l = n\frac{c}{z}, \tag{1.4}$$

where $n$ is the common factor used to reduce the indices to the smallest integers possible. Solving for $x$, $y$, and $z$ from (1.4) and substituting into (1.3), one obtains

$$d_{hkl} = \frac{n}{\left[\frac{h^2}{a^2} + \frac{k^2}{b^2} + \frac{l^2}{c^2}\right]^{1/2}}, \tag{1.5}$$

which is the required formula. Thus the interplanar distance of the (111) planes in a simple cubic crystal is $d = na\sqrt{3}$, where $a$ is the cubic edge.

## 1.7 EXAMPLES OF SIMPLE CRYSTAL STRUCTURES

In order to gain an appreciation of actual crystals, let us familiarize ourselves with a few of the better-known structures, and with the sizes of their unit cells. The cumulative knowledge obtained over the years on the structures of various crystals is truly enormous, but here we shall touch on only the few simple and better-known examples which we shall meet repeatedly in this book.

### Face-centered and body-centered cubic

Many of the common metals crystallize in one or the other of these two lattices.

Thus the most familar metals—Ag, Al, Au, Cu, Co($\beta$), Fe($\gamma$), Ni($\beta$), Pb, and Pt—all crystallize in the fcc structure (Fig. 1.12a). The unit cell contains four atoms: one from the eight corner atoms which it shares with other cells, and three from the six surface atoms it shares with other cells.

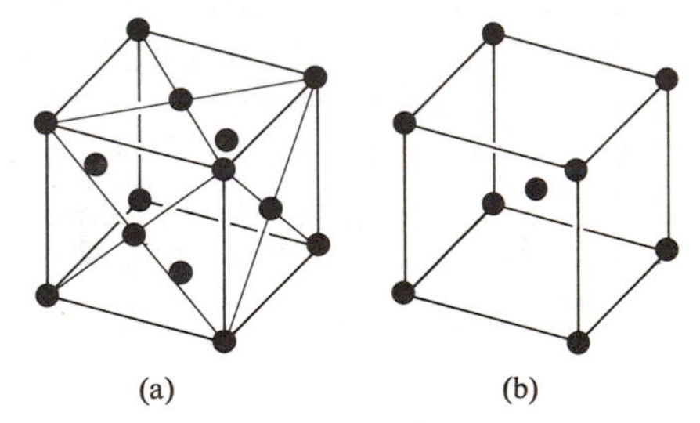


**Fig. 1.12** (a) An fcc unit cell. (b) A bcc unit cell.

Some of the metals which crystallize in the bcc structure are: Fe($\alpha$), and the alkalis Li, Na, K, Rb, and Cs (Fig. 1.12b). Here the unit cell has two atoms. One is from the shared corner atoms and the other is the central atom, which is not shared.

### The sodium chloride structure

This is the structure assumed by ordinary table salt, NaCl. The structure is cubic, and is such that, along the three principal directions (axes), there is an alternation of Na and Cl atoms, as shown in Fig. 1.13(a). In three dimensions the unit cell appears as shown in Fig. 1.13(b). That is, the cell is a face-centered cubic one. The positions of the four Na atoms are 000, $\frac{1}{2}\frac{1}{2}0$, $\frac{1}{2}0\frac{1}{2}$, $0\frac{1}{2}\frac{1}{2}$, while those of

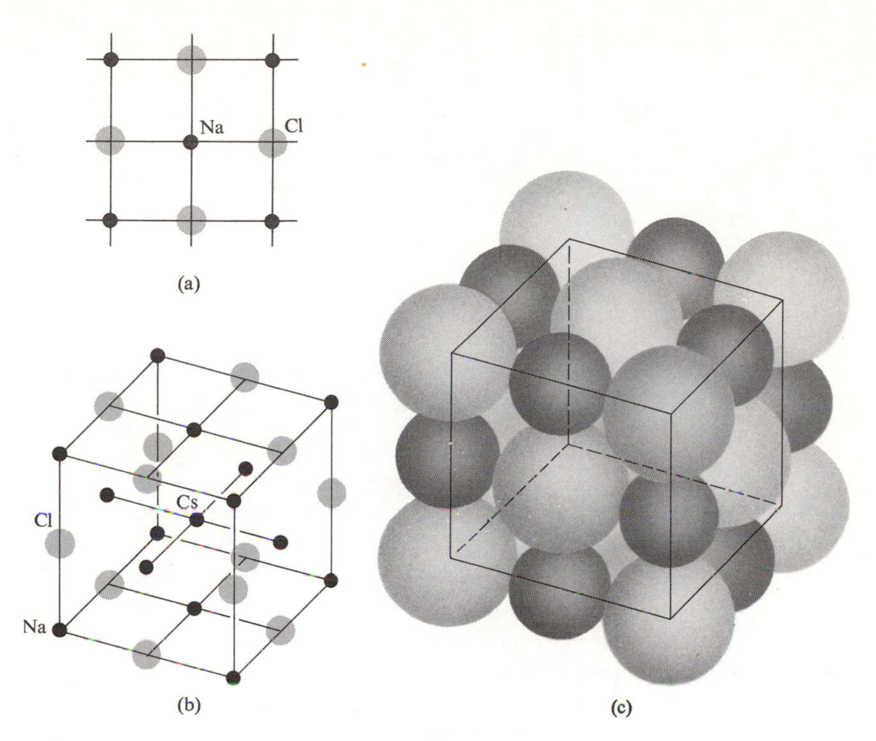


**Fig. 1.13** (a) A two-dimensional view of the NaCl structure. (b) The NaCl structure in three dimensions. The Na atoms form an fcc structure which is interlocked with another fcc structure composed of the Cl atoms. (c) The NaCl structure drawn close to scale, with the ions nearly touching. The sodium atoms, small solid spheres, reside in the octahedral voids between the chlorine atoms.

the four Cl atoms are located at $\frac{1}{2}\frac{1}{2}\frac{1}{2}$, $00\frac{1}{2}$, $\frac{1}{2}00$, $0\frac{1}{2}0$ (the numbers refer to coordinates given in fractions of the cubic edge).

We summarize this by saying that NaCl is a non-Bravais structure composed of two interpenetrating fcc sublattices; one made up of Na atoms and the other of Cl atoms, and the two sublattices are displaced relative to each other by $\frac{1}{2}\mathbf{a}$.

Many ionic crystals such as KCl and PbS also have this structure. For a more complete list, including the lattice constants, refer to Table 1.2.

### The cesium chloride structure

This again is a cubic crystal, but here the cesium and chlorine atoms alternate on lines directed along the four diagonals of the cube. Thus the unit cell is a bcc one, as shown in Fig. 1.14. There are, per unit cell, one Cs atom located at the point 000 and one Cl atom located at $\frac{1}{2}\frac{1}{2}\frac{1}{2}$. Therefore this is a non-Bravais lattice composed of two sc (simple cubic) lattices which are displaced relative to each other along the diagonal by an amount equal to one-half the diagonal. For a list of certain ionic compounds crystallizing in this structure, see Table 1.2.

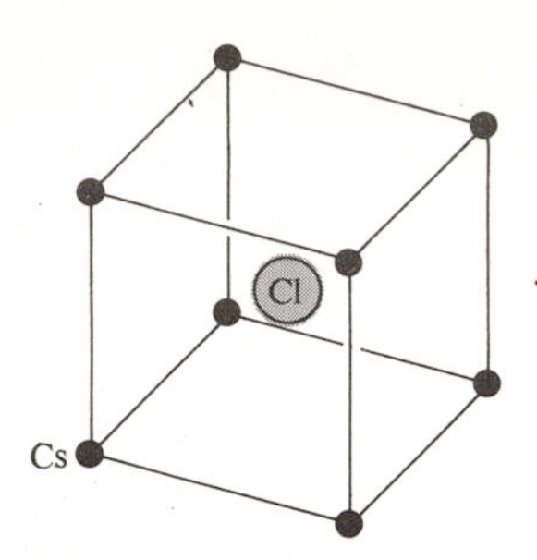


**Fig. 1.14** Structure of cesium chloride. The Cs atoms form an sc lattice interlocked with another sc lattice formed by the Cl ions.

**Table 1.2**

Structures and Cell Dimensions of Some Elements and Compounds

| Element or compound | Structure | $a$, Å | $c$, Å |
|---|---|---|---|
| Al | fcc | 4.04 | |
| Be | hcp | 2.27 | 3.59 |
| Ca | fcc | 5.56 | |
| C | Diamond | 3.56 | |
| Cr | bcc | 2.88 | |
| Co | hcp | 2.51 | 4.07 |
| Cu | fcc | 3.61 | |
| Ge | Diamond | 5.65 | |
| Au | fcc | 4.07 | |
| Fe | bcc | 2.86 | |
| K | fcc | 3.92 | |
| Si | Diamond | 5.43 | |
| Ag | fcc | 4.08 | |
| Na | bcc | 4.28 | |
| Zn | hcp | 2.66 | 4.94 |
| LiH | Sodium chloride | 4.08 | |
| NaCl | Sodium chloride | 5.63 | |
| AgBr | Sodium chloride | 5.77 | |
| MnO | Sodium chloride | 4.43 | |
| CsCl | Cesium chloride | 4.11 | |
| TlBr | Cesium chloride | 3.97 | |
| CuZn ($\beta$-brass) | Cesium chloride | 2.94 | |
| CuF | Zincblende | 4.26 | |
| AgI | Zincblende | 6.47 | |
| ZnS | Zincblende | 5.41 | |
| CdS | Zincblende | 5.82 | |

### The diamond structure

The unit cell for this structure is an fcc cell with a basis, where the basis is made up of two carbon atoms associated with each lattice site. The positions of the two basis atoms are 000 and $\frac{1}{4}\frac{1}{4}\frac{1}{4}$. A two-dimensional view of the cell is shown in Fig. 1.15(a), and the whole cell in three dimensions is shown in Fig. 1.15(b). There are eight atoms per unit cell.

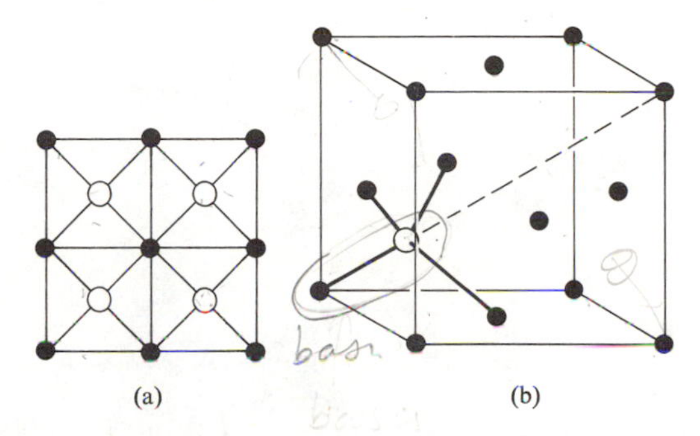


**Fig. 1.15** The diamond structure. (a) Projection of the atoms on the base of the cube. One dark circle plus an adjacent white circle form a basis for the structure. (b) A simplified three-dimensional view. Only one of the 14 white spheres is shown, together with the tetrahedral coordination.

Note that the present structure is such that each atom finds itself surrounded by four nearest atoms, which form a regular tetrahedron whose center is the atom in question. Such a configuration is common in semiconductors, and is referred to as a *tetrahedral bond.* This structure occurs in many semiconductors, for example, Ge, Si, etc. Table 1.2 contains a few examples, with appropriate numerical values.

### The zinc sulfide† (ZnS) structure

This structure, named after the compound ZnS, is closely related to the diamond structure discussed above, the only difference being that the two atoms forming the basis are of different kinds, e.g., Zn and S atoms. Here each unit cell contains four ZnS molecules, and each Zn (or S) atom finds itself at the center of a tetrahedron formed by atoms of the opposite kind.

Many of the compound semiconductors—such as InSb, GaSb, GaAs, etc.—do crystallize in this structure (Table 1.2).

### The hexagonal close-packed structure

This is another structure that is common, particularly in metals. Figure 1.16 demonstrates this structure. In addition to the two layers of atoms which form

† Also known as the *zincblende* structure.

the base and upper face of the hexagon, there is also an intervening layer of atoms arranged such that each of these atoms rests over a depression between three atoms in the base. The atoms in a hexagonal close-packed (hcp) structure are thus packed tightly together, which explains why this structure is so common in metals, where the atoms tend to assemble very close to each other. Examples of hcp crystals are Be, Mg, Ca, Zn, and Hg—all divalent metals.

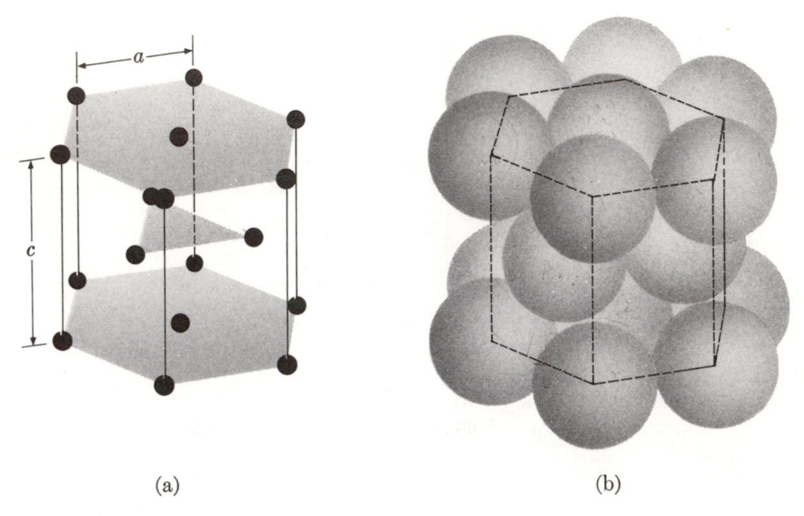


**Fig. 1.16** (a) Hexagonal close-packed structure. (b) The hcp when the atoms are nearly touching, as in the actual situation.

## 1.8 AMORPHOUS SOLIDS AND LIQUIDS

Amorphous solids have received increasing attention in recent years, particularly as a result of the discovery of the electrical properties of amorphous semiconductors (Section 13.2). It behooves us, therefore, to glance at least briefly at the structure of these solids.

The most familiar example of an amorphous solid is ordinary window glass. Chemically the substance is a silicon oxide. Structurally it has no crystal structure at all; the silicon and oxygen are simply distributed in what appears to be a *random* fashion.

Another familiar case of an amorphous structure is that of a liquid. Here again the system has no crystal structure, and the atoms appear to have a random distribution. As time passes, the atoms in the liquid drift from one region to another, but their random distribution persists.

This suggests a strong similarity between liquids and amorphous solids, even though the atoms in the latter are fixed in space and do not drift as they do in liquids. This is why amorphous solids, such as glass, are sometimes referred to as *supercooled* liquids. In fact, if one could take an instantaneous picture of the atoms in a liquid, the result would be the same as, and indistinguishable from, that of an amorphous solid. The same mathematical formalism may therefore be employed to describe both types of substance.

Even a liquid does actually have a certain kind of "order" or structure, even though this structure is not crystalline. Consider the case of mercury, for instance. This metal crystallizes in the hcp structure. When the substance is in the solid state, below the melting point, all the atoms are in their regular positions, and each atom is surrounded by a certain number of nearest neighbors, next-nearest neighbors, etc., all of which are positioned at exactly defined distances from the central atom. When the metal is heated and melts, the atoms no longer hold to their regular positions, and the crystal structure as such is destroyed. Yet as we view the system from the vantage point of the original atom, we discover that insofar as the number of nearest and next-nearest neighbors and their distances is concerned, the situation in the liquid state remains substantially the same as it was in the solid state. Of course, when we speak of the "number of nearest neighbors" in the liquid state, we actually mean the average number, since the actual number is constantly changing as a result of the motion of the atoms.

It is apparent, therefore, that a liquid has a structure, and that this structure is quite evident from x-ray diffraction pictures of liquids. The important point, however, is that the order in a liquid is restricted only to the few shells of neighbors surrounding the central atom. As one goes to farther and farther atoms, their distribution relative to the central atom becomes entirely random. This is why we say that a liquid has only a *short-range order*. *Long-range order* is absent. Contrast this with the case of a crystal. In a crystal, the positions of all atoms, even the farthest ones, are exactly known once the position of the central atom is given. A crystal therefore has both short-range and long-range orders, i.e., perfect order.

It is not surprising that some order should exist, even in the liquid state. After all, the interatomic forces responsible for the crystallinity of a solid remain operative even after the solid melts and becomes a liquid. Furthermore, since the expansion of volume that is concomitant with melting is usually small, the average interatomic distances and hence the forces remain of the same magnitude as before. The new element now entering the problem is that the thermal kinetic energy of the atoms, resulting from heating, prevents them from holding to their regular positions, but the interatomic forces are still strong enough to impart a certain partial order to the liquid.

To turn now to the mathematical treatment: We take a typical atom and use it as a central atom in order to study the distribution of other atoms in the system relative to it. We draw a spherical shell of radius $R$ and thickness $\Delta R$ around this atom. The number of atoms in this shell is given by

$$\Delta N(R) = n(R) 4\pi R^2 \, \Delta R, \tag{1.6}$$

where $n(R)$ is the concentration of atoms in the system. Note that the quantity $4\pi R^2 \Delta R$ is the volume of the spherical shell, which, when we multiply it by the concentration, yields the number of particles. Note also that, since a liquid is isotropic, we need not be concerned with any angular variation of the concentration. Only the radial dependence is relevant here.

The structural properties of the liquid are now contained entirely in the concentration $n(R)$. Once this quantity and its variation with the radial distance $R$ are determined, the structure of the liquid is completely known.

The concentration $n(R)$ versus $R$ in liquid mercury as revealed by x-ray diffraction is shown in Fig. 1.17. The curve has a primary peak at $R \simeq 3$Å, beyond which it oscillates a few times before reaching a certain constant value. The concentration vanishes for $R \lesssim 2.2$ Å.

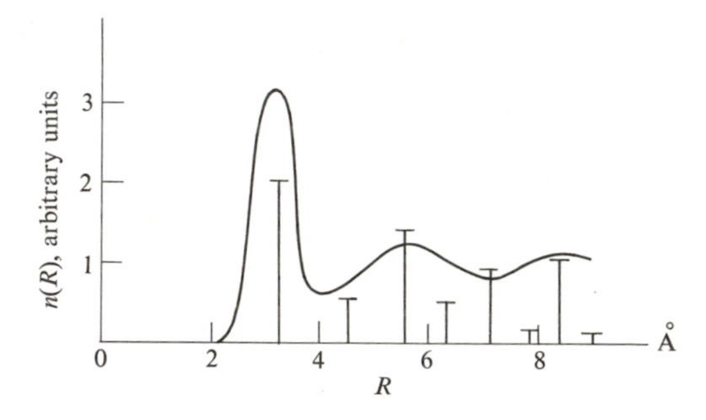


**Fig. 1.17** The atomic concentration $n(R)$ in liquid mercury. Vertical lines indicate the atomic distribution in crystalline mercury.

These features can be made quite plausible on the basis of interatomic forces. The vanishing of $n(R)$ at small values of $R$ is readily understandable; as other atoms approach the central one very closely, strong repulsive forces arise which push these atoms away (see the following two sections). These repulsive forces therefore prevent the other atoms from overlapping the central atom, which explains why $n(R) = 0$ at small $R$. One expects the value of $R$ where $n(R) = 0$ to be nearly equal to the diameter of the atom.

The reason for the major peak (Fig. 1.17) is closely related to the attractive interatomic force. We shall explain below that, except at very short distances, atoms attract each other. This force therefore tends to pull other atoms toward the center, resulting in a particularly large density at a certain specific distance. The other oscillations in the curve arise from an interplay between the force of the central atom and the forces of the near neighbors acting on neighbors still farther away.

At large values of $R$, the concentration $n(R)$ approaches a constant value $n_0$, which is actually equal to the average concentration in the system. We expect this result because we have seen that a liquid does not have a long-range order; thus at large $R$ the distribution of the atoms is completely random, and independent of the position of the central atom, i.e., independent of $R$.

Instead of $n(R)$, it is customary to express the correlation between atoms by introducing the so-called *pair distribution function* $g(R)$. This is defined as

$$g(R) = \frac{n(R)}{n_0}. \tag{1.7}$$

Thus this function has the meaning of a relative density, or probability. Since $n_0$ is a constant, the shape of $g(R)$ is the same as that of $n(R)$, that is, the same as in Fig. 1.17. Note in particular that $g(R) \to 1$ as $R \to \infty$, which is the situation corresponding to the absence of correlation between atoms.

As alluded to above, the pair function $g(R)$ is determined by x-ray diffraction. We shall discuss this in Section 2.8.

## 1.9 INTERATOMIC FORCES

Solids are stable structures, e.g., a crystal of NaCl is more stable than a collection of free Na and Cl atoms. Similarly, a Ge crystal is more stable than a collection of free Ge atoms. This implies that the Ge atoms attract each other when they get close to each other, i.e., an attractive interatomic force exists which holds the atoms together. This is the force responsible for crystal formation.

This also means that the energy of the crystal is lower than that of the free atoms by an amount equal to the energy required to pull the crystal apart into a set of free atoms. This is called the *binding energy* (also the cohesive energy) of the crystal.

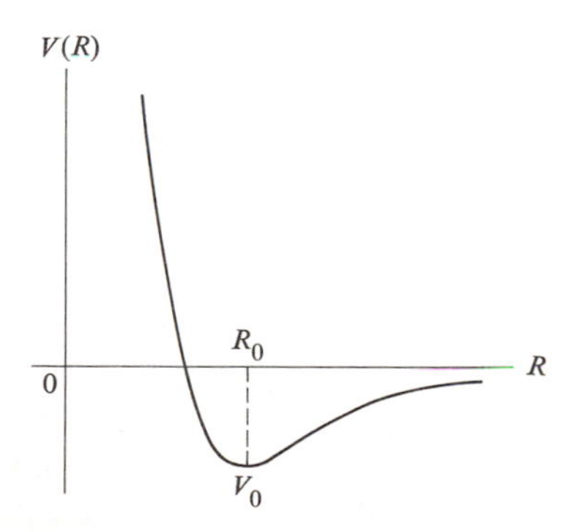


**Fig. 1.18** Interatomic potential $V(R)$ versus interatomic distance.

The potential energy representing the interaction between two atoms varies greatly with the distance between the atoms. A typical curve of this pair potential, shown in Fig. 1.18, has a minimum at some distance $R_0$. For $R > R_0$, the potential increases gradually, approaching 0 as $r \to \infty$, while for $R < R_0$ the potential increases very rapidly, approaching $\infty$ at small radius.

Because the system—the atom pair—tends to have the lowest possible energy, it is most stable at the minimum point $A$, which therefore represents the *equilibrium position*; the equilibrium interatomic distance is $R_0$, and the binding energy $-V_0$. Note that, since $V_0 < 0$, the system is stable, inasmuch as its energy is lower than that state in which two atoms are infinitely far apart (free atoms).

A typical value for the equilibrium radius $R_0$ is a few angstroms, so the forces under consideration are, in fact, rather short-range. The decay of the potential with distance is so rapid that once this exceeds a value of, say, 10 or 15 Å, the force may be disregarded altogether, and the atoms may then be treated as free, noninteracting particles. This explains why the free-atom model holds so well in gases, in which the average interatomic distance is large.

The interatomic force $F(R)$ may be derived from the potential $V(R)$. It is well known from elementary physics that

$$F(R) = -\frac{\partial V(R)}{\partial R}. \tag{1.8}$$

That is, the force is the negative of the potential gradient. If we apply this to the curve of Fig. 1.18, we see that $F(r) < 0$ for $R_0 < R$. This means that in the range $R_0 < R$ the force is *attractive*, tending to pull the atoms together. On the other hand, the force $F(R) > 0$ for $R_0 < R$. That is, when $R < R_0$, the force is *repulsive*, and tends to push the atoms apart.

It follows from this discussion that the interatomic force is composed of two parts: an attractive force, which is the dominant one at large distances, and a repulsive one, which dominates at small distances. These forces cancel each other exactly at the point $R_0$, which is the point of equilibrium.

We shall discuss the nature of the attractive and repulsive forces in the following section.

## 1.10 TYPES OF BONDING

The presence of attractive interatomic forces leads to the bonding of solids. In chemist's language, one may say that these forces form *bonds* between atoms in solids, and it is these bonds which are responsible for the stability of the crystal.

There are several types of bonding, depending on the physical origin and nature of the bonding force involved. The three main types are: *ionic bonding*, *covalent bonding*, and *metallic bonding*. Let us now take these up one by one, and also consider secondary types of bonding which are important in certain special cases.

### The ionic bond

The most easily understood type of bond is the ionic bond. Take the case of

NaCl as a typical example. In the crystalline state, each Na atom loses its single valence electron to a neighboring Cl atom, resulting in an ionic crystal containing both positive and negative ions. Thus each $Na^+$ ion is surrounded by six $Cl^-$ ions, and vice versa, as pointed out in Section 7.

If we examine a pair of Na and Cl ions, it is clear that an attractive electrostatic coulomb force, $e^2/4\pi\epsilon_0 R^2$, exists between the pairs of oppositely charged ions. It is this force which is responsible for the bonding of NaCl and other ionic crystals.

It is more difficult, however, to understand the origin of the repulsive force at small distances. Suppose the ions in NaCl were brought together very closely by a (hypothetical) decrease of the lattice constant. Then a repulsive force would begin to operate at some point. Otherwise the ions would continue to attract each other, and the crystal would simply collapse—which is, of course, not in agreement with experiment. We cannot explain this repulsive force on the basis of coulomb attraction; therefore it must be due to a new type of interaction.

A qualitative picture of the origin of the repulsive force may be drawn as follows: When the $Na^+$ and $Cl^-$ ions approach each other closely enough so that the orbits of the electrons in the ions begin to overlap each other, then the electrons begin to repel each other by virtue of the repulsive electrostatic coulomb force (recall that electrons are all negatively charged). Of course, the closer together the ions are, the greater the repulsive force, which is in qualitative agreement with Fig. 1.18 in the region $R < R_0$.

There is yet another equally important source which contributes to the repulsive force: the *Pauli exclusion principle*. As ions approach each other, the orbits of the electrons begin to overlap, i.e., some electrons attempt to occupy orbits already occupied by others. But this is forbidden by the exclusion principle, inasmuch as both the $Na^+$ and $Cl^-$ ions have outermost shells that are completely full. To prevent a violation of the exclusion principle, the potential energy of the system increases very rapidly, again in agreement with Fig. 1.18, in the range $R < R_0$.

The ionic bond is strong when compared with other bonds, a typical value for the binding energy of a pair of atoms being about 5 eV. This strength is attributed to the strength of the coulomb force responsible for the bonding. Experimentally, this strength is characterized by the high melting temperatures associated with ionic crystals. Thus the melting temperature for the ionic crystal NaCl is 801°C, while the melting temperatures for the Na and K metals are 97.8°C and 63°C, respectively.

Ionic bonding is most likely to exist when the elements involved are of widely differing electronegativities. Example: an electropositive alkali atom plus an electronegative halogen atom, as in NaCl.

**The covalent bond**

To introduce this bond, we consider a well-known covalent solid: diamond.

Recall from Section 7 that this crystal is formed from carbon atoms arranged in a certain type of fcc structure in which each atom is surrounded by four others, forming a regular tetrahedron. We cannot invoke the ionic bond to explain the bonding in diamond, because here each atom retains its own electrons, i.e., there is no transfer of electrons between the atoms, and in consequence no ions are formed. This is evident from the fact that all the atoms are identical. Hence no reason exists for an electron to transfer from one atom to another.

Instead, the bonding in diamond takes place in the following manner: Each atom has four valence electrons, and it forms four bonds with its four nearest neighbors (Fig. 1.19). The bond here is composed of two electrons, one contributed by each of the two atoms. This double-electron bond is well known in chemistry and physics. It is referred to as a *covalent bond.* As an indication of the appropriateness of this bond in the case of diamond, we note that as a result of electron sharing, each C atom now has 8 electrons surrounding it, resulting in a complete—and hence stable—shell structure for the valence shell at hand (in this case the familiar p shell).

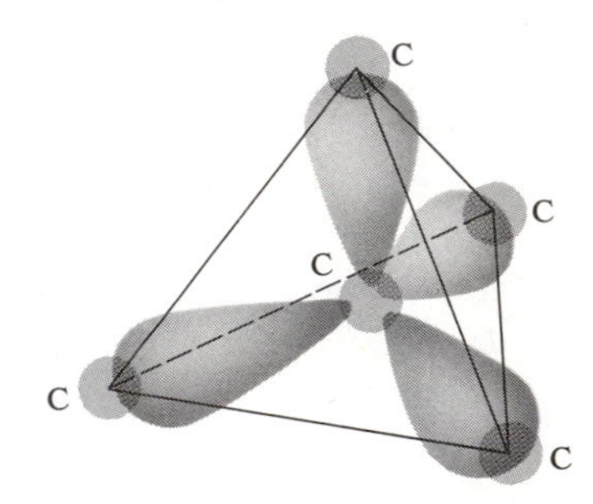


**Fig. 1.19** The tetrahedral covalent bond in diamond. Each elongated region represents the charge distribution of the two electrons forming the corresponding bond.

This plausible account still does not explain just why a double-electron arrangement produces a bond, i.e., an attractive interatomic force. The explanation of the covalent bond can be given only through quantum mechanics. The simplest known example of the covalent bond occurs in the hydrogen molecule ($H_2$), in which the two atoms are held together by just this bond, i.e., they share their two electrons. We discuss the quantum explanation of bonding in $H_2$ in Section A.7, and its adaptation to the tetrahedral bond (as in diamond) in Section A.8. Refer to these sections for further details.

The covalent bond is also strong, as attested to by the unusual hardness of diamond, and its high melting point ($> 3000°C$). A typical value for covalent-bond binding energy is a few electron volts per bond.

The covalent bond is particularly important for those elements in column IV of the periodic table. We have already mentioned diamond (C). Other elements are Si, Ge, and Sn, all of which crystallize in the diamond structure and are covalent crystals. The elements silicon (Si) and germanium (Ge) hold special interest, since both are among the best known semiconductors. We shall study them in considerable detail in Chapters 6 and 7, which concern the semiconducting properties of solids.

Covalent crystals tend to be hard and brittle, and incapable of appreciable bending. These facts are understandable in terms of the underlying atomic forces. Since the bonds have well-defined directions in space, attempts to alter them are strongly resisted by the crystal.

In our discussion of bonding, we have considered only pure ionic or pure covalent bonds. There are, however, many crystals in which the bond is not pure, but a mixture of ionic and covalent. A good example is the case of the semiconductor GaAs. Here a charge transfer does take place, but the transfer is not complete; only about 0.46 of an electron is transferred on the average from the Ga to the As atom. This transfer accounts for part of the binding force in GaAs, but the major part is due to a covalent—or electron-sharing—bond between the Ga and neighboring As atoms.

**The metallic bond**

Most elements are metals. Metals are characterized by high electrical conductivity and mechanical strength, and also by the property of being highly ductile. We are particularly interested here in the bonding mechanism, but the model we shall invoke for this purpose is capable of accounting for the above-mentioned properties as well. (These properties will be the subject of ample discussions later; see Chapters 4, 5, and 11.)

To understand the metallic bond, let us consider a typical simple metal: sodium. How can an assembly of Na atoms, brought together to form a crystal, attract each other and form a stable solid? Quick reflection tells us that neither the ionic nor the covalent bond can account for the interatomic attraction in sodium (why not?). The correct explanation is this: Each free Na atom has a single valence electron which is only loosely bound to the atom. When a crystal is formed, the valence electron detaches from its own atom and becomes an essentially free electron, capable of moving throughout the crystal. This picture of free valence electrons (in metals these are referred to as *valence* or *conduction* electrons) is drastically different from the valence electrons in ionic and covalent solids, in which the electrons are tightly bound to their atoms; it is, in fact, the primary feature distinguishing metals from these latter crystals.

The model we now have of sodium metal is an assemblage of positive $Na^+$ ions forming a bcc lattice, immersed in a gas of free electrons. The question confronting us now is: Why is the energy of such a system lower than that of free Na atoms? First, it is clear that the $Na^+$ ions would tend to repel each other as a result of the electrostatic coulomb force. But this force, which acts against stability, is largely ineffective, because free electrons strongly screen ions from each other, resulting in essentially neutralized noninteracting ions, much as in the case of free atoms. But the great reduction in energy needed for the bonding can be explained only in quantum terms: It follows from quantum considerations that when a particle is restricted to move in a small volume, it must by necessity have a large kinetic energy. This energy is proportional to $V^{-2/3}$, where $V$ is the volume of confinement (see Section A.3). The origin of this energy is entirely quantum in nature, and is intimately related to the *Heisenberg uncertainty principle*.

We now apply this interesting idea to the case of metals. When the Na atoms are in the gaseous state, their valence electrons have large kinetic energies because they are restricted to move in the very small atomic volumes. But, in the crystalline state, the electrons are free to wander throughout the volume of the crystal, which is very large. This results in a drastic decrease of their kinetic energies, and thus an appreciable diminution in the total energy of the system, which is the source of the metallic bonding. (Figuratively speaking, the free electrons, which are of course negative, act as a glue that holds the positive ions together.)

The metallic bond is somewhat weaker than the ionic and covalent bonds (for instance, the melting point of Na is only 97.8°C), but is still far from being small or negligible.

To account briefly for the other metallic properties listed earlier, we note that the high *electrical conductivity* is due to the ability of the valence electrons to move readily under the influence of an electric field, resulting in a net electrical current in the field direction. A similar explanation may be given for the high *thermal conductivity*. The high *density* is due to the fact that the metallic ions may be packed together tightly, even though the free electrons produce a strong and effective screening between them. The high *ductility* is a consequence of the fact that the metallic bond is nondirectional, so that if an external bending torque is applied and the ions change positions to accommodate this torque, the electrons, being very small and highly mobile, readily adapt themselves to the new deformed situation.

This metallic bonding model works well in the simple metals, particularly the alkalis. More complicated metals—especially the transition elements such as Fe, Ni, etc.—require more complex models, as one would expect. Thus in Fe and Ni the 3d electrons have well-localized properties, and hence they tend to form covalent bonds with their neighbors. This covalent bonding is in addition to the contribution of the 4s valence electrons, which produce a metallic bonding.

**Secondary bonds**

In addition to the three primary bonds discussed above (ionic, covalent, and metallic), there are other, weaker bonds which often play important roles in explaining some of the "fine-scale" bonding properties. For example, the ice crystal ($H_2O$). First, consider the bonding in a single water molecule. A covalent bond is formed between the oxygen atom and each of the two hydrogen atoms (Fig. 1.20a); the electron sharing makes it possible for the oxygen atom to have 8 valence electrons, i.e., a stable shell structure. Thus the atoms in an $H_2O$ molecule are stongly bonded.

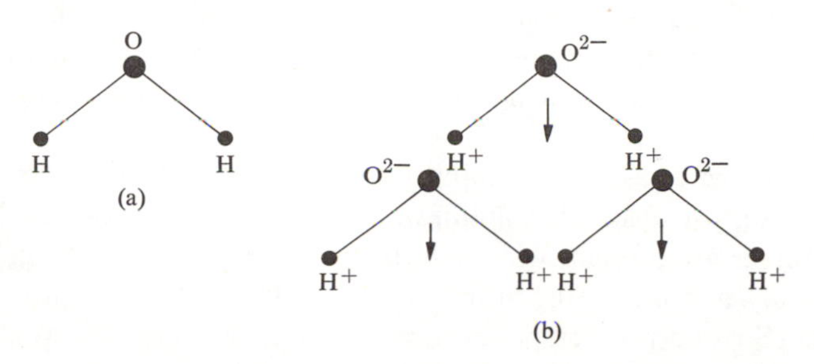


**Fig. 1.20** (a) Water molecule. (b) Arrangement of water molecule as a result of hydrogen bond. Arrows represent electric dipole moments of the molecules.

But when we consider the bonding between the water molecules themselves to form ice, we find that the bonding strength is much weaker, e.g., the melting point of water is only 0°C. The explanation of this is that, although each $H_2O$ molecule is, on the whole, electrically neutral, the distribution of internal charge is such as to produce an interaction between the molecules. Thus in describing the electron sharing in the H-O bond, we should also mention the fact that the electrons are actually pulled more strongly toward the oxygen atom, resulting in a net negative charge on the oxygen atom and a corresponding positive charge on the hydrogen atom (Fig. 1.20b). This produces a so-called *electric dipole* in the water molecule, as indicated by the vector in the figure. Now electric dipoles attract each other. Thus water molecules are attracted to each other, forming a crystal (Fig. 1.20b). (We can also appreciate the dipole attraction on a more elementary level by noting that the negative oxygen atom in one water molecule is attracted toward that corner in another water molecule which contains a positive hydrogen atom.)

The bond described here is referred to as the *hydrogen bond*—sometimes also known as the hydrogen bridge—because of the important role played by the small hydrogen nucleus (which is a proton).

Another bond which plays an especially important role in inert-gas solids is the *van der Waals bond.* You undoubtedly recall from basic chemistry that the inert-gas elements—i.e., those that occur in column VIII of the periodic table (He, Ne, Ar, etc.)—display extremely small attraction toward each other, or other elements. So these elements do not usually participate in chemical reactions (hence the name inert), and they form *monatomic* gases rather than *diatomic* ones such as $H_2$, $O_2$, or other polyatomic gases. The weakness of the interatomic forces in the inert-gas solids is also illustrated by their low melting points: $-272.2$, $-248.7$

and $-189.2°C$ for He, Ne, and Ar, respectively. In other words, He remains in the liquid state down to a temperature of only about one degree from absolute zero!†

If one uses the principles of quantum theory, it is not difficult to explain the weakness of interatomic forces in the inert gases. In each of these gases, the atom has an outer shell that is completely full. Consequently an atom has very little predilection to exchange or share electrons with other atoms. This rules out any ionic and covalent forces, and likewise rules out any metallic-bonding forces in inert-gas crystals.

Yet even the inert-gas atoms exhibit interatomic forces, albeit very weak ones. The fact that Ne, for instance, solidifies at $-248.7°C$ indicates clearly that some interatomic forces are present, which are responsible for the freezing; by contrast, a system of truly noninteracting atoms would remain gaseous down to the lowest temperature. So our problem is not so much to explain the weakness of the forces, but rather to account for their presence in the first place.

Without becoming embroiled in physical and mathematical complexities, we may present the following model for the attraction in inert-gas elements. Consider two such atoms. Each contains a number of orbital electrons, which are in a continuous state of rotation around the nucleus. If their motion were such that their charge was always symmetric around the nucleus, then the effect would be to screen the nucleus completely from an adjacent atom, and the two atoms would not interact. This supposition is not quite correct, however. Although the distribution of the electrons is essentially symmetric, and is certainly so on the average, as time passes there are small fluctuations, whose effect is to produce a fluctuating electric dipole on each of the atoms. The dipoles tend to attract each other (as mentioned in connection with the hydrogen bond), and this is the source of the van der Waals force. The resulting potential is found to decrease with distance as $1/R^6$, far more rapidly than the ionic potential, which decreases only as $1/R$.

Two reasons may be given to account for the smallness of the van der Waals force (also known as the *London* force): (a) The fluctuating atomic dipoles are small, and (b) the dipoles on the different atoms are not synchronized with each other, a fact which tends to cancel their attractive effects. Following the various steps in detail, however, one arrives at a net attractive force.

## REFERENCES

### Crystals

L. V. Azaroff, 1960, *Introduction to Solids*, New York: McGraw-Hill

C. S. Barrett and T. Massalski, 1966, *Structure of Metals*, third edition, New York: McGraw-Hill

† Actually He solidifies only if an external pressure of at least 26 atms is applied to it.

F. C. Phillips, 1963, *An Introduction to Crystallography*, third edition, New York: John Wiley

G. Weinreich, 1965, *Solids*, New York: John Wiley

J. Wulff *et al.*, 1963, *Structures and Properties of Materials*, Vol. I, Cambridge, Mass.: MIT Press

R. W. G. Wykoff, 1963, *Crystal Structures*, second edition, New York: John Wiley/Interscience

### Interatomic bonds

A. Holden, 1970, *Bonds Between Atoms*, Oxford: Oxford University Press

W. Hume-Rothery, 1955, *Atomic Theory for Students in Metallurgy*, London: Institute of Metals

Linus Pauling, 1964, *General Chemistry*, San Francisco: Freeman

Linus Pauling, 1948, *The Nature of the Chemical Bond*, Ithaca, N.Y.: Cornell University Press

B. L. Smith, "Inert Gas Crystals," *Contem. Phys.* **11**, 125, 1970

J. Wulff *et al.*, 1963, *Structures and Properties of Materials*, Vol. I, Cambridge, Mass.: MIT Press

## QUESTIONS

1. What is the reason for the fact that the tetrahedral bond is the dominant bond in carbon compounds?
2. Estimate the strength of the hydrogen bond in water (in electron volts per bond).
3. Show that two parallel electric dipoles attract each other.
4. Estimate the strength of the van der Waals bond for neon.

## PROBLEMS

1. Given that the primitive basis vectors of a lattice are $\mathbf{a} = (a/2)(\mathbf{i} + \mathbf{j})$, $\mathbf{b} = (a/2)(\mathbf{j} + \mathbf{k})$ and $\mathbf{c} = (a/2)(\mathbf{k} + \mathbf{i})$, where **i**, **j**, and **k** are the usual three unit vectors along cartesian coordinates, what is the Bravais lattice?
2. Using Table 1.2 and the data below, calculate the densities of the following solids: Al, Fe, Zn, and Si, whose atomic weights are respectively 26.98, 55.85, 65.37, and 28.09.
3. Show that in an ideal hexagonal-close-packed (hcp) structure, where the atomic spheres touch each other, the ratio $c/a$ is given by

$$\frac{c}{a} = \left(\frac{8}{3}\right)^{1/2} = 1.633.$$

(The hcp structure is discussed in Section 7.)

4. The *packing ratio* is defined as the fraction of the total volume of the cell that is filled by atoms. Determine the maximum values of this ratio for equal spheres located at the points of simple-cubic, body-centered-cubic, and face-centered-cubic crystals.

5. Repeat Problem 4 for simple hexagonal, and rhombohedral lattices.
6. Repeat Problem 4 for an hcp structure.
7. Consider a face-centered-cubic cell. Construct a primitive cell within this larger cell, and compare the two. How many atoms are in the primitive cell, and how does this compare with the number in the original cell?
8. a) Show that a two-dimensional lattice may not possess a 5-fold symmetry.
   b) Establish the fact that the number of two-dimensional Bravais lattices is five: Oblique, square, hexagonal, simple rectangular, and body-centered rectangular. (The proof is given in Kittel, 1970.)
9. Demonstrate the fact that if an object has two reflection planes intersecting at $\pi/4$, it also possesses a 4-fold axis lying at their intersection.
10. Sketch the following planes and directions in a cubic unit cell: (122), [122], $(1\bar{1}2)$, $[1\bar{1}2]$.
11. a) Determine which planes in an fcc structure have the highest density of atoms.
    b) Evaluate this density in atoms/cm$^2$ for Cu.
12. Repeat Problem 11 for Fe, which has a bcc structure.
13. Show that the maximum packing ratio in the diamond structure is $\pi\sqrt{3}/16$. [*Hint*: The structure may be viewed as two interpenetrating fcc lattices, arranged such that each atom is surrounded by four other atoms, forming a regular tetrahedron.]
14. A quantitative theory of bonding in ionic crystals was developed by Born and Meyer along the following lines: The total potential energy of the system is taken to be

$$E = N\frac{A}{R^n} - N\frac{\alpha e^2}{4\pi\epsilon_0 R},$$

where $N$ is the number of positive–negative ion pairs. The first term on the right represents the repulsive potential, where $A$ and $n$ are constants determined from experiments. The second term represents the attractive coulomb potential, where $\alpha$, known as the *Madelung constant*, depends only on the crystal structure of the solid.

a) Show that the equilibrium interatomic distance is given by the expression

$$R_0^{n-1} = \frac{4\pi\epsilon_0 A}{\alpha e^2} n.$$

b) Establish that the bonding energy at equilibrium is

$$E_0 = -\frac{\alpha N e^2}{4\pi\epsilon_0 R_0}\left(1 - \frac{1}{n}\right).$$

c) Calculate the constant $n$ for NaCl, using the data in Table 1.2 and the fact that the measured binding energy for this crystal is 1.83 kcal/mole (or 7.95 eV/molecule). The constant $\alpha$ for NaCl is 1.75.

# CHAPTER 2 X-RAY, NEUTRON, AND ELECTRON DIFFRACTION IN CRYSTALS



---

*All things visible and invisible.*

The Book of Common Prayer

## 2.1 INTRODUCTION

In this chapter we shall discuss the determination of crystal structures. One can determine the structure of a crystal by studying the diffraction pattern of a beam of radiation incident on the crystal. Beam diffraction takes place only in certain specific directions, much as light is diffracted by a grating. By measuring the directions of the diffraction and the corresponding intensities, one obtains information concerning the crystal structure responsible for the diffraction.

Three types of radiation are used: x-rays, neutrons, and electrons. The treatment of these three types is quite similar; therefore we shall examine in detail only the x-ray case. After a brief discussion of the generation and absorption of x-rays, we shall give a simple derivation of Bragg's law. We shall then proceed to show that this law follows also from a more sophisticated treatment utilizing the concepts of scattering theory. In this connection we shall discuss the reciprocal lattice, and also experimental aspects of the determination of crystal structure by x-rays. We shall then talk about neutron and electron diffractions along the same lines, and point out their advantages.

Determining the structure of a liquid is discussed in Section 2.8, in which it is shown how one can obtain the pair distribution by measuring the so-called liquid structure factor.

Some of the concepts presented here, particularly diffraction and the reciprocal lattice, will be found useful in the discussion of lattice vibrations in Chapter 3, and of electron states in a crystal, in Chapter 5.

## 2.2 GENERATION AND ABSORPTION OF X-RAYS

X-rays are electromagnetic waves whose wavelengths are in the neighborhood of 1 Å. Except for the fact that their wavelength is so short, they have the same physical properties as other electromagnetic waves, such as optical waves. The wavelength of an x-ray is thus of the same order of magnitude as the lattice constants of crystals, and it is this which makes x-rays useful in the analysis of crystal structures. The energy of an x-ray photon is given by the Einstein relation $E = h\nu$, where $h$ is Planck's constant and $\nu$ is the frequency (Section A.1). Substituting $h = 6.6 \times 10^{-27}$ erg-s and $\lambda = 1$ Å (recall that $\nu = c/\lambda$), one finds an energy $E \simeq 10^4$ eV, which is a typical value.

The basic experimental arrangement for generating an x-ray beam is sketched in Fig. 2.1. Electrons emitted from the cathode of a vacuum tube are accelerated by a large potential acting across the tube. The electrons thus acquire high kinetic energy, and when they impinge on a metallic target, forming the anode at the end of the tube, bursts of x-rays are emitted from the target. Some of the x-ray radiation is then extracted from the tube and used for the intended purpose. The emitted radiation has a wide continuous spectrum, on which is superimposed a series of discrete lines. The continuous spectrum is due to emission of radiation by the incident electrons as they are deflected by the nuclear charges in the target, while the discrete lines are due to the emission by atoms in the target after they are excited by the incident electrons. The maximum frequency of the continuous spectrum $\nu_0$ is related to the accelerating potential by $eV = h\nu_0$, since the maximum energy of a photon cannot exceed the kinetic energy of the incident electron. The corresponding wavelength $\lambda_0$ is given by

$$\lambda_0 = \frac{12.3}{V} \text{ Å}, \tag{2.1}$$

where $V$ is in kilovolts.

When an x-ray beam passes through a material medium it is partially absorbed. The intensity of the beam is attenuated according to the relation

$$I = I_0 e^{-\alpha x}, \tag{2.2}$$

where $I_0$ is the initial intensity at the surface of the medium and $x$ the distance traveled. The parameter $\alpha$ is known as the *absorption coefficient*. The attenuation of the intensity expressed by (2.2) is due to the scattering and absorption of the beam by the atoms of the medium.

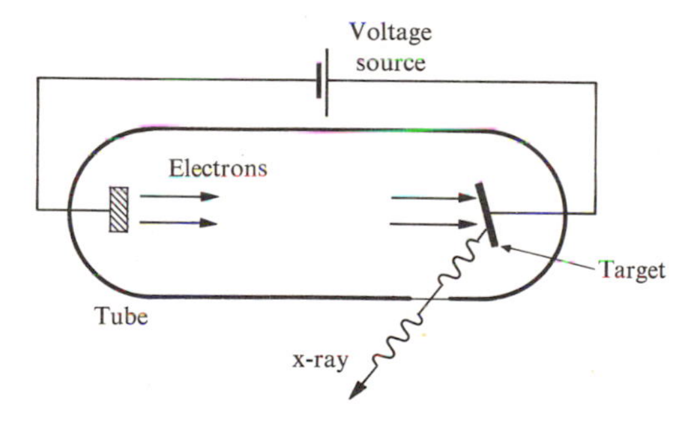


**Fig. 2.1** Generation of x-rays.

## 2.3 BRAGG'S LAW

When a monochromatic x-ray beam is incident on the surface of a crystal, it is reflected. However, the reflection takes place only when the angle of incidence has certain values. These values depend on the wavelength and the lattice constants of the crystal, and consequently it seems reasonable to attempt to explain the selective reflectivity in terms of interference effects, as in physical optics. The model is illustrated in Fig. 2.2(a), where the crystal is represented by a set of parallel planes, corresponding to the atomic planes. The incident beam is reflected partially at each of these planes, which act as mirrors, and the reflected rays are then collected simultaneously at a distant detector. The reflected rays interfere at the detector and, according to physical optics, the interference is constructive only if the

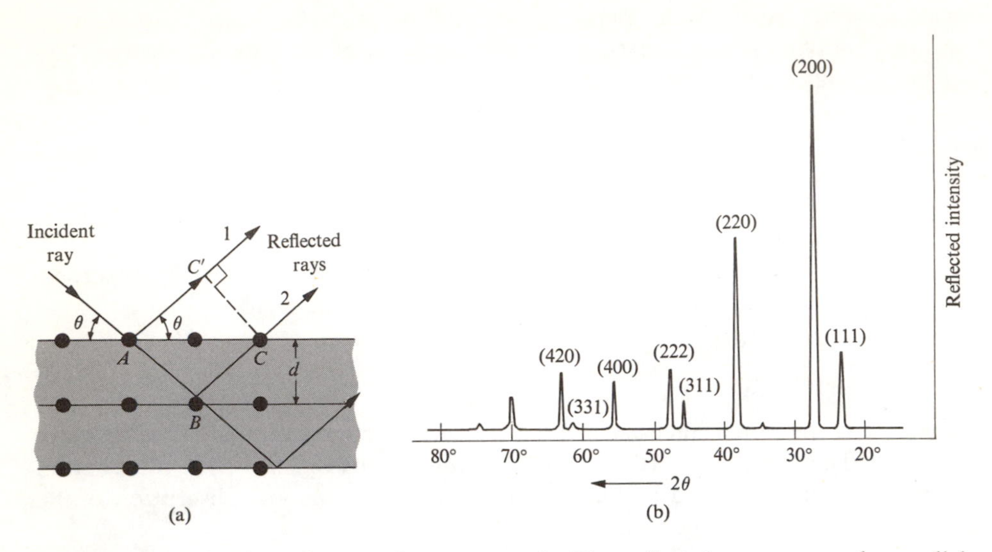


**Fig. 2.2** (a) Reflection of x-rays from a crystal. The reflected rays are nearly parallel because the detector is positioned far from the crystal. (b) Reflected intensity from a KBr crystal. The reflecting planes for the various peaks are indicated.

difference between the paths of any two consecutive rays is an integral multiple of the wavelength. That is,

$$\text{Path difference} = n\lambda, \qquad n = 1, 2, 3, \ldots, \tag{2.3}$$

where $\lambda$ is the wavelength and $n$ a positive integer. The path difference $\Delta$ between rays 1 and 2 in the figure is

$$\Delta = \overline{AB} + \overline{BC} - \overline{AC'} = 2\overline{AB} - \overline{AC'}.$$

In equating $\overline{AB}$ and $\overline{BC}$, we have assumed that the reflection is *specular*, i.e., that the angles of incidence equal the angles of reflection. When the interplanar distance is denoted by $d$, it follows from the figure that

$$\overline{AB} = d/\sin\theta \qquad \text{and} \qquad \overline{AC'} = \overline{AC}\cos\theta = (2d/\tan\theta) \times \cos\theta,$$

where $\theta$ is the *glancing angle* between the incident beam and the reflecting planes. Substituting these into (2.3) and performing some trigonometric manipulation, we arrive at the following condition for constructive interference:

$$2d\sin\theta = n\lambda. \tag{2.4}$$

This is the celebrated *Bragg's law*. The angles determined by (2.4), for a given $d$ and $\lambda$, are the only angles at which reflection takes place. At other angles the reflected rays interfere with each other destructively, and consequently the reflected beam disappears, i.e., the incident beam passes through the crystal undisturbed. The reflections corresponding to $n = 1, 2$, etc., are referred to as first order, second order, etc., respectively. The intensity of the reflected beam decreases as the order increases.† It is actually more appropriate to think of the reflection taking place here as a diffraction, as the concept of interference is an essential part of the process.

The basic idea underlying the use of Bragg's law in studying crystal structures is readily apparent from (2.4). Since $\lambda$ can be determined independently, and since $\theta$ can be measured directly from the reflection experiment (it is half the angle between the incident and diffracted beams, as shown in the figure), one may employ (2.4) to calculate the interplanar distance $d$. Note that, according to (2.4), diffraction is possible only if $\lambda < 2d$, which shows why optical waves, for example, cannot be used here. Note also that if the crystal is rotated, a new diffracted beam may appear corresponding to a new set of planes. Figure 2.2(b) shows the Bragg reflection from KBr.

The model we have used in arriving at Bragg's law is oversimplified. In view of the fact that the scattering of the x-ray beam is caused by the discrete atoms themselves, one may object to representing the atomic planes by a set of continuous reflecting mirrors. The proper treatment should consider the diffracted beam to be due to the interference of partial rays scattered by all atoms in the lattice. That is, one should treat the lattice as a three-dimensional diffraction grating. In adding the contributions of the partial rays, one must pay particular attention to the phases of these rays, as in the optical analog. This program, which is developed in the following sections, leads us back to Bragg's law, but we shall gain a much deeper appreciation of the diffraction process along the way.

## 2.4 SCATTERING FROM AN ATOM

The diffraction process can be divided naturally into two stages: (1) scattering by individual atoms, and (2) mutual interference between the scattered rays. Since the two stages are distinct from each other, we shall treat them independently, for convenience.

Why does an atom scatter the x-ray beam? Well, any atom is surrounded by electrons which undergo acceleration under the action of the electric field associated with the beam. Since an accelerated charge emits radiation (a fact well known from electromagnetism), so do the atomic electrons. In effect the electrons absorb energy from the beam, and scatter it in all directions. But the electrons form a charge cloud surrounding the atom, so when we are considering scattering from the atom as a whole, we must take into account the phase differences between the rays scattered from the different regions of the charge cloud. We do this as follows:

---

† In the remainder of this chapter and in the problem section, we shall consider only first-order reflections.

Consider a single electron, as shown in Fig. 2.3(a). A plane-wave field given by

$$u = A\, e^{i(\mathbf{k}_0 \cdot \mathbf{r} - \omega t)} \tag{2.5}$$

is incident on the electron, where $A$ is the amplitude, $\mathbf{k}_0$ the wave vector ($k_0 = 2\pi/\lambda$), and $\omega$ the angular frequency. The scattered field is an outgoing spherical wave represented by

$$u' = f_e \frac{A}{D} e^{i(kD - \omega t)}, \tag{2.6}$$

where $f_e$ is a parameter known as the *scattering length* of the electron, and $D$ is the radial distance from the electron to the point at which the field is evaluated. The quantity $k$ is the wave number of the scattered wave, and has the same magnitude as $k_0$. Note that the amplitude of the scattered wave decreases with distance as $1/D$, a property shared by all spherical waves.

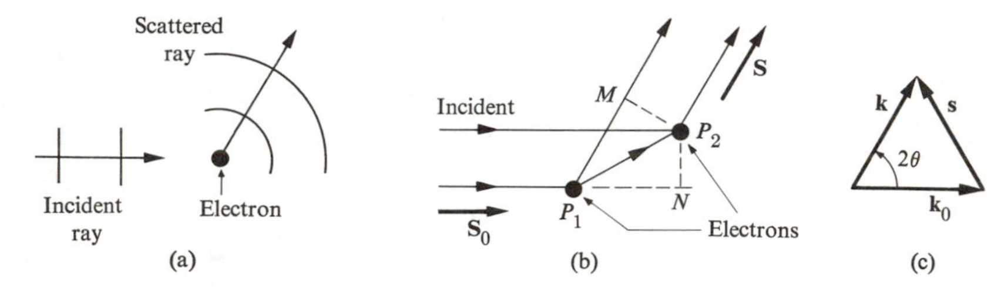


**Fig. 2.3** Scattering from (a) a single electron, (b) two electrons. (c) The scattering vector $\mathbf{s}$. Note that the vectors $\mathbf{k}_0$, $\mathbf{k}$, and $\mathbf{s}$ form an isosceles triangle.

Suppose now that the incident wave acts on two electrons, as in Fig. 2.3(b). In this case, both electrons emit spherical waves, and the scattered field observed at a distant point is the sum of the two partial fields, where their phase difference has to be taken into account. Thus we have

$$u' = f_e \frac{A}{D} \left[ e^{ikD} + e^{i(kD + \delta)} \right] \tag{2.7}$$

where $\delta$ is the phase lag of the wave from electron 1 behind that of electron 2.† (The time factor has been omitted for the sake of brevity, but its presence is implied.) Referring to the figure, we may write

$$\delta = (\overline{P_1 N} - \overline{P_1 M}) 2\pi/\lambda = (\mathbf{r} \cdot \mathbf{S} - \mathbf{r} \cdot \mathbf{S}_0) k,$$

† The distance $D$ to the field point is assumed to be large, otherwise the denominator $D$ in (2.7) would not be the same for the two electrons. This condition simplifies the calculations, and is the reason why the detector is usually placed far from the crystal.

where $\mathbf{r}$ is the vector radius of electron 2 relative to electron 1, and $\mathbf{S}_0$ and $\mathbf{S}$ are the unit vectors in the incident and scattered directions, respectively. The expression for $\delta$ can be set forth in the form

$$\delta = \mathbf{s} \cdot \mathbf{r}, \tag{2.8}$$

where the *scattering vector* $\mathbf{s}$ is defined as

$$\mathbf{s} = k(\mathbf{S} - \mathbf{S}_0) = \mathbf{k} - \mathbf{k}_0. \tag{2.9a}$$

As seen from Fig. 2.3(c), the magnitude of the scattering vector is given by

$$s = 2k \sin \theta, \tag{2.9b}$$

where $\theta$ is half of the scattering angle. Substituting the expression (2.8) for $\delta$ into (2.7), one finds

$$u' = f_e \frac{A}{D} e^{ikD} \left[1 + e^{i\mathbf{s}\cdot\mathbf{r}}\right]. \tag{2.10}$$

In deriving this we have chosen the origin of our coordinates at electron 1. But it is now more convenient to choose the origin at an arbitrary point, and in this manner treat the two electrons on an equal footing. The ensuing expression for the scattered field is then

$$u' = f_e \frac{A}{D} e^{ikD} \left[e^{i\mathbf{s}\cdot\mathbf{r}_1} + e^{i\mathbf{s}\cdot\mathbf{r}_2}\right], \tag{2.11}$$

where $\mathbf{r}_1$ and $\mathbf{r}_2$ are the position vectors of the two electrons relative to the new origin. Equation (2.10) is a special case of (2.11), where $\mathbf{r}_1 = 0$, that is, where the origin is chosen at electron 1, as pointed out above. The generalization of (2.11) to an arbitrary number of scatterers is now immediate, and the result is

$$u' = f_e \frac{A}{D} e^{ikD} \sum_l e^{i\mathbf{s}\cdot\mathbf{r}_l}, \tag{2.12}$$

where $\mathbf{r}_l$ is the position of the $l$th electron, and the sum is carried out over all the electrons. By analogy with the case of the single electron, Eq. (2.6), the scattering length for the system as a whole is now given by the sum

$$f = f_e \sum_l e^{i\mathbf{s}\cdot\mathbf{r}_l}. \tag{2.13}$$

That is, the total scattering length is the sum of individual lengths with the phases taken properly into account. The intensity $I$ of the scattered beam is proportional to the square of the magnitude of the field, and therefore

$$I \sim |f|^2 = f_e^2 |\sum e^{i\mathbf{s}\cdot\mathbf{r}}|^2. \tag{2.14}$$

Results (2.13) and (2.14) are the basic equations in the treatment of scattering and diffraction processes, and we shall use them time and again in the following pages.

We may digress briefly to point out an important aspect of the scattering process: the *coherence* property involved in the scattering. This property means that the scatterers maintain definite phase relationships with each other. Consequently we can speak of interference between the partial rays. By contrast, if the scatterers were to oscillate randomly, or incoherently, the partial rays would not interfere, and the intensity at the detector would be simply the sum of the partial intensities, that is,

$$I \sim N f_e^2, \tag{2.15}$$

where $N$ is the number of scatterers. Note the marked difference between this result and that of coherent scattering in (2.14).

The scattering length of the electron is well known, and can be found in books on electromagnetism. Its value is

$$f_e = [(1 + \cos^2 2\theta)/2]^{1/2} r_e,$$

where $r_e$, the so-called classical radius of the electron, has a value of about $10^{-15}$ m.†

We can now apply these results to the case of a single free atom. In attempting to apply (2.13), where the sum over the electrons appears, we note that the electrons do not have discrete positions, but are spread as a continuous charge cloud over the volume of the atom. It is therefore necessary to convert the discrete sum to the corresponding integral. This readily leads to

$$f_e \sum_l e^{i\mathbf{s}\cdot\mathbf{r}_l} \rightarrow f_e \int \rho(\mathbf{r})\, e^{i\mathbf{s}\cdot\mathbf{r}} d^3r,$$

where $\rho(\mathbf{r})$ is the density of the cloud (in electrons per unit volume), and the integral is over the atomic volume. The *atomic scattering factor* $f_a$ is defined as the integral appearing in the above expression, i.e.,

$$f_a = \int d^3\rho\,(\mathbf{r})\, e^{i\mathbf{s}\cdot\mathbf{r}}. \tag{2.16}$$

(Note that $f_a$ is a dimensionless quantity.) The integral can be simplified when the density $\rho(\mathbf{r})$ is spherically symmetric about the nucleus, because then the integration over the angular part of the element of volume can be readily per-

† For the sake of visual thinking, consider the electron to be in the form of a sphere whose radius is roughly equal to the scattering length $f_e$. Thus the electron "appears" to the radius as a circular obstacle of cross section $\pi f_e^2$.

formed (see the problems at the end of this chapter). The resulting expression is

$$f_a = \int_0^R 4\pi r^2 \rho(r) \frac{\sin sr}{sr}\, dr, \tag{2.17}$$

where $R$ is the radius of the atom (the nucleus being located at the origin). As seen from (2.17), the scattering factor $f_a$ depends on the scattering angle (recall that $s = 2k \sin \theta$), and this comes about from the presence of the oscillating factor $(\sin sr)/sr$ in the integrand. The wavelength of oscillation is inversely proportional to $s$ in Fig. 2.4(a), and the faster the oscillation—i.e., the shorter the wavelength—the smaller is $f_a$, due to the interference between the partial beams scattered by different regions of the charge cloud. Recalling that $s = 2k \sin \theta$, Eq. (2.9), we see that as the scattering angle $2\theta$ increases, so also does $s$, and this results in a decreasing scattering factor $f_a$.

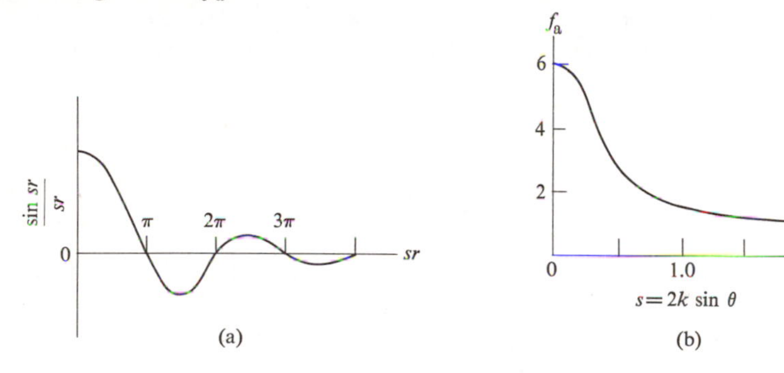


**Fig. 2.4** (a) Oscillating factor $\sin(sr)/sr$. (b) Atomic scattering factor for a carbon atom as a function of the scattering angle (after Woolfson).

To proceed further with the evaluation of $f_a$, we need to know the electron density $\rho(\mathbf{r})$ for the atom in question. For this information we have to turn to the literature on atomic physics. Figure 2.4(b) shows the scattering factor for carbon.

There is one special direction for which $f_a$ can be evaluated at once, namely the forward direction. In this case, $\theta = 0$, $s = 0$, and hence the oscillating factor $(\sin sr)/sr$ reduces to unity (recall that $\sin 0/0 = 1$). Equation (2.17) then becomes

$$f_a = \int_0^R 4\pi r^2 \rho(r)\, dr,$$

and the integral is simply equal to the total number of electrons in the atom, i.e., the atomic number Z. We may therefore write

$$f_a\,(\theta = 0) = Z. \tag{2.18}$$

Thus for carbon $f_a\,(\theta = 0) = 6$ in agreement with Fig. 2.4(b). The physical interpretation of (2.18) is quite apparent: When one looks in the forward direction all the partial rays are in phase, and hence they interfere constructively.

## 2.5 SCATTERING FROM A CRYSTAL

Our primary aim in this chapter is, of course, to investigate the scattering from a crystal, and we shall now proceed to apply Eq. (2.13) to this situation. By analogy with the atomic case, we define the *crystal scattering factor* $f_{cr}$ as

$$f_{cr} = \sum_l e^{i\mathbf{s}\cdot\mathbf{r}_l}, \tag{2.19}$$

where the sum here extends over all the electrons in the crystal. To make use of the atomic scattering factor discussed in the previous section, we may split the sum (2.19) into two parts: First we sum over all the electrons in a single atom, and then sum over all the atoms in the lattice. The double summation then amounts to the sum over all the electrons in the crystal, as required by (2.19). Since the first of the above sums leads to the atomic scattering factor, Eq. (2.19) may thus be written in the form

$$f_{cr} = \sum_l f_{al}\, e^{i\mathbf{s}\cdot\mathbf{R}_l}, \tag{2.20}$$

where $\mathbf{R}_l$ is the position of the $l$th atom, and $f_{al}$ the corresponding atomic factor.

It is now convenient to rewrite (2.20) as a product of two factors, one involving a sum over the unit cell, and the other the sum over all unit cells in the crystal. Thus we define the *geometrical structure factor* $F$ as

$$F = \sum_j f_{aj}\, e^{i\mathbf{s}\cdot\boldsymbol{\delta}_j}, \tag{2.21}$$

where the summation is over all the atoms in the unit cell, and $\boldsymbol{\delta}_j$ is the relative position of the $j$th atom. Similarly we define the *lattice structure factors* as

$$S = \sum_l e^{i\mathbf{s}\cdot\mathbf{R}_l^{(c)}}, \tag{2.22}$$

where the sum extends over all the unit cells in the crystal, and $\mathbf{R}_l^{(c)}$ is the position of the $l$th cell. To express $f_{cr}$ in terms of $F$ and $S$, we return to (2.20), write $\mathbf{R}_l = \mathbf{R}_l^{(c)} + \boldsymbol{\delta}_j$, and then use (2.21) and (2.22). The result is evidently

$$f_{cr} = FS. \tag{2.23}$$

Note that the lattice factor $S$ depends only on the crystal system involved, while $F$ depends on the geometrical shape as well as the contents of the unit cell. In the special case of a simple lattice, where the unit cell contains a single atom, the factor $F$ becomes equal to $f_a$. The factorization of $f_{cr}$ as shown in

(2.23) merits some emphasis: We have separated the purely structural properties of the lattice, which are contained in $S$, from the atomic properties contained in $F$. Great simplification is achieved thereby, because the two factors may now be treated independently. Since the factor $F$ involves a sum over only a few atomic factors, it can be easily evaluated in terms of the atomic factors, as discussed in the previous section. We shall therefore not concern ourselves with this straightforward task for the moment, but press on and consider the evaluation of the lattice factor $S$.

### The lattice structure factor

The lattice structure factor $S$, defined in (2.22), is of vital importance in the discussion of x-ray scattering. Let us now investigate its dependence on the scattering vector $\mathbf{s}$, and show that the values of $\mathbf{s}$ for which $S$ does not vanish form a discrete set, which is found to be related to Bragg's law.

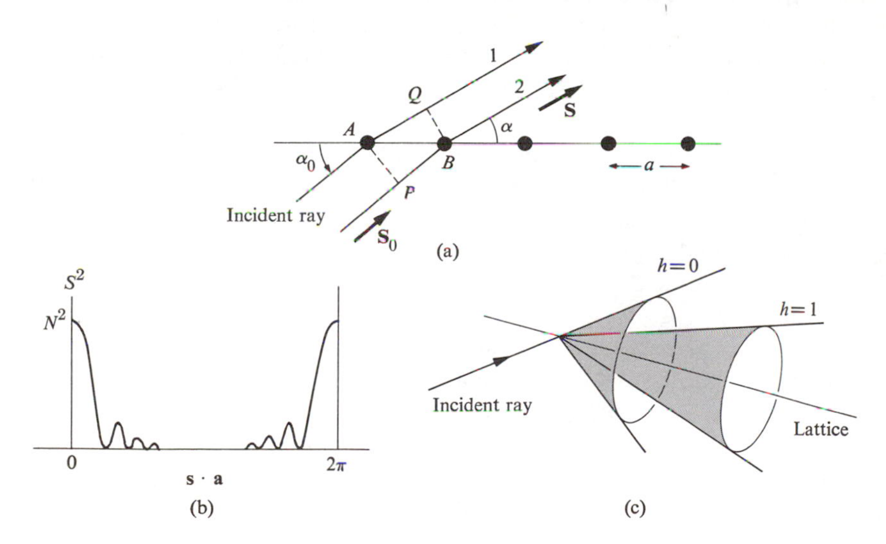


**Fig. 2.5** (a) Scattering from a one-dimensional lattice. (b) Diffraction maxima. (c) Diffraction cones for first order ($h = 0$) and second order ($h = 1$) maxima.

We start with the simplest possible situation, an x-ray beam scattered from a one-dimensional monatomic lattice, as illustrated in Fig. 2.5(a). When we denote the basis vector of the lattice by $\mathbf{a}$, the structure factor becomes

$$S = \sum_{l=1}^{N} e^{i\mathbf{s}\cdot l\mathbf{a}}, \tag{2.24}$$

where we have substituted $\mathbf{R}_l^{(c)} = l\mathbf{a}$, and $N$ is the total number of atoms. The series in (2.24) is a geometric progression, the common ratio being $e^{i\mathbf{s}\cdot\mathbf{a}}$, and can readily be evaluated. The result is

$$S = \frac{\sin[(\frac{1}{2})N\mathbf{s}\cdot\mathbf{a})]}{\sin[(\frac{1}{2})\mathbf{s}\cdot\mathbf{a})]}. \tag{2.25}$$

Physically, it is more meaningful to examine $S^2$ than $S$, since this is the quantity which enters directly into calculations of intensity. It is given by

$$S^2 = \frac{\sin^2[(\frac{1}{2})N\mathbf{s}\cdot\mathbf{a}]}{\sin^2[(\frac{1}{2})\mathbf{s}\cdot\mathbf{a}]}. \tag{2.26}$$

We now wish to see how this function depends on the scattering vector **s**. As we see from (2.26), $S^2$ is the ratio of two oscillating functions having a common period $\mathbf{s}\cdot\mathbf{a} = 2\pi$, but, because $N$ is much larger than unity in any practical case, the numerator oscillates far more rapidly than the denominator. Note, however, that for the particular value $\mathbf{s}\cdot\mathbf{a} = 0$, both the numerator and denominator vanish simultaneously, but the limiting value of $S^2$ is equal to $N$, a very large number. Similarly the value of $S^2$ at $\mathbf{s}\cdot\mathbf{a} = 2\pi$ is equal to $N^2$, as follows from the periodicity of $S^2$, mentioned above. The function $S^2$ is sketched versus $\mathbf{s}\cdot\mathbf{a}$ in Fig. 2.5(b), for the range $0 < \mathbf{s}\cdot\mathbf{a} \leqslant 2\pi$. It has two primary maxima, at $\mathbf{s}\cdot\mathbf{a} = 0$ and $\mathbf{s}\cdot\mathbf{a} = 2\pi$, separated by a large number of intervening subsidiary maxima, the latter resulting from the rapid oscillations of the numerator in (2.26). Calculations (see the problem section) show that when the number of cells is very large, as it is in actual cases, these subsidiary maxima are negligible compared with the primary ones. For instance, the peak of the highest subsidiary maximum is only 0.04 that of a primary maximum. It is therefore a good approximation to ignore all the subsidiary maxima, and take the function $S^2$ to be nonvanishing only in the immediate neighborhoods of the primary maxima. Furthermore, it can also be demonstrated that the width of each primary maximum decreases rapidly as $N$ increases, and that this width vanishes in the limit as $N \to \infty$. Therefore $S^2$ is nonvanishing only at the values given exactly by $\mathbf{s}\cdot\mathbf{a} = 0, 2\pi$. But because $S^2$ is periodic, with a period of $2\pi$, it is also finite at all the values

$$\mathbf{s}\cdot\mathbf{a} = 2\pi h, \qquad h = \text{any integer}. \tag{2.27}$$

At these values $S^2$ is equal to $N^2$, and hence $S = N$.

Equation (2.26) determines all the directions in which $S$ has a nonzero value, and hence the directions in which diffraction takes place. The physical interpretation of this equation is straightforward. Recalling the definition of **s**, Eq. (2.9), and referring to Fig. 2.5, we obtain

$$\mathbf{s}\cdot\mathbf{a} = \frac{2\pi}{\lambda}(S - S_0)\cdot\mathbf{a} = \frac{2\pi}{\lambda}(\overline{AQ} - \overline{PB}),$$

which is the phase difference between the two consecutive scattered rays. Thus

Eq. (2.27) is the condition for constructive interference, i.e., the lattice scattering factor survives only in these directions, which is hardly surprising.

For a given $h$, the condition (2.27) does not actually determine a single direction, but rather an infinite number of directions forming a cone whose axis lies along the lattice line. To see this, we can write (2.27) as

$$\frac{2\pi a}{\lambda}(\cos\alpha - \cos\alpha_0) = 2\pi h, \tag{2.28}$$

where $\alpha_0$ is the angle between the incident beam and the lattice line and $\alpha$ is the corresponding angle for the diffracted beam. Thus for a given $h$ and $\alpha_0$, the beam diffracts along all directions for which $\alpha$ satisfies (2.28). These form a cone whose axis lies along the lattice, and whose half angle is equal to $\alpha$. The case $h = 0$ is a special one; its cone includes the direction of forward scattering. Diffraction cones corresponding to several values of $h$ are shown in Fig. 2.5(c).

In treating the lattice-structure factor, we have so far confined ourselves to the case of a one-dimensional lattice. Now let us extend the treatment to the real situation of a three-dimensional lattice. Referring to (2.22) and substituting for the lattice vector,

$$\mathbf{R}^{(c)} = l_1\mathbf{a} + l_2\mathbf{b} + l_3\mathbf{c},$$

where **a**, **b**, and **c** are the basis vectors, we find for the structure factor

$$S = \sum_{l_1,l_2,l_3} e^{i\mathbf{s}\cdot(l_1\mathbf{a}+l_2\mathbf{b}+l_3\mathbf{c})}, \tag{2.29}$$

where the triple summation extends over all the unit cells in the crystal. We can separate this sum into three partial sums,

$$S = \left(\sum_{l_1} e^{i\mathbf{s}\cdot l_1\mathbf{a}}\right)\left(\sum_{l_2} e^{i\mathbf{s}\cdot l_2\mathbf{b}}\right)\left(\sum_{l_3} e^{i\mathbf{s}\cdot l_3\mathbf{c}}\right), \tag{2.30}$$

and in this manner we factor out $S$ into a product of one-dimensional factors, and we can therefore use the results we developed earlier. The condition for constructive interference now is that each of the three factors must be finite individually, and this means that **s** must satisfy the following three equations simultaneously:

$$\begin{aligned} \mathbf{s}\cdot\mathbf{a} &= h2\pi \\ \mathbf{s}\cdot\mathbf{b} &= k2\pi \\ \mathbf{s}\cdot\mathbf{c} &= l2\pi \end{aligned} \tag{2.31}$$

where $h$, $k$, and $l$ are any set of integers. Written in terms of the angles made by **s** with the basis vectors, in analogy with (2.27), these equations become respectively

$$a(\cos \alpha - \cos \alpha_0) = h\lambda$$
$$b(\cos \beta - \cos \beta_0) = k\lambda \qquad (2.32)$$
$$c(\cos \gamma - \cos \gamma_0) = l\lambda$$

where $\alpha_0$, $\beta_0$, and $\gamma_0$ are the angles which the incident beam makes with the basis vectors, while $\alpha$, $\beta$, and $\gamma$ are the corresponding angles for the diffracted beam. Equations (2.31) and (2.32) are known as the *Laue equations*, after the physicist who first derived them.

The question is how to determine the values of the scattering vector **s** which satisfy the diffraction condition (2.31). We shall show in the next section that these values form a discrete set which corresponds to Bragg's law.

## 2.6 THE RECIPROCAL LATTICE AND X-RAY DIFFRACTION

Starting with a lattice whose basis vectors are **a**, **b**, and **c**, we can define a new set of basis vectors **a***, **b***, and **c*** according to the relations

$$\mathbf{a}^* = \frac{2\pi}{\Omega_c}(\mathbf{b} \times \mathbf{c}), \qquad \mathbf{b}^* = \frac{2\pi}{\Omega_c}(\mathbf{c} \times \mathbf{a}), \qquad \text{and} \quad \mathbf{c}^* = \frac{2\pi}{\Omega_c}(\mathbf{a} \times \mathbf{b}), \qquad (2.33)$$

where $\Omega_c = \mathbf{a} \cdot (\mathbf{b} \times \mathbf{c})$, the volume of a unit cell. We can now use the vectors **a***, **b***, and **c*** as a basis for a new lattice whose vectors are given by

$$\mathbf{G}_n = n_1\mathbf{a}^* + n_2\mathbf{b}^* + n_3\mathbf{c}^*, \qquad (2.34)$$

where $n_1$, $n_2$, and $n_3$ are any set of integers. The lattice we have just defined is known as the *reciprocal lattice*, and **a***, **b***, and **c*** are called the *reciprocal basis vectors*.†

The relation of the reciprocal basis vectors **a***, **b***, and **c*** to the direct basis vectors **a**, **b**, **c** is shown in Fig. 2.6. The vector **a***, for instance, is normal to the

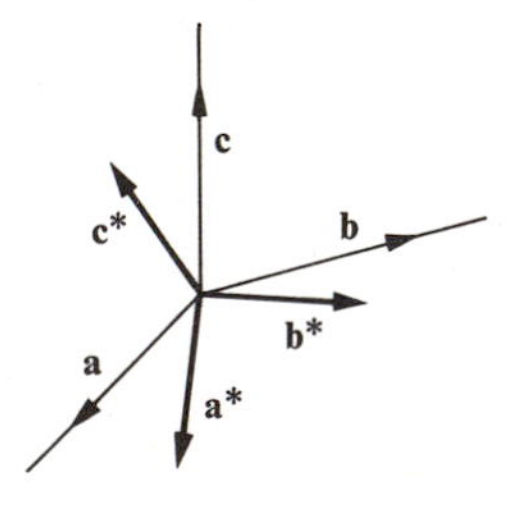


**Fig. 2.6.** Reciprocal basis vectors.

† For the construction of the reciprocal lattice to be valid, the real basis vectors **a**, **b**, and **c** must form a primitive basis; in other words, the cell in the real lattice must be primitive (Section 1.3).

plane defined by the vectors **b** and **c**, and analogous statements apply to **b*** and **c***. Also note that if the direct basis vectors **a**, **b**, **c** form an orthogonal set, then **a***, **b***, and **c*** also form another orthogonal set with **a*** parallel to **a**, **b*** parallel to **b**, and **c*** parallel to **c**. In general, of course, neither set is orthogonal.

The following mathematical relations are useful in dealing with the reciprocal lattice:

$$\begin{aligned} \mathbf{a}^* \cdot \mathbf{a} &= 2\pi, \qquad & \mathbf{a}^* \cdot \mathbf{b} &= \mathbf{a}^* \cdot \mathbf{c} = 0, \\ \mathbf{b}^* \cdot \mathbf{b} &= 2\pi, \qquad & \mathbf{b}^* \cdot \mathbf{a} &= \mathbf{b}^* \cdot \mathbf{c} = 0, \\ \mathbf{c}^* \cdot \mathbf{c} &= 2\pi, \qquad & \mathbf{c}^* \cdot \mathbf{a} &= \mathbf{c}^* \cdot \mathbf{b} = 0. \end{aligned} \tag{2.35}$$

The first row of equations, for instance, can be established as follows: To prove the first of the equations, we substitute for **a*** from (2.33) and find that

$$\mathbf{a}^* \cdot \mathbf{a} = \frac{2\pi}{\Omega_c} (\mathbf{b} \times \mathbf{c}) \cdot \mathbf{a}.$$

But $(\mathbf{b} \times \mathbf{c}) \cdot \mathbf{a}$ is also equal to the volume of the unit cell $\Omega_c$, and hence $\mathbf{a}^* \cdot \mathbf{a} = 2\pi$, as required. The second two equations in the first row reflect the fact, already mentioned, that **a*** is perpendicular to the plane formed by **b** and **c**. The remainder of the equation in (2.35) can be established in a similar manner.

Examples of reciprocal lattices are shown in Fig. 2.7. Figure 2.7(a) shows a direct one-dimensional lattice and its reciprocal. Note that in this case **a*** is parallel to **a**, and that $\mathbf{a}^* = 1/a$. Figure 2.7(b) shows a plane rectangular lattice and its reciprocal.† Three-dimensional examples are more complex, but the procedure for finding them is straightforward. One employs (2.33) to find the basis **a***, **b***, **c***, and then uses (2.34) to locate all the lattice points. It is evident, for instance, that the reciprocal of an sc lattice of edge $a$ is also an sc lattice with a cube edge equal to $2\pi/a$ (Fig. 2.8).

We can similarly establish that the reciprocal of a bcc is an fcc lattice, and vice versa (see the problem section). One may extend the argument to other crystal systems. When we realize that the reciprocal lattice is a lattice in its own right, and that it possesses the same rotational symmetry as the direct lattice, we see that the reciprocal lattice always falls in the same crystal system as its direct lattice (see Table 1.1). Thus the reciprocals for monoclinic, triclinic,... and hexagonal lattices are also monoclinic, triclinic,... and hexagonal, respectively. (Note, however, that the two lattices need not have the same Bravais structure within the same system; see the bcc and fcc examples above.)

---

† In one and two dimensions, Eq. (2.33), which defines the reciprocal lattice, does not apply because the vector cross product is defined only in three dimensions. Therefore in dealing with one- and two-dimensional lattices, we use instead (2.35) to define the reciprocal lattice.

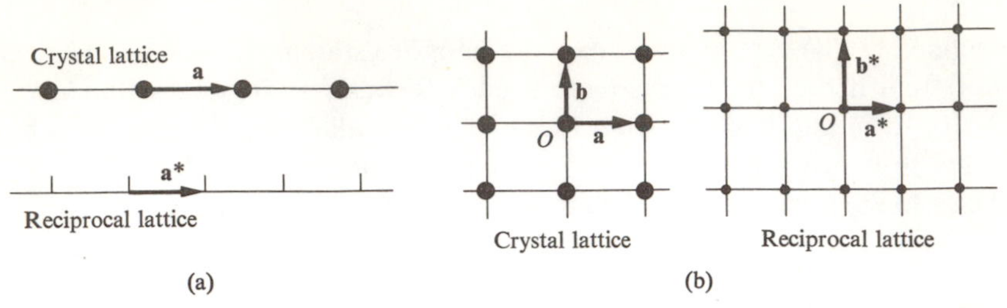


**Fig. 2.7** (a) Reciprocal lattice for a one-dimensional crystal lattice. (b) Reciprocal lattice for a two-dimensional lattice.

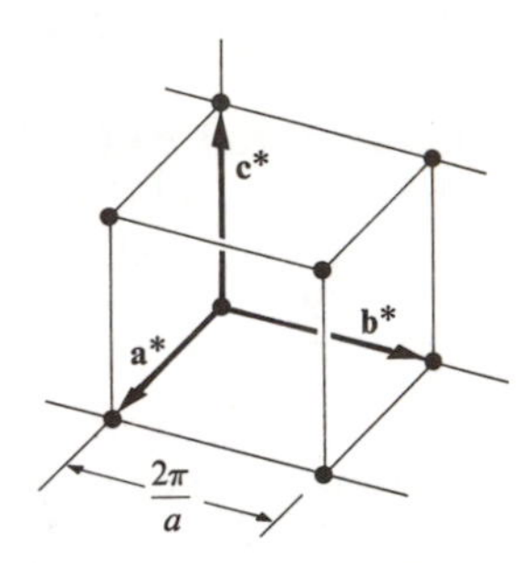


**Fig. 2.8** A part of the reciprocal lattice for an sc lattice.

The unit cell of the reciprocal is chosen in a particular manner. For the rectangular lattice of Fig. 2.9, let $O$ be the origin point, and draw the various lattice vectors connecting the origin with the neighboring lattice points. Then draw the straight lines which are perpendicular to these vectors at their midpoints. The smallest area enclosed by these lines, the rectangle $A$ in the figure, is the unit cell we are seeking, and is called the *first Brillouin zone*. This Brillouin zone (BZ) is an acceptable unit cell because it satisfies all the necessary requirements. It also has the property that its corresponding lattice point falls precisely at the cell center, unlike the case of the direct lattice, in which the lattice points usually lie at the corners of the cell. If the first BZ is now translated by all the reciprocal vectors $\mathbf{G}_n$, then the whole reciprocal lattice space is covered, as it must be, since the BZ is a true unit cell.

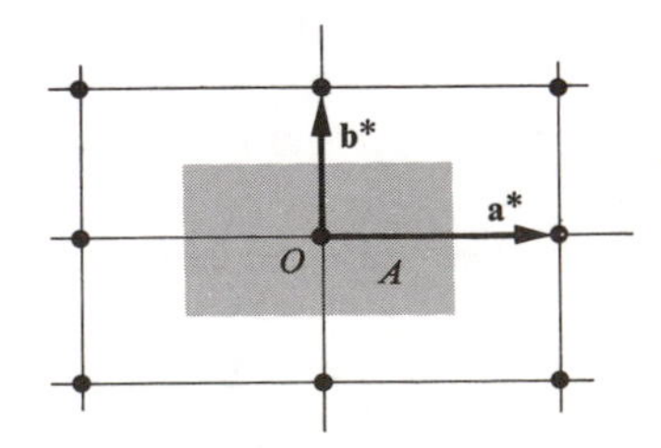


**Fig. 2.9** The first Brillouin zone for a rectangular lattice.

The Brillouin zone for a 3-dimensional lattice can be constructed in a similar manner, but note that in this case the lattice vectors are bisected by perpendicular planes, and that the first BZ is now the smallest *volume* enclosed by these planes. In the simplest case, the sc lattice, the BZ is a cube of edge $2\pi/a$, centered at the origin. The BZ's for the other cubic lattices are more complicated in shape, and we shall defer discussion of these and other lattices to a later section.

Sometimes one also uses higher-order Brillouin zones, which correspond to vectors joining the origin to farther points in the reciprocal lattice, but we shall not discuss these here, as they will not be needed. We shall find that the concept of the Brillouin zone is very important in connection with lattice vibrations (Chapter 3), and electron states in a crystal (Chapter 5).

Having defined the reciprocal lattice and discussed some of its properties, let us now proceed to demonstrate its usefulness. One important application lies in its use in the evaluation of lattice sums, and this rests on the following mathematical statement:

$$\sum_{l=1}^{N} e^{i\mathbf{A}\cdot\mathbf{R}_l} = N\,\delta_{\mathbf{A},\mathbf{G}_n}. \tag{2.36}$$

Here $\mathbf{A}$ is an arbitrary vector, the summation is over the direct lattice vectors,† and $N$ is the total number of cells in the direct lattice. Because of the delta symbol, the meaning of (2.36) is that the lattice sum on the left vanishes whenever the vector $\mathbf{A}$ is not equal to some reciprocal lattice vector $\mathbf{G}_n$. When it is equal to some $\mathbf{G}_n$, however, the lattice sum becomes equal to $N$. To establish the validity of (2.36), we shall first treat the case $\mathbf{A} = \mathbf{G}_n$; to evaluate the exponent $\mathbf{A}\cdot\mathbf{R}_l$ on the left of (2.36), we substitute $\mathbf{A} = \mathbf{G}_n = n_1\mathbf{a}^* + n_2\mathbf{b}^* + n_3\mathbf{c}^*$ and $\mathbf{R}_l = l_1\mathbf{a}_1 + l_2\mathbf{a}_2 + l_3\mathbf{a}_3$, and the result is

$$\begin{aligned}\mathbf{A}\cdot\mathbf{R}_l = \mathbf{G}_n\cdot\mathbf{R}_l &= (n_1\mathbf{a}^* + n_2\mathbf{b}^* + n_3\mathbf{c}^*)\cdot(l_1\mathbf{a} + l_2\mathbf{b} + l_3\mathbf{c})\\ &= (n_1l_1 + n_2l_2 + n_3l_3)\,2\pi,\end{aligned} \tag{2.37}$$

where in evaluating the scalar products of the basis vectors we used (2.35). For example, $\mathbf{a}^*\cdot\mathbf{a} = 2\pi$, $\mathbf{a}^*\cdot\mathbf{b} = 0$, etc. Each term in the sum in (2.36) is therefore of the form $e^{im2\pi}$, where $m$ is an integer and is consequently equal to unity. The total sum is then equal to $N$, as demanded by (2.36). In the case $\mathbf{A} \neq \mathbf{G}_n$, we can follow the same procedure employed in evaluating (2.24), and the result is the same as before, namely, that for large $N$ the sum vanishes except for certain values of $\mathbf{A}$. The exceptional values are, in fact, those singled out above, that is, $\mathbf{A} = \mathbf{G}_n$.

As a final point, we shall now show that the vectors of the reciprocal lattice are related to the crystal planes of the direct lattice. In this manner, the somewhat

† To distinguish the real lattice from the reciprocal lattice, we shall refer to the former as the *direct lattice*.

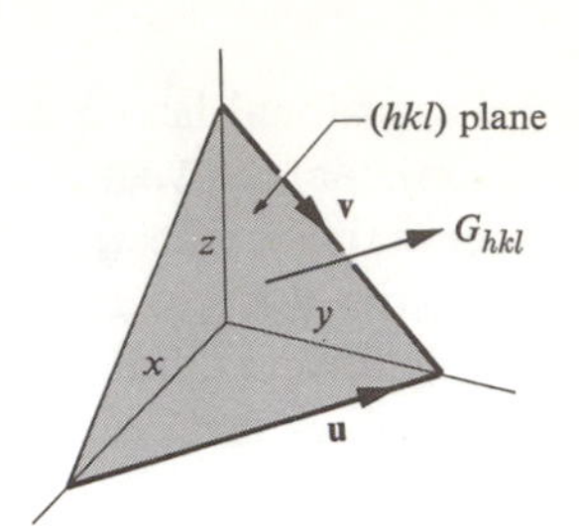


**Fig. 2.10** The reciprocal lattice vector $\mathbf{G}_{hkl}$ is normal to the plane $(hkl)$.

abstract reciprocal vectors will acquire a concrete meaning. Consider the set of crystal planes whose Miller indices are $(hkl)$ and the corresponding reciprocal lattice vector $\mathbf{G}_{hkl} = h\mathbf{a}^* + k\mathbf{b}^* + l\mathbf{c}^*$, where the numbers $h$, $k$, and $l$ are a set of integers. We shall now establish the following properties:

i) The vector $\mathbf{G}_{hkl}$ is normal to the $(hkl)$ crystal planes.

ii) The interplanar distance $d_{hkl}$ is related to the magnitude of $\mathbf{G}_{hkl}$ by

$$d_{hkl} = 2\pi/G_{hkl}. \tag{2.38}$$

To establish these relations, we refer to Fig. 2.10, where we have drawn one of the $(hkl)$ planes. The intercepts of the plane with the axes are $x$, $y$, $z$ and they are related to the indices by

$$(h, k, l) \sim \left(\frac{1}{x}, \frac{1}{y}, \frac{1}{z}\right), \tag{2.39}$$

where use is made of the definition of the Miller indices (Section 1.6). Note also the vectors $\mathbf{u}$ and $\mathbf{v}$ which lie along the lines of intercepts of the plane with the $xy$ and $yz$ planes, respectively. According to the figure, these vectors are given by $\mathbf{u} = x\,\mathbf{a} - y\,\mathbf{b}$, and $\mathbf{v} = y\,\mathbf{b} - z\,\mathbf{c}$. In order to prove relation (i) above, we need only prove that $\mathbf{G}_{hkl}$ is orthogonal to both $\mathbf{u}$ and $\mathbf{v}$. We have

$$\mathbf{u} \cdot \mathbf{G}_{hkl} = (x\mathbf{a} - y\mathbf{b}) \cdot (h\mathbf{a}^* + k\mathbf{b}^* + l\mathbf{c}^*) = 2\pi(xh - yk) = 0,$$

where we have used (2.35) to establish the second equality; the last equality follows from (2.39). In the same manner we can also show that $\mathbf{G}_{hkl}$ is orthogonal to $\mathbf{v}$, and this establishes property (i).

In order to prove (2.38), one observes that $d_{hkl}$, the interplanar distance, is equal to the projection of $x\mathbf{a}$ along the direction normal to the $(hkl)$ planes; this direction can be represented by the unit vector $\hat{\mathbf{G}}_{hkl} = \mathbf{G}_{hkl}/G_{hkl}$, since we have already established that $\mathbf{G}_{hkl}$ is normal to the plane. Therefore

$$d_{hkl} = x\mathbf{a} \cdot \hat{\mathbf{G}}_{hkl} = (x\mathbf{a} \cdot \mathbf{G}_{hkl})/G_{hkl}. \tag{2.40}$$

We now note that $x\mathbf{a} \cdot \mathbf{G}_{hkl} = 2\pi hx$, and this is equal to $2\pi$, because, according to (2.39), $xh = 1$. This completes the proof of (2.38).

The connection between reciprocal vectors and crystal planes is now quite clear. The vector $\mathbf{G}_{hkl}$ is associated with the crystal planes $(hkl)$, which are, in fact, normal to it, and the separation of these planes is $2\pi$ times the inverse of the length $G_{hkl}$ in the reciprocal space. The crystallographer prefers to think in terms of the crystal planes, which have a physical reality, and their Miller indices, while the solid-state physicist prefers the reciprocal lattice, which is mathematically more elegant; the two approaches are, however, equivalent, and one can change from one to the other by using the relations connecting the two. Of the two approaches, we shall mostly use the reciprocal lattice in this book.

## 2.7 THE DIFFRACTION CONDITION AND BRAGG'S LAW

We shall now employ the concept of the reciprocal lattice to evaluate the lattice-structure factor $S$, which is involved in the x-ray scattering process. This factor is given in (2.22). Comparing this with (2.36), we see that $S$ vanishes for every value of $\mathbf{s}$ except where

$$\mathbf{s} = \mathbf{G}_{hkl}. \tag{2.41}$$

The condition for diffraction is therefore that the scattering vector $\mathbf{s}$ is equal to a reciprocal lattice vector. Equation (2.41) implies that $\mathbf{s}$ is normal to the $(hkl)$ crystal planes [property (i) of Section 6], as shown in Fig. 2.11. The equation can be rewritten in a different form. Recalling that $s = 2(2\pi/\lambda) \sin \theta$, $G_{hkl} = 2\pi/d_{hkl}$, and substituting into (2.41), we find that

$$2d_{khl} \sin \theta = \lambda. \tag{2.42}$$

This is exactly the same form as Bragg's law, Eq. (2.4), which is seen to follow from the general treatment of scattering theory. It is therefore physically meaningful to use the Bragg model (Section 3), and speak of reflection from atomic planes. This manner of viewing the diffraction process is conceptually simpler than that of scattering theory.

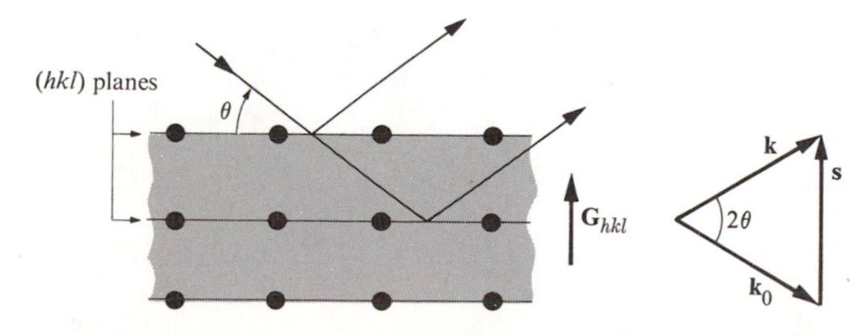


**Fig. 2.11** The scattering vector is equal to a reciprocal lattice vector.

When the condition (2.41) is satisfied, the structure factor is nonzero, and its value is equal to $N$ as seen from (2.36). Thus

$$S_{hkl} = N. \tag{2.43}$$

Substituting this into (2.23), we find the crystal scattering factor $f_{cr}$ to be

$$f_{cr,\,hkl} = NF_{hkl}, \tag{2.44}$$

and the intensity $I$ is then

$$I_{hkl} \sim |f_{cr,\,hkl}|^2 \sim |F_{hkl}|^2. \tag{2.45}$$

The scattered intensity vanishes in all directions except those in which the structure factor $S$ is nonvanishing. These latter directions are therefore the directions of diffraction: they are the ones which satisfy the condition of constructive interference. When the Bragg condition is satisfied, then the incident beam is diffracted into a single beam (neglecting higher orders), which is recorded at the detector as a single spot on a film. This spot represents the whole set of reflecting planes ($hkl$). When the crystal is rotated so that a new set of planes again satisfies the Bragg condition, then this new set appears as a new spot on the film at the detector. Therefore each spot on the film represents a whole set of crystalline planes, and from the arrangement of these spots one can determine the structure of the crystal, as discussed in Section 9.

According to our statements, each diffracted beam can be associated with a set of planes of certain Miller indices; this is evident from (2.45). It is experimentally observed, however, that diffraction from certain planes may be missing. This is due to the geometrical structure factor $F_{hkl}$, which depends on the shape and contents of the unit cell. Thus, if $F_{hkl}$ is zero for certain indices, then the intensity vanishes according to (2.45), even though the corresponding planes satisfy the Bragg condition. To evaluate $F_{hkl}$, we return to (2.21). We assume the atoms to be identical, and take $\boldsymbol{\delta}_j = \mathbf{U}_j\mathbf{a} + \mathbf{V}_j\mathbf{b} + \mathbf{W}_j\mathbf{c}$, where $\delta_j$ is the position of the $j$th atom. Furthermore, we take

$$\mathbf{s} = \mathbf{G}_{hkl} = h\mathbf{a}^* + k\mathbf{b}^* + l\mathbf{c}^*.$$

Therefore

$$F_{hkl} = f_a \sum_j e^{i2\pi(hu_j + kv_j + lw_j)}. \tag{2.46}$$

Consider, for instance, the bcc lattice. The unit cell has two atoms whose coordinates are $(u_j, v_j, w_j) = (0, 0, 0)$, and $(\frac{1}{2}, \frac{1}{2}, \frac{1}{2})$. Using (2.46), one has

$$F_{hkl} = f_a(1 + e^{i\pi(h+k+l)}).$$

This expression can take only two values; when $(h + k + l)$ is even, $F_{hkl} = 2f_a$, while $F_{hkl} = 0$ when $(h + k + l)$ is odd. Thus for the bcc lattice, the diffraction is absent for all those planes in which the sum $(h + k + l)$ is odd, and is present for the planes in which $(h + k + l)$ is even. We leave it as a problem to show that in an fcc lattice the allowed reflections correspond to the cases in which $h$, $k$, $l$ are either all even or all odd. Note that the missing planes give direct information concerning the symmetry of the unit cell.†

Equation (2.41) can be rewritten in still another form. We recall from (2.9a) that $\mathbf{s} = \mathbf{k} - \mathbf{k}_0$, where $\mathbf{k}_0$ and $\mathbf{k}$ are the vectors of the incident and diffracted beams. Substituting into (2.41), one finds

$$\mathbf{k} = \mathbf{k}_0 + \mathbf{G}. \tag{2.47}$$

Multiplication of both sides by $\hbar$ leads to

$$\hbar\mathbf{k} = \hbar\mathbf{k}_0 + \hbar\mathbf{G}.$$

But the quantity $\hbar\mathbf{k}$ is the momentum of the x-ray photon associated with the beam (see the deBroglie relation, Section A.1). Thus the above equation may be viewed as momentum conservation, and the diffraction process as a collision process between the x-ray photon and the crystal. In the collision the photon recoils and gains a momentum $\hbar\mathbf{G}$. Conversely, the crystal recoils in the opposite direction with a momentum $-\hbar\mathbf{G}$. The recoil energy of the crystal is very small because the motion is that of a rigid-body displacement, and therefore the kinetic energy is $(\hbar G)^2/2M$, where $M$ is the total mass of the crystal. Since $M$ is extremely large compared with the mass of the atom, the recoil energy is very small, and may be neglected. Therefore the collision process may be regarded as elastic; this has been implicitly assumed throughout, of course, since we have taken $k$ to be equal to $k_0$.

## 2.8 SCATTERING FROM LIQUIDS

X-ray scattering is also used in the investigation of liquid structure. By observing the pattern of the scattered beam, one can determine the pair-distribution function of the liquid (see Section 1.8). Returning to the general result (2.20), we write for the liquid scattering factor

$$f_{lq} = f_a \sum_l e^{i\mathbf{s}\cdot\mathbf{R}_l}, \tag{2.48}$$

where $f_a$ is the atomic factor and the summation is over all the atoms in the liquid; we have assumed a monotonic liquid. But in a liquid the atoms are continually moving from one region to another, unlike the case for a solid, in which

† The formula governing the missing planes is referred to as the *extinction rule*.

they are restricted to certain sites, and the sum in (2.48) is therefore difficult to evaluate. This can be mitigated by dealing instead with the scattered intensity, which is, after all, the quantity recorded experimentally. The intensity is proportional to $|f_{lq}|^2$ which, with the use of (2.48), can be written as

$$|f_{lq}|^2 = f_a^2 \sum_{j,l} e^{i\mathbf{s}\cdot(\mathbf{R}_l - \mathbf{R}_j)}. \tag{2.49}$$

The liquid structure factor $S_{lq}$ is now defined as the double sum in this equation. That is,

$$S_{lq} = \sum_{j,l} e^{i\mathbf{s}\cdot(\mathbf{R}_l - \mathbf{R}_j)}, \tag{2.50}$$

which is analogous to the lattice-structure factor $S$ of (2.22). The sum can be split into two different types of terms: Those for which $j = l$, that is, the indices $j$ and $l$ referring to the same atom, and those for which $j \neq l$. The former type is readily seen to add up to $N$—there being $N$ terms in all—and the latter can be expressed in terms of the pair-distribution function. The result is

$$S_{lq}/N = 1 + n_0 \int d^3R e^{i\mathbf{s}\cdot\mathbf{R}} g(R), \tag{2.51}$$

where $n_0$ is the average atomic density and $g(R)$ the pair function (Section 1.8). The integration is over the volume of the liquid. We note, however, that only the deviation of $g(R)$ from unity contributes to scattering because the remainder, $g(R) = 1$, corresponds to a uniform distribution which would allow the beam to pass through without any scattering. Thus we may rewrite (2.51) as

$$S_{lq}/N = 1 + n_0 \int d^3R e^{i\mathbf{s}\cdot\mathbf{R}} [g(R) - 1]. \tag{2.52}$$

The integral is now extended over all space, since $[g(R) - 1]$ decays rapidly at

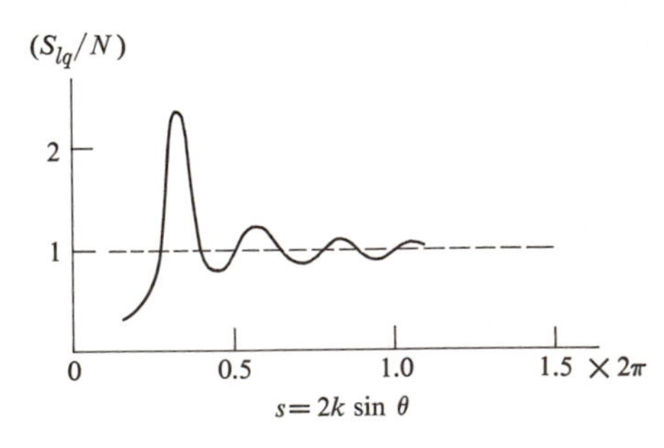


**Fig. 2.12** The structure factor for liquid mercury (after Guinier)

large $R$ (Section 1.8), and hence no appreciable error is introduced by extending the integration range to include the whole space.

Equation (2.52) allows the computation of $S_{lq}$ if $g(R)$ is given, but the problem we usually face is the converse one. That is, $S_{lq}$ can be measured and $g(R)$ must be inferred from this measurement. We must therefore invert (2.52). This inversion can be carried out using the Fourier-transform theorem. Examining (2.52), we note that $(S_{lq} - 1)/n_0$ is simply the Fourier transform of $g(R) - 1$. Therefore, using the Fourier-transform theorem, one writes at once

$$g(R) - 1 = \frac{1}{(2\pi)^3}\frac{1}{n_0}\int d^3s e^{-i\mathbf{s}\cdot\mathbf{R}}\left[\frac{S_{lq}(\mathbf{s}) - 1}{N}\right], \tag{2.53}$$

the integral now being over the whole space of the scattering vector $\mathbf{s}$. Figure 2.12 shows the structure factor of liquid mercury as determined by x-ray scattering techniques. Another scattering technique which is increasingly used in the study of liquid structures is that of neutron scattering, which is discussed in Section 11.

## 2.9 EXPERIMENTAL TECHNIQUES

In this section we review the experimental techniques used in collecting x-ray diffraction data. Our discussion is brief, and covers primarily the physical principle underlying the methods used. This is not the place to discuss each method in detail, nor the various practical difficulties or corrections attendant on each method. Anyone interested in further details may refer to the books by Woolfson, Cullity, Buerger, and Guinier, and the bibliographies given therein.

There are essentially three methods: The *rotating-crystal method*, the *Laue method*, and the *powder method*. Regardless of the method used, the quantities measured are essentially the same.

i) The scattering angle $2\theta$ between the diffracted and incident beams. By substituting $\sin\theta$ into Bragg's law, one determines the interplanar spacing as well as the orientation of the plane responsible for the diffraction.

ii) The intensity $I$ of the diffracted beam. This quantity determines the cell-structure factor, $F_{hkl}$, and hence gives information concerning the arrangement of atoms in the unit cell.

### The rotating-crystal method

This method is used for analysis of the structure of a single crystal. The experimental arrangement is shown in Fig. 2.13. The crystal is usually about 1 mm in diameter, and is mounted on a spindle which can be rotated. A photographic film is placed on the inner side of a cylinder concentric with the axis of

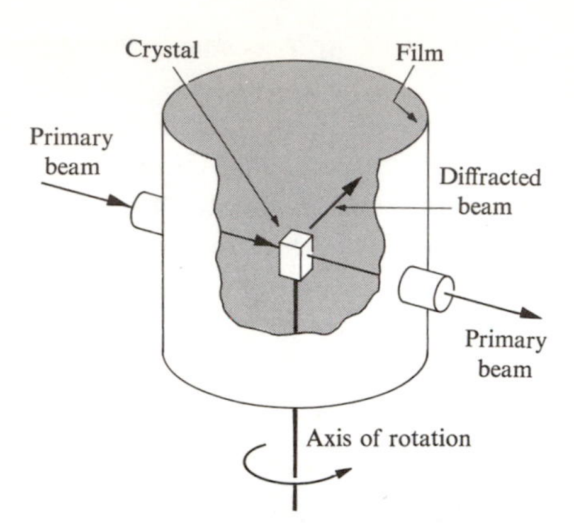


**Fig. 2.13** Experimental arrangement for the rotating-crystal method.

rotation. A monochromatic incident beam of wavelength $\lambda$ is collimated and made to impinge on the crystal. The specimen is then rotated, if necessary, until a diffraction condition obtains, that is, $\lambda$ and $\theta$ satisfy Bragg's law. When this occurs, a diffracted beam (or beams) emerges from the crystal and is recorded as a spot on the film.

By recording the diffraction patterns (both angles and intensities) for various crystal orientations, one can determine the shape and size of the unit cell as well as the arrangement of atoms inside the cell.

### The Laue method

This method can be used for a rapid determination of the symmetry and orientation of a single crystal. The experimental arrangement is shown in Fig. 2.14(a). A *white* x-ray beam—i.e., one with a spectrum of continuous wavelength— is made to fall on the crystal, which has a fixed orientation relative to the incident beam. Flat films are placed in front of and behind the specimen. Since $\lambda$ covers a continuous range, the crystal selects that particular wavelength which satisfies Bragg's law at the present orientation, and a diffracted beam emerges at the corresponding angle. The diffracted beam is then recorded as a spot on the film. But since the wavelength corresponding to a spot is not measured, one cannot determine the actual values of the interplanar spacings—only their ratios. Therefore one can determine the shape but not the absolute size of the unit cell. A typical Laue photograph is shown in Fig. 2.14(b).

Note that if the direction of the beam is an axis of symmetry of the crystal, then the diffraction pattern should exhibit this symmetry. Figure 2.14(b) shows the 6-fold symmetry of the symmetry axis in Mg, which has the hexagonal structure.

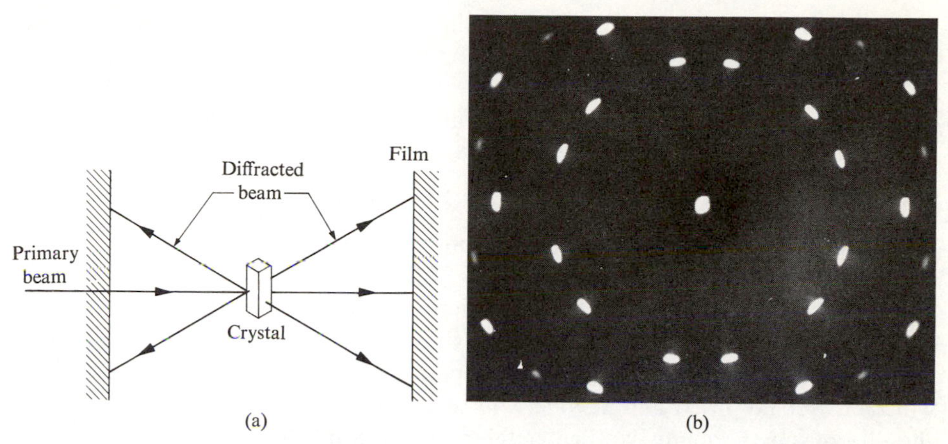


**Fig. 2.14** The Laue method: (a) Experimental arrangement. (b) Laue pattern for an Mg crystal, with the x-ray beam parallel to the 6-fold symmetry axis. [After Barrett (1966)]

### The powder method

This method is used to determine the crystal structure even if the specimen is not a single crystal. The sample may be made up of fine-grained powder packed into a cylindrical glass tube, or it may be polycrystalline, in which case it is made up of a large number of small crystallites oriented more or less randomly. A monochromatic beam impinges on the specimen, and the diffracted beams are recorded on a cylindrical film surrounding it.

Because of the large number of crystallites which are randomly oriented, there is always enough of these which have the proper orientation relative to the incident monochromatic beam to satisfy Bragg's law, and hence a diffracted beam emerges at the corresponding angle (Fig. 2.15). Since both $\lambda$ and $\theta$ are measurable, one can determine the interplanar spacing.

Other sets of planes lead to other diffracted beams corresponding to different planar spacing for the same wavelength. Thus one can actually determine the

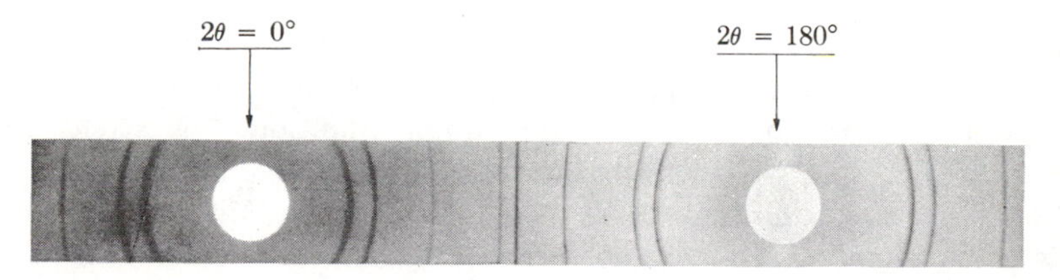


**Fig. 2.15** The x-ray powder diffraction pattern for Cu. $2\theta$ is the scattering angle. [After Cullity (1956)]

lattice parameters quite accurately, particularly if the crystal structure is already known. Note also that, since the specimen is symmetric under rotation around the incident beam as an axis, the diffracted beam corresponding to each scattering angle $2\theta$ fans out along a cone whose axis lies along the incident beam.

## 2.10 OTHER X-RAY APPLICATIONS IN SOLID-STATE PHYSICS

The x-ray diffraction technique, in addition to its major use in analysis of crystal structures, finds a multitude of other applications in solid-state physics. Indeed much of our knowledge of the microscopic world has been derived from the persistent use of x-ray techniques. To illustrate the usefulness of this powerful tool, we shall consider some of the applications here. However, our discussion will be extremely brief, and anyone who seeks more than a cursory acquaintance should consult the many excellent references listed in the bibliography.

The most ambitious goal would be to plot the distribution of electrons inside the solid, i.e., to draw an electron-density map, for this would be tantamount to "seeing" the electronic cloud in the solid. In principle, it should be possible to achieve this through the use of x-ray diffraction, since it is the electrons which are responsible for the diffraction process. In order to see how this may be done, refer to Eq. (2.21), in which we note that the electron density is contained in $f_a$. If the crystal-scattering factor is measured, then this equation can be inverted. That is, one finds the electron density $\rho(\mathbf{r})$ in terms of $f_{cr}(\mathbf{s})$. The mathematical procedure involves the use of the Fourier transform, in a manner somewhat similar to that used in liquids, as indicated in the previous section.

Another important application is in the study of lattice imperfections, such as foreign impurities, dislocations, regions of strain, etc. In the presence of such imperfections, the diffraction pattern no longer corresponds to that of a perfect crystal, and by studying the deviation one can obtain information concerning the type of imperfections and their distribution in the crystal. Such techniques are in common use by chemists, metallurgists, and materials scientists.

In our study of pure crystals we have assumed perfect periodicity (except in the previous paragraph). We have assumed, in other words, that each atom is located at its lattice site at all times. However, it is well known that when the temperature is above absolute zero, atoms undergo some vibration around their sites as a result of the thermal excitations. The presence of these *lattice vibrations* leads to a modification in the x-ray diffraction pattern. In particular, some diffraction is observed along directions which do not satisfy the Bragg condition; this is referred to as *diffuse scattering*. This type of scattering has long been used in the study of lattice vibrations, and has contributed greatly to our understanding of this important subject. We shall examine lattice vibrations in considerable detail in the following chapter.

Finally, x-ray diffraction is used to determine the structure of biological molecules. Many of the recent great strides in our knowledge of molecular biology have been accomplished in this manner. The discovery of the double-helical structure of the DNA molecule is but one example.

## 2.11 NEUTRON DIFFRACTION

We have already indicated that other forms of radiation, in addition to x-rays, can also be used in the investigation of crystal structures and other related problems. The primary requirements are these: First, the radiation should possess a wave property so that the scattered waves can superpose coherently, thereby revealing the structure of the scattering medium. Second, the wavelength of the radiation should be of the same order of magnitude as the lattice constant. Neutron radiation satisfies these requirements.

The neutron (and other particles) has wave properties, as you recall from elementary physics (see also Section A.1). The wavelength, also known as the deBroglie wavelength, is given by the relation $\lambda = h/p$, where $p$ is the momentum of the neutron. The wavelength can also be expressed in terms of the energy $E = p^2/2m$, where $m$ is the mass. Substituting the mass value appropriate to the neutron, one finds

$$\lambda = \frac{0.28}{E^{1/2}}, \tag{2.54}$$

where $\lambda$ is in angstroms and $E$ in electron volts. To be useful in structure analysis, $\lambda$ must be about 1 Å, which, when substituted into (2.54), yields an energy of about 0.08 eV. This energy is of the same order of magnitude as the thermal energy $kT$ at room temperature, 0.025 eV, and for this reason we speak of *thermal neutrons*.

The scattering mechanism for the neutron is the interaction between it and the atomic nuclei present in the crystal. This interaction is referred to as the *strong interaction*; it is the reaction responsible for holding the nucleons (neutrons and protons) together in the nucleus. Being electrically neutral, the neutron does not interact with the electrons in the crystal. Thus, unlike the x-ray, which is scattered entirely by electrons, the neutron is scattered entirely by nuclei (see below for an exception).

Since the details of neutron diffraction are precisely the same as those for x-rays, we need not go into it further here. The only difference lies in the fact that the neutron analog to (2.6) now contains the scattering length of the neutron instead of that of the electron. The results of interest to us here—e.g., Bragg's law, Laue equations, etc.—are exactly the same as before. All these are direct consequences of the structure factor which, being a lattice sum, depends only on the lattice structure and not on the atomic scattering factor; the type of radiation used is irrelevant.

Neutron diffraction has several advantages over its x-ray counterpart.

a) Light atoms such as hydrogen are better resolved in a neutron pattern because, having only a few electrons to scatter the x-ray beam, they do not contribute significantly to the x-ray diffracted pattern.

b) A neutron pattern distinguishes between different atomic isotopes, whereas an x-ray pattern does not.

c) Neutron diffraction has made important contributions to the studies of magnetic materials. In magnetic crystals the electrons of the atomic orbitals have a net spin, and hence a net magnetic moment. The relative orientations of these moments may be either random or parallel, or antiparallel, depending on the range of temperature of the crystal. One can use neutron diffraction to reveal the crystalline magnetic pattern because the neutron does interact with the moments. The interaction results from the fact that the neutron also has a magnetic moment of its own (it is a tiny magnet), which feels the field generated by the moments of the electrons. Examples of the application of neutron diffraction to this important branch of magnetism are given in Sections 9.9 and 9.14.

d) The technique of neutron diffraction is far superior to that of x-rays in the studies of lattice vibrations, which will be discussed in the following chapter.

The disadvantages of neutron diffraction techniques are:

a) The necessity for using nuclear reactors, which are not commonly available. Furthermore, even the most powerful neutron sources have intensities of only about $10^{-5}$ the intensity available from common x-ray sources. Because of this, large crystals are used in neutron diffraction, and the exposure time is made as long as possible.

b) Neutrons, being electrically neutral, are harder to detect than the ionizing x-rays. Therefore neutrons are converted first into ionizing radiation through their reaction with, e.g., boron nuclei.

## 2.12 ELECTRON DIFFRACTION

A beam of electrons incident on a crystal suffers Bragg diffraction in a manner similar to the x-ray and neutron diffraction discussed previously. The electron, like the neutron, possesses wave properties, and the wavelength is also given by $\lambda = h/p$. Writing $p$ in terms of the energy $E$, and the lattice in terms of the accelerating potential $V$, that is, $E = eV$, and inserting the values appropriate to the electron, one finds

$$\lambda = \sqrt{150/V}, \tag{2.55}$$

where $\lambda$ is in angstroms and $V$ in electron volts. For a $\lambda = 1$ Å, the potential is $V = 150$ V, or $E = 150$ eV.

The mechanism responsible for the electron scattering is the electric field associated with the atoms in the solid. This field is produced both by the nucleus and by the orbital electrons in each atom. It is large at the nucleus, but decreases rapidly away from the nucleus. In the latter region the nucleus is screened by the orbital electrons.

Calculations show that the scattering length associated with the scattering of the electron from an atom is large. This means that an electron beam is strongly scattered, and hence has a short stopping distance. This distance is only about 50 Å for $V = 50$ kV, for example. Even though the electron beam is restricted to a rather small depth near the surface, this depth does nonetheless include a number of atomic layers, so that a crystal diffraction pattern obtains (Fig. 2.16). It also follows that the electron diffraction pattern is particularly sensitive to the physical properties of the surface, which explains its wide use in the study of surfaces, e.g., oxide layers forming on the surface of solids, thin films, and so forth.†

**Fig. 2.16** Continuous rotation electron diffraction pattern of a single crystal of silver. The axis of rotation is normal to the paper. [After Leighton]

† For a readable elementary review of the subject, see K. A. R. Mitchell, *Contemp. Physics*, **14,** 251 (1973). The article discusses how the currently popularly LEED (low-energy electron diffraction) technique may be used in studies of surface crystallography and surface chemistry, and their bearing on understanding the interatomic bonds of surface atoms, as well as such technologically important topics as surface catalysis, corrosion, and epitaxial crystal growth.

We have been concerned here only with the diffraction of external electrons, but internal electrons also suffer the same type of diffraction as they move through the crystal. We shall find this concept to be very helpful in our discussion of the electron states in crystals (Chapter 5).

Finally, a point of historical interest. The wave properties of material particles were first demonstrated in connection with electron diffraction. In 1927, Davisson and Germer observed the scattering of an electron beam from the surface of a nickel crystal. In obtaining a diffraction pattern, they confirmed the wave properety of the electron, as postulated earlier by deBroglie. In recognition of this work, Davisson was awarded the Nobel prize in 1937.

## SUMMARY

The crystal structure is determined from the diffraction pattern observed when the crystal is irradiated with an x-ray beam. The fundamental result is the *Bragg law*,

$$2d \sin \theta = n\lambda,$$

where $d$ is the interplanar distance, $\theta$ the glancing angle, and $\lambda$ the wavelength of the beam. By measuring $\theta$ and $\lambda$, one may determine $d$, and, eventually, the crystal structure.

A more rigorous treatment of the diffraction process considers the crystal to be composed of discrete electrons. The scattering factor is

$$f = \sum_l e^{i\mathbf{s}\cdot\mathbf{r}_l},$$

where the sum is over all the electrons in the system, and $\mathbf{s}$ is the scattering vector,

$$\mathbf{s} = \mathbf{k} - \mathbf{k}_0.$$

Applying the result to a single atom leads to the *atomic scattering factor*,

$$f_a = \int_0^R 4\pi r^2 \rho(r) \frac{\sin sr}{sr}\, dr.$$

The factor $f_a$ decreases as the scattering angle $2\theta$ increases, because of the interference between the various shells of the electronic cloud in the atom.

The scattering factor for a crystal may be written as a product

$$f_{cr} = FS,$$

where $F$ is the *geometrical structure factor* and $S$ the *lattice-structure factor*. These are given, respectively, by

$$F = \sum_j f_{aj} e^{i\mathbf{s}\cdot\boldsymbol{\delta}_j},$$

the summation being over all the atoms in a unit cell, and

$$S = \sum_{l} e^{i\mathbf{s}\cdot\mathbf{R}_l^{(c)}},$$

this summation being over all unit cells in the crystal. The factor $F$ depends only on the atomic properties and the shape of the unit cell, and $S$ depends only on the lattice structure. The factorization of $f_{cr}$ into $F$ and $S$ is useful because it enables us to treat the atomic and lattice properties of the crystal independently.

An examination of the lattice factor $S$ shows that it vanishes, except when

$$\mathbf{s} = \mathbf{G}.$$

That is, the scattering vector is equal to a reciprocal-lattice vector. This is the same condition as Bragg's law for reflection from the atomic planes normal to **G**.

Liquid structures can also be studied by x-ray diffraction. By measuring the liquid structure factor, one may evaluate the pair-distribution function for atoms in the liquid.

The x-ray diffraction pattern is recorded on a film, which is sensitized by the diffracted beams emerging from the crystal. Each beam represents a reflection from a set of atomic planes in the crystal, and is recorded as a spot on the film. The position and symmetry of the spot pattern contain the information needed to decipher the crystal structure.

A neutron beam may also be used to determine the crystal structure. The same formulas developed above apply here also, provided the deBroglie wavelength

$$\lambda = h/p$$

is used. The energy of the neutron is very small, about 0.1 V, and we speak of thermal neutrons. The scattering of neutrons is accomplished by their interaction with the nuclei of the crystal, and not their interactions with the electrons, as in x-rays.

Electron diffraction has also been used in the analysis of crystal structure. Since electrons interact very strongly with the atoms in a crystal, the stopping distance of the electron is very short—only about 50 Å. Consequently, electron diffraction is employed primarily in the study of surface phenomena.

## REFERENCES

### X-ray diffraction

C. S. Barrett and T. B. Massalksi, 1966, *Structure of Metals*, third edition, New York: McGraw-Hill

J. M. Buerger, 1960, *Crystal Structure Analysis*, New York: John Wiley

B. D. Cullity, 1972, *X-Ray Diffraction*, New York: Freeman; Reading, Mass.: Addison-Wesley

A. Guinier and D. L. Dexter, 1963, *X-Ray Studies of Materials*, New York: John Wiley/Interscience (These last two references discuss scattering by crystals and liquids.)
C. Kittel, 1966, *Introduction to Solid State Physics*, New York: John Wiley
T. Kovacs, 1969, *Principles of X-Ray Metallurgy*, New York: Plenum
M. M. Woolfson, 1970, *X-Ray Crystallography*, Cambridge: Cambridge University Press. (An excellent treatment which had great influence on the presentation of this chapter.)

### Neutron diffraction

G. E. Bacon, 1962, *Neutron Diffraction*, Oxford: Oxford University Press
P. A. Egelstaff, editor, 1965, *Thermal Neutron Scattering*, New York: Academic Press

### Electron diffraction

B. K. Vainstein, 1964, *Structure Analysis by Electron Diffraction*, London, Pergamon Press

### Liquids

P. A. Egelstaff, 1967, *An Introduction to the Liquid State*, New York: Academic Press

## QUESTIONS

1. What is the justification for drawing the scattered rays in Fig. 2.2(a) as nearly parallel?
2. In the scattering of x-rays by electrons, there is a small probability that the photon may suffer *Compton scattering* by the electron—this in addition to the scattering considered in this chapter, which is known as *Thompson scattering*. Compton scattering is inelastic, and the photon loses some of its energy to the electron; the energy loss depends on the scattering angle. Would you expect Compton scattering to produce a diffraction pattern? Why or why not?
3. It was stated following Eq. (2.6) that the amplitude of the wave decreases as the inverse of the radial distance from the scattering center. Justify this on the basis of energy conservation.
4. The crystal scattering factor $f_{cr}$ of (2.19) is a complex number. What is the advantage of using complex representation?
5. Diamond and silicon have the same type of lattice structure, an fcc with a basis, but different lattice constants. Is the lattice structure factor $S$ the same for both substances?
6. A reciprocal-lattice vector has a dimension equal to the reciprocal of length, for example, $cm^{-1}$. Is it meaningful to compare the magnitudes of a direct-lattice vector **R** with a reciprocal-lattice vector **G**? Is it meaningful to compare their directions? If the latter answer is yes, find the angle between **R** and **G** in terms of their components in a cubic crystal. What is the angle between **R** = [111] and **G** = [110]?
7. Does a real lattice vector have a corresponding unique reciprocal vector?
8. Draw a figure illustrating momentum conservation in the Bragg reflection considered as a photon–crystal collision. Why is this collision elastic? Justify your answer with numerical estimates.

9. Why is the energy of a neutron so much smaller than that of an electron in radiation beams employed in crystal diffraction?
10. Can a light beam be used in the analysis of crystal structure? Estimate the lattice constant for a crystal amenable to analysis by visible light.
11. Why is the neutron more useful than the proton in structure analysis?

## PROBLEMS

1. The minimum wavelength observed in x-ray radiation is $\lambda = 1.23$ Å. What is the kinetic energy, in eV, of the primary electron hitting the target?
2. The edge of a unit cell in a cubic crystal is $a = 2.62$ Å. Find the Bragg angle corresponding to reflection from the planes (100), (110), (111), (200), (210) and (211), given that the monochromatic x-ray beam has a wavelength $\lambda = 1.54$ Å.
3. A Cu target emits an x-ray line of wavelength $\lambda = 1.54$ Å.
   a) Given that the Bragg angle for reflection from the (111) planes in Al is 19.2°, compute the interplanar distance for these planes. Recall that aluminum has an fcc structure.
   b) Knowing that the density and atomic weight of Al are, respectively, 2.7 g/cm$^3$ and 27.0, compute the value of Avogadro's number.
4. a) The Bragg angle for reflection from the (110) planes in bcc iron is 22° for an x-ray wavelength of $\lambda = 1.54$ Å. Compute the cube edge for iron.
   b) What is the Bragg angle for reflection from the (111) planes?
   c) Calculate the density of bcc iron. The atomic weight of Fe is 55.8.
5. Establish the validity of (2.11) for an arbitrary origin.
6. Prove the result of (2.17).
7. Establish the result (2.20).
8. Establish the fact that Eq. (2.23) follows from (2.20) and the definitions (2.21) and (2.22).
9. The electron density in a hydrogen atom in its ground state is spherically symmetric, and given by

$$\rho(r) = e^{-2r/a_0}/\pi a_0^3,$$

   where $a_0$, the first Bohr radius, has the value 0.53 Å. Compute the atomic scattering factor $f_a$ for hydrogen, and plot it as a function of $s = 2k \sin\theta = 4\pi \sin\theta/\lambda$. Explain physically why the scattering factor is small for back reflection ($\theta = \pi/2$).
10. The crystal-structure factor $f_{cr}$ depends on the origin of the coordinate system. Show that the intensity, which is the observed quantity, is independent of the choice of origin.
11. Evaluate the first subsidiary minimum of $S^2$ (Fig. 2.5b), and show that it is equal to $0.04N^2$, in the limit of large $N$.
12. The geometrical structure factor $F_{hkl}$ for a bcc lattice was evaluated in the text by assuming the cell to contain one atom at a corner and another at the center of the unit cell. Show that the same result is obtained by taking the cell to contain one-eighth of an atom at each of its eight corners, plus one atom at the center.
13. Evaluate the geometrical structure factor $F_{hkl}$ for reflection from the $(hkl)$ planes in an fcc lattice, and show that the factor vanishes unless the numbers $h$, $k$, and $l$ are all even or all odd.

14. Which of the following reflections would be missing in a bcc lattice: (100), (110), (111), (200), (210), (220), (211)? Answer a similar question for an fcc lattice.
15. Diamond has an fcc structure in which the basis is composed of two identical atoms, one at the lattice point, and another at a point ($a/4$, $a/4$, $a/4$) relative to the first atom, where $a$ is the edge of the cube (see Fig. 2.15). Find the geometrical structure factor for diamond, and express it in terms of the factor corresponding to an fcc Bravais lattice. Which of the reflections in Problem 14 are missing in diamond?
16. Cesium chloride (CsCl) crystallizes in the bcc structure, in which one type of atom is located at the corners and the other at the center of the cell. Calculate the geometrical structure factor $F_{100}$, assuming that $f_{Cs} = 3f_{Cl}$. Explain why the extinction rule derived in the text is violated here.
17. Repeat Problem 15 for GaSb, which crystallizes in the zincblende structure (see Section 1.7), assuming that $f_{Sa} = 2f_{Ga}$.
18. Show that the volume of the reciprocal cell is equal to the inverse of the real cell.
19. Construct the reciprocal lattice for a two-dimensional lattice in which $a = 1.25$ Å, $b = 2.50$ Å, and $\gamma = 120°$.
20. A unit cell has the dimensions $a = 4$ Å, $b = 6$ Å, $c = 8$ Å, $\alpha = \beta = 90°$, $\gamma = 120°$. Determine:
    a) $a^*$, $b^*$, and $c^*$ for the reciprocal cell.
    b) The volume of the real and reciprocal unit cells.
    c) The spacing between the (210) planes.
    d) The Bragg angle $\theta$ for reflection from the above planes.
21. Show that if the crystal undergoes volume expansion, then the reflected beam is rotated by the angle

$$\delta\theta = -\frac{\gamma}{3}\tan\theta,$$

where $\gamma$ is the volume coefficient of expansion and $\theta$ the Bragg angle.
22. Discuss the variation of the intensity with the half scattering angle $\theta$. Include the effects of the lattice-structure factor, the geometrical-structure factor, and the electron scattering length.
23. Write an essay on the experimental aspects of x-ray diffraction.
24. Prove the result (2.52).
25. A beam of 150-eV electrons falls on a powder nickel sample. Find the two smallest Bragg angles at which reflection takes place, recalling that Ni has an fcc lattice with a cube edge equal to 3.25 Å.

# CHAPTER 3 LATTICE VIBRATIONS: THERMAL, ACOUSTIC, AND OPTICAL PROPERTIES



---

*We are no other than a moving row*
*Of visionary shapes that come and go*
*Round with this Sun-illumin'd Lantern held*
*In Midnight by the Master of the Show.*

Omar Khayyam

## 3.1 INTRODUCTION

In studying crystal structures in the last two chapters, we have assumed that the atoms were at rest at their lattice sites. Atoms, however, are not quite stationary, but oscillate around their equilibrium positions as a result of thermal energy. Let us now discuss these lattice vibrations in detail, and include their influence on the thermal, acoustic, and optical properties of crystals.

In this chapter we shall first consider crystal vibrations in the elastic long-wavelength limit, in which the crystal may be treated as a continuous medium, and we shall compare the various models used to explain specific heat. It is found that agreement with experiment is achieved only through the use of quantum concepts. Thus we shall introduce the phonon, the quantum unit of sound waves. This is followed by a discussion of lattice vibrations, taking into account the discrete nature of the lattice, and we shall also take up the conduction of heat by the lattice.

Direct observations of lattice waves, by the scattering of radiation (such as x-rays) are also discussed. This is followed by a section on some interesting aspects of lattice waves in the microwave region. Finally we shall discuss the reflection and absorption of infrared light by lattice vibrations in ionic crystals.

## 3.2 ELASTIC WAVES

A solid is composed of discrete atoms, and this discreteness must be taken into account in the discussion of lattice vibrations. However, when the wavelength is very long, one may disregard the atomic nature and treat the solid as a continuous medium. Such vibrations are referred to as *elastic waves*.

Let us now examine the propagation of an elastic wave in a sample in the shape of a long bar (Fig. 3.1). Suppose that the wave is longitudinal, and denote the elastic displacement at the point $x$ by $u(x)$. The *strain* is defined as

$$e = \frac{du}{dx}, \tag{3.1}$$

which is the change of length per unit length. The *stress* $S$ is defined as the force per unit area, and is also a function of $x$. According to *Hooke's law*, the stress is proportional to the strain. That is,

$$S = Ye, \tag{3.2}$$

where the elastic constant $Y$ is known as *Young's modulus*.

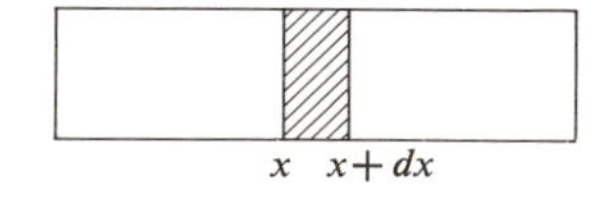


**Fig. 3.1** Elastic wave in a bar.

To examine the dynamics of the bar, we choose an arbitrary segment of length $dx$, as shown in the figure. Using Newton's second law, we can write for the motion of this segment,

$$(\rho A' dx)\frac{\partial^2 u}{\partial t^2} = [S(x + dx) - S(x)]\, A', \tag{3.3}$$

where $\rho$ is the mass density and $A'$ the cross-sectional area of the bar. The term on the left is simply the mass times the acceleration, while that on the right is the net force resulting from the stresses at the ends of the segment. Writing $S(x + dx) - S(x) = \partial S/\partial x\, dx$ for a short segment, substituting for $S$ from (3.2), and then using (3.1) for the strain, one can rewrite the dynamical equation (3.3) as

$$\frac{\partial^2 u}{\partial x^2} - \frac{\rho}{Y}\frac{\partial^2 u}{\partial t^2} = 0, \tag{3.4}$$

which is the well-known *wave equation in one dimension.*

We now attempt a solution in the form of a propagating plane wave

$$u = Ae^{i(qx - \omega t)}, \tag{3.5}$$

where $q$, of course, is the wave number ($q = 2\pi/\lambda$), $\omega$ the frequency of the wave, and $A$ is its amplitude. Substitution in (3.4) leads to

$$\omega = v_s q, \tag{3.6}$$

where

$$v_s = \sqrt{Y/\rho}. \tag{3.7}$$

The relation (3.6) connecting the frequency and wave number is known as the *dispersion relation.* Since the velocity of the wave is equal to $\omega/q$, a fact well known from wave theory, it follows that the constant $v_s$ in (3.6) is equal to this velocity. It is expressed in terms of the properties of the medium by (3.7). The wave under discussion is the familiar sound wave.

Figure 3.2 shows the dispersion relation for the elastic wave. It is a straight line whose slope is equal to the velocity of the sound. This type of dispersion relation, where $\omega$ is related linearly to $q$, is satisfied by other familiar waves. For example, an optical wave traveling in vacuum has a dispersion relation $\omega = cq$, where $c$ is the speed of light. Sound waves in liquids and gases satisfy similar relations.

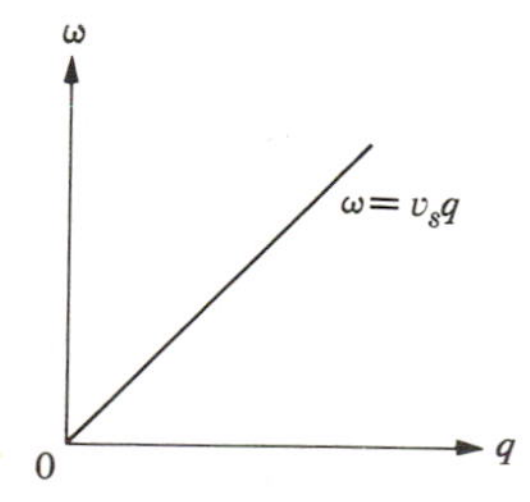


**Fig. 3.2** Dispersion curve of an elastic wave.

Deviations from the linear relationship are often observed, however, and this is known as *dispersion*. We shall see in Section 6, for instance, that the effect of lattice discreteness is to introduce a significant amount of dispersion into the dispersion curve of Fig. 3.2, particularly when the wavelength is so short as to be comparable to the interatomic distance.

Equation (3.7) can be used to evaluate Young's modulus. Measurements show that typical values in solids are $v_s = 5 \times 10^5$ cm/s and $\rho = 5$ g/cm$^3$, which leads to $Y = 5 \times (5 \times 10^5)^2 = 1.25 \times 10^{12}$ g/cm s$^2$.

We have treated a longitudinal wave here, but the same type of analysis also applies to a transverse, or shear wave. This introduces a shear elastic constant, analogous to Young's modulus, and the velocity of the shear wave is related to it by an equation similar to (3.7). The two elastic constants can then be used to describe the propagation of an arbitrary elastic wave in the solid.

It has been tacitly assumed that the solid is isotropic. However, crystals are, in fact, anisotropic, and the effect of anisotropy on the elastic properties is readily demonstrated. This leads in general to the introduction of many more elastic constants than the two needed for the isotropic solid. Considerations of symmetry show, however, that many of these constants are interrelated, a fact which results in a substantial decrease in the number of independent elastic constants. For instance, in the important case of a cubic crystal, it can be shown that only three independent constants are required. They are denoted by $C_{11}$, $C_{12}$, and $C_{44}$. The constant $C_{11}$ relates the compression stress and strain along the [100] direction, e.g., the $x$-axis, while $C_{44}$ relates the shear stress and strain in the same direction. The constant $C_{12}$ relates the compression stress in one direction to the strain in another; these may, for instance, be the $x$- and $y$-directions. The three constants $C_{11}$, $C_{12}$, and $C_{44}$ are determined by measuring the sound velocities in certain directions in the crystal. It can be shown, for example, that the velocities of longitudinal and shear waves along the [100] direction are, respectively, $\sqrt{C_{11}/\rho}$ and $\sqrt{C_{44}/\rho}$, which is expected on the basis of (3.2). The constant $C_{12}$ can be determined from the velocity of the longitudinal wave in the [111] direction, which is found to be $\sqrt{(C_{11} + 2C_{12} + 4C_{44})/3\rho}$. Anyone interested in the further discussion of this topic should read the excellent treatment in Kittel's book.†

## 3.3 ENUMERATION OF MODES; DENSITY OF STATES OF A CONTINUOUS MEDIUM

Consider the elastic waves in the long bar of Fig. 3.1, in which the wave travels in one dimension only. The solution has already been written in (3.5). That is,

$$u = A\, e^{iqx}, \tag{3.8}$$

† References used most frequently in solid state physics are listed at the end of the book.

where we have omitted the temporal factor, since it is not relevant to the present discussion. We shall now consider the effects of the boundary conditions on the solution (3.8). These boundary conditions are determined by the external constraints applied to the ends of the bar. For example, the ends might be clamped as the interior of the bar vibrates, or they might be free to vibrate with the rest of the bar. The type of boundary condition which we shall find most convenient, and which is used throughout this book, is known as the *periodic boundary condition*. By this we mean that the right end of the bar is constrained in such a way that it is always in the same state of oscillation as the left end. It is as if the bar were deformed into a circular shape so that the right end joined the left. Given that the length of the bar is $L$, if we take the origin as being at the left end, the periodic condition means that

$$u(x = 0) = u(x = L), \tag{3.9}$$

where $u$ is the solution given in (3.8). If we substitute (3.8) into (3.9), we find that

$$e^{iqL} = 1. \tag{3.10}$$

This equation imposes a condition on the admissible values of $q$; only those values which satisfy (3.10) are *allowed*. Noting that $e^{in2\pi} = 1$ for any integer $n$, we conclude from (3.10) that the allowed values are

$$q = n\frac{2\pi}{L}, \tag{3.11}$$

where $n = 0, \pm 1, \pm 2$, etc. When these values are plotted along a $q$-axis, they form a one-dimensional mesh of regularly spaced points, as shown in Fig. 3.3. The spacing between the points is $2\pi/L$. When the bar length is large, the spacing becomes small and the points form a quasi-continuous mesh.

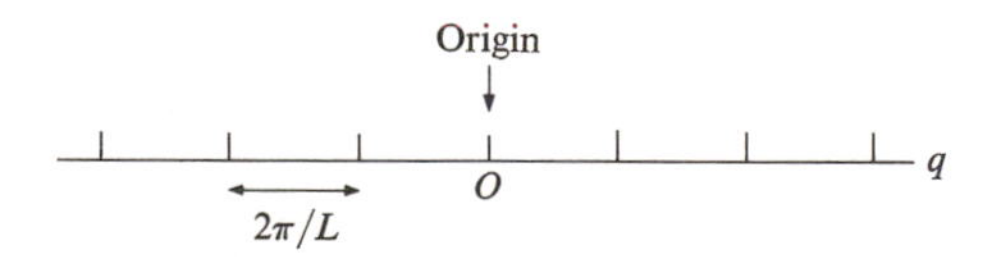


**Fig. 3.3** Allowed values of $q$.

Each $q$-value of (3.11), or each point in Fig. 3.3, represents a *mode* of vibration.† Suppose we choose an arbitrary interval $dq$ in $q$-space, and look for the number of modes whose $q$'s lie in this interval. We assume here that $L$ is large, so that the points are quasi-continuous; this is true, of course, for the macroscopic

† Note that $q = 2\pi/\lambda$, where $\lambda$ is the wavelength of the wave. Thus "quantization" of $q$ in (3.11) is equivalent to quantizing the wavelengths of the allowed waves in the bar.

objects with which we are dealing. Since the spacing between the points is $2\pi/L$, the number of modes is

$$\frac{L}{2\pi}\,dq. \tag{3.12}$$

But $q$ and the frequency $\omega$ are interrelated via the dispersion relation, and we may well seek the number of modes in the frequency range $d\omega$ lying between $(\omega, \omega + d\omega)$. The *density of states* $g(\omega)$ is defined such that $g(\omega)\,d\omega$ gives this number. Comparing this definition with (3.12), one may write $g(\omega)\,d\omega = (L/2\pi)\,dq$, or $g(\omega) = (L/2\pi)/(d\omega/dq)$. We note from Fig. 3.4, however, that in calculating $g(\omega)$ we must include the modes lying in the negative $q$-region as well as in the positive region. The former represent waves traveling to the left, and the latter waves traveling to the right. The effect is to multiply the above expression for $g(\omega)$ by a factor of two. That is,

$$g(\omega) = \frac{L}{\pi}\frac{1}{d\omega/dq}. \tag{3.13}$$

This is a general result for the one-dimensional case, and we see that the density of states $g(\omega)$ is determined by the dispersion relation. For the linear relation Eq. (3.6), $d\omega/dq = v_s$, and therefore

$$g(\omega) = \frac{L}{\pi}\frac{1}{v_s}, \tag{3.14}$$

which is a constant independent of $\omega$.

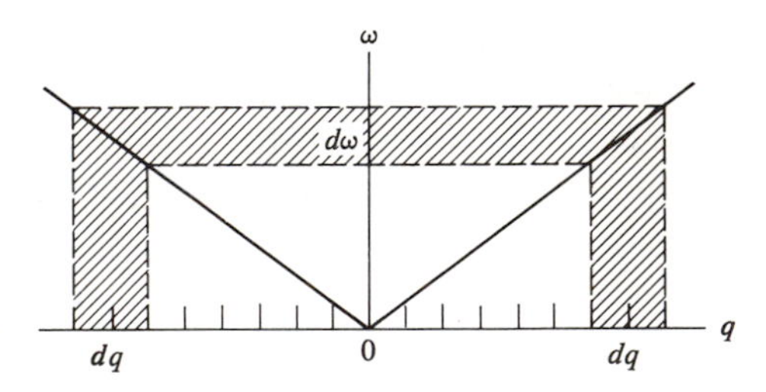


**Fig. 3.4** The enumeration of modes. The dispersion curve is composed of two segments: $\omega = v_s q$ and $\omega = -v_s q$. The former represents waves traveling to the right, the latter waves traveling to the left.

Now let us extend the results to the three-dimensional case. The wave solution analogous to (3.8) is now

$$u = Ae^{i[q_x x + q_y y + q_z z]} = Ae^{i\mathbf{q}\cdot\mathbf{r}}, \tag{3.15}$$

where the propagation is described by the wave vector $\mathbf{q}$, whose direction specifies that of the propagation, and whose magnitude is proportional to the inverse wavelength. Here again we need to inquire into the effects of the boundary conditions. For the sake of simplicity, let us assume a cubic sample whose edge is $L$. By imposing the periodic boundary conditions, one finds that the allowed values of $\mathbf{q}$ must satisfy the condition

$$e^{i(q_xL+q_yL+q_zL)} = 1.$$

That is, the values are given by

$$(q_x, q_y, q_z) = \left(n\frac{2\pi}{L}, m\frac{2\pi}{L}, l\frac{2\pi}{L}\right), \tag{3.16}$$

where $n$, $m$, and $l$ are any three integers.

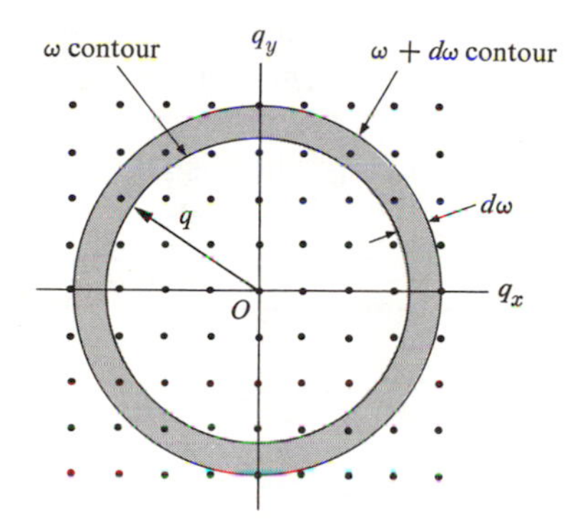


**Fig. 3.5** Allowed values of $\mathbf{q}$ for a wave traveling in 3 dimensions. (Only the cross section in the $q_xq_y$-plane is shown.) The shaded circular shell is used for counting the modes.

If we plot these values in a $\mathbf{q}$-space, as in Fig. 3.5, we obtain a three-dimensional cubic mesh. The volume assigned to each point in this $\mathbf{q}$-space is $(2\pi/L)^3$.

Each point in Fig. 3.5 determines one mode. Suppose we now wish to find the number of modes inside a sphere whose radius is $q$. The volume of this sphere is $(4\pi/3)q^3$, and since the volume per point is $(2\pi/L)^3$, it follows that the number we seek is

$$\left(\frac{L}{2\pi}\right)^3 \frac{4\pi}{3} q^3 = \frac{V}{(2\pi)^3} \frac{4\pi}{3} q^3, \tag{3.17}$$

where $V = L^3$ is the volume of the sample.

This equation gives the number of all the allowed waves whose $q$ is less than a certain value, and which travel in all directions. If we differentiate (3.17) with respect to $q$, we obtain

$$\frac{V}{(2\pi)^3} 4\pi q^2 \, dq, \tag{3.18}$$

which therefore gives the number of modes, or points, in the spherical shell between the radii $q$ and $q + dq$ in Fig. 3.5.

We recall that the density of states $g(\omega)$ is defined such that $g(\omega)\,d\omega$ is the number of modes whose frequencies lie in the interval $(\omega, \omega + d\omega)$. This number can be obtained from (3.18) by making a change of variable from $q$ to $\omega$, which may be accomplished by the use of the dispersion relation. Using the relation $\omega = v_s q$, Eq. (3.6), one finds

$$g(\omega)\,d\omega = \frac{V}{(2\pi)^3}\,4\pi\left(\frac{\omega}{v_s}\right)^2 \frac{d\omega}{v_s}.$$

This expression gives the number of points between the surface of constant frequency at $\omega$ and a similar one at $\omega + d\omega$. Plotted in the **q**-space, these surfaces are spheres, and the volume between them is the spherical shell shown in Fig. 3.5. The above expression for $g(\omega)\,d\omega$ is the number of points inside the shell.

According to the above equation, the density of states $g(\omega)$ is thus given by

$$g(\omega) = \frac{V}{2\pi^2}\frac{\omega^2}{{v_s}^3}. \tag{3.19}$$

This function is plotted versus $\omega$ in Fig. 3.6, where we see that $g(\omega)$ increases as $\omega^2$, unlike the one-dimensional case in which $g(\omega)$ was a constant. The increase in the present case is a reflection of the fact that the volume of the spherical shell in Fig. 3.5 increases as $q^2$, and hence as $\omega^2$, since $\omega$ is proportional to $q$.

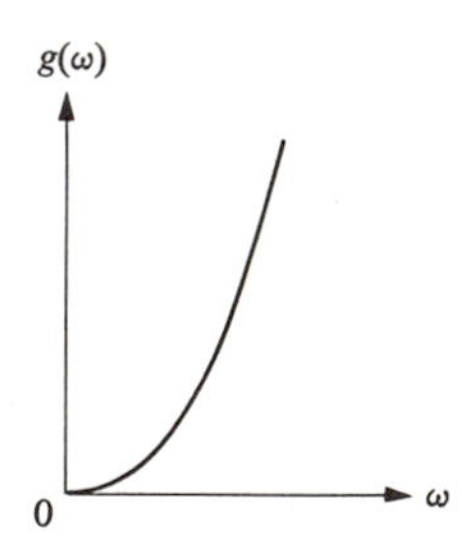


**Fig. 3.6** Density of modes, or states, in an elastic medium.

One last modification is necessary. In the above discussion we have associated a single mode with each value of **q**. This is not quite true for the 3-dimensional case, however, because for each **q** the wave may be either longitudinal or transverse. There are actually three different modes, one longitudinal and two transverse, associated with the same value of **q**. The dispersion relations for the longitudinal and transverse waves are different, since they have different velocities, but if we ignore this difference and assume a common velocity, we may obtain the total density of states from (3.19) by simply multiplying it by a factor of three. That is,

$$g(\omega) = \frac{3V}{2\pi^2} \frac{\omega^2}{v_s^{\,3}}. \tag{3.20}$$

We shall make use of this formula shortly in connection with the Debye theory of specific heat. Note incidentally that $g(\omega)$ is proportional to $V$, the volume of the specimen. We shall often conveniently omit this factor by taking our volume to be equal to unity.

A remark concerning the choice of the periodic boundary conditions: It can be shown that, when the wavelengths of the modes are small compared with the dimensions of the sample, the density-of-states function $g(\omega)$ is independent of the choice of boundary conditions. In using the periodic conditions, we have made the choice which is mathematically most convenient for our purposes.

## 3.4 SPECIFIC HEAT: MODELS OF EINSTEIN AND DEBYE

The specific heat per mole is defined as

$$C = \frac{\Delta Q}{\Delta T},$$

where $\Delta Q$ is the heat required to raise the temperature of one mole by an amount equal to $\Delta T$. If the process is carried out at constant volume, then $\Delta Q = \Delta E$, where $\Delta E$ is the increase in the internal energy of the system. The specific heat at constant volume $C_v$ is therefore given by

$$C_v = \left(\frac{\partial E}{\partial T}\right)_V. \tag{3.21}$$

The specific heat depends on the temperature in the manner shown in Fig. 3.7. At high temperatures the value of $C_v$ is close to $3R$, where $R$ is the universal gas constant. Since $R \simeq 2$ cal/°K mole, at high temperatures $C_v \simeq 6$ cal/°K mole.

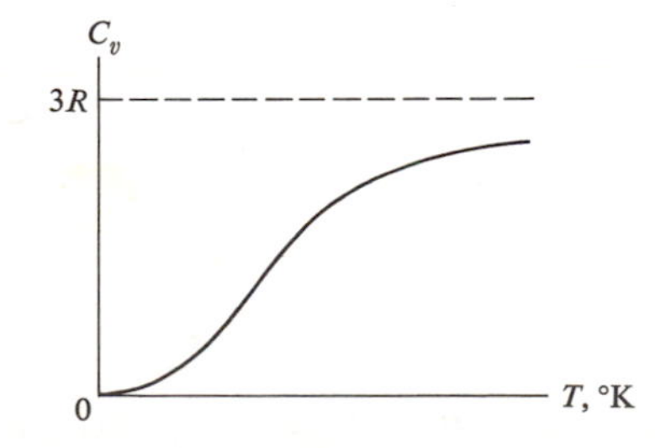


**Fig. 3.7** Dependence of specific heat of solids on temperature.

This range usually includes room temperature. The fact that $C_v$ is nearly equal to $3R$ at high temperatures regardless of the substance described is called the *Dulong–Petit law*.

The deviation from this law in low-temperature regions is strikingly demonstrated by the figure. As $T$ decreases, $C_v$ also decreases, and vanishes entirely at absolute zero. Another observation (which will be relevant to future discussions) is that near absolute zero the specific heat $C_v$ is proportional to $T^3$.

Let us now evaluate $C_v$ theoretically, and compare the value so obtained with the experimental result. First, the so-called *classical* theory: The model used to describe the solid is one in which each atom is bound to its site by a *harmonic* force. When the solid is heated, the atoms vibrate around their sites like a set of harmonic oscillators. The energy associated with this motion is the energy $E$ which appears in Eq. (3.21). Recall from elementary physics that the average energy $\bar{\epsilon}$ for a one-dimensional oscillator is equal to $kT$,† where $k$ is the *Boltzmann constant*. That is,

$$\bar{\epsilon} = kT. \tag{3.22}$$

Therefore the average energy per atom, regarded as a three-dimensional oscillator, is $3kT$, and consequently the energy per mole is

$$E = 3N_A kT = 3RT, \tag{3.23}$$

where $N_A$ is Avogadro's number. We have used the relation $R = N_A k$. When we substitute (3.23) into (3.21), we find, when we have effected the differentiation, that

$$C_v = 3R. \tag{3.24}$$

This result is certainly in agreement with experiment at high temperatures, but it fails completely at low temperatures. Although it predicts a constant value for $C_v$, the actual value, as we have seen, decreases as $T$ decreases, and, in fact, vanishes entirely as $T \to 0°K$. This discrepancy between theory and experiment was one of the outstanding paradoxes in physics until 1905, when it was resolved by Einstein, when he used the then-new quantum mechanics.

**The Einstein model**

In this model, the atoms are treated as independent oscillators, but the energy of the oscillator is given by quantum mechanics rather than by the classical result (3.22). According to quantum mechanics, the energy of an isolated oscillator is restricted to the values

$$\epsilon_n = n\hbar\omega, \tag{3.25}$$

† See, for instance, M. Alonso and E. J. Finn, 1968, *Fundamental University Physics*, Volume III, Reading, Mass.: Addison-Wesley.

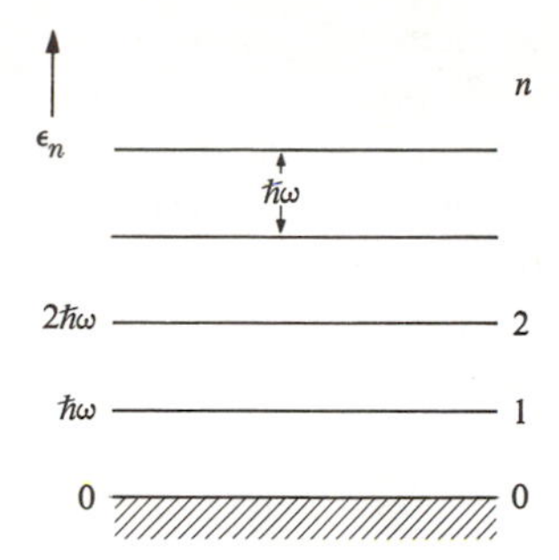


**Fig. 3.8** Spectrum of a one-dimensional oscillator, according to quantum mechanics.

where $n$ is a positive integer or zero.† That is, $n = 0, 1, 2, \ldots$. The constant $\omega$ is the frequency of the oscillator. Thus the energy of the oscillator is *quantized*. The ground state, corresponding to $n = 0$, has an energy $\epsilon_0 = 0$, and the excited states form a discrete, uniformly spaced spectrum, as shown in Fig. 3.8, with an interlevel spacing equal to $\hbar\omega$.

Equation (3.25) refers to an isolated oscillator, but the atomic oscillators in a solid are not isolated. They are continually exchanging energy with the ambient thermal bath surrounding the solid. The energy of the oscillator is therefore continually changing, but its average value at thermal equilibrium is given by

$$\bar{\epsilon} = \sum_{n=0}^{\infty} \epsilon_n e^{-(\epsilon_n/kT)} \Bigg/ \sum_{n=0}^{\infty} e^{-\epsilon_n/kT}.$$

The exponential $e^{-\epsilon_n/kT}$ is the well-known *Boltzmann factor*, which gives the probability that the energy state $\epsilon_n$ is occupied, and the sum in the denominator is inserted for correct normalization.‡ When we substitute from (3.25) into the above equation and evaluate the series involved, we find the simple result§

$$\bar{\epsilon} = \frac{\hbar\omega}{e^{\hbar\omega/kT} - 1}. \tag{3.26}$$

† Actually the exact expression is $\epsilon_n = (n + \frac{1}{2})\hbar\omega$. The lowest state, $n = 0$, is the ground state, while the higher states are the excited states. This shows that the oscillator executes some motion even in the lowest possible state. This is referred to as *zero-point motion*, and its energy as *zero-point energy*. Zero-point motion, since it is irrelevant to the discussion of specific heat, may be disregarded here.

‡ See Alonso and Finn, *op. cit.*

§ The above expression for the average energy may be written as

$$\bar{\epsilon} = -\frac{\partial}{\partial(1/kT)} \ln \left[\sum_{n=0}^{\infty} e^{-\epsilon_n/kT}\right].$$

When expression (3.25) is substituted for $\epsilon_n$, the summation inside the logarithm becomes an infinite geometric series. Summing the series and carrying out the differentiation leads to (3.26).

In Fig. 3.9, which plots the energy $\bar{\epsilon}$ versus temperature, we see that at high temperature the energy $\bar{\epsilon} \to kT$, which is the same as the classical value given above. But as the temperature decreases, the energy $\bar{\epsilon}$ decreases, and continues to decrease until $T = 0°K$, at which point the energy $\bar{\epsilon}$ vanishes entirely. This behavior of $\bar{\epsilon}$ at low temperature is a consequence of the quantum nature of the motion, and is responsible for the classically unexpected decrease in specific heat in the low-temperature region.

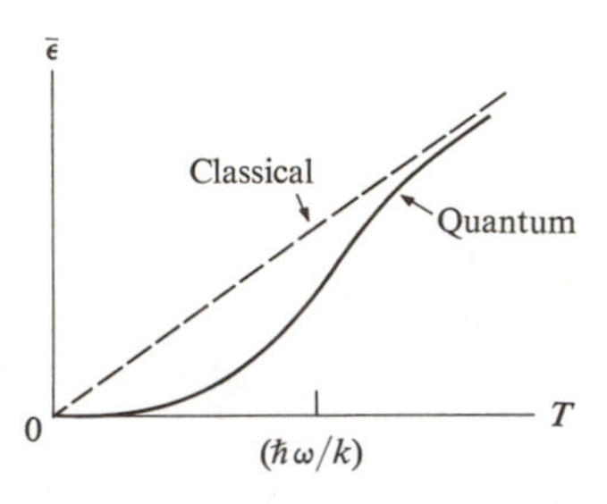


**Fig. 3.9** Energy of the average oscillator versus temperature. The dashed curve is the classical result $\bar{\epsilon} = kT$. Note that the quantum value for $\bar{\epsilon}$ is much less than the classical value at low temperatures.

The behavior shown in the figure may also be understood from the following qualitative argument: An oscillator coupled to a thermal bath exchanges with it an amount of energy which is on the average equal to $kT$. At high temperature, we have $kT \gg \hbar\omega$, which means that the oscillator is in a highly excited quantum state. Since the energy $kT$ is much larger than the quantum step $\hbar\omega$, the quantum nature of the spectrum becomes unimportant, and one expects to obtain the classical result $\bar{\epsilon} = kT$. By contrast, at low temperature, $kT \ll \hbar\omega$, and the energy of exchange $kT$ is not sufficient to lift the oscillator to the first excited state. In this case the energy of the oscillator is much less than $kT$, and is, in fact, very close to zero. as we have found above. Here the quantum nature of the motion plays the dominant role.

Equation (3.26) is the same formula used by Planck in his theory of blackbody radiation. It was there that the concept of the quantization of energy was postulated for the first time. In fact, Einstein's treatment of specific heat closely parallels Planck's theory of blackbody radiation.

We can now find the energy of the solid by noting that each atom is equivalent to three oscillators, so that there is a total of $3N_A$ such oscillators. The total energy is, therefore,

$$E = 3N_A \frac{\hbar\omega_E}{e^{\hbar\omega_E/kT} - 1}, \tag{3.27}$$

where we used $\omega_E$, the *Einstein frequency*, to denote the common frequency of the oscillators. The specific heat, found by differentiating this expression in accordance with (3.21), is

$$C_v = 3R\left(\frac{\hbar\omega_E}{kT}\right)^2 \frac{e^{\hbar\omega_E/kT}}{(e^{\hbar\omega_E/kT}-1)^2}. \tag{3.28}$$

This equation may be simplified by introducing the Einstein temperature $\theta_E$, where $k\theta_E = \hbar\omega_E$. Expression (3.28) then reduces to

$$C_v = 3R\left(\frac{\theta_E}{T}\right)^2 \frac{e^{\theta_E/T}}{(e^{\theta_E/T}-1)^2}. \tag{3.29}$$

If we now plot $C_v$ versus $T$ using this equation, we obtain a curve of the same general shape as Fig. 3.7, which indicates that the theory is now in agreement with experiment, at least qualitatively, over the entire temperature range. Note in particular that $C_v \to 0$ as $T \to 0°K$, a new and important feature of (3.29) which was lacking in the classical theory.

The temperature $\theta_E$ is an adjustable parameter chosen to produce the best fit to the measured values over the whole temperature range. Figure 3.10

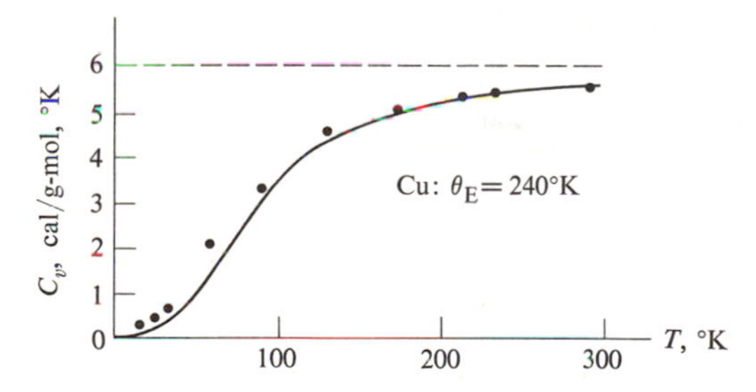


**Fig. 3.10** Specific heat of copper versus temperature. The dots represent experimental values, and the curve is given by the Einstein expression.

illustrates the procedure for copper, where $\theta_E$ is found to be 240°K. The fact that such a good agreement is obtained over such a wide temperature range by adjusting only one parameter is indeed impressive.

We can calculate the Einstein frequency $\omega_E$ once we have determined the temperature $\theta_E$. Thus, for $\theta_E = 240°K$, the frequency $\omega_E = k\theta_E\hbar$ is about $2.5 \times 10^{13}\ s^{-1}$, which is in the infrared region.

Let us now examine the behavior of $C_v$, as given by (3.29) in extreme temperature limits. In the high-temperature limit, where $T \gg \theta_E$, one may expand the exponential $e^{\theta_E/T}$ in a power series of $\theta_E/T$. Carrying this out and retaining only the largest terms in the series, one finds that $C_v \simeq 3R$, which is the classical result. This is to be expected, of course, because in the high-temperature region the quantum aspects of the problem become irrelevant, as we have previously indicated.

In the low-temperature range, in which $T \ll \theta_E$, the exponent $e^{\theta_E/T}$ in (3.29) is much larger than unity. The expression for $C_v$ then reduces to

$$C_v \simeq 3R\left(\frac{\theta_E}{T}\right)^2 e^{-\theta_E/T} \simeq B(T)\, e^{-\theta_E/T}, \tag{3.30}$$

where $B(T)$ is a function relatively insensitive to temperature. Because of the exponential $e^{-\theta_E/T}$, the specific heat approaches zero very rapidly—exponentially, in fact—and vanishes at $T = 0°K$. Although the fact that $C_v \to 0$ as $T \to 0°K$ agrees with experiment, the manner in which this is approached is not. Equation (3.30) indicates that $C_v$ approaches zero exponentially, while experiments show that $C_v$ approaches zero as $T^3$. The decrease predicted by (3.30) is much faster than warranted by experiment, and this, as we shall see, is the basic weakness of the Einstein model.

The Einstein model may be summarized as follows: At high temperature, the oscillator is fully excited, acquiring an average energy equal to $kT$, which leads to a molar specific heat $C_v \simeq 3R$. On the other hand, at low temperature, the oscillator is essentially unexcited, and hence $C_v = 0$; in other words, the oscillator is "frozen" in its ground state. This "freezing" is also the reason why the vibrational modes in diatomic molecules, such as $H_2$, do not contribute to specific heat, except at high temperatures.

In most respects, the Einstein model has been a remarkable success; its results are in good agreement with experiment over most of the temperature range. Nevertheless, the model is incorrect at very low temperatures, at which it predicts a specific heat that is much smaller than the observed value. This disagreement is removed by the Debye model, to which we now turn.

### The Debye model

The atoms in the Einstein model were assumed to oscillate independently of each other. Actually, the idea of independence here is not a viable one because, since the atoms *do* interact with each other, the motion of one atom affects its neighbors. The motion of these in turn affects their neighbors, and so forth, so that the motion of one atom anywhere in the solid, in fact, affects all other atoms present. Thus we need to consider the motion of the lattice as a whole, and not a single independent atom. That is, we must consider the collective lattice modes.

The most familiar example of such collective modes is the sound waves in solids, which were discussed in Section 3.3. When a sound wave propagates in a solid, the atoms do not oscillate independently; their motions are orchestrated in such a manner that they all move with the same amplitude and with a fixed phase relationship.

Using sound waves as a prototype of lattice modes, Debye assumed that all these modes have a character similar to sound waves, i.e., they obey the same dispersion relation given in (3.6),

$$\omega = v_s q. \tag{3.6}$$

We shall see shortly how this may be used to evaluate specific heat. Note that in the Debye model the frequency of the lattice vibration covers a wide range of values since, as $q$ [or the wavelength in (3.6)] varies, so does $\omega$. This is unlike the Einstein model, in which only a single frequency was assumed. The lowest frequency in the Debye model is $\omega = 0$, corresponding to $q = 0$, or an infinite wavelength; the highest allowed frequency is determined by a procedure which will be discussed below.

The assumption that the sound-dispersion relation (3.6) holds for lattice waves is an approximation, inasmuch as it ignores the discreteness of the lattice. The approximation is expected to hold well for those waves of long wavelength, or low frequency, where the consequences of discreteness are unimportant. But when the wavelength is short enough to be comparable to interatomic spacing, the Debye approximation (3.6) will certainly break down. The manner in which (3.6) fails in the short-wavelength region will be discussed in detail in Section 3.5.

Now let us calculate specific heat on the basis of the Debye model. In finding the energy of vibration, we note that each mode is equivalent to a single harmonic oscillator whose average energy is, therefore, given by expression (3.26). The total energy of vibration for the entire lattice is now given by the expression

$$E = \int \bar{\epsilon}(\omega)\, g(\omega)\, d\omega, \tag{3.31}$$

where the integration is effected over all the allowed frequencies. Here $g(\omega)$ is the density-of-states function (Section 3.3), and Eq. (3.31) follows from noting that $g(\omega)\, d\omega$ is the number of modes in the range $(\omega, \omega + d\omega)$, and the energy of each of these modes is equal to $\bar{\epsilon}(\omega)$. In other words, we are treating the vibrating lattice as a set of collective modes which vibrate independently of each other.† In evaluating (3.31), we substitute for $\bar{\epsilon}(\omega)$ from (3.26). The density of states $g(\omega)$ is substituted from (3.20) because, in the Debye approximation, the lattice vibrates as a continuous medium, as we pointed out above in connection with Eq. (3.6). The ensuing expression for the total energy is

$$E = \frac{3V}{2\pi^2 v_s^3} \int \omega^2 \frac{\hbar\omega}{e^{\hbar\omega/kT} - 1}\, d\omega. \tag{3.32}$$

† We may treat the modes as independent of each other, but the atoms themselves must interact. Thus two sound waves in a solid may propagate independently, but then atoms have to interact with each other for any wave to propagate at all.

Before we can evaluate the integral in (3.32), we need to know its limits, namely, the lower and upper ends of the frequency spectrum. The lower limit is evidently $\omega = 0$. The upper cutoff frequency was determined by Debye, by requiring that the total number of modes included must be equal to the number of degrees of freedom for the entire solid. Since this number is equal to $3N_A$, because each atom has three degrees of freedom, the above condition may be expressed in terms of the density of states as

$$\int_0^{\omega_D} g(\omega)\, d\omega = 3N_A, \tag{3.33}$$

where the cutoff frequency, denoted by $\omega_D$, is called the *Debye frequency*. Figure 3.11 shows graphically the manner in which this cutoff is accomplished. It may

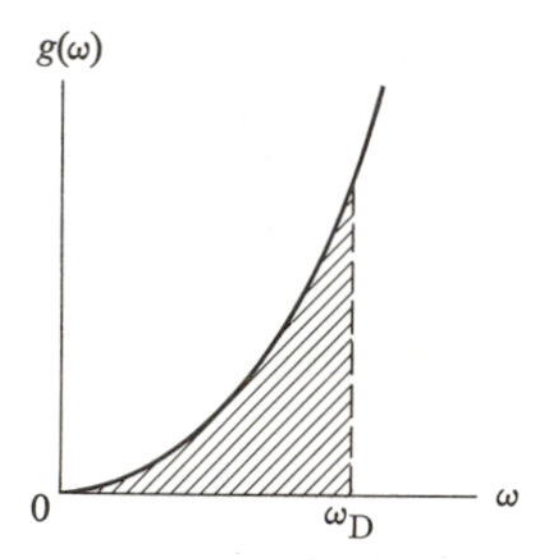


**Fig. 3.11** The Debye cutoff procedure. The shaded area is equal to the number of modes, which is $3N_A$.

be remarked, in justification of the Debye cutoff procedure, that it is a most reasonable one. One must introduce some upper limiting frequency because otherwise many difficulties would arise. For instance, if the upper frequency in (3.32) were allowed to go to infinity, such an infinite energy would not make any sense physically. (The presence of an upper cutoff frequency will arise naturally when we discuss lattice waves in Section 3.6.)

The Debye frequency can be determined by substituting for $g(\omega)$ from (3.20) into (3.33), and carrying out the indicated integration. The result is readily found to be

$$\omega_D = v_s(6\pi^2 n)^{1/3}, \tag{3.34}$$

where $n = N_A/V$ is the concentration of atoms in the solid.

Equation (3.34) may also be derived geometrically. If we draw the contour corresponding to the frequency $\omega = \omega_D$ in the **q**-space of Fig. 3.5, we obtain a sphere enclosing a number of **q**-points equal to $N_A$, as shown in Fig. 3.12. (Recall that each point represents three modes, one longitudinal and two transverse, so the number of modes is equal to $3N_A$, the number of degrees of freedom.) We shall call this surface the *Debye sphere*, and its radius the *Debye radius* $q_D$. Since the number of points inside the sphere is

$$\frac{V}{(2\pi)^3}\frac{4\pi}{3}q_D^3,$$

it follows that the radius $q_D$ must be such that

$$\frac{V}{(2\pi)^3}\frac{4\pi}{3}q_D^3 = N_A.$$

Solving for $q_D$ from this equation, one finds that

$$q_D = (6\pi^2 n)^{1/3}. \tag{3.35}$$

The Debye frequency $\omega_D$ is now found by substituting this value for $q_D$ into the dispersion relation (3.6), and the result is readily seen to lead to (3.34).

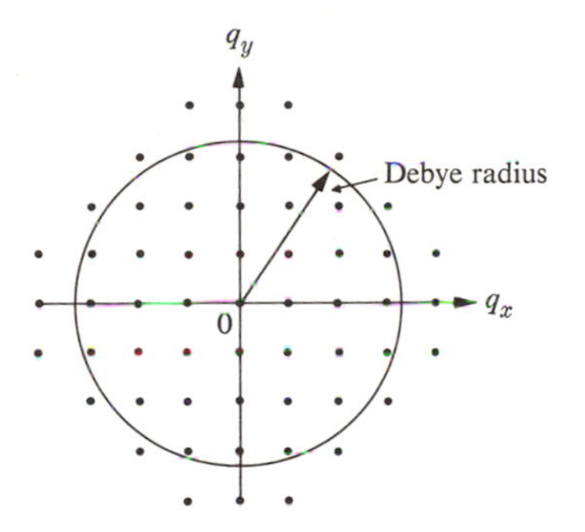


**Fig. 3.12** The Debye sphere.

Returning to (3.32), the total energy is now given by

$$E = \frac{3V}{2\pi^2 v_s^3}\int_0^{\omega_D}\frac{\hbar\omega^3}{e^{\hbar\omega/kT}-1}\,d\omega, \tag{3.36}$$

and the specific heat $C_v$, which is found by differentiating this equation with respect to $T$, is

$$C_v = \frac{3V}{2\pi^2 v_s^3}\frac{\hbar^2}{kT^2}\int_0^{\omega_D}\frac{\omega^4 e^{\hbar\omega/kT}}{(e^{\hbar\omega/kT}-1)^2}\,d\omega. \tag{3.37}$$

We can simplify the appearance of this equation by changing to a dimensionless variable $x = \hbar\omega/kT$, and by defining the *Debye temperature* $\theta_D$ as $k\theta_D = \hbar\omega_D$. Equation (3.37) then takes the form

$$C_v = 9R\left(\frac{T}{\theta_D}\right)^3\int_0^{\theta_D/T}\frac{x^4 e^x}{(e^x-1)^2}\,dx, \tag{3.38}$$

where the velocity of sound $v_s$ has been eliminated by using (3.34). Equation (3.38), which is the specific heat in the Debye model, is the result we have sought.

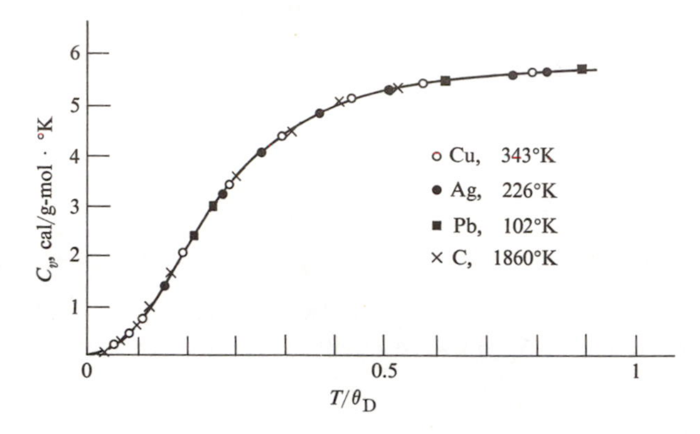


**Fig. 3.13** Specific heats versus reduced temperature for four substances. Numbers refer to Debye temperatures. Note the high Debye temperature for diamond.

**Table 3.1**

Debye Temperatures

| Element | $\theta_D$, °K | Compound | $\theta_D$, °K |
|---|---|---|---|
| Li | 335 | NaCl | 280 |
| Na | 156 | KCl | 230 |
| K | 91.1 | $CaF_2$ | 470 |
| Cu | 343 | Lif | 680 |
| Ag | 226 | $SiO_2$ (quartz) | 255 |
| Au | 162 | | |
| Al | 428 | | |
| Ga | 325 | | |
| Pb | 102 | | |
| Ge | 378 | | |
| Si | 647 | | |
| C | 1860 | | |

To compare (3.38) with experimental results, one must know the Debye temperature $\theta_D$. This is determined by choosing the value which, when substituted into (3.38), yields the best fit over the whole temperature range. We see from (3.38) that if $C_v$ is plotted versus a *reduced* temperature $T/\theta_D$, then the same curve should obtain for all substances. That is, there is a *universal* curve for specific heat. This observation is tested experimentally in Fig. 3.13 for four widely different substances, where we see that it holds true to a remarkable degree. The values for $\theta_D$ have been tabulated in the literature, and a short list is given in Table 3.1.

The value of $\theta_D$ depends on the given substance, but, as Table 3.1 shows, a typical value is about 300°K. The corresponding Debye frequency $\omega_D = k\theta_D/\hbar$ is about $3 \times 10^{13}\ \mathrm{s}^{-1}$, which lies in the infrared region of the spectrum. This frequency may also be evaluated from (3.34). Substitution of the typical values $v_s = 5 \times 10^5$ cm/s and $n = 10^{22}$ atoms/cm$^3$ yields $\omega_D \simeq 4 \times 10^{13}\ \mathrm{s}^{-1}$, which is of the same order as the value calculated above. We note that $\theta_D \sim \omega_D \sim v_s n^{1/3}$, and consequently

$$\theta_D \sim \sqrt{\frac{Y}{\rho}}\, n^{1/3} \sim \sqrt{\frac{Y}{M}}, \tag{3.39}$$

where $Y$ is Young's modulus and $M$ the atomic mass (using $\rho = nM$). Therefore $\theta_D$ depends primarily on the elastic constant of the substance $Y$ and on the atomic mass $M$. The stiffer the crystal and the smaller $M$, the higher is $\theta_D$. This explains, for instance, why $\theta_D$ is high for carbon (1860°K), which is stiff and light, and low for lead (102°K), which is soft and heavy.

It may be demonstrated that $C_v$ as given by (3.38) has the correct values in the appropriate temperature limits. At high temperature, $T \gg \theta_D$, so the upper limit of the integral is very small. Thus $x$ in the integrand is small over its entire range, and we may make the approximation $e^x \simeq 1 + x$. In the first approximation, the integral reduces to $\int_0^{\theta_D} x^2\, dx = \frac{1}{3}(\theta_D/T)^3$, which leads to $C_v = 3R$, in agreement with the Dulong–Petit law. This can also be seen from the following intuitive argument: When $T \gg \theta_D$, every mode of oscillation is completely excited, and has an energy equal to the classical value $kT$. That is, $\bar{\epsilon} = kT$. When we insert this into (3.31), we find that $E = kT \int g(\omega)\, d\omega = 3N_A kT = 3RT$, which leads to $C_v = 3R$. Note also that this saturation in specific heat is already evident when $T \simeq \theta_D$. Since $\theta_D$ is typically close to room temperature, we see that $C_v$ begins to approach its classical value $R$ for $T$ equal to room temperature and above.

The low-temperature limit is more interesting. Here $T \ll \theta_D$, and hence the upper limit of the integral (3.38) approaches $\infty$. The ensuing integral $\int_0^\infty [x^4/(e^x - 1)^2]\, dx$, which may be evaluated analytically, has the value $4\pi^4/15$. The specific heat is now given by

$$C_v = \frac{12\pi^4}{5} R(T/\theta_D)^3. \tag{3.40}$$

This shows the $T^3$-dependence referred to earlier.

The cubic dependence may also be appreciated from the following qualitative argument: At low temperature, only a few of the modes are excited. These are the modes whose quantum energy $\hbar\omega$ is less than $kT$. The number of these modes may be estimated by drawing a sphere in the **q**-space whose frequency $\omega = kT/\hbar$, and counting the number of points inside, as shown in Fig. 3.14. This sphere may be called the *thermal sphere*, in analogy with the Debye sphere discussed above. The number of modes inside the thermal sphere is proportional to $q^3 \sim \omega^3 \sim T^3$. Each mode is fully excited and has an average energy equal to $kT$. Therefore the total energy of excitation is proportional to $T^4$, which leads to a specific heat proportional to $T^3$, in agreement with (3.40).

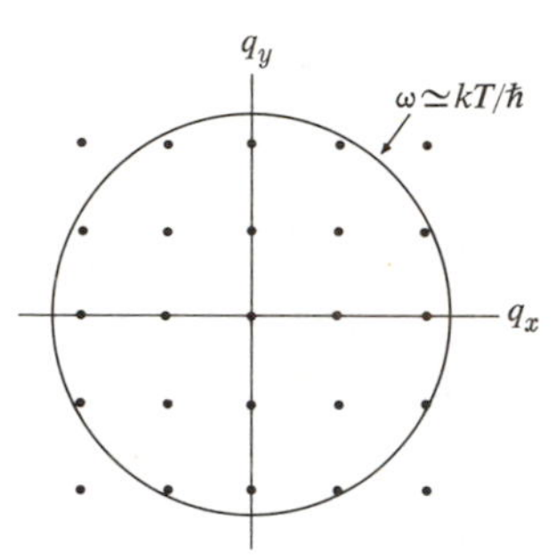


**Fig. 3.14** The thermal sphere which is the frequency contour $\omega = kT/\hbar$.

The reason for the error in the Einstein model at low temperature is now evident. This model ignores the presence of the very low-frequency, long-wavelength modes which can absorb heat even at very low temperature, because their energies of quantization are very small. The exponential freezing of the modes does not actually occur, and the specific heat has a finite value, small though it may be.

Despite its impressive success, the Debye model also remains only an approximation. The nature of the approximation, as pointed out previously, lies in assuming the continuum dispersion relation to hold true for all possible modes of excitation. Experimentally, the approximate nature of the Debye model is shown by plotting $\theta_D$ versus $T$ over a wide temperature range, where $\theta_D$ is found at each temperature by matching the experimental value for $C_v$ with (3.38) at that temperature. If the Debye model were strictly valid, the value of $\theta_D$ so obtained should be independent of $T$. Instead one finds that $\theta_D$ varies with $T$, the variation reaching as much as 10% or even more in some cases. In order to improve on the Debye model, one needs to remove the long-wavelength approximation and use, instead, the correct dispersion relation and the corresponding density of states. This will be taken up in the following sections, beginning with Section 3.6.

## 3.5 THE PHONON

Implicit in the Debye theory is a very important and far-reaching concept. We have seen that the energy of each mode is quantized, the unit of quantum energy being $\hbar\omega$. Since the modes are elastic waves, we have, in fact, quantized the elastic energy of sound waves. The procedure is closely analogous to that used in quantizing the energy of an electromagnetic field, in which the corpuscular nature of the field is expressed by introducing the *photon*. In the present case, the particle-like entity which carries the unit energy of the elastic field in a particular mode is called a *phonon*. The energy of the phonon is therefore given by

$$\epsilon = \hbar\omega. \tag{3.41}$$

Since the phonon also represents a traveling wave, it carries a momentum of its own. By analogy with the photon (see also the deBroglie relation, Section A.2), the momentum of the phonon is given by $p = h/\lambda$, where $\lambda$ is the wavelength. Writing $\lambda = 2\pi/q$, where $\mathbf{q}$ is the wave vector, we obtain for the momentum of the phonon

$$\mathbf{p} = \hbar\mathbf{q}. \tag{3.42}$$

Just as we think of an electromagnetic wave as a stream of photons, we now view an elastic sound wave as a stream of phonons which carry the energy and momentum of the wave. The speed of travel of the phonon is equal to the velocity of sound in the medium.

The number of phonons in a mode at thermal equilibrium can be found from inspection of Eq. (3.26). Since the energy per phonon is equal to $\hbar\omega$, and since the average energy of phonons in the mode is given by $\bar{\epsilon}$ in (3.26), it follows that the average number of phonons in the mode is given by

$$\bar{n} = \frac{1}{e^{\hbar\omega/kT} - 1}. \tag{3.43}$$

This number depends on the temperature; at $T = 0$, $\bar{n} = 0$, but as $T$ increases, $\bar{n}$ also increases, eventually reaching the value $\bar{n} \simeq kT/\hbar$ at high temperatures. Here we see an interesting point: Phonons are created simply by raising the temperature, and therefore the number of them in the system is not conserved. This is unlike the case of the more familiar elementary particles of physics—e.g., electrons or protons—in which the number is conserved.

The concept of the phonon is an extremely important one in solid-state physics, and we shall encounter it time and again in this book. For instance, in Section 3.10, we shall study the interaction of the phonon with other forms of radiation, such as x-rays, neutrons, and light. These interactions will not only validate Eqs. (3.41) and (3.42) for the energy and momentum of the phonon, but will also furnish valuable information on the state of vibration of the solid.

## 3.6 LATTICE WAVES

In discussing waves in solids, we have thus far treated the solid as a continuum, in which the discreteness of the lattice played no significant role. In this section we shall seek to relax this serious oversimplification by treating a solid as the entity composed of discrete atoms which we know it to be.

Of primary importance in our consideration of this point is the form of the dispersion relation $\omega = \omega(q)$. We have already seen that in the long wavelength limit, corresponding to $q \to 0$, the linear relation $\omega = v_s q$ holds good, because there the interatomic spacing is so much smaller than the wavelength that the medium may be treated as a continuum. However, as the wavelength decreases and $q$ increases, the discreteness of the lattice becomes more significant because the atoms begin to scatter the wave. The effect of this scattering is to impede the propagation by decreasing the velocity of the wave. As $q$ increases further, the scattering becomes greater, since the strength of scattering increases as the wavelength decreases (a fact well known in wave physics), and the velocity decreases even further. The effect of this on the dispersion curve is to bend it downward, as indicated in Fig. 3.15, because, as we shall see shortly, the slope of the curve gives the velocity of the wave. Let us now show that the dispersion curve, obtained by solving the equation of motion of the lattice, does indeed have the general shape of Fig. 3.15.

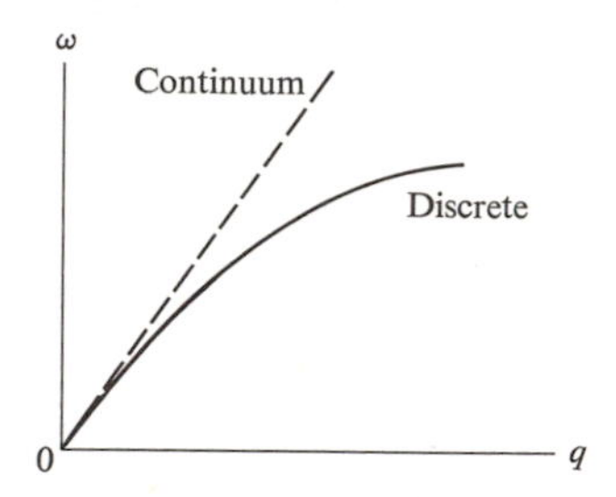


**Fig. 3.15** Expected dispersion curve of a discrete lattice. The dashed line is the continuum model approximation. Note that the two curves coincide at $q = 0$.

For the sake of simplicity, we shall begin the quantitative discussion with the one-dimensional lattice.

### The one-dimensional monatomic lattice

Figure 3.16 shows a one-dimensional monatomic lattice with a lattice constant equal to $a$. When the lattice is at equilibrium, each atom is positioned exactly at its lattice site. Now suppose that the lattice begins to vibrate, so that each atom is displaced from its site by a small amount. Because the atoms interact with each other, the various atoms move simultaneously, so that we must consider the motion of the entire lattice.

Consider the $n^{\text{th}}$ atom. The force exerted on it as a result of its interaction with the $(n + 1)^{\text{th}}$ atom is given by $-\alpha(u_{n+1} - n_n)$, where $u_n$ and $u_{n+1}$ are the displacements of the $n^{\text{th}}$ and $(n + 1)^{\text{th}}$ atoms, respectively, and $(u_{n+1} - u_n)$ is the relative displacement of the atoms. The parameter $\alpha$ is known as the interatomic *force constant.* The assumption that force is proportional to relative displacement is known as the *harmonic approximation*, and it is expected to hold well, provided the displacements are small. This approximation is equivalent to the well-known Hooke's law, familiar from elementary elastic theory [see also Eq. (3.2)]. It is as though the atoms were interconnected by elastic springs. The force exerted on the $n^{\text{th}}$ atom by the $(n-1)^{\text{th}}$ atom is similarly found to be $-\alpha(u_{n+1} - u_n)$. Applying Newton's second law to the motion of the $n^{\text{th}}$ atom, we have therefore

$$M \frac{d^2 u_n}{dt^2} = -\alpha(u_{n+1} - u_n) - \alpha(u_{n-1} - u_n) = -\alpha(2u_n - u_{n+1} - u_{n-1}), \tag{3.44}$$

where $M$ is the mass of the atom.

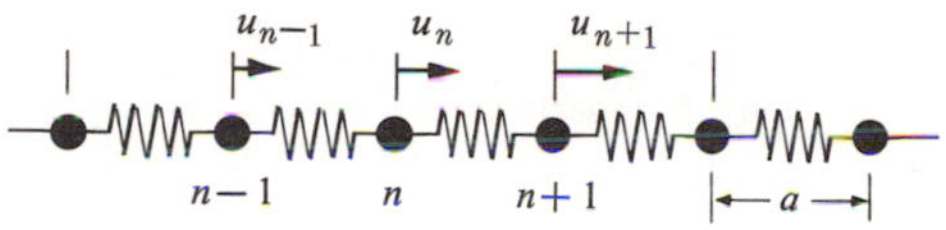

**Fig. 3.16** A segment of a one-dimensional lattice. The arrows represent atomic displacements from equilibrium positions (displacements are exaggerated for illustrative purposes). Springs represent elastic forces between the atoms.

Note that we have neglected the interaction of the $n^{\text{th}}$ atom with all but its nearest neighbors. Although these neglected interactions are small, as the force decreases rapidly with distance, they are not negligible, and must be taken into account in any realistic calculation. The simplified approximation of (3.44) will suffice, however, to illustrate the new physical concepts without involving cumbersome mathematical complexities.

In attempting to solve (3.44), we note that the motion of the $n^{\text{th}}$ atom is coupled to those of the $(n+1)^{\text{th}}$ and $(n-1)^{\text{th}}$ atoms. Similarly the motion of the $(n+1)^{\text{th}}$ atom is found to be related to those of *its* two neighbors, and so forth. Mathematically speaking, one has to write an equation of motion similar to (3.8) for each atom in the lattice, resulting in $N$ coupled differential equations to be solved simultaneously, where $N$ is the total number of the atoms. In addition, the boundary conditions applied to the end atoms of the lattice must also be taken into account.

Let us now attempt a solution of the form

$$u_n = Ae^{i(qX_n - \omega t)}, \tag{3.45}$$

where $X_n$ is the equilibrium position of the $n^{\text{th}}$ atom, that is, $X_n = na$. This equation represents a traveling wave, in which all atoms oscillate with the same frequency $\omega$ and the same amplitude $A$. As expected of such a wave, the phases of the atoms are interlocked such that the phase increases regularly from one atom to the next by an amount $qa$.

Note that a solution of the form (3.45) is possible only because of the translational symmetry of the lattice, i.e., the presence of equal masses at regular intervals. If, on the other hand, the masses had random values, or if they were distributed randomly along the line, then the solution would be expected to be a strongly attenuated wave. In extreme cases, a propagating solution may not even be possible at all. In the discussion of extended systems a mode of vibration such as (3.44), in which all elements of the system oscillate with the same frequency, is referred to as a *normal mode*. In the case of the lattice, the normal mode is a propagating wave.

If we substitute (3.45) into (3.44) and cancel the common quantities (amplitude and time factors), we find

$$M(-\omega^2)\, e^{iqna} = -\alpha[2e^{iqna} - e^{iq(n+1)a} - e^{iq(n-1)a}].$$

This equation can be further simplified by canceling the common factor $e^{iqna}$, and making use of the Euler formula $e^{iy} + e^{-iy} = 2\cos y$. After a simple trigonometric manipulation, we can write the result as

$$\omega = \omega_m |\sin(qa/2|, \tag{3.46}$$

where $\omega_m = (4\alpha/M)^{1/2}$, and where we have restricted $\omega$ to positive values only because of the physical meaning of the frequency. Equation (3.46), which is the dispersion relation for the one-dimensional lattice, is the result we have been seeking. It is sketched in Fig. 3.17, in which the dispersion curve is seen to be a sinusoid with a period equal to $2\pi/a$ in $q$-space, and a maximum frequency equal to $\omega_m$.

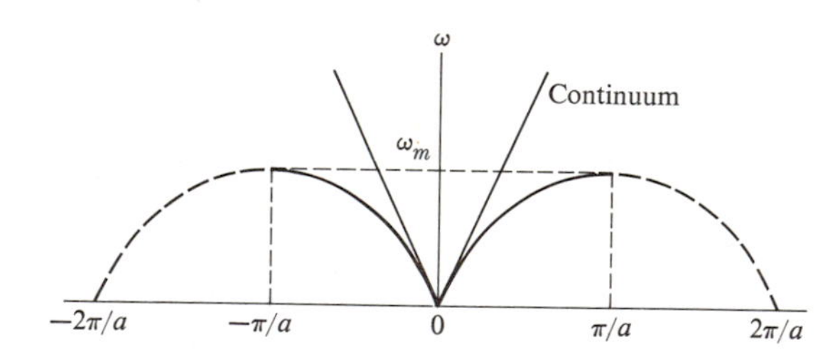


**Fig. 3.17** The dispersion curve, $\omega$ versus $q$, for a one-dimensional lattice with nearest-neighbor interaction. The curve is periodic, but is drawn as a dashed line outside the region $-\pi/a < q < \pi/a$ (see text).

The dispersion relation (3.46) has several important and intriguing properties, which we now examine in some detail, as they apply not only to one- but to two- and three-dimensional lattices as well.

i) *The long-wavelength limit*

Since the dispersion curve is periodic and symmetric around the origin, we may confine our attention for the moment to the range $0 < q < \pi/a$. We see that the frequencies cover the continuous range $0 < \omega < \omega_m$. These frequencies, and only these, will be transmitted by the lattice, while other frequencies will be strongly attenuated. The lattice therefore acts as a *low-pass mechanical filter*.

In the long-wavelength limit as $q \to 0$, the dispersion relation (3.46) may be approximated by

$$\omega = \left(\frac{\omega_m a}{2}\right) q, \tag{3.47}$$

which is a linear relation between $\omega$ and $q$. This result is expected because, in this limit, the lattice behaves as an elastic continuum. The velocity of sound $v_s$ is given by $\omega_m a/2$. We can use (3.47) to relate the interatomic force constant $\alpha$ to the Young's modulus $Y$ of Section 3.2.

Consider the cubic lattice shown in Fig. 3.18. The vibration of the atomic planes satisfies the same equations as that of the one-dimensional lattice. Equating the velocities of sound obtained from (3.47) and (3.7), one finds that $\omega_m a/2 = \sqrt{Y/\rho}$. The substitution of $\omega_m = (4\alpha/M)^{1/2}$ and $\rho = M/a^3$ in this equation leads to

$$\alpha = aY, \tag{3.48}$$

a useful relation for estimating $\alpha$. Inserting typical values for $a$ and $Y$, one obtains $\alpha = (5 \times 10^{-8})(10^{11}) = 5 \times 10^3$ dynes/cm, a typical value.

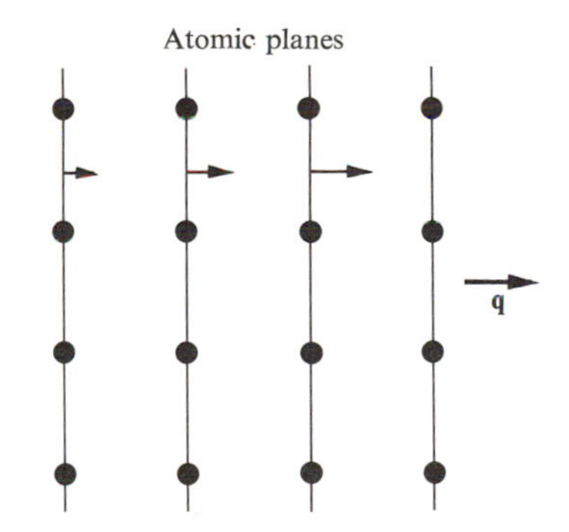


**Fig. 3.18** Motion of atomic planes.

Note, however, that as $q$ increases, the dispersion curve begins to deviate from the straight line, and bends downward, as predicted in Fig. 3.15. Eventually the curve saturates at $q = \pi/a$ with a maximum frequency equal to $\omega_m$, which we found to be

$$\omega_m = \left(\frac{4\alpha}{M}\right)^{1/2}. \tag{3.49}$$

The dependence of this frequency on the force constant and the atomic mass is as one would expect for a harmonic oscillator. In particular, $\omega_m$ is inversely proportional to $M^{1/2}$. The value of $\omega_m$ may be estimated. Substituting $\alpha = 5 \times 10^3$ dynes/cm and $M = 2 \times 10^{-24}$ g (for hydrogen), one finds $\omega_m \simeq 2 \times 10^{13}\ \mathrm{s}^{-1}$, which is in the infrared region.

The above results for the behavior of the dispersion curve in the range $0 < q < \pi/a$ may also be understood from the following qualitative argument. For small $q$, $\lambda \gg a$, and the atoms move essentially in phase with each other, as indicated in Fig. 3.19(a). The restoring force on the atom due to its neighbors is therefore small, which is the reason why $\omega$ is also small. In fact for $q = 0$, $\lambda = \infty$, and the whole lattice moves as a rigid body, which results in the vanishing of the restoring force. This explains why $\omega = 0$ at $q = 0$. The opposite limit occurs at $q = \pi/a$ (Fig. 19b), where $\lambda = 2a$. As we see from the figure, the neighboring atoms are now out of phase, and consequently the restoring force and the frequency are at a maximum.

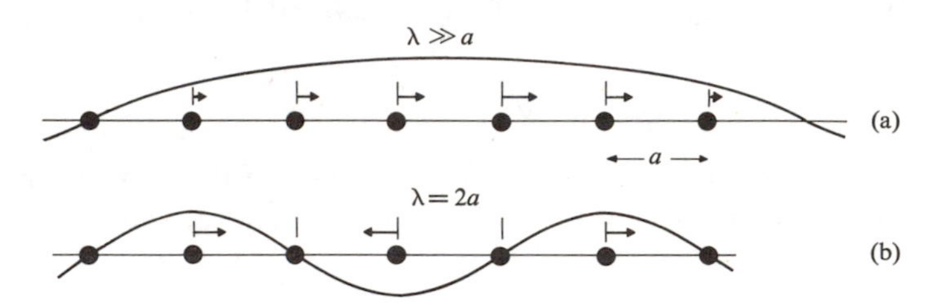


**Fig. 3.19** (a) Atomic displacements in long-wavelength limit. (b) Atomic displacements at wavelength $\lambda = 2a$, which corresponds to $q = \pi/a$.

This discussion may be made quantitative by introducing a force constant which is $q$-dependent: If we return to (3.44), and substitute $u_{n+1} = u_n e^{\pm iqa}$, as follows from (3.45), the former equation reduces to

$$M \frac{d^2 u_n}{dt^2} = -[4\alpha \sin^2 (qa/2]u_n, \tag{3.50}$$

which corresponds to a harmonic oscillator of force constant

$$\alpha(q) = 4\alpha \sin^2 (qa/2). \tag{3.51}$$

This force constant depends on $q$, or $\lambda$, because the motions of the atoms are correlated. The frequency of the oscillator described by (3.51) is given by the familiar harmonic oscillator formula $\omega = \sqrt{\alpha(q)/M}$, which leads precisely to the dispersion relation (3.46) found earlier.

ii) *Phase and group velocity*

What is the velocity of the lattice wave? In wave theory, a distinction is made between two kinds of velocities: *phase velocity* and *group velocity*. For an arbitrary dispersion relation, phase velocity is given by

$$v_p = \frac{\omega}{q}, \tag{3.52}$$

and group velocity by

$$v_g = \frac{\partial \omega}{\partial q}. \tag{3.53}$$

The physical distinction between these velocities is that $v_p$ is the velocity of propagation for a pure wave of an exactly specified frequency $\omega$ and a wave vector $\mathbf{q}$, while $v_g$ describes the velocity of a wave pulse whose average frequency and wave vector are specified by $\omega$ and $q$. Since energy and momentum are transmitted, in practice, via pulses rather then by pure waves, group velocity is physically the more significant.

Let us now examine the behavior of $v_g$ for the discrete lattice. In the long-wavelength limit, in which $\omega = v_s q$, $v_g$ is equal to $v_p$ and both are equal to the velocity of sound $v_s$. In this limit the lattice behaves as a continuum, and no dispersion takes place. But as $q$ increases, it is seen from Fig. 3.17 that $v_g$, being the slope of the dispersion curve [see (3.53)], decreases steadily and reaches a value $v_g = 0$ at the point $q = \pi/a$. The reason for this decrease is that, as $q$ increases, the scattering of the wave by discrete atoms becomes more pronounced, as mentioned earlier in this section.

A particularly interesting situation arises at $q = \pi/a$, in Fig. 3.17, where the group velocity $v_g$ is found to vanish. What is unique about this value of $q$ that leads to the vanishing of $v_g$?

At this value of $q$, the wavelength $\lambda = 2a$. Consequently, as seen in Fig. 3.20, the wavelets scattered from the neighboring atoms are out of phase by an amount $\pi$. But when the wavelet reflected from $B$ reaches that reflected from $A$, the two are in phase. Since this applies to other wavelets as well, it follows that, at $q = \pi/a$, all the scattered wavelets interfere constructively, and consequently the reflection is at a maximum. The situation is obviously the same as the Bragg condition of Section 2.3, but applied here to elastic waves. We now understand physically why $v_g = 0$ at $q = \pi/a$; it is because the reflected wave is so strong that, when combined with the incident wave, it leads to a *standing wave*, which, of course, has a vanishing group velocity.

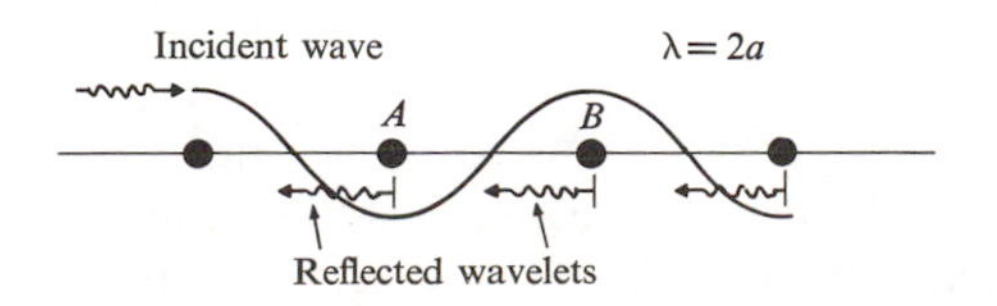


**Fig. 3.20** Bragg reflection of lattice waves.

The reappearance of the Bragg condition in the present context is an important feature of lattice-dispersion curves. It is not surprising that such a condition arises here, inasmuch as it is a consequence of the wave nature of the incident field and the periodicity of the lattice; the particular nature of the field—be it electromagnetic or acoustic—is irrelevant.

iii) *Symmetry in q-space: the first Brillouin zone*

The dispersion curve, Fig. 3.17, has some interesting symmetry properties: It is periodic in $q$-space, and is symmetric with respect to reflection around the origin $q = 0$. We shall now show that these symmetries are not accidental, but follow directly from the translational symmetry of the real lattice.

Consider first the periodic symmetry, which is perhaps the most intriguing property of all. The dispersion relation (3.46) shows that $\omega(q)$ is periodic in $q$-space, with a period equal to $2\pi/a$. That is,

$$\omega(q + 2\pi/a) = \omega(q). \tag{3.54}$$

The physical origin of this becomes clear from the following simple example: Consider the points $q = \pi/2a$ and $q' = q + 2\pi/a$. The wavelengths corresponding to these values are, respectively, $\lambda = 4a$ and $\lambda' = 4a/5$, and are drawn in Fig. 3.21(a). Note from the figure that these two waves represent exactly the *same* physical motion. The shorter wave has more oscillations, but as far as the motions of the atoms themselves (the only entities available for observation), are concerned, the two waves in the figure are physically identical. The two modes must therefore have the same frequency. The same conclusion may be drawn about any two points $q$ and $q'$, where $q' = q + n(2\pi/a)$ for any integer $n$. This explains why the frequency $\omega$ is a periodic function of $q$ with a period $2\pi/a$.

This discussion shows that, in a discrete lattice, the wavelength associated with a certain wave is not a unique quantity, because to this wave many equivalent $q$'s may be assigned, which are related to each other by translations in $q$-space equal to $n2\pi/a$. To each of these $q$'s there is a corresponding wavelength. In order to make a unique representation, one must therefore choose a certain interval in $q$-space whose length is equal to the period, which is, of course, $2\pi/a$. In principle, the choice is entirely arbitrary; however, the one we shall find most convenient to take is $q$ in the range

$$-\frac{\pi}{a} < q < \frac{\pi}{a}.$$

In making this choice, we specify a wave by a unique $q$ and hence a unique $\lambda$. The choice is such that $\lambda$ has the largest possible value consistent with a given set of atomic displacements. The wavelengths corresponding to additional, unobservable oscillations between the atoms have been eliminated. Figure

3.21(b) is a plot of the lattice dispersion curve confined to the chosen interval. Figure 3.21(c) indicates some of the regions which are equivalent to the interval $0 < q < \pi/a$, and others that are equivalent to the interval $-\pi/a < q < 0$. Note that the intervals $0 < q < \pi/a$ and $-\pi/a < q < 0$ are not equivalent, however, because they cannot be related by a translation equal to $n2\pi/a$.

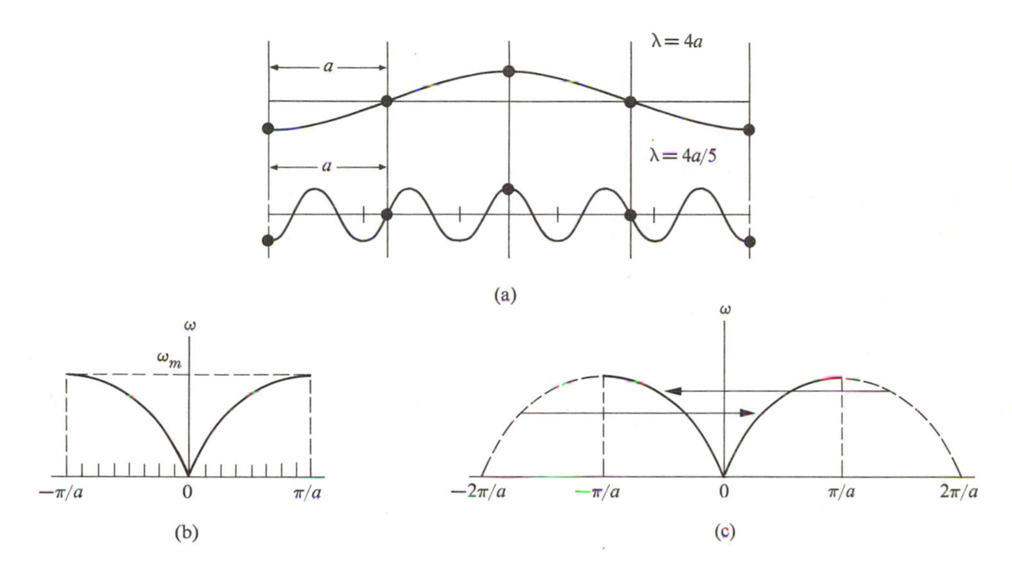


**Fig. 3.21** (a) Transverse waves corresponding to $q = \pi/2a$ and $q' = q + 2\pi/a$, or $\lambda = 4a$ and $\lambda' = 4a/5$, respectively. (b) The range $-\pi/a < q < \pi/a$ is sufficient to give a unique wavelength for all physical oscillations in a one-dimensional lattice. (c) The regions in $q$-space connected by arrows are physically equivalent.

Note that the interval $-\pi/a < q < \pi/a$ is, in fact, the first Brillouin zone for the one-dimensional lattice (see Section 2.7). It follows that we may confine our consideration of $q$-space to the first zone only, disregarding thereby the higher zones, which we have shown to be equivalent to the first zone. This is a mathematical convenience which we shall also use in the three-dimensional lattice, as well as in later discussions on electron states in crystals. Note also that the Bragg condition is satisfied at the ends of the zone, that is, $\pm\pi/a$, another feature which will also be found to hold true in higher-dimensional lattices.

We turn next to the reflection symmetry in $q$-space; that is,

$$\omega(-q) = \omega(q). \tag{3.55}$$

To prove this, note that a mode $q$ represents a wave traveling in the lattice toward the right [see (3.45)], provided $q > 0$. The mode $-q$ represents a wave of the same wavelength, but traveling to the left. Since the lattice is equivalent in these two directions, it responds in the same fashion to the two waves, and the corresponding frequencies must be identical, as indicated by (3.55).

According to this discussion, symmetry properties should hold true in general, regardless of the type of interaction between the atoms, because these properties follow from the symmetry of the real lattice. For instance, if other interactions, besides those of nearest neighbors, were included, the dispersion relation would be more complicated than (3.46), but the translational and reflection symmetries in $q$-space would remain valid.

iv) *The number of modes in the first zone*

We have yet to consider the effects of the boundary conditions on the vibration of the discrete lattice. As in the case of the continuous line (Section 3.3), we shall assume periodic boundary conditions, which means that the first and last atoms have exactly the same oscillation. Applying this to the solution (3.45), one finds, again in analogy with the continuum case, that the only allowed values of $q$ are

$$q = n\frac{2\pi}{L}, \tag{3.56}$$

where $n = 0, \pm 1, \pm 2$, etc. This leads to a uniform mesh of $q$-values, as marked in Fig. 3.21(b), with a spacing equal to $2\pi/L$. When $L$ is large, as would be the case for any lattice of macroscopic size that we would meet in practice, the allowed points come close together, their distribution along the $q$-axis becoming quasi-continuous. The total number of points inside the first zone is $(2\pi/a)/(2\pi/L) = L/a = N$, where $N$ is the total number of atoms, or unit cells in the lattice. This is an important result which holds good in general: The number of allowed $q$-points is equal to the number of unit cells in the lattice.

This conclusion is expected, because the values of $q$ inside the zone uniquely describe all the vibration modes of the lattice. Therefore the number of these values must be equal to the number of degrees of freedom in the lattice, which is $N$.

Finally, let us mention a more general type of lattice motion than has hitherto been considered. Namely, when several waves propagate simultaneously in the lattice, then an atom vibrates with all the corresponding frequencies at the same time. By superposing all the normal modes, it is possible to produce any arbitrary motion in the lattice. This assertion can be established through Fourier analysis, in a manner analogous to that used in the discussion of the vibrating string.†

**The one-dimensional diatomic lattice**

Now consider the one-dimensional *diatomic* lattice. In addition to having the properties of the monatomic lattice, the diatomic lattice also exhibits important features of its own. Figure 3.22 shows a diatomic lattice in which the

† See any textbook on mathematical physics.

unit cell is composed of two atoms of masses $M_1$ and $M_2$, and the distance between two neighboring atoms is $a$. For example, in NaCl, the two masses are those of the sodium and chlorine atoms.

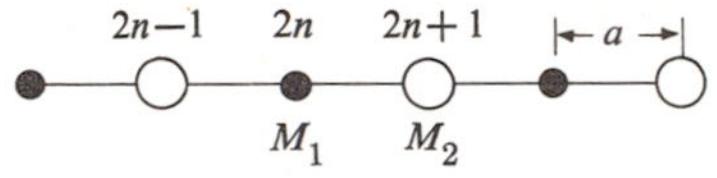

**Fig. 3.22** A one-dimensional diatomic lattice. The unit cell has a length $2a$.

The motion of this lattice can be treated in a manner similar to the motion of the monatomic lattice. Since there are two different types of atoms, we shall write two equations of motion. By analogy with (3.44), we have

$$M_1 \frac{d^2 u_{2n+1}}{dt^2} = -\alpha(2u_{2n+1} - u_{2n} - u_{2n+2}),$$
$$M_2 \frac{d^2 u_{2n+2}}{dt^2} = -\alpha(2u_{2n+2} - u_{2n+1} - u_{2n+3}), \tag{3.57}$$

where $n$ is an integral index, and the subscripts on the displacements are such that all atoms with mass $M_1$ are labeled as odd and those with mass $M_2$ as even. The two equations in (3.57) are coupled. By writing a similar set for each cell in the crystal, we have a total of $2N$ coupled differential equations that have to be solved simultaneously ($N$ is the number of unit cells in the lattice.) To proceed with the solution, we rely on the discussion of the monatomic lattice, and look for a normal mode for the diatomic lattice. Thus we attempt a solution in the form of a traveling wave,

$$\begin{bmatrix} u_{2n+1} \\ u_{2n} \end{bmatrix} = \begin{bmatrix} A_1 e^{iqX_{2n+1}} \\ A_2 e^{iqX_{2n+2}} \end{bmatrix} e^{-i\omega t}, \tag{3.58}$$

which is written in an obvious matrix form. Note that all the atoms of mass $M_1$ have the same amplitude $A_1$, and all those of mass $M_2$ have amplitude $A_2$. If we now substitute (3.58) into (3.57), and make some straightforward simplifications, we find

$$\begin{bmatrix} 2\alpha - M_1\omega^2 & -2\alpha\cos(qa) \\ -2\alpha\cos(qa) & 2\alpha - M_2\omega^2 \end{bmatrix} \begin{bmatrix} A_1 \\ A_2 \end{bmatrix} = 0, \tag{3.59}$$

which is a matrix equation equivalent to a set of two simultaneous equations (write these out) in the unknowns $A_1$ and $A_2$. Since the equations are homogeneous, a nontrivial solution exists only if the determinant of the matrix in (3.59) vanishes. This leads to the *secular equation*,

$$\begin{vmatrix} 2\alpha - M_1\omega^2 & -2\alpha\cos(qa) \\ -2\alpha\cos(qa) & 2\alpha - M_2\omega^2 \end{vmatrix} = 0. \tag{3.60}$$

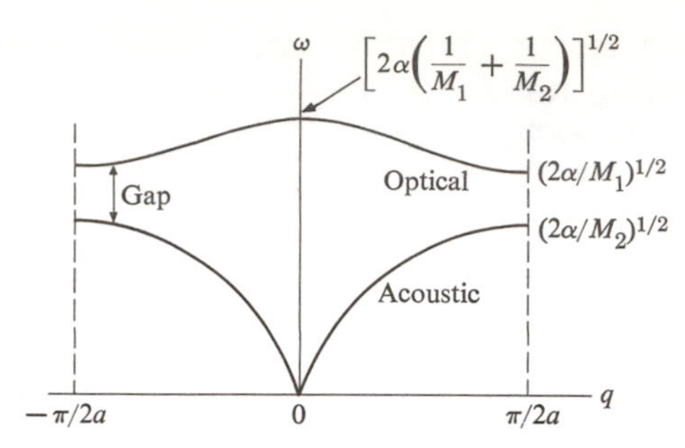


**Fig. 3.23** The two dispersion branches of a diatomic lattice ($M_1 < M_2$), showing frequency gap.

This is a quadratic equation in $\omega^2$, which can be readily solved. Its two roots are

$$\omega^2 = \alpha\left(\frac{1}{M_1}+\frac{1}{M_2}\right) \pm \alpha\sqrt{\left(\frac{1}{M_1}+\frac{1}{M_2}\right)^2 - \frac{4\sin^2(qa)}{M_1M_2}}. \tag{3.61}$$

Corresponding to the two signs in (3.61), there are thus two dispersion relations, and consequently two dispersion curves, or *branches*, associated with the diatomic lattice.

Figure 3.23 shows these curves. The lower curve, corresponding to the minus sign in (3.61), is the *acoustic branch*, while the upper curve is the *optical branch*. The acoustic branch begins at the point $q = 0$, $\omega = 0$. As $q$ increases, the curve rises, linearly at first (which explains why this branch is called acoustic), but then the rate of rise decreases. Eventually the curve saturates at the value $q = \pi/2a$, as can be seen from (3.61), at a frequency $(2\alpha/M_2)^{1/2}$. It is assumed that $M_1 < M_2$. As for the optical branch, it begins at $q = 0$ with a finite frequency

$$\omega = \left[2\alpha\left(\frac{1}{M_1}+\frac{1}{M_2}\right)\right]^{1/2},$$

and then decreases slowly, saturating at $q = \pi/2a$ with a frequency $(2\alpha/M_1)^{1/2}$. The frequency of this branch does not vary appreciably over the entire $q$-range, and, in fact, it is often taken to be approximately a constant.

The frequency range between the top of the acoustic branch and the bottom of the optical branch is forbidden, and the lattice cannot transmit such a wave; waves in this region are strongly attenuated. One speaks here of a *frequency gap*. Therefore the diatomic lattice acts as a *band-pass mechanical filter*.

The dynamic distinction between the acoustic and optical branches can be seen most clearly by comparing them at the value $q = 0$ (infinite wavelength). We may use (3.59) to find the ratio of the amplitude $A_2/A_1$. Inserting $\omega = 0$, for the acoustic branch, one finds that the equation is satisfied only if

$$A_1 = A_2. \tag{3.62}$$

Thus for this branch the two atoms in the cell, or molecule, have the same amplitude, and are also in phase.† In other words, the molecule (and indeed the whole lattice) oscillates as a rigid body, with the center of mass moving back and forth, as shown in Fig. 3.24(a). As $q$ increases, the two atoms in the molecule no longer satisfy (3.62) exactly, but they still move approximately in phase with each other.

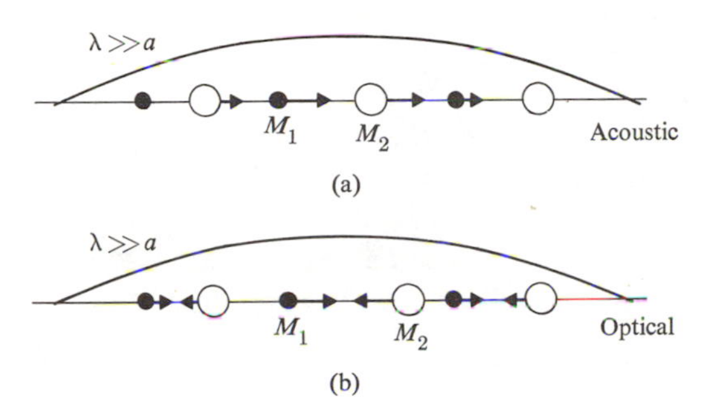


**Fig. 3.24** (a) Atomic displacements in the acoustic mode at infinite wavelength ($q = 0$). (b) Atomic displacements in the optical mode at infinite wavelength.

On the other hand, if we substitute

$$\omega = \left[2\alpha\left(\frac{1}{M_1} + \frac{1}{M_2}\right)\right]^{1/2}$$

for the optical branch, we find that

$$M_1A_1 + M_2A_2 = 0. \tag{3.63}$$

This means that the optical oscillation takes place in such a way that the center of mass of the cell remains fixed. The two atoms move $\pi$ out of phase with each other, and the ratio of their amplitudes $A_2/A_1 = -M_1/M_2$. This type of oscillation around the center of mass is well known in the study of molecular vibrations. As $q$ increases beyond zero, the frequency of the diatomic vibration decreases, but the decrease is not large because the atoms continue to oscillate approximately $\pi$ out of phase with each other throughout the entire $q$-range.

The reasons for referring to the upper branch as *optical* are: First, the frequency of this branch is given approximately by $(2\alpha/M)^{1/2}$, which has a typical value of about $(2 \times 5 \times 10^3/10^{-23})^{1/2} \simeq 3 \times 10^{13}\ \text{s}^{-1}$, using typical values for $\alpha$ and $M$. This frequency lies in the infrared region. Furthermore, if the atoms are charged, as in NaCl, the cell carries a strong electric dipole moment as the lattice oscillates in the optical mode, and this results in a strong reflection and absorption of the infrared light by the lattice, as we shall see in Section 3.12.

† The diatomic lattice may be viewed as an array of diatomic molecules.

Finally, we note that the dispersion curve for the diatomic lattice satisfies the same symmetry properties in $q$-space discussed in connection with the one-dimensional lattice. For example, the dispersion wave is periodic with a period $\pi/2a$, and has a reflection symmetry about $q = 0$. Note that here the first Brillouin zone lies in the range $-\pi/2a < q < \pi/2a$, since the period of the real lattice is $2a$ and not $a$. These assertions concerning symmetry can be established by referring either to (3.61) or Fig. 3.23. It can also be shown, using the periodic boundary conditions, that the number of allowed $q$-values inside the first zone is $N$, and consequently the total number of modes inside this zone is $2N$, since two modes—one acoustic and the other optical—correspond to each $q$. Therefore the total number of modes inside the first zone is equal to the number of degrees of freedom in the lattice, as must be the case.

This suggests that we may confine our attention to the first zone only, as in the monatomic lattice, a procedure we have already followed implicitly.

### Three-dimensional lattice

Let us now extend our discussion to the three-dimensional lattice. To avoid mathematical details, which become quite involved here, we shall present an essentially qualitative discussion. However, the results follow smoothly and logically from the one-dimensional case treated previously.

Consider first the monatomic Bravais lattice, in which each unit cell has a single atom. The equation of motion of each atom can be written in a manner similar to that of (3.44). Here also the atoms are coupled together because of their mutual interactions. In attempting a normal-mode solution, we write

$$\mathbf{u}_n = \mathbf{A}e^{i(\mathbf{q}\cdot\mathbf{r}-\omega t)}, \tag{3.64}$$

where the wave vector $\mathbf{q}$ specifies both the wavelength and direction of propagation. A vector is necessary here because propagation takes place in three dimensions. The vector $\mathbf{A}$ specifies the amplitude as well as the direction of vibration of the atoms. Thus this vector specifies the *polarization* of the wave, i.e., whether the wave is longitudinal ($\mathbf{A}$ parallel to $\mathbf{q}$) or transverse ($\mathbf{A} \perp \mathbf{q}$). (In general the wave in a lattice is neither purely longitudinal nor purely transverse, but a mixture of both.)

When we substitute (3.64) into the equation of motion, we obtain three simultaneous equations involving $A_x$, $A_y$, and $A_z$, the components of $\mathbf{A}$. These equations are coupled together and are equivalent to a $3 \times 3$ matrix equation. Writing the secular equation for this matrix, we arrive at a $3 \times 3$ determinantal equation, analogous to (3.60), which is cubic in $\omega^2$. The roots of this equation lead to three different dispersion relations, or three dispersion curves, as shown in Fig. 3.25(a). All three branches pass through the origin, which means that in this lattice all the branches are acoustic. This is of course to be expected, since we are dealing with a monatomic Bravais lattice.

Note that, in this three-dimensional situation, the dispersion relations are not necessarily isotropic in **q**-space, and the dispersion curve in Fig. 3.25 represents only the "profile" of the dispersion in a certain **q**-direction. If the dispersion relations are plotted in another direction, a new profile will result which may look quite different from the previous one. In the three-dimensional case, therefore, a complete representation of the dispersion relations requires giving the frequencies for points throughout the three-dimensional **q**-space. This is often accomplished by plotting the frequency contours in this space.

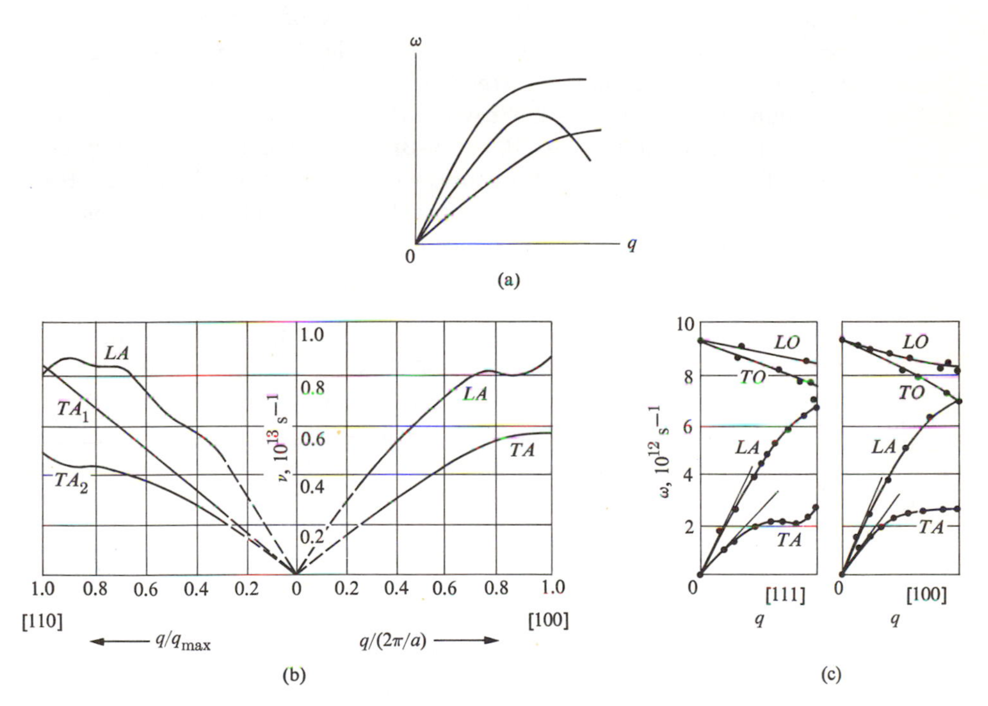


**Fig. 3.25** (a) The three acoustic branches in a three-dimensional Bravais lattice. (b) Dispersion curves for Al in [100] direction (right portion) and in [111] direction, left portion. The TA branch in the [100] direction actually represents two coincident, or degenerate, branches. (Note that because each branch is individually symmetric relative to the origin, only half of each branch is plotted.) (c) Dispersion curves of Ge in the [100] and [111] directions.

The three branches in Fig. 3.25 differ in their polarization. When **q** lies along a direction of high symmetry—for example, the [100] or [110] directions—these waves may be classified as either pure longitudinal or pure transverse waves. In that case, two of the branches are transverse and one is longitudinal.† We

† Usually the longitudinal branch is higher than the transverse branches because the restoring forces associated with longitudinal oscillations are greater.

usually refer to these as the TA (*transverse acoustic*) and LA (*longitudinal acoustic*) branches, respectively. However, along nonsymmetry directions the waves may not be pure longitudinal or pure transverse, but have a mixed character. One may still refer to the branches as TA or LA on the basis of their polarizations along the directions of high symmetry. Figure 3.25(b) shows the dispersion curves for Al in the [100] and [111] directions. Note that in certain high-symmetry directions, such as the [100] in Al, the two transverse branches coincide. The branches are then said to be *degenerate*.

As we have seen, the polarization and degeneracy of dispersion curves are intimately related to the crystal symmetry relative to the direction of propagation. We could carry the subject much further by using the elegant mathematical methods of group theory, but that lies beyond the scope of this book.

We turn our attention now to the non-Bravais three-dimensional lattice. Here the unit cell contains two or more atoms. If there are $r$ atoms per cell, then on the basis of our previous experience we conclude that there are $3r$ dispersion curves. Of these, three branches are acoustic, and the remaining $(3r - 3)$ are optical. The mathematical justification for this assertion is as follows: We write the equation of motion for each atom in the cell, which results in $r$ equations. Since these are vector equations, they are equivalent to $3r$ scalar equations, or to a single matrix equation of the order of $3r \times 3r$. Therefore the secular equation is of degree $3r$ in $\omega^2$, and has three roots, leading to $3r$ branches. It can be shown that three of the roots always vanish at $q = 0$, which results in three acoustic branches. The remaining $(3r - 3)$ roots, therefore, belong to the optical branches, as stated above.

The acoustic branches may be classified, as before, by their polarizations as $TA_1$, $TA_2$, and LA. The *optical* branches can also be classified as longitudinal or transverse when $q$ lies along a high-symmetry direction, and one speaks of LO and TO branches. As in the one-dimensional case, one can also show that, for an optical branch, the atoms in the unit cell vibrate out of phase relative to each other. As an example of a non-Bravais lattice, the dispersion curves for Ge are shown in Fig. 3.25(c). Since there are two atoms per unit cell in germanium, there are six branches: three acoustic and three optical. Note that the two transverse branches are degenerate along the [100] direction, as indicated earlier.

Lattice-dispersion curves are measured by inelastic x-ray or neutron-scattering methods, as we shall see in Section 3.10. These curves can also be calculated theoretically by a procedure similar to that employed in the one-dimensional case. We assume force constants corresponding to the interaction of the atom with its various neighbors. Substitution in the equations of motion, and solution of the corresponding secular equations, leads to the dispersion curves. We then compare these with those measured experimentally, and the force constants are chosen so as to achieve agreement between the experimental and theoretical results (de Launey, 1956).

**Symmetry in $q$-space: the first Brillouin zone (three-dimensional)**

The dispersion relation in the case of a three-dimensional lattice is of the form

$$\omega = \omega_j(q), \tag{3.65}$$

where the index $j$ specifies the branch of interest. The dispersion relation for each individual branch satisfies symmetry properties similar to those discussed in connection with the one-dimensional lattice. In the following discussion, therefore, we shall omit the mathematical details, inasmuch as they are quite similar to those for the one-dimensional case.

First $\omega_j(q)$ satisfies the periodic property

$$\omega_j(\mathbf{q} + \mathbf{G}) = \omega_j(\mathbf{q}), \tag{3.66}$$

where **G** is any reciprocal lattice vector. This means that we may confine our attention to the first BZ (Brillouin zone) only. Also the inversion symmetry

$$\omega_j(-q) = \omega_j(q) \tag{3.67}$$

holds true. Note again that these symmetries, following directly from the translational symmetry of the real lattice, are always satisfied regardless of the solid under consideration.

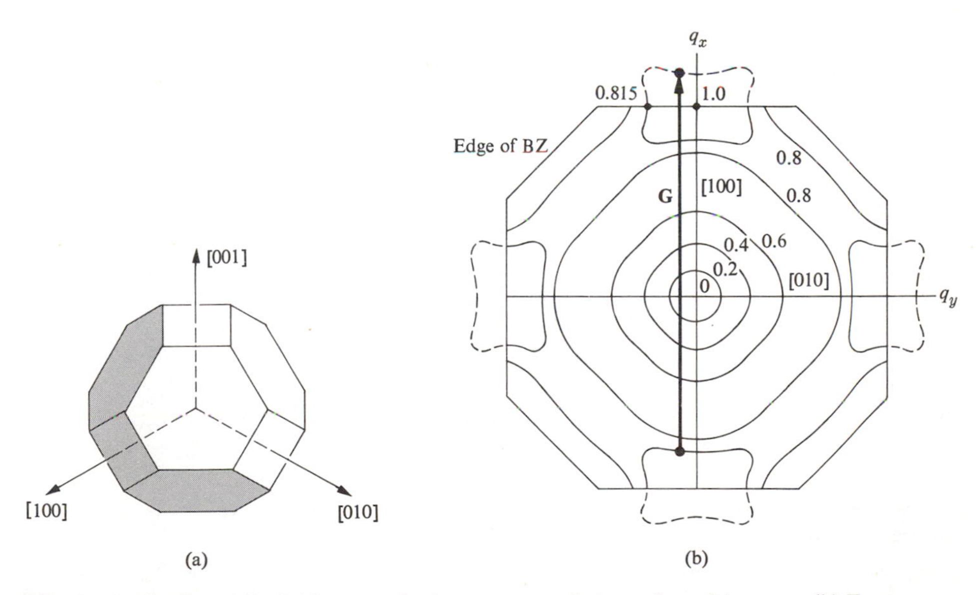


**Fig. 3.26** The first BZ of Al: a tetrahedron truncated along the cubic axes. (b) Frequency ($\omega$) contours for the LA branch in Al (numbers are in units of $2\pi \times 10^{13}\ \text{s}^{-1}$). Note that only a cross section in the $q_xq_y$-plane is shown. (After Walker.)

In addition, the dispersion relation exhibits any rotational symmetry possessed by the real lattice. For instance, in a cubic crystal, each dispersion relation $\omega_j(q)$ exhibits cubic symmetry.

The various symmetries referred to are illustrated in Fig. 3.26. Figure 3.26(a) shows the BZ for Al, which has an fcc lattice, and Fig. 3.26(b) is a plot of the frequency contours in this zone. It is readily seen from this figure that the periodic, inversion, and rotational symmetries are all fulfilled.

In addition to their esthetic value, these symmetries are also important in a practical sense. Thus we usually need to determine the dispersion curve in a small region of the BZ only, and the remainder of the zone can then be completed by using symmetry. Thus in a cubic crystal the dispersion curve need be determined only in $1/48^{\text{th}}$ of the BZ (the cubic rotational group has 48 elements).

Finally, note that the aforementioned symmetries apply to each dispersion branch individually. They do not relate the different branches to each other.

## 3.7 DENSITY OF STATES OF A LATTICE

The *density of states* $g(\omega)$ is defined, as before, such that $g(\omega)\,d\omega$ gives the number of modes in the frequency range $(\omega, \omega + d\omega)$. This function plays an important role in most phenomena involving lattice vibrations, particularly specific heat. We have previously calculated this function for the continuous solid (Section 3.3), and used it in connection with the Debye model of specific heat. Here we shall derive the appropriate function for the discrete lattice, and then use the result in the following section, devoted to the exact theory of specific heat.

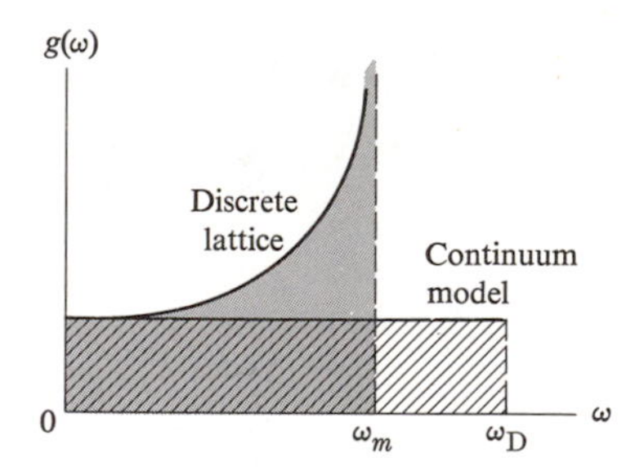


**Fig. 3.27** Density of states for a one-dimensional lattice. For comparison, the density of states for the continuum model is also shown.

Consider first the one-dimensional case. We derived the general formula for the density of states previously, in Eq. (3.13); that is,

$$g(\omega) = \frac{L}{\pi}\frac{1}{d\omega/dq}. \tag{3.68}$$

We see that $g(\omega)$ is calculated by using the dispersion relation. Thus, for the continuous line, the dispersion relation $\omega = v_s q$ leads to $g(\omega) = L/\pi v_s$, while the lattice dispersion relation (3.46) leads to

$$g(\omega) = \frac{2L}{\pi a \omega_m} [\cos (qa/2)]^{-1}. \qquad (3.69)$$

This latter equation is plotted versus $\omega$ in Fig. 3.27. Starting at a finite value at $\omega = 0$, it increases as $\omega$ increases, and reaches an infinite value at $\omega = \omega_m$. For $\omega > \omega_m$, the density $g(\omega)$ vanishes, because this corresponds to a region outside the BZ.

The area under the curve, the stippled region, is equal to the total number of modes, which is $N$. [This can be demonstrated by integrating $g(\omega)$ of (3.69); see the problem section at the end of this chapter.] The figure also shows, for comparison, the density of states for the continuous line in which the upper frequency $\omega_D$ is the Debye frequency, i.e., the cross-hatched area is equal to $N$. Note the structure in $g(\omega)$ for the lattice case, particularly the singularity at $\omega_m$. This is due to the fact that at $\omega = \omega_m$ the dispersion curve, Fig. 3.17, is flat, and consequently a large number of $q$-values—i.e., modes—are included even in a very small frequency interval.

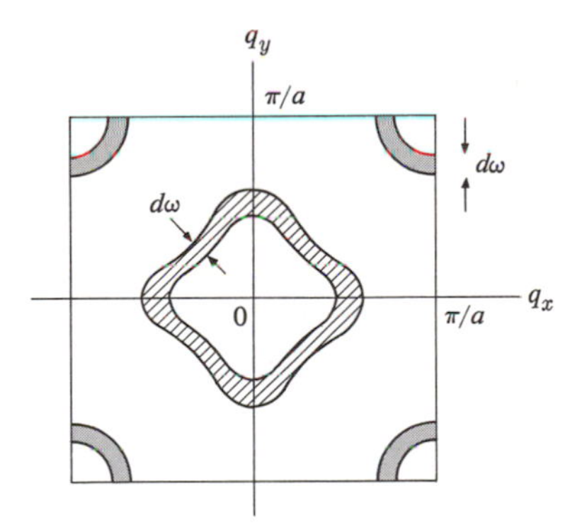


**Fig. 3.28** Counting the number of modes. The cross-hatched region represents a shell well inside the BZ, while the shaded region illustrates the situation when the frequency is so high that the frequency contours intersect the boundaries of the BZ.

To find $g(\omega)$ for the three-dimensional lattice, we follow the same general procedure used in Section 3.3. Consider the $j^{\text{th}}$ branch; we plot the frequency contours $\omega_j(\mathbf{q}) = \omega$ and $\omega_j(\mathbf{q}) = \omega + d\omega$, as shown in Fig. 3.28, and then count the number of modes enclosed between these surfaces. This number is equal to $g_j(\omega)\, d\omega$, and in this manner determine $g_j(\omega)$.

Figure 3.29 illustrates the general features of $g_j(\omega)$. At low frequencies $g_j(\omega)$ increases as $\omega^2$, because the modes involved there are long-wavelength acoustic modes. As $\omega$ increases further, however, $g_j(\omega)$ exhibits some structure determined by the actual dispersion relation, which in turn determines the shape of the shell in Fig. 3.28. (The dispersion relation is, of course, determined by the interatomic force constants, and hence it depends on the crystal in question.) At some frequency, the density $g_j(\omega)$ begins to decrease rapidly, and eventually it vanishes entirely, as shown in the figure. This can be understood by referring to Fig. 3.28. At some frequency the shell begins to intersect the boundaries of the BZ, and when this occurs the number of modes inside the shell decreases (the modes outside the BZ are not counted). When the radius of the shell is sufficiently large for the shell to lie completely outside the zone, the density of states $g_j(\omega)$ vanishes entirely.

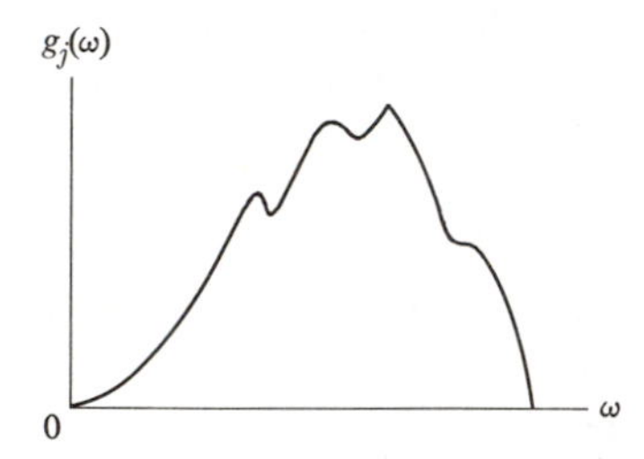


**Fig. 3.29** A typical density-of-states curve.

To find the total density of states, one sums the individual densities of all the branches. That is,

$$g(\omega) = \sum_j g_j(\omega). \tag{3.70}$$

The total density $g(\omega)$ shows the same type of behavior as in Fig. 3.29, except that the structure is even more complicated because of the interference of the various branches. Figure 3.30 shows, for example, the density of states for copper.

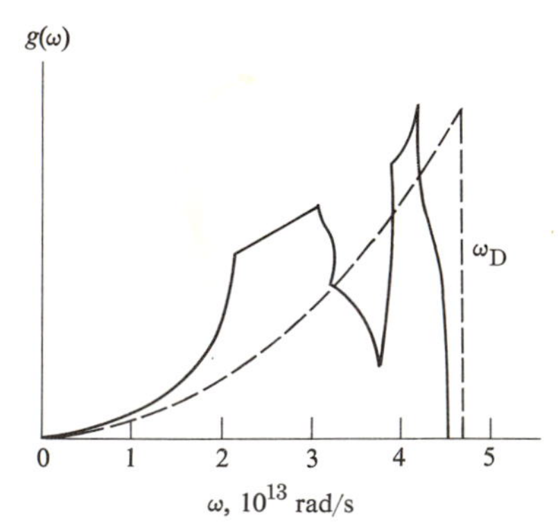


**Fig. 3.30** Total density of states for Cu, as deduced from data on neutron scattering. Dashed curve is the Debye approximation, which has the same area (under the curve) as the solid curve.

## 3.8 SPECIFIC HEAT: EXACT THEORY

In Section 3.4 we discussed the Debye model of specific heat. Recall that the approximation was made there using the linear dispersion relation to describe all vibration modes. In light of our discussion of lattice vibrations, this approximation is justified only near the center of the BZ. In the remainder of the zone, and particularly near the edges, the approximation breaks down entirely because the effects of dispersion there are especially severe.

Also, the treatment of the optical modes lies completely outside the scope of the above model, because their frequencies are essentially independent of $q$, and are, in fact, more appropriately described by the Einstein model.

Now we can dispose of the long-wavelength approximation, and write a general expression for the specific heat involving the actual density of states of the lattice. The general expression for the thermal lattice energy is given by Eq. (3.31). That is,

$$E = \int \bar{\epsilon}(\omega)\, g(\omega)\, d\omega, \tag{3.71}$$

and the specific heat is found by differentiating this expression with respect to temperature, which yields

$$C_v = k\int \left(\frac{\hbar\omega}{kT}\right)^2 e^{\hbar\omega/kT}\, (e^{\hbar\omega/kT} - 1)^{-2}\, g(\omega)\, d\omega, \tag{3.72}$$

since only $\bar{\epsilon}(\omega)$ in (3.71) depends on $T$.

To evaluate $C_v$, we must now substitute the actual density of states function $g(\omega)$. For instance, the specific heat of copper may be evaluated by substituting $g(\omega)$ from Fig. 3.30, and carrying out the integral (3.72) numerically. The agreement with experiment obtained in this manner is decidedly better than that given by the Debye model, particularly in the intermediate-temperature region.

It can readily be established that Eq. (3.72) reduces to the correct values in the appropriate temperature limits. Thus at high temperature all modes are excited, and one can show that $C_v \simeq 3R$, while at low temperature only the long-wavelength phonons are excited. That is, we may take $g(\omega) \sim \omega^2$, which leads to the $T^3$ behavior discussed earlier. The exact theory and the Debye model have the same value at the extreme temperature limits. Where the two theories diverge—i.e., in the intermediate range—the Debye model may be viewed as a good, simple interpolation.

## 3.9 THERMAL CONDUCTIVITY

When the two ends of a sample of a given material are at two different temperatures, $T_1$ and $T_2$ ($T_2 > T_1$), heat flows down the thermal gradient, i.e., from the hotter to the cooler end, as shown in Fig. 3.31. Observations show that the heat current density $Q$ (current per unit area) is proportional to the temperature gradient $(\partial T/\partial x)$. That is,

$$Q = -K\frac{\partial T}{\partial x}. \tag{3.73}$$

The proportionality constant $K$, known as the *thermal conductivity*, is a measure of the ease of transmission of heat across the bar (the minus sign is included so that $K$ is a positive quantity).

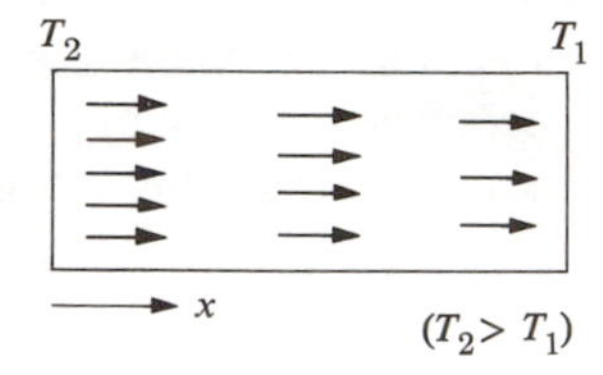


**Fig. 3.31** Thermal conduction by lattice waves (phonons). Arrows represent phonons.

Heat may be transmitted in the material by several independent agents. In metals, for example, the heat is carried both by electrons and by lattice waves, or phonons, although the contribution of the electrons is much the larger. In insulators, on the other hand, heat is transmitted entirely by phonons, since there are no mobile electrons in these substances. Only transmission by phonons will be considered in this section (conduction of heat by electrons is considered in Section 4.6).

When we discuss transmission of heat by phonons, it is convenient to think of these as forming a phonon gas, as shown in Fig. 3.31. In every region of space there are phonons traveling randomly in all directions, corresponding to all the **q**'s in the Brillouin zone, much like the molecules in an ordinary gas.† The advantage of using this gas model is that many of the familiar concepts of the kinetic theory of gases can also be applied here. In particular, thermal conductivity is given by

$$K = \tfrac{1}{3} C_v v l, \tag{3.74}$$

where $C_v$ is the specific heat per unit volume, $v$ the speed of the particle, and $l$ its mean free path. In the present case, $v$ and $l$ refer, of course, to the speed and mean free path of the phonon, respectively. (More explicitly, $v$ and $l$ are *average*

† The process of conduction may be viewed as follows: Since the left end of the bar is hotter, the atoms are moving more violently there than on the right end. Thus the concentration of phonons is greater on the left, and since phonon gas is inhomogeneous, phonons flow from the left to the right, i.e., diffuse down the temperature gradient, carrying heat energy with them.

quantities over all the occupied modes in the Brillouin zone.) Table 3.2 lists the thermal conductivities and mean free paths for a few substances.

**Table 3.2**

Thermal Conductivities and Phonon Mean Free Paths

| | ($T$ = 273°K) | | ($T$ = 20°K) | |
|---|---|---|---|---|
| | $K$, watt/m · °K | $l$, Å | $K$, watt/m · °K | $l$, cm |
| $SiO_2$ (quartz) | 14 | 97 | 760 | $7.5\times10^{-3}$ |
| $CaF_2$ | 11 | 72 | 85 | $1.0\times10^{-3}$ |
| NaCl | 6.4 | 67 | 45 | $2.3\times10^{-4}$ |
| Si | 150 | 430 | 4200 | $4.1\times10^{-2}$ |
| Ge | 70 | 330 | 1300 | $4.5\times10^{-3}$ |

Values of $l$ are calculated from (3.74) by substituting observed values of $K$ and $v$.

Let us now investigate the dependence of the thermal conductivity on temperature. This can be understood by examining (3.74): The dependence of $C_v$ on temperature has already been studied in detail (see Section 3.4), while the velocity $v$ is found to be essentially insensitive to temperature. The mean free path $l$ depends strongly on temperature, as will be seen from the following argument. Recall the analogy from kinetic theory, in which length $l$ is the average distance the phonon travels between two successive collisions. Therefore $l$ is determined by the collision processes operating in the solid. Three important mechanisms may be distinguished: (a) The collision of a phonon with other phonons, (b) the collision of a phonon with imperfections in the crystal, such as impurities and dislocations, and (c) the collision of a phonon with the external boundaries of the sample.

Consider a collision of type (a): When one phonon "sees" another phonon in the crystal, the two scatter from each other, due to the *anharmonic* interaction between them. In our treatment thus far, we have considered phonons to be independent of each other, a conclusion based on the harmonic approximation introduced in Section 3.6. This approximation becomes inadequate, however, when the atomic displacements become appreciable, and this gives rise to anharmonic coupling between the phonons, causing their mutual scattering. It follows that phonon–phonon collision becomes particularly important at high temperature, at which the atomic displacements are large. In this region, the corresponding mean free path is inversely proportional to the temperature, that is, $l \sim 1/T$. This is reasonable, since the larger $T$ is, the greater the number of phonons participating in the collision.

Crystal imperfections, such as impurities and defects, also scatter phonons because they partially destroy the perfect periodicity which is at the very basis of the concept of a freely propagating lattice wave [see the discussion following Eq. (3.45)]. For instance, a substitutional point impurity having a mass different from that of the host atom causes scattering of the wave at the impurity. The greater the difference in mass and the greater the density of impurities, the greater is the scattering, and the shorter the mean free path.

At very low temperature (say below 10°K), both phonon–phonon and phonon–imperfection collisions become ineffective, because, in the former case, there are only a few phonons present, and in the latter the few phonons which are excited at this low temperature are long-wavelength ones. These are not effectively scattered by objects such as impurities, which are much smaller in size than the wavelength.† In the low-temperature region, the primarys cattering mechanism is the external boundary of the specimen, which leads to the so-called *size* or *geometrical effects*. This mechanism becomes effective because the wavelengths of the excited phonons are very long—comparable, in fact, to the size of the specimen. The mean free path here is $l \simeq D$, where $D$ is roughly equal to the diameter of the specimen, and is therefore independent of temperature. The general behavior of the mean free path as a function of temperature is therefore as shown in Fig. 3.32(a). At low temperature, $l$ is a constant $\simeq D$, while at high temperature it decreases as $1/T$. Values of $l$ are given in Table 3.2,

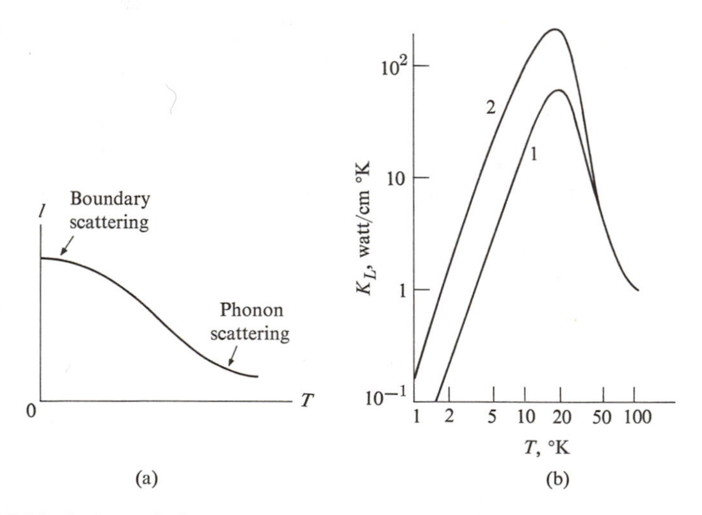


**Fig. 3.32** (a) Variation of phonon mean free path with temperature. (b) Thermal conductivity versus temperature for LiF. Curve 1 is for a bar of cross section $0.755 \times 0.697 \text{cm}^2$. (After Berman.)

† It is well known in wave physics that the strength of scattering of a wave by an object depends on the ratio of the diameter of the object to the wavelength. The smaller this ratio—i.e. the longer the wavelength—the weaker the scattering.

where it is seen that $l$ decreases by several orders of magnitude as $T$ increases from 20°K to, say, room temperature.

Figure 3.32(b) illustrates the temperature dependence of thermal conductivity $K$. At low temperature $K \sim T^3$, the dependence resulting entirely from the specific heat $C_v$ [see (3.74)], while at high temperature $K \sim 1/T$, the dependence now being entirely due to $l$. These conclusions are in agreement with the experimental results of Fig. 3.32(b).

In our discussion of phonon–phonon collision, we glossed over an important but subtle point. Suppose that two phonons of vectors $\mathbf{q}_1$ and $\mathbf{q}_2$ collide, and produce a third phonon of vector $\mathbf{q}_3$. Since momentum must be conserved, it follows that $\mathbf{q}_3 = \mathbf{q}_1 + \mathbf{q}_2$. Although both $\mathbf{q}_1$ and $\mathbf{q}_2$ lie inside the Brillouin zone, $\mathbf{q}_3$ may not do so. If it does, then the momentum of the system before and after collision is the same. Such a process has no effect at all on thermal resistivity, as it has no effect on the flow of the phonon system as a whole. It is called a *normal* process.†

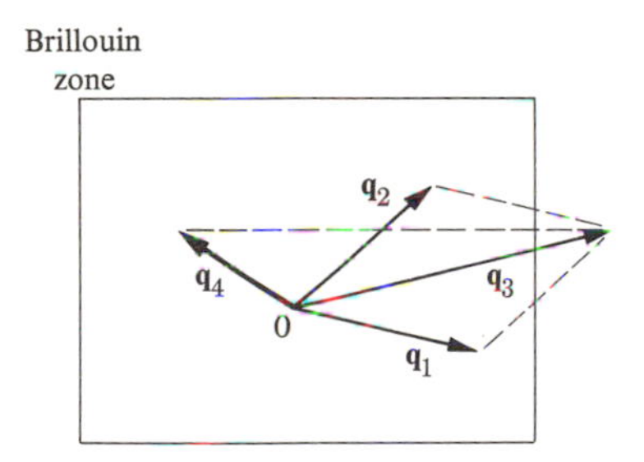


**Fig. 3.33** The umklapp process.

By contrast, if $\mathbf{q}_3$ lies outside the BZ, an interesting new factor enters the picture (Fig. 3.33). Since such a vector is not physically meaningful according to our convention, we reduce it to its equivalent $\mathbf{q}_4$ inside the first zone, where $\mathbf{q}_3 = \mathbf{q}_4 + \mathbf{G}$ (the vector $\mathbf{G}$ is the appropriate reciprocal lattice vector). We see that the effective phonon vector $\mathbf{q}_4$ produced by the collision travels in a direction almost opposite to either of the original phonons $\mathbf{q}_1$ and $\mathbf{q}_2$. (The difference in momentum is transferred to the center of mass of the lattice.) This type of process is thus highly effective in changing the momentum of the phonon, and is responsible for the mean free path of the phonon at high temperature. It is known as the *umklapp process* (German for "flipping over"). It is clear that the umklapp

† Thermal resistivity is simply the inverse of conductivity. What we are saying is that a normal process conserves momentum, and consequently does not contribute to resistivity. In other words, if the normal process was the only process taking place, then the resistivity would be zero, and the thermal conductivity would be infinite. Thus resistivity is due entirely to other collision processes.

process can be effective only at high temperature, where many phonons near the boundaries of the BZ are excited.

## 3.10 SCATTERING OF X-RAYS, NEUTRONS, AND LIGHT BY PHONONS

The dispersion curves of phonons in crystals are determined by the inelastic scattering of x-rays or neutrons from these materials. We have already discussed how the elastic scattering of these radiations is employed to determine the crystalline structure of substances (Chapter 2). There we said that, because of the crystalline arrangement of atoms in a solid, an incident beam underwent a Bragg diffraction, and that this arrangement of atoms was determined by examining the angle and intensity of the diffracted beam. Analogously, when the lattice is in a mode of vibration, the incident beam may be scattered from this mode, and examination of the scattered beam should yield information concerning the mode.

### Inelastic x-ray scattering

Consider first the x-ray scattering process. Figure 3.34 shows an incident beam scattered from a lattice wave whose wave vector is $\mathbf{q}$. Viewing the situation from the quantum vantage point, one concludes that the incident photon absorbs a phonon, and is consequently scattered in a new direction. The law of conservation of momentum requires that

$$\mathbf{k} = \mathbf{k}_0 + \mathbf{q}, \tag{3.75}$$

where $\mathbf{k}_0$ and $\mathbf{k}$ are the wave vectors for incident and scattered x-ray photons, respectively. That is, the momentum transferred to the photon is equal to the momentum of the absorbed phonon. The same equation also holds good if the

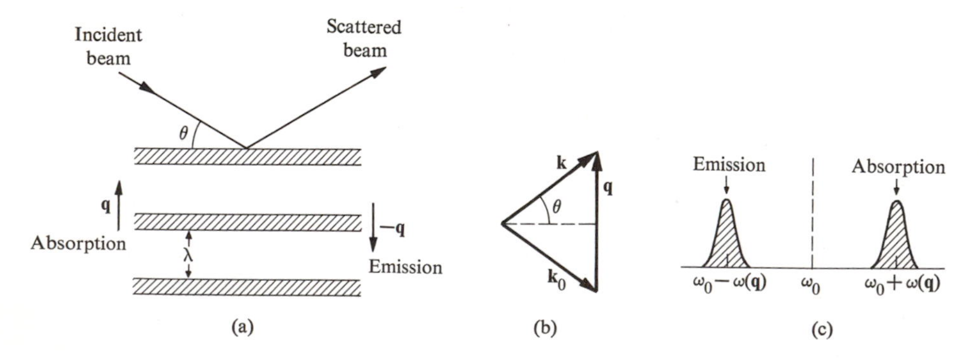


**Fig. 3.34** Scattering of x-rays by phonons. (a) The vibrating lattice acts as a set of planes at spacing equal to $\lambda$. Absorption of a phonon $\mathbf{q}$ and emission of a phonon $-\mathbf{q}$ lead to the same momentum conservation, and hence the two processes are observed simultaneously at the detector. Their frequencies are different, however. (b) Conservation of momentum for x-ray photon–phonon collision. (c) Shifted x-ray frequencies.

x-ray photon has instead emitted a phonon of wave vector $-\mathbf{q}$. This is represented by a lattice wave traveling in the opposite direction, as shown in Fig. 3.34(a). The conservation of momentum (3.75) is illustrated graphically in Fig. 3.34(b).

Energy is also conserved in the scattering process, which requires that

$$\omega = \omega_0 \pm \omega(\mathbf{q}), \tag{3.76}$$

where $\omega_0$ and $\omega$ are the frequencies of the incident and scattered phonon, respectively, and $\omega(\mathbf{q})$ the frequency of the phonon involved. The positive sign in (3.76) refers to the phonon-absorption case, while the minus sign refers to the phonon-emission case [recall that $\omega(-\mathbf{q}) = \omega(\mathbf{q})$; see (3.67)].

The spectrum of the scattered beam, when analyzed at the detector, reveals therefore two lines which are shifted from the incident frequency $\omega_0$ by amounts equal to the frequency of the phonon involved. The positively shifted line at $\omega_0 + \omega(\mathbf{q})$ corresponds to the phonon absorption, and the line at $\omega_0 - \omega(\mathbf{q})$ to the phonon emission. The two shifted lines are situated symmetrically about the unshifted frequency $\omega_0$. The frequency of the phonons can thus be determined from spectral analysis.

The phonon wave vector $\mathbf{q}$ can be determined from Fig. 3.34(b). The magnitude of $\mathbf{q}$ is given by

$$q = 2k_0 \sin\theta = 2n\frac{\omega_0}{c}\sin\theta, \tag{3.77}$$

where $n$ is the index of refraction of the medium and $\theta$ half the scattering angle. In deriving (3.77), we have assumed that $\omega(\mathbf{q}) \ll \omega_0$, which is an excellent approximation because usually $\hbar\omega_0 \simeq 10^4$ eV, while $\hbar\omega(\mathbf{q}) \simeq 0.03$ eV. [Usually the frequency $\omega_0$ is in the visible range, while $\omega(\mathbf{q})$ is in the infrared region, or lower.]†

By measuring the frequency shift and the scattering angle, one can therefore determine both $\mathbf{q}$ and $\omega(\mathbf{q})$, and this determines one point on the dispersion curve of the lattice. By rotating the detector (or the crystal), thereby allowing different phonons to enter the picture, one can sample other points in the Brillouin zone, and by repeating this procedure as often as necessary one can cover the whole zone. The x-ray technique is a standard method for measuring dispersion curves in solids; the dispersion curve for Al shown in Fig. 3.26, for example, was obtained in this manner.

The main disadvantage of the x-ray technique in the study of lattice vibrations lies in the accurate determination of the frequency shift. The photon frequency

---

† The incident frequency $\omega_0$ does not appear at the detector because the angle $\theta$ usually does not satisfy the Bragg condition. Thus only the shifted frequencies are observed. This type of x-ray scattering, which violates the Bragg condition, is referred to as *diffused scattering*. At those angles at which the Bragg condition is satisfied, the incident frequency $\omega_0$ appears together with the shifted frequencies.

$\omega_0$ is so much greater than the phonon frequency $\omega(\mathbf{q})$, typically $\omega_0/\omega(\mathbf{q}) \simeq 10^5$, that a considerable effort must be expended to achieve the needed resolution. This difficulty is overcome by the use of neutron scattering, as will be discussed shortly.

*N.B.*: The scattering of x-rays by phonons, treated above from a quantum point of view, may also be viewed as a classical process in which the electromagnetic wave is diffracted from the acoustic wave. The lattice wave, in producing regions of compression and rarefication in the medium, acts as a set of atomic planes from which the x-ray beam suffers Bragg diffraction, the interplanar spacing being equal to the wavelength. From this vantage point, the momentum equation (3.75) is simply the Bragg condition for constructive interference, Eq. (3.9). The energy equation (3.76) follows from the fact that, since the wave is moving, the x-ray beam should suffer a Doppler shift in its frequency. In the case of phonon absorption, the wave is traveling toward the x-ray beam and the shift is positive, while in the process of phonon emission, the wave travels away from the beam and the shift is negative. When the Doppler shift is treated quantitatively, it leads precisely to (3.76), as you may convince yourself.

### Inelastic neutron scattering

Inelastic scattering of neutrons by phonons may be discussed along the same lines as x-ray scattering, and the details therefore will not be repeated. In particular, the conditions of conservation of momentum and energy [(3.75) and (3.76)] hold true here also. In (3.76), however, the frequency of the neutron is the so-called Einstein frequency, which is related to the energy by $\omega = E/\hbar$, where $E$ is the energy of the neutron ($E = p^2/2m$, where $p$ is the momentum of the neutron).

Just as one can use x-rays, one can use neutron scattering to determine dispersion curves. The important advantage of the neutron technique over the x-ray technique is that the energy of a thermal neutron is only about 0.08 eV (see Section 2.11), which is of the same order as the phonon frequency. The relative shift in frequency is now appreciable, and can therefore be determined with great accuracy. For this reason, the neutron technique is preferable to the x-ray technique, and is used whenever suitable sources of neutrons are available. The dispersion curve for Ge shown in Fig. 3.26(b) was determined by this method.

### Light scattering: Brillouin and Raman

Light waves (or visible photons) may also be scattered by phonons, and this can be used to study the dispersion curves of solids. When the phonon involved is acoustic, the process is known as *Brillouin scattering*; when the phonon is optical, it is *Raman scattering*.

Let us consider first Brillouin scattering. According to the law of conservation of energy, (3.76), the spectrum of the scattered beam should reveal two lines displaced relative to the incident frequency by an amount equal to the phonon frequency, as in the x-ray case. The two lines, shown in Fig. 3.35, are known as *Brillouin wings*. The central unshifted line is not produced by phonon scattering, of course, but by Rayleigh scattering caused by static impurities in the sample, and is called the *Rayleigh line*.

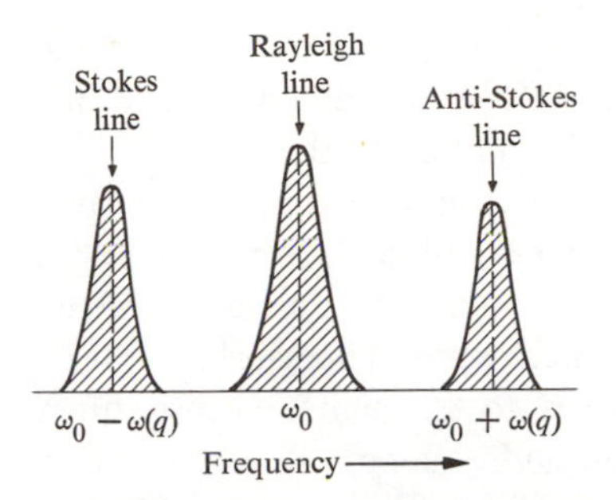


**Fig. 3.35** Raman spectrum, showing undisplaced Rayleigh line, as well as Stokes and anti-Stokes lines.

The lower Brillouin wing, arising from phonon emission, is called the *Stokes line*, and the upper wing, which arises from phonon absorption, is known as the *anti-Stokes line*.

Let us now calculate the Brillouin shift in terms of the scattering angle. This is simplified by the observation that for visible photons, the wave vector $k$ is very small, unlike the x-ray case, in which $k$ is large. To see this, we note that $k = 2\pi/\lambda$, and for the typical value $\lambda = 5000$ Å, $k \simeq 10^5\ \text{cm}^{-1}$. This value is to be compared with the radius of the BZ which, being of the order of $\pi/a$, is about $10^8\ \text{cm}^{-1}$. Therefore $k$ is smaller than the BZ radius by a factor of about $10^{-3}$. Since the $q$ for the phonons involved in the scattering is of the same order as $k$, as seen from Fig. 3.24(b), (which should apply here also), it follows that $q$ is also small, and only long-wavelength phonons participate in light scattering. In other words, this type of scattering probes only that region lying very close to the center of the zone, unlike x-rays or neutrons, which probe the entire zone.

The long-wavelength approximation $\omega(\mathbf{q}) = v_s q$ holds true near the center of the BZ. Using this fact, and Eq. (3.77), one obtains for the Brillouin shift

$$\Delta\omega = \pm 2n\omega_0\left(\frac{v_s}{c}\right)\sin\theta. \tag{3.78}$$

The shift increases with the scattering angle $\theta$. Measurements are often made at right angle to the incident beam—that is, at $\theta = \pi/2$—in order to avoid any interference from this beam.

Note also, from (3.78), that $\Delta\omega/\omega_0 \sim v_s/c$, which is the ratio of the velocity of sound to the velocity of light. One can readily appreciate this if one views the Brillouin scattering as a Doppler-shifted Bragg diffraction, as indicated earlier in connection with x-rays. Since $v_s/c \simeq 10^{-5}$, one sees that the relative shift $\Delta\omega/\omega_0$ is very small (hence $\Delta\omega \simeq 10^{-5}\,\omega_0 \simeq 10^{11}\ \text{s}^{-1}$), and special painstaking techniques are needed for accurate measurements. The task is greatly facilitated by the use of laser sources, in which the frequency can be controlled very accurately.

One can also see from (3.78) that the velocity of sound $v_s$ can be determined from the Brillouin shift. Note that here the sound waves need not be generated externally, as in usual velocity measurements, since the waves are already present in the solid, by virtue of thermal excitations.†

Much of the above holds true for Raman scattering, in which optical phonons are involved. Again Stokes and anti-Stokes lines are observed, and probing is restricted to the region very close to the center of the BZ. There are, however, two primary points in which Raman scattering differs from Brillouin scattering. (1) Raman scattering leads to a much larger frequency shift, since $\Delta\omega$, being equal to the optical-phonon frequency, is of the order of $10^{13}\ \mathrm{s}^{-1}$, compared to about $10^{11}\ \mathrm{s}^{-1}$ or less for Brillouin scattering. (2) Inasmuch as the frequency of the optical phonon is essentially independent of $q$, the Raman shift does not depend on the scattering angle to any significant extent.

Figure 3.36 shows the Raman shifts in ZnSe, the two lines shown corresponding to the LO and TO phonons.

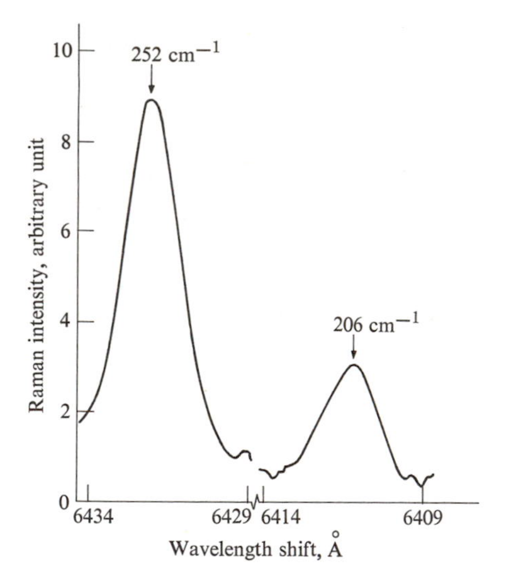


**Fig. 3.36** Raman spectrum of ZnSe, showing scattering from both longitudinal (252 $\mathrm{cm}^{-1}$) and transverse (206 $\mathrm{cm}^{-1}$) optical phonons. (After Mitra.)

Research on Raman and Brillouin scattering in solids has proliferated in recent years, particularly since the advent of the laser. Such a source provides

† The linewidth of the Brillouin wing may be used to determine the lifetime of the phonon. According to the uncertainty relation (Section A.1), the linewidth $\Delta\omega$ and the lifetime $\tau$ are related by the equation $\Delta\omega\tau \simeq 1$. Thus the phonon lifetime is given by $\tau \simeq 1/\Delta\omega$.

a beam which is both intense and monochromatic. The first property is needed for the observation of a sufficiently strong signal, since both Raman and Brillouin scattering, being nonlinear effects, are generally very weak. The high monochromaticity is needed for good resolution of the scattered signal. (Conversely, the phenomenon of light scattering by sound has made beneficial contributions to laser technology. Thus Brillouin scattering is employed for light-beam deflection in Q-switching, a technique for generating high laser pulses.)

Note also that these scatterings can be used to provide sources of tunable coherent radiation. If the optical beam is coherent, which is the case for a laser source, then the phonons emitted are phase-locked to the incident beam, and consequently the scattered Stokes radiation is also coherent. (It is assumed that the temperature is sufficiently low for the anti-Stokes radiation, which is incoherent, to be suppressed.) Phonon beams have been generated in this manner extending from 100 kHz up to several GHz. In this respect, the lattice acts as a parametric amplifier.

## 3.11 MICROWAVE ULTRASONICS

Some of the most interesting phenomena associated with lattice waves are exhibited in the study of those acoustic waves which lie in the microwave region, where $\omega \simeq 10^{10}\ \text{s}^{-1}$; these are known as *ultrasonic* waves. We shall also see that these phenomena hold the promise of providing extremely useful electronic devices. This explains, at least in part, the considerable practical attention directed toward these waves in recent years.

It should be noted at the outset that, for $\omega \simeq 10^{10}\ \text{s}^{-1}$, the wavelength $\lambda$ is about $10^{-5}$ cm, which is still much larger than the interatomic spacing. The continuum approximation may therefore be employed in the ultrasonic region.

Figure 3.37 shows a common method for generating an ultrasonic wave in a sample. The sample is bonded to a quartz crystal rod, which in turn is coupled to an electromagnetic microwave cavity. As the cavity is excited, its electromagnetic field couples to the end of the quartz rod through piezoelectric interaction,† and causes the end to oscillate. The oscillation is then propagated as an elastic wave down the quartz rod, and then through the sample. In effect, part of the electromagnetic energy in the cavity has been converted into elastic energy.

The quartz rod acts as a *transducer*, which is a convenient intermediary for coupling the electromagnetic field to the sample. Were it not for this, the coupling between the sample and the field would be weak, since the former is not usually piezoelectric.

The efficiency of conversion decreases rapidly with the frequency, but is typically of the order of $10^{-4}$. Therefore, to obtain an appreciable amount of

---

† Piezoelectricity is discussed in Section 8.11.

elastic energy, the cavity is usually pulsed at high power, of the order of several watts, for a short period of about 1 $\mu$s.

In this manner one can generate a coherent ultrasonic phonon beam, which can then be employed to study physical processes in the solid. Because one can control the direction, frequency, and polarization of such a beam, it is more amenable to accurate measurement than the ultrasonic phonons excited thermally; these cannot be conveniently controlled, inasmuch as they are excited in all directions, with all possible polarizations, and over a large frequency range.

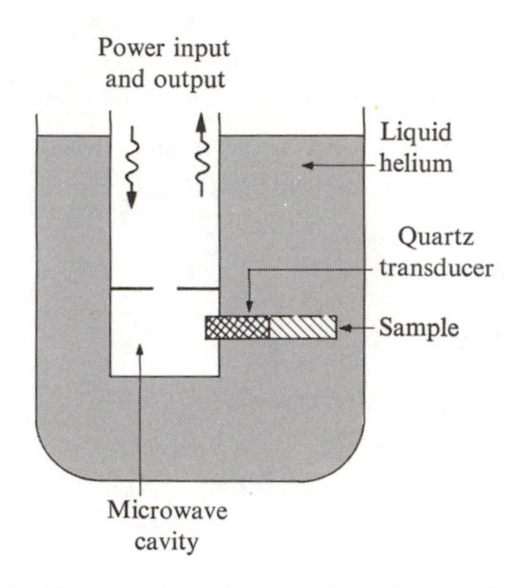


**Fig. 3.37** Experimental setup for ultrasonic studies.

The amount of information obtainable increases rapidly with the frequency of the ultrasonic beam. Most of the work in the area has been performed in the range 1–16 GHz, although frequencies up to 100 GHz have been generated and studied.

The physical quantities measured are the velocity of sound and the attenuation coefficient in the sample. From these one obtains information on the internal structure of the solid, as we shall see. The velocity is determined by measuring the travel time of the sound pulse as it propagates back and forth in the sample, while the attenuation coefficient is determined from the amplitude of the reflected echo pulse. The attenuation is often quite large, and increases rapidly with frequency and temperature. To mitigate excessive attenuation, experiments are often made at very low temperatures, near that of liquid helium.

One of the many applications of ultrasonic waves in electronics is in the design of microwave acoustic delay lines and delay line amplifiers. In such devices, an electromagnetic signal is fed into one end of the sample, where it is converted into an acoustic wave. The wave propagates down the sample, amplified if necessary, and is then reconverted to an electromagnetic signal at the other end of the sample. Note, however, that the wave, being acoustic, travels in the sample with the slow speed of sound $v_s$. Had the signal not been converted, it would have traveled with a much greater velocity, closer to the speed of light $c$. Since $v_s/c \simeq 10^{-5}$, the signal is, in fact, delayed significantly. The same delay can be achieved acoustically as that achieved by purely electromagnetic means, using a cable $10^5$ times the length of the sample, e.g., a sample 5 cm long is equivalent to a cable 5 km long. The size reduction is very striking indeed.

Many other applications are anticipated, and it is hoped that many of the functions of microwave cavities will one day be accomplished by the use of ultrasonic devices, at a great reduction in cost and size.

Let us now talk about some of the physical processes which take place when a coherent beam of phonons travels along a crystalline sample. The coherent beam of phonons is scattered by thermal phonons and by imperfections, of course (as we discussed in connection with thermal conductivity in Section 3.39), and also by conduction electrons in the case of metals. By measuring the effect of this scattering on the coherent phonons, one obtains information about the thermal phonons, and also about the imperfections.

N. S. Shiren has also studied the interaction between two coherent phonons. He caused two waves of frequencies 16.45 GHz and 8.5 GHz to be propagated in an MgO sample. These two waves coupled by anharmonic nonlinear interaction, and Shiren found that by pumping at the higher frequency, he could also increase the intensity of the lower frequency. Here the lattice acted as a parametric acoustic amplifier.

A coherent beam of phonons may also be used in the study of spin–phonon interaction. If the sample contains paramagnetic impurities—for example, $Mn^{2+}$ in quartz or $Cr^{3+}$ in MgO—the energy level of the impurity splits in the presence of a magnetic field, as shown in Fig. 3.38.†

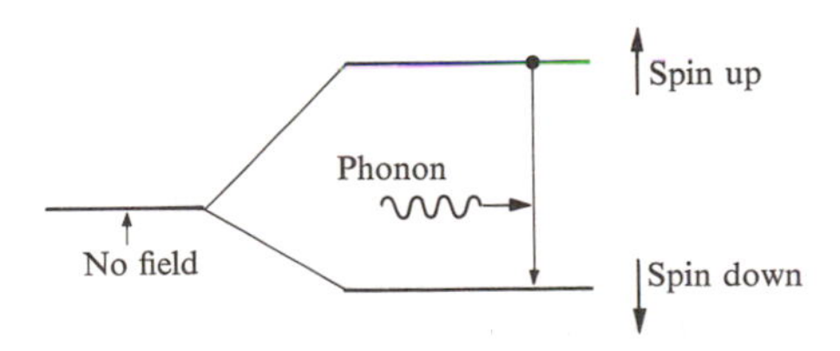


**Fig. 3.38** Absorption of a phonon by a magnetic impurity.

If the frequency of the phonon is such that $\hbar\omega$ equals the energy split between the spin levels, then the phonon is strongly absorbed by the spin system; for each phonon absorbed, an atom in the system flips its spin. By studying the phonon absorption, one therefore obtains information about the energy structure of the impurities, and the strength of their coupling to the phonon. (The process

† Magnetic impurities are discussed in Section 9.6.

discussed here is analogous to the more familiar electron spin resonance, in which the spin flip results from photon absorption. See Section 9.12.)

In the above example, the phonon beam suffers some attenuation. This is the normal situation. Under some circumstances, however, the beam may actually grow as it travels down the sample, and in that case the sample then acts as a *phonon amplifier*. Many types of such amplifiers have been produced, but the one which has received the most attention is the one that involves the acoustoelectric effect in piezoelectric semiconductors. The physical principle underlying the operation is as follows.

A semiconductor contains many free electrons which, under the application of a suitably large electric field, can be made to drift down the sample at a high velocity, as shown in Fig. 3.39(a). Suppose now that an acoustic wave also travels down the sample. The wave then couples to the drifting electrons, the coupling being particularly strong in piezoelectric materials. It can be shown that, if the wave velocity $v_s$ is slightly less than the drift velocity $v$, then energy is transferred from the electron beam to the wave, and hence the wave is amplified.† We can appreciate the physical process if we refer to Fig. 3.39(b). Because of the wave, the electrons find themselves effectively in a periodic electrostatic potential, with more electrons on the leading side of the wave trough. The electrons, therefore, tend to slide down the slope to the bottom of the trough, and the energy lost thereby is then converted into an elastic energy in the wave. Useful acousto-elastic delay line amplifiers have been built using CdS and ZnO up to a frequency of about 14 GHz. An amplification up to 100 dB/cm has been achieved for a frequency of 1 to 2 GHz.

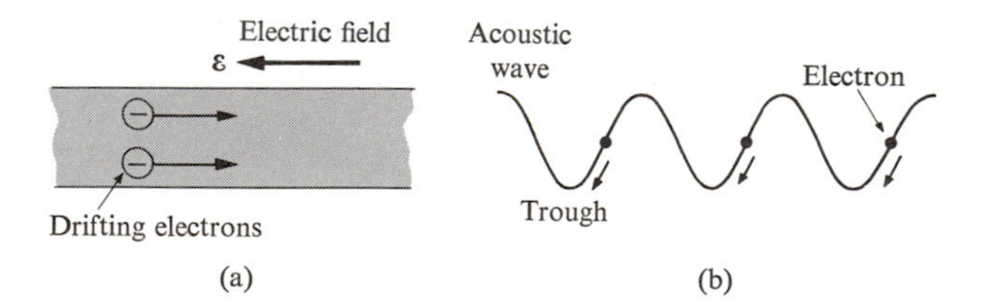


**Fig. 3.39** The principle of the acoustoelectric amplifier. (a) Electrons are set adrift at high velocity by application of a large electric field. (b) Electrons slide down the wave trough, thereby releasing some of their energy to the wave, which is thus amplified.

Perhaps the most promising of all the microwave ultrasonic devices are those employing *surface* lattice waves. These waves, also known as *Rayleigh waves*, travel strictly along the surface of the sample, the amplitude damping out completely within a distance roughly equal to one wavelength from the surface. The

† A familiar analog is that of wind blowing over a wave in the sea. If the wind is faster than the wave, then the wave is amplified by the transference of energy from the wind to the wave.

velocity of the surface waves is approximately the same as that of the bulk waves. The former, however, are much more easily coupled to an external circuit either at the input or output ends. Figure 3.40 shows the basic design of a surface-wave delay line. The applications of surface waves will undoubtedly have a major impact on electronic technology in the microwave region in the coming years.†

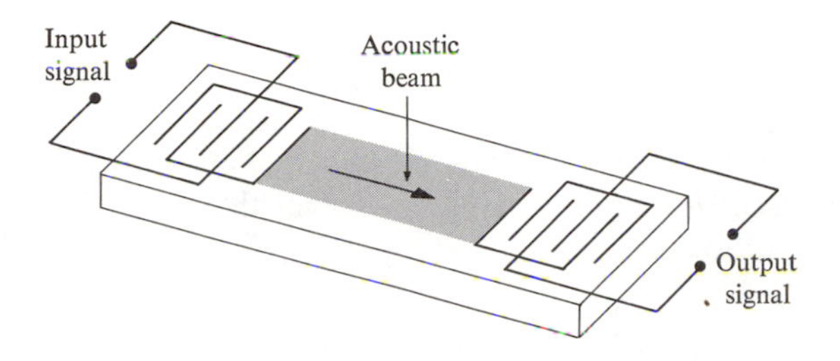


**Fig. 3.40** Basic design of a surface-wave delay line.

## 3.12 LATTICE OPTICAL PROPERTIES IN THE INFRARED

Ionic crystals exhibit interesting optical properties associated with optical phonons. The frequency range here lies in the infrared region of the spectrum, the region in which optical phonons are active, as we said in Section 3.6. Characteristically, an ionic crystal in the infrared region exhibits strong optical reflection, concomitant with a strong absorption. Let us now relate these observations to the optical vibrations of the lattice in ionic crystals.

We can approach the discussion most conveniently via the dielectric function of the medium. This quantity, denoted by $\epsilon$, is defined by

$$D = \epsilon\, \mathscr{E} = \epsilon_0\, \mathscr{E} + P, \tag{3.79}$$

where $D$ is the electric displacement, $\mathscr{E}$ the electric field, and $P$ the polarization in the medium; $\epsilon_0$ is the familiar dielectric constant associated with the vacuum. We shall soon use this equation to evaluate the dielectric constant for an ionic crystal. It may also be pointed out that, in writing (3.79) we assumed that $\mathscr{E}$ and $P$ lie in the same direction, and consequently $D$ also points in the same direction. In other words, the crystal is treated as an isotropic medium, for the sake of simplicity, and the vector symbol may be deleted when writing these quantities.

Once the dielectric function $\epsilon$ is known, it may be used to study the optical properties of the medium. The procedure is as follows: The relative dielectric function $\epsilon_r$, defined as $\epsilon_r = \epsilon/\epsilon_0$, can be written as

$$\epsilon_r = (n + i\kappa)^2, \tag{3.80}$$

† For an interesting discussion of these waves and their applications in electronics, and optics, see "Acoustic Wave Amplifiers" by G. S. Kino and J. Shaw, *Scientific American*, October 1972, page 50. The photographs in this article are excellent.

where $n$ is the optical index of refraction and $\kappa$ the extinction coefficient. The quantities $n$ and $\kappa$ may now be used to calculate the *reflectivity* and the absorption of the medium according to the relations†

$$R = \frac{(n-1)^2 + \kappa^2}{(n+1)^2 + \kappa^2} \tag{3.81}$$

and

$$\alpha_{\mathrm{ab}} = 2\kappa q. \tag{3.82}$$

Here $R$ is the reflectivity, evaluated at normal incidence to the surface, and $\alpha_{\mathrm{ab}}$ is the *absorption coefficient*; $q$ is the wave vector of the wave. We shall now apply this procedure to an ionic crystal.

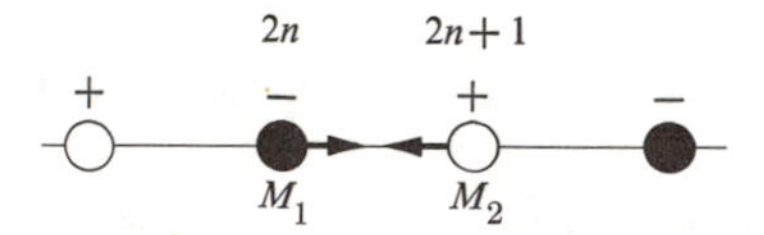


**Fig. 3.41** An ionic diatomic lattice.

Figure 3.41 shows a diatomic crystal in which the two atoms of the unit cell have masses $M_1$ and $M_2$ and electrical charges $e^*$ and $-e^*$. (The quantity $e^*$, called the *effective charge*, is smaller than the charge on the electron $e$ because the transfer of the electron—in the alkali halides, for example—from the alkali atom to the halogen atom is not complete; in NaCl, $e^* = 0.74\,e$.) When an alternating electric field $\mathscr{E}$ is applied to the crystal, the equations of motion for the two ions may be written as

$$M_1 \frac{\partial^2 u_{2n+1}}{\partial t^2} = -\alpha[2u_{2n+1} - u_{2n} - u_{2n+2}] + e^* \mathscr{E}, \tag{3.83}$$

$$M_2 \frac{\partial^2 u_{2n}}{\partial t^2} = -\alpha[2u_{2n} - u_{2n-1} - u_{2n+1}] - e^* \mathscr{E}. \tag{3.84}$$

In each of these equations, the first term on the right represents the short-range elastic restoring force due to the interaction between the atoms, as used in connection with (3.57), while the second term represents the force due to the electric field. In comparing the present situation with that of Section 3.6, we note that here we are discussing the *forced* vibration of the lattice, while the earlier discussion was concerned with the *free* vibration. The forced term, of course, arises from the electric field.

---

† See any intermediate-level textbook on optics or electromagnetism, e.g., G. R. Fowles, 1968, *Introduction to Modern Optics*, Holt, Rinehart, and Winston, New York.

In solving the above equation, we take the field $\mathscr{E}$ to be a propagating plane wave,

$$\mathscr{E} = \mathscr{E}_0 e^{i(qx-\omega t)}. \tag{3.85}$$

Also, for the sake of simplicity, we assume that the wavelength is very large compared to the interatomic distance, so that we may use the infinite wavelength limit $q = 0$. In that case, all similar atoms have the same displacement, e.g., atoms of mass $M_1$ have the displacement $u_+$, and those of mass $M_2$ the displacement $u_-$, the positive and negative signs being used to label the positive and negative ions. These displacements in the steady states have forms similar to the forcing field (3.85). That is,

$$u_+ = u_{0+}e^{-i\omega t}, \qquad u_- = u_{0-}e^{-i\omega t}, \tag{3.86}$$

where $u_{0+}$ and $u_{0-}$ are the amplitudes, and where we have set $q = 0$, in accordance with our approximation. Substitution of (3.85) and (3.86) into the equations of motion (3.83) and (3.84) leads to the determination of the ionic displacements

$$u_{0+} = \frac{e^*}{M_1(\omega_t^2 - \omega^2)} \mathscr{E}_0, \tag{3.87}$$

$$u_{0-} = -\frac{e^*}{M_2(\omega_t^2 - \omega^2)} \mathscr{E}_0, \tag{3.88}$$

where $\omega_t^2 = 2\alpha(1/M_1 + 1/M_2)$. Referring to Section 3.6, we note that $\omega_t$ is the frequency of the transverse optical phonon at $q = 0$.

The ionic polarization $P_i$ of the medium is defined as the electric dipole moment per unit volume, which may therefore be written as

$$P_i = n_m e^*(u_{0+} - u_{0-}), \tag{3.89}$$

where $n_m$ is the number of molecules, or cells, per unit volume. Equation (3.89) follows from noting that the electric dipole moment per molecule is $e^*(u_{0+} - u_{0-})$. In addition to the ionic polarization, there is also an electronic polarization due to the fact that the electrons in the atomic shells of the ions also respond to, and polarize in, the electric field. This polarization will be denoted by $P_e$.

The ionic polarization (3.89) may be evaluated by using (3.87) and (3.88), and when the result is substituted into (3.79) and the common factor $\mathscr{E}$ canceled, we find

$$\epsilon_r(\omega) = 1 + \frac{P_e}{\epsilon_0 \mathscr{E}} + \frac{n_m e^{*2}}{\epsilon_0 \omega_t^2 \mu} \frac{1}{1 - \omega^2/\omega_t^2}, \tag{3.90}$$

where $\mu = M_1 M_2/(M_1 + M_2)$ is the reduced mass of the two ions. On the right side, the second term represents the electronic contribution, and the third term the ionic contribution. For $\omega \ll \omega_t$, both terms contribute, resulting in the familiar *static* dielectric function $\epsilon_r(0)$. At the opposite end of the spectrum, where $\omega \gg \omega_t$, it is seen from (3.90) that the ionic contribution vanishes, because the frequency there is too high for the ions to follow the oscillation of the field. In that range the dielectric constant is denoted by $\epsilon_r(\infty)$, and contains only the electronic contribution. We may now rewrite (3.90) in the convenient form

$$\epsilon_r(\omega) = \epsilon_r(\infty) + \frac{\epsilon_r(0) - \epsilon_r(\infty)}{1 - \omega^2/\omega_t^2}, \tag{3.91}$$

where the ionic contribution is contained entirely in the second term on the right side. In this manner, the dielectric function is conveniently expressed in terms of quantities which are directly measurable, that is $\epsilon_r(0)$, $\epsilon_r(\infty)$, and $\omega_t$.

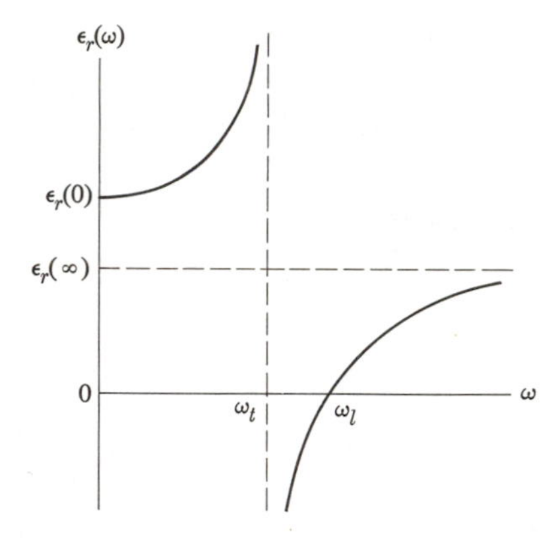


**Fig. 3.42** Dielectric function $\epsilon_r(\omega)$ versus frequency. The function is singular at the transverse frequency $\omega_t$ and vanishes at the longitudinal frequency $\omega_l$. The former condition represents resonance.

Figure 3.42 sketches the dielectric function $\epsilon_r\omega)$ versus $\omega$ over the entire frequency range. An important feature of this figure is that $\epsilon_r(\omega)$ is negative in the frequency range $\omega_t < \omega < \omega_l$, where $\omega_l$ is the frequency at which $\epsilon_r(\omega)$ vanishes, as shown. This frequency can be determined from the expression (3.91), and is readily found to be

$$\omega_l = \left(\frac{\epsilon_r(0)}{\epsilon_r(\infty)}\right)^{1/2} \omega_t. \tag{3.92}$$

We shall shortly explain the physical significance of $\omega_l$, but for the moment let us continue our discussion of the dielectric function. Since $\epsilon_r(\omega)$ is negative in the range $\omega_t < \omega < \omega_l$, it follows from (3.80) that $n = 0$ and $\kappa \neq 0$ which, when substituted into (3.81), shows that the reflectivity $R = 1$. That is, an incident wave whose frequency lies in the range $\omega_t < \omega < \omega_l$ suffers *total reflection*. The wave in this range does not propagate inside the crystal, and we speak of a *forbidden* gap. The dependence of the reflectivity on the frequency, as determined by (3.81), is illustrated in Fig. 3.43(a). Compare this with the experimental curve for NaCl

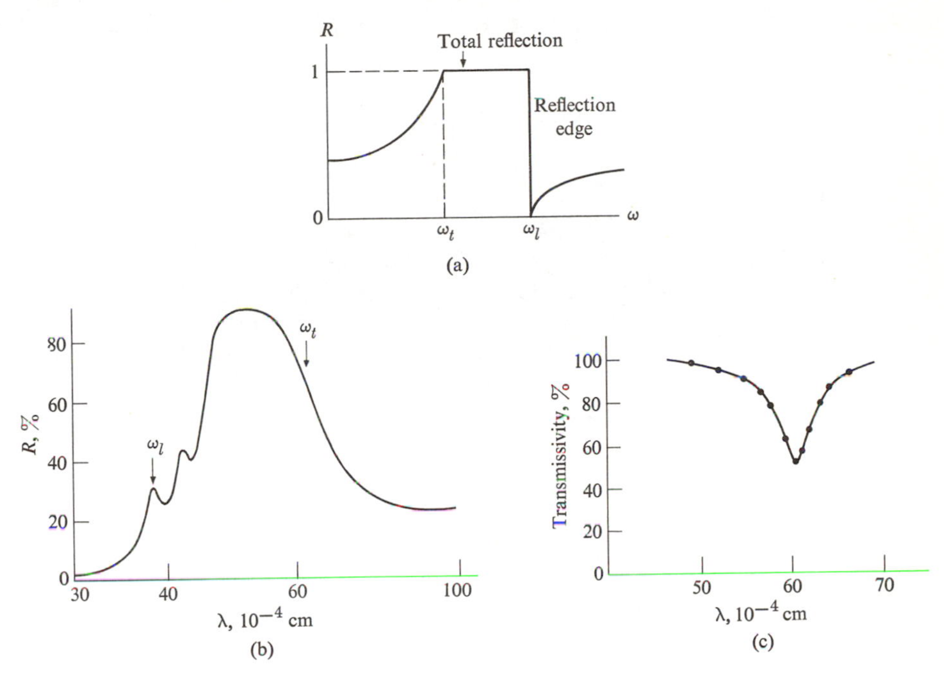


**Fig. 3.43** (a) Reflectivity versus frequency for an ideal crystal. (b) Infrared reflectivity versus wavelength for NaCl at room temperature. The frequencies $\omega_t$ and $\omega_l$ correspond to $\lambda = 61$ and $38 \times 10^{-4}$ cm, respectively. (c) Infrared transmissivity versus wavelength for an NaCl thin film (of thickness $0.17 \times 10^{-4}$ cm). Dip is at frequency $\omega_t$.

given in Fig. 3.43(b). Note that the sharp edges of the reflectivity are rounded off in the experimental curve. This can be explained partly by introducing a damping term in the lattice equation of motion (3.83) and (3.84). Such a damping may be due to any of the phonon-collision mechanisms discussed in Section 3.9. The primary mechanism is the anharmonic phonon-phonon collision, which explains why the shape of the reflectivity depends to some extent on the temperature.

Figure 3.43(c) shows the observed infrared absorption in a thin film of NaCl. As we have indicated previously, the absorption coefficient may be found from (3.82). The reason for using a thin film is the strong reflection incurred in this region. The point of maximum absorption marks the transverse frequency $\omega_t$ (recall that at $\omega_t$ the function $\epsilon_r(\omega) \rightarrow \infty$, and hence $\kappa$ and $\alpha$ have their maximum values).

The phenomena of strong infrared reflection and absorption by the lattice are sometimes referred to as *reststrahlen* (German for "residual rays").

The physical significance of $\omega_l$ is that it is the frequency of the longitudinal optical photon, and it is to be contrasted with $\omega_t$, the frequency of the transverse optical phonon. The reason for the distinction between these two types of vibrations in an ionic crystal can be understood from the following argument: If we use the Maxwell equation $\nabla \cdot \mathbf{D} = 0$, there being no net charge, and (3.1), we may write

$$\nabla \cdot \mathscr{E} = -\frac{1}{\epsilon_0} \nabla \cdot \mathbf{P}, \tag{3.93}$$

where we have solved for the field $\mathscr{E}$ in terms of the polarization of the medium **P**. Now for a transverse wave the divergence $\nabla \cdot \mathbf{P}$ vanishes and this, in conjunction with (3.93), indicates that $\nabla \cdot \mathscr{E} = 0$. The field associated with this wave is therefore a constant, and may be taken to be zero. By contrast, however, the divergence $\nabla \cdot \mathbf{P} \neq 0$ for a longitudinal wave, which means that such a wave has an associated electric field. This conclusion is also evident from Fig. 3.44,

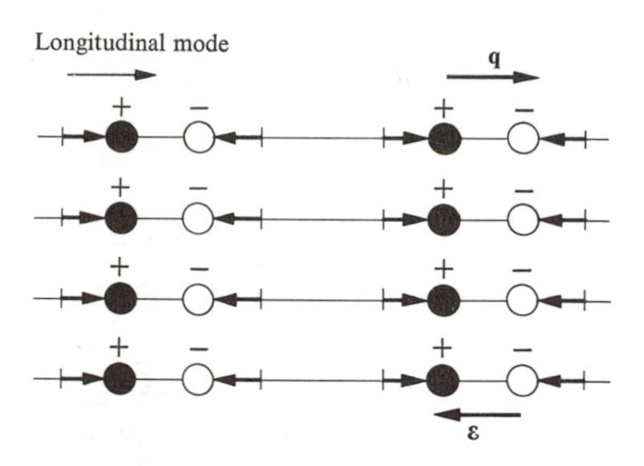


**Fig. 3.44** Bunching of charges in the longitudinal mode.

where we see that the *bunching* together of electric charges, associated with the longitudinal mode, leads to the creation of an electric field. The effect of this field is to increase the restoring force beyond that of the short-range interaction, and this makes the longitudinal frequency larger than the transverse frequency.

We have seen why the longitudinal frequency is larger than the transverse frequency, but we have yet to show that the former is given by $\omega_l$ of Eq. (3.92). To demonstrate this, we return to the equation $\nabla \cdot \mathbf{D} = 0$, which we write, with the assistance of (3.79), as

$$\epsilon_r(\omega)\, \nabla \cdot \mathscr{E} = 0. \tag{3.94}$$

This condition must hold true whether the wave is transverse or longitudinal, but the manner in which this is accomplished is very different in the two cases. In the former, $\nabla \cdot \mathscr{E} = 0$, and the condition (3.94) is thus satisfied. But in the longitudinal case $\nabla \cdot \mathscr{E} \neq 0$, and the only way in which (3.94) may be satisfied is if $\epsilon_r(\omega) = 0$. In other words, the frequency of the longitudinal mode is equal to the *root* of the dielectric constant. Since $\omega_l$ was determined by putting $\epsilon_r(\omega)$ equal to zero, it follows that $\omega_l$ is equal to the frequency of the longitudinal mode. Equation (3.91), relating $\omega_l$ to $\omega_t$, is known as the *LST* (Lyddane–Sachs–Teller) relation.

Table 3.3 gives the optical parameters of some common ionic crystals. The values for $\epsilon_r(0)$, $\epsilon_r(\omega)$, and $\omega_t$ are determined experimentally, while those of $\omega_l$ are calculated from the LST relation. The effective charge ratio $e^*/e$ is determined by comparing (3.91) with (3.90).

**Table 3.3**

Infrared Lattice Data for Ionic Crystals

| | $\epsilon_r(0)$ | $\epsilon_r(\infty)$ | $\omega_t$, $10^{13}$ rad/s | $\omega_l$, $10^{13}$ rad/s | $e^*/e$ |
|---|---|---|---|---|---|
| LiF | 8.9 | 1.9 | 5.8 | 12 | 0.87 |
| NaF | 5.3 | 1.75 | 4.4 | 7.8 | 0.93 |
| NaCl | 5.62 | 2.25 | 3.08 | 5.0 | 0.74 |
| NaBr | 5.99 | 2.62 | 2.55 | 3.9 | 0.69 |
| KCl | 4.68 | 2.13 | 2.71 | 4.0 | 0.80 |
| KBr | 4.78 | 2.33 | 2.18 | | 0.76 |
| KI | 4.94 | 2.69 | 1.91 | 2.64 | 0.69 |
| RbCl | 5 | 2.19 | 2.24 | | 0.84 |
| RbBr | 5 | 2.33 | 1.69 | | 0.82 |
| AgCl | 12.3 | 4.04 | 1.94 | 3.4 | 0.78 |
| AgBr | 13.1 | 4.62 | 1.51 | 2.5 | 0.73 |
| CsCl | 7.20 | 2.60 | 1.87 | 3.1 | 0.85 |
| CsBr | 6.51 | 2.87 | 1.39 | | 0.78 |

The following remarks concerning the distinction between transverse and longitudinal modes may be helpful. When an electromagnetic wave impinges on the surface of the crystal at normal incidence, it excites TO phonons inside the crystal if the frequency is matched correctly to that of the phonons. This is so because both waves are transverse, and consequently they couple together. On the other hand, LO phonons may not be excited in this manner, since these phonons, being longitudinal, do not couple to the incident wave, which is transverse. Other means must be used to excite these phonons. For example, if the incident wave falls obliquely onto the surface, there is a longitudinal electric field at the surface which then acts to excite the LO phonons.

The other point to note is that for a nonionic crystal, the frequencies $\omega_t$ and $\omega_l$ are identical (at $q = 0$), as we can see from Fig. 3.25b for germanium. The reason for this is that in such crystals the vibrating entities are electrically neutral, and hence the ionic polarization is zero.

**The polariton**

Interesting effects arise when one considers explicitly the influence of optical phonons on a *transverse* electromagnetic wave actually propagating in the crystal. This influence can be taken into account via the dielectric function $\epsilon_r(\omega)$ of the medium. The dispersion relation for the electromagnetic wave, which is $\omega = cq$ in vacuum, is now modified to $\omega = cq/\sqrt{\epsilon_r(\omega)}$, where $\sqrt{\epsilon_r(\omega)}$, being equal to the index of refraction, introduces the effects of the medium on the velocity of the wave in the usual manner. By substituting $\epsilon_r(\omega)$ from (3.90) into the above equation, squaring both sides, and rearranging terms, one finds the dispersion relation

$$\omega^2\left[\varepsilon_r(\infty) + \frac{\epsilon_r(0) - \epsilon_r(\infty)}{1 - \omega^2/\omega_t^2}\right] = c^2 q^2. \qquad (3.95)$$

Equation (3.95) contains not one, but *two* different dispersion relations. We can see this algebraically by noting that, for a given $q$, the equation, being quadratic in $\omega^2$, has two frequency roots. Thus when we vary $q$, the two roots trace two separate dispersion curves, as shown in Fig. 3.45. These results are particularly interesting because the dispersion curves obtained do not conform either to the photon, where $\omega \sim q$, or to the phonon, where $\omega$ is independent of $q$. And in fact the modes described here are neither pure photons nor pure phonons, but a photon–phonon mixture, which is given the name of *polariton* (referring to polar or ionic crystals). The reason for the photon–phonon mixing is that, in ionic crystals, there is a strong coupling between the two pure modes, and because of this the pure modes are modified to new coupled modes. Thus, starting at each $q$ with a photon mode and a phonon mode, we find two new polariton modes.

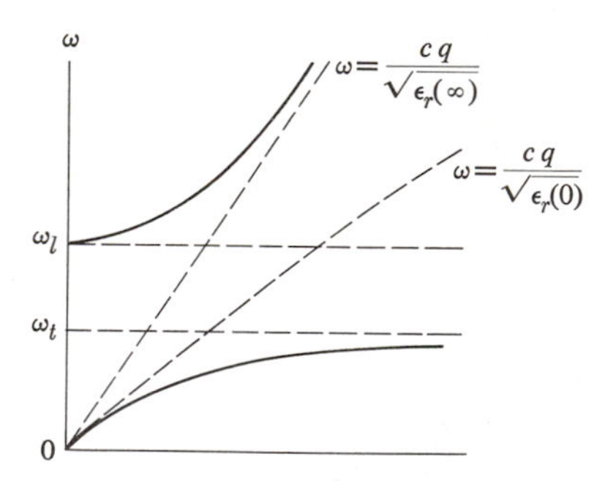


**Fig. 3.45** Dispersion curves for the polariton. Dashed curves represent free modes, while solid curves describe interacting modes—the polariton.

There is a familiar analogy to the coupling scheme found above. Consider two harmonic oscillators with frequencies $\omega_1$ and $\omega_2$. Without mutual coupling between them, the oscillators vibrate independently, each with its own frequency.

However, if they are connected by a spring, the two oscillators no longer vibrate independently. They vibrate together, in two different modes, whose frequencies, no longer equal to $\omega_1$ and $\omega_2$, can be expressed in terms of $\omega_1$, $\omega_2$, and the coupling strength. In this analogy, the pure photon and pure phonon represent two independent oscillators, which couple in an ionic crystal to produce new modes, the polaritons.

We see from Fig. 3.45 that the coupling is strongest in the region $\omega \simeq \omega_t$, where the frequencies of the pure modes are nearly equal to the crossover point. (This is the region of intersection of the dashed lines, representing the pure modes.) This is expected because, in the oscillator analogy above, the two oscillators are most strongly affected by the coupling if the frequencies $\omega_1$ and $\omega_2$ are nearly equal. Conversely, in the region far from the intersection, the two mixed modes reduce to essentially pure modes. Consider, for example, the lower polariton curve in Fig. 3.45: At $q \simeq 0$, the dispersion relation is $\omega = \left(c/\sqrt{\epsilon_r(0)}\right)q$, and the mode is essentially a pure photon mode. Since $\omega$ is much lower than the lattice mode $\omega_t$, the lattice vibrations are not dynamically evident, and the crystal merely acts as a rigid medium of dielectric constant $\epsilon_r(0)$. In the opposite limit, where $q$ is large, $\omega \simeq \omega_t$, and is independent of $q$; then the lower polariton mode becomes almost a pure transverse phonon. The electric field associated with the wave becomes very small here, and the energy is almost entirely mechanical. However, in the intermediate $q$-region, the polariton is a mixture of both electromagnetic and mechanical fields, and has the intermediate behavior described above. Analogous comments can be made concerning the upper polariton curve.

Note also that no mode can propagate in the frequency interval $\omega_t < \omega < \omega_l$, which is the frequency gap encountered previously.

The reason for our interest in the polariton from a fundamental point of view is twofold: (a) It results from the coupling of two collective modes, and (b) it is a collective mode in its own right. The subject of collective modes in solids has received a great deal of attention in recent years. Many other examples of such modes, both free and coupled, will be found throughout this text.

## SUMMARY

This chapter concerned lattice vibrations and their influence on the thermal, acoustic, and optical properties of solids.

### The continuum model

In the long-wavelength limit, a crystal may be treated as a continuous medium because the wavelength is much greater than the lattice constant. The dispersion relation is linear,

$$\omega = v_s q,$$

where $v_s$ is the velocity of sound. Consideration of boundary conditions shows that the modes of oscillations—the sound waves—have discrete values of $q$ and $\omega$. The density of modes is

$$g(\omega) = \frac{v}{(2\pi)^3} \frac{2}{v_s^3}.$$

**Specific heat**

The atoms in the lattice are regarded as a set of harmonic oscillators, and the thermal energy is the average energy of these oscillations. According to classical theory, the average energy for a one-dimensional oscillator is

$$\bar{\epsilon} = kT.$$

Thus the total thermal energy per mole is $\bar{E} = 3N_A kT$, where $N_A$ is Avogadro's number, and the molar specific heat $C_v \equiv \partial \bar{E}/\partial T$ is given by

$$C_v = 3R,$$

where $R = N_A k$ is the universal gas constant. This result, known as the Dulong–Petit law, asserts that $C_v$ is a constant independent of temperature. This law is found to be valid only at high temperatures; at low temperatures, specific heat decreases and then vanishes at $T = 0°\text{K}$.

Einstein rectified this discrepancy by treating the oscillator quantum-mechanically. The average thermal energy for the oscillator is then given by

$$\bar{\epsilon} = \frac{\hbar\omega}{e^{\hbar\omega/kT} - 1},$$

which approaches the classical value $kT$ only at high temperatures. At low temperatures, the quantum energy decreases very rapidly because of the "freezing" of the motion. Treating the atoms as independent oscillators, vibrating with a common frequency, Einstein found that the specific heat is

$$C_v = 3R\left(\frac{\theta_E}{T}\right)^2 \frac{e^{\theta_E/T}}{(e^{\theta_E/T} - 1)^2},$$

where $\theta_E$ is the Einstein temperature. Specific heat approaches the classical value $3R$ at high temperatures, and vanishes at $T \to 0°\text{K}$. Both these facts are in accord with experiment.

Careful measurements show that the decrease in $C_v$ near absolute zero is slower than predicted by Einstein. Debye explained this by treating the atoms not as independent oscillators, but as coupled oscillators vibrating collectively as sound waves. Making the long-wavelength approximation, he found that the specific heat is given by

$$C_v = 9R\left(\frac{T}{\theta_D}\right)^3 \int_0^{\theta_D/T} \frac{x^4 e^x}{(e^x - 1)^2}\, dx,$$

where $\theta_D$ is the Debye temperature. This expression for $C_v$ approaches the classical value $3R$ at high temperatures, but at low temperature $C_v \sim T^3$. This latter result, known as the Debye $T^3$-law, is in agreement with observation.

The agreement of Debye's model with experiment is good throughout the entire range of temperature. Better agreement can be achieved only by removing the long-wavelength approximation and treating the crystal as a discrete lattice.

**The phonon**

The elastic energy of sound waves in solids is quantized, and the quantum unit is the phonon. The phonon carries an energy

$$\epsilon = \hbar\omega,$$

and a momentum

$$\mathbf{p} = \hbar\mathbf{q},$$

where $\omega$ and $\mathbf{q}$ are the frequency and wave vector of the wave, respectively.

**Lattice vibrations**

The dispersion relation for a one-dimensional monatomic lattice, with nearest-neighbor interaction, is

$$\omega = \omega_m \sin(aq/2),$$

where the cutoff frequency $\omega_m = (4\alpha/M)^{1/2}$. The quantities $\alpha$ and $M$ are, respectively, the interatomic force constant and the atomic mass. The dispersion curve is linear near $q = 0$, the long-wavelength regime, and saturates at large values of $q$. The lattice acts as a low-pass filter: Only waves whose frequencies are lower than $\omega_m$ are transmitted; modes with frequencies exceeding $\omega_m$ are heavily attenuated.

The dispersion curves for a one-dimensional diatomic lattice consist of two branches: the lower one the acoustic, and the upper the optical branch. The character of the acoustic branch is similar to that of the monatomic case, while the optical is essentially flat throughout the $q$-space. There is a frequency gap between the two branches, and thus the lattice acts as a band-pass filter.

The dispersion relation in a three-dimensional lattice is an extension of the one-dimensional case. The wave vector $\mathbf{q}$ is now a three-dimensional vector, and the frequency is a function of both the magnitude and direction of $\mathbf{q}$. Thus the dispersion has the form

$$\omega = \omega_j(q).$$

The subscript $j$ is the branch index. A Bravais lattice has three acoustic branches—one longitudinal and two transverse. A non-Bravais lattice, with $r$ atoms per unit cell, has $3r$ branches, three of which are acoustic and the remainder optical.

The dispersion curve exhibits symmetry properties in $\mathbf{q}$-space. The translational symmetry $\omega_j(\mathbf{q} + \mathbf{G}) = \omega_j(\mathbf{q})$ enables us to restrict consideration to the first Brillouin zone only, while the inversion symmetry $\omega_j(-\mathbf{q}) = \omega_j(\mathbf{q})$ and the rotation symmetry establish the relation between various regions of the Brillouin zone.

The density of modes is found by counting the number of modes in each branch, using the actual dispersion relation for that branch. The total density is found by summing the densities of modes of all branches. When this density is employed in calculations of specific heat, good agreement with experiment is obtained.

**Thermal conductivity**

The conduction of heat in insulators is accomplished by lattice waves, or phonons. Treating the phonons as a gas, and using results from kinetic theory, we find that thermal conductivity is given by

$$K = \tfrac{1}{3} C_v v_s l,$$

where $l$ is the mean free path of the phonon.

The mean free path is determined by the scattering of a phonon by other phonons, or by defects in the solid. At low temperatures, the scattering is due to the boundaries of the sample, and $l \simeq D$, where $D$ is the diameter of the sample. Scattering at high temperatures is due to anharmonic interaction between a phonon and other phonons in the solid, and the mean free path is then found to vary inversely with temperature. That is, $l \simeq 1/T$. Impurities in the lattice also contribute to scattering.

**The scattering of radiation by phonons**

Radiation—x-rays, neutrons, and light—may be scattered by phonons, and scattering is used to measure the dispersion curves of the lattice. The law of conservation of momentum requires that

$$\mathbf{k} = \mathbf{k}_0 + \mathbf{q},$$

where $\mathbf{k}$ and $\mathbf{k}_0$ are the wave vectors of the incident and scattered particles, and $\mathbf{q}$ is the wave vector of the phonon involved in the scattering process. The law of conservation of energy requires that

$$\omega = \omega_0 \pm \omega(\mathbf{q}),$$

where $\omega$ and $\omega_0$ are the frequencies of the incident and scattered beams, respectively, and $\omega(\mathbf{q})$ is that of the phonon. The plus sign refers to phonon absorption, and the minus to a phonon emission process.

The scattered frequency is thus shifted from the incident frequency, and by measuring this shift as a function of the wave vector $\mathbf{q}$, one can determine the dispersion curve of the lattice. X-rays and neutrons, being of short wavelength, can be used to determine the dispersion curve throughout the zone. Light waves, on the other hand, have much greater wavelength, and sample only the region near the center of the BZ.

**Ultrasonic waves**

Ultrasonic waves are important in research and applications. By employing coherent ultrasonic phonons of carefully controlled frequency and polarization, one may investigate several basic properties of solids, such as anharmonic interaction and phonon–spin interaction. An important application is the acousto-electric amplifier, in which the acoustic wave is amplified by absorbing energy from high-velocity electrons. Such an amplifier is particularly useful in the design of acoustic delay lines.

**Infrared optical properties**

Optical phonons in ionic crystals interact strongly with light, which leads to strong reflection and absorption in the infrared region. Reflectivity and absorption can be expressed in terms of the dielectric function of the lattice. This function is frequency dependent, as

$$\epsilon_r(\omega) = \epsilon_r(\infty) + \frac{\epsilon_r(0) - \epsilon_r(\infty)}{1 - (\omega^2/\omega_t{}^2)},$$

where $\epsilon_r(0)$ is the static dielectric constant and $\epsilon_r(\infty)$ the high-frequency dielectric constant [$\epsilon_r(\infty) = n^2$, where $n$ is the optical index of refraction]. The quantity $\omega_t$ is the frequency of the transverse phonon. As $\omega$ increases and crosses $\omega_t$, the function $\epsilon_r(\omega)$ decreases from $\epsilon_r(0)$ to $\epsilon_r(\infty)$, due to the fact that the ions are no longer able to follow the field at high frequencies. Consequently the vibrations of the optical phons are suppressed.

The longitudinal phonon in an ionic crystal has a higher frequency than the transverse phonon, due to the bunching of charges associated with longitudinal oscillations. Longitudinal frequency is given by the relation

$$\omega_l = \left(\frac{\epsilon_r(0)}{\epsilon_r(\infty)}\right)^{1/2} \omega_t.$$

A lattice exhibits total reflection in the frequency range $\omega_t$ to $\omega_l$. Thus light in this range cannot propagate through the crystal, resulting in a frequency gap.

**REFERENCES**

**Specific heat**

H. M. Rosenberg, 1963, *Low Temperature Solid State Physics*, Oxford: Oxford University Press

**Lattice vibrations (also optical properties and scattering of radiation)**

T. A. Bak, editor, 1964, *Phonons and Phonon Interaction*, New York: Benjamin

M. Born and K. Huang, 1954, *Dynamical Theory of Crystal Lattices*, Oxford: Oxford University Press

L. Brillouin, 1953, *Wave Propagation in Periodic Structures*, New York: Dover Press

A. A. Maradudin, *et al.*, 1963, "Theory of Lattice Dynamics in the Harmonic Approximation," Supplement 3 of *Solid State Physics*

R. W. H. Stevenson, editor, 1966, *Phonons*, London: Oliver and Boyd

### Thermal conductivity

P. G. Klemens, 1958, "Thermal Conductivity and Lattice Vibration Modes," *Solid State* Physics, **1**

H. M. Rosenberg, listed above under Specific Heat

J. M. Ziman, 1960, *Electrons and Phonons*, Oxford: Oxford University Press

## QUESTIONS

1. Equation (3.11) gives the allowed values of $q$ in a continuous line under periodic boundary conditions. Plot a few of the corresponding wavelengths, and compare with results from elementary physics for, say, a vibrating string.
2. Determine the density of states for a two-dimensional continuous medium using periodic boundary conditions.
3. In the Einstein model, atoms are treated as independent oscillators. The Debye model, on the other hand, treats atoms as coupled oscillators vibrating collectively. However, the collective modes are regarded here as independent. Explain the meaning of this independence, and contrast it with that in the Einstein model.
4. Would you expect to find sound waves in small molecules? If not, how do you explain the propagation of sound in gaseous substances?
5. Explain qualitatively why the interatomic force constant diminishes rapidly with distance.
6. Show that the total number of allowed modes in the first BZ of a one-dimensional diatomic lattice is equal to $2N$, the total number of degrees of freedom.
7. Suppose that we allow two masses $M_1$ and $M_2$ in a one-dimensional diatomic lattice to become equal. What happens to the frequency gap? Is this answer expected? Compare the results with those of the monatomic lattice.
8. Derive an expression for the specific heat of a one-dimensional diatomic lattice. Make the Debye approximation for the acoustic branch, and assume that the optical branch is flat.
9. Figure 3.25(b) shows that the TA branches, as well as the TO branches, in Ge are degenerate in the [111] direction. Explain this qualitatively on the basis of symmetry.
10. Convince yourself that the BZ of an fcc lattice has the shape given in Fig. 3.26(b).
11. Give a physical argument to support the plausibility of (3.74) for thermal conductivity.
12. Explain the dependence of thermal conductivity on temperature as displayed in Fig. 3.32(b).
13. In the microwave generator of a miniature semiconductor, a considerable amount of undesirable heat is generated in the conversion of dc to ac power. Explain why diamond is being increasingly used as a heat sink to transport the heat away from the device.

14. Discuss two experimental techniques for measuring the mean free paths of phonons in solids.
15. Verify (3.91).
16. Verify (3.92).
17. Draw a figure for a transverse oscillation in an ionic crystal and show that, unlike the case of longitudinal oscillations, no charge bunching takes place.

## PROBLEMS

1. The longitudinal and transverse velocities of sound in diamond, a cubic crystal, along the [100] direction are, respectively, 1.76 and $1.28 \times 10^6$ cm/s. The longitudinal velocity in the [111] direction is $1.86 \times 10^6$ cm/s. From these data, and the fact that the density is 3.52 g/cm$^3$, calculate the elastic constants $C_{11}$, $C_{12}$, and $C_{44}$ for diamond.
2. In deriving (3.19) for the density of states for a continuous medium, it was assumed that the longitudinal and transverse velocities $v_l$ and $v_t$ were equal. Derive the density of states for a case in which this assumption is no longer true.
3. It is more convenient in practice to measure the specific heat at constant pressure, $C_p$, than the specific heat at constant volume, $C_v$, but the latter is more amenable to theoretical analysis.
   a) Using a thermodynamic argument, show that the two specific heats are related by
   $$C_p - C_v = \alpha^2 TV/K,$$
   where $\alpha$ is the volume coefficient of thermal expansion and $K$ the compressibility.
   b) Show that $C_p - C_v = R$ for an ideal gas.
   c) Show that $C_p \simeq C_v$ for a solid at room temperature. (Look up the needed parameters in appropriate reference works, e.g., the *Handbook of Chemistry and Physics*.)
4. Using the Maxwell–Boltzmann distribution, show that the average energy of a one-dimensional oscillator at thermal equilibrium is $\bar{\epsilon} = kT$.
5. Prove the result (3.26) for the average energy of a quantum oscillator.
6. a) If the classical theory of specific heat were valid, what would be the thermal energy of one mole of Cu at the temperature $T = \theta_D$? The Debye temperature for Cu is 340°K.
   b) Calculate the actual thermal energy according to the Debye theory (use Fig. 3.13), and compare with the classical value obtained above. (For the purpose of this calculation, you may approximate the Debye curve by a straight line joining the origin to the point on the Debye curve at $T = \theta_D$.)
   c) What is the order of magnitude of the maximum displacement of a Cu atom at the Debye temperature? Compare this displacement with the interatomic distance.
7. It was stated in the text that the Debye temperature $\theta_D$ is proportional to $(Y/M)^{1/2}$, where $Y$ is Young's modulus and $M$ the atomic mass. For solids of similar chemical and structural characteristics, the parameters $Y$ are nearly equal, and thus $\theta_D \sim 1/M^{1/2}$. Plot $\theta_D$ versus $M^{-1/2}$ for the alkali metals (Li, Na, K, Rb, Cs), the noble metals (Cu, Ag, Au), the covalent crystals (C, Si, Ge, Sn), and discuss how well this prediction is satisfied.

8. Verify the mathematical reasoning between (3.36) and (3.38).
9. a) Derive the expression for the specific heat of a linear continuous chain according to the Debye theory. Discuss the high and low temperature limits.
   b) Repeat part (a) for a continuous sheet.
10. Determine the phase and group velocities for a monatomic lattice. Plot the results versus the wave vector $q$ and give a brief discussion of their significance.
11. When the frequency of a wave in a one-dimensional lattice is greater than the cutoff frequency $\omega_m$, the wave is heavily attenuated. Assuming that the solution may still be expressed in the form (3.55), but with $q$ being an imaginary number, calculate the attenuation coefficient, i.e., the coefficient governing the exponential decay of the intensity, and plot the result as a function of the frequency. [*Hint*: Use the formula $\sin iy = i \sinh y$.]
12. Verify (3.62) and (3.63).
13. Using the optical mode frequency for NaCl (Table 3.3), calculate the interatomic force constant and Young's modulus for this substance. From these data and the density (2.18 g/cm$^3$), calculate the velocity of sound in NaCl.
14. What is the minimum wavelength for a wave traveling in the [100] direction in an fcc structure? In the [111] direction? Use Fig. 3.26(a), and assume that the cube edge of the real unit cell is $a = 5$ Å.
15. Using the density of states (3.69) for a one-dimensional monatomic crystal, show that the total number of states is equal to $N$.
16. Using data on thermal conductivity, calculate the velocity of sound in NaCl at $T = 20$°K and $T = 300$°K. Compare your answer with that for Problem 13.
17. In discussing the behavior of the phonon's mean free path we treated the various collision processes separately. However, in most situations, several of these processes act simultaneously to scatter the phonon. Show that the effective path in that case is given by $1/l = \sum_i 1/l_i$, where the $l_i$'s refer to the mean free paths of the individual collision mechanisms. [*Hint*: You may use a probabilistic argument. A similar approach is employed in Section 4.5 in connection with the scattering of electrons in metals.]
18. The text stated that the equations for the conservation of momentum and energy for the scattering of a photon by a phonon, Eqs. (3.75) and (3.76), may also be derived by treating the scattering process as a Doppler-shifted Bragg reflection. Prove this statement.
19. Brillouin scattering of a monochromatic light beam, $\lambda_0 = 6328$ Å, from water at room temperature leads to a Brillouin sideband whose shift from the central line is $\Delta\nu = 4.3 \times 10^9$ Hz at scattering angle of 90°. Knowing that the refractive index of water is 1.33, what is the velocity of sound in this substance at room temperature?
20. Fill in the entries left vacant in Table 3.3.
21. Solve for the two polariton dispersion relations from (3.86), and show that the dispersion curves are as shown in Fig. 3.45.

# CHAPTER 4 METALS I: THE FREE-ELECTRON MODEL



---

*Freedom has a thousand charms to show,*
*That slaves, howe'er contented, never know.*

William Cowper

## 4.1 INTRODUCTION

Metals are of great importance in our daily lives. Iron is used in automobiles, copper in electrical wiring, silver and gold as jewelry, to give only a few examples. These and other metals have played an exceedingly important role in the growth of our technological, industrial world from early historical times to the present, and will continue to do so in the future.

Metals are characterized by common physical properties: great physical strength, high density, good electrical and thermal conductivities, and high optical reflectivity, which is responsible for their characteristic lustrous appearance. The explanation of these properties is important to the physicist who is interested in understanding the microscopic structure of materials, and also to the metallurgist and engineer who wish to use metals for practical purposes.

In this chapter we shall see that these properties are intimately related. They can all be explained by assuming that a metal contains a large concentration of essentially free electrons which are able to move throughout the crystal. In the introductory sections we develop the concept of the free-electron model. We then describe how electrons can carry a current in the presence of an electric field. After that we shall calculate the specific heat of electrons, and show that agreement with experiment can be obtained only if the electrons obey the Pauli exclusion principle. This introduces the important concepts of the Fermi level and Fermi surface, which are then employed to develop a more refined description of electrical and thermal conduction in metals.

The effects of a magnetic field on the motion of free electrons will also be discussed. We shall point out, in particular, how cyclotron resonance and measurements of the Hall effect can yield basic information on metals.

Some of the most interesting properties are associated with metals when studied in the optical frequency range. We shall discuss these in some detail, and show that the free-electron model is capable of explaining most of the observed properties. We shall also discuss thermionic emission of electrons from metals. Then, finally, we shall criticize the free-electron model, and discuss its limitations.

## 4.2 CONDUCTION ELECTRONS

What are the conduction electrons? Let us answer this question by an example, using the simplest metal, Na, as illustration. Consider first an Na gas, which is a collection of free atoms, each atom having 11 electrons orbiting around the nucleus. In chemistry these electrons are grouped into two classes: The 10 core electrons which comprise the stable structure of the filled first and second shells (Bohr orbits), and a *valence electron* loosely bound to the rest of the system. This valence electron, which occupies the third atomic shell, is the electron which is responsible for most of the ordinary chemical properties of Na. In chemical reactions the Na atom usually loses this valence electron—it being loosely bound—and an $Na^+$ ion is formed. This is what happens, for example, in NaCl, in which the electron is transferred from the Na to the Cl atom. The radius of the third shell in Na is 1.9 Å.

Let us now bring the Na atoms together to form a metal. In the metallic state, Na has a bcc structure (Section 1.7), and the distance between nearest neighbors is 3.7 Å. We see from Fig. 4.1 that in the solid state two atoms overlap slightly. From this observation it follows that a valence electron is no longer attached to a particular ion, but belongs to both neighboring ions at the same time. This idea can be carried a step further: A valence electron really belongs to the whole crystal, since it can move readily from one ion to its neighbor, and then the neighbor's neighbor, and so on. This mobile electron, which is called a valence electron in a free atom, becomes a *conduction electron* in a solid.

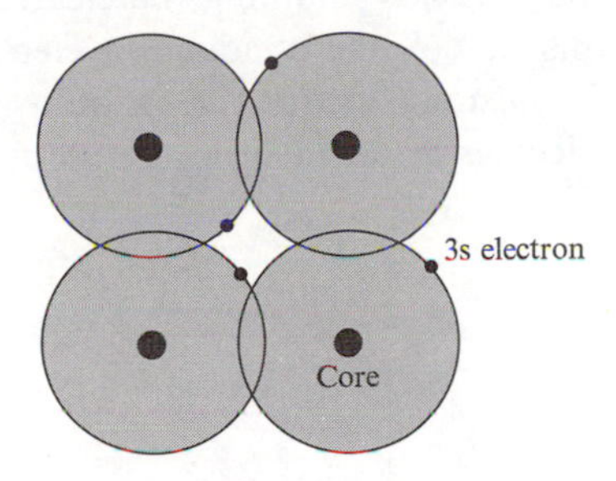


**Fig 4.1** Overlap of the 3s orbitals in solid sodium.

Of course, each atom contributes its own conduction electron, and each of these electrons belongs to the whole crystal. These are called conduction electrons because they can carry an electric current under the action of an electric field. The conduction is possible because each conduction electron is spread throughout the solid (delocalized) rather than being attached to any particular atom. On the contrary, well localized electrons do not carry a current. For example, the *core electrons* in metallic Na—i.e., those centered around the nuclei at the lattice sites—do not contribute anything to the electric current. The states of these electrons in the solid differ little from those in the free atom.

In summary: When free atoms form a metal, all the valence electrons become conduction electrons and their states are profoundly modified, while the core electrons remain localized and their character remains essentially unchanged. Just as valence electrons are responsible for chemical properties, so conduction electrons are responsible for most of the properties of metals, as we shall see.

One can calculate the number of conduction electrons from the valence of the metal and its density. Thus in Na the number of conduction electrons is the same as the number of atoms, and the same is true for K, and also for the noble metals Cu, Ag, Au, all of which are monovalent. In divalent metals—such as Be, Mg, Zn, and Cd—the number of electrons is twice the number of atoms, and so on. If the density of the substance is $\rho_m$, then the atom concentration is

$(\rho_m/M')N_A$, where $M'$ is the atomic weight and $N_A$ is Avogadro's number. Denoting the atomic valence by $Z_v$, one finds the electron concentration†

$$N = Z_v \frac{\rho_m N_A}{M'}. \tag{4.1}$$

## 4.3 THE FREE-ELECTRON GAS

In the free-electron model, which is the basis of this chapter, the conduction electrons are assumed to be completely *free*, except for a potential at the surface (see Fig. 4.2), which has the effect of confining the electrons to the interior of the specimen. According to this model, the conduction electrons move about inside the specimen without any collisions, except for an occasional reflection from the surface, much like the molecules in an ideal gas. Because of this, we speak of a *free-electron gas*.

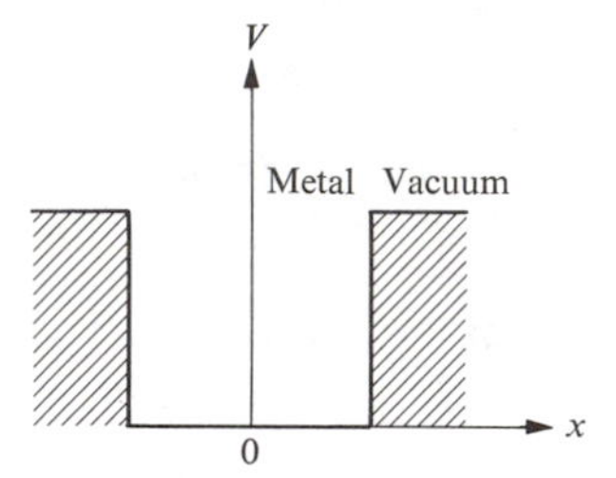


**Fig. 4.2** The potential in the free-electron model.

Let us look at the model a little more closely. It is surprising that it should be valid at all, because, at first sight, one expects the conduction electrons to interact with the ions in the background, and also with each other. These interactions are strong, and hence the electrons ought to suffer frequent collisions; a picture of a highly nonideal gas should therefore emerge. Why then does the free-electron model work? The answer to this fundamental question was not known to the workers who first postulated the model. We now know the answer, but since it requires the use of quantum mechanics, we shall postpone the discussion to Chapter 5. Only a brief qualitative statement is offered here.

The reason why the interaction between the ions appears to be weak is as follows. Although the electron does interact with an ion through coulomb attraction, quantum effects introduce an additional *repulsive* potential, which tends to cancel the coulomb attraction. The net potential—known as the *pseudopotential*—turns out to be weak, particularly in the case of alkali metals. Another way of approaching this is to note that, when an electron passes an ion, its velocity

† In this chapter we use the symbol $N$ for electron concentration. The symbol $n$ will be reserved for the optical index of refraction, discussed in Section 4.11.

increases rather rapidly in the ion's neighborhood (Fig. 4.3), due to the decrease in the potential. Because of this, the electron spends only a small fraction of its time near the ion, where the potential is strong. Most of the time the electron is far away in a region in which the potential is weak, and this is why the electron behaves like a free particle, to a certain approximation.† We shall talk about the electron–ion interaction again in Section 5.3, and the pseudopotential in Section 5.9.

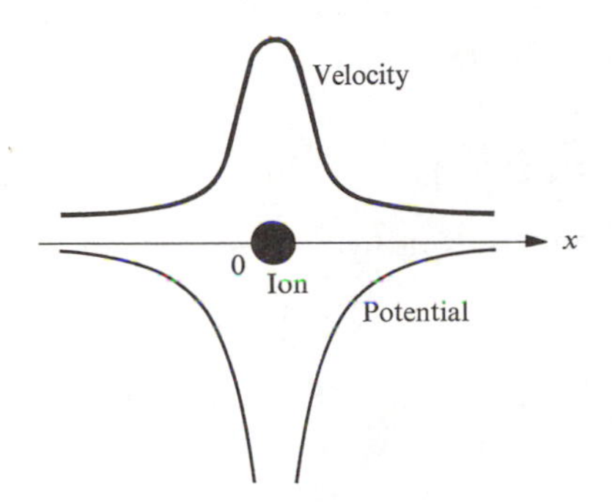


**Fig. 4.3** Variation of the local velocity of electrons in space.

We come now to the interaction between the conduction electrons themselves, and the reason for the weakness of this interaction. There are actually two reasons: First, according to the Pauli exclusion principle, electrons of parallel spins tend to stay away from each other. Second, even if their spins are opposite, electrons tend to stay away from each other, in order to minimize the energy of the system. If two electrons come very close to each other, the coulomb potential energy becomes exceedingly large, and this violates the tendency of the electron system to have the lowest possible energy. When these two considerations are carried out mathematically, the following situation results: Each electron is surrounded by a (spherical) region which is deficient of other electrons. This region, called a *hole*, has a radius of about 1 Å (the exact value depends on the concentration of electrons). As an electron moves, its hole—sometimes known as a *Fermi hole*—moves with it. We see now why the interaction between electrons is weak. If we examine the interaction between two particular electrons, we find that other electrons distribute themselves in such a manner that our two electrons are screened from each other. Consequently there is very little interaction between them.

Free-electron gas in metals differs from ordinary gas in some important respects. First, free-electron gas is charged (in ordinary gases the molecules are mostly

† Note that the interaction between the electron and ion is very weak when the distance between them is large because the ions are screened by other electrons. This means that the interaction has the form of a short-range *screened* coulomb potential rather than a long-range pure coulomb potential.

neutral). Free-electron gas is thus actually similar to a *plasma*. Second, the concentration of electrons in metals is large: $N \simeq 10^{29}$ electrons-m$^{-3}$. By contrast, the ordinary gas has about $10^{25}$ molecules-m$^{-3}$. We may thus think of free-electron gas in a metal as a dense plasma.

Our model of the electron (sometimes called the *jellium* model) corresponds to taking metallic positive ions and smearing them uniformly throughout a sample. In this way there is a positive background which is necessary to maintain charge *neutrality*. But, because of the uniform distribution, the ions exert zero field on the electrons; the ions form a uniform *jelly* into which the electrons move.

## 4.4 ELECTRICAL CONDUCTIVITY

The law of electrical conduction in metals—*Ohm's law*— is

$$I = V/R, \tag{4.2}$$

where $I$ is the current, $V$ the potential difference, and $R$ the resistance of the wire. We want to express this law in a form which is independent of the length and cross section of the wire, since these factors are, after all, irrelevant to the basic physics of conduction. Suppose that $L$ and $A$ are, respectively, the length and cross section of the wire; then

$$J = \frac{I}{A}, \qquad \mathscr{E} = \frac{V}{L}, \quad \text{and} \quad R = \frac{L\rho}{A}, \tag{4.3}$$

where $J$ is the *current density* (current per unit area), $\mathscr{E}$ the electric field, and $\rho$ the *electrical resistivity*. The inverse of the resistivity is called the *conductivity*, denoted by $\sigma$. That is,

$$\sigma = \frac{1}{\rho}. \tag{4.4}$$

When we substitute (4.3) and (4.4) into (4.2), we arrive at

$$J = \sigma\mathscr{E}, \tag{4.5}$$

which is the form of Ohm's law which we shall use. Since the dimension of $\rho$ is ohm-m, $\sigma$ has the dimension ohm$^{-1}$-m$^{-1}$. Now we want to express $\sigma$ in terms of the microscopic properties pertaining to the conduction electrons.

The current is due to the motion of the conduction electrons under the influence of the field. Because these particles are charged, their motion leads to an electrical current; the motion of neutral particles does not lead to an electrical current. We say that it is the conduction electrons which are responsible for the current because the ions are attached to and vibrate about the lattice sites. They have no net translational motion, and hence do not contribute to the current. Let us now treat the motion of the conduction electrons in an electric field.

Consider one typical electron: The field exerts on the electron a force $-e\mathscr{E}$. There is also a *friction force* due to the collision of the electron with the rest of the medium. Let us assume that this friction force has the form $-m^*v/\tau$, where $v$ is the velocity of the electron and $\tau$ is a constant called the *collision time*. Using Newton's law, we have

$$m^* \frac{dv}{dt} = -e\mathscr{E} - m^* \frac{v}{\tau}, \tag{4.6}$$

where $m^*$ is the *effective mass* of the electron.† We see that the effect of the collision, as usual in friction or viscous forces, tends to reduce the velocity to zero. We are interested in the steady-state solution; that is, where $dv/dt = 0$. The appropriate solution of (4.6) in this case is

$$v = -\frac{e\tau}{m^*}\mathscr{E}. \tag{4.7}$$

This, then, is the *steady-state velocity* of the electron (in discussions of friction it is usually called the *terminal velocity*). It is opposite to $\mathscr{E}$ because the charge on the electron is negative.

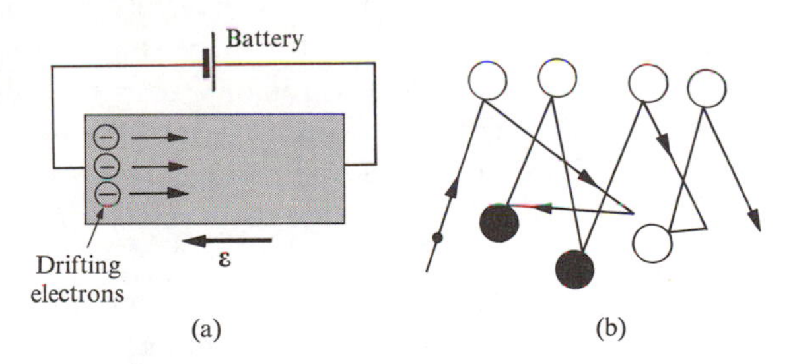


**Fig. 4.4** (a) An electric field applied to a metallic wire. (b) Random versus drift motion of electrons. Circles represent scattering centers.

We should make a distinction here between the two different velocities associated with the electron: The velocity appearing in (4.7) is called the *drift velocity*. This is superimposed on a much higher velocity or speed, known as the *random velocity*, due to the random motion of the electron. Just as in an ordinary gas, the electrons have random motion even in the absence of the field. This is due to the fact that the electrons move about and occasionally scatter and change direction. The random motion, which contributes zero current, exists also in the

† The effective mass of the electron in a metal, denoted by $m^*$, is in general different from the free-electron mass, usually denoted by $m$ or $m_0$. This difference is due to the interaction of the electron with the lattice, as will be discussed in Section 5.15. The effective masses in various metals are listed in Table 4.1.

presence of a field; but in that case there is an additional net velocity opposite to the field, as given by (4.7). The distinction between random and drift motions is shown in Fig. 4.4. We shall denote the two velocities by $v_r$ and $v_d$; it will be shown later that $v_d \ll v_r$.

The current density $J$ can be calculated from (4.7). Since there is a charge $(-Ne)$ per unit volume, and since each electron has a drift velocity given by (4.7), it follows that the amount of charge crossing a unit area per unit time is

$$J = (-Ne)\, v_d = (-Ne)\left(-\frac{e\tau}{m^*}\mathscr{E}\right) = \frac{Ne^2\tau}{m^*}\mathscr{E}. \tag{4.8}$$

The current is *parallel* to the field. Comparing (4.8) with Ohm's law, (4.5), one finds the following expression for the conductivity,

$$\sigma = \frac{Ne^2\tau}{m^*}, \tag{4.9}$$

which is the expression we have been seeking. We see that $\sigma$ increases as $N$ increases. This is reasonable because, as $N$ (or the concentration) increases, there are more current carriers. The conductivity $\sigma$ is inversely proportional to $m^*$, which is again expected, sense the larger $m^*$ is, the more sluggish the particle, and the harder it is for it to move. The proportionality to $\tau$ follows because $\tau$ is actually the time between two consecutive collisions, i.e., the *mean free lifetime.* Therefore the larger $\tau$ is, the more time the electron has to be accelerated by the field between collisions, and hence the larger the drift velocity (4.7), and also the larger $\sigma$ is.

We can evaluate the conductivity $\sigma$ if we know the quantities on the right of (4.7). We shall take $m^*$ to be the same as the free mass $m_0 = 9.1 \times 10^{-31}$ kg. Then we can calculate $N$ as discussed in Section 4.2. There remains the collision time $\tau$; this is a quantity which is difficult to calculate from first principles, so we shall postpone discussing it until Section 4.5. For the time being, we can use (4.8) and the measured value of $\sigma$ to calculate $\tau$. Table 4.1 gives a list of $\sigma$, $N$, $\tau$ and other related quantities for various common metals. Note that $\sigma$ is about $5 \times 10^7$ (ohm-m)$^{-1}$. Note in particular that $\tau$ has a value of about $10^{-14}$ s. This is an extremely small time interval on the common time scale, and we shall see later that important conclusions may be drawn from this.

The time $\tau$ is also called the *relaxation time*. To see the reason for this, let us suppose that an electric field is applied, long enough for a drift velocity $v_{d,0}$ to be established. Now let the field be suddenly removed at some instant. The drift velocity after this instant is governed by

$$m^* \frac{dv}{dt} = -m^* \frac{v}{\tau},$$

which follows from (4.6) for $\mathscr{E} = 0$. The solution appropriate to the initial condition is now

$$v_d(t) = v_{d,0}e^{-t/\tau}, \tag{4.10}$$

showing that $v_d(t)$ approaches zero exponentially with a characteristic time $\tau$. This behavior is called a *relaxation process*. Since we found above that $\tau$ is very short, it follows that $v_d(t)$ relaxes to zero very rapidly.

**Table 4.1**

Electrical Conductivities and Other Transport Parameters for Metals

| Element | $\sigma$, $\text{ohm}^{-1}\,\text{m}^{-1}$ | $N$, $\text{m}^{-3}$ | $\tau$, s | $v_F$, m/s | $l$, Å | $E_F$, eV | $E_F$(obs.), eV | $m^*/m_0$ |
|---|---|---|---|---|---|---|---|---|
| Li | $1.07\times10^7$ | $4.6\times10^{28}$ | $0.9\times10^{-14}$ | $1.3\times10^6$ | 110 | 4.7 | 3.7 | 1.2 |
| Na | 2.11 | 2.5 | 3.1 | 1.1 | 350 | 3.1 | 2.5 | 1.2 |
| K | 1.39 | 1.3 | 4.3 | 0.85 | 370 | 2.1 | 1.9 | 1.1 |
| Rb | 0.80 | 1.1 | 2.75 | 0.80 | 220 | 1.8 | — | — |
| Cs | 0.50 | 0.85 | — | 0.75 | 160 | 1.5 | — | — |
| Cu | 5.88 | 8.45 | 2.7 | 1.6 | 420 | 7.0 | 7.0 | 1.0 |
| Ag | 6.21 | 5.85 | 4.1 | 1.4 | 570 | 5.5 | — | — |
| Au | 4.55 | 5.90 | 2.9 | 1.4 | 410 | 5.5 | — | — |
| Zn | 1.69 | 13.10 | — | 1.82 | — | 9.4 | 11.0 | 0.85 |
| Cd | 1.38 | 9.28 | — | 1.62 | — | 7.5 | — | — |
| Hg | 0.10 | — | — | — | — | — | — | — |
| Al | 3.65 | 18.06 | — | 2.02 | — | 11.6 | 11.8 | — |
| Ga | 0.67 | 15.30 | — | 1.91 | — | 10.3 | — | — |
| In | 1.14 | 11.5 | — | 1.74 | — | 8.6 | — | — |

Values quoted are for metals at room temperature. The concentration is found by using the usual chemical valences. The Fermi velocity $v_F$ and $E_F$ are evaluated by using $m^* = m_0$ and the appropriate equation from Section 4.6. The Fermi energy $E_F$ (observed) is the experimentally determined value as discussed in Chapter 6. The effective mass $m^*$ is determined by using the experimental value $E_F$ (observed) and the relation $E_F = (\hbar/2m^*)(3\pi^2N)^{2/3}$, Eq. (4.34).

We shall now rewrite (4.9) in a form which brings out some aspects of the physics more clearly. Since $\tau$ is the time between two successive collisions, it may be expressed as

$$\tau = \frac{l}{v_r}, \tag{4.11}$$

where $l$ is the *distance* between two successive collisions and $v_r$ is the random velocity. In terms of these, $\sigma$ becomes

$$\sigma = \frac{Ne^2 l}{m^* v_r}. \tag{4.12}$$

Let us compare the results of applying this formula to metals and semiconductors. For the former, $\sigma \simeq 5 \times 10^7$ (ohm-m)$^{-1}$, as we have seen, while for the latter, $\sigma \simeq 1$ (ohm-m)$^{-1}$. The difference can be accounted for by (4.12). First, in semiconductors, $N \simeq 10^{20}$ m$^{-3}$, as compared with $N \simeq 10^{29}$ m$^{-3}$ in metals. This reduces $\sigma$ by a factor of $10^{-9}$ for semiconductors. Second, $v_r$ in metals is of the order of the *Fermi velocity* (Section 4.7), which is about $10^6$ m-s$^1$, while it is only about $10^4$ m-s$^{-1}$ in semiconductors.† If we include the effects of both $N$ and $v_r$, we find the conductivity to be the right order of magnitude for semiconductors.

Let us compare the magnitudes of $v_r$ and $v_d$. The former has a value of about $10^6$ m-s$^{-1}$; on the other hand, $v_d$ can be evaluated from (4.7). When we substitute for $e$, $\tau$, and $m^*$ in (4.7) their values: $e \simeq 10^{-19}$ coul, $\tau = 10^{-14}$ s, $m^* \simeq 10^{-30}$ kg, and $\mathscr{E} \simeq 10$ V/m, we find that $v_d \simeq 10^{-2}$ m-s$^{-1}$. Thus $v_d/v_r \simeq 10^{-8}$, a very small ratio indeed.

We can also find the microscopic expression for the joule heat. The power dissipated as joule heat must be equal to the power absorbed by the electron system from the field. Recalling from elementary physics that the power absorbed by a particle from force $F$ is $Fv$, where $v$ is the velocity of the particles, we see that the power absorbed by the electron system per unit volume is

$$P = NFv_d = N(-e\mathscr{E})\left(-\frac{e\tau\mathscr{E}}{m^*}\right)$$

$$= \frac{Ne^2\tau}{m^*}\mathscr{E}^2. \tag{4.13}$$

**The origin of collision time**

We have introduced $\tau$ as collision time due to some friction force, the source of which was not discussed. It seems natural to assume that the friction force is caused by the collision of electrons with ions. According to this particular model of collision, an electron, as it moves in the lattice, collides with ions, which has the effect of slowing down the electron's momentum. This model turns out to be untenable because it leads to many points of disagreement with experiment. To cite only one: The mean free path $l$ can be calculated from (4.9). If we substitute the values $\tau \simeq 10^{-14}$s and $v_r \simeq 10^6$ m-s$^{-1}$, we find that $l \simeq 10^{-8}$ m $\simeq 10^2$ Å. This means that, between two collisions, the electron travels a distance of more than

† In semiconductors, random velocity is given by the usual expression $v_r = (3kT/m^*)^{1/2}$ due to thermal motion. If we substitute $T = 300$°K and $m^* = m_0/5$, a typical value for the effective mass in semiconductors, we find that $v_r = 10^4$ m $\cdot$ s$^{-1}$.

20 times the interatomic distance. This is much larger than one would expect if the electron really *did* collide with the ions whenever it passed them. Especially in close-packed structures, in which the atoms are densely packed, it is difficult to see how the electrons could travel so far between collisions.

This paradox can be explained only by the use of quantum concepts. The essence of the argument is as follows: We saw in Section 2.12 that, according to quantum mechanics, an electron has a wave character. The wavelength of the electron in the lattice is given by the deBroglie relation (Section A.1),

$$\lambda = \frac{h}{m^* v_r}. \tag{4.14}$$

It is well known from the theory of wave propagation in discrete structures† that, when a wave passes through a periodic lattice, it continues propagating indefinitely without scattering. The effect of the atoms in the lattice is to absorb energy from the wave and radiate it back, so that the net result is that the wave continues without modification in either direction or intensity. The *velocity* of propagation, however, *is* modified. This is what happens in the case of an electron wave in a regular lattice, except that in this case we are dealing with a matter wave.

We discussed the mathematical reason why a regular lattice does not scatter a wave in some detail in Chapter 2. There we saw that the wave—be it x-ray, neutron, or electron—does not scatter or diffract except when the Bragg condition is satisfied. Save under this special condition, the conduction electron should not be scattered by a regular lattice of ions at all.

There is a familiar example in optics: A light wave traveling in a crystal is not scattered at all. The only effect the crystal has is to introduce the index of refraction $n$ so that the velocity in the medium is $c/n$. Therefore we see that, if the ions form a perfect lattice, there is no collision at all—that is, $l = \infty$—and hence $\tau = \infty$, which in turn leads to infinite conductivity. It has been shown, however, that the observed $l$ is about $10^2$ Å. The finiteness of $\sigma$ must thus be due to the deviation of the lattice from perfect periodicity; this happens either because of thermal vibration of the ions, or because of the presence of imperfections or foreign impurities, as we shall see in the next section.

## 4.5 ELECTRICAL RESISTIVITY VERSUS TEMPERATURE

The electrical conductivity of a metal varies with the metal's temperature in a characteristic manner. This variation is usually discussed in terms of the behavior of the resistivity $\rho$ versus $T$. Figure 4.5 shows the observed curve for Na. At $T \simeq 0°\text{K}$, $\rho$ has a small *constant* value; above that, $\rho$ increases with $T$, slowly at first, but afterward $\rho$ increases linearly with $T$. The linear behavior continues essen-

† See, e.g., L. Brillouin, 1953, *Wave Propagation in Periodic Structures*, New York: Dover Press.

tially until the melting point is reached. This pattern is followed by most metals (except as noted below), and usually room temperature falls into the linear range. The linear behavior is readily verified experimentally, as you may recall from elementary physics.

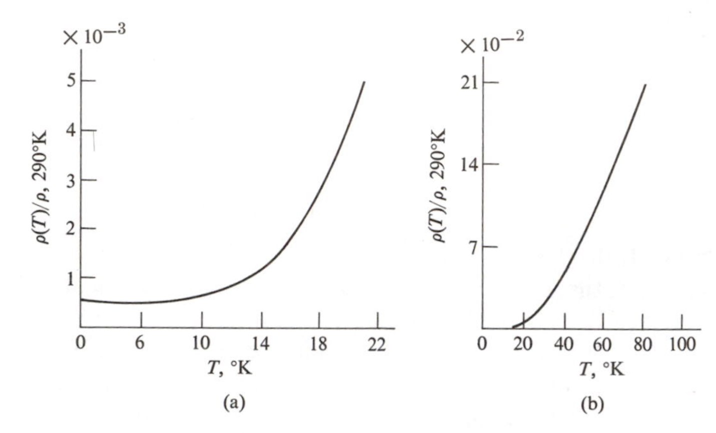


**Fig. 4.5** The normalized resistivity $\rho(T)/\rho(290°\text{K})$ versus $T$ for Na in the low-temperature region (a), and at higher temperatures (b). $\rho(290) \simeq 2.10 \times 10^{-8}$ Ω-m.

We want to explain this behavior of $\rho$ in terms of the formulas developed in Section 4.4. Recalling that $\rho = \sigma^{-1}$, and using (4.9), we have

$$\rho = \frac{m^*}{Ne^2}\frac{1}{\tau}. \tag{4.15}$$

We note from the interpretation of $\tau$ in the last section that $1/\tau$ is actually equal to the probability of the electron suffering a scattering per unit time. Thus, if $\tau = 10^{-14}$s, then the electron undergoes $10^{14}$ collisions in one second. But in Section 4.4 we saw that the electron undergoes a collision only because the lattice is not perfectly regular. We group the deviations from a perfect lattice into two classes.

a) Lattice vibrations (phonons) of the ions around their equilibrium position due to thermal excitation of the ions.

b) All static imperfections, such as foreign impurities or crystal defects. Of this latter group we shall take foreign impurities as an example. Now the probabilities of electrons being scattered by phonons and by impurities are additive, since these two mechanisms are assumed to act independently. Therefore we may write

$$\frac{1}{\tau} = \frac{1}{\tau_{\text{ph}}} + \frac{1}{\tau_i}, \tag{4.16}$$

where the first term on the right is due to phonons and the second is due to impurities. The former is expected to depend on $T$ and the latter on impurities, but not on $T$. When (4.16) is substituted into (4.15), we readily find

$$\rho = \rho_i + \rho_{\text{ph}}(T) = \frac{m^*}{Ne^2}\frac{1}{\tau_i} + \frac{m^*}{Ne^2}\frac{1}{\tau_{\text{ph}}}. \tag{4.17}$$

We note that $\rho$ has split into two terms: a term $\rho_i$ due to scattering by impurities (which is independent of $T$), called the *residual resistivity*. Added to this is another term $\rho_{\text{ph}}(T)$ due to scattering by phonons; hence it is temperature dependent, and is called the *ideal resistivity*, in that it is the resistivity of a pure specimen.

At very low $T$, scattering by phonons is negligible because the amplitudes of oscillation are very small; in that region $\tau_{\text{ph}} \to \infty$, $\rho_{\text{ph}} \to 0$, and hence $\rho = \rho_i$, a constant. This is in agreement with Fig. 4.5. As $T$ increases, scattering by phonons becomes more effective, and $\rho_{\text{ph}}(T)$ increases; this is why $\rho$ increases. When $T$ becomes sufficiently large, scattering by phonons dominates and $\rho \simeq \rho_{\text{ph}}(T)$. In the high-temperature region, $\rho_{\text{ph}}(T)$ increases linearly with $T$, as we shall shortly show. This is again in agreement with experiment, as shown in Fig. 4.5. The statement that $\rho$ can be split into two parts, one of which is independent of $T$, is known as the *Matthiessen rule*. This rule is embodied in (4.17).

We expect that $\rho_i$ should increase with impurity concentrations, and indeed it will be shown that for small concentrations $\rho_i$ is proportional to the impurity concentration $N_i$. We also remark that, for small impurity concentration, $\rho_{\text{ph}} \gg \rho_i$, except at very low $T$. Let us now derive approximate expressions for $\tau_i$ and $\tau_{\text{ph}}$, using arguments from the kinetic theory of gases. We shall assume, for simplicity, that the collision is of the hard-spheres (billiard-ball) type.

Consider first the collision of electrons with impurities. We write

$$\tau_i = \frac{l_i}{v_r}, \tag{4.18}$$

after (4.11), where $l_i$ is the mean free path for collision with impurities. Given that the *scattering cross section* of an impurity is $\sigma_i$—which is the area an impurity atom presents to the incident electron—then, using an argument familiar from the kinetic theory of gases, one may write

$$l_i \sigma_i N_i = 1$$

or

$$l_i = \frac{1}{N_i \sigma_i}. \tag{4.19}$$

It is expected that $\sigma_i$ is of the same magnitude as the actual geometrical area of the impurity atom. That is, that $\sigma_i \simeq 1\text{Å}^2$. (Calculations of the exact value of $\sigma_i$ require quantum scattering theory.) By substituting from (4.18) and (4.19) into (4.17), one can find $\rho_i$. One then sees that $\rho_i$ is proportional to $N_i$, the concentration of impurities.

Calculating $\tau_{ph}$ is more difficult, but equations similar to (4.18) and (4.19) still hold. In particular, one may write

$$l_{ph} = \frac{1}{N_{ion}\sigma_{ion}}, \tag{4.20}$$

where $N_{ion}$ is the concentration of metallic ions in the lattice and $\sigma_{ion}$ is the scattering cross section per ion. We should note here that $\sigma_{ion}$ has no relation to the geometrical cross section of the ion. Rather it is the area presented by the thermally fluctuating ion to the passing electron. Suppose that the distance of deviation from equilibrium is $x$; then the average scattering cross section is about

$$\sigma_{ion} \simeq \pi\langle x^2\rangle, \tag{4.21}$$

where $\langle x^2\rangle$ is the average of $x^2$. The value $\langle x^2\rangle$ can be estimated as follows: Since the ion is a harmonic oscillator (Section 3.4), the average of its potential energy is equal to half the total energy. Thus

$$\tfrac{1}{2}k\langle x^2\rangle = \langle E\rangle = \frac{\hbar\omega}{e^{\hbar\omega/kT} - 1}, \tag{4.22}$$

where we used the formula for the energy of a quantum oscillator (Section 3.4). The frequency $\omega$ is either the Einstein or the Debye frequency, because in this rough argument we can ignore the difference between these two frequencies. We may introduce the Debye temperature $\theta$ so that $\hbar\omega = k\theta$. When we make these substitutions into (4.17), we find that $\rho_{ph}(T)$ can be written as

$$\rho_{ph}(T) = \left(\frac{\pi\hbar}{k\theta M}\right)\frac{1}{e^{\theta/T} - 1}, \tag{4.23}$$

where $M$ is the mass of the ion. In the range $T \gg \theta$, this can be written as

$$\rho_{ph} \simeq \frac{\pi\hbar^2}{k\theta M}\frac{T}{\theta}, \tag{4.24}$$

which is linear in $T$, as promised, and in agreement with experiment.

In the low-temperature range, Eq. (4.23) predicts that $\rho_{ph}(T)$ will decrease exponentially as $e^{-\theta/T}$. However, the observed decrease is as $T^5$. The reason for this discrepancy is that we used the Einstein model, in which the motion of the neighboring ions was treated independently. When the correlation between ionic motions is taken into account, as in the Debye theory of lattice vibrations, one obtains the $T^5$ behavior.

Deviations from Matthiessen's rule are often observed, the best known being the *Kondo effect*. When some impurities of Fe, for example, are dissolved in Cu, $\rho$ does not behave as in Fig. 4.5 at low $T$. Instead $\rho$ has a minimum at low $T$. This anomalous behavior is due to an additional scattering of electrons by the magnetic moments on the impurity centers. Also, deviations from Matthiessen's rule attributable to complications in the band structure of the conduction electrons have been reported. We see from these two examples that the behavior of $\rho$ versus $T$ at very low $T$ may be much more complex than that implied by the simple statement of Mathiessen's rule.

## 4.6 HEAT CAPACITY OF CONDUCTION ELECTRONS

In the *free-electron model* the conduction electrons are treated as free particles which obey the classical laws of mechanics, electromagnetism, and statistical mechanics. We have already pointed out the difficulty of treating collisions in this model, and also how one must appeal to quantum concepts in order to salvage the model. Another difficulty arises in connection with the heat capacity of the conduction electrons.

Let us calculate the heat capacity per mole for the conduction electrons on the basis of the Drude–Lorentz model. It is well known from the kinetic theory of gases that a free particle in equilibrium at temperature $T$ has an average energy of $\frac{3}{2}\,kT$. Therefore the average energy per mole is

$$\langle \bar{E} \rangle = N_A(\tfrac{3}{2}kT) = \tfrac{3}{2}RT, \tag{4.25}$$

where $N_A$ is Avogadro's number and $R = N_A k$. The electrons' heat capacity $C_e = \partial[\bar{E}]/\partial T$. Therefore

$$C_e = \tfrac{3}{2}R \simeq 3 \text{ cal/mole °K}. \tag{4.26}$$

The total heat capacity in metals, including phonons, should then be

$$C = C_{ph} + C_e, \tag{4.27}$$

which, at high temperature, has the value

$$C = 3R + \tfrac{3}{2}\,R = 4.5R \simeq 9 \text{ cal/mole °K}. \tag{4.28}$$

Experiments on heat capacity in metals show, however, that $C$ is very nearly equal to $3R$ at high $T$, as is the case for insulators. Accurate measurements in which the contributions of electrons to total heat capacity are isolated show that $C_e$ is smaller than the classical value $\frac{3}{2}R$ by a factor of about $10^{-2}$. To explain this discrepancy, we must once again turn to quantum concepts.

The energy of the electron in a metal is quantized according to quantum mechanics. Figure 4.6(a) shows the quantum energy levels. The electrons in the metal occupy these levels. In doing so, they follow a very important quantum principle, the *Pauli exclusion principle*, according to which an energy level can accommodate at most two electrons, one with spin up, and the other with spin down. Thus in filling the energy levels, two electrons occupy the lowest level, two more the next level, and so forth, until all the electrons in the metal have been accommodated, as shown in Fig. 4.6(a). The energy of the highest occupied level is called the *Fermi energy* (or simply the *Fermi*) *level.* We shall evaluate the Fermi level in Section 4.7. A typical value for the Fermi energy in metals is about 5 eV.

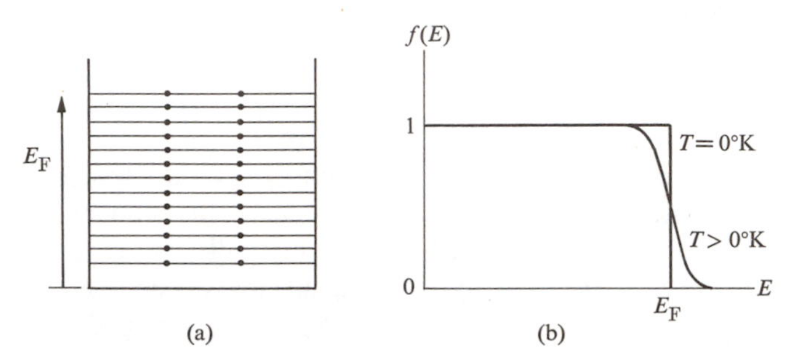


**Fig. 4.6** (a) Occupation of energy levels according to the Pauli exclusion principle. (b) The distribution function $f(E)$ versus $E$, at $T = 0°K$ and $T > 0°K$.

The situation described obtains in metals at $T = 0°K$. Even at the lowest possible temperature, the electron system has a considerable amount of energy, by virtue of the exclusion principle. If it were not for this principle, all the electrons would fall into the lowest level, and the total energy of the system would be negligible. This corresponds to the assertion, usually made in classical mechanics, that as $T \rightarrow 0°K$ all motion ceases, and the energy vanishes. This assertion clearly does not apply to the conduction electrons.

The distribution of electrons among the levels is usually described by the *distribution function*, $f(E)$, which is defined as the probability that the level $E$ is occupied by an electron. Thus if the level is certainly empty, then $f(E) = 0$, while if it is certainly full, then $f(E) = 1$. In general, $f(E)$ has a value between zero and unity.

It follows from the preceding discussion that the distribution function for electrons at $T = 0°K$ has the form

$$f(E) = \begin{cases} 1, & E < E_F \\ 0, & E_F < E \end{cases} \tag{4.29}$$

That is, all levels below $E_F$ are completely filled, and all those above $E_F$ are completely empty. This function is plotted in Fig. 4.6(b), which shows the discontinuity at the Fermi energy.

We have thus far restricted our treatment to the temperature at absolute zero. When the system is heated ($T > 0°K$), thermal energy excites the electrons. But this energy is not shared equally by all the electrons, as would be the case in the classical treatment, because the electrons lying well below the Fermi level $E_F$ cannot absorb energy. If they did so, they would move to a higher level, which would be already occupied, and hence the exclusion principle would be violated.

Recall in this context that the energy which an electron may absorb thermally is of the order $kT$ ( $= 0.025$ eV at room temperature), which is much smaller than $E_F$, this being of the order of 5 eV. Therefore only those electrons close to the Fermi level can be excited, because the levels above $E_F$ are empty, and hence when those electrons move to a higher level there is no violation of the exclusion principle. Thus only these electrons—which are a small fraction of the total number—are capable of being thermally excited, and this explains the low electronic specific heat (or heat capacity).

The distribution function $f(E)$ at temperature $T \neq 0°K$ is given by

$$f(E) = \frac{1}{e^{(E-E_F)/kT} + 1}. \tag{4.30}$$

This is known as the *Fermi–Dirac distribution.*† This function is also plotted in Fig. 4.6(b), which shows that it is substantially the same as the distribution at $T = 0°K$, except very close to the Fermi level, where some of the electrons are excited from below $E_F$ to above it. This is, of course, to be expected, in view of the above discussion.‡

One can use the distribution function (4.30) to evaluate the thermal energy and hence the heat capacity of the electrons, but this is a fairly tedious undertaking, so instead we shall attempt to obtain a good approximation with a minimum of mathematical effort. Since only electrons within the range $kT$ of the Fermi level are excited, we conclude that only a fraction $kT/E_F$ of the electrons is affected. Therefore the number of electrons excited per mole is about $N_A(kT/E_F)$, and since each electron absorbs an energy $kT$, on the average, it follows that the thermal energy per mole is given approximately by

$$\bar{E} = \frac{N_A(kT)^2}{E_F},$$

and the specific heat $C_e = \partial\bar{E}/\partial t$ is

$$C_e = 2R\frac{kT}{E_F}. \tag{4.31}$$

We see that the specific heat of the electrons is reduced from its classical value, which is of the order of $R$, by the factor $kT/E_F$. For $E_F = 5$ eV and $T = 300°K$, this

† For a derivation see, for example, M. Alonso and E. J. Finn, 1968, *Fundamental University Physics,* Volume III, Reading Mass.: Addison-Wesley.

‡ Note that, in the energy range far above the Fermi energy, $(E - E_F)/kT \gg 1$, and hence the Fermi–Dirac distribution function has the form $f(E) = e^{E_F/kT}e^{-E/kT} = \text{constant} \times e^{-E/kT}$, which is the classical—or Maxwell–Boltzmann—distribution. Thus in the high energy range, i.e., in the tail of the Fermi–Dirac distribution, electrons may be treated by classical statistical mechanics.

factor is equal to 1/200. This great reduction is in agreement with experiment, as pointed out previously.

The so-called *Fermi temperature* $T_F$, which is sometimes used in this context, is defined as $E_F = kT_F$, and the specific heat may now be written as

$$C_e = 2R\frac{T}{T_F}.$$

A typical value for $T_F$, corresponding to $E_F = 5$ eV, is 60,000°K. Thus in order for the specific heat of the electrons in a solid to reach its classical value, the solid must be heated to a temperature comparable to $T_F$. But this is not possible, of course, as the solid would long since have melted and evaporated! At all practical temperatures, therefore, the specific heat of electrons is far below its classical value.

Another interesting conclusion from (4.32) is that the heat capacity $C_e$ of the electrons is a linear function of temperature. This is unlike the lattice heat capacity $C_L$, which is constant at high temperature, and proportional to $T^3$ at low temperature.

An exact evaluation of the electronic heat capacity yields the value

$$C_e = \frac{\pi^2}{2}R\frac{kT}{E_F}, \qquad (4.32)$$

which is clearly of the same order of magnitude as the approximate expression (4.31).

## 4.7 THE FERMI SURFACE

The electrons in a metal are in a continuous state of random motion. Because these electrons are considered to be free particles, the energy of an electron is entirely kinetic, and one may therefore write

$$E = \tfrac{1}{2}m^* v^2,$$

where $v$ is the speed of the particle. Now let us introduce the concept of *velocity space*, whose axes are $v_x$, $v_y$, and $v_z$. Each point in this space represents a unique velocity—both in magnitude and direction.

Consider the conduction electrons in this velocity space. These electrons have many different velocities, and since these velocities are random, the points representing them fill the space uniformly, as shown in Fig. 4.7. Note, however, that there is a sphere outside which all points are empty. The radius of this sphere is the *Fermi speed* $v_F$, which is related to the Fermi energy by the usual relation

$$E_F = \tfrac{1}{2}m^* v_F^2. \qquad (4.33)$$

The reason why all points outside the sphere are empty is that they correspond to energies greater than $E_F$, which are unoccupied at $T = 0°K$, as discussed in Section 4.6. All the points inside the sphere are completely full. This sphere is known as the *Fermi sphere*, and its surface as the *Fermi surface*.

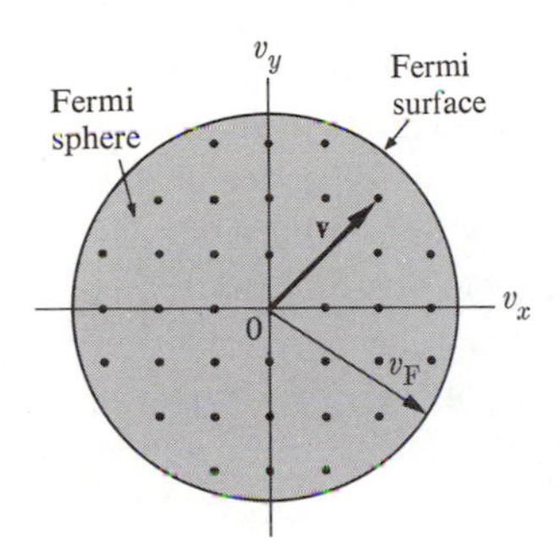


**Fig. 4.7** The Fermi surface and the Fermi sphere.

The Fermi surface (FS), which is very significant in many solid-state phenomena—for example, transport properties—is not affected appreciably by temperature. When the temperature is raised, only relatively few electrons are excited from the inside to the outside of the Fermi surface, and these have very little effect, as we have seen. Thus the FS has an independent, permanent identity, and should be regarded as a real physical characteristic of the metal.

The Fermi speed $v_F$ is very large. If we substitute $E_F = 5$ eV in (4.33) and calculate $v_F$, we find that $v_F = (2E_F/m^*)^{1/2} \simeq (2 \times 5 \times 1.6 \times 10^{-19}/9 \times 10^{-31})^{1/2} \simeq 10^6 \text{ m}\cdot\text{s}^{-1}$, which is about one-hundredth of the speed of light. Thus electrons at the FS are moving very fast. Furthermore, the Fermi speed, like the Fermi surface, is independent of temperature.

The value of the Fermi energy is determined primarily by the electron concentration. The greater the concentration, the higher the topmost energy level required to accommodate all the electrons (refer to Fig. 4.6a), and hence the higher the $E_F$. Section 5.12 will show that $E_F$ is given by

$$E_F = \frac{\hbar^2}{2m^*}(3\pi^2 N)^{2/3}. \tag{4.34}$$

If one substitutes the typical value $N = 10^{28} \text{ m}^{-3}$, one finds that $E_F \simeq 5$ eV, in agreement with our earlier statements. Table 4.1 lists the Fermi energies for various metals.

The Fermi surface will be discussed in much greater detail in Section 5.12, where the interaction of the electrons with the lattice is taken into account. We shall find there that the FS may be distorted from the simple spherical shape considered here, this distortion being engendered by the electron–lattice interaction. For the time being, however, the free-electron model and its FS satisfy our purpose.

## 4.8 ELECTRICAL CONDUCTIVITY; EFFECTS OF THE FERMI SURFACE

We discussed electrical conductivity in Section 4.3, in which we treated electrons on a classical basis. How are the results modified when the FS is taken into account?

Let us refer to Fig. 4.8. In the absence of an electric field, the Fermi sphere is centered at the origin (Fig. 4.8a). The various electrons are all moving—some at very high speeds—and they carry individual currents. But the total current of the system is zero, because, for every electron at velocity **v** there exists another electron with velocity $-\mathbf{v}$, and the sum of their two currents is zero. Thus the total current vanishes due to pairwise cancellation of the electron currents.

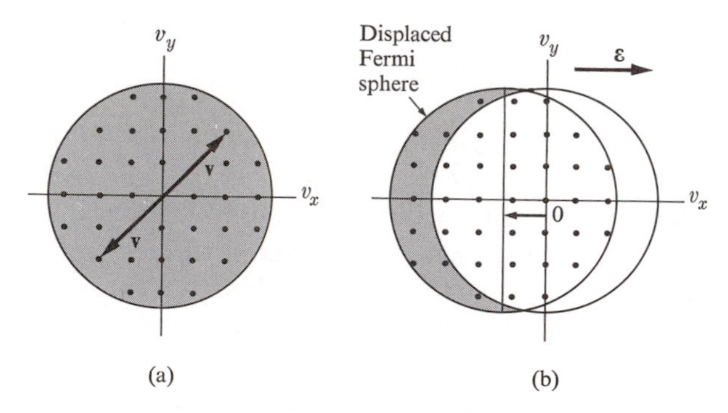


**Fig. 4.8** (a) The Fermi sphere at equilibrium. (b) Displacement of the Fermi sphere due to an electric field.

The situation changes when a field is applied. If the field is in the positive $x$-direction, each electron acquires a drift velocity $v_d = -(e\tau/m^*)\mathscr{E}$, as given by (4.7). Thus the whole Fermi sphere is displaced to the left, as shown in Fig. 4.8(b). Although the displacement is very small, and although the great majority of the electrons still cancel each other pairwise, some electrons—in the shaded crescent in the figure—remain uncompensated. It is these electrons which produce the observed current.

Let us estimate the current density: The fraction of electrons which remain uncompensated is approximately $v_d/v_F$. The concentration of these electrons is therefore $N(v_d/v_F)$, and since each electron has a velocity of approximately $-v_F$, the current density is given by

$$J \simeq -e\,N(v_d/v_F)(-v_F) = N\,e\,v_d,$$

which, on substitution of $v_d = -(e\tau/m^*)\mathscr{E}$, yields

$$J = \frac{N\,e^2\tau_F}{m^*}\,\mathscr{E},$$

where $\tau_F$ is the collision time of an electron at the FS. The resulting electrical conductivity is therefore

$$\sigma = \frac{N\,e^2\tau_F}{m^*}. \tag{4.35}$$

This is precisely the same as the result obtained classically, except that $\tau$ is replaced by $\tau_F$. The expression (4.35), which is only an approximate derivation, can be corroborated by a more detailed and accurate statistical analysis.

The actual picture of electrical conduction is thus quite different from the classical one envisaged in Section 4.4, in which we assumed that the current is carried equally by all electrons, each moving with a very small velocity $v_d$. The current is, in fact, carried by very few electrons only, all moving at high velocity. Both approaches lead to the same result, but the latter is the more accurate. This can be seen from the fact that only the collision time for electrons at the FS, $\tau_F$, appears in expression (4.35) for $\sigma$.

If we substitute $\tau_F = l_F/v_F$ into (4.35), we find that

$$\sigma = \frac{Ne^2 l_F}{m^* v_F}.$$

The only quantity on the right side which depends on temperature is the mean free path $l_F$. Since $l_F \sim 1/T$ at high temperature, as we saw in Section 4.5, it follows that $\sigma \sim 1/T$ or $\rho \sim T$, in agreement with our previous discussion of electrical resistivity.

The importance of the FS in transport phenomena is now clear. Since the current is transported by electrons lying close to the Fermi surface, these phenomena are very sensitive to the properties, shape, etc., of this surface. The inner electrons are irrelevant so far as conduction processes are concerned.

The fact that essentially the same answer may be obtained classically as quantum mechanically (with proper adjustment of the collision time) encourages us to use the simpler classical procedure. This we shall do wherever feasible in the following sections.

## 4.9 THERMAL CONDUCTIVITY IN METALS

When the ends of a metallic wire are at different temperatures, heat flows from the hot to the cold end. (Recall our discussion in Section 3.9 on thermal conductivity in insulators.) The basic experimental fact is that the heat current $Q$—that is, the amount of thermal energy crossing a unit area per unit time—is proportional to the temperature gradient,

$$Q = -K\frac{dT}{dx},$$

where $K$ is the thermal conductivity. In insulators, heat is carried entirely by phonons, but in metals heat may be transported by both electrons and phonons. The conductivity $K$ is therefore equal to the sum of the two contributions,

$$K = K_e + K_{ph},$$

where $K_e$ and $K_{ph}$ refer to electrons and phonons, respectively. In most metals, the contribution of the electrons greatly exceeds that of the phonons, because of the great concentration of electrons; typically $K_{ph} \simeq 10^{-2} K_e$. This being so, the conductivity of the phonons will henceforth be ignored in this section.

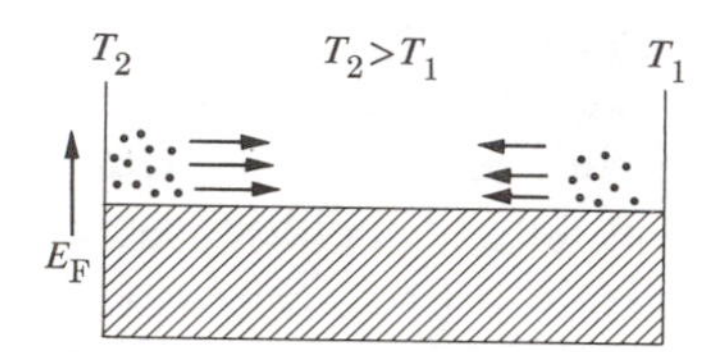


**Fig. 4.9** The physical basis for thermal conductivity. Energetic electrons on the left carry net energy to the right.

The physical process by which heat conduction takes place via electrons is illustrated in Fig. 4.9. Electrons at the hot end (to the left) travel in all directions, but a certain fraction travel to the right and carry energy to the cold end. Similarly, a certain fraction of the electrons at the cold end (on the right) travel to the left, and carry energy to the hot end. These oppositely traveling electron currents are equal, but because those at the hot end are more energetic on the average than those on the right, a net energy is transported to the right, resulting in a current of heat. Note that heat is transported almost entirely by those electrons near the Fermi levels, because those well below this level cancel each other's contributions. Once more it is seen that the electrons at the FS play a primary role in transport phenomena.

To evaluate the thermal conductivity $K$ quantitatively, we use the formula $K = \frac{1}{3} cvl$, used in Section 3.9 in treating heat transport in insulators. We recall that $C_v$ is the specific heat per unit volume, $v$ the speed, and $l$ the mean free path of the particles involved. In the present case, where electrons are involved, $C_v$ is the electronic specific heat and should be substituted from (4.32); also $R$ should be replaced by $Nk$, since we are dealing here with a unit volume rather than a mole. In addition, $v$ and $l$ should be replaced by $v_F$ and $l_F$, since only electrons at the

Fermi levels are effective. Thus

$$K = \frac{1}{3}\left(\frac{\pi^2 N k^2 T}{2E_F}\right) v_F\, l_F.$$

Noting that

$$E_F = \tfrac{1}{2} m^* v_F^2$$

and that $l_F/v_F = \tau_F$, we can simplify this expression for $K$ to

$$K = \frac{\pi^2 N\, k^2 T \tau_F}{3m^*}, \tag{4.36}$$

which expresses thermal conductivity in terms of the electronic properties of the metal. Substituting the usual values of the electrons' parameters, one finds $K \simeq 50$ cal/m °K-s. Table 4.2 gives the measured values of $K$ for some metals, and shows that theory is in basic agreement with experiment.

**Table 4.2**

Thermal Conductivities and Lorenz Numbers (Room Temperature)

| Element | Na | Cu | Ag | Au | Al | Cd | Ni | Fe |
|---|---|---|---|---|---|---|---|---|
| $K$, cal/m °K · s | 33 | 94 | 100 | 71 | 50 | 24 | 14 | 16 |
| L, cal · ohm/s · °K | $5.2 \times 10^{-9}$ | 5.4 | 5.6 | 5.9 | 4.7 | 6.3 | 3.7 | 5.5 |

Many of the parameters appearing in the expression for $K$ were also included in the expression for electrical conductivity $\sigma$. Recalling that $\sigma = Ne^2\tau_F/m^*$, we readily establish that the ratio $K/\sigma T$ is given by

$$\mathrm{L} = \frac{1}{3}\left(\frac{\pi k}{e}\right)^2. \tag{4.37}$$

This *Lorenz number* L, because it depends only on the universal constants $k$ and $e$, should be the same for *all* metals. Its numerical value is $5.8 \times 10^{-9}$ cal-ohm/s °K$^2$. This conclusion suggests that the electrical and thermal conductivities are intimately related, which is to be expected, since both electrical and thermal current are carried by the same agent: electrons.

Table 4.2 lists Lorenz numbers for widely differing metals, and we see that they are close to the predicted values. The fact that the agreement is not exact stems from (a) the use of the rather simple free-electron model, and (b) the simplified treatment used in calculating the transport coefficients $\sigma$ and $K$. A more refined treatment shows that L does indeed depend on the metal under discussion.

## 4.10 MOTION IN A MAGNETIC FIELD: CYCLOTRON RESONANCE AND THE HALL EFFECT

The application of a magnetic field to a metal gives rise to several interesting effects arising from the conduction electrons. The *cyclotron resonance* and the *Hall effect* are two which we shall use to investigate the properties of conduction electrons.

### Cyclotron resonance

Figure 4.10 illustrates the phenomenon of cyclotron resonance. A magnetic field applied across a metallic slab causes electrons to move in a counterclockwise circular fashion in a plane normal to the field. The frequency of this *cyclotron* motion, known as the *cyclotron frequency*, is given by

$$\omega_c = \frac{eB}{m^*}. \tag{4.38}$$

If we substitute the value of the free-electron mass, we find that

$$\nu_c = \omega_c/2\pi = 2.8B \text{ GHz},$$

where $B$ is in kilogauss. Thus for $B = 1$ kG, the cyclotron frequency is $\nu_c = 2.8$ GHz, which is in the microwave range.

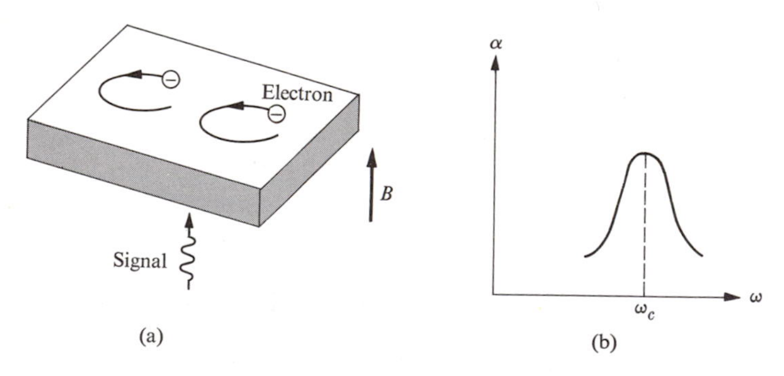


**Fig. 4.10** (a) Cyclotron motion. (b) The absorption coefficient $\alpha$ versus $\omega$.

Suppose now that an electromagnetic signal is passed through the slab in a direction parallel to **B**, as shown in the figure. The electric field of the signal acts on the electrons, and some of the energy in the signal is absorbed. The rate of absorption is greatest when the frequency of the signal is exactly equal to the frequency of the cyclotron:

$$\omega = \omega_c. \tag{4.39}$$

This is so because, when this condition holds true, each electron moves synchronously with the wave throughout the cycle, and therefore the absorption continues all through the cycle. Thus Eq. (4.39) is the condition for cyclotron resonance. On the other hand, when Eq. (4.39) is not satisfied, the electron is in phase with the wave through only a part of the cycle, during which time it absorbs energy from the wave. In the remainder of the cycle, the electron is out of phase and returns energy to the wave. The shape of the absorption curve as a function of the frequency is shown in Fig. 4.10(b).†

Cyclotron resonance is commonly used to measure the electron mass in metals and semiconductors. The cyclotron frequency is determined from the absorption curve, and this value is then substituted in (4.38) to evaluate the effective mass. The accuracy with which $m^*$ is determined depends on the accuracy of $\omega_c$ and $B$. One can measure the cyclotron frequency $\omega_c$ very accurately, particularly if one uses a laser beam, and therefore the accuracy of measurement of $m^*$ is limited only by the accuracy of measurement of the magnetic field and its homogeneity across the sample.

**The Hall effect**

The physical process underlying the Hall effect is illustrated in Fig. 4.11. Suppose that an electric current $J_x$ is flowing in a wire in the $x$-direction, and a magnetic field $B_z$ is applied normal to the wire in the $z$-direction. We shall show that this leads to an additional electric field, normal to both $J_x$ and $B_z$, that is, in the $y$-direction.

To see how this comes about, let us first consider the situation before the magnetic field is introduced. There is an electric current flowing in the positive $x$-direction, which means that the conduction electrons are drifting with a velocity $\mathbf{v}$ in the negative $x$-direction. When the magnetic field is introduced, the Lorentz force $\mathbf{F} = e(\mathbf{v} \times \mathbf{B})$ causes the electrons to bend downward, as shown in the figure. As a result, electrons accumulate on the lower surface, producing a net negative charge there. Simultaneously a net positive charge appears on the upper surface, because of the deficiency of electrons there. This combination of positive and

† If the peak of the absorption curve is to be clearly discernible, and hence the cyclotron frequency accurately determined, the condition $\omega_c\tau \gg 1$ must be satisfied. This means that the electron can execute many cyclotron cycles during the time it takes to make a single collision. If this condition is not fulfilled, the curve of the collision time is so broad that no unique frequency $\omega_c$ is distinguishable.

To make the quantity $\omega_c\tau$ as large as possible, one raises the frequency $\omega_c$ by using very high magnetic fields—about 50 kG— and increases the collision time by cooling the sample to low temperatures, e.g., 10°K.

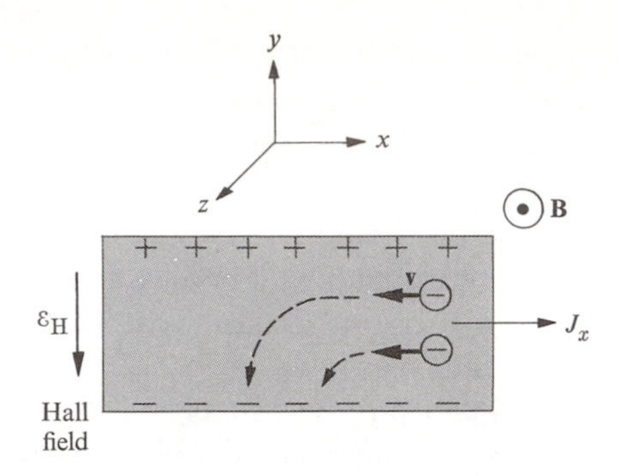


**Fig. 4.11** Origin of the Hall field and Hall effect.

negative surface charges creates a downward electric field, which is called the *Hall field*.

Let us evaluate this Hall field. The Lorentz force L which produces the charge accumulation in the first place is in the negative $y$-direction, and has the value

$$F_L = ev_x\mathrm{B},$$

where the sign is properly adjusted so that $F_L$ is negative, in accordance with the figure (recall that $v_x$, being to the left, is negative). Now the field created by the surface charges produces a force which opposes this Lorentz force. The accumulation process continues until the Hall force completely cancels the Lorentz force. Thus, in the steady state, $F_H = F_L$:

$$-e\mathscr{E}_H = -e\,v_x B \qquad \text{or} \qquad \mathscr{E}_H = v_x B,$$

which is the Hall field. It is convenient to express this in terms of measurable quantities, and for this purpose the velocity $v_x$ is expressed in terms of the current density $J_x = N(-e)v_x$. This leads to

$$\mathscr{E}_H = -\frac{1}{Ne} J_x B. \tag{4.40}$$

The Hall field is thus proportional both to the current and to the magnetic field. The proportionality constant—that is, $\mathscr{E}_H/J_xB$—is known as the *Hall constant*, and is usually denoted by $R_H$. Therefore

$$R_H = -\frac{1}{Ne}. \tag{4.41}$$

The result (4.41) is a very useful one in practice. Since $R_H$ is inversely proportional to the electron concentration $N$, it follows that we can determine $N$ by measuring the Hall field. In fact, this is the standard technique for determining electron concentration. The technique is particularly valuable because, apart from $N$, the only other quantity on which $R_H$ depends is the charge on the electron, $-e$, which is a fundamental physical constant whose value is known very accurately. Table 4.3 gives Hall constants for some of the common metals.

**Table 4.3**

Hall Constants (in volt $m^3$/amp weber at Room Temperature)

| Li | Na | Cu | Ag | Au | Zn | Cd | Al |
|---|---|---|---|---|---|---|---|
| $-1.7\times 10^{-10}$ | $-2.50$ | $-0.55$ | $-0.84$ | $-0.72$ | $+0.3$ | $+0.6$ | $-0.30$ |

Another useful feature of the Hall constant is that its sign depends on the sign of the charge of the current carriers. Thus electrons, being negatively charged, lead to a negative Hall constant. By contrast, we shall see in Chapter 5 that the Hall coefficient due to conduction by holes (which are positively charged) is positive.† Thus the sign of $R_H$ indicates the sign of the carriers involved, which is very valuable information, particularly in the case of semiconductors. For example, the Hall constants for both Zn and Cd are positive (see Table 4.3), indicating that the current in these substances is carried by holes.

The above analysis shows another interesting aspect of the transport process in the presence of a magnetic field: The current itself, flowing in the $x$-direction, is uninfluenced by the field. Therefore electrical resistance is independent of magnetic field. This result, even though it is a negative one, is interesting because it is somewhat unexpected. The Lorentz force of the field, which tends to influence $J_x$, is canceled by the Hall force, so that the electrons flow horizontally through the specimen, oblivious of the field.

## 4.11 THE AC CONDUCTIVITY AND OPTICAL PROPERTIES

We discussed static electrical conductivity in Section 4.4. Now let us consider electrical conductivity in the presence of an alternating-current field. This is intimately related to the optical properties, as we shall see shortly; the term "optical" here covers the entire frequency range, and is not restricted to the visible region only.

Consider a transverse EM wave, propagating in the $x$-direction and polarized in the $y$-direction. Its electric field may thus be expressed as

$$\mathscr{E}_y = \mathscr{E}_0\, e^{i(qx-\omega t)}. \tag{4.42}$$

The equation of motion of a conduction electron in the presence of this ac field

† These holes, which are different from the Fermi holes mentioned in Section 4.3, will be introduced in Section 5.17, and discussed at length in Chapter 6 on semiconductors.

is the same as Eq. (4.6), which yields the steady-state solution

$$v_y = -\frac{e\tau}{m^*}\frac{1}{1-i\omega\tau}\mathscr{E}. \tag{4.43}$$

The current density $J_y = N(-e)v_y$, which, in light of Eq. (4.43), leads to the ac conductivity,

$$\tilde{\sigma} = \frac{\sigma_0}{1-i\omega\tau}, \tag{4.44}$$

where $\sigma_0 = Ne^2\tau/m^*$ is the familiar static conductivity. The conductivity is now a complex quantity $\tilde{\sigma} = \sigma' + i\sigma''$, whose real and imaginary components are

$$\sigma' = \frac{\sigma_0}{1+\omega^2\tau^2}, \qquad \sigma'' = \frac{\sigma_0\omega\tau}{1+\omega^2\tau^2}. \tag{4.45}$$

The real part $\sigma'$ represents the in-phase current which produces the resistive joule heating, while $\sigma''$ represents the $\pi/2$ out-of-phase inductive current. An examination of $\sigma'$ and $\sigma''$ as functions of the frequency shows that in the low-frequency region, $\omega\tau \ll 1$, $\sigma'' \ll \sigma'$. That is, the electrons exhibit an essentially resistive character. Since $\tau \simeq 10^{-14}$ s, this spans the entire familiar frequency range up to the far infrared. In the high-frequency region, $1 \ll \omega\tau$, however, which corresponds to the visible and ultraviolet regimes, $\sigma' \ll \sigma''$, and the electrons evince an essentially inductive character. No energy is absorbed from the field in this range, and no joule heat appears.

Let us look at the response of the electrons from another point of view. We recall one of the Maxwell equations

$$\nabla \times \mathbf{H} = \epsilon_L \frac{\partial \mathscr{E}}{\partial t} + \mathbf{J}, \tag{4.46}$$

where the first term on the right represents the *displacement* current associated with the polarization of the ion cores (subscript $L$ for lattice), while the second term, $\mathbf{J}$, is the *convective* current of the conduction electrons. We may group the two currents together thus: Writing $\mathbf{J} = \tilde{\sigma}\mathscr{E} = (\tilde{\sigma}/-i\omega)\partial\mathscr{E}/\partial t$ for an ac field, we rewrite Eq. (4.46) as

$$\nabla \times \mathbf{H} = \tilde{\epsilon}\frac{\partial \mathscr{E}}{\partial t}, \tag{4.47}$$

where $\tilde{\epsilon}$ is the total dielectric constant,

$$\tilde{\epsilon} = \epsilon_L + i\frac{\tilde{\sigma}}{\omega}. \tag{4.48}$$

We now view the conduction electrons as part of the dielectric medium, which is plausible, since they merely oscillate around their equilibrium positions without a net translational motion. Substitution of $\tilde{\sigma}$ from (4.45) into (4.48) yields, for the relative dielectric constant, $\tilde{\epsilon}_r = \tilde{\epsilon}/\epsilon_0$,

$$\tilde{\epsilon}_r = \epsilon_r' + i\epsilon_r'' = \left(\epsilon_{L,r} - \frac{\sigma_0 \tau}{\epsilon_0(1 + \omega^2\tau^2)}\right) - i\frac{\sigma_0}{\epsilon_0\omega(1 + \omega^2\tau^2)}. \tag{4.49}$$

The complex index of refraction of the medium $n$ is defined as

$$\tilde{n} = \tilde{\epsilon}_r^{1/2} = n + i\kappa, \tag{4.50}$$

where $n$ is the usual refractive index and $\kappa$ the *extinction coefficient*. In optical experiments, one does not usually measure $n$ and $\kappa$ directly, however, but rather the reflectivity $R$ and the absorption coefficient $\alpha$. It can be shown† that these are related to $n$ and $\kappa$ by the expressions

$$R = \frac{(n-1)^2 + \kappa^2}{(n+1)^2 + \kappa^2}, \tag{4.51}$$

$$\alpha = \frac{2\omega}{c}\kappa, \tag{4.52}$$

where $c$ is the velocity of light in vacuum. Equations (4.49) through (4.52) describe the behavior of the electrons in the entire frequency range, but their physical contents can best be understood by examining their implications in the various frequency regions.

a) *The low-frequency region* $\omega\tau \ll 1$. The above equations show that $\tilde{\epsilon}_r$ reduces to the imaginary value $\tilde{\epsilon}_r \simeq i\epsilon_r''$ in this region, and hence

$$|n| \simeq |\kappa| = \left(\frac{\epsilon_r''}{2}\right)^{1/2} = \left(\frac{\sigma_0}{2\epsilon_0\omega}\right)^{1/2}. \tag{4.53}$$

The inverse of the absorption coefficient $\delta = 1/\alpha$ is known as the *skin depth*. [Recall that the intensity $I = I_0 e^{-\alpha x}$, and hence $1/\alpha$, is a measure of the distance of penetration of the optical beam into the medium before the beam is dissipated.] We can now evaluate $\delta$ as

$$\delta = \left(\frac{\epsilon_0 c^2}{2\sigma_0\omega}\right)^{1/2}. \tag{4.54}$$

In practice, $\delta$ has a very small value (for Cu at $\omega = 10^7\ \text{s}^{-1}$, $\delta = 100\mu$), indicating that an optical beam incident on a metallic specimen penetrates only a short distance below the surface.

---

† See any textbook on optics. Also note that Eq. (4.51) gives the reflectivity at normal incidence.

b) *The high-frequency region* $1 \ll \omega\tau$. This region covers the visible and ultraviolet ranges. Equation (4.49) shows that $\tilde{\epsilon}_r$ reduces to the real value

$$\epsilon_r = \epsilon_{L,r}\left(1 - \frac{\omega_p^2}{\omega^2}\right), \tag{4.55}$$

where

$$\omega_p^2 = \frac{Ne^2}{\epsilon_L m^*}, \tag{4.56}$$

and where we have made use of the relation $\sigma_0 = ne^2\tau/m^*$. The frequency $\omega_p$ is known as the *plasma frequency*; its significance will be revealed shortly. We can see from Eq. (4.55) that the high-frequency region can now be divided into two subregions: In the subregion $\omega < \omega_p$, $\epsilon_r < 0$, and consequently, from (4.50), $n = 0$. In view of (4.51), this leads to $R = 1$. That is, the metal exhibits *perfect reflectivity*. In the higher subregion $\omega_p < \omega$, however, $0 < \epsilon_r$, and hence, by similar reasoning, $\kappa = 0$. In this range, therefore, $\alpha = 0$, $0 < R < 1$, and the metallic medium acts like a nonabsorbing transparent dielectric, e.g., glass.

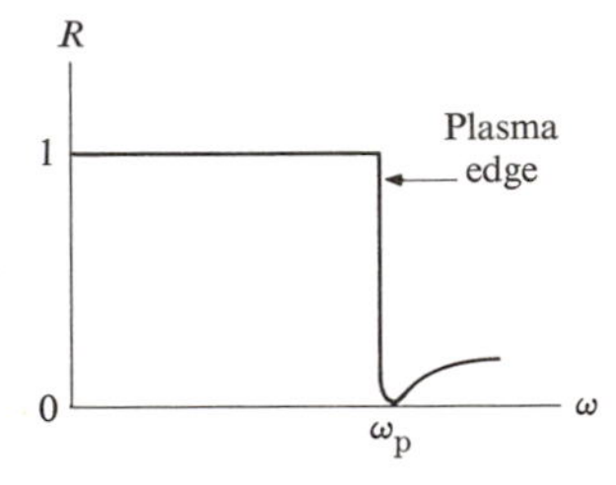


**Fig. 4.12** The plasma reflection edge.

Figure 4.12 illustrates the dependence of reflectivity on frequency, exhibiting the dramatic *discontinuous* drop in $R$ at $\omega = \omega_p$, which has come to be known as the *plasma reflection edge*. The frequency $\omega_p$ as seen from (4.56) is proportional to the electron density $N$. In metals, the densities are such that $\omega_p$ falls into the high visible or ultraviolet range (Table 4.4).

**Table 4.4**

Reflection Edges (Plasma Frequencies) and Corresponding Wavelengths for Some Metals

| | Li | Na | K | Rb |
|---|---|---|---|---|
| $\omega_p$ | $1.22 \times 10^{16}\ \mathrm{s}^{-1}$ | 0.89 | 0.593 | 0.55 |
| $\lambda_p$ | 1550 Å | 2100 | 3150 | 3400 |

Another significant property of $\omega_p$ can be deduced from the Maxwell equation

$$\nabla \cdot \mathbf{D} = \epsilon \nabla \cdot \mathscr{E} = 0, \tag{4.57}$$

where $\mathbf{D} = \epsilon \mathscr{E}$ is the familiar electric displacement field (see also Section 8.2). [Note that, since the conduction electrons have been included in the dielectric treatment, the so-called free charge has been set equal to zero]. This equation admits the existence of a *longitudinal* mode, for which $\nabla \cdot \mathscr{E} \neq 0$, provided only that

$$\epsilon = \epsilon_0 \epsilon_r = 0. \tag{4.58}$$

It may be seen from (4.55) that $\epsilon_r$ vanishes only at $\omega = \omega_p$. This mode, known as the *plasma mode*, has been observed in metals, and received much attention in the 1950's and 1960's.

Note that, of the two components of the dielectric constant, the real part $\epsilon_r'$ represents the polarization of the charges induced by the field, while $\epsilon_r''$ represents the absorption of energy by the system. We can see this because Eqs. (4.48) and (4.49) imply that $\epsilon_r'' \sim \sigma'$, and the latter quantity is related to the energy absorption, as pointed out earlier in this section.

## 4.12 THERMIONIC EMISSION

When a metal is heated, electrons are emitted from its surface, a phenomenon known as *thermionic emission*. This property is employed in vacuum tubes, in which the metallic cathode is usually heated in order to supply the electrons required for the operation of the tube.

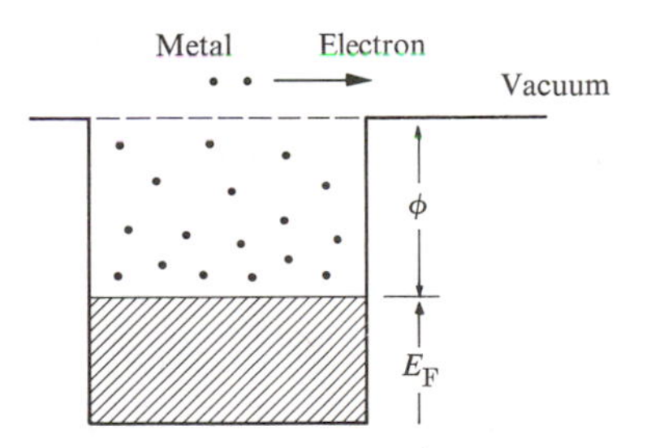


**Fig. 4.13** Thermionic emission.

Figure 4.13 shows the energy-level scheme for electrons in metals, according to the free-electron model. At $T = 0°K$, all levels are filled up to the Fermi level $E_F$, above which all levels are empty. Note also that an electron at $E_F$ cannot escape from the metal because of the presence of an energy barrier at the surface. The height of this barrier, denoted by $\phi$, is known as the *work function*. This function varies from one metal to another, but generally falls in the range 1.5–5 eV.

At $T = 0°K$, no electrons can escape from the metal. But as the temperature is raised, the levels above $E_F$ begin to be occupied because of the transfer of electrons from below $E_F$. Even the levels above the barrier—i.e., at energies higher than $(E_F + \phi)$—become populated to some extent. The electrons in these latter levels now have enough energy to overcome the barrier, and they are the ones responsible for the observed emission from the surface.

Let us now evaluate the current density for the emitted electrons, taking the metal surface to be normal to the $x$-direction. Consider the number of electrons whose velocity components fall in the range $(v_x, v_y, v_z)$ to $(v_x + dv_x, v_y + dv_y, v_z + dv_z)$. Their concentration is given by

$$d^3N = N\left(\frac{m^*}{2\pi kT}\right)^{3/2} e^{-m^*(v_x^2+v_y^2+v_z^2)/2kT}\, dv_x dv_y dv_z. \tag{4.59}$$

We have used the Maxwell–Boltzmann distribution because the electrons involved in the emission process are all so high above the Fermi level that they can be described very accurately by this distribution (Section 4.6). The density of the emitted current due to these electrons is given by

$$dJ_x = -e\, v_x\, d^3N, \tag{4.60}$$

as follows from the reasoning used in writing (4.8). To find the current density due to all the electrons, we must sum over all the velocities involved. Thus

$$J_x = \int dJ_x$$

$$= -eN\left(\frac{m^*}{2\pi kT}\right)^{3/2} \iiint v_x e^{-m^*(v_x^2+v_y^2+v_z^2)}\, dv_x dv_y dv_z.$$

When we carry out this integration over all the velocities, the ranges for $v_y$ and $v_z$ are both $(-\infty, \infty)$, but the range for $v_x$ is such that $\frac{1}{2}m^*v_x^2 \geqslant E_F + \phi$, because only these electrons have sufficient velocity in the relevant direction to escape from the surface. We have therefore

$$J_x = -eN\left(\frac{m^*}{2\pi kT}\right)^{1/2} \int_{v_{xo}}^{\infty} v_x e^{-m^*v_x^2/2kT}\, dv_x,$$

where $v_{x0} = [2(E_F + \phi)/m^*]^{1/2}$. The integration may be readily effected, and leads to

$$J_x = AT^2 e^{-\phi/kT}, \tag{4.61}$$

where $A = m^*ek^2/2\pi^2\hbar^3$. The numerical value of $A$ is 120 amp/cm$^2 \cdot$°K$^2$. The result (4.61), known as the *Richardson–Dushman equation*, holds good experimentally. It shows that the current density increases very rapidly with temperature. Since $\phi \gg kT$ for the usual range of temperature, the current density essentially increases exponentially with temperature. Table 4.5 gives the work functions for some metals, as determined from measurements of thermionic emission.

**Table 4.5**

Work Functions, eV

| W | Ta | Ni | Ag | Cs | Pt |
|---|---|---|---|---|---|
| 4.5 | 4.2 | 4.6 | 4.8 | 1.8 | 5.3 |

## 4.13 FAILURE OF THE FREE-ELECTRON MODEL

We have discussed the free-electron model in great detail to show how invaluable it is in accounting for observed metallic properties. Nevertheless, the model is only an approximation, and as such has its limitations. Consider the following points.

a) The model suggests that, other things being equal, electrical conductivity is proportional to electron concentration, according to (4.9). No definite conclusion to the contrary can be drawn from the data (Table 4.1), since we do not know the other quantities in the formula (because these were determined from $\sigma$), but it is surprising that the divalent metals (Be, Cd, Zn, etc.), and even trivalent metals (Al, In), are consistently less conductive than the monovalent metals (Cu, Ag, and Au) despite the fact that the former have higher concentrations of electrons.

b) A far more damaging testimony against the model is the fact that some metals exhibit positive Hall constants, for example, Be, Zn, Cd (Table 4.3). The free-electron model always predicts a negative Hall constant.

c) Measurements of the Fermi surface indicate that it is often nonspherical in shape (Section 5.12). This contradicts the model, which predicts a spherical FS.

These difficulties, and others which need not be enumerated here, can be resolved by a more sophisticated theory which takes into account the interaction of the electrons with the lattice. We shall take up this subject in the following chapter.

## SUMMARY

### Conduction electrons

When atoms are packed together to form a metal, the *valence electrons* detach from their own atoms and move throughout the crystal. These delocalized electrons are the *conduction electrons*. Their concentration is given by

$$N = Z_v \frac{\rho_m N_A}{M'_{mol}},$$

where $Z_v$ is the atomic valence, and other symbols have their usual meanings.

### Electrical conductivity

The electrical conductivity of conduction electrons, treated as free particles with a collision time $\tau$, is

$$\sigma = \frac{Ne^2\tau}{m^*}.$$

Comparing this result with experimental values shows that the collision time is extremely short, of the order of $10^{-14}$ s at room temperature.

When one evaluates the collision time, one finds that a perfect lattice produces no scattering. Only lattice vibrations or imperfections lead to scattering, and hence determine the collision time. Treating lattice vibrations and static impurities in the crystal as independent collision mechanisms, one finds that the electrical resistivity $\rho$ is

$$\rho = \rho_{ph}(T) + \rho_i,$$

where $\rho_{ph} \sim T$ is the resistivity due to collisions caused by lattice vibrations or phonons, and $\rho_i$ is the residual resistivity due to collision of the electrons with impurities within the crystal.

### Thermal conductivity

The thermal conductivity of metals is given by the expression

$$K = \mathrm{L} T \sigma,$$

where L, a constant known as the *Lorenz number*, is

$$\mathrm{L} = \frac{\pi^2}{3}\left(\frac{k}{e}\right)^2.$$

### Heat capacity

Experiments show that the heat capacity of conduction electrons is much smaller than predicted by classical statistical mechanics. This is explained on the basis of the exclusion principle. All the energy levels up to the Fermi level are occupied, and when the system is heated, only those electrons near the Fermi level are excited. The electron heat capacity per mole is

$$C_e = \frac{\pi^2}{2} R \frac{kT}{E_F}.$$

**The Fermi energy**

The Fermi energy is determined by the concentration of electrons. Its value is

$$E_F = \frac{\hbar^2}{2m^*}(3\pi^2 N)^{2/3}.$$

**Cyclotron resonance and the Hall effect**

When a magnetic field is applied to a solid, the electrons execute a circular cyclotron motion. The cyclotron frequency is

$$\omega_c = \frac{eB}{m^*},$$

and its measurement enables one to determine the effective mass of the electron.

When a magnetic field is applied to a current-carrying wire, it produces an electric field normal to both the current and the magnetic field. This electric field, or *Hall field*, has the form $\mathscr{E}_H = RBJ$, where the *Hall constant* is

$$R_H = -\frac{1}{Ne}.$$

Measuring $R$ yields the electron concentration $N$.

**Optical properties**

The complex conductivity of conduction electrons is

$$\tilde{\sigma} = \frac{\sigma_0}{1 + i\omega\tau},$$

where $\sigma_0$ is the static conductivity. The form of $\tilde{\sigma}$ indicates that the electrons have a mixed resistive–inductive character. The resistive character dominates in the low-frequency region $\omega < 1/\tau$, while the inductive character dominates in the high-frequency region $\omega > 1/\tau$. Because $\tau$ is very short, the former region includes all frequencies up to and including microwaves.

The dielectric constant for the whole crystal, including both the lattice and the electrons, is

$$\tilde{\epsilon}(\omega) = \epsilon_L + \frac{i\tilde{\sigma}}{\omega}.$$

Once we know the dielectric constant, we can determine the reflective and absorptive properties of the crystal. The following frequency regimes can be delineated.

a) *Low-frequency region*, $\omega \ll 1/\tau$. The wave penetrates the metal a short distance, known as the *skin depth*, whose value is

$$\delta = \left(\frac{2\epsilon_0 c^2}{\sigma_0 \omega}\right)^{1/2}$$

The reflectivity in this frequency range is very close to unity.

b) *Intermediate-frequency region* $1/\tau \ll \omega < \omega_p$. The wave is evanescent in this region, and the metal exhibits total reflection.

c) *High-frequency region* $\omega_p < \omega$. The metal acts as a regular dielectric, through which the wave propagates without attenuation.

### The plasma mode

This mode refers to the longitudinal oscillation of the electron system. Its frequency is equal to the plasma frequency $\omega_p = (Ne^2/\epsilon_L m^*)^{1/2}$.

### Thermionic emission

When a metal is heated, some electrons at the tail end of the Fermi distribution acquire sufficient energy to escape from the surface of the metal. The thermionic current density is

$$J = AT^2 e^{-\phi/kT},$$

where $A$ is a constant and $\phi$ is the work function of the metal.

## REFERENCES

### Transport properties

There are a great many references which treat the transport properties of metals in considerable detail. Among these are the following.

F. J. Blatt, 1968, *Physics of Electron Conduction in Solids*, New York: McGraw-Hill

B. Donovan, 1967, *Elementary Theory of Metals*, New York: Pergamon Press

N. F. Mott and H. Jones, 1958, *Theory of the Properties of Metals and Alloys*, New York: Dover Press

H. M. Rosenberg, 1963, *Low-Temperature Solid State Physics*, Oxford: Oxford University Press

F. Seitz, 1940, *Modern Theory of Solids*, New York: McGraw-Hill

J. M. Ziman, 1960, *Electrons and Phonons*, Oxford: Oxford University Press

### Optical properties

B. Donovan, *op. cit.*

F. Stern, "Elementary Theory of the Optical Properties of Metals," *Solid State Physics* **15**, 1963

## QUESTIONS

1. Explain the distinction between localized and delocalized (or core) electrons in solids. Describe one experimental method of testing the difference between the two types.
2. The text said that the conduction electrons are better described as a plasma than an ordinary gas. In what essential ways does a plasma differ from a gas?
3. Trace the steps which show that the electrical current of the electrons is in the same direction as the field, even though the particles are negatively charged.
4. Assuming that the conduction electrons in Cu are a classical gas, calculate the rms value of the electron speed, and compare the value obtained with the Fermi velocity (see Problem 1).
5. Explain why electrons carry a net energy but not a net current in the case of thermal conduction.
6. Show that if the random velocity of the electrons were due to the thermal motion of a classical electron gas, the electrical resistivity would increase with the temperature as $T^{3/2}$.
7. In a cyclotron resonance experiment, part of the signal is absorbed by the electrons. What happens to this energy when the system is in a steady-state situation?
8. Explain qualitatively why the Hall constant $R_H$ is inversely proportional to the electron concentration $N$.
9. Demonstrate qualitatively that the Hall constant for a current of positive charges is positive.
10. Equation (4.54) shows that the skin depth $\delta$ becomes infinite at zero frequency. Interpret this result.
11. Describe the variation of skin depth with temperature.
12. According to the discussion in Section 4.11, free electrons make a negative contribution to the dielectric constant, while bound electrons make a positive contribution. Explain this difference in electron behavior.

## PROBLEMS

1. Copper has a mass density $\rho_m = 8.95\ \text{g/cm}^3$, and an electrical resistivity $\rho = 1.55 \times 10^{-8}$ ohm-m at room temperature. Assuming that the effective mass $m^* = m_0$, calculate:
   a) The concentration of the conduction electrons
   b) The mean free time $\tau$
   c) The Fermi energy $E_F$
   d) The Fermi velocity $v_F$
   e) The mean free path at the Fermi level $l_F$
2. Derive Eq. (4.19) for the mean free path.
3. The residual resistivity for 1 atomic percent of As impurities in Cu is $6.8 \times 10^{-8}$ ohm-m. Calculate the cross section for the scattering of an electron by one As impurity in Cu.
4. Sodium has a volume expansion coefficient of $15 \times 10^{-5}\,^\circ\text{K}^{-1}$. Calculate the percentage change in the Fermi energy $E_F$ as the temperature is raised from $T = 0^\circ\text{K}$ to $300^\circ\text{K}$. Comment on the magnitude of the change.
5. Repeat Problem 4 for silver, whose volume coefficient of expansion is $18.6 \times 10^{-5}$ $^\circ\text{K}^{-1}$.

6. Calculate the Fermi temperatures $T_F$ for Cu, Na, and Ag. Also calculate the ratio $T/T_F$ in each case for $T = 300°K$. The effective masses of Cu and Na are 1.0 and 1.2 times $m_0$.
7. Estimate the fraction of electrons excited above the Fermi level at room temperature for Cu and Na.
8. Calculate the ratio of electrons to lattice heat capacities for Cu at $T = 0.3°$, 4°, 20°, 77°, and 300°K. The lattice heat capacity of Cu is given in Fig. 3.10.
9. Plot the Fermi–Dirac function $f(E)$ versus the energy ratio $E/E_F$ at room temperature $T = 300°K$. (Assume $E_F$ independent of temperature.) If $E_F = 5$ eV, determine the energy values at which $f(E) = 0.5$, 0.7, 0.9, and 0.95.
10. Cyclotron resonance has been observed in Cu at a frequency of 24 GHz. Given that the effective mass of Cu is $m^* = m_0$, what is the value of the applied magnetic field?
11. Using Table 4.4, giving the Hall constants, calculate the electron concentrations in Na, Cu, Cd, Zn, Al, and In. Compare these results with those given in Table 1.1.
12. a) Using the appropriate values of $\epsilon_{L,r}$, $\sigma_0$, and $\tau$ for Ag at room temperature, calculate the refractive index $n$ and the extinction coefficient $\kappa$ for Ag, and plot these versus $\omega$ on the logarithmic scale.
    b) Evaluate the optical reflectivity and plot it versus $\omega$. (Data on Ag are found in Table 4.1.) The values of $\omega$ may be confined to the range $\omega < 10^{16}\ s^{-1}$.
13. Evaluate the skin depth for Cu at room temperature, and plot the results versus the frequency on a logarithmic scale. (Data are given in Table 4.1.) The value of $\omega$ may be confined to $\omega < 10^{13}\ s^{-1}$.
14. Carry out the integration which leads to (4.61).
15. Calculate the density of the thermionic emission current in Cs at 500, 1000, 1500, and 2000°K.

# CHAPTER 5 METALS II: ENERGY BANDS IN SOLIDS



---

*On the surface there is infinite variety*
*of things; at base a simplicity of cause.*

Ralph Waldo Emerson

## 5.1 INTRODUCTION

In Chapter 4 we talked about the motion of electrons in solids, using the free-electron model. This model is oversimplified, however, because the crystal potential is neglected. But this potential cannot be entirely disregarded if one is to explain the experimental results quantitatively. In addition, some effects cannot be explained at all without taking this potential into account, as we pointed out at the end of Chapter 4. The present chapter therefore treats the influence of the crystal potential on the electronic properties of solids.

In the first part of the chapter we shall consider the energy spectrum of an electron in a crystal. We shall see that the spectrum is composed of continuous *bands*, unlike the case for atoms, in which the spectrum is a set of discrete levels. We shall discuss the properties and the corresponding wave functions of these bands in detail, and develop a useful criterion for distinguishing metals from insulators in this band model. Then we shall deal with the density of states and the Fermi surface, which serve as useful characteristics of a solid.

The electrons in a crystal are in a constant state of motion. Formulas are developed for calculating the velocity of an electron, and its effective mass. We shall study the effects of an electric field on the motion of an electron, and then derive an expression for the electron's electrical conductivity. Although this expression reduces to the one derived previously in Chapter 4 under the appropriate circumstances, the form we shall develop here is more general, and brings out more clearly the physical factors influencing conductivity.

Cyclotron resonance and the Hall effect will also be discussed again and we shall show how these phenomena may be used to obtain information on a solid.

The last section will deal with the limitations of the energy-band model, and the metal–insulator transition.

## 5.2 ENERGY SPECTRA IN ATOMS, MOLECULES, AND SOLIDS

The primary purpose of this section is to describe qualitatively the energy spectrum of an electron moving in a crystalline solid. It is helpful, however, to begin the discussion by considering the spectrum of a free atom, and see how this spectrum is gradually modified as atoms are assembled to form the solid.

Let us take lithium as a concrete example. Consider a free lithium atom: The electron moves in a potential well, as shown in Fig. 5.1(a). When we solve the Schrödinger equation, we obtain a series of discrete energy levels, as shown. As in the case of the hydrogen atom, these levels are denoted by 1s, 2s, 2p, etc. The lithium atom contains three electrons, two of which occupy the 1s shell (completely full), and the third the 2s subshell.

Now consider the situation in which two lithium atoms assemble to form the lithium molecule $Li_2$. The potential "seen" by the electron is now the double well shown in Fig. 5.1(b). The energy spectrum here is comprised of a set of discrete *doublets*: Each of the atomic levels—that is, the 1s, 2s, 2p, etc.—has split into two closely spaced levels. Because of the close generic relation between the atomic and molecular levels, we may also speak of the 1s, 2s, 2p, etc., molecular

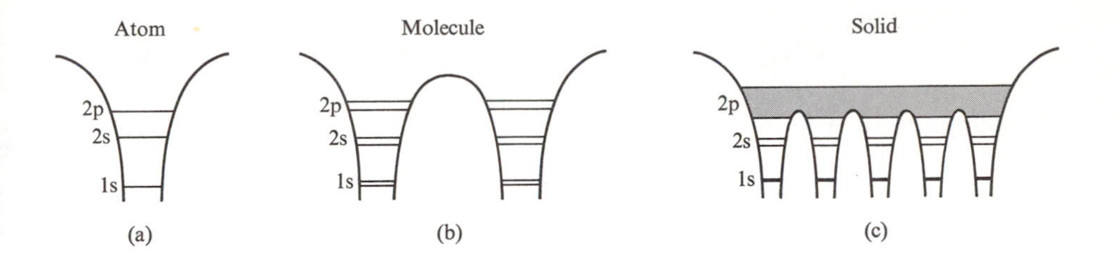


**Fig. 5.1** The evolution of the energy spectrum of Li from an atom (a), to a molecule (b), to a solid (c).

energy levels, recognizing that each of these is, in fact, composed of two sublevels.

We can see why the atomic level splits into two, and only two, sublevels in a diatomic molecule from our treatment of the hydrogen molecule ion $H_2^+$ (Section A.7). The reason is essentially as follows: When the two Li atoms are far apart, the influence of one atom on an electron in the other atom is very small, and may be treated as a perturbation. In this approximation, the unperturbed levels 1s, 2s, etc., are each doubly degenerate, because an electron in a 1s level, for instance, may occupy that level in either atom; and since there are two atoms, the energy is thus doubly degenerate. This degeneracy is strictly valid only if the interaction between the atoms is neglected entirely. When this interaction is included, the double degeneracy is lifted, and each level is split into its two sublevels. The molecular orbitals corresponding to these sublevels are usually taken to be the symmetric and antisymmetric combinations of the corresponding atomic orbitals, as in the case of $H_2^+$ (Section A.7).

Each molecular level can accommodate at most two electrons, of opposite spins, according to the exclusion principle. The $Li_2$ molecule has six electrons; four occupy the 1s molecular doublet, and the other two the lower level of the 2s doublet.

According to this discussion, the amount of splitting depends strongly on the internuclear distance of the two atoms in the molecule. The closer the two nuclei, the stronger the perturbation and the larger the splitting. The splitting also depends on the atomic orbital: The splitting of the 2p level is larger than that of the 2s level, which is larger still than that of the 1s level. The reason is that the radius of the ls orbital, for instance, is very small, and the orbital is therefore tightly bound to its own nucleus. It is not greatly affected by the perturbation. The same is not true for the 2s and 2p orbitals, which have larger radii and are only loosely bound to their own nuclei. It follows that, generally speaking, the higher the energy, the greater the splitting incurred.

The above considerations may be generalized to a polyatomic Li molecule of an arbitrary number of atoms. Thus in a 3-atom molecule, each atomic level is split into a triplet, in a 4-atom molecule into a quadruplet, and so forth. The lithium solid may then be viewed as the limiting case in which the number of atoms has become very large, resulting in a gigantic lithium molecule. What has happened to the shape of the energy spectrum? We can answer this on the basis of the above discussion: Each of the atomic levels is split into $N$ closely spaced sublevels, where $N$ is the number of atoms in the solid. But since $N$ is so very large, about $10^{23}$, the sublevels are so extremely close to each other that they coalesce, and form an *energy band*. Thus the 1s, 2s, 2p levels give rise, respectively, to the 1s, 2s, and 2p bands, as shown in Fig. 5.1(c).

To illustrate how close to each other the sublevels lie within the bands, consider the following numerical example. Suppose that the width of the band is 5 eV (a typical value). The energy interval between two adjacent levels is therefore of the order $5/10^{23} = 5 \times 10^{-23}$ eV. Since this is an extremely small value, the individual sublevels are indistinguishable, so we can consider their distribution as a continuous energy band.

To recapitulate, the spectrum in a solid is composed of a set of energy bands. The intervening regions separating these bands are energy *gaps*—i.e., regions of forbidden energy—which cannot be occupied by electrons. Contrast this situation with that of a free atom or a molecule, in which the allowed energies form a set of discrete levels. This broadening of discrete levels into bands is one of the most fundamental properties of a solid, and one we shall use often throughout this book.

The width of the band varies, but in general the higher the band the greater its width, because, as we recall from the case of molecules, a high energy state corresponds to a large atomic radius, and hence a strong perturbation, which is the cause of the level broadening in the first place. By contrast, low energy states correspond to tightly bound orbitals, which are affected but slightly by the perturbation.

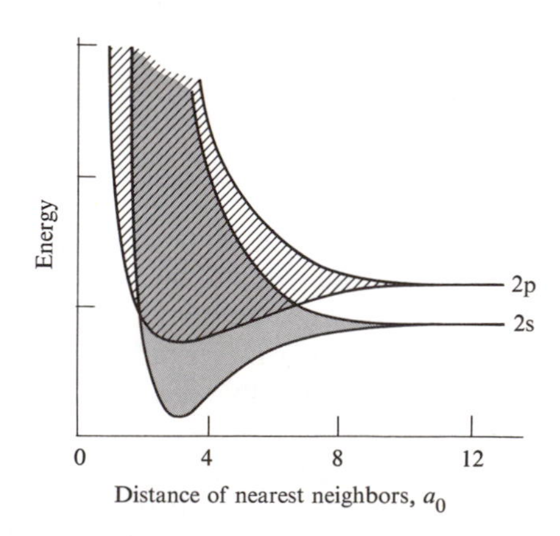


**Fig. 5.2** The broadening of the 2s and 2p levels into energy bands in a lithium crystal ($a_0$ is the Bohr radius, 0.53 Å).

Figure 5.2 shows 2s and 2p bands for metallic lithium plotted as functions of the lattice constants $a$. Note that the band widths increase as $a$ decreases, as is to be expected, since the smaller the interatomic distance the greater the perturbation. Note also that, for $a < 6a_0$, the 2s and 2p bands broaden to the point at which they begin to overlap, and the gap between them vanishes entirely.

The *crystal orbitals*—i.e., the wave functions describing the electronic states in the bands—extend throughout the solid, unlike the atomic orbitals, which are *localized* around particular atoms, and decay exponentially away from those atoms. In this sense, we refer to solid wave functions as *delocalized orbitals*. We shall see shortly that these orbitals actually describe electron waves traveling in the solid. The concept of delocalization is a basic one. It is responsible for all electronic transport phenomena in solids, e.g., electrical conduction.

We have already presented many concepts related to electronic states in a crystalline solid. In the following sections we shall place these concepts on a firmer, more mathematical basis by writing the Schrödinger equation and discussing the properties of its solution. This will also lead to many interesting and novel concepts which we shall discuss as we go along.

## 5.3 ENERGY BANDS IN SOLIDS; THE BLOCH THEOREM

### The Bloch function

The behavior of an electron in a crystalline solid is determined by studying the appropriate Schrödinger equation. This may be written as (Section A.2),

$$\left[ -\frac{\hbar^2}{2m} \nabla^2 + V(\mathbf{r}) \right] \psi(\mathbf{r}) = E\psi(\mathbf{r}), \tag{5.1}$$

where $V(\mathbf{r})$ is the crystal potential "seen" by the electron, and $\psi(\mathbf{r})$ and $E$ are, respectively, the state function and energy of this electron. The potential $V(\mathbf{r})$ includes the interaction of the electron with all atoms in the solid, as well as its interaction with other electrons (we will get back to this later). At this point we make the important observation that the potential $V(\mathbf{r})$ is periodic. It has the same

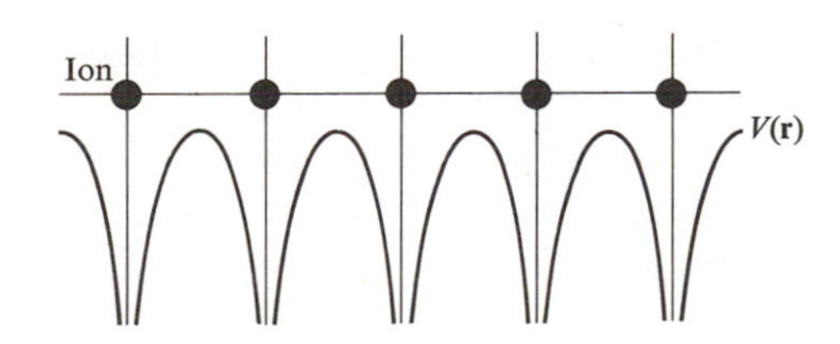


**Fig. 5.3** The crystal potential seen by the electron.

translational symmetry as the lattice, that is,

$$V(\mathbf{r} + \mathbf{R}) = V(\mathbf{r}), \tag{5.2}$$

where **R** is a lattice vector. Such a potential is shown schematically in Fig. 5.3.

According to the *Bloch theorem*, the solution of (5.1) for a periodic potential $V(\mathbf{r})$ has the form

$$\psi_{\mathbf{k}}(\mathbf{r}) = e^{i\mathbf{k}\cdot\mathbf{r}} u_{\mathbf{k}}(\mathbf{r}), \tag{5.3}$$

where the function $u_{\mathbf{k}}(\mathbf{r})$ has the same translational symmetry as the lattice, that is,

$$u_{\mathbf{k}}(\mathbf{r} + \mathbf{R}) = u_{\mathbf{k}}(\mathbf{r}). \tag{5.4}$$

The vector **k** is a quantity related to the momentum of the particle, as we shall see.

We shall now give a physical proof of the Bloch theorem. Anyone interested may pursue the more rigorous treatment in the references cited in the bibliography, e.g., Seitz (1940). The proof presented here is chosen to bring out the physical concepts with a minimum of mathematical detail. Returning to Eq. (5.1), it is always possible to write its solution as

$$\psi(\mathbf{r}) = f(\mathbf{r})\, u(\mathbf{r}),$$

where $u(\mathbf{r})$ is periodic, as in (5.4), and where the function $f(\mathbf{r})$ is to be determined. However, since the potential $V(\mathbf{r})$ is periodic, one requires that all observable quantities associated with the electron also be periodic. In particular, the quantity $|\psi(\mathbf{r})|^2$, which gives the electron probability, must also be periodic.† This imposes the following condition on $f(\mathbf{r})$:

$$|f(\mathbf{r} + \mathbf{R})|^2 = |f(\mathbf{r})|^2.$$

The only function which satisfies this requirement for all **R**'s is one of the exponential form $e^{i\mathbf{k}\cdot\mathbf{r}}$. This demonstrates that the solution of the Schrödinger equation has the Bloch form (5.3), as we set out to prove.

The state function $\psi_{\mathbf{k}}$ of the form (5.3), known as the *Bloch function*, has several interesting properties.

a) It has the form of a traveling plane wave, as represented by the factor $e^{i\mathbf{k}\cdot\mathbf{r}}$, which implies that the electron propagates through the crystal like a free particle. The effect of the function $u_{\mathbf{k}}(\mathbf{r})$ is to modulate this wave so that the amplitude oscillates periodically from one cell to the next, as shown in Fig. 5.4, but this does not affect the basic character of the state function, which is that of a traveling wave.

---

†It is well known in quantum mechanics that the quantity $|\psi(\mathbf{r})|^2$ is the probability density, and as such is physically measurable. However, the wave function $\psi(\mathbf{r})$ itself is *not* physically measurable.

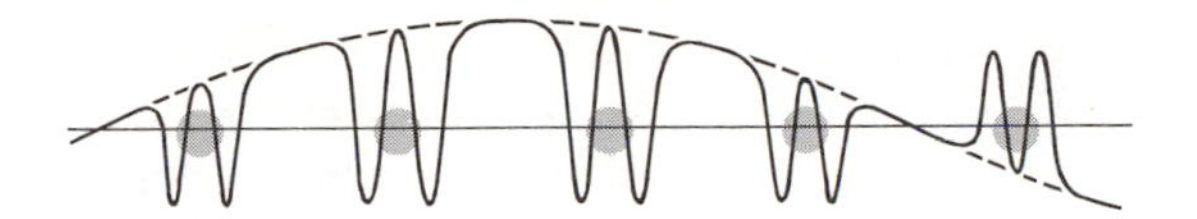

**Fig. 5.4** The Bloch function or wave. The smooth curve represents the wave $e^{i\mathbf{k}\cdot\mathbf{r}}$ which is modulated by the atomic-like "wiggly" function $u_{\mathbf{k}}(\mathbf{r})$.

If the electron were indeed entirely free, the state function $\psi_{\mathbf{k}}$ would be given by $(1/V^{1/2})\, e^{i\mathbf{k}\cdot\mathbf{r}}$, that is, the function $u_{\mathbf{k}}(\mathbf{r})$ is a constant. But the electron is not free, since it interacts with the lattice, and this interaction determines the special character of the periodic function $u_{\mathbf{k}}$.

b) Because the electron behaves as a wave of vector **k**, it has a deBroglie wavelength $\lambda = 2\pi/k$, and hence a *momentum*

$$\mathbf{p} = \hbar\mathbf{k}, \tag{5.5}$$

according to the deBroglie relation. We shall call the vector the *crystal momentum* of the electron, and discuss its properties in later sections.

c) The Bloch function $\psi_{\mathbf{k}}$ is a crystal orbital, as it is delocalized throughout the solid, and not localized around any particular atom. Thus the electron is shared by the whole crystal. This is, of course, consistent with property (a) above, in which we described the electron as a traveling wave. Note also that the function $\psi_{\mathbf{k}}$ is so chosen that the electron probability distribution $|\psi_{\mathbf{k}}|^2$ is periodic in the crystal.

In the above discussion, we have stressed the analogy between a *crystalline* electron and a *free* one; this is very helpful in understanding the properties of electrons in crystals. One should not, however, jump to the conclusion that the two are identical in their behavior. The Bloch-function electron exhibits many intriguing properties not shared by a free electron, properties which result from the interaction of the electron with the lattice.

### Energy bands

The discussion has thus far centered on the state function; nothing has been said about energy. We now turn to the energy spectrum which results from solving the Schrödinger equation (5.1). Toward this end, we rewrite this equation in a different form. Substituting for $\psi_{\mathbf{k}}$ from the Bloch form (5.3), and eliminating the factor $e^{i\mathbf{k}\cdot\mathbf{r}}$, after performing the necessary operations, we arrive at

$$\left[ -\frac{\hbar^2}{2m}(\nabla + i\mathbf{k})^2 + V(\mathbf{r}) \right] u_{\mathbf{k}}(\mathbf{r}) = E_{\mathbf{k}}\, u_{\mathbf{k}}(\mathbf{r}), \tag{5.6}$$

which is actually the wave equation for the periodic function $u_{\mathbf{k}}(\mathbf{r})$. This is an eigenvalue equation, like the Schrödinger equation, and can therefore be solved in a similar manner. Note that the operator in the brackets is an explicit function of **k**, and hence both the eigenfunctions and eigenvalues depend on **k**, a fact we have already used explicitly by labeling them with the vector **k**. An eigenvalue equation leads, however, not to one but to many solutions. For each value of **k**, therefore, we find a large number of solutions, giving a set of *discrete* energies $E_{1,\mathbf{k}}, E_{2,\mathbf{k}}, \ldots$, as shown in Fig. 5.5.† Since these energies depend on **k**, they vary continuously as **k** is varied over its range of values. Each level leads to an energy band, as shown in the figure. We shall henceforth write the energy eigenvalue as $E_n(\mathbf{k})$, and refer to the subscript $n$ as the *band index*, for obvious reasons.

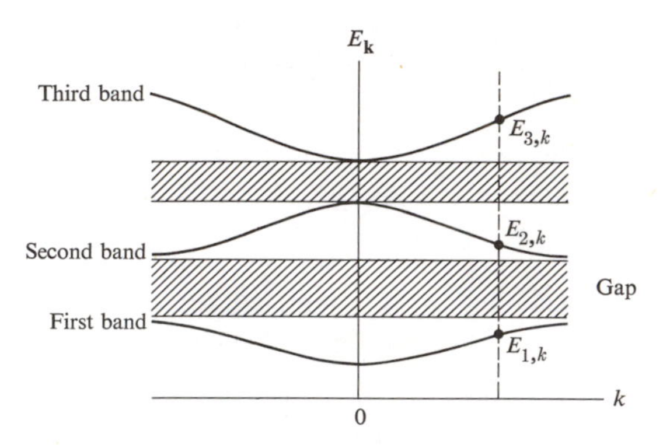


**Fig. 5.5** Energy bands and gaps. The cross-hatched regions indicate energy gaps.

The number of bands is large—usually infinite—but only the lowest ones are occupied by electrons. Each band covers a certain energy range, extending from the lowest to the highest value it takes when plotted in **k**-space. The energy intervals interspersed between the bands constitute the energy gaps, which are forbidden energies that cannot be occupied by electrons.

Note also that, since **k** is a vector quantity, a diagram such as Fig. 5.5 is a plot of the energy bands in only one particular direction in **k**-space. If these bands were plotted in a different **k**-direction, their appearance would change, in general. A complete representation of the bands therefore requires one to specify the energy values throughout the **k**-space. Often this is accomplished, at least partially, by drawing the energy contours in **k**-space for the various bands, as we shall do in the following sections. We shall also show that the bands satisfy certain important symmetry relations that enable us to restrict our considerations to relatively small regions in **k**-space.

The energy bands which have emerged from this analysis are the same as those discussed in the previous section, and in fact we can establish a one-to-one correspondence between the energy bands and the atomic levels from which they arise. The particular significance of the present results is that here we can classify

† In other words, the energy is a multivalued function of **k**.

the electron states within the band according to their momentum as given by **k**. Such a classification, which we shall find extremely useful, was not evident from the last section.

### The crystal potential

We turn now to the crystal potential $V(\mathbf{r})$ which acts on the electron. This potential is composed of two parts: the interaction of the electron with the ion cores, forming the lattice, and its interaction with other Bloch electrons moving through the lattice. In metallic sodium, for example, an electron in the 3s band interacts with the $Na^+$ ions forming the bcc structure, as well as with other electrons in this band. We may therefore write $V(\mathbf{r})$ as the sum

$$V(\mathbf{r}) = V_i(\mathbf{r}) + V_e(\mathbf{r}), \tag{5.7}$$

where the first term on the right represents the interaction with the ion cores and the second the interaction with the electrons.

The ionic part may be written as

$$V_i(\mathbf{r}) = \sum_j v_i(\mathbf{r} - \mathbf{R}_j), \tag{5.8}$$

where $v_i(\mathbf{r} - \mathbf{R}_j)$ is the potential of an ion located at the lattice vector $\mathbf{R}_j$, as in Fig. 5.6(a). and the summation is over all the ions. The potential $V_i(\mathbf{r})$ obviously has the same periodicity as that of the lattice.

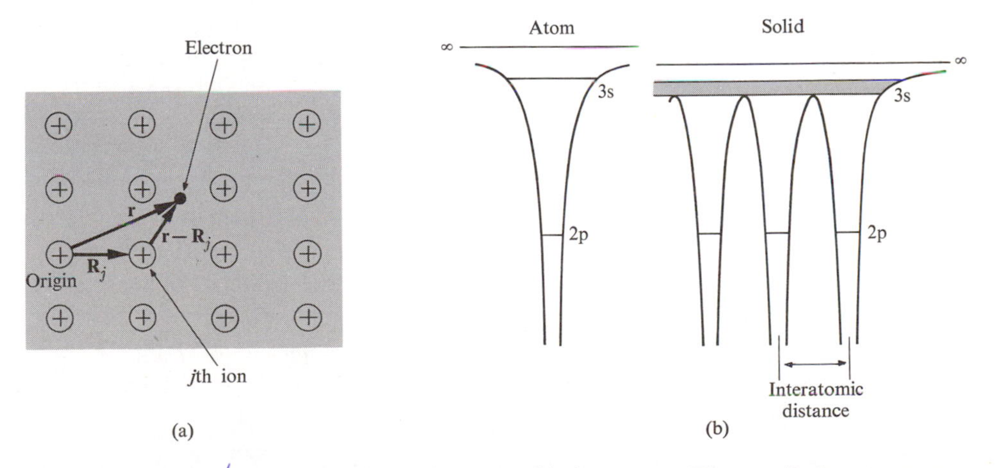


**Fig. 5.6** (a) The interaction of an electron with ion cores. The small dots represent electrons. (The spatial distribution of the electrons is not shown accurately. They actually tend to be positioned primarily around the ions.) (b) The spectrum of an Na atom (left), and an Na solid (right). [After J. C. Slater, *Physics Today* **21,** 43 (1968)]. Note the broadening of the 3s level into a 3s band in the solid, and that this band lies almost entirely above the potential barriers of the atoms, which facilitates the delocalization of the electrons in this band. By contrast, electrons in the 2p level or band are so highly constrained by the barriers that they are localized.

The electronic potential $V_e(\mathbf{r})$, the so-called *electron–electron interaction*, presents several hurdles which make its treatment very difficult. First, we can evaluate this term only if we know the states for all other electrons, but these states are not given in advance. In fact, they are the very states we are trying to find. Second, the potential $V_e(\mathbf{r})$ is not strictly periodic, since the electrons are in constant motion through the lattice. Third, a proper treatment should really consider the dynamics of all the electrons simultaneously, not one electron at a time, as we have done above. This is a typical example of the *many-body problems* which are often encountered in solid-state physics.

In view of these difficulties, it is fortunate that the electron–electron interaction turns out to be quite weak, for the reason given in Section 4.3, because this fact makes the above difficulties far less serious than they could otherwise be. The major effect of this interaction is that the electrons distribute themselves primarily around the ions, so that they *screen* these ions from other electrons. This has the additional effect of making the electron–ion interaction weak even at long range, which is another fortunate circumstance.

So we can write an approximate expression for the potential as

$$V(\mathbf{r}) = \sum_j v_s(\mathbf{r} - \mathbf{R}_j), \tag{5.9}$$

where $v_s(\mathbf{r} - \mathbf{R}_j)$ is the potential of the screened ion located at the lattice point $\mathbf{R}_j$. And precisely because this potential *is* once again periodic, it satisfies the requirements of the Bloch theorem. Figure 5.6(b) shows the crystal potential for Na.

In discussing the crystal potential, we have so far tacitly assumed that the atoms are at rest at their lattice sites. However, they are not in fact stationary. They are in a constant state of oscillation as a result of their thermal excitation, as discussed in Chapter 3. Clearly, then, our assumption of a stationary lattice is an approximation, and the question now is: How good is our approximation? One may answer this pragmatically by pointing out that band structures calculated on the basis of a stationary lattice are usually in good agreement with experiment, except at temperatures close to the melting point of the solid. The reason the stationary-lattice approximation seems to hold so well is that amplitudes of lattice vibrations are much smaller than the interatomic distance at all temperatures, even up to the melting point.† Therefore the distortion of the lattice, as seen by the electron, is not appreciable.

## 5.4 BAND SYMMETRY IN k-SPACE; BRILLOUIN ZONES

The energy eigenvalues $E_n(\mathbf{k})$ for the bands have many useful symmetry properties when these bands are plotted in **k**-space. Before broaching this subject, however, let us say a few words about the Brillouin zones.

---

† The average amplitude of the atomic oscillation due to thermal excitation at the melting point is typically about 5% of the interatomic distance.

## Brillouin zones

We first encountered Brillouin zones in our discussion of Bragg diffraction of x-rays in Section 2.6. When one draws the normal planes which bisect the reciprocal lattice vectors, the regions enclosed between these planes form the various Brillouin zones.

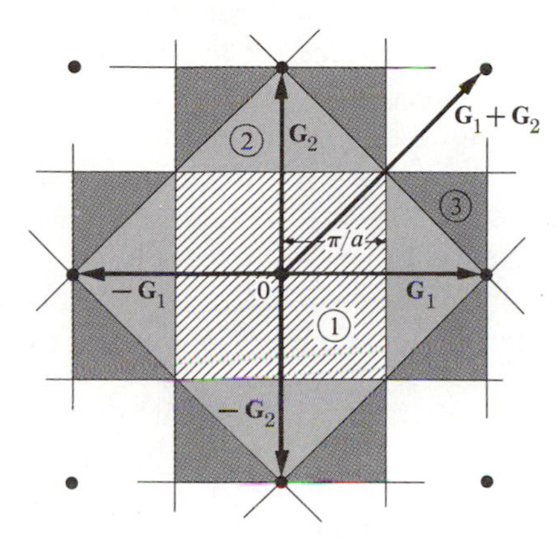


**Fig. 5.7** The first three Brillouin zones of the square lattice: First zone (cross-hatched), second zone (shaded) and third zone (screened). Numbers indicate indices of zones.

Consider, for instance, the square lattice whose reciprocal—also a square lattice of edge equal to $2\pi/a$—is shown, in Fig. 5.7, which also shows the reciprocal vectors $\mathbf{G}_1$, $-\mathbf{G}_1$, $\mathbf{G}_2$, and $-\mathbf{G}_2$, etc., as well as the corresponding normal bisectors. The smallest enclosed region centered around the origin (the cross-hatched area) is the first zone. The shaded area (composed of four separate half-

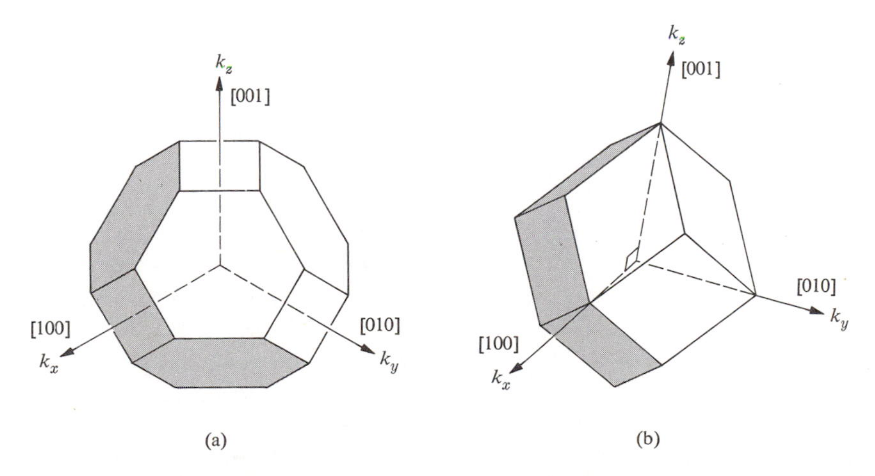


**Fig. 5.8** The first Brillouin zone for (a) an fcc lattice, and (b) a bcc lattice.

diamond-shaped pieces enclosed between the normal bisectors to $\mathbf{G}_1$, $\mathbf{G}_2$, and $\mathbf{G}_1 + \mathbf{G}_2$, etc.) forms the second zone. Similarly, the screened area (eight parts) forms the third zone. As higher-order bisectors are included, higher-order zones are also formed, which may have quite complicated shapes.

However, *all the zones have the same area*, regardless of the complexity of the zone. Thus we can see in the figure that the second zone has the same area as the first, that is, $(2\pi/a)^2$. The same is true for the third zone, and this can also be shown to hold true for all zones. This equality of the areas of the Brillouin zones holds true for all plane lattices, not just for square lattices.

In three dimensions, the zones are three-dimensional volumes. Figure 5.8 shows the first zone for fcc (a truncated octahedron) and bcc (a regular rhombic dodecahedron) lattices. Higher-order zones in these lattices are somewhat complicated in appearance and difficult to visualize; they will not concern us further here.

Let us now discuss the relation of the Brillouin zones to the band structure.

### Symmetry properties

It can be shown that each energy band $E_n(\mathbf{k})$ satisfies the following symmetry properties.

i) $$E_n(\mathbf{k} + \mathbf{G}) = E_n(\mathbf{k}) \tag{5.10}$$

ii) $$E_n(-\mathbf{k}) = E_n(\mathbf{k}) \tag{5.11}$$

iii) $E_n(\mathbf{k})$ has the same rotational symmetry as the real lattice.

Note that these properties are the same as those obeyed by the dispersion relations of lattice vibrations (Section 3.6), and can be proved in a similar manner—i.e., by invoking the symmetry properties of the real lattice—as will be discussed later in this section.

Property (i) indicates that $E_n(\mathbf{k})$ is periodic, with a period equal to the reciprocal lattice vector. In other words, any two points in $\mathbf{k}$-space related to each other by a displacement equal to a reciprocal lattice vector have the same energy. For instance, in Fig. 5.9(a), the energy is the same at points $P_1$, $P_2$, and $P_3$, because $P_2$ is related to $P_1$ by a translation equal to $-\mathbf{G}_2$, $P_3$ is related to $P_1$ by a translation $-\mathbf{G}_1$, and both $-\mathbf{G}_1$ and $-\mathbf{G}_2$ are reciprocal lattice vectors.

Figure 5.9(b) illustrates how, by using this translational symmetry, the various pieces of the second zones may be translated by reciprocal lattice vectors to fit precisely over the first zone. Each two areas connected by an arrow are *equivalent*. The first and second zones are equivalent. Similarly, higher-order zones can be appropriately translated to fit over the first zone. It follows, therefore, that we may confine our attention to the first zone only, since this contains all the necessary information.

The inversion property (ii) shows that the band is symmetric with respect to inversion around the origin $\mathbf{k} = 0$. Thus, in Fig. 5.9(a), the energy at point $P_1'$ is equal to that at $P_1$.

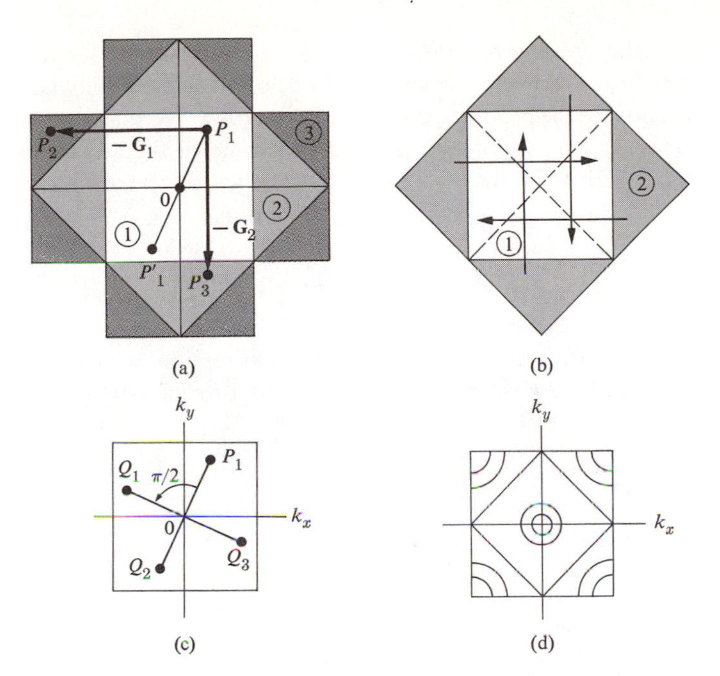


**Fig. 5.9** (a) Translational symmetry of the energy $E(\mathbf{k})$ in **k**-space for a square lattice. (b) Mapping of the second zone into the first. (c) Rotational symmetry of $E(\mathbf{k})$ in **k**-space for a square lattice. (d) Energy contours in the first zone.

Property (iii) asserts that the band has the same rotational symmetry as the real lattice. For instance, in a square lattice, the energy should exhibit the rotational symmetry of the square. Since this is symmetric with respect to a rotation by $\pi/2$ (and its multiples), it follows that in Fig. 5.9(c) the energies at points $Q_1$, $Q_2$, and $Q_3$ are equal to that at point $P_1$, because these points may be obtained from $P_1$ by symmetry rotations. [Note that $Q_2$ is the same as $P'_1$ of Fig. 5.9(a); this is so for a square lattice, but it does not hold good for other lattices.]

In Fig. 5.9(d) energy contours are sketched for a band in the first zone of a square lattice. This figure satisfies the various symmetry properties described above.

The symmetry properties are particularly important because we can use them to reduce the labor involved in determining energy bands. For example, with inversion symmetry, we need to know the band in only half of the first zone, and rotational symmetry usually enables us to reduce this even further. In the case of a square lattice, for example, only one-eighth of the zone need be specified independently, as you may see, and the remainder of the zone can then be completed by using symmetry properties.

The labor-saving is even greater in three-dimensional cases. Thus, in the case of a cubic lattice, the band need be specified independently in only 1/48th of the first zone.

Note that the symmetry properties discussed above refer to the same band. They hold for every band separately, but do not relate one band to another.

Let us turn now to the proofs of the above properties. We shall only outline these proofs here, leaving you to pursue the details in some of the advanced references listed at the end of the chapter. Consider first the translational property (i): The Bloch function at the point $\mathbf{k} + \mathbf{G}$ may be written as

$$\psi_{\mathbf{k}+\mathbf{G}} = e^{i(\mathbf{k}+\mathbf{G})\cdot\mathbf{r}} u_{\mathbf{k}+\mathbf{G}} = e^{i\mathbf{k}\cdot\mathbf{r}}(e^{i\mathbf{G}\cdot\mathbf{r}} u_{\mathbf{k}+\mathbf{G}}). \tag{5.12}$$

Note that the factor inside the brackets of the last expression, which may be denoted by $v(\mathbf{r})$, is periodic in the $\mathbf{r}$-space with a period equal to the lattice vector. That is,

$$v(\mathbf{r} + \mathbf{R}) = e^{i\mathbf{G}\cdot(\mathbf{r}+\mathbf{R})} u_{\mathbf{k}+\mathbf{G}}(\mathbf{r} + \mathbf{R}) = e^{i\mathbf{G}\cdot\mathbf{r}} u_{\mathbf{k}+\mathbf{G}}(\mathbf{r}) = (v)\mathbf{r}.$$

This follows from the fact that $u_{\mathbf{k}+\mathbf{G}}$ is periodic, and $e^{i\mathbf{G}\cdot\mathbf{R}} = 1$, since $\mathbf{G}\cdot\mathbf{R} = n2\pi$, where $n$ is some integer. The expression in the brackets in (5.12) has, therefore, the same behavior as $u_{\mathbf{k}}(\mathbf{r})$ in Eq. (5.3). We have thus shown that the state function $\psi_{\mathbf{k}+\mathbf{G}}$ has the same form as $\psi_{\mathbf{k}}$, and consequently the two functions have the same energy, since there is no physical basis for distinguishing between them.

Property (ii) may be established by noting that the Schrödinger equation analogous to (5.6), which corresponds to the point $-\mathbf{k}$, is the same as the equation obtained by writing the complex conjugate equation of (5.6). This means that the corresponding eigenvalues are equal, that is, that $E_n(-\mathbf{k}) = E_n^*(\mathbf{k})$. Since the energy $E_n(\mathbf{k})$ is a real number, however, it follows that $E_n(-\mathbf{k}) = E_n(\mathbf{k})$, which is property (ii).

Property (iii) is derived by noting that if the real lattice is rotated by a symmetry operation, the potential $V(\mathbf{r})$ remains unchanged, and hence the new state function obtained must have the same energy as the original state function. One may show further that these new states correspond to rotations in $\mathbf{k}$-space, and this leads to the desired property.

## 5.5 NUMBER OF STATES IN THE BAND

We denoted the Bloch function by $\psi_{n,\mathbf{k}}$, which indicates that each value of the band index $n$ and the vector $\mathbf{k}$ specifies an electron state, or orbital. We shall now show that *the number of orbitals in a band inside the first zone is equal to the number of unit cells in the crystal.* This is much the same as the statement made in connection with the number of lattice vibrational modes (Section 3.3), and is proved in a like manner, by appealing to the boundary conditions.

Consider first the one-dimensional case, in which the Bloch function has the form

$$\psi_k(x) = e^{ikx} u_k(x). \tag{5.13}$$

If we impose the periodic boundary condition on this function, it follows that the only allowed values of $k$ are given by

$$k = n\frac{2\pi}{L}, \tag{5.14}$$

where $n = 0, \pm 1, \pm 2$, etc. [Note that $u_k(x)$ is intrinsically periodic, so the condition $u_k(x + L) = u_k(x)$ is automatically satisfied.] As in Section 3.3, the allowed values of $k$ form a uniform mesh whose unit spacing is $2\pi/L$. The number of states inside the first zone, whose length is $2\pi/a$, is therefore equal to

$$(2\pi/a)\,(2\pi/L) = L/a = N,$$

where $N$ is the number of unit cells, in agreement with the assertion made earlier.

A similar argument may be used to establish the validity of the statement in two- and three-dimensional lattices.

It has been shown that each band has $N$ states inside the first zone. Since each such state can accommodate at most two electrons, of opposite spins, in accordance with the Pauli exclusion principle, it follows that the maximum number of electrons that may occupy a single band is $2N$. This result is significant, as it will be used in a later section to establish the criterion for predicting whether a solid is going to behave as a metal or an insulator.

## 5.6 THE NEARLY-FREE-ELECTRON MODEL

In Section 5.3 and 5.4 we studied the general properties of the state functions, and of the energies of an electron moving in a crystalline solid. To obtain explicit results, however, we must solve the Schrödinger equation (5.1) for the actual potential $V(\mathbf{r})$ in the particular solid of interest. But the process of solving the Schrödinger equation for any but the simplest potentials is an arduous and time-consuming task, inundated with mathematical details. Although this is essential for obtaining results that may be compared with experiments, it is preferable to start the discussion of explicit solutions by using rather simplified potentials. The advantage is that we can solve the Schrödinger equation with only minimal mathematical effort and thus concentrate on the new physical concepts involved.

In the present section we shall treat the *nearly-free-electron* (NFE) *model*, in which it is assumed that the crystal potential is so weak that the electron behaves essentially like a free particle. The effects of the potential are then treated by the use of perturbation methods, which should be valid inasmuch as the potential is weak. This model should serve as a rough approximation to the valence bands in the simple metals, that is, Na, K, Al, etc.

In the following section, we shall treat the *tight-binding model*, in which the atomic potentials are so strong that the electron moves essentially around a single atom, except for a small interaction with neighboring atoms, which may then be treated as a perturbation. This model lies at the opposite end from the NFE model in terms of the strength of crystal potential involved, and should serve as a rough approximation to the narrow, inner bands in solids, e.g., the 3d band in transition metals.

### The empty-lattice model

The starting point for the NFE model is the solution of the Schrödinger equation for the case in which the potential is exactly zero, i.e., the electron is entirely free. However, we also require that the solutions satisfy the symmetry properties of Section 5.4, which are imposed by the translational symmetry of the real lattice. This leads to the so-called *empty-lattice model.*

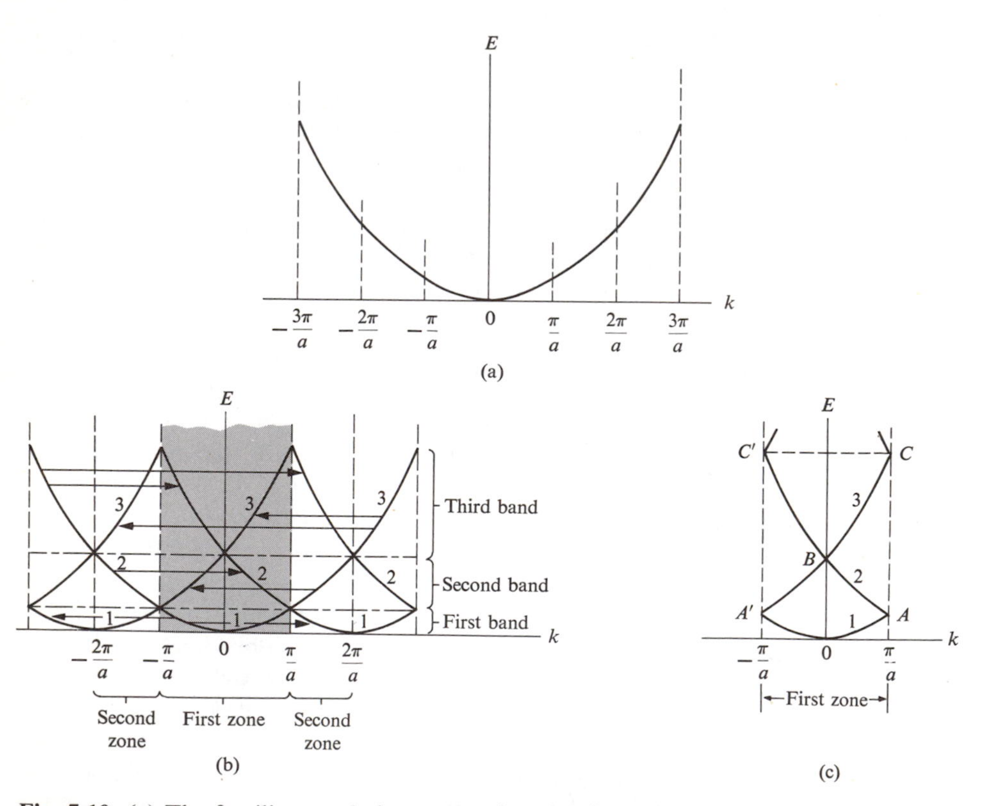


**Fig. 5.10** (a) The familiar parabola representing the dispersion curve for a free particle, $E = \hbar^2k^2/2m_0$. (b) The dispersion curves for the same particle in the empty-lattice model, showing translational symmetry and the various bands. (c) Dispersion curves in the empty-lattice model (first zone only).

For a one-dimensional lattice, the state functions and energies for the empty-lattice model are

$$\psi_k^{(0)} = \frac{1}{L^{1/2}} e^{ikx}, \tag{5.15}$$

and

$$E_{(k)}^{(0)} = \frac{\hbar^2 k^2}{2m_0}, \tag{5.16}$$

where the superscript 0 indicates that the solutions refer to the unperturbed state (Section A.7). The energy $E_{(k)}^{(0)}$ which is plotted versus $k$ in Fig. 5.10(a) exhibits a curve in the familiar *parabolic* shape. Figure 5.10(b) shows the result of imposing the symmetry property (i) of Section 5.4. Segments of the parabola of Fig. 5.10(a) are cut at the edges of the various zones, and are translated by multiples of $G = 2\pi/a$ in order to ensure that the energy is the same at any two equivalent points. Figure 5.10(c) displays the shape of the energy spectrum when we confine our consideration to the first Brillouin zone only. [Conversely, Fig. 5.10(b) may be viewed as the result of translating Fig. 5.10(c) by multiples of $G$.]

The type of representation used in Fig. 5.10(c) is referred to as the *reduced-zone scheme*. Because it specifies all the needed information, it is the one we shall find most convenient. The representation of Fig. 5.10(a), known as the *extended-zone scheme*, is convenient when we wish to emphasize the close connection between a crystalline and a free electron. However, Fig. 5.10(b) employs the *periodic-zone* scheme, and is sometimes useful in topological considerations involving the **k**-space. All these representations are strictly equivalent; the use of any particular one is dictated by convenience, and not by any intrinsic advantages it has over the others.

**The nearly-free-electron model**

How is the energy spectrum of Fig. 5.10(c) altered when the crystal potential is taken into account, or "turned on?" Figure 5.11(a) shows this. The first and second bands, which previously touched at the point $A$ (and $A'$) in Fig. 5.10(c) are now split, so that an energy gap is created at the boundary of the Brillouin zone. A similar gap is created at the center of the zone, where bands 2 and 3 previously intersected (point $B$ in Fig. 5.10c) and also at point $C$, where bands 3 and 4 previously intersected. Thus, in general, in the empty-lattice model, energy gaps are created in **k**-space wherever bands intersect, which occurs either at the center or the boundaries of the BZ. At these points the shape of the spectrum is strongly modified by the crystal potential, weak as this may be. (In effect, what the crystal potential has accomplished is to smooth over the sharp "corners" present in the band structure of the empty lattice.)

In the remainder of the zone, however, the shape of the spectrum is affected very little by the crystal potential, since this is assumed to be weak. In that region

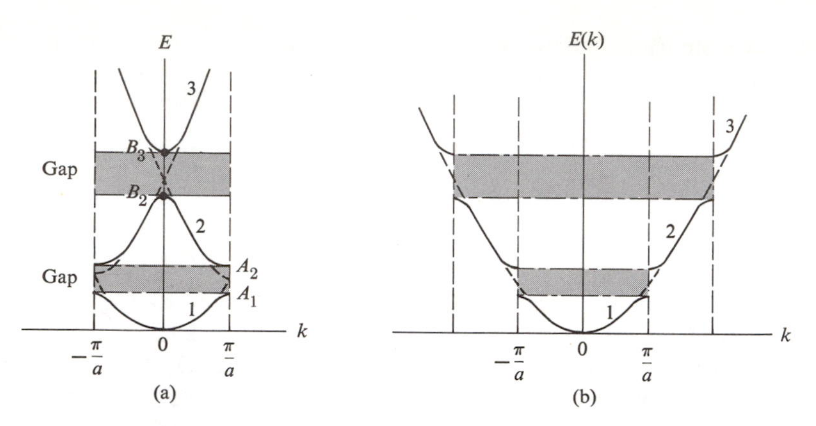


**Fig. 5.11** (a) Dispersion curves in the nearly-free-electron model, in the reduced-zone scheme. (b) The same dispersion curves in the extended-zone scheme.

of the **k**-space the bands essentially retain their parabolic shape inherited from the empty-lattice model of Fig. 5.10(c), and the electron there behaves essentially like a free electron.

By comparing Fig. 5.10(c) and Fig. 5.11(a), one notes that a hint of a band structure is almost present even in the empty-lattice model, except that the gaps there vanish, since the bands touch at the zone boundaries. This vanishing is foreseen, of course, since no energy gaps are expected to appear in the spectrum of a free particle. The point is that even a weak potential leads to the creation of gaps, in agreement with the results of Sections 5.2 and 5.3.

Figure 5.11(b) shows the band structure for the NFE model, represented according to the extended-zone scheme, which should be compared with Fig. 5.10(a). Note that, except at the zone boundaries at which gaps are created, the dispersion curve is essentially the same as the free-electron curve.

We made the above assertions without proofs; we shall now outline proofs on the basis of the perturbation method of Section A.7. Suppose, for instance, that we seek to find the influence of the crystal potential on the first band in Fig. 5.10(c). When we treat the potential $V(x)$ as a perturbation, the perturbed energy $E_1(k)$ up to the second order of the potential is given by

$$E_1(k) = E_1^{(0)}(k) + \langle \psi_{1,k}^{(0)} \mid V \mid \psi_{1,k}^{(0)} \rangle + \sum_{k',n}{}' \frac{|\langle|\langle n, k'|V|1, k\rangle|^2}{E_1^{(0)}(k) - E_n^{(0)}(k')}. \tag{5.17}$$

Here the subscript 1 refers to the first band, which is the one of interest, and the superscript 0 refers to the empty-lattice model of Eqs. (5.15) and (5.16). The second term on the right side of (5.17), which is the first-order correction, is the average value of the potential. The third term, giving the second-order correction, involves summing over all states $n$, $k$, except where these indices are equal to the state 1, $k$ under investigation.

First we note that the first-order correction is equal to

$$\langle \psi_{1,k}^{(0)} \mid V \mid \psi_{1,k}^{(0)} \rangle = \frac{1}{L} \int e^{-ikx} V(x)\, e^{ikx}\, dx = \frac{1}{L} \int V(x)\, dx,$$

which is the average value of the potential over the entire lattice. It is independent of $k$, and hence it is merely a constant. Its effect on the spectrum of Fig. 5.10(c) is simply to displace it rigidly by a constant amount, without causing any change in the shape of the energy spectrum. Since this term does not lead to anything of interest to us here, it will be set equal to zero, which can be accomplished by shifting the zero energy level.

We must therefore consider the second-order correction in Eq. (5.17). We first assert that the quantity $\langle n, k' \mid V \mid 1, k \rangle$ can be shown to vanish except when $k' = k$, where both $k$ and $k'$ are restricted to the first zone. That is, the only states which are coupled to the $1, k$ state by the perturbation are those lying directly above this state, as shown in Fig. 5.12.

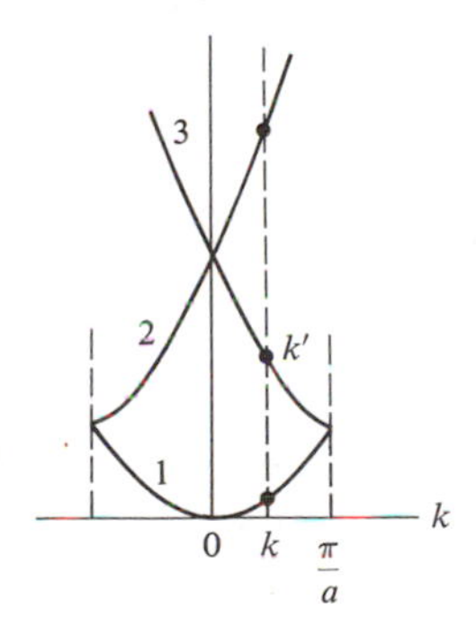


**Fig. 5.12** Only those states lying directly above the state $\psi_{1,k}^{(0)}$ in **k**-space are coupled to it by the perturbation.

This assertion rests on the translational symmetry of the crystal potential $V(x)$.† Furthermore, since the energy difference in the denominator of the third

---

† An arbitrary potential $V(x)$ can always be expanded as a Fourier series

$$V(x) = \textstyle\sum_k V_k\, e^{ikx},$$

where the summation is over all the allowed $k$'s. The Fourier coefficiemt $V_k$ is given by

$$V_k = (1/L) \int_0^L V(x)\, e^{-ikx}\, dx.$$

But if $V(x)$ is periodic, as is the case in a crystal, then only the values $k = G$ contribute to the above summation; that is, $V_k = 0$ for $k \neq G$. A periodic potential therefore has the expansion

$$V(x) = \textstyle\sum_G V_G\, e^{iGx}.$$

It can be shown that the bracket in the numerator of (5.17) is the Fourier coefficient $V_{k'-k}$, and hence this bracket vanishes except for $k' - k = G$.

term in (5.17) increases rapidly as the band $n$ rises, the major effect on band 1 arises from its coupling to band 2. We may therefore write

$$E_1(k) \simeq E_1^{(0)}(k) + \frac{|V_{-2\pi/a}|^2}{E_1^{(0)}(k) - E_2^{(0)}(k)}, \tag{5.18}$$

where $V_{-2\pi/a}$ is the Fourier component of the potential, that is,

$$V_{-2\pi/a} = \frac{1}{L}\int V(x)\, e^{+i(2\pi/a)x}\, dx.$$

An explicit expression for $E_1(k)$ can be obtained by substituting the values for $E_1^{(0)}(k)$ and $E_2^{(0)}(k)$: namely $E_1^{(0)}(k) = \hbar^2 k^2/2m$, and $E_2^{(0)}(k) = \hbar\,(k - 2\pi/a)^2\ 2m$. [Note that if $0 < k < \pi/a$, then the second band is obtained by translating that part of the free-electron curve lying in the interval $-2\pi/a < k < -\pi/a$, as seen in Fig. 5.10(b), and hence the above expression for $E_2^{(0)}(k)$.] But this is not really necessary, because if the potential is weak, then $|V_{-2\pi/a}|^2$ is very small, and the second term in (5.18) is negligibly small compared with the first. In other words, $E_1(k) \simeq E_1^{(0)}(k)$, and the effect of the lattice potential is negligible.

There is, however, one point in $k$-space at which the above conclusion breaks down: the point $k = \pi/a$ at the zone edge. At this point the energies $E_1^{(0)}(k)$ and $E_2^{(0)}(k)$ are equal [recall that bands 1 and 2 touch there; see Fig. 5.10(c)], the denominator of the perturbation term in (5.18) vanishes, and hence the perturbation correction becomes very large. Since the above perturbation theory presumes the smallness of the correction, it follows that this theory cannot hold true in the neighborhood of the zone edge. In this neighborhood, one should instead invoke the *degenerate* perturbation theory, in which both bands 1 and 2 are treated simultaneously, and on an equal footing. The resulting energy values are (Ziman, 1963),

$$E_\pm(k) = \tfrac{1}{2}\{E_1^{(0)}(k) + E_2^{(0)}(k) \pm [(E_2^{(0)}(k) - E_1^{(0)}(k))^2 + 4|V_{-2\pi/a}|^2]^{1/2}\}, \tag{5.19}$$

where the plus sign corresponds to the deformed upper band—i.e., band 2—near the edge of the zone, and the minus sign refers to the deformed lower band—i.e., band 1.

Now let us substitute the values of $E_1^{(0)}(k)$ and $E_2^{(0)}(k)$ into (5.19) and plot $E_+(k)$ and $E_-(k)$ in the neighborhood of the zone edge. We obtain the spectrum shown in Fig. 5.11(a). In particular, the energy gap $E_g$ is equal to the difference $E_+(k) - E_-(k)$ evaluated at the point $k = \pi/a$. Using (5.19), we readily find that

$$E_g = 2\,|V_{-2\pi/a}|. \tag{5.20}$$

That is, the energy gap is equal to twice the Fourier component of the crystal potential. In effect, band 1 has been depressed by an amount equal to $|V_{-2\pi/a}|$ and band 2 has been raised by the same amount, leading to an energy gap given by (5.20).

The same formula (5.19) may also be used to find the energy gap that arises at the center of the zone, at the intersection between bands 2 and 3, except that we now replace $E_1^{(0)}(k)$, $E_2^{(0)}(k)$ by $E_2^{(0)}(k)$ and $E_3^{(0)}(k)$, respectively. We also replace the potential term by $V_{-4\pi/a}$. This leads to the splitting of bands 2 and 3, as shown in Fig. 5.11(a), with an energy gap of $2|V_{-4\pi/a}|$. Obviously the procedure can be used to find both the splitting of the bands and the corresponding gaps at all appropriate points.

In addition to the above results, two qualitative conclusions emerge from the analysis. First, the higher the band, the greater its width; this is evident from referring back to the empty lattice model in Fig. 5.10(a), since the energy there increases as $k^2$. Second, the higher the energy, the narrower the gap; this follows from the fact that the gap is proportional to a certain Fourier component of the crystal potential, but note that the order of the component increases as the energy rises (from $V_{-2\pi/a}$ to $V_{-4\pi/a}$ in our discussion above). Since the potential is assumed to be well behaved, the components decrease rapidly as the order increases, and this leads to a decrease in the energy gap. It follows therefore that, as we move up the energy scale, the bands become wider and the gaps narrower; i.e., the electron behaves more and more like a free particle. This agrees with the qualitative picture drawn in Section 5.2.

Since the greatest effect of the crystal potential takes place near the points in $k$-space at which two bands touch, let us examine the behavior there more closely. If one applies the degenerate perturbation formula (5.17) to the splitting of bands 2 and 3 at the center of the zone, one finds that, for small $k$ ($k \ll \pi/a$),

$$E_3(k) = E_B + |V_{-4\pi/a}| + \frac{\hbar^2}{2m_0}\alpha k^2, \tag{5.21}$$

and

$$E_2(k) = E_B - |V_{-4\pi/a}| - \frac{\hbar^2}{2m_0}\alpha k^2, \tag{5.22}$$

where the parameter $\alpha$ is given by

$$\alpha = 1 + \frac{4E_B}{E_g} \tag{5.23}$$

and $E_B = \hbar^2(2\pi/a)^2/2m_0$ is the energy of point $B$ in Fig. 5.10(c). These results are very interesting for several reasons.

a) Equation (5.21) shows that, for an electron near the bottom of the third band, $E \sim k^2$ (ignoring the first two terms on the right, since they are simply constants), which is similar to the dispersion relation of a free electron. In other words, the electron there behaves like a free electron, with an effective mass $m^*$ given by

$$m^* = m_0/\alpha,$$

which is different from the free mass. Referring to (5.23), one sees that the effective mass increases as the energy gap $E_g$ increases. Such a relationship between $m^*$ and $E_g$ is familiar in the study of semiconductors.

b) Equation (5.22) shows that, for an electron near the top of the second band, $E \sim -k^2$, which is like a free electron, except for the surprising fact that the effective mass is negative. Such behavior is very unlike that of a free electron, and its cause lies, of course, in the crystal potential. The phenomenon of a negative effective mass near the top of the band is a frequent occurrence in solids, particularly in semiconductors, as we shall see later (Chapter 6).

We have thus far confined ourselves to a one-dimensional lattice, but we may extend this treatment to two- and three-dimensional lattices in a straightforward fashion. We find again, as expected, that starting with the empty-lattice model, the "turning on" of the crystal potential leads to the creation of energy gaps. Furthermore, these gaps occur at the boundaries of the Brillouin zone.

## 5.7 THE ENERGY GAP AND THE BRAGG REFLECTION

In discussing the NFE model, we focused on energy values. But perturbation also modifies state functions, and we shall now study this modification. If we apply the perturbation theory to the one-dimensional empty lattice, we find that the state function of the first band in Fig. 5.11(a) is given by

$$\psi_{1,k} = \psi_{1,k}^{(0)} + \frac{V_{-2\pi/a}}{E_1^{(0)}(k) - E_2^{(0)}(k)}\,\psi_{2,k}^{(0)}, \tag{5.24}$$

where—again because of the form of the potential and also the energy difference in the denominator—the perturbation summation has been reduced to one term only, involving the state function of the second band $\psi_{2,k}^{(0)}$.

The state functions $\psi_{1,k}^{(0)}$ and $\psi_{2,k}^{(0)}$ refer to a free electron; $\psi_{2,k}^{(0)} \sim e^{ikx}$ represents a wave traveling to the right, while $\psi_{2,k}^{(0)} \sim e^{i(k-2\pi/a)x}$ represents a wave traveling to the left (note that $|k| < \pi/a$). The effect of the lattice potential is then to introduce a new left-traveling wave in addition to the incident free wave. This new wave is generated by the scattering of the electron by the crystal potential. If $k$ is not close to the zone edge, however, the coefficient of $\psi_{2,k}^{(0)}$ in (5.24) is negligible. That is,

$$\psi_{1,k} \simeq \psi_{1,k}^{(0)} = \frac{1}{L^{1/2}}\,e^{ikx}, \tag{5.25}$$

and the electron behaves like a free electron. The effects of the potential are negligible there, which is in agreement with the conclusions reached in Section 5.6.

Near the zone edge, however, the energy denominator in the correction term in (5.24) becomes very small, and the perturbation term large, which means that the form (5.24) becomes invalid. As stated in Section 5.6, one must then use the degenerate perturbation theory, in which the state functions $\psi_{1,k}^{(0)}$ and $\psi_{2,k}^{(0)}$ are treated on an equal footing. One finds that, at the zone edge itself,

$$\psi_{\pm}(x) = \frac{1}{\sqrt{2}}\,[\psi_{1,\pi/a}^{(0)}(x) \pm \psi_{2,\pi/a}^{(0)}(x)] = \frac{1}{\sqrt{2L}}\,(e^{i(\pi/a)x} \pm e^{-i(\pi/a)x}). \tag{5.26}$$

The function $\psi_{+}(x) \sim \cos(\pi/a)x$, and hence the probability is proportional to $|\psi_{+}(x)|^2 \sim \cos^2(\pi/a)x$. Such a state function distributes the electron so that it is piled predominantly at the nuclei (recall that the origin $x = 0$ is at the center of an ion) [see Fig. 5.13], and since the potential is most negative there, this

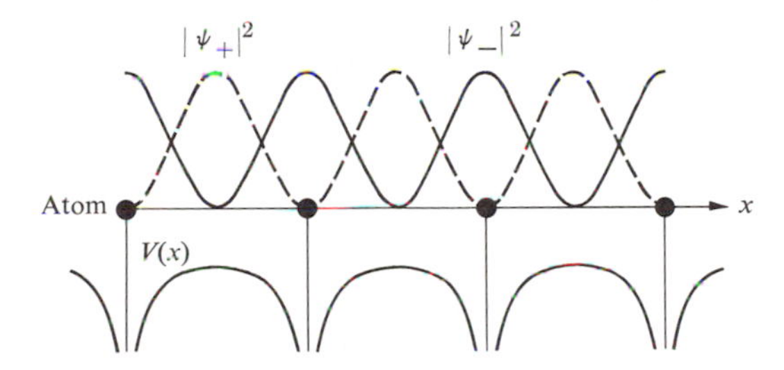


**Fig. 5.13** Spatial distributions of electrons described by the functions $\psi_{+}$ and $\psi_{-}$.

distribution has a low energy. The function $\psi_{+}(x)$ therefore corresponds to the energy at the top of band 1, that is, point $A$ in Fig. 5.11(a).

By contrast, the function $\psi_{-}(x) \sim \sin \pi/ax$, depositing its electron mostly between the ions (as shown in Fig. 5.13), corresponds to the bottom of band 2 in Fig. 5.11(a), that is, point $A_2$. The gap arises, therefore, because of the two different distributions for the same value $k = \pi/a$, the distributions having different energies.

Scrutinizing (5.26) from the viewpoint of scattering, we see that at the zone edge, $k = \pi/a$, the scattering is so strong that the reflected wave has the same amplitude as the incident wave. As found above, the electron is represented there by a *standing* wave, $\cos \pi/ax$ or $\sin \pi/ax$, very unlike a free particle. An interesting result of this is that the electron, as a standing wave, has a zero velocity at $k = \pi/a$. This is a general result which is valid at all zone boundaries, and one which we shall encounter often in the following sections.

We have seen that the periodic potential causes strong scattering at $k = \pi/a$. Recall from Section 3.6 on lattice vibrations that this strong scattering arises as a result of the Bragg diffraction at the zone edge. In the present situation, the wave diffracted is the electron wave, whose wavelength is $\lambda = 2\pi/k$.

In higher-dimension lattices, the Bragg condition is satisfied along all boundaries of the Brillouin zone, as discussed in Section 2.6, and this results in the creation of energy gaps along these boundaries, in agreement with the conclusions of the last section.

## 5.8 THE TIGHT-BINDING MODEL

In the tight-binding model, it is assumed that the crystal potential is strong, which is the same as saying that the ionic potentials are strong. It follows, therefore, that when an electron is captured by an ion during its motion through the lattice, the electron remains there for a long time before leaking, or tunneling, to the next ion [see Fig. 5.14(a), which also shows that the energy of the electron is appreciably

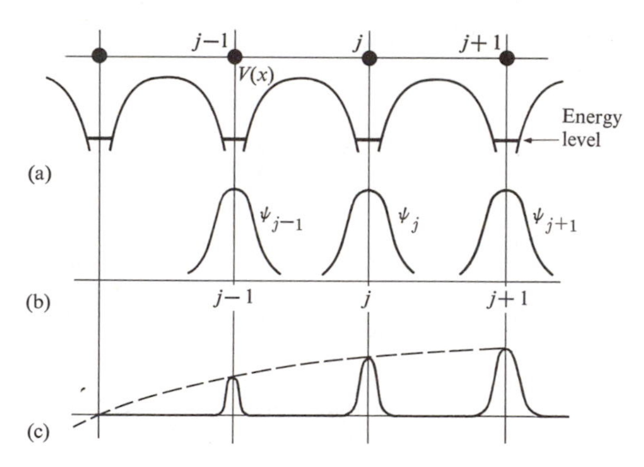


**Fig. 5.14** The tight-binding model. (a) The crystal potential. (b) The atomic wave functions. (c) The corresponding Bloch function.

lower than the top of the potential barrier]. During the capture interval, the electron orbits primarily around a single ion, i.e., its state function is essentially that of an atomic orbital, uninfluenced by other atoms. Most of the time the electron is tightly bound to its own atom. The mathematical analysis to be developed must reflect this important fact.

As we said in Section 5.6, the TB (tight-binding) model is primarily suited to the description of low-lying narrow bands for which the shell radius is much smaller than the lattice constant. Here the atomic orbital is modified only slightly by the other atoms in the solid. An example is the 3d band, so important in transition metals.

Let us begin, then, with an atomic orbital, $\phi_v(x)$, whose energy in a free atom is $E_v$. We wish to examine the effects of the presence of other atoms in the solid. The index $v$ characterizes the atomic orbital (for the atomic shell of interest).

First, the one-dimensional case: It is necessary to choose a suitable Bloch function, and while the choice is not unique, the following offers a reasonable form.

$$\psi_k(x) = \frac{1}{N^{1/2}} \sum_{j=1}^{N} e^{ikX_j} \phi_v(x - X_j), \tag{5.27}$$

where the summation extends over all the atoms in the lattice. The coordinate $X_j$ specifies the position of the $j^{\text{th}}$ atom. That is, $X_j = ja$, where $a$ is the lattice constant. The function $\phi_v(x - X_j)$ is the atomic orbital centered around the $j^{\text{th}}$ atom; it is large in the neighborhood of $X_j$, but decays rapidly away from this point, as shown in Fig. 5.14(b). By the time the neighboring site at $X_{j+1}$ (or $X_{j-1}$) is reached, the function $\phi_v(x - X_j)$ has decayed so much that it has become almost negligible. In other words, there is only a little overlap between neighboring atomic orbitals. This is the basic assumption of the TB model. The factor $N^{1/2}$ is included in (5.27) to ensure that the function $\psi_k$ is normalized to unity (if the atomic orbital $\phi_v$ is so normalized).

Let us turn now to the properties of the function $\psi_k(x)$, as defined by (5.27), First, it is necessary to ascertain that this function is a Bloch function, namely, that it can be written in the form (5.3). This can be established by rewriting (5.27) in the form

$$\psi_k(x) = \frac{1}{N^{1/2}} e^{ikx} \sum_{j=1}^{N} e^{-i\,k(x-X_j)} \phi_v(x - X_j),$$

where it is now readily recognized that the factor defined by the summation is periodic, with a period equal to the lattice constant $a$. Thus the function $\psi_k(x)$ has indeed the desired Bloch form, i.e., it describes a propagating electron wave, as shown in Fig 5.14(c).

Note also that near the center of the $j^{\text{th}}$ ion, the function $\psi_k(x)$ reduces to

$$\psi_k(x) \simeq e^{ikX_j} \phi_v(x - X_j) \sim \phi_v(x - X_j). \tag{5.28}$$

That is, the Bloch function is proportional to the atomic orbital. Thus in the neighborhood of the $j^{\text{th}}$ ion, the crystal orbital behaves much like an atomic orbital, in agreement with the basic physical assumption of the TB model.

The function $\psi_k(x)$ therefore satisfies both the mathematical requirement of the Bloch theorem and the basic assumption of the TB model, and as such is a suitable crystal orbital. It will be used now to calculate the energy of the band.

The energy of the electron described by $\psi_k$ is given, according to quantum mechanics, by

$$E(k) = \langle \psi_k \,|\, H \,|\, \psi_k \rangle, \tag{5.29}$$

where $H$ is the Hamiltonian of the electron†. Substituting for $\psi_k$ from (5.27), one has

$$E(k) = \frac{1}{N}\sum_{j,j'} e^{ik(X_j - X_{j'})}\langle \phi_v(x - X_{j'}) \mid H \mid \phi_v(x - X_j)\rangle, \tag{5.30}$$

where the double summation over $j$ and $j'$ extends over all the atoms in the lattice. Note that each term in the summation is a function of the difference $X_j - X_{j'}$, and not of $X_j$ and $X_{j'}$ individually. Therefore, for each particular choice of $j'$, the sum over $j$ yields the same result, and since $j'$ can take $N$ different values, one obtains $N$ equal terms, which thus leads to

$$E(k) = \sum_{j=-N/2}^{(N-1)/2} e^{ikX_j}\langle \phi_v(x) \mid H \mid \phi_v(x - X_j)\rangle, \tag{5.31}$$

where we have arbitrarily put $X_{j'} = 0$ in (5.30). By splitting the term $j = 0$ from the others, one may write the above expression as

$$E(k) = \langle\phi_v(x) \mid H \mid \phi_v(x)\rangle + \sum_j{}' e^{ikX_j}\langle\phi_v(x) \mid H \mid \phi_v(x - X_j)\rangle, \tag{5.32}$$

The first term gives the energy the electron would have if it were indeed entirely localized around the atom $j = 0$, while the second term includes the effects of the electron tunneling to the various other atoms. The terms in the summation are expected to be appreciable only for nearest neighbors—that is, $j = 1$ and $j = -1$—because as $j$ increases beyond that point, the overlap between the corresponding functions and the state function at the origin becomes negligible (Fig. 5.14b). Note also that, since the property of electron delocalization is included entirely in the second term of (5.32), it is this term which is responsible for the band structure, and as such is of particular interest to us here.

To proceed with the evaluation of $E(k)$, according to (5.32), we need to examine the Hamiltonian $H$ more closely. The expression for this quantity is given by

$$H = -\frac{\hbar^2}{2m_0}\frac{d^2}{dx^2} + V(x), \tag{5.33}$$

where $V(x)$ is the crystal potential. Writing this potential as a sum of atomic potentials, one has

$$V(x) = \sum_j v(x - X_j). \tag{5.34}$$

† The Hamiltonian $H$ is simply the quantum operator which represents the total energy of the particle. Thus $H = -(\hbar^2/2m_0)\nabla^2 + V(\mathbf{r})$, where the first term on the right represents kinetic energy and the second term potential energy. The expression (5.29) for the energy is very plausible, since the term on the right is the average value of the energy in quantum mechanics.

In using this to evaluate the first term in (8.32), we shall find it convenient to split $V(x)$ into a sum of two terms

$$V(x) = v(x) + V'(x), \tag{5.35}$$

where $v(x)$ is the atomic potential due to the atom at the origin and $V'(x)$ is that due to all the other atoms. These potentials are plotted in Figs. 5.15(a) and (b),

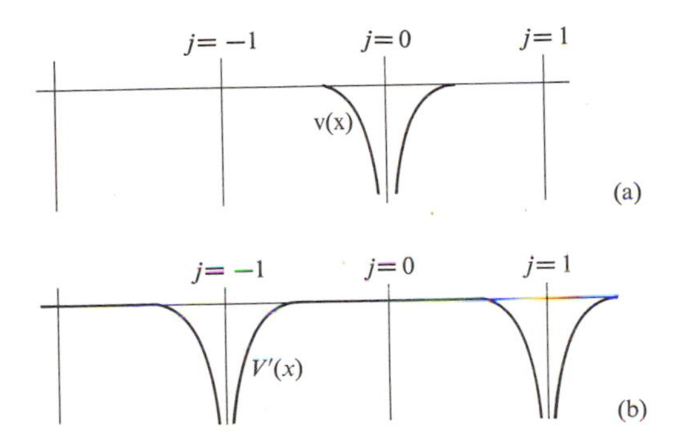


**Fig. 5.15** The splitting of the crystal potential into (a) an atomic potential and (b) the remainder of the crystal potential.

respectively. Note in particular that $V'(x)$ is small in the neighborhood of the origin. The first term in (5.32) may now be written as

$$\langle \phi_v(x) \mid H \mid \phi_v(x) \rangle = \left\langle \phi_v(x) \middle| \left[ -\frac{h^2}{2m}\frac{d^2}{dx^2} + v(x) \right] \middle| \phi_v(x) \right\rangle + \langle \phi_v(x) \mid V'(x) \mid \phi_v(x) \rangle. \tag{5.36}$$

The first term on the right is equal to $E_v$, the atomic energy, since the operator involved is the Hamiltonian for a free atom. The second term is an integral which can be evaluated, and will be denoted by the constant $-\beta$. Explicitly,

$$\beta = -\int \phi_v^*(x)\, V'(x) \phi_v(x)\, dx, \tag{5.37}$$

where the minus sign is introduced so that $\beta$ is a positive number.† Note that $\beta$ is a small quantity, since the function $\phi_v(x)$ is appreciable only near the origin, whereas $V'(x)$ is small there. Collecting the two terms above, we have

$$\langle \phi_v(x) \mid H \mid \phi_v(x) \rangle = E_v - \beta. \tag{5.38}$$

---

† The integral in (5.37) is negative because $V'(x)$ is negative (Fig. 5.15b).

Let us now turn to the interaction term, i.e., the summation in (5.32). The term involving interaction with the nearest neighbor at $X_1 = a$ involves an integral which may be written as

$$\langle \phi_v(x) \mid H \mid \phi_v(x-a) \rangle = \langle \phi_v(x) \mid -\frac{h^2}{2m_0} \frac{d^2}{dx^2}$$

$$+ v(x-a) \mid \phi_v(x-a) \rangle + \langle \phi_v(x) \mid V'(x-a) \mid \phi_v(x) \rangle. \tag{5.39}$$

The first term on the right is equal to $E_v \langle \phi_v(x) \mid \phi_v(x-a) \rangle$, which is a negligible quantity, since the two functions $\phi_v(x)$ and $\phi_v(x-a)$, being centered at two different atoms, do not overlap appreciably. The second term on the right of (5.39) is a constant which we shall call $-\gamma$, that is,

$$\gamma = -\int \phi_v^*(x)\, V'(x-a)\, \phi_v(x-a)\, dx. \tag{5.40}$$

Note that $\gamma$, though small, is still nonvanishing because $V'(x-a)$ is appreciable near the origin, that is, $x = 0$ (although not at $x = a$). The parameter $\gamma$ is called the *overlap integral*, since it is dependent on the overlap between orbitals centered at two neighboring atoms.

The integral arising from the term $j = -1$ in the sum in (5.32), which is due to the atom on the left side of the origin, yields the same result as (5.39) because the atomic functions are symmetric.

Substituting the above results into (5.32), and restricting the sum to nearest neighbors only, one finds

$$E(k) = E_v - \beta - \gamma \sum_{j=-1}^{1} e^{ikX_j}, \tag{5.41}$$

which may thus be written as

$$E(k) = E_v - \beta - 2\gamma \cos ka. \tag{5.42}$$

This is the expression we have been seeking. It gives band energy as a function of $k$ in terms of well-defined parameters which we can evaluate from our knowledge of atomic energy and atomic orbitals.

Equation (5.42) may be rewritten more conveniently as

$$E(k) = E_0 + 4\gamma \sin^2\left(\frac{ka}{2}\right), \tag{5.43}$$

where

$$E_0 = E_v - \beta - 2\gamma. \tag{5.44}$$

The energy $E(k)$ is plotted versus $k$ in Fig. 5.16, where $k$ is restricted to the first zone [although $E(k)$ is obviously periodic in $k$, in agreement with property (i) of Section 5.4]. We see, as expected, that the original atomic level $E_v$ has broadened into an energy band. The bottom of the band, located at $k = 0$, is equal to $E_0$, and its *width* is equal to $4\gamma$.

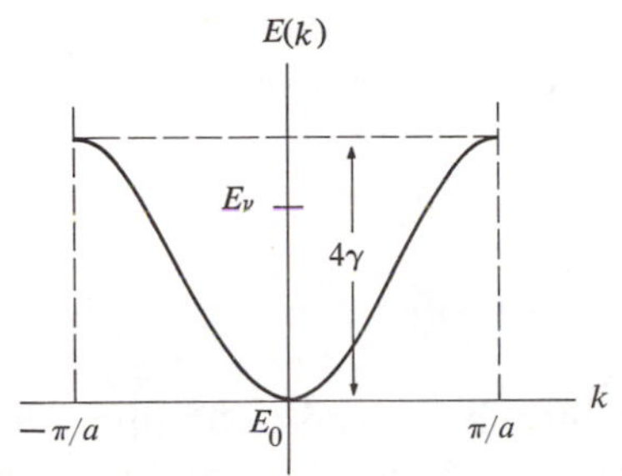


**Fig. 5.16** The dispersion curve in the tight-binding model.

Note that the bottom of the band $E_0$ is lower than the atomic energy $E_v$, which is to be expected, since one effect of the presence of the other atom is to depress the potential throughout the system (refer to Fig. 5.14a). In addition to $E_0$, the electron has an amount of energy given by the second term in (5.43). This is a *kinetic energy*, arising from the fact that the electron is now able to move through the crystal.

Note also that the bandwidth, $4\gamma$, is proportional to the overlap integral. This is reasonable, because, as we saw in Section 5.2, the greater the overlap the stronger the interaction, and consequently the wider the band.

When the electron is near the bottom of the band, where $k$ is small, one may make the approximation $\sin(ka/2) \sim ka/2$, and hence

$$E(k) - E_0 = \gamma a^2 k^2, \tag{5.45}$$

which is of the same form as the dispersion relation of a free electron. An electron in that region of $k$-space behaves like a free electron with an effective mass

$$m^* = \frac{\hbar^2}{2a^2}\frac{1}{\gamma}. \tag{5.46}$$

It is seen that the effective mass is inversely proportional to the overlap integral $\gamma$. This is intuitively reasonable, since the greater the overlap the easier it is for the electron to tunnel from one atomic site to another, and hence the smaller is the inertia (or mass) of the electron. Conversely, a small overlap leads to a large mass, i.e., a sluggish electron. Of course, in the TB model, the overlap is supposed to be small, implying a large effective mass.

Note, however, that an electron near the top of the band shows unusual behavior. If we define $k' = \pi/a - k$, and expand the energy $E(k)$ near the maximum point, using (5.43), we arrive at

$$E(k') - E_{max} = -\frac{a^2}{2}\gamma k'^2, \tag{5.47}$$

which shows that the electron behaves like a particle of *negative* effective mass

$$m^* = -\frac{\hbar^2}{a^2\gamma}. \tag{5.48}$$

This, you recall, is in agreement with the results obtained on the basis of the NFE model.

The above treatment can be extended to three dimensions in a straightforward manner. Thus for a sc lattice, the band energy is given by

$$E(k) = E_0' + 4\gamma\left[\sin^2\left(\frac{k_x a}{2}\right) + \sin^2\left(\frac{k_y a}{2}\right) + \sin^2\left(\frac{k_z a}{2}\right)\right]. \tag{5.49}$$

where $E_0'$ is the energy at the bottom of the band. The energy contours for this band, in the $k_x - k_y$ plane, are shown in Fig. 5.17(a), and the dispersion curves along the [100] and [111] directions are shown in Fig. 5.17(b). The bottom of the band is at the origin $k = 0$, and the electron there behaves as a free particle with an effective mass given by (5.46). The top of the band is located at the corner of the zone along the [111] direction, that is, at $[\pi/a, \pi/a, \pi/a]$; the electron there has a negative effective mass given by (5.48). The width of the band is equal to $12\gamma$.

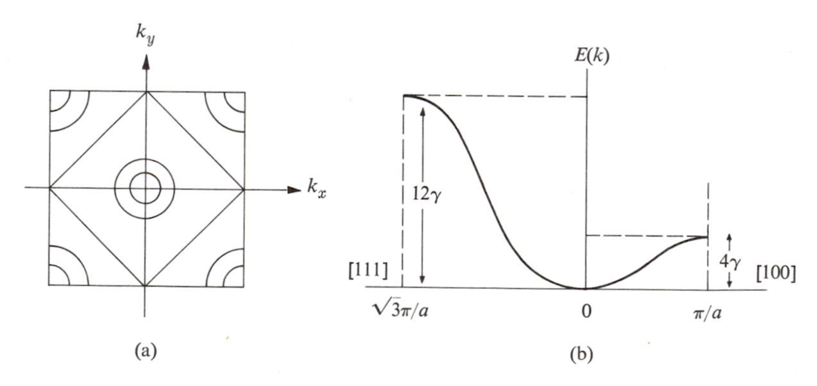


**Fig. 5.17** (a) Energy contours for an sc lattice in the tight-binding model. (b) Dispersion curves along the [100] and [111] directions for an sc lattice in the TB model.

In this treatment of the TB model, we have seen how an atomic level *broadens* into a band as a result of the interaction between atoms in the solid. In this manner, each atomic level leads to its own corresponding band, and each band reflects the character of the atomic level from which it has originated.

In conclusion, we see that both the NFE and TB models lead to the same qualitative results, although the models start from opposite points of view. The principal results arrived at in both models are: (a) Energy gaps appear at zone boundaries. (b) An electron near the bottom of the band behaves like a free particle with a positive effective mass. (c) An electron near the top of the band behaves like a free particle with a negative effective mass.

## 5.9 CALCULATIONS OF ENERGY BANDS

In the last few sections we have discussed some methods of calculating energy bands. However, these methods—the NFE and TB models—are too crude to be useful in calculations of actual bands which are to be compared with experimental results. In this section we shall consider therefore some of the common methods employed in calculations of actual bands. Because this subject is an advanced one, requiring a considerable background in quantum mechanics, as well as meticulous attention to almost endless mathematical details, our discussion will be brief, primarily qualitative, and somewhat superficial. We shall nevertheless try to give the reader a glimpse of this fundamental subject in the hope that he may pursue it further, if he so desires, by referring to books listed in the bibliography at the end of the chapter.

Several different schemes for calculating energy bands have been used. Let us now discuss them individually.

### The cellular method

The cellular method was the earliest method employed in band calculations (Wigner and Seitz, 1935). It was applied with success to the alkali metals, particularly Na and K; we shall use Na as an example.

The Schrödinger equation whose solution we seek is

$$\left[-\frac{\hbar^2}{2m_0}\nabla^2 + V(\mathbf{r})\right]\psi_{\mathbf{k}} = E(\mathbf{k})\,\psi_{\mathbf{k}}, \tag{5.50}$$

where $V(\mathbf{r})$ is the crystal potential and $\psi_{\mathbf{k}}$ the Bloch function. Here we are interested only in the 3s band. It is at once evident that this equation cannot be solved analytically. We must therefore use an approximation procedure.

When we use the cellular method, we divide the crystal into unit cells; each atom is centered at the middle of its cell, as shown in Fig. 5.18. Such a cell, known as the *Wigner–Seitz* (WS) *cell*, is constructed by drawing bisecting planes normal to the lines connecting an atom $A$, say, to its neighbors, and "picking out" the volume enclosed by these planes. (The procedure for constructing the WS cell, you may note, is analogous to that used in constructing the Brillouin zone in **k**-space.) For Na, which has a bcc structure, the WS cell has the shape of a regular dodecahedron (similar to Fig. 5.8b, but in real space).

In order to solve (5.50), we now assume that the electron, when in a particular cell, say $A$, is influenced by the potential of the ion in that cell only. The ions in other cells have a negligible effect on the electron in cell $A$ because each of these cells is occupied, on the average, by another conduction electron which tends to screen the ion, thereby reducing its potential drastically. To ensure that the function $\psi_{\mathbf{k}}$ satisfies the Bloch form, it is necessary that $u_{\mathbf{k}}$—where $\psi_{\mathbf{k}} = e^{i\mathbf{k}\cdot\mathbf{r}}u_{\mathbf{k}}$—be periodic, that is, $u_{\mathbf{k}}$ has the same points on opposite faces of the cell, e.g., points $P_1$ and $P_2$ in Fig. 5.18(a).

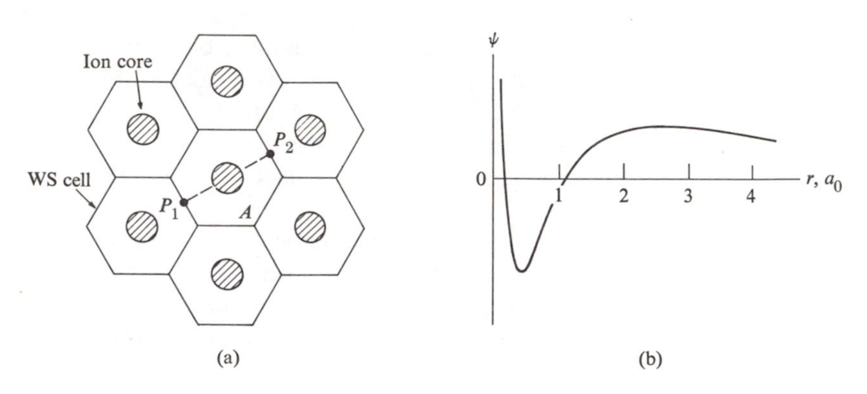


**Fig. 5.18** (a) The WS cell. (b) The wave function $\psi_0$ at the bottom of the 3s band in Na versus the radial distance, in units of the Bohr radius.

The procedure is now clear in principle: We attempt to solve (5.50) in a single cell, using for $V(\mathbf{r})$ the potential of a *free* ion, which can be found from atomic physics. In Na, for instance, $V(\mathbf{r})$ is the potential of the ion core $Na^+$. It is still very difficult, however, to impose the requirements of periodicity on the function for the actual shape of the cell (the truncated octahedron), and to overcome this hurdle Wigner and Seitz replaced the cell by a WS *sphere* of the same volume as the actual cell, i.e., one employs a *WS sphere*. Using these simplifying assumptions concerning the potential and the periodic conditions, one then solves the Schrödinger equation numerically, since an analytical solution cannot usually be found. The resulting wave function $\psi_0$ at the bottom of the band, $k = 0$, is shown in Fig. 5.18(b). The wave functions at other values of $\mathbf{k}$ near the bottom of the band may then be approximated by

$$\psi_{\mathbf{k}} \simeq \frac{1}{V^{1/2}} e^{i\mathbf{k}\cdot\mathbf{r}} \psi_0, \tag{5.51}$$

which has the Bloch form.

The procedure is also capable of yielding the energy $E(\mathbf{k})$. The energy $E_0$ of the bottom of the band is obtained from the same calculations which give $\psi_0$, and the energy at any other point $\mathbf{k}$ is obtained by using

$$E(\mathbf{k}) = \left\langle \psi_{\mathbf{k}} \left| -\frac{\hbar^2}{2m_0} \nabla^2 + V(\mathbf{r}) \right| \psi_{\mathbf{k}} \right\rangle, \tag{5.52}$$

where the wave function $\psi_k$ is substituted from (5.51). The energy found in this manner was used by Wigner and Seitz to evaluate the cohesive energy, and the results are in satisfactory agreement with experiment.

One noteworthy feature of these results is the shape of the wave function in Fig. 5.18(b). The wave function oscillates at the ion core, but once outside the core the function is essentially a constant. This constancy of the wave function holds true for almost 90% of the cell volume. Thus the wave function behaves like a plane wave, as seen from (5.51), over most of the cell, and hence over most of the crystal. Looking at this in terms of the potential, we see that where the function is a plane wave, the potential must be a constant. Thus the *effective* potential acting on the electron is essentially a constant, except in the region at the ion core itself. Viewing the motion of the electron in the crystal as a whole, we conclude that the electron moves in a region of constant potential throughout most of the crystal; only at the cores themselves does the electron experience any appreciable potential. This surprising result explains why the conduction electrons in Na, for example, may be regarded as essentially free electrons. Mathematically, it is a consequence of the periodic conditions imposed on the wave function in the cell, and this is particularly apparent when one realizes that the wave function for the 3s electron in a free Na atom is very unlike $\psi_0$ outside the ion core. The flatness of $\psi_0$ is thus due to the imposition of the periodic conditions, and not to any special property of the ionic potential.† The effect of the periodic condition is to cancel out the ionic potential outside the core, and thus render the potential a constant. We shall find this result very useful in the development of other methods of band calculation.

Despite its usefulness, the cellular method is greatly oversimplified, and is not currently much in use. One of its chief disadvantages is that when one replaces the WS cell by a sphere, one ignores the crystal structure entirely. All anisotropic effects, for instance, are completely masked out.

### The augmented-plane wave (APW) method

The APW method (Slater, 1937) uses the results of the cellular method, but is so formulated as to avoid its shortcomings. Since the effective crystal potential was found to be constant in most of the open spaces between the cores, the APW method begins by assuming such a potential (Fig. 5.19), which is referred to as the

† The boundary conditions require that the derivative of the function $\psi_0$ vanish at the surface of the WS sphere (why?). Thus the function is flat near the surface of this sphere, as shown in Fig. 5.18(b).

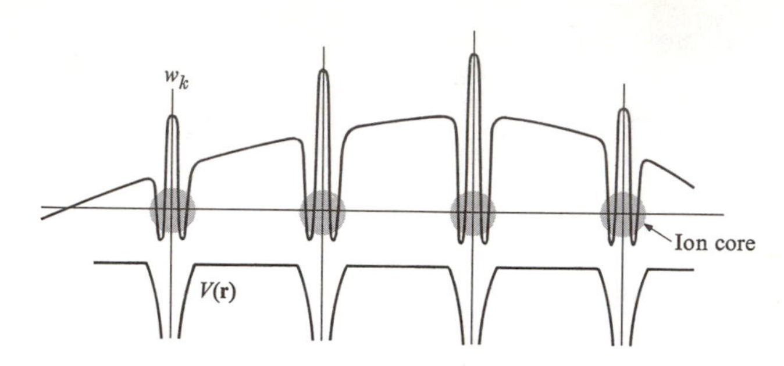


**Fig. 5.19** The potential and wave function in the APW method.

*muffin-tin potential*. The potential is that of a free ion at the core, and is strictly constant outside the core. The wave function for the wave vector **k** is now taken to be

$$w_{\mathbf{k}} = \begin{cases} \dfrac{1}{V^{1/2}} e^{i\mathbf{k}\cdot\mathbf{r}}, & r > r_s, \\ \text{atomic function}, & r < r_s, \end{cases} \tag{5.53}$$

where $r_s$ is the core radius. Outside the core the function is a plane wave because the potential is constant there. Inside the core the function is atom-like, and is found by solving the appropriate free-atom Schrödinger equation. Also, the atomic function in (5.53) is chosen such that it joins continuously to the plane wave at the surface of the sphere forming the core; this is the boundary condition here.

The function $w_{\mathbf{k}}$ does not have the Bloch form, but this can be remedied by forming the linear combination

$$\psi_{\mathbf{k}} = \sum_{\mathbf{G}} a_{\mathbf{k}+\mathbf{G}}\, w_{\mathbf{k}+\mathbf{G}}, \tag{5.54}$$

where the sum is over the reciprocal lattice vectors, which has the proper form. The coefficients $a_{\mathbf{k}+\mathbf{G}}$ are determined by requiring that $\psi_{\mathbf{k}}$ minimize the energy.† In practice the series in (5.54) converges quite rapidly, and only four or five terms—or even less—suffice to give the desired accuracy.

The APW method is a sound one for calculating the band structure in metals, and has been used a great deal in the past few years. It incorporates the essential features of the problem in a straightforward and natural fashion.

### The pseudopotential method

Yet another method popular among solid-state physicists for calculating band structure in solids is the pseudopotential method, which is distinguished by the manner

† The "best" linear combination (5.54) is that which makes the energy as low as possible.

in which the wave function is chosen. We seek a function which oscillates rapidly inside the core, but runs smoothly as a plane wave in the remainder of the open space of the WS cell. Such a function was chosen in the APW method according to (5.53), but this is not the only choice possible. Suppose we take

$$w_{\mathbf{k}} = \phi_{\mathbf{k}} - \sum_i a_i v_i \tag{5.55}$$

where $\phi_{\mathbf{k}}$ is a plane wave and $v_i$ an atomic function. The sum over $i$ extends over all the atomic shells which are occupied. For example, in Na, the sum extends over the 1s, 2s, and 2p shells. The coefficients $a_i$ are chosen such that the function $w_{\mathbf{k}}$, representing a 3s electron, is orthogonal to the core function $v_i$.† By requiring this orthogonality, we ensure that the 3s electron, when at the core, does not occupy the other atomic orbitals already occupied. Thus we avoid violating the Pauli exclusion principle.

The function $w_{\mathbf{k}}$ has the features we are seeking: Away from the core, the atomic functions $v_i$ are negligible, and thus $w_{\mathbf{k}} \simeq \phi_{\mathbf{k}}$, a plane wave. At the core, the atomic functions are appreciable, and act so as to induce rapid oscillations, as shown in Fig. 5.20.

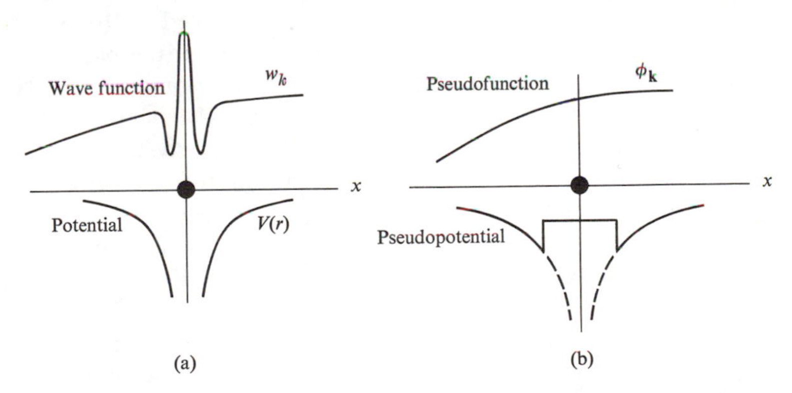


**Fig. 5.20** The pseudopotential concept. (a) The actual potential and the corresponding wave function, as seen by the electron. (b) The corresponding pseudopotential and pseudofunction.

If one now substitutes $w_{\mathbf{k}}$ into the Schrödinger equation

$$\left[-\frac{\hbar^2}{2m_0}\nabla^2 + V\right] w_{\mathbf{k}} = E(\mathbf{k}) w_{\mathbf{k}}, \tag{5.56}$$

† Two functions $\psi_1$ and $\psi_2$ are said to be *orthogonal* if the integral $\int \psi_1^* \psi_2 \, d^3r = 0$. This concept of orthogonality is very useful in quantum mechanics. The atomic functions in the various atomic shells are all mutually orthogonal.

and rearranges the terms, one finds that the equation may be written in the form

$$\left[-\frac{\hbar^2}{2m}\nabla^2 + V'\right]\phi_{\mathbf{k}} = E(\mathbf{k})\,\phi_{\mathbf{k}}, \tag{5.57}$$

where

$$V' = V - \sum_i b_i \langle v_i | V | v_i \rangle. \tag{5.58}$$

These results are very interesting: Equation (5.57) shows that the effective potential is given by $V$, while (5.58) shows that $V'$ is weaker than $V$, because the second term on the right of (5.58) tends to cancel the first term. This cancellation of the crystal potential by the atomic functions is usually appreciable, often leading to a very weak potential $V'$. This is known as the *pseudopotential*. Since $V'$ is so weak, the wave function as seen from (5.57) is almost a plane wave, given by $\phi_{\mathbf{k}}$, and is called the *pseudofunction*.

The pseudopotential and pseudofunction are illustrated graphically in Fig. 5.20(b). Note that the potential is quite weak, and, in particular, the singularity at the ion core is entirely removed. Correspondingly, the rapid "wiggles" in the wave function have been erased, so that there is a smooth plane-wave-like function.

Now we can understand one point which has troubled us for some time: why the electrons in Na, for instance, seem to behave as free particles despite the fact that the crystal potential is very strong at the ionic cores. Now we see that, when the exclusion principle is properly taken into account, the effective potential is indeed quite weak. The free-particle behavior, long taken to be an empirical fact, is now borne out by quantum-mechanical calculations. The explanation of this basic paradox is one of the major achievements of the pseudopotential method. This method has also been used to calculate band structure in many metals and semiconductors (Be, Na, K, Ge, Si, etc.) with considerable success.

The APW and pseudopotential methods, as well as other related systems, require much numerical work which can feasibly be carried out only by modern electronic computers. It often takes a whole year or more to develop the necessary program and perform the calculations for one substance on a large computer!

## 5.10 METALS, INSULATORS, AND SEMICONDUCTORS

Solids are divided into two major classes: *Metals* and *insulators*. A metal—or conductor—is a solid in which an electric current flows under the application of an electric field. By contrast, application of an electric field produces no current in an insulator. There is a simple criterion for distinguishing between the two classes on the basis of the energy-band theory. This criterion rests on the following statement: *A band which is completely full carries no electric current, even in the presence of an electric field.* It follows therefore that a solid behaves as a metal only when some of the bands are partially occupied. The proof of this statement will be supplied later (Section 5.13), but we shall accept it for the time being as an established fact.

Let us now apply this statement to Na, for example. Since the inner bands 1s, 2s, 2p are all fully occupied, they do not contribute to the current. We may therefore concern ourselves only with the topmost occupied band, the *valence band.* In Na, this is the 3s band. As we saw in Section 5.5, this band can accommodate $2N_c$ electrons, where $N_c$ is the total number of unit cells. Now in Na, a Bravais bcc lattice, each cell has one atom, which contributes one valence (or 3s) electron. Therefore the total number of valence electrons is $N_c$, and as these electrons occupy the band, only half of it is filled, as shown in Fig. 5.21(a). Thus sodium behaves like a metal because its valence band is only partially filled.

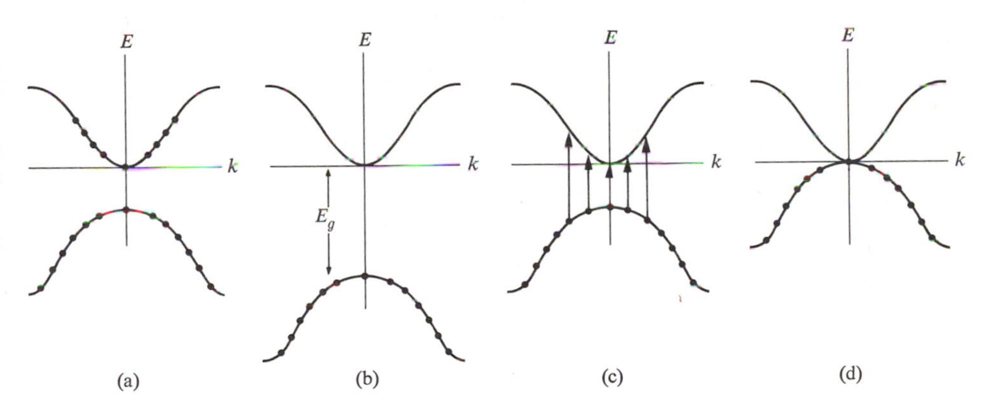


**Fig. 5.21** The distribution of electrons in the bands of (a) a metal, (b) an insulator, (c) a semiconductor, and (d) a semimetal.

In a similar fashion, we conclude that the other alkalis, Li, K, etc., are also metals because their valence bands—the 2s, 4s, etc., respectively—are only partially full. The noble metals, Cu, Ag, Au, are likewise conductors for the same reason. Thus in Cu the valence band (the 4s band) is only half full, because each cell in its fcc structure contributes only one valence electron.

As an example of a good insulator, we mention diamond (carbon). Here the top band originates from a hybridization of the 2s and 2p atomic states (Section A.10), which gives rise to two bands split by an energy gap (Fig. 5.21b.) Since these bands arise from s and p states, and since the unit cell here contains two atoms, each of these bands can accommodate $8N_c$ electrons. Now in diamond each atom contributes 4 electrons, resulting in 8 valence electrons per cell. Thus the valence band here is completely full, and the substance is an insulator, as stated above.†

There are substances which fall in an intermediate position between metals and insulators. If the gap between the valence band and the band immediately above it is small, then electrons are readily excitable thermally from the former to the latter band. Both bands become only partially filled and both contribute to the electric condition. Such a substance is known as a *semiconductor*. Examples are Si and Ge, in which the gaps are about 1 and 0.7 eV, respectively. By contrast, the gap in diamond is about 7 eV. Roughly speaking, a substance behaves as a semiconductor at room temperature whenever the gap is less than 2 eV.

The conductivity of a typical semiconductor is very small compared to that of a metal, but it is still many orders of magnitude larger than that of an insulator. It is justifiable, therefore, to classify semiconductors as a new class of substance, although they are, strictly speaking, insulators at very low temperatures.

In some substances the gap vanishes entirely, or the two bands even overlap slightly, and we speak of *semimetals* (Fig. 5.21d). The best-known example is Bi, but other such substances are As, Sb, and white Sn.

An interesting problem is presented in this connection by the divalent elements, for example, Be, Mg, Zn, etc. For instance, Be crystallizes in the hcp structure, with one atom per cell. Since there are two valence electrons per cell, the 2s band should completely fill up, resulting in an insulator. In fact, however, Be is a metal—although a poor one, in that its conductivity is small. The reason for the apparent paradox is that the 2s and 2p bands in Be overlap somewhat, so that electrons are transferred from the former to the latter, resulting in incompletely filled bands, and hence a metal. The same condition accounts for the metallicity of Mg, Ca, Zn, and other divalent metals.

A substance in which the number of valence electrons per unit cell is *odd* is necessarily a metal, since it takes an even number of electrons to fill a band completely. But when the number is even, the substance may be either an insulator or a metal, depending on whether the bands are disparate or overlapping.

---

† The case of hydrogen is of special interest. Although it is gaseous at atmospheric pressure, hydrogen solidifies at high pressure. But the familiar solid hydrogen is an insulator, having two atoms per unit cell, which causes the complete filling of the ls band. Theory predicts, however, that at very high pressure ($\simeq 2$ megabars), solid hydrogen undergoes a crystal structure transformation and a concomitant change to a metallic state. Many experimenters are currently attempting to observe this transformation, and tentative successes have been reported, but definitive results are still lacking at the time of writing. Even diamond has been reported to undergo transition to the metallic state at high pressure ($\simeq 1.5$ megabars). Simultaneously a structural phase transformation to a body-centered tetragonal structure occurs. The decrease in the lattice constant caused by the pressure is about 17%.

## 5.11 DENSITY OF STATES

The *density of states* for electrons in a band yields the number of states in a certain energy range. This function is important in electronic processes, particularly in transport phenomena. When we denote the density-of-states function by $g(E)$, it is defined by the relation

$$g(E)\,dE = \text{number of electron states per unit volume in the energy range } (E, E + dE). \tag{5.59}$$

This definition of $g(E)$ is analogous to that of the phonon density of states $g(\omega)$, so our discussion here parallels that presented in connection with $g(\omega)$. (See Sections 3.3 and 3.7; particularly 3.7.) To evaluate $g(E)$ one applies the definition (5.59): One draws a shell in **k**-space whose inner and outer surfaces are determined by the energy contours $E(\mathbf{k}) = E$ and $E(\mathbf{k}) = E + dE$, respectively, as shown in Fig. 5.22. The number of allowed **k** values lying inside this shell then gives the number of states which, when divided by the thickness of the shell $dE$, yields the desired function $g(E)$.

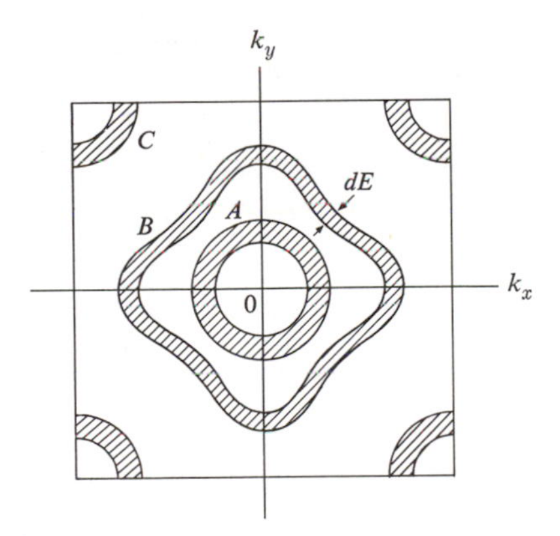


**Fig. 5.22** Concentric shells in **k**-space used to evaluate the density of states $g(E)$.

It is evident that $g(E)$ is intimately related to the shape of the energy contours, and hence the band structure. The complexities of this structure are reflected in the form taken by $g(E)$. Let us first evaluate $g(E)$ for the case in which the dispersion relation for electron energy has the standard form

$$E = \frac{\hbar^2 k^2}{2m^*}. \tag{5.60}$$

As we have seen earlier, such a dispersion relation often holds true for those states lying close to the bottom of the band near the origin of the Brillouin zone. The energy contours corresponding to (5.60) are clearly concentric spheres surrounding the origin. The resulting density-of-states shell is then spherical in shape, as illustrated by shell $A$ in Fig. 5.22, and since this is spherical, its volume is given by $4\pi k^2\, dk$, where $k$ is the radius and $dk$ the thickness of the shell. Recalling from Section 3.3 that the number of allowed **k** values per unit volume of **k**-space is $1/(2\pi)^3$, it follows that the number of states lying in the shell—i.e., in the energy range $(E, E + dE)$—is

$$\text{Number of states} = \frac{1}{(2\pi)^3}\, 4\pi k^2\, dk. \tag{5.61}$$

We may convert the right side by writing it in terms of $E$, the energy, rather than in terms of $k$, by using (5.60). We then find that

$$\text{Number of states} = \frac{1}{4\pi^2}\left(\frac{2m^*}{\hbar^2}\right)^{3/2} E^{1/2} dE.$$

Comparing this result with the definition (5.59), we infer that

$$g(E) = \frac{1}{4\pi^2}\left(\frac{2m^*}{\hbar}\right)^{3/2} E^{1/2}. \tag{5.62}$$

In order to take into account the spin degeneracy—i.e., the fact that each **k** state may accommodate two electrons of opposite spins—we multiply this expression by 2, which yields

$$g(E) = \frac{1}{2\pi^2}\left(\frac{2m^*}{\hbar^2}\right)^{3/2} E^{1/2}. \tag{5.63}$$

This shows that $g(E) \sim E^{1/2}$, which means that the curve $g(E)$ has a parabolic shape (Fig. 5.23). The function $g(E)$ increases with $E$ because, as we see from Fig. 5.22, the larger the energy the greater the radius, and hence the volume of the shell,

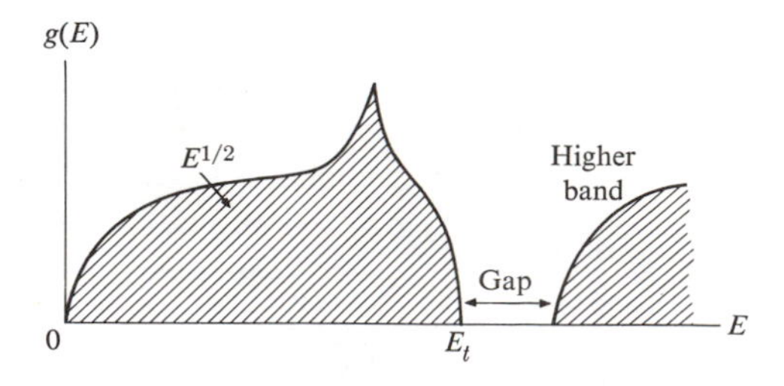


**Fig. 5.23** The density of states.

and consequently the larger the number of states lying within it. Also note that $g(E) \sim m^{*3/2}$. That is, the larger the mass the greater the density of states.

The result (5.63) is very useful, and will be used repeatedly in subsequent discussions, but note that its validity is restricted to that region in **k**-space in which the standard dispersion relation (5.60) is satisfied. As the energy increases, a point is reached at which the energy contours become nonspherical—e.g., shell $B$ in Fig. 5.22, in which region Eq. (5.63) no longer holds. One must then resort to a more complicated formula to evaluate $g(E)$. As a result, the shape of $g(E)$ is no longer parabolic at large energy, as shown in Fig. 5.23, the actual shape being determined by the dispersion relation $E = E(\mathbf{k})$ of the band. Note also that, at sufficiently large energies, the shell begins to intersect the boundaries of the zone, e.g., shell $C$ in Fig. 5.22, in which case the volume of the shell begins to shrink, with a concomitant decrease in the number of states. The density of states of the shell plummets, and continues to decrease as the energy increases, until it vanishes completely when the shell lies entirely outside the zone, as shown in Fig. 5.23. The energy at which $g(E)$ vanishes marks the top of the valence band. The density of states remains zero for a certain energy range beyond that, this range marking the energy gap, until a new energy band appears, with its own density of states.

In simple metals, such as alkalis and noble metals, the standard form (5.60) holds true for most of the zone until the energy contours come close to the boundaries of the zone. It follows therefore that for these substances the expression (5.63) applies throughout most of the energy band, except close to the top of the band.

It is sometimes useful to have an expression for the density of states in the energy range lying close to the top of the band. This can be derived readily if the band there can be represented by a negative effective mass, as is usually the case (see Section 3.6). We may then show, by following a procedure analogous to that in deriving (5.63), that

$$g(E) = \frac{1}{2\pi^2}\left(\frac{2\,|m^*|}{\hbar^2}\right)^{3/2}(E_t - E)^{1/2}, \tag{5.64}$$

where $E_t$ is the top of the valence band (note that here $E < E_t$). Thus the density function $g(E)$ has an inverted parabolic shape, where the parabola is at the top of the band. (See Fig. 5.23.).

Figure 5.24 illustrates situations in which bands overlap each other. Figure 5.24(a) represents a circumstance typical of divalent metals, in which the top of a band is at higher energy than the bottom of the next-higher band. Figure 5.24(b) shows the overlap of the 4s and 3d bands in transition metals. The 3d band, narrow and high, lies in the midst of the wide and flat 4s band.

According to definition, the quantity $g(E)\,dE$ gives the number of states lying in the energy range $(E, E + dE)$. The number of electrons actually occupying this

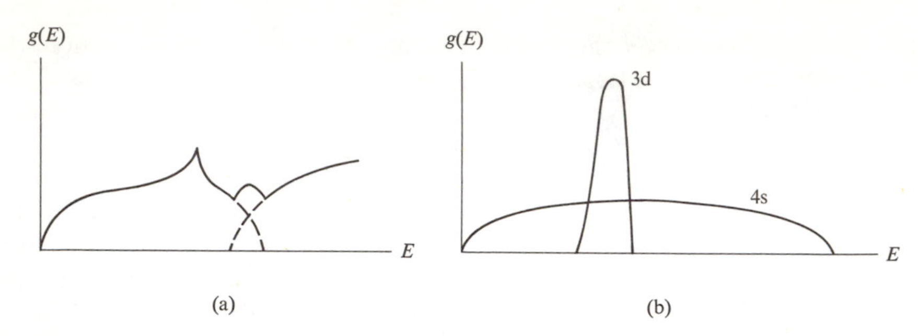


**Fig. 5.24** (a) The shape of the density of states when two bands overlap each other as, e.g., in divalent metals. (b) The overlap of the 3d and 4s bands in transition metals.

range of energy is then given by

$$dn(E) = f(E)\,g(E)\,dE, \tag{5.65}$$

where $f(E)$ is the Fermi–Dirac distribution function, $f(E) = (1 + e^{(E-E_F)/k_BT})^{-1}$, discussed in Section 4.6. Expression (5.65) follows from the fact that since $g(E)\,dE$ gives the number of available states, and $f(E)$ the probability that each of these is occupied by an electron, then the product $f(E)\,g(E)\,dE$ must give the number of electrons present in that energy range.

## 5.12 THE FERMI SURFACE

In Section 4.7 we discussed the *Fermi surface* (FS) in connection with the free-electron model. There we saw that the significance of this surface in solid-state physics derives from the fact that only those electrons lying near it participate in thermal excitations or transport processes. Here we shall consider the Fermi surface again, and now we shall incorporate the effects of the crystal potential. The significance of the FS remains unchanged, but its shape, in some cases, may be considerably more complicated than the spherical shape of the free-electron model. We shall now consider the effects of the crystal potential on the shape of the FS, while in later sections we shall see how this change may influence the physical properties of the crystal. Experimental determination of the FS will be considered in Section 5.19.

As we recall, the FS is defined as the surface in **k**-space inside which all the states are occupied by valence electrons.† All the states lying outside the surface are

† In Section 4.7 we discussed the FS in velocity space. However, for a free-like electron, the velocity is given by $\mathbf{v} = \hbar\mathbf{k}/m^*$. Thus **v** and **k** are proportional to each other, and one could equally well speak of the FS in **k**-space, provided an appropriate change in scale were made.

empty. The definition is strictly valid only at absolute zero, $T = 0°K$, but, as we saw in Section 4.6, the effect of temperature on the FS is very slight, and the surface remains sharp even at room temperature or higher. The shape of the FS is determined by the geometry of the energy contours in the band, since the FS is itself an energy contour, where $E(\mathbf{k}) = E_F$, $E_F$ being the Fermi energy. (Because of this, the FS should display the same rotational symmetry as the lattice.)

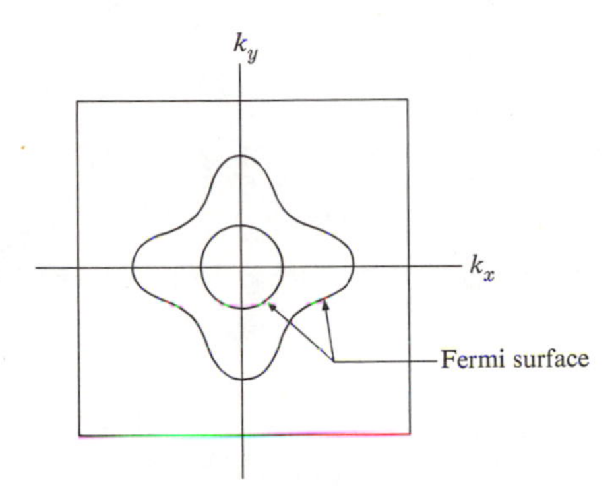


**Fig. 5.25** The evolution of the shape of the FS as the concentration of valence or conduction electrons increases.

Figure 5.25 illustrates the evolution of the shape of the FS as the concentration of valence electrons increases. For small $n$, only those states lying near the bottom of the band at the center of the zone are populated, and the occupied volume is a sphere in **k**-space, which is therefore bounded by a spherical FS. As $n$ increases and more states are populated, the "Fermi volume" expands, and so does the FS. This surface, which is spherical near the origin, begins to deform gradually as $n$ increases, following the distortion in the contours at large energies (as discussed previously) as seen in Fig. 5.25. The distortion in the shape of the FS may become quite pronounced, particularly as the FS approaches the boundaries of the zone. The distortion is even greater when the surface intersects the boundaries, as will be discussed later in this section.

The alkali metals Li, Na, and K crystallize in the bcc structure, whose Brillouin zone is a truncated octahedron (Fig. 5.9a). As we saw in Section 5.6, the valence band is half filled. The FS is still far from the boundaries, and since the standard dispersion relation holds well throughout most of the zone, it follows that the FS in these substances is essentially spherical in shape. Experiments confirm this, showing that in Na and K the distortion of the FS from sphericity is of the order of $10^{-3}$.

The noble metals Cu, Ag, and Au crystallize in the fcc structure. The shape of the BZ here is that of a truncated octahedron (Fig. 5.26). Here again the valence band is only half-filled, and consequently the FS, being far from the zone

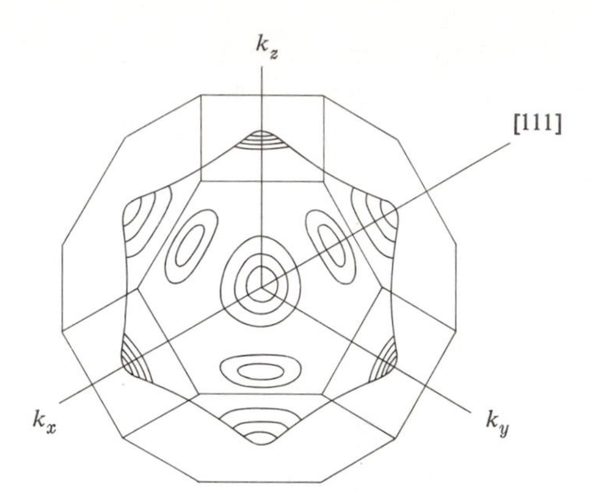


**Fig. 5.26** The FS in noble metals. The surface protrudes toward the zone faces in the [111] directions.

boundaries, should be essentially spherical, which is substantially true for most of the FS. However, along the ⟨111⟩ directions, the FS comes close to the zone boundaries, because of the shape of the zone, and as a result the surface suffers strong distortion in that region. As seen in Fig. 5.26, the FS protrudes along the ⟨111⟩ directions so much as to touch the zone face. In effect the zone boundaries have "pulled" the FS, giving it the shape shown in the figure—a sphere with eight "necks" protruding in the ⟨111⟩ directions. In this respect the FS in the noble metals is quite different from that in the alkali metals.

The position of the Fermi level $E_F$ for various classes of solids is illustrated in Fig. 5.27. Figure 5.27(a) illustrates the density of states and the position of $E_F$

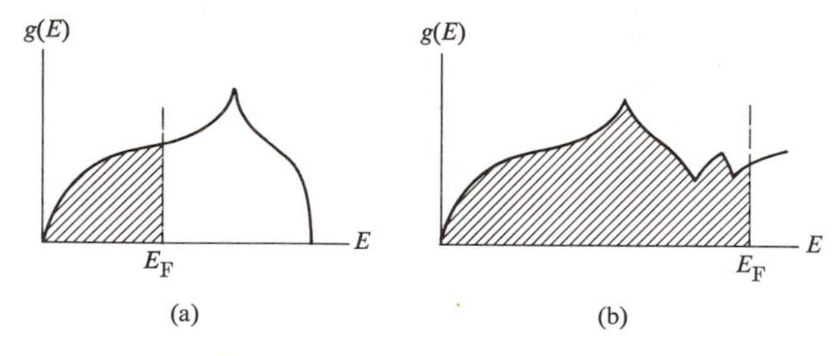


**Fig. 5.27** The position of the Fermi energy in (a) a monovalent metal, and (b) a divalent metal.

for a typical monovalent metal, where only half the band is filled, and the substance acts as a conductor. Figure 5.27(b) shows a divalent metal. Here the bands overlap to some extent, and the number of valence electrons is so large that the FS spills over into the higher band. Figure 5.28 shows an insulator, in which the

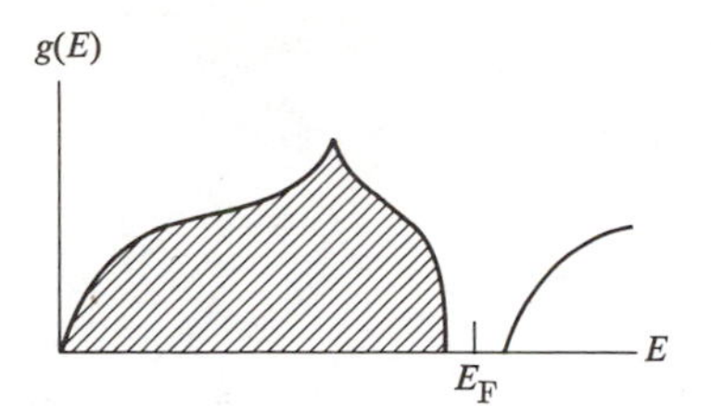


**Fig. 5.28** The position of the Fermi energy in an insulator.

valence band is completely filled and the Fermi level lies somewhere in the energy gap.

We shall now determine the Fermi energy $E_F$ for the case in which the standard form (5.16) holds. As seen above, this applies to the alkali metals and, to a lesser extent, to the noble metals as well. By its very definition, the Fermi energy satisfies the relation (at $T = 0°K$)

$$\int_0^{E_F} g(E)\, dE = n. \tag{5.66}$$

because the integral on the left gives the number of states from the bottom of the band, $E = 0$, right up to the Fermi level. This number must be equal to the number of electrons, which is the meaning of (5.66). If we substitute for $g(E)$ from (5.63), perform the necessary integration (which can be readily accomplished), and solve for $E_F$, we find that

$$E_F = \frac{\hbar^2}{2m^*}(3\pi^2 n)^{2/3}, \tag{5.67}$$

which is the result quoted previously in the case of the free-electron model (Section 4.7). Refer to Table 4.1 for a list of Fermi levels, and note that $E_F$ is typically of the order of a few electron volts.

Let us now turn to the FS in polyvalent metals. Suppose that the number of valence electrons is sufficiently large so that the FS intersects the boundaries of the zone, as shown in Fig. 5.29(a). In constructing the FS here, we used the empty-lattice model, so the crystal potential is set equal to zero. The FS is now seen to extend over two zones. The part of the FS lying in the first zone is repeated in Fig. 5.29(b). Note that it is composed of the four sides of a diamond-shaped figure. Figure 5.29(c) replots the part of the FS lying in the second zone using the reduced-zone scheme. We see that it is composed of the sides of four half-bubble-shaped figures. When viewed in the various individual zones, the shape of the FS appears quite complicated, even for a free electron, belying its original simplicity. Of course, if one uses the extended-zone scheme, the original spherical shape of Fig. 5.29(a) may be reconstructed, but this is not immediately apparent from Fig. 5.29(b) or (c) individually. If we now turn on a weak crystal potential, the

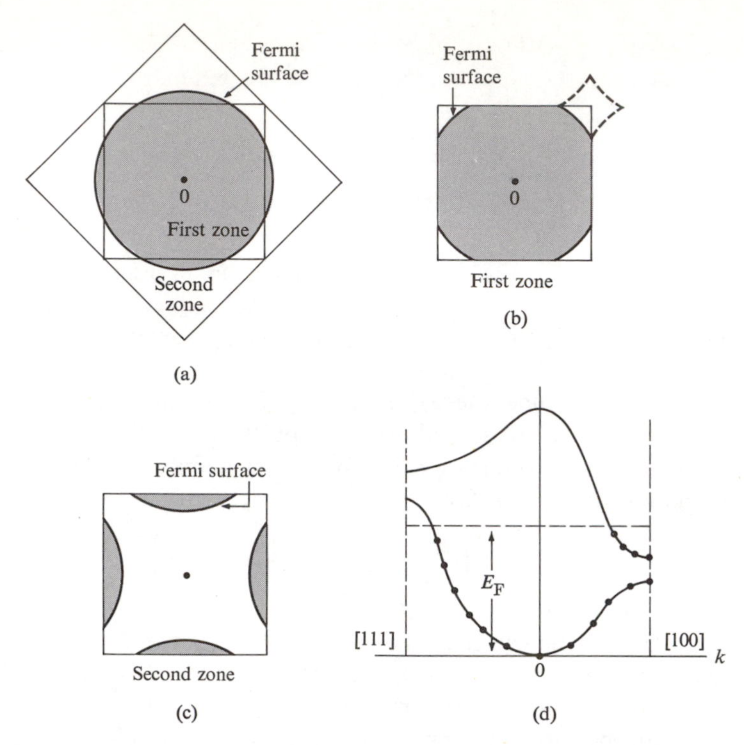


**Fig. 5.29** The Harrison construction. (a) The FS in the empty-lattice model using the extended-zone scheme. (b) The FS in the first zone. (c) The FS in the second zone. (d) Band overlap.

shape of the FS in the two zones is affected only slightly, the effect being primarily to round off the sharp corners. The point here is that the complicated FS's usually observed in polyvalent metals are not necessarily the result of strong crystal potentials (as was once thought to be the case). They may be due largely to the crossing of the zone and the piecing together of the various parts of the FS. (The procedure for reconstructing Fermi surfaces on the basis of the empty-lattice model is known as the *Harrison construction.*)

Figure 5.29(d) shows the energy bands in the two zones plotted in two different directions. The two bands overlap. The top of the first band along the [111] direction is higher than the bottom of the second band in the [100] direction. The Fermi level crosses both bands, and both contribute to the conduction process.

It is important to note here that the Fermi level crosses the lower band (on the left in Fig. 5.29d) in a region in which the curvature of the band is downward, i.e., a region of negative effective mass. As we shall see in Section 5.17, such a situation is best described in terms of *holes*.

Figure 5.29(d) illustrates what is known as the *two-band model* for a metal.

The electric current is transported by carriers in two bands: Electrons in the higher band, holes in the lower. We shall exploit this model to full advantage in Section 5.18.

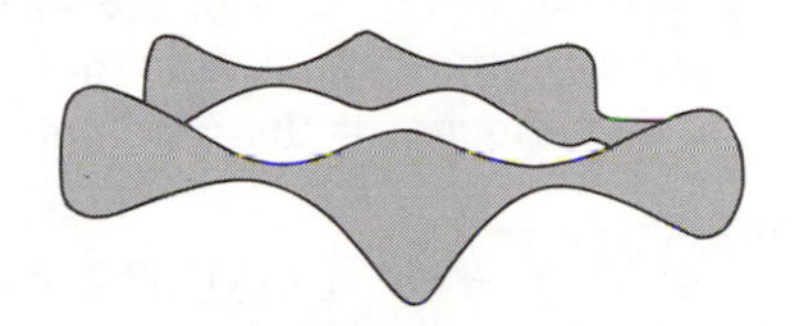

**Fig. 5.30** The Fermi surface of beryllium.

Finally, Fig. 5.30 shows the FS for Be (known also as the Be *coronet*). Complicated as this appears to be, the surface is quite similar to the shape obtained using the Harrison construction. Note the hexagonal symmetry, expected as a consequence of the hexagonal crystal structure of Be.

## 5.13 VELOCITY OF THE BLOCH ELECTRON

Now let us study the motion of the Bloch electrons in solids. An electron in a state $\psi_{\mathbf{k}}$ moves through the crystal with a velocity directly related to the energy of that state. Consider first the case of a free particle. The velocity is given by $\mathbf{v} = \mathbf{p}/m_0$, where $\mathbf{p}$ is the momentum. Since $\mathbf{p} = \hbar\mathbf{k}$, it follows that, for a free electron, the velocity is given by

$$\mathbf{v} = \frac{\hbar \mathbf{k}}{m_0}, \tag{5.68}$$

i.e., the velocity is proportional to and parallel to the wave vector $\mathbf{k}$, as shown in Fig. 5.31(a).

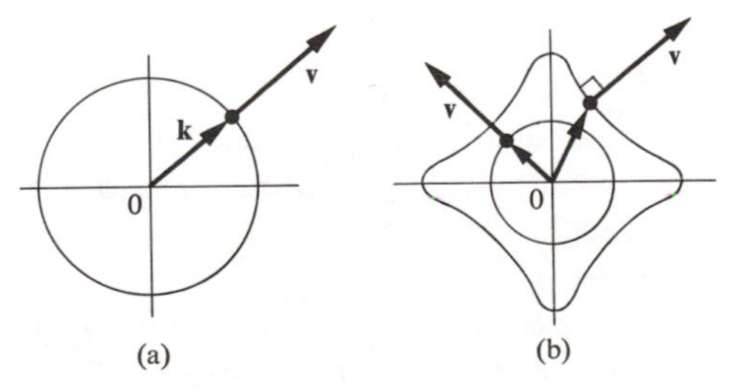


**Fig. 5.31** The velocity of (a) a free electron, and (b) a Bloch electron.

For a Bloch electron, the velocity is also a function of $\mathbf{k}$, but the functional relationship is not as simple as (5.68). To derive this relationship, we use a well-known formula in wave propagation. That is, the group velocity of a wave packet is given by

$$\mathbf{v} = \nabla_{\mathbf{k}}\, \omega(\mathbf{k}), \tag{5.69}$$

where $\omega$ is the frequency and **k** the wave vector of the wave packet. Applying this equation to the electron wave in the crystal, and noting the Einstein relation $\omega = E/\hbar$, we may write for the velocity of the Bloch electron

$$\mathbf{v} = \frac{1}{\hbar}\, \nabla_{\mathbf{k}}\, E(\mathbf{k}), \tag{5.70}$$

which states that the velocity of an electron in state **k** is proportional to the gradient of the energy in **k**-space. [Equation (5.70) can also be derived more rigorously by writing the quantum expression for the velocity of the probability wave associated with the Bloch electron and finding the quantum expectation value; see Mott (1936).] We assume implicitly that we are dealing here with the valence band, and hence the band index has been suppressed, although it should be clear from the derivation that (5.70) is valid in any band.

Since the gradient vector is perpendicular to the contour lines, a fact well known from vector analysis, it follows that the velocity **v** at every point in **k**-space is normal to the energy contour passing through that point, as shown in Fig. 5.31(b). Because these contours are in general nonspherical, it follows that the velocity is not necessarily parallel to the wave vector **k**, unlike the situation of a free particle.

Note, however, that near the center of the zone, where the standard dispersion relation $E = \hbar^2 k^2/2m^*$ is expected to hold true, the relation (5.70) leads to

$$\mathbf{v} = \frac{\hbar \mathbf{k}}{m^*}, \tag{5.71}$$

which is of the same form as the relation for a free particle, (5.68), except that $m_0$ has been replaced by $m^*$, the effective mass. This is to be expected, of course, since we have often stated that a Bloch electron behaves in many respects like a free electron, except for the difference in mass. It follows that near the center of the zone **v** is parallel to **k**, and points radially outward, as shown in Fig. 5.31(b). It is near the zone boundaries at which the energy contours are so distorted that this simple relationship between **v** and **k** is destroyed, and so one must resort to the more general result (5.70).

Note also that when an electron is in a certain state $\psi_{\mathbf{k}}$, it remains in that state forever, provided only that the lattice remains periodic. Thus as long as this situation persists, the electron will continue to move through the crystal with the same velocity **v**, unhampered by any scattering from the lattice.† In other words,

† See the remarks about the propagation of waves in periodic lattices (Section 4.5).

the velocity of the electron is a constant. Any effect the lattice may exert on the propagation velocity has already been included in (5.70) through the energy $E(\mathbf{k})$.

Deviations in the periodicity of the lattice would, of course, cause a scattering of the electron, and hence a change in its velocity. For example, an electron moving in a vibrating lattice suffers numerous collisions with phonons, resulting in a profound influence being exerted on the velocity. Also, external fields—electric or magnetic—lead to change in the velocity of the electron. We shall discuss these effects in the following sections.

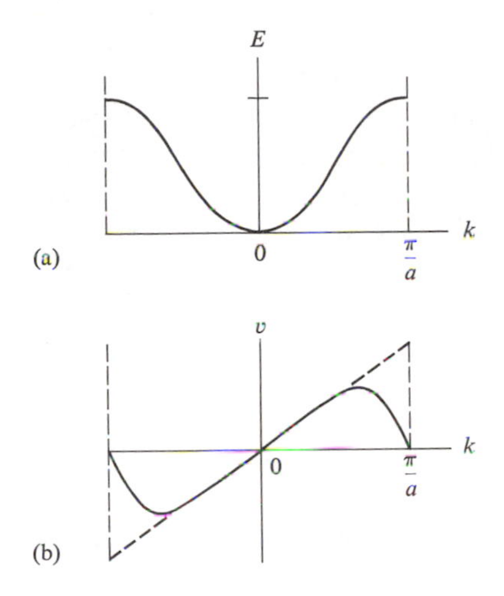


**Fig. 5.32** (a) The band structure, and (b) the corresponding electron velocity in a one-dimensional lattice. The dashed line in (b) represents the free-electron velocity.

Figure 5.32(a) shows a typical one-dimensional band structure, and Fig. 5.32(b) shows the corresponding velocity, which in this case reduces to

$$v = \frac{1}{\hbar}\frac{\partial E}{\partial k}, \tag{5.72}$$

that is, the velocity is proportional to the slope of the energy curve. We see that as $k$ varies from the origin to the edge of the zone, the velocity increases at first linearly, reaches a maximum, and then decreases to zero at the edge of the zone. We wish now to explain this behavior on the basis of the NFE model, particularly the seemingly anomalous decreases in the velocity near the edge of the zone. The following discussion is closely related to the discussion in Section 5.7.

Near the zone center, the electron may be adequately represented by a single plane wave $\psi_k \sim e^{ikx}$, and hence $\mathbf{v} = \hbar k/m_0$, explaining the linear region of Fig. 5.32(b). However, as $\mathbf{k}$ increases, the scattering of the free wave by the lattice introduces a new left-traveling wave whose wave vector $k' = k - 2\pi/a$, and which is to be superimposed on the original right-traveling wave $k$. Therefore the electron is now represented by the wave mixture

$$\psi_k \simeq e^{ikx} + be^{-i(2\pi/a-k)x}, \tag{5.73}$$

where the coefficient $b$ is found from perturbation theory (Eq. 5.24). The velocity of this wave, according to quantum mechanics, is given by

$$v = \frac{\hbar k}{m_0} - |b|^2 \frac{\hbar}{m_0}\left(\frac{2\pi}{a} - k\right), \tag{5.74}$$

where the first term on the right is the contribution of the right-traveling wave, while the second term is the contribution of the left-traveling wave. At small $k$, the coefficient $b$ is small, and $v$ is given essentially by $\hbar k/m_0$, as stated above. As $k$ increases, however, the coefficient of the scattered wave increases, and so the second term in (5.74) becomes appreciable. Since the second term is negative ($k < 2\pi/a$), its effect tends to cancel the first term. Near the zone boundaries, the coefficient $b$ is so large that the resulting cancellation is greater than the increase in the first term, which leads to a net decrease in the velocity, as we have seen.

At the zone boundary itself ($k = \pi/a$), the scattered wave becomes equal to the incident wave as a result of the strong Bragg reflection, that is, $b = 1$, which, when substituted into (5.74), yields $v = 0$, in agreement with Fig. 5.32(b). We anticipated this result in Section 5.7, in which we found that at the zone edge the electron is represented by a standing wave.

Similar applications of the NFE model in two and three dimensions explain why the relationship between **v** and **k** near the zone boundaries differs considerably from that for a free particle (see the problem section at the end of this chapter).

Now we shall derive a result which was used earlier in Section 5.10, namely, that a completely filled band carries no electric current. To establish this, we note that according to (5.70)

$$\mathbf{v}(-\mathbf{k}) = \mathbf{v}(\mathbf{k}), \tag{5.75}$$

where $\mathbf{v}(\mathbf{k})$ and $\mathbf{v}(-\mathbf{k})$ are the velocities of electrons in the Bloch states **k** and $-\mathbf{k}$, respectively (see Fig. 5.33). This equation follows from the symmetry relation

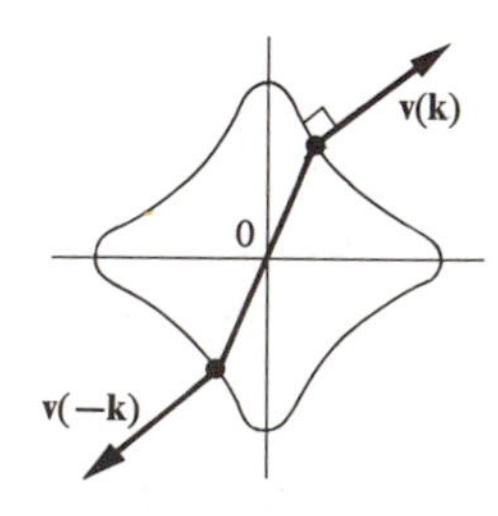


**Fig. 5.33** $\mathbf{v}(-\mathbf{k}) = -\mathbf{v}(\mathbf{k})$.

$E(-\mathbf{k}) = E(\mathbf{k})$, which was established in Section 5.4. The current density due to all electrons in the band is given by

$$\mathbf{J} = \frac{1}{V}(-e)\sum_{\mathbf{k}} \mathbf{v}(\mathbf{k}), \tag{5.76}$$

where $V$ is the volume, $-e$ the electronic charge, and the sum is over all states in the band. But as a consequence of (5.75), the sum over a whole band is seen to vanish, that is, $\mathbf{J} = 0$, with the electrons' velocities canceling each other out in pairs.

## 5.14 ELECTRON DYNAMICS IN AN ELECTRIC FIELD

When an electric field is applied to the solid, the electrons in the solid are accelerated. We can study their motions most easily in $k$-space. Suppose that an electric field $\mathscr{E}$ is applied to a given crystal. As a result, an electron in the crystal experiences a force $\mathbf{F} = -e\mathscr{E}$, and hence a change in its energy. The rate of absorption of energy by the electron is

$$\frac{dE(\mathbf{k})}{dt} = -e\mathscr{E}\cdot\mathbf{v}, \tag{5.77}$$

where the term on the right is clearly the expression for the power absorbed by a moving object. If we write

$$\frac{dE(\mathbf{k})}{dt} = \nabla_{\mathbf{k}} E(\mathbf{k})\cdot\frac{d\mathbf{k}}{dt},$$

and use the expression (5.70) for $\mathbf{v}$, then substitute these into (5.77), we find the surprisingly simple relation

$$\hbar\frac{d\mathbf{k}}{dt} = -e\mathscr{E} = \mathbf{F}. \tag{5.78}$$

This shows that the rate of change of $\mathbf{k}$ is proportional to—and lies in the same direction as—the electric force $\mathbf{F}$ (i.e., opposite to the field $\mathscr{E}$, by virtue of the negative electron charge). This relation is a very important one in the dynamics of Bloch electrons, and is known as the *acceleration theorem*.

Equation (5.78) is not totally unexpected. We have already noted the fact that the vector $\hbar\mathbf{k}$ behaves like the momentum of the Bloch electron (Section 5.3). In that context, Eq. (5.78) simply states that the time rate of change of the momentum is equal to the force, which is Newton's second law.

Let us now consider the consequences of the acceleration theorem, starting with the one-dimensional case. Equation (5.78) may be written in the form

$$\frac{dk}{dt} = \frac{F}{\hbar}, \tag{5.79}$$

showing that the wave vector $k$ increases uniformly with time. Thus, as $t$ increases, the electron traverses the $k$-space at a uniform rate, as shown in Fig. 5.34. If we

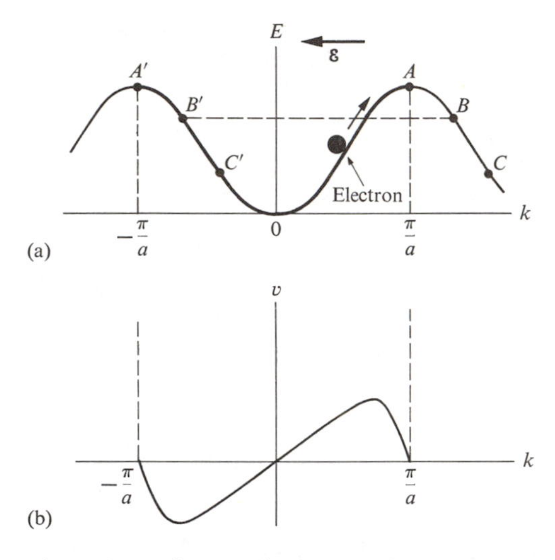


**Fig. 5.34** (a) The motion of an electron in $k$-space in the presence of an electric field (directed to the left). (b) The corresponding velocity.

use the repeated-zone scheme, the electron, starting from $k = 0$, for example, moves up the band until it reaches the top (point $A$) and then starts to descend along the path $BC$. If we use the reduced-zone scheme, then once the electron passes the zone edge at $A$, it immediately reappears at the equivalent point $A'$, then continues to descend along the path $A'B'C'$. Recall that, according to the translational-symmetry property of Section 5.4, the points $B'$, $C'$ are respectively equivalent to the points $B$, $C$, so that we may use either of the two schemes.

Note that, in the presence of an electric field, the electron is in constant motion in $k$-space; it is never at rest.

Also note that the motion in $k$-space is periodic in the reduced-zone scheme, since after traversing the zone once, the electron repeats the motion. The *period* of the motion is readily found, on the basis of (5.79), to be

$$T = \frac{2\pi\hbar}{Fa} = \frac{2\pi\hbar}{e\mathscr{E}a}. \tag{5.80}$$

Figure 5.34(b) shows the velocity of the electron as it traverses the $k$-axis. Starting at $k = 0$, as time passes, the velocity increases, reaches a maximum, decreases. and then vanishes at the zone edge. The electron then turns around and acquires a negative velocity, and so forth. The velocity we are discussing is the velocity in real space, i.e., the usual physical velocity. It follows that a Bloch electron, in the presence of a static electric field, executes an oscillatory periodic motion in real space, very much unlike a free electron. This is one of the surprising conclusions of electron dynamics in a crystal.

Yet the oscillatory motion described above has not been observed, and the reason is not hard to come by. The period $T$ of (5.80) is about $10^{-5}$ s for usual values of the parameters, compared with a typical electron collision time $\tau = 10^{-14}$ s at room temperature. Thus the electron undergoes an enormous number of collisions, about $10^9$, in the time of one cycle. Consequently the oscillatory motion is completely "washed out."†

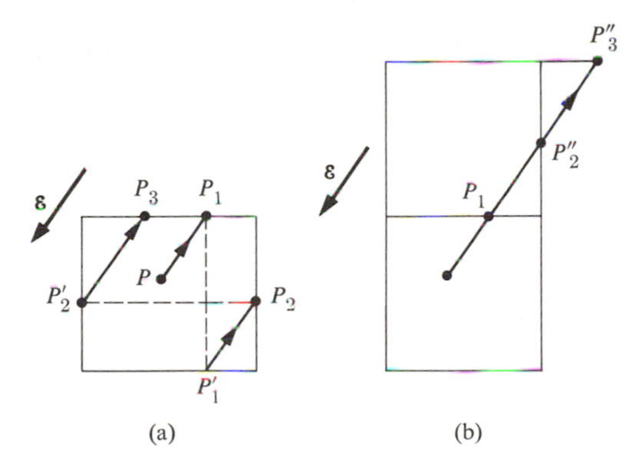


**Fig. 5.35** The motion of an electron in a two-dimensional lattice in the presence of an electric field (a) according to the reduced-zone scheme, (b) according to the repeated-zone scheme.

Let us now consider the situation in two dimensions (Fig. 5.35). When an electric force **F** is applied, the electron, starting at some arbitrary point $P$, moves in a straight line in **k**-space, according to (5.78). As it reaches the zone edge at point $P_1$, it reappears at $P_1'$, continues on to $P_2$, and reappears at $P_2'$. It follows the crisscross path shown in Fig. 5.35(a). If we used the repeated-zone scheme instead (Fig. 5.35b), then the path of the electron in **k**-space would simply be the straight line $P\,P_1\,P_2''\,P_3''$ (note that $P_2''$ is equivalent to $P_2$, $P_3''$ to $P_3$, etc.). This is one situation in which the repeated-zone scheme proves to be more convenient than the extended-zone scheme.

## 5.15 THE DYNAMICAL EFFECTIVE MASS

When an electric field is applied to a crystal, the Bloch electron undergoes an

†Leo Esaki and his collaborators are currently attempting to build a device for which $T \lesssim \tau$, by growing highly pure superlattices for which $a \simeq 50 - 100$Å. Such a *Bloch oscillator* may be used as an oscillator or amplifier.

acceleration. This can be calculated as follows: Since acceleration is the time derivative of velocity, we have

$$a = \frac{dv}{dt}, \tag{5.81}$$

where we have chosen to treat the one-dimensional case first. But velocity is a function of the wave vector $k$, and consequently the above equation may be rewritten as

$$a = \frac{dv}{dk}\frac{dk}{dt},$$

which, when we substitute for the velocity from (5.72), and for $dk/dt$ from (5.78), yields

$$a = \frac{1}{\hbar}\frac{d^2E}{dk^2}F. \tag{5.82}$$

This has the same form as Newton's second law, provided we define a *dynamical effective mass* $m^*$ by the relation

$$m^* = \hbar^2 \bigg/ \left(\frac{d^2E}{dk^2}\right). \tag{5.83}$$

Thus, insofar as the motion in an electric field is concerned, the Bloch electron behaves like a free electron whose effective mass is given by (5.83).

The mass $m^*$ is inversely proportional to the curvature of the band; where the curvature is large—that is, $d^2E/dk^2$ is large—the mass is small; a small curvature implies a large mass (Fig. 5.36).

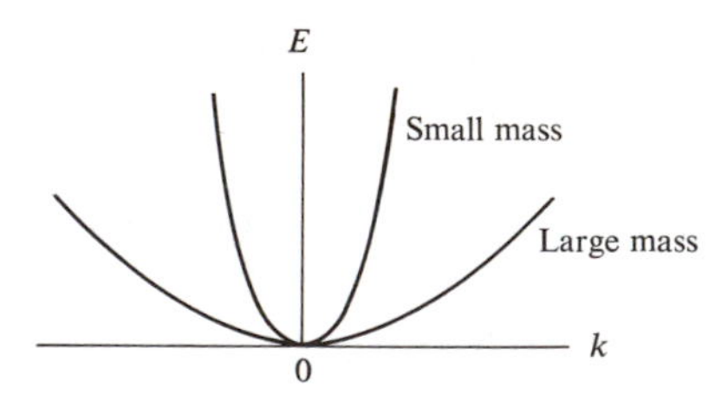


**Fig. 5.36** The inverse relationship between the mass and the curvature of the energy band.

We have previously used the concept of effective mass (Sections 5.6 and 5.8). Those situations are now superseded by—and are in fact special cases of—the general relation (5.83). Thus, if the energy is quadratic in $k$,

$$E = \alpha k^2, \tag{5.84}$$

where $\alpha$ is a constant. Then Eq. (5.83) yields

$$m^* = \hbar^2/2\alpha, \tag{5.85}$$

which is equivalent to rewriting (5.84) as $E = \hbar^2 k^2/2m^*$, the standard form.

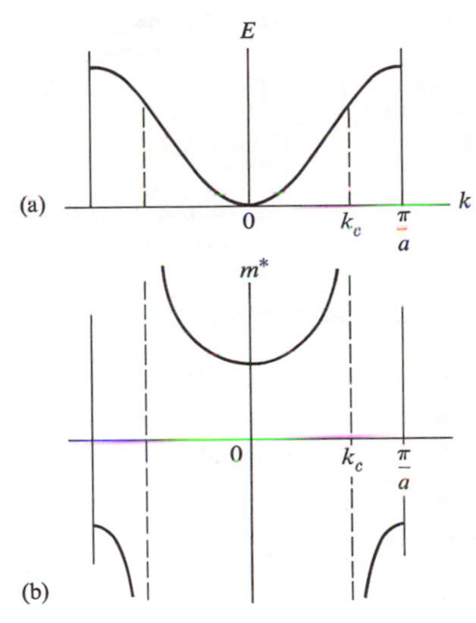


**Fig. 5.37** (a) The band structure, and (b) the effective mass $m^*$ versus $k$.

Figures 5.37(a) and (b) show, respectively, the band structure and the effective mass $m^*$, the latter calculated according to (5.83). Near the bottom of the band, the effective mass $m^*$ has a constant value which is positive, because the quadratic relation (5.84) is satisfied near the bottom of the band. But as $k$ increases, $m^*$ is no longer a strict constant, being now a function of $k$, because the quadratic relaion (5.84) is no longer valid.

Note also that beyond the *inflection point* $k_c$ the mass $m^*$ becomes negative, since the region is now close to the top of the band, and a negative mass is to be expected (Sections 5.6 and 5.8).

The negative mass can be seen dynamically by noting that, according to Fig. 5.34, the velocity decreases for $k > k_c$. Thus the acceleration is negative, i.e., opposite to the applied force, implying a negative mass. This means that in this region of $k$-space the lattice exerts such a large retarding (or braking) force on the electron that it overcomes the applied force and produces a negative acceleration.

The above results may be extended to three dimensions. The acceleration is now

$$\mathbf{a} = \frac{d\mathbf{v}}{dt}.$$

If we write this in cartesian coordinates, and use (5.70) and (5.78), we find that

$$a_i = \sum_j \frac{1}{\hbar^2} \frac{\partial^2 E}{\partial k_i \partial k_j} F_j, \qquad i, j = x, y, z,$$

which leads to the definition of effective mass as

$$\left(\frac{1}{m^*}\right)_{ij} = \frac{1}{\hbar^2} \frac{\partial^2 E}{\partial k_i \partial k_j}, \qquad i, j = x, y, z. \tag{5.86}$$

The effective mass is now a second-order tensor which has nine components.

When the dispersion relation can be written as†

$$E(\mathbf{k}) = (\alpha_1 k_x^2 + \alpha_2 k_y^2 + \alpha_3 k_z^2), \tag{5.87}$$

then using (5.86) leads to an effective mass with three components: $m_{xx}^* = \hbar^2/2\alpha_1$, $m_{yy}^* = \hbar^2/2\alpha_2$, and $m_{zz}^* = \hbar^2/2\alpha_3$. In this case the mass of the electron is *anisotropic*, and depends on the direction of the external force. When the force is along the $k_x$-axis, the electron responds with a mass $m_{xx}^*$, while a force in the $k_y$-direction elicits an effective mass $m_{yy}^*$. A relation of the type (5.87), corresponding to ellipsoidal contours, is a common occurrence in semiconductors, e.g., Si and Ge. Note that in this case, unlike the free-electron case, the acceleration is not, in general, in the same direction as the applied force.

It may also happen that one of the $\alpha_i$'s in (5.87) is negative. This means that the mass in the corresponding direction is negative, while the other directions exhibit positive masses. This again is vastly different from the behavior of the free electron.

The concept of effective mass is very useful, in that it often enables us to treat the Bloch electron in a manner analogous to a free electron. Nonetheless, the Bloch electron exhibits many unusual properties which are alien to those of a free electron.

## 5.16 MOMENTUM, CRYSTAL MOMENTUM, AND PHYSICAL ORIGIN OF THE EFFECTIVE MASS

We have said on several occasions that a Bloch electron in the state $\psi_{\mathbf{k}}$ behaves as if it had a momentum $\hbar\mathbf{k}$. Basically, there are three different reasons to support this statement.

† This is possible near a point at which the energy has a minimum, a maximum, or a saddle point.

a) The Bloch function has the form

$$\psi_{\mathbf{k}} = e^{i\mathbf{k}\cdot\mathbf{r}} u_{\mathbf{k}}, \tag{5.89}$$

which, since $u_{\mathbf{k}}$ is periodic, appears essentially as a plane wave of wavelength $\lambda = 2\pi/k$. This, combined with the deBroglie relation, leads to a momentum $\hbar\mathbf{k}$.

b) When an electric field is applied, the wave vector varies with time according to

$$\frac{d(\hbar\mathbf{k})}{dt} = \mathbf{F}_{\text{ext}}, \tag{5.90}$$

again indicating that $\hbar\mathbf{k}$ acts as a momentum. Here $\mathbf{F}_{\text{ext}}$ refers to the external force applied to the crystal.

c) In collision processes involving a Bloch electron, the electron contributes a momentum equal to $\hbar\mathbf{k}$.

These reasons are sufficiently important to warrant identification of $\hbar\mathbf{k}$ with the momentum. The fact is, nevertheless, that $\hbar\mathbf{k}$ is *not* equal to the actual momentum of the Bloch electron. To make the distinction clear, let us denote the vector $\hbar\mathbf{k}$ by $\mathbf{p}_c$. That is,

$$\mathbf{p}_c = \hbar\mathbf{k}. \tag{5.91}$$

We shall refer to this as the *crystal momentum*.

The actual momentum of the electron $\mathbf{p}$ can be evaluated using quantum methods. According to quantum mechanics the average momentum is given by

$$\mathbf{p} = \langle \psi_{\mathbf{k}} | -i\hbar\nabla | \psi_{\mathbf{k}} \rangle, \tag{5.92}$$

where $-i\hbar\nabla$ is the momentum operator and $\psi_{\mathbf{k}}$ is the Bloch function. If one evaluates this integral, using the properties of the wave function $\psi_{\mathbf{k}}$ (see the problem section at the end of this chapter), one finds that

$$\mathbf{p} = m_0\mathbf{v}, \tag{5.93}$$

where $m$ is the mass of the *free* electron and $\mathbf{v}$ is the velocity as given by (5.70). Thus the true momentum of the electron is equal to the true mass $m$ times the actual velocity $\mathbf{v}$, which seems to be a plausible result.

In retrospect, one may have suspected the original identification of $\mathbf{p}_c$ with the actual momentum from the outset. Since the function $u_{\mathbf{k}}$ in (5.89) is not a constant, the Bloch function $\psi_k$ is not quite a plane wave, and correspondingly the vector $\hbar\mathbf{k}$ is not quite equal to the momentum. Also, if $\mathbf{p}_c = \hbar\mathbf{k}$ were the true momentum, then the force appearing on the right of (5.90) should have been the *total* force, and not just the *external* force. As we shall see, there is a force exerted by the lattice, yet this force does not appear to influence $\mathbf{p}_c$.

The above ideas may now be assembled to give a physical interpretation of the effective mass. Since the vector $\mathbf{p} = m_0\mathbf{v}$ is equal to the true momentum, one may write

$$m_0 \frac{dv}{dt} = F_{\text{tot}} = F_{\text{ext}} + F_L, \tag{5.94}$$

where $F_{\text{tot}}$ and $F_L$ are, respectively, the total force and the lattice force acting on the electron. By lattice force, we mean the force exerted by the lattice on the electron as a result of its interaction with the crystal potential. The left side in (5.94) can be readily expressed in terms of the effective mass, namely

$$m_0 \frac{dv}{dt} = m_0 \frac{F_{\text{ext}}}{m^*}, \tag{5.95}$$

as we can see by referring to Eqs. (5.81) through (5.83). Substituting this into (5.94), and solving for $m^*$, one finds

$$m^* = m_0 \frac{F_{\text{ext}}}{F_{\text{ext}} + F_L}. \tag{5.96}$$

Now we see that the reason why $m^*$ is different from $m_0$, the free mass, lies in the presence of the lattice force $F_L$. If $F_L$ were to vanish, the effective mass would become equal to the true mass.

The effective mass $m^*$ may be smaller or larger than $m_0$, or even negative, depending on the lattice force. Suppose that the electron is "piled up" primarily near the top of the crystal potential, as shown in Fig. 5.38(a). When an

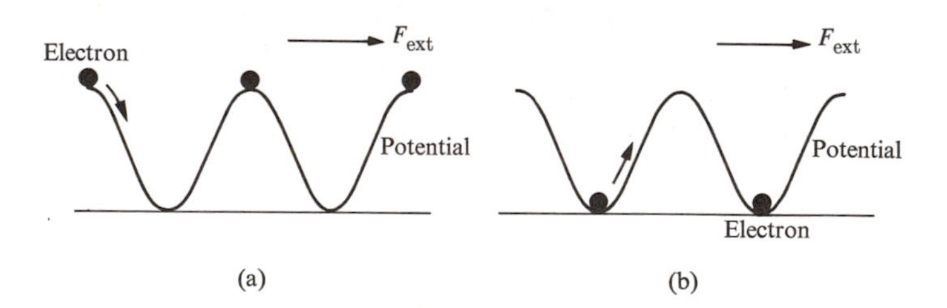


**Fig. 5.38** (a) Electron spatial distribution leading to an effective mass $m^*$ smaller than $m_0$. (b) A distribution leading to $m^* > m_0$.

external force is applied, it causes the electron to "roll downhill" along the potential curve. As a result, a positive lattice force becomes operative and hence, according to (5.96), $m^* < m_0$. This is what happens in alkali metals, for instance, and in the conduction band in semiconductors. Here $m^*$ is less than $m_0$ because the lattice force assists the external force.

On the other hand, when the electron is piled mainly near the bottom of the potential curve (Fig. 5.38b), then clearly the lattice force tends to oppose the external force, resulting in $m^* > m_0$. This is the situation in the alkali halides, for instance. If the potential wave is sufficiently steep, then $F_L$ becomes larger than $F_{\text{ext}}$, and $m^*$ becomes negative.

Note that the lattice force $F_L$, which appears in (5.94), is a force induced by the external force. Thus if $F_{\text{ext}} = 0$, then the velocity is constant (Section 5.13), and hence $F_L = 0$, according to (5.94). It is true that the lattice also exerts a force on an otherwise-free electron even in the absence of $F_{\text{ext}}$, but that force has already been included in the solution of the Schrödinger equation, and hence in the properties of the state $\psi_{\mathbf{k}}$. That force (as we stated in Sections 5.13 and 4.4) does not scatter the wave $\psi_{\mathbf{k}}$.

However, the crystal momentum $\mathbf{p}_c = \hbar\mathbf{k}$ is still a very useful quantity. In problems of electron dynamics in external fields, crystal momentum is much more useful than true momentum, since it is easier to follow motion in $k$-space than in real space. Therefore we shall continue to use $\mathbf{p}_c$ and refer to it as the momentum, when there is no ambiguity, and even drop the subscript $c$.

In other words, the effective mass $m^*$ and the crystal momentum $\hbar\mathbf{k}$ are artifices which allow us—formally at least—to ignore the lattice force and concentrate on the external force only. This is very useful, because lattice force is not known *a priori*, nor is it easily found and manipulated as is the external force.

## 5.17 THE HOLE

A *hole* occurs in a band that is completely filled except for one vacant state. Figure 5.39 shows such a hole. When we consider the dynamics of the hole in an

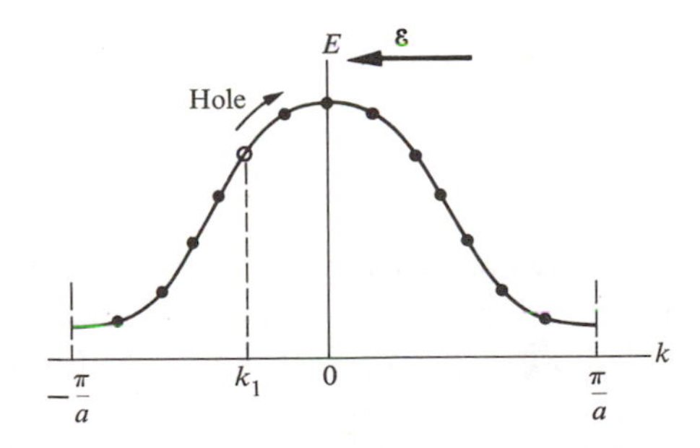


**Fig. 5.39** The hole and its motion in the presence of an electric field.

external field, we find it far more convenient to focus on the motion of the vacant site than on the motion of the enormous number of electrons filling the band. The concept of the hole is an important one in band theory, particularly in semiconductors, in which it is essential to the operation of many valuable devices, e.g., the transistor.

Suppose the hole is located at the wave vector $k_1$, as shown in Fig. 5.39. The current density of the whole system is

$$J_h = \frac{-e}{V} \sum_k{}' v_e(k), \tag{5.97}$$

where the sum is over all the electrons in the band, with the prime over the summation sign, indicating that the state $k_1$ is to be excluded, since that state is vacant. Since the sum over the filled band is zero, the current density (5.97) is also equal to

$$J_h = \frac{e}{V} v_e(k_1). \tag{5.98}$$

That is, the current is the same as if the band were empty, except for an electron of *positive* charge $+e$ located at $k_1$.

When an electric field is now applied to the system, and directed to the left (Fig. 5.39), all the electrons move uniformly to the right, in $k$-space, and at the same rate (Section 5.14). Consequently the vacant site also moves to the right, together with the rest of the system. The change in the hole current in a time interval $\delta t$ can be found from (5.98):

$$\delta J_h = \frac{e}{V} \left( \frac{dv_e}{dk} \right)_{k_1} \frac{dk}{dt} \delta t,$$

which, when we use (5.70), (5.83), and (5.78), can be transformed into

$$\delta J_h = \frac{e}{V} \frac{1}{m^*(k_1)} F \, \delta t = \frac{1}{V} \left( \frac{-e^2}{m^*(k_1)} \right) \mathscr{E} \, \delta t, \tag{5.99}$$

where $m^*(k_1)$ is the mass of an electron occupying state $k_1$.

This equation gives the electric current of the hole, induced by the electric field, which is the observed current.† Since the hole usually occurs near the top of the band—due to thermal excitation of the electron to the next-higher band, where the mass $m^*(k_1)$ is negative—it is convenient to define the mass of a hole as

$$m_h^* = -m^*(k_1), \tag{5.100}$$

† In practice a band contains not a single hole but a large number of holes, and in the absence of an electric field the net current of these holes is zero because of the mutual cancelation of the contributions of the various holes, i.e., the sum of the expression (5.98) over the holes vanishes. When a field is applied, however, induced currents are created, and since these are additive, as seen from (5.99), a nonvanishing net current is established.

which is a positive quantity, and write (5.99) as

$$\delta J_{\mathrm{h}} = \frac{1}{V} \frac{e^2}{m_{\mathrm{h}}^*} \mathscr{E}\, \delta t. \tag{5.101}$$

Note that the hole current, like the electron current, is in the same direction as the electric field.

By examining (5.98) and (5.101), we can see that the motion of the hole, both with and without an electric field, is the same as that of a *particle with a positive charge e and a positive mass* $m_{\mathrm{h}}^*$. Viewing the hole in this manner results in a great simplification, in that the motion of all the electrons in the band has been reduced to that of a single "particle." This representation will be used frequently in the following discussions.

We may note, incidentally, that according to (5.99), if the hole were to lie near the bottom of the band, where $m^*(k_1) > 0$, then the current would be opposite to the field. This means that the system would act as an amplifier, with the field absorbing energy from the system. This situation is not likely to occur, however, because the hole usually lies near the top of the band.†

## 5.18 ELECTRICAL CONDUCTIVITY

We discussed electrical conductivity previously in connection with the free-electron model (Sections 4.4 and 4.8), in which we obtained the result

$$\sigma = \frac{ne^2\tau_{\mathrm{F}}}{m^*}. \tag{5.102}$$

The quantity $n$ is the concentration of the conduction—or valence—electrons and $\tau_{\mathrm{F}}$ is the collision time for an electron at the Fermi surface. Now let us derive the corresponding expression for electrical conductivity within the framework of band theory.

When the system is at equilibrium—i.e., when there is no electric field—the FS is centered exactly at the origin, as shown in Fig. 5.40(a). Consequently the net current is zero, because the velocities of the electrons cancel in pairs. That is, for every electron in state **k** whose velocity is **v**(**k**), another electron exists in state −**k** whose velocity **v**(−**k**) = −**v**(**k**) is simply the reverse of the former. This result, found in the free-electron model, also holds good in band theory, and accounts for the vanishing of the current at equilibrium.

When an electric field is applied, each electron travels through **k**-space at a

† A proposal for an amplifier operating on essentially the same principle was advanced by H. Kroemer, *Phys. Rev.* **109**, 1856 (1955).

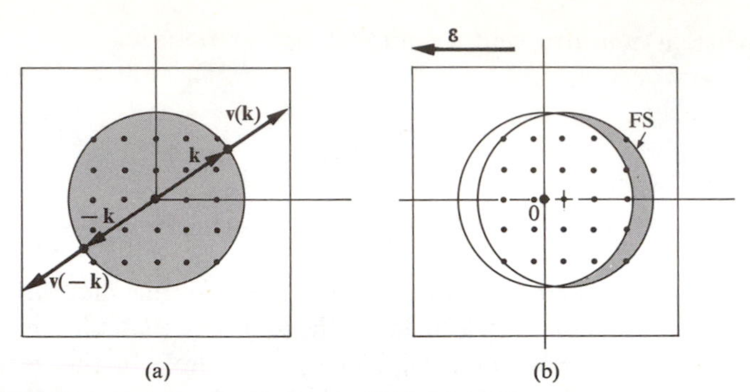


**Fig. 5.40** (a) In the absence of an electric field the FS is centered at the origin, and the electron currents cancel in pairs. (b) In the presence of an electric field, the FS is displaced and a net current results.

uniform rate, as discussed in Section 5.14. That is,

$$\delta k_x = -\frac{e\mathscr{E}}{\hbar}\,\delta t,$$

where $\delta k_x$ is the displacement in a time interval $\delta t$. Since an electron usually "lives" for an interval equal to the collision time $\tau$, the average displacement is

$$\delta k_x = -\frac{e\mathscr{E}}{\hbar}\,\tau. \tag{5.103}$$

Consequently the FS is displaced rigidly by this amount, as shown in Fig. 5.40(b). There are now some electrons which are *not* compensated—i.e., canceled—by other electrons, and which are indicated by the cross-hatched crescent-shaped region. They contribute a *net* current.

The density of this current can be calculated as follows: It is given by

$$\begin{aligned} J_x &= -\,e\,\bar{v}_{F,x} \times \text{concentration of uncompensated electrons} \\ &= -\,e\bar{v}_{F,x}g(E_F)\,\delta E \\ &= -\,e\bar{v}_{F,x}g(E_F)\left(\frac{\partial E}{\partial k_x}\right)_{E_F}\delta k_x, \end{aligned} \tag{5.104}$$

where $\bar{v}_{F,x}$ is the component of the Fermi velocity in the $x$-direction and the bar indicates an average value.

Note that $g(E_F)\,\delta E$ gives the concentration of uncompensated electrons, $g(E_F)$ being the density of states at the FS and $\delta E$ the energy absorbed by the electron from the field. Noting that $\partial E/\partial k_x = \hbar v_{F,x}$, and substituting for $\delta k_x$ from (5.103). one obtains

$$J_x = e^2\bar{v}_{F,x}^2\tau_F g(E_F)\mathscr{E}, \tag{5.105}$$

where the collision time has been designated as $\tau_F$, inasmuch as we are clearly dealing with electrons lying at the FS. Note that the current is in the same direction as the field.

For a spherical FS, there is a spherical symmetry, and hence one may write $\bar{v}_{F,x}^2 = \frac{1}{3}v_F^2$ which, when substituted into (5.105), leads finally to the following expression for the electrical conductivity:

$$\sigma = \tfrac{1}{3}\, e^2 v_F^2 \tau_F g(E_F), \tag{5.106}$$

which is the expression we have been seeking.

Note that $\sigma$ depends on the Fermi velocity and the collision time, but also note the dependence on the density of states at the FS, $g(E_F)$. Often this is the predominant factor in determining the conductivity, as we shall see shortly.

Expression (5.106) is more general than the free-electron formula (5.102), and far more meaningful. Equation (5.102) implies that conductivity is controlled primarily by $n$, the electron concentration. However, conductivity is, in fact, controlled primarily by the density of states $g(E_F)$ instead. In the appropriate limit, expression (5.106) reduces to (5.102) as a special case, as it must. To establish this, we use the relation $g(E_F) = \frac{1}{2}\pi^2(2m^*/\hbar^2)^{3/2}E_F^{1/2}$ [see (5.63)], $E_F = \frac{1}{2}m^* v_F^2$, and $E_F = (\hbar^2/2m^*)(3\pi^2 n)^{2/3}$ [from (5.67)], which we find reduce (5.106) to (5.102).

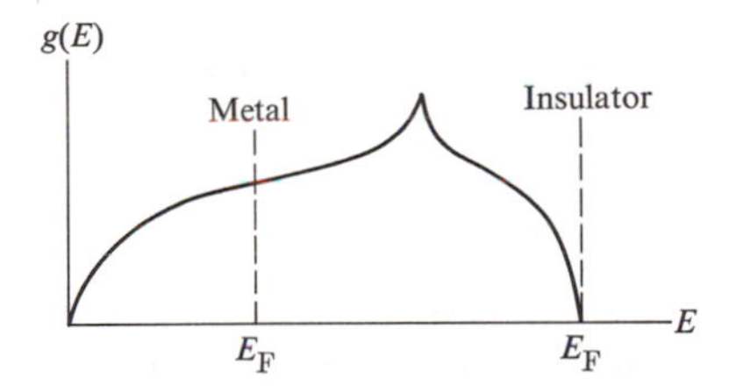


**Fig. 5.41** Position of the Fermi energy level in a monovalent metal and in an insulator. In the former, $g(E_F)$ is large, while in the latter, $g(E_F) = 0$.

Figure 5.41 shows the density of states for a typical solid, indicating the position of the Fermi level for a monovalent metal, and also for an insulator. In the metal, the level $E_F$ is located near the middle of the band where $g(E_F)$ is large, leading to a large conductivity, according to (5.106). In the insulator, the level $E_F$ is right at the top of the band, where $g(E_F) = 0$. Thus the conductivity is zero, despite the fact that the Fermi velocity, which also appears in (5.106), is very large.

The expression (5.106), though restricted to the case in which the FS is spherical, is useful in unraveling the important role played by the density of states. The results may be generalized to include the effects of more complex FS shapes (as you will find by referring to the bibliography), which often lead to unwieldy expressions.

Another important aspect of the electrical conduction process—and of transport phenomena in general—is that they enable us to calculate the collision time $\tau_F$. We discussed this subject in a semiclassical fashion in Section 4.4 for the free-electron model, but a more rigorous treatment involves the use of quantum methods (see Appendix A), and perturbation theory in particular. The scattering mechanisms are the same as those discussed in connection with the free-election model (Section 4.5)—scattering by lattice vibrations, impurities, and other lattice defects—but the details of the calculation are highly complicated (Ziman, 1960), and will not be given here.

## 5.19 ELECTRON DYNAMICS IN A MAGNETIC FIELD: CYCLOTRON RESONANCE AND THE HALL EFFECT

We discussed electron dynamics in a magnetic field in Section 4.10 with respect to the free-electron model, where we also treated cyclotron resonance and the Hall effect. Here we shall discuss the way in which this is modified for a Bloch electron, taking into account the interaction with the crystal potential. This subject is more useful in practice, as the magnetic field is often used in studies of band structure.

### Cyclotron resonance

The basic equation of motion describing the dynamics in a magnetic field is

$$\hbar \frac{d\mathbf{k}}{dt} = -e\,[\mathbf{v}(\mathbf{k}) \times \mathbf{B}], \tag{5.107}$$

where the left side is the time derivative of the crystal momentum, and the right side the well-known Lorentz force due to the magnetic field. This equation is a plausible one in light of the discussion in Sections 5.14 and 5.16, in which we concluded that the momentum of the crystal usually acts as the familiar momentum, provided only the external force is included. [The equation (5.107) may also be derived from detailed quantum calculations.]

According to (5.107), the change in $\mathbf{k}$ in a time interval $\delta t$ is given by

$$\delta\mathbf{k} = -(e/\hbar)\,[\mathbf{v}(\mathbf{k}) \times \mathbf{B}]\,\delta t, \tag{5.108}$$

which shows that the electron moves in $\mathbf{k}$-space in such a manner that its displacement $\delta\mathbf{k}$ is perpendicular to the plane defined by $\mathbf{v}$ and $\mathbf{B}$. Since $\delta\mathbf{k}$ is perpendicular to $\mathbf{B}$, this means that the electron trajectory lies in a plane normal to the magnetic field. In addition, $\delta\mathbf{k}$ is perpendicular to $\mathbf{v}$ which, inasmuch as $\mathbf{v}$ is normal to the energy contour in $\mathbf{k}$-space, means that $\delta\mathbf{k}$ lies along such a contour. Putting these two bits of information together, we conclude that the electron rotates along

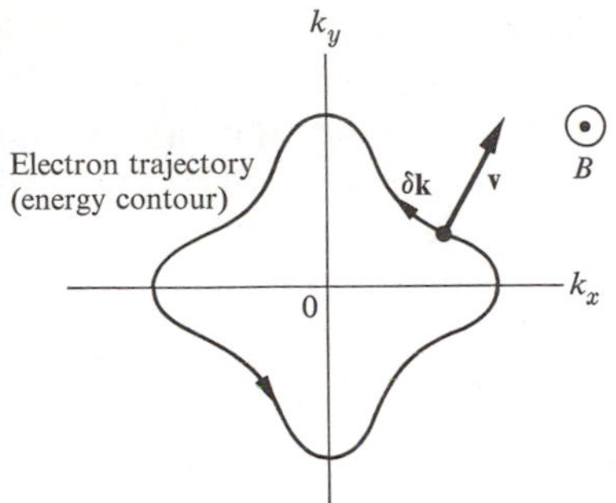


**Fig. 5.42** Trajectory of the electron in **k**-space in the presence of a magnetic field.

an energy contour normal to the magnetic field (Fig. 5.42), and in a counterclockwise fashion.

Also note that, because the electron moves along an energy contour, no energy is absorbed from, or delivered to, the magnetic field, in agreement with the well-known facts concerning the interaction of electric charges with a magnetic field.

As Fig. 5.42 shows, the motion of the electron in **k**-space is *cyclic*, since, after a certain time, the electron returns to the point from which it started. The period $T$ for the motion is, according to (5.108), given by

$$T = \oint \delta t = \frac{\hbar}{eB} \oint \frac{\delta k}{v(\mathbf{k})}, \tag{5.109}$$

where the circle on the integration sign denotes that this integration is to be carried out over the complete cycle in **k**-space, i.e., a closed orbit. In (5.109), the differential $\delta k$ is taken along the perimeter of the orbit, while $v(\mathbf{k})$ is the magnitude of the electron velocity normal to the orbit. Also note that in deriving (5.109) from (5.108), we have used the fact that $\mathbf{v}$ is normal to $\mathbf{B}$, since the electron trajectory lies in a plane normal to $\mathbf{B}$.

The angular frequency $\omega_c$ associated with the motion is $\omega_c = 2\pi/T$, which, in light of (5.109), is given by

$$\omega_c = (2\pi eB/\hbar) \bigg/ \oint \frac{\delta k}{v(\mathbf{k})}. \tag{5.110}$$

This is the *cyclotron frequency* for the Bloch electron. It is the generalization of the cyclotron frequency (4.38) derived for the free-electron model.

We conclude that the motion of a Bloch electron in a magnetic field is a natural generalization of the motion of a free electron in the same field. A free electron executes circular motion in velocity space along an energy contour with a frequency $\omega_c = eB/m^*$. A Bloch electron executes a cyclotron motion along an energy contour with a frequency given by (5.110). The energy contour in this latter case may, of course, be very complicated.

When the standard form $E = \hbar^2k^2/2m^*$ is applicable, the frequency $\omega_c$ in (5.110) may be readily calculated. The cyclotron orbit is circular in this case, and in evaluating the integral we note that $v(\mathbf{k}) = \hbar k/m^*$, which is a constant along the orbit, since the magnitude $k$ of the wave vector is constant along this contour trajectory. Thus

$$\oint \frac{\delta k}{v(\mathbf{k})} = \frac{1}{(\hbar k/m^*)} \oint \delta k = \frac{2\pi k}{(\hbar k/m^*)} = \frac{2\pi m^*}{\hbar},$$

which, when substituted into (5.110), produces

$$\omega_c = eB/m^*.$$

This, as expected, agrees with the result for the free-electron model.

But, of course, Eq. (5.110) is more general than the free-electron result, and applies to a contour of arbitrary shape, although evaluating the integral may become very tedious. In the problem section at the end of this chapter, you will be asked to evaluate $\omega_c$ for contours which, although more complicated than those in the free-electron model, are still simple enough to render the integral in (5.110) tractable.

In discussing the above cyclotron motion, we have disregarded the effects of collision. Of course, if this cyclotron motion is to be observed at all, the electron must complete a substantial fraction of its orbit during one collision time; that is, $\omega_c\tau \gtrsim 1$. This necessitates the use of very pure samples at low temperature under a very strong magnetic field.

**The Hall effect**

When we were discussing the Hall effect in the free-electron model (Section 4.10), we found that the Hall constant is given by

$$R_e = -\frac{1}{n_e e}, \tag{5.111}$$

where $n_e$ is the electron concentration. The negative sign is due to the negative charge of the electron. The general treatment of the Hall effect for Bloch electrons becomes quite complicated for arbitrary FS, requiring considerable mathematical effort (Ziman, 1960). However, we can obtain some important results quite readily.

Suppose that only holes were present in the sample. Then we could apply to the holes the same treatment used for electrons in Section 4.10, and would obtain a Hall constant

$$R_h = \frac{1}{n_h e}, \tag{5.112}$$

where $R$ is now positive because of the positive charge on the hole ($n_h$ is the hole concentration).

Actually, in metals, holes are not present by themselves; there are always some electrons present. Thus when two bands overlap with each other, electrons are present in the upper band and holes in the lower. The expression for the Hall constant when both electrons and holes exist simultaneously is given by (see the problem section)

$$R = \frac{R_e\sigma_e^2 + R_h\sigma_h^2}{(\sigma_e + \sigma_h)^2}, \tag{5.113}$$

where $R_e$ and $R_h$ are the contributions of the individual electrons and holes, as given above, and $\sigma_e$ and $\sigma_h$ are the conductivities of the electrons and holes ($\sigma_e = n_e e^2 \tau_e / m_e^*$ and $\sigma_h = n_h e^2 \tau_h / m_h^*$).

Equation (5.113) shows that the sign of the Hall constant $R$ may be either negative or positive depending on whether the contribution of the electrons or the holes dominates. If we take $n_e = n_h$, which is the case in metals, then $|R_e| = |R_h|$ and the sign of $R$ is determined entirely by the relative magnitudes of the conductivities $\sigma_e$ and $\sigma_h$. Thus if $\sigma_e > \sigma_h$—that is, if the electrons have small mass and long lifetime—the electrons' contribution dominates and $R$ is negative. And when the opposite condition prevails, the holes' contribution dominates, and $R$ is positive. We can now understand why some polyvalent metals—e.g., Zn and Cd—exhibit positive Hall constants (see Table 4.3).

## 5.20 EXPERIMENTAL METHODS IN DETERMINATION OF BAND STRUCTURE

Now let us discuss some of the experimental techniques used to determine the band structure in metals. For example, how did physicists determine the Fermi energies in Table 4.1, or the Fermi surfaces shown in Fig. 5.26 for Cu and Fig. 5.30 for Be? This field of solid-state physics is a wide one, and has been expanding at a rapid pace. Our discussion here will therefore be rather sketchy, leaving it to the reader to pursue the subject in greater detail by referring to the entries in the bibliography.

One can determine the Fermi energy by the method of *soft x-ray emission*. When a metal is bombarded by a beam of high-energy electrons, electrons from the inner K shell† are knocked out, leaving empty states behind. Electrons in the valence band now move to fill these vacancies, undergoing downward transitions, as shown in Fig. 5.43(a). The photons emitted in the transition, usually lying in the soft x-ray region—about 200 eV—are recorded and their energies measured. Figure 5.43(b) shows the intensity of the x-ray spectrum as well as the energy

† The atomic shells $n = 0, 1, 2$, etc., are usually referred to as the K, L, M, etc., shells, respectively.

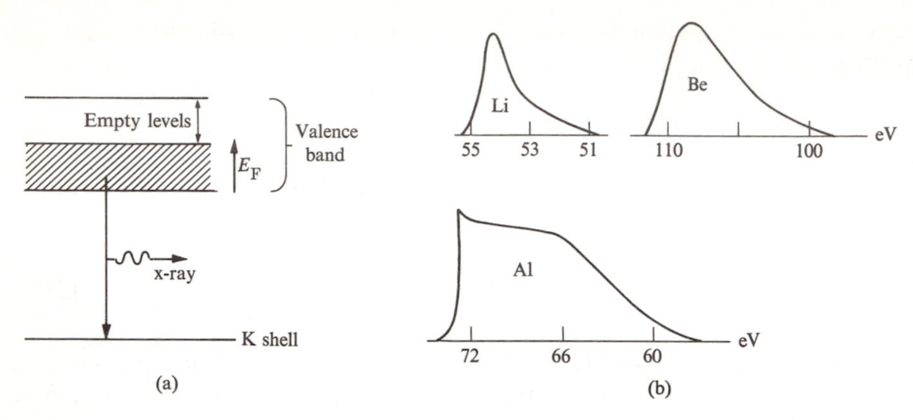


**Fig. 5.43** (a) Emission of soft x-rays. (b) Intensity of the spectrum of x-ray emission versus energy for Li, Be, and Al.

range for several metals. Since the K shell is very narrow, almost to the point of being a discrete level, the width of the range shown in Fig. 5.43(b) is due entirely to the spread of the occupied states in the valence band, i.e., the width is equal to the Fermi level. One can also extract information from Fig. 5.43(b) on the shape of the density of states. In fact, the shape of the curve is determined primarily by the density of states of the valence band.

Let us now turn to the determination of the FS, and discuss one of the many methods in common use: the *Azbel–Kaner cyclotron resonance* (AKCR) technique. A semi-infinite metallic slab is placed in a strong static magnetic field $\mathbf{B}_0$, which is parallel to the surface (Fig. 5.44). As a result, electrons in the

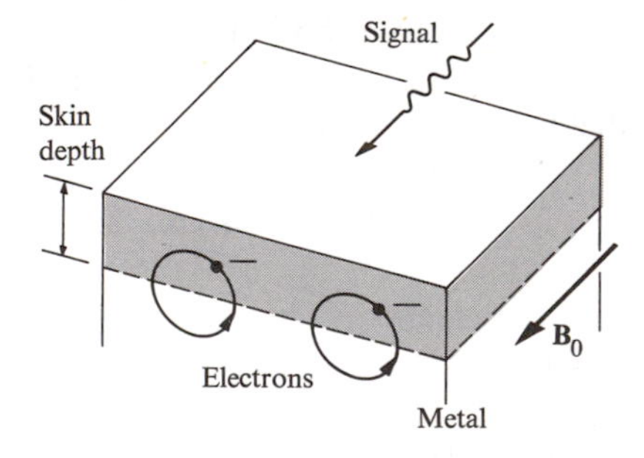


**Fig. 5.44** Physical setup for Azbel–Kaner cyclotron resonance.

metal begin to execute a cyclotron motion, with a cyclotron frequency $\omega_c$. Now an alternating electromagnetic signal of frequency $\omega$, circularly polarized in a counterclockwise direction, is allowed to travel parallel to the surface and along the direction of the static field $\mathbf{B}_0$. This signal penetrates the metal only to a small extent, equal to the skin depth (see Section 4.11), and so is confined to a short distance from the surface. Only electrons in this region are affected by the signal.

The electrons near the surface feel the field of the signal and absorb energy from it. This absorption is greatest when the condition

$$\omega = \omega_c \tag{5.114}$$

is satisfied, because the electron then remains in phase with the signal field throughout the cycle. This is the *resonance* condition.

During a part of its cycle, the electron actually penetrates the metal beyond the skin depth, where the signal field vanishes. A resonance condition is still satisfied, provided only that, when the electron returns to the region at the surface, it is again in phase with the field. In general, therefore, the condition for resonance is

$$\omega = l\omega_c, \tag{5.115}$$

where $l = 1, 2, 3$, etc., at all harmonics of the cyclotron frequency $\omega_c$. The AKCR for Cu is shown in Fig. 5.45. (Usually the frequency $\omega$ is held fixed and the field is varied until the resonance condition is satisfied.)

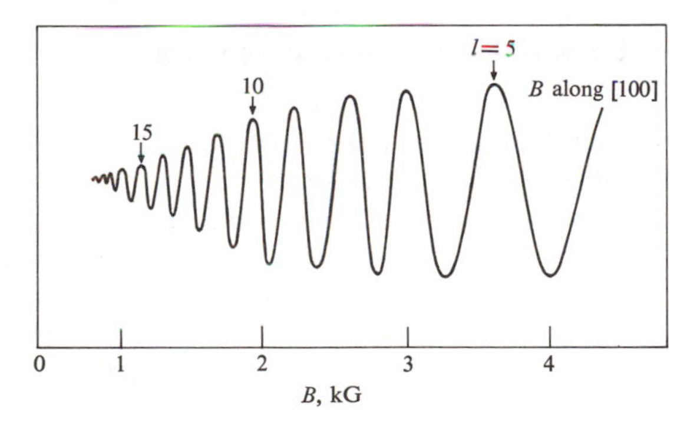


**Fig. 5.45** AKCR spectrum in Cu at $T = 4.2°K$. The crystal surface (upper surface) is cut along the (100) plane. The ordinate of the curve represents the derivative of the surface resistivity with respect to the field. [After Haüssler and Wells, *Phys. Rev.*, **152,** 675, 1966]

Not only is the method capable of determining $\omega_c$ (and hence the effective mass $m^*$), but also the actual shape of the FS. In general, electrons in different regions of the surface have different cyclotron frequencies, but the frequency which is most pronounced in the absorption is the frequency appropriate to the *extremal* orbit, i.e., where the FS cross section perpendicular to $\mathbf{B}_0$ is greatest, or smallest. Therefore, by varying the orientation of $\mathbf{B}_0$, one can measure the extremal sections in various directions, and reconstruct the FS.

The experiment is usually performed at very low temperatures, that is, $T \simeq 4°K$, on very pure samples, and at very strong fields—about 100 kG. Under these conditions, the collision time $\tau$ is long enough, and the cyclotron frequency $\omega_c$ high enough, so that the high-field condition $\omega_c \tau \gg 1$ is satisfied. In this limit, the electron executes many cycles in a single collision time, leading to a sharp, well-resolved resonance. The frequency $\omega_c$ usually falls in the microwave range.

Optical ultraviolet techniques are also used in determining band structure. Figure 5.46 shows the principle of the method. When a light beam impinges on a

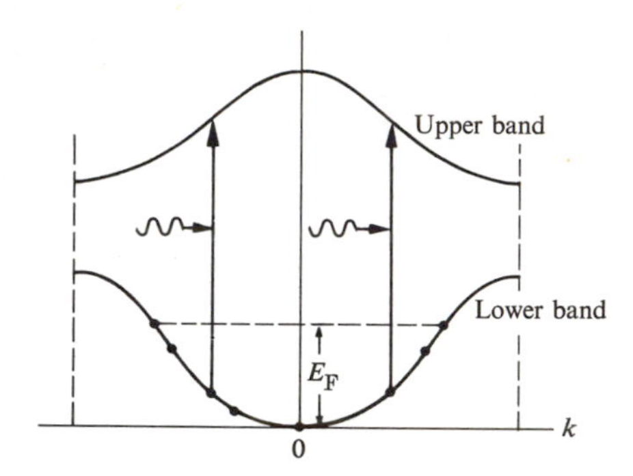


**Fig. 5.46** Interband optical absorption.

metal, electrons are excited from below the Fermi level into the next-higher band. This *interband absorption* may be observed by optical means—i.e., reflectance and absorption techniques, which give information concerning the shape of the energy bands. In this case, two bands are involved simultaneously, and the results cannot be expressed in terms of the individual bands separately. But if the shape of one of these is known, the shape of the other may be determined. For further discussion of the optical properties of metals in the ultraviolet region—which is where the frequencies happen to lie in the case of most metals—refer to Section 8.9.

## 5.21 LIMIT OF THE BAND THEORY; METAL–INSULATOR TRANSITION

So far in this chapter we have based our discussion entirely on the so-called *band model* of solids. This model has been of immense value to us; it is capable of explaining all the observed properties of metals, and is the basis of the semiconductor properties to be discussed in Chapters 6 and 7. Yet this model has a limitation which we now wish to probe.

Consider, for example, the case of Na. This substance is a conductor because the 3s band is only partially filled—half filled, to be exact. Suppose that we cause the Na to expand by some means, so that the lattice constant $a$ can be increased arbitrarily. Would the material then remain a conductor for any arbitrary value of $a$? The answer must be yes, if one is to believe the band model, because, regardless of the value of $a$, the 3s band would always be half full. It is true (the model predicts further) that the conductivity $\sigma$ decreases as $a$ increases, but the decrease is gradual, as shown in Fig. 5.47.

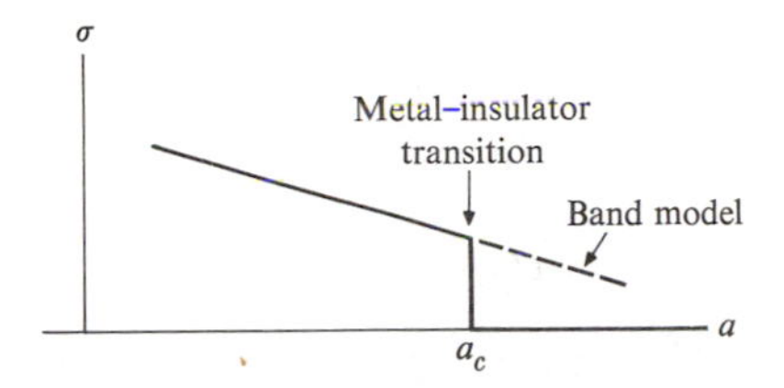


**Fig. 5.47** Electrical conductivity $\sigma$ versus lattice constant $a$.

In fact, however, this is not correct. As $a$ increases, a critical value $a_c$ is reached at which the conductivity drops to zero abruptly, rendering the solid an *insulator*, and it remains so for all values $a > a_c$. Thus for a sufficiently large lattice constant, the metal is transformed into an insulator, and we speak of the *metal–insulator transition* (also known as the *Mott transition*).

To explain this transition, we need to recall some of the fundamental concepts underlying band theory. In this theory, Bloch electrons are assumed to be *delocalized*, extending throughout the crystal, and it is this delocalization which is responsible for metallic conductivity. As a delocalized particle, the Bloch electron spends a fraction of its time ($1/N$, to be exact), at each atom. The interaction between the various Bloch electrons is taken into account only in an average manner, i.e., the interaction between individual electrons is neglected.

However, as $a$ increases, the bandwidth decreases (recall the TB model, Section 5.8), until it becomes quite small at sufficiently large $a$. In that case, the band model breaks down because it allows the presence of two or more electrons at the same lattice site, which cannot happen because of the Coulomb repulsion between electrons. When the band is wide, this is not serious, because electrons can readjust their kinetic energies to compensate for the increase in the coulomb potential energy. But for a narrow band the kinetic energy is, at best, quite small, and this readjustment is not possible.

In effect, for very large $a$, the proper electronic orbitals in a crystal are not of the Bloch type. They are localized orbitals centered around their respective sites, which mitigates the large coulomb energy. Since the orbitals are localized, as in the case of free atoms, conductivity vanishes, as depicted in Fig. 5.47.

Note that the above conclusion holds true even though the energy levels still form a band, and even though the band is only half full. The point is that electronic orbitals become localized, and hence nonconducting.

The metal–insulator transition has been observed in $VO_2$ (vanadium oxide) and other oxide materials. Although $VO_2$ is normally an insulator, it is transformed into a metallic material at sufficiently high pressure.

## SUMMARY

### The Bloch theorem and energy bands in solids

The wave function for an electron moving in a periodic potential, as in the case of a crystal, may be written in the *Bloch form*,

$$\psi_{\mathbf{k}}(\mathbf{r}) = e^{i\mathbf{k}\cdot\mathbf{r}} u_{\mathbf{k}}(\mathbf{r}),$$

where the function $u_{\mathbf{k}}(\mathbf{r})$ has the same periodicity as the potential. The function $\psi_{\mathbf{k}}$ has the form of a plane wave of vector $\mathbf{k}$, which is modulated by the periodic function $u_{\mathbf{k}}$. Although the function $\psi_{\mathbf{k}}$ itself is nonperiodic, the electron probability density $|\psi_{\mathbf{k}}|^2$ is periodic; i.e., the electron is delocalized, and is deposited periodically throughout the crystal.

The energy spectrum of the electron is comprised of a set of continuous *bands*, separated by regions of forbidden energies which are called *energy gaps*. The electron energy is commonly denoted by $E_n(\mathbf{k})$, where $n$ is the band index.

Regarded as a function of the vector $\mathbf{k}$, the energy $E(\mathbf{k})$ satisfies several symmetry properties. First, it has translational symmetry

$$E(\mathbf{k} + \mathbf{G}) = E(\mathbf{k}),$$

which enables us to restrict our consideration to the first Brillouin zone only. The energy function $E(\mathbf{k})$ also has inversion symmetry, $E(-\mathbf{k}) = E(\mathbf{k})$, and rotational symmetry in $\mathbf{k}$-space.

### The NFE and TB models

In the NFE model the crystal potential is taken to be very weak. Solving the Schrödinger equation shows that the electron behaves essentially as a free particle, except when the wave vector $\mathbf{k}$ is very close to, or at, the boundaries of the zone. In these latter regions, the potential leads to the creation of energy gaps. The first gap is given by

$$E_g = 2\,|V_{-2\pi/a}|,$$

where $V_{-2\pi/a}$ is a *Fourier component* of the crystal potential.

The wave functions at the zone boundaries are described by standing waves, which result from strong Bragg reflection of the electron wave by the lattice.

The TB model, in which the crystal potential is taken to be strong, leads to the same general conclusions as the NFE model, i.e., the energy spectrum is composed of a set of continuous bands. The TB model shows that the width of the band increases and the mobility of the electron becomes greater (the mass lighter) as the overlap between neighboring atomic functions increases.

### Metals versus insulators

If the valence band of a given substance is only partially full, the substance acts like a metal or conductor because an electric field produces an electric current in the material. If the valence band is completely full, however, no current is produced, regardless of the field, and the substance is an insulator.

When the gap between the valence band and the band immediately above it is small, electrons may be thermally excited across the gap. This gives rise to a small conductivity, and the metal is called a *semiconductor*.

### Velocity of the Bloch electron

An electron in the Bloch state $\psi_{\mathbf{k}}$ moves through the crystal with a velocity

$$\mathbf{v} = \frac{1}{\hbar}\,\nabla_{\mathbf{k}} E(\mathbf{k}).$$

This velocity remains constant so long as the lattice remains perfectly periodic.

### Electron dynamics in an electric field

In the presence of an electric field, an electron moves in **k**-space according to the relation

$$\dot{\mathbf{k}} = -\,(e/\hbar)\mathscr{E}.$$

The motion is uniform, and its rate proportional to the field. One obtains this relation at once if one regards the electron as having a momentum $\hbar\mathbf{k}$.

### Effective mass

The effective mass of a Bloch electron is given by

$$m^* = \hbar^2/(d^2E/dk^2).$$

The mass is positive near the bottom of the band, where the curvature is positive. But near the top, where the band curvature is negative, the effective mass is also negative. The fact that the effective mass is different from the free mass is due to the effect of the lattice force on the electron.

### The hole

A hole exists in a band which is completely full, with one vacant state. The hole acts as a particle of positive charge $+e$. When the hole lies near the top of the band, which is the usual situation, the hole also behaves as if it has a positive effective mass.

### Electrical conductivity

Electrical conductivity is given by

$$\sigma = \tfrac{1}{3}\,e^2 v_F^2 \tau_F g(E_F).$$

This expression is a particularly sensitive function of $g(E_F)$, the density of states at the Fermi energy. In monovalent metals, $\sigma$ is large because $g(E_F)$ is large, while the opposite is true for polyvalent metals. In insulators, the electrical conductivity vanishes because $g(E_F) = 0$.

Under appropriate circumstances, the above expression for $\sigma$ reduces to the familiar form $\sigma = ne^2\tau_F/m^*$ of the free-electron model.

### Cyclotron resonance and the Hall effect

The motion of a Bloch electron in a magnetic field is governed by

$$\hbar \frac{d\mathbf{k}}{dt} = -e(\mathbf{v} \times \mathbf{B}).$$

The electron moves along an energy contour in a trajectory perpendicular to the field $\mathbf{B}$, and the motion is referred to as cyclotron motion.

The *cyclotron frequency* is found to be

$$\omega_c = (2\pi eB/h) \Big/ \oint \frac{\delta k}{v},$$

where the integral in the denominator is taken over a closed contour. Measuring this frequency gives information about the shape of the contour, and hence about the shape of the band. The above expression reduces to the familiar form $\omega_c = eB/m^*$ for the case of a standard band.

When both electrons and holes are present in the metal, they both contribute to the Hall constant. The resulting expression is

$$R = \frac{R_e\sigma_e^2 + R_h\sigma_h^2}{(\sigma_e + \sigma_h)^2}.$$

When the electron term dominates, the Hall constant $R$ is negative; when the hole term dominates, the Hall constant $R$ is positive.

### REFERENCES

J. Callaway, 1963, *Energy Band Theory*, New York: Academic Press

J. F. Cochran and R. R. Haering, editors, 1968, *Electrons in Metals*, London: Gordon and Breach

W. A. Harrison and M. B. Webb, editors, 1968, *The Fermi Surface*, New York: Wiley

W. A. Harrison, 1970, *Solid State Theory*, New York: McGraw-Hill

N. F. Mott and H. Jones, 1936, *Theory of the Properties of Metals and Alloys*, Oxford: Oxford University Press; also Dover Press (reprint)

A. B. Pippard, 1965, *Dynamics of Conduction Electrons*, London: Gordon and Breach

F. Seitz, 1940, *Modern Theory of Solids*, New York: McGraw-Hill

D. Schoenberg, "Metallic Electrons in Magnetic Fields," *Contemp. Phys.* **13**, 321, 1972

J. C. Slater, 1965, *Quantum Theory of Molecules and Solids*, Volume II, New York: McGraw-Hill

A. H. Wilson, 1953, *Theory of Metals*, second edition, Cambridge: Cambridge University Press

## QUESTIONS

1. It was pointed out in Sections 6.3 and 4.3 that an electron spends only a little time near an ion, because of the high speed of the electron there. At the same time it was claimed that the ions are "screened" by the electrons, implying that the electrons are so distributed that most of them are located around the ions. Is there a paradox here? Explain.
2. Figure 5.10(c) is obtained from Fig. 5.10(a) by cutting and displacing various segments of the free-electron dispersion curve. Is this rearrangement justifiable for a truly free electron? How do you differentiate between an empty lattice and free space?
3. Explain why the function $\psi_0$ in Fig. 5.18(b) is flat throughout the Wigner–Seitz cell except close to the ion, noting that this behavior is different from that of an atomic wave function, which decays rapidly away from the ion. This implies that the coulomb force due to the ion in cell $A$ is much weakened in the flat region. What is the physical reason for this?
4. *Band overlap* is important in the conductivity of polyvalent metals. Do you expect it to take place in a one-dimensional crystal? You may invoke the symmetry properties of the energy band.

## PROBLEMS

1. Figure 5.7 shows the first three Brillouin zones of a square lattice.
   a) Show that the area of the third zone is equal to that of the first. Do this by appropriately displacing the various fragments of the third zone until the first zone is covered completely.
   b) Draw the fourth zone, and similarly show that its area is equal to that of the first zone.
2. Draw the first three zones for a two-dimensional rectangular lattice for which the ratio of the lattice vectors $a/b = 2$. Show that the areas of the second and third zones are each equal to the area of the first.
3. Convince yourself that the shapes of the first Brillouin zones for the fcc and bcc lattices are those in Fig. 5.8.
4. Show that the number of allowed **k**-values in a band of a three-dimensional sc lattice is $N$, the number of unit cells in the crystal.
5. Repeat Problem 4 for the first zone of an fcc lattice (zone shown in Fig. 5.8a).
6. Derive Eqs. (5.21) and (5.22).
7. Show that the first three bands in the empty-lattice model span the following energy ranges.

$$E_1:\quad 0 \text{ to } \pi^2\hbar^2/2m_0a^2; \qquad E_2:\quad \pi^2\hbar^2/m_0a^2 \text{ to } 2\pi^2\hbar^2/m_0a^2;$$

$$E_3:\quad 2\pi^2\hbar^2/m_0a^2 \text{ to } 9\pi^2\hbar^2/2m_0a^2.$$

8. a) Show that the octahedral faces of the first zone of the fcc lattice (Fig. 5.8a) are due to Bragg reflection from the (111) atomic planes, while the other faces are due to reflection from the (200) planes.
   b) Show similarly that the faces of the zone for the bcc lattice are associated with Bragg reflection from the (110) atomic planes.

9. Suppose that the crystal potential in a one-dimensional lattice is composed of a series of rectangular wells which surround the atom. Suppose that the depth of each well is $V_0$ and its width $a/5$.
   a) Using the NFE model, calculate the values of the first three energy gaps. Compare the magnitudes of these gaps.
   b) Evaluate these gaps for the case in which $V_0 = 5$ eV and $a = 4$ Å.
10. Prove that the wave function used in the TB model, Eq. (5.27), is normalized to unity if the atomic function $\phi_v$ is so normalized. [*Hint*: For the present purpose you may neglect the overlap between the neighboring atomic functions.]
11. The energy of the band in the TB model is given by

$$E(\mathbf{k}) = E_v - \beta - \gamma \sum_j e^{i\mathbf{k}\cdot\mathbf{x}_j},$$

   where $\beta$ and $\gamma$ are constants, as indicated in the text, and $\mathbf{x}_j$ is the position of the $j$th atom relative to the atom at the origin.
   a) Find the energy expression for a bcc lattice, using the nearest-neighbor approximation. Plot the energy contours in the $k_x - k_y$ plane. Determine the width of the energy band.
   b) Repeat part (a) for the fcc lattice.
12. a) Using the fact that the allowed values of $k$ in a one-dimensional lattice are given by $k = n(2\pi/L)$, show that the density of electron states in the lattice, for a lattice of unit length, is given by

$$g(E) = \frac{1}{2\pi} \Big/ \left(\frac{dE}{dk}\right).$$

   b) Evaluate this density of states in the TB model, and plot $g(E)$ versus $E$.
13. Calculate the density of states for the first zone of an sc lattice according to the empty-lattice model. Plot $g(E)$, and determine the energy at which $g(E)$ has its maximum. Explain qualitatively the behavior of this curve.
14. a) Using the free-electron model, and denoting the electron concentration by $n$, show that the radius of the Fermi sphere in **k**-space is given by

$$k_F = (3\pi^2 n)^{1/3}.$$

   b) As the electron concentration increases, the Fermi sphere expands. Show that this sphere begins to touch the faces of the first zone in an fcc lattice when the electron-to-atom ratio $n/n_a = 1.36$, where $n_a$ is the atom concentration.
   c) Suppose that some of the atoms in a Cu crystal, which has an fcc lattice, are gradually replaced by Zn atoms. Considering that Zn is divalent while Cu is monovalent, calculate the atomic ratio of Zn to Cu in a CuZn alloy (brass) at which the Fermi sphere touches the zone faces. Use the free-electron model. (This particular mixture is interesting because the solid undergoes a structural phase change at this concentration ratio.)
15. a) Calculate the velocity of the electron for a one-dimensional crystal in the TB model, and prove that the velocity vanishes at the zone edge.
   b) Repeat (a) for a square lattice. Show that the velocity at a zone boundary is parallel to that boundary. Explain this result in terms of the Bragg reflection.

c) Repeat for a three-dimensional sc lattice, and show once more that the electron velocity at a zone face is parallel to that face. Explain this in terms of Bragg reflection. Can you make a general statement about the direction of the velocity at a zone face?

16. Suppose that a static electric field is applied to an electron at time $t = 0$, at which instant the electron is at the bottom of the band. Show that the position of the electron in real space at time $t$ is given by

$$x = x_0 + \frac{1}{F} E(k = Ft/\hbar),$$

where $x_0$ is the initial position and $F = -e\mathscr{E}$ is the electric force. Assume a one-dimensional crystal, and take the zero-energy level at the bottom of the band. Is the motion in real space periodic? Explain.

17. a) Using the TB model, evaluate the effective mass for an electron in a one-dimensional lattice. Plot the mass $m^*$ versus $k$, and show that the mass is independent of $k$ only near the origin and near the zone edge.
b) Calculate the effective mass at the zone center in an sc lattice using the TB model.
c) Repeat (b) at the zone corner along the [111] direction.

18. Prove Eq. (5.18).

19. a) Calculate the cyclotron frequency $\omega_c$ for an energy contour given by

$$E(\mathbf{k}) = \frac{\hbar^2}{2m_1^*} k_x^2 + \frac{\hbar^2}{2m_2^*} k_y^2,$$

where the magnetic field is perpendicular to the plane of the contour.

$$\left[ \textit{Answer}: \ \omega_c = \sqrt{\frac{e^2}{m_1^* m_2^*}} B. \right]$$

b) Repeat (a) for an ellipsoidal energy surface

$$E(\mathbf{k}) = \frac{\hbar^2}{2m_1^*} (k_x^2 + k_y^2) + \frac{\hbar^2}{2m_3^*} k_z^2,$$

where the field $\mathbf{B}$ makes an angle $\theta$ with the $k_z$-axis of symmetry of the ellipsoid.

$$\left[ \textit{Answer}: \ \omega_c = \left[ \left( \frac{eB}{m_1^*} \right)^2 \cos^2\theta + \frac{e^2 B^2}{m_1^* m_3^*} \sin^2\theta \right]^{1/2}. \right]$$

20. In Section 5.19 we discussed the motion of a Bloch electron in **k**-space in the presence of a magnetic field. The electron also undergoes a simultaneous motion in **r**-space. Discuss this motion, and in particular show that the trajectory in **r**-space lies in a plane parallel to that in **k**-space, that the shapes of the two trajectories are the same except that the one in **r**-space is rotated by an angle of $-\pi/2$ relative to the other, and expanded by a linear scale factor of $(\hbar/eB)$. [*Hint:* Use Eq. (5.108) to relate the electron displacements in **r**- and **k**-space.]

21. Prove Eq. (5.113) for the Hall constant of an electron–hole system.

# CHAPTER 6 SEMICONDUCTORS I: THEORY



---

*There be none of Beauty's daughters*
*With a magic like thee.*

Lord Byron

## 6.1 INTRODUCTION

As a group, semiconductors are among the most interesting and useful substances of all classes of solids. They exhibit a wide spectrum of phenomena, covering the entire range from the strictly metallic to the insulator, and they are extremely versatile in terms of applications. The wide variety of physical devices employing semiconductors is truly impressive.

Although semiconductors have been studied for a long time—since the 1920's—they actually came into their own only after Shockley, Bardeen, and Brattain invented the transistor in the late 1940's. Because of this invention, and because of the ensuing development of other related devices, semiconductors have become the most actively studied substances in solid-state physics. And in the process, we have learned a great deal about the basic properties of these solids, and how to utilize them in designing still newer and more efficient devices.

Because of the wide coverage given to semiconductors here, their study has been divided into two chapters. The present chapter is devoted to the basic physical properties, and the following one explores the use of these properties in the operation of important practical devices.

This chapter begins with the bonding forces in semiconductors, followed by the band structure of these substances. Then we present a procedure for evaluating the number of electrons and holes, the particles responsible for transporting the electric current. Semiconductors are rarely used in a pure form, but are usually doped with foreign impurities, and so we shall discuss the effects of impurities on the supply of electrons and holes.

The most important property of a semiconductor is its electrical conductivity; therefore this topic is considered in some detail. We then show how the cyclotron-resonance and Hall-effect techniques are employed in measuring important parameters. The special effects arising from high electric fields are examined in connection with hot electrons and the Gunn effect.

The optical properties of semiconductors are also discussed, and the intimate relationship between these properties and band structure is indicated. These properties find applications in the phenomena of photoconductivity and luminescence.

The chapter closes with a discussion of diffusion, a phenomenon which arises whenever the electron (or hole) distribution is spatially nonuniform. Since such a distribution obtains in many devices—in all devices, in fact, which contain junctions between positive and negative materials—the diffusion process plays a decisive part in many applications.

## 6.2 CRYSTAL STRUCTURE AND BONDING

Semiconductors include a large number of substances of widely different chemical and physical properties. These materials are grouped into several classes of similar behavior, the classification being based on the position in the periodic table of the elements.

The best-known class is the Group IV semiconductors—C (diamond), Si, Ge, and $\alpha$–Sn (gray tin)—all of which lie in the fourth column of the periodic table. The semiconducting character of these elements was recognized quite early in the history of solid-state research, and they have been studied intensively, particularly Si and Ge, which have found many applications in electronic devices.

The elemental semiconductors all crystallize in the diamond structure. (See Section 1.7 and Fig. 1.15.) The diamond structure has an fcc lattice with a basis composed of two identical atoms, and is such that each atom is surrounded by four neighboring atoms, forming a regular tetrahedron. Figure 6.1 gives a planar view of this coordinational environment in Si, with the three-dimensional tetrahedron projected on a plane.

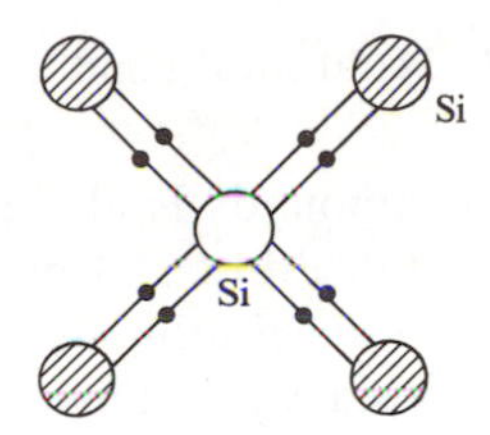


**Fig. 6.1** Tetrahedral bond in Si. Small solid circles represent electrons forming covalent bonds. (See also Fig. 1.19).

Group IV semiconductors are covalent crystals, i.e., the atoms are held together by covalent bonds. These bonds (see Section A.7) consist of two electrons of opposite spins distributed along the line joining the two atoms. Thus, in Fig. 6.1, each of the four bonds joining an Si atom to its neighbors is a double-electron covalent bond. Each of the two atoms on the extremities contributes one electron to the bond. Also the covalent electrons forming the bonds are hybrid $sp^3$ atomic orbitals (see Section A.8). These remarks on Si apply equally well to other Group IV elements.

The picture which has emerged of a covalent crystal is one in which the positive ion cores occupy the lattice sites, and are interconnected by an intricate net of covalent bonds. The total charge on each atom is zero, because the ionic charge is compensated by the covalent electrons for every atom.

Another important group of semiconductors is the Group III–V compounds, so named because each contains two elements, one from the third and the other from the fifth column of the periodic table. The best-known members of this group are GaAs and InSb, but the list also contains compounds such as GaP, InAs, GaSb, and many others.

These substances crystallize in the zincblende structure. As may be recalled from Section 1.7, this is the same as the diamond structure, except that the two atoms forming the basis of the lattice are now different. Thus, in GaAs, the basis of the fcc lattice consists of two atoms, Ga and As. Because of this structure, each atom is surrounded by four others of the opposite kind, and these latter atoms form a regular tetrahedron, just as in the diamond structure. Figure 6.2 shows this for the case of GaAs.

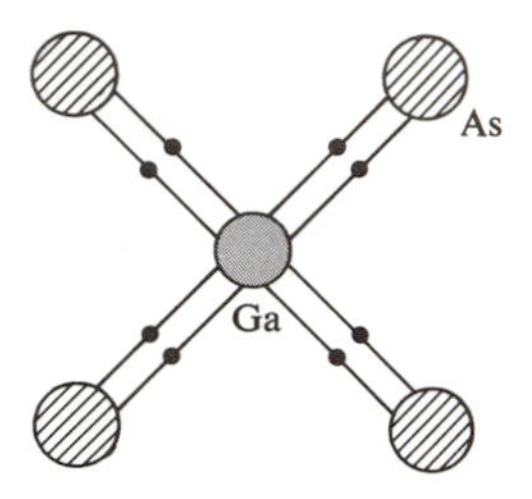


**Fig. 6.2** Tetrahedral bond in GaAs.

The bonding in the III–V compounds is also primarily covalent. The eight electrons required for the four tetrahedral covalent bonds are supplied by the two types of atoms, the trivalent atom contributing its three valence electrons, and the pentavalent five electrons. One would expect the bonding in these substances to be covalent because of the crystal structure, since the tetrahedral bond is usually associated with covalent bonding.

The bonding in this group, however, is not entirely covalent. Because the two elements in the compound are different, the distribution of the electrons along the bond is not symmetric, but is displaced toward one of the atoms. As a result, one of the atoms acquires a net electric charge. Such a bond is called *heteropolar*, in contrast to the purely covalent bond in the elemental semiconductors, which is called *homopolar*.

The distribution of electrons in the bond is displaced toward the atom of higher *electronegativity*. In GaAs, for instance, the As atom has a higher electronegativity than the Ga, and consequently the As atom acquires a net negative charge, whose value is $-0.46e$ per atom (a typical value in Group III–V compounds). The Ga atom correspondingly acquires a net positive charge of $0.46e$. The transferred charge per atom is known as the *effective charge*.

Charge transfer leads to an ionic contribution to the bonding in Group III–V compounds. Their bonding is therefore actually a mixture of covalent and ionic components, although covalent ones predominate in most of these substances.

The III–V compounds possess a *polar* character. Because of the opposite charges on the ions, the lattice may be polarized by the application of an electric field. Thus, in these substances, the ions' displacement contributes to the dielectric constant. A particularly interesting manifestation of this is the strong dispersion in the infrared region due to the interaction of light with the optical phonons (Section 3.12).

Another class of substances which has received much attention lately is the II–VI semiconductor, such as CdS and ZnS. Most of these compounds also crystallize in the zincblende structure, indicating that the bonding is primarily covalent in nature. This is so, but the charge transfer here is greater than in the III–V compounds (a typical value is 0.48*e*). Hence in the II–VI compounds the ionic contribution to the bonding is greater and the polar character stronger.

And finally there is the important group of lead salts which form the IV–VI compounds, for example, PbTe.

## 6.3 BAND STRUCTURE

A semiconductor was defined in Section 5.10 as a solid in which the highest occupied energy band, the *valence band*, is completely full at $T = 0°K$, but in which the gap above this band is also small, so that electrons may be excited thermally at room temperature from the valence band to the next-higher band, which is known as the *conduction band*.† Generally speaking, the number of excited electrons is appreciable (at room temperature) whenever the energy gap $E_g$ is less than 2 eV. The substance may then be classified as a semiconductor. When the gap is larger, the number of electrons is negligible, and the substance is an insulator.

When electrons are excited across the gap, the bottom of the conduction band (CB) is populated by electrons, and the top of the valence band (VB) by holes. As a result, both bands are now only partially full, and would carry a current if an electric field were applied. The conductivity of the semiconductor is small compared with the conductivities of metals of the small number of electrons and holes involved, but this conductivity is nonetheless sufficiently large for practical purposes.

Only the CB and VB are of interest to us here, because only these two bands contribute to the current. Bands lower than the VB are completely full, and those higher than the CB completely empty, so that neither of these groups of bands contribute to the current; hence they may be ignored so far as semiconducting properties are concerned. In characterizing a semiconductor, therefore, we need describe only the CB and VB.

The simplest band structure of a semiconductor is indicated in Fig. 6.3. The energy of the CB has the form

$$E_c(\mathbf{k}) = E_g + \frac{\hbar^2 k^2}{2m_e^*}, \tag{6.1}$$

where $\mathbf{k}$ is the wave vector and $m_e^*$ the effective mass of the electron. The energy $E_g$ represents the energy gap. The zero-energy level is chosen to lie at the top of the VB.

† A word of caution concerning terminology: When we are discussing metals, the words "valence" and "conduction" are used interchangeably. Thus the delocalized electrons in metals are called either valence electrons or conduction electrons. When we are dealing with semiconductors, however, the words "valence" and "conduction" refer to two distinctly different electrons or bands.

We have used the standard band form to describe the CB, because we are primarily interested in the energy range close to the bottom of the band, since it is this range which contains most of the electrons. Recall from Section 5.6 that the standard-band form holds true near the bottom of the band.

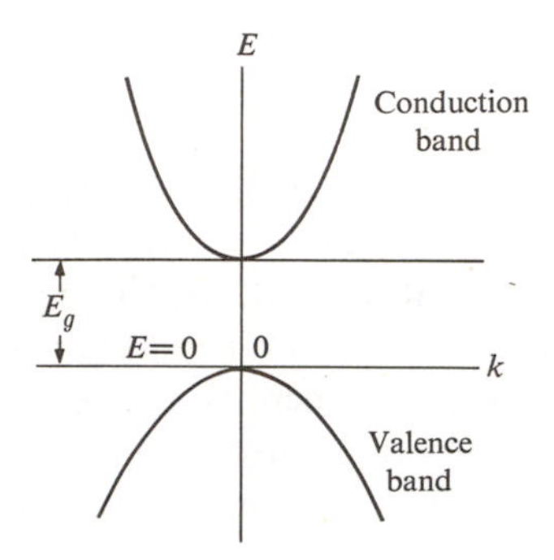


**Fig. 6.3** Band structure in a semiconductor.

The energy of the VB (Fig. 6.3) may be written as

$$E_v(\mathbf{k}) = -\frac{\hbar^2 k^2}{2m_h^*}, \tag{6.2}$$

where $m_h^*$ is the effective mass of the hole. (Recall from Section 5.1 that, because of the inverted shape of the VB, the mass of an electron at the top of the VB is negative, equal to $-m_h^*$, but the mass of a hole is positive.) The VB is again represented by the standard inverted form because we are interested only in the region close to the top of the band, where most of the holes lie.

The primary band-structure parameters are thus the electron and hole masses $m_e$ and $m_h$ (the asterisks have been dropped for convenience), and the band gap $E_g$. Table 6.1 gives these parameters for various semiconductors. Note that the masses differ considerably from—and are often much smaller than—the free-electron mass, and that the energy gaps range from 0.18 eV in InSb to 3.7 eV in ZnS. The table also shows that the wider the gap, the greater the mass of the electron. We have already alluded to this property in the discussion of the NFE model (see the remark following Eq. (5.23)].

The energy gap for a semiconductor varies with temperature, but the variation is usually slight. That a variation with temperature should exist at all can be appreciated from the fact that the crystal, when it is heated, experiences a volume expansion, and hence a change in its lattice constant. This, in turn, affects the band structure, which, as we found in Chapter 5, is a sensitive function of the lattice constant.

It also follows that the gap may be varied by applying pressure, as this too induces a change in the lattice constant. Studies of semiconductors under high pressure have, in fact, proved very helpful in elucidating some of their properties.

**Table 6.1**

Parameters for Band Structure of Semiconductors (Room Temperature)

| Group | Crystal | $E_g$, eV | Effective mass, $m/m_0$ Electrons | Holes |
|---|---|---|---|---|
| IV | C | 5.3 | | |
| | Si | 1.1 | $m_l = 0.97, m_t = 0.19$ | 0.5, 0.16 |
| | Ge | 0.7 | $m_l = 1.6, m_t = 0.08$ | 0.3, 0.04 |
| | $\alpha$ Sn | 0.08 | | |
| III–V | GaAs | 1.4 | 0.07 | 0.09 |
| | GaP | 2.3 | 0.12 | 0.50 |
| | GaSb | 0.7 | 0.20 | 0.39 |
| | InAs | 0.4 | 0.03 | 0.02 |
| | InP | 1.3 | 0.07 | 0.69 |
| | InSb | 0.2 | 0.01 | 0.18 |
| II–VI | CdS | 2.6 | 0.21 | 0.80 |
| | CdSe | 1.7 | 0.13 | 0.45 |
| | CdTe | 1.5 | 0.14 | 0.37 |
| | ZnS | 3.6 | 0.40 | 5.41 |
| | ZnSe | 2.7 | 0.10 | 0.60 |
| | ZnTe | 2.3 | 0.10 | 0.60 |
| IV–VI | PbS | 0.4 | 0.25 | 0.25 |
| | PbSe | 0.3 | 0.33 | 0.34 |
| | PbTe | 0.3 | 0.22 | 0.29 |

*Note*: $m_l$ and $m_t$ refer to longitudinal and transverse masses, respectively, of ellipsoidal energy surfaces. When there is more than one value for hole mass, the values refer to heavy and light holes (see Section 6.9).

The conduction and valence bands in semiconductors are related to the atomic states. The discussion of the hydrogen molecule (Section A.7) states that, when two hydrogen atoms are brought together to form a molecule, the atomic 1s state splits into two states: a low-energy bonding state and a high-energy antibonding state. In solid hydrogen, these states broaden into bonding and antibonding energy bands, respectively. In like fashion, the valence and conduction bands in semiconductors are, respectively, the bonding and antibonding bands of the corresponding atomic valence states. Thus the VB and CB in Si, for example, result from the bonding and antibonding states of the hybrid $3s^1 3p^3$ (see Section A.8). Similar remarks apply to the bands in Ge, C, and other semiconductors.

The band structure in Fig. 6.3 is the simplest possible structure. Band structures of real semiconductors are somewhat more complicated, as we shall see, but for the present the simple structure will suffice for our purposes.

## 6.4 CARRIER CONCENTRATION; INTRINSIC SEMICONDUCTORS

In the field of semiconductors, electrons and holes are usually referred to as *free carriers*, or simply *carriers*, because it is these particles which are responsible for carrying the electric current. The number of carriers is an important property of a semiconductor, as this determines its electrical conductivity. In order to determine the number of carriers, we need some of the basic results of statistical mechanics.

The most important result in this regard is the Fermi–Dirac (FD) distribution function

$$f(E) = \frac{1}{e^{(E-E_F)/k_B T} + 1}. \tag{6.3}$$

This function,† which we encountered in Section 4.6, gives the probability that an energy level $E$ is occupied by an electron when the system is at temperature $T$. The function is plotted versus $E$ in Fig. 6.4.

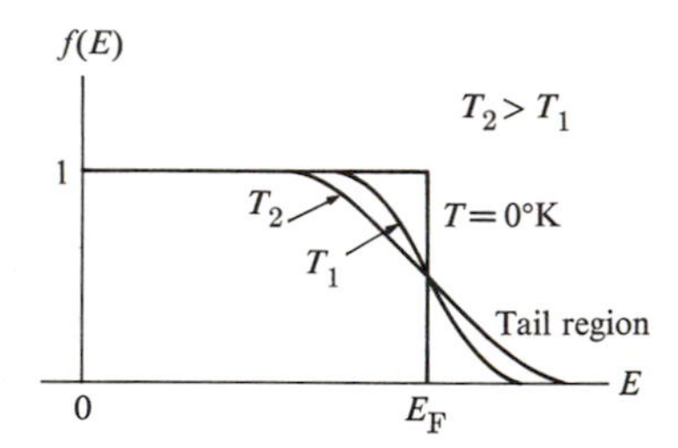


**Fig. 6.4** The Fermi–Dirac distribution function.

Here we see that, as the temperature rises, the unoccupied region below the Fermi level $E_F$ becomes longer, which implies that the occupation of high energy states increases as the temperature is raised, a conclusion which is most plausible, since increasing the temperature raises the overall energy of the system. Note also that $f(E) = \frac{1}{2}$ at the Fermi level ($E = E_F$) regardless of the temperature. That is, the probability that the Fermi level is occupied is always equal to one-half.

† In this chapter as well as in the following one, the Boltzmann constant is denoted by $k_B$ rather than the usual $k$ in order to avoid confusion, because the latter symbol has been used to denote the wave vector in $k$-space in band theory. In the remainder of the book, however, this confusion does not arise and the Boltzmann constant will therefore be denoted by $k$, as usual.

In semiconductors it is the tail region of the FD distribution which is of particular interest. In that region the inequality $(E - E_F) \gg k_B T$ holds true, and one may therefore neglect the term unity in the denominator of (6.1). The FD distribution then reduces to the form

$$f(E) = e^{E_F/k_B T} e^{-E/k_B T}, \tag{6.4}$$

which is the familiar Maxwell–Boltzmann, or classical, distribution. This simple distribution therefore suffices for the discussion of electron statistics in semiconductors.

We can calculate the concentration of electrons in the CB in the following manner. The number of states in the energy range $(E, E + dE)$ is equal to $g_e(E)\,dE$, where $g_e(E)$ is the density of electron states (Section 5.11). Since each of these states has an occupation probability $f(E)$, the number of electrons actually found in this energy range is equal to $f(E)g_e(E)\,dE$. The concentration of electrons throughout the CB is thus given by the integral over the band

$$n = \int_{E_{c1}}^{E_{c2}} f(E) g_e(E)\,dE, \tag{6.5}$$

where $E_{c1}$ and $E_{c2}$ are the bottom and top of the band, respectively, as shown in Fig. 6.5(a).

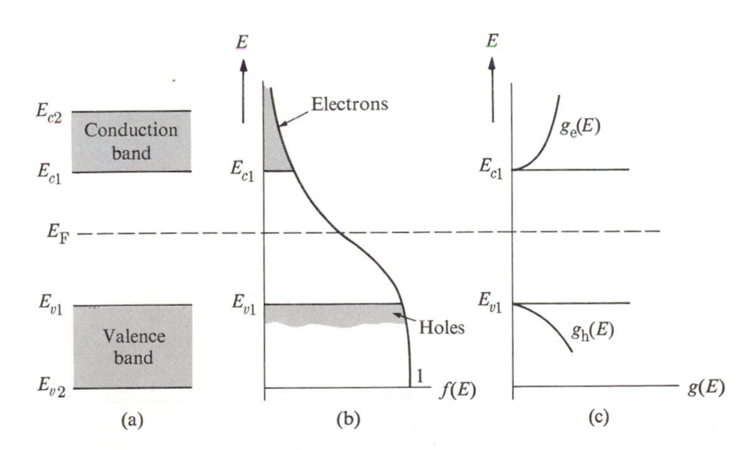


**Fig. 6.5** (a) Conduction and valence bands. (b) The distribution function. (c) Density of states for electrons and holes: $g_e(E)$ and $g_h(e)$.

The distribution function is shown in Fig. 6.5(b). Note that the entire CB falls in the tail region. Thus we may use the Maxwell–Boltzmann function for $f(E)$ in (6.5). (Proof of this statement will come later, when we show that the Fermi energy lies very near the middle of the energy gap.)

We calculated the density of states in Section 5.11, where the expression appropriate to the standard band form is, according to Eq. (5.63), given by

$$g_e(E) = \frac{1}{2\pi^2}\left(\frac{2m_e}{\hbar^2}\right)^{3/2}(E - E_g)^{1/2}, \tag{6.6}$$

where the zero-energy level has been chosen to lie at the top of the VB. Thus $g_e(E)$ vanishes for $E < E_g$, and is finite only for $E_g < E$, as shown in Fig. 6.5(c).

When we substitute for $f(E)$ and $g_e(E)$ into (6.5), we obtain

$$n = \frac{1}{2\pi^2}\left(\frac{2m_e}{\hbar^2}\right)^{3/2} e^{E_F/k_BT}\int_{E_g}^{\infty}(E - E_g)^{1/2}e^{-E/k_BT}dE. \tag{6.7}$$

For convenience, the top of the CB has been set equal to infinity. Since the integrand decreases exponentially at high energies, the error introduced by changing this limit from $E_{c2}$ to $\infty$ is quite neglibible. By changing the variable, and using the result

$$\int_0^{\infty} x^{1/2}e^{-x}dx = \frac{\pi^{1/2}}{2},$$

one can readily evaluate the integral in (6.7). The electron concentration then reduces to the expression

$$n = 2\left(\frac{m_e k_B T}{2\pi\hbar^2}\right)^{3/2} e^{E_F/k_BT}\, e^{-E_g/k_BT}. \tag{6.8}$$

The electron concentration is still not known explicitly because the Fermi energy $E_F$ is so far unknown. This can be calculated in the following manner. Essentially the same ideas employed above may also be used to evaluate the number of holes in the VB. The probability that a hole occupies a level $E$ in this band is equal to $1 - f(E)$, since $f(E)$ is the probability of electron occupation. Thus the probability of hole occupation $f_h$ is

$$f_h = 1 - f(E). \tag{6.9}$$

Since the energy range involved here is much lower than $E_F$, the FD function of (6.3) must be used rather than (6.4). Thus

$$f_h = 1 - \frac{1}{e^{(E-E_F)/k_BT} + 1} = \frac{1}{e^{(E_F-E)/k_BT} + 1} \simeq e^{-E_F/k_BT}e^{E/k_BT}, \tag{6.10}$$

where the approximation in the last expression follows as a result of the inequality $(E_F - E) \gg k_BT$. The validity of this inequality in turn can be seen by referring to Fig. 6.5(b), which shows that $E_F - E$ is of the order of $E_g/2$, which is much larger than $k_BT$ at room temperature.

The density of states for the holes is

$$g_{\mathrm{h}}(E) = \frac{1}{2\pi^2}\left(\frac{2m_{\mathrm{h}}}{\hbar^2}\right)^{3/2}(-E)^{1/2}, \tag{6.11}$$

which is appropriate for an inverted band [see also Eq. (5.64)]. Note that the term $(-E)$ in this equation is positive, because the zero-energy level is at the top of the VB, and the energy is measured positive upward and negative downward from this level.

The hole concentration is thus given by

$$p = \int_{-\infty}^{0} f_{\mathrm{h}}(E)\, g_{\mathrm{h}}(E)\, dE. \tag{6.12}$$

When we substitute for $f_{\mathrm{h}}(E)$ and $g_{\mathrm{h}}(E)$ from the above equations and carry out the integral as in the electron case, we obtain

$$p = 2\left(\frac{m_{\mathrm{h}} k_{\mathrm{B}} T}{2\pi\hbar^2}\right)^{3/2} e^{-E_{\mathrm{F}}/k_{\mathrm{B}}T}. \tag{6.13}$$

The electron and hole concentrations have thus far been treated as independent quantities. The two concentrations are, in fact, equal, because the electrons in the CB are due to excitations from the VB across the energy gap, and for each electron thus excited a hole is created in the VB. Therefore

$$n = p. \tag{6.14}$$

If we substitute $n$ and $p$ from (6.8) and (6.13), respectively, into (6.14), we obtain an equation involving the only unknown, $E_{\mathrm{F}}$. The solution of this equation is

$$E_{\mathrm{F}} = \tfrac{1}{2} E_g + \tfrac{3}{4} k_{\mathrm{B}} T \log\left(\frac{m_{\mathrm{h}}}{m_{\mathrm{e}}}\right). \tag{6.15}$$

Since $k_{\mathrm{B}}T \ll E_g$ under usual circumstances, the second term on the right of (6.15) is very small compared with the first, and the energy level is close to the middle of the energy gap. This is consistent with earlier assertions that both the bottom of the CB and the top of the VB are far from the Fermi level.†

The concentration of electrons may now be evaluated explicitly by using the above value of $E_{\mathrm{F}}$. Substitution of (6.15) into (6.8) yields

$$n = 2\left(\frac{k_{\mathrm{B}}T}{2\pi\hbar^2}\right)^{3/2} (m_{\mathrm{e}} m_{\mathrm{h}})^{3/4} e^{-E_g/2k_{\mathrm{B}}T}. \tag{6.16}$$

The important feature of this expression is that $n$ increases very rapidly—exponentially—with temperature, particularly by virtue of the exponential factor.

† The fact that the Fermi level falls in the energy gap—the forbidden region—poses no difficulties. This level is a theoretical concept and no electrons need be present there.

Thus as the temperature is raised, a vastly greater number of electrons is excited across the gap. (This can be visualized by recalling that as the temperature is raised, the tail of the FD distribution in the CB becomes longer, and more states are occupied in this band.)

Figure 6.6 is a plot of log $n$ versus $1/T$. The curve is a straight line of slope equal to $(-E_g/2k_B)$. [The $T^{3/2}$-dependence in (6.16) is so weak in comparison with the exponential dependence that the former may be disregarded for the purpose of this discussion.)

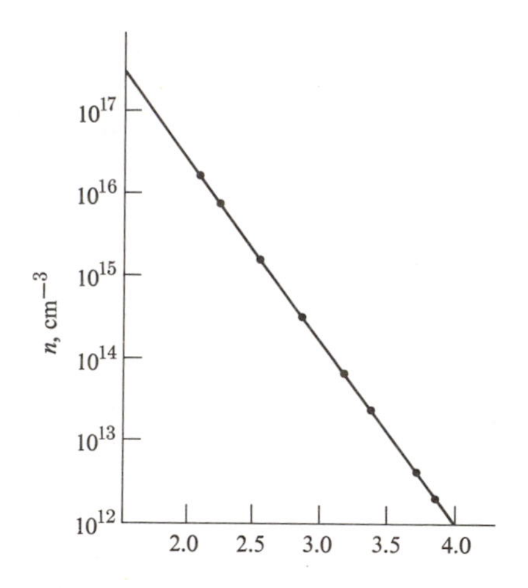


**Fig. 6.6** Electron concentration $n$ versus $1/T$ in Ge. [After Morin and Morita, *Phys. Rev.* **96**, 28, 1954]

One can estimate the numerical value of $n$ by substituting the values $E_g = 1$ ev, $m_e = m_h = m_0$, and $T = 300°$K. One finds $n \simeq 10^{15}$ electrons/cm$^3$, a typical value of carrier concentration in semiconductors.

Note that the expression (6.16) also gives the hole concentration, since $n = p$.

Our discussion of carrier concentration in this section is based on the premise of a pure semiconductor. When the substance is *im*pure, additional electrons or holes are provided by the impurities, as will be seen in Section 6.5. In that case, the concentrations of electrons and holes may no longer be equal, and the amount of each depends on the concentration and type of impurity present. When the substance is sufficiently pure so that the concentrations of electrons and holes are equal, we speak of an *intrinsic semiconductor*. That is, the concentrations are determined by the intrinsic properties of the semiconductor itself. On the other hand, when a substance contains a large number of impurities which supply most of the carriers, it is referred to as an *extrinsic semiconductor*.

## 6.5 IMPURITY STATES

A pure semiconductor has equal numbers of both types of carriers, electrons and holes. In most applications, however, one needs specimens which have one type of carrier only, and none of the other. (This will be seen in Chapter 7 when we discuss, for example, the junction transistor.) By doping the semiconductor with appropriate impurities, one can obtain samples which contain either electrons only or holes only.

Consider, for instance, a specimen of Si which has been *doped* by As. The As atoms (the impurities) occupy some of the lattice sites formerly occupied by the Si host atoms. The distribution of the impurities is random throughout the lattice. But their presence affects the solid in one very important respect: The As atom is pentavalent (while Si is tetravalent). Of the five electrons of As, four participate in the tetrahedral bond of Si, as shown in Fig. 6.7. The fifth electron cannot enter the bond, which is now saturated, and hence this electron detaches from the impurity and is free to migrate through the crystal as a conduction electron, i.e., the electron enters the CB. The impurity is now actually a positive ion, $As^+$ (since it has lost one of its electrons), and thus it tends to capture the free electron, but we shall show shortly that the attraction force is very weak, and not enough to capture the electron in most circumstances.

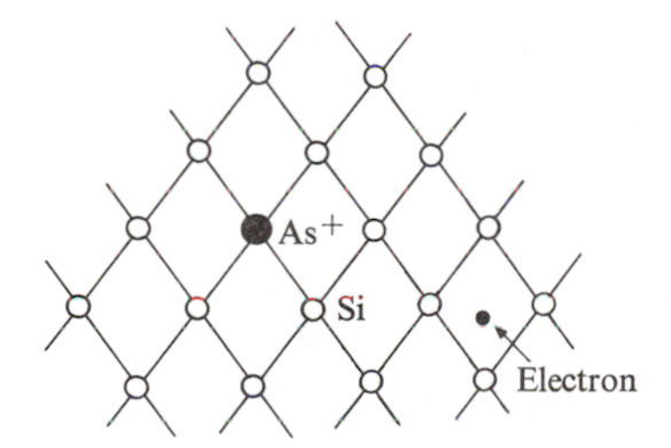


**Fig. 6.7** An As impurity in a Si crystal. The extra electron migrates through the crystal.

The net result is that the As impurities contribute electrons to the CB of the semiconductors, and for this reason these impurities are called *donors*. Note that the electrons have been created *without* the generation of holes.

When an electron is captured by an ionized donor, it orbits around the donor much like the situation in hydrogen (Fig. 6.8). We can calculate the binding energy by using the familiar Bohr model. However, we must take into account the fact that the coulomb interaction here is weakened by the screening due to the presence of the semiconductor crystal, which serves as a medium in which both the donor and ion reside. Thus the coulomb potential is now given by

$$V(r) = -\frac{e^2}{4\pi\epsilon_r\epsilon_0 r}, \tag{6.17}$$

where $\epsilon_r$ is the reduced dielectric constant of the medium. The dielectric constant $\epsilon_r = 11.7$ in Si, for example, showing a substantial decrease in the interaction force. It is this screening which is responsible for the small binding energy of the electron at the donor site.

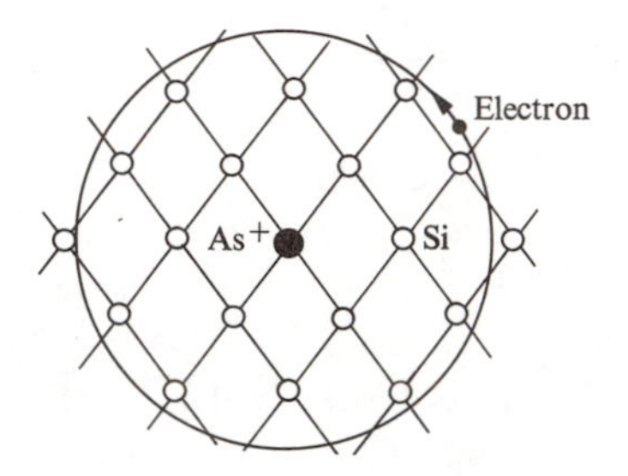


**Fig. 6.8** Orbit of an electron around a donor.

When one used this potential in the Bohr model, one finds the binding energy, corresponding to the ground state of the donor, to be

$$E_d = +\frac{1}{\epsilon_r^2}\left(\frac{m_e}{m_0}\right)\left[\frac{e^4 m_0}{2(4\pi\epsilon_0\hbar)^2}\right]. \tag{6.18}$$

Note that the effective mass $m_e$ has been used rather than the free mass $m_0$. [The mass $m_0$ in (6.18) actually cancels out, and is inserted only for convenience.] The last factor on the right in (6.18) is the binding energy of the hydrogen atom, which is equal to 13.6 eV. The binding energy of the donor is therefore reduced by the factor $1/\epsilon_r^2$, and also by the mass factor $m_e/m_0$, which is usually smaller than unity. If we used the typical values $\epsilon_r \simeq 10$ and $m_e/m_0 \simeq 0.2$, we would see that the binding energy of the donor is about 1/500th as much as the hydrogen energy, i.e., about 0.01 eV. This is indeed the order of the observed values.

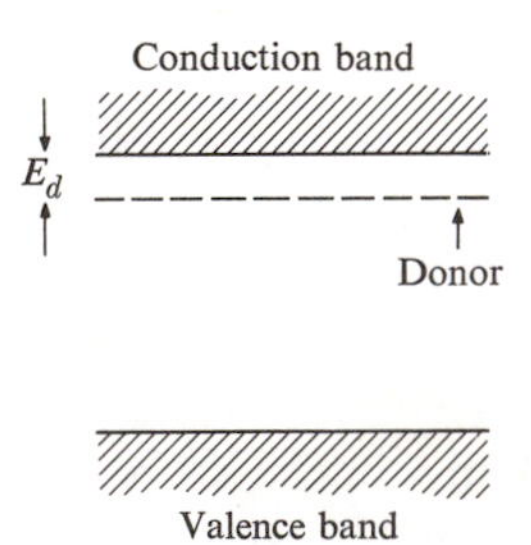


**Fig. 6.9** The donor level in a semiconductor.

The donor level lies in the energy gap, very slightly below the conduction band, as shown in Fig. 6.9. Because the level is so close to the CB, almost all the donors are ionized at room temperature, their electrons having been excited into the CB. (Recall that the thermal energy $k_B T = 0.025$ eV at room temperature.)

Table 6.2 lists the binding energies of various crystals.

**Table 6.2**

Ionization Energies of Donors and Acceptors in Si and Ge (in Electron Volts)

| Impurity | Si ($\epsilon_r = 11.7$) | Ge ($\epsilon_r = 16.0$) |
|---|---|---|
| *Donors* | | |
| Li | 0.033 | — |
| P | 0.044 | 0.012 |
| AS | 0.049 | 0.013 |
| Sb | 0.039 | 0.096 |
| Bi | 0.069 | |
| *Acceptors* | | |
| B | 0.045 | 0.010 |
| Al | 0.057 | 0.010 |
| Ga | 0.065 | 0.011 |
| In | 0.16 | 0.011 |

It is instructive to evaluate the Bohr radius of the donor electron. Straightforward adaptation of the Bohr result leads to

$$r_d = \epsilon_r \left(\frac{m_0}{m_e}\right) a_0, \tag{6.19}$$

where $a_0$ is the Bohr radius, equal to 0.53 Å. The radius of the orbit is thus much larger than $a_0$, by a factor of 50, if we use the previous values for $\epsilon_r$ and $m_e$. A typical radius is thus of the order of 30 Å. Since this is much greater than the interatomic spacing, the orbit of the electron encloses a great many host atoms (Fig. 6.8), and our picture of the lattice acting as a continuous, polarizable dielectric is thus a plausible one.

Since the donors are almost all ionized, the concentration of electrons is nearly equal to that of the donors. Typical concentrations are about $10^{15}$ cm$^{-3}$. But sometimes much higher concentrations are obtained by heavy doping of the sample, for example, $10^{18}$ cm$^{-3}$ or even more.

**Acceptors**

An appropriate choice of impurity may produce holes instead of electrons. Suppose that the Si crystal is doped with Ga impurity atoms. The Ga impurity resides at a site previously occupied by an Si atom, but since Ga is trivalent, one of the electron bonds remains vacant (Fig. 6.10). This vacancy may be filled by an electron moving in from another bond, resulting in a vacancy (or hole) at this latter bond. The hole is then free to migrate throughout the crystal. In this manner, by introducing a large number of trivalent impurities, one creates an appreciable concentration of holes, which lack electrons.

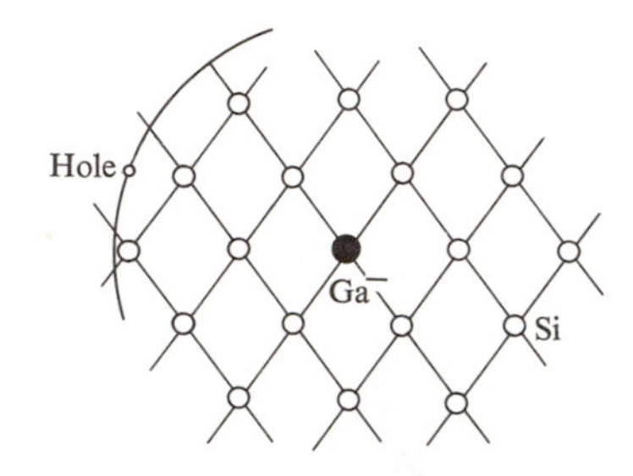


**Fig. 6.10** A Ga impurity in a Si crystal. The extra hole migrates through the crystal.

The trivalent impurity is called an *acceptor*, because it accepts an electron to complete its tetrahedral bond.

The acceptor is negatively charged, by virtue of the additional electron it has entrapped. Since the resulting hole has a positive charge, it is attracted by the acceptor. We can evaluate the binding energy of the hole at the acceptor in the same manner followed above in the case of the donor. Again this energy is very small, of the order of 0.01 eV. (See Table 6.2 for a list.) Thus essentially all the acceptors are ionized at room temperature.

The acceptor level lies in the energy gap, slightly above the edge of the VB, as shown in Fig. 6.11. This level corresponds to the hole being captured by the acceptor. When an acceptor is ionized (an electron excited from the top of the VB to fill this hole), the hole falls to the top of the VB, and is now a free carrier. Thus the ionization process, indicated by upward transition of the electron on the energy scale, may be represented by a downward transition of the hole on this scale.

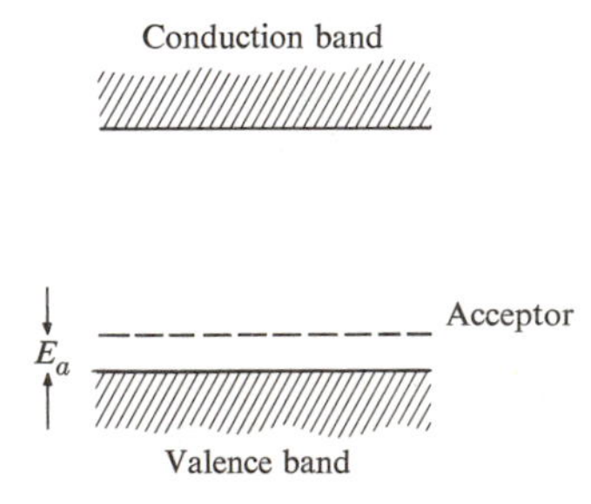


**Fig. 6.11** The acceptor level in a semiconductor.

We have just been saying that the energy levels of both donors and acceptors have been found to lie in the energy gap of the crystal. Yet in Chapter 5 when we discussed the band model we emphasized that the energy range of the gap is forbidden, and that no electron states could exist there. There is no contradiction, however, because the discussion in Chapter 5 was concerned with perfect crystals, while the donor and acceptor levels are related to *impurity* states, and thus to *imperfections* in the crystal. Another manifestation of this difference is that impurity states, representing bound states, are localized, not delocalized, as are Bloch electrons. Thus impurity states are nonconducting.

## 6.6 SEMICONDUCTOR STATISTICS

Semiconductors usually contain both donors and acceptors. Electrons in the CB can be created either by interband thermal excitation or by thermal ionization of the donors. Holes in the VB may be generated by interband excitation, or by thermal excitation of electrons from the VB into the acceptor level. And in addition, electrons may fall from the donor levels to the acceptor level. Figure 6.12 indicates these various processes.

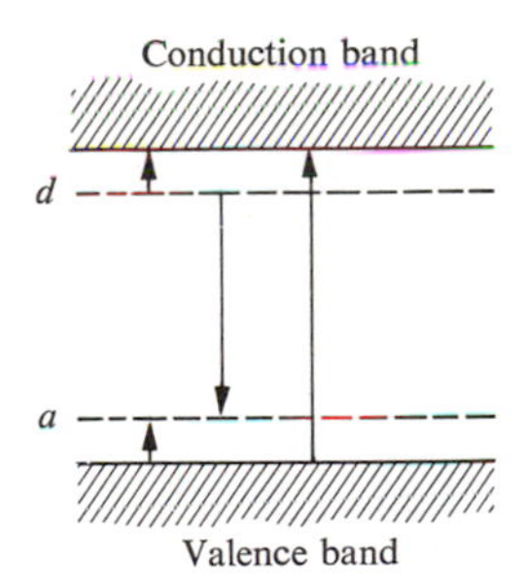


**Fig. 6.12** The various electronic processes in a semiconductor (see text).

Finding the concentrations of carriers—both electrons and holes—under the most general of circumstances, taking all these processes into account, is quite complicated. Instead of giving the details of such general calculations here, we shall treat a few special cases instead, ones which are often encountered in practice. Two regions may be distinguished, depending on the physical parameters involved: The *intrinsic* and the *extrinsic* regions.

### The intrinsic region

The concentration of carriers in the intrinsic region is determined primarily by thermally induced interband transitions. Consequently we have, to a good approximation,

$$n = p. \tag{6.20}$$

In that case, we find the carrier concentrations as we did in Section 6.4, namely

$$n = p = n_i = 2\left(\frac{k_B T}{2\pi\hbar^2}\right)^{3/2}(m_e m_h)^{3/4} e^{-E_g/2k_B T}. \tag{6.21}$$

This is known as the *intrinsic concentration*, denoted by $n_i$.

The intrinsic region obtains when the impurity doping is small. When we denote the concentrations of donors and acceptors by $N_d$ and $N_a$, the requirement for the validity of the intrinsic condition is

$$n_i \gg (N_d - N_a). \tag{6.22}$$

The reason for this condition is readily understandable. There are $N_d$ electrons at the donor level, but of these a number $N_a$ may fall into the acceptors, leaving only $N_d - N_a$ electrons to be excited from the donor level into the conduction band. When condition (6.22) is satisfied, the ionization of all these remaining impurities is not sufficient to appreciably affect the number of electrons excited thermally from the VB. The semiconductor may then be treated as a pure sample, and the influence of impurities disregarded. This is precisely what we did in obtaining (6.21).

Since $n_i$ increases rapidly with temperature, the intrinsic condition becomes more favorable at higher temperatures. All semiconductors, in fact, become intrinsic at sufficiently high temperatures (unless the doping is unusually high).

**The extrinsic region**

Quite often the intrinsic condition is not satisfied. For the common dopings encountered, about $10^{15}$ cm$^{-3}$, the number of carriers supplied by the impurities is large enough to change the intrinsic concentration appreciably at room temperature. The contribution of impurities, in fact, frequently exceeds those carriers that are supplied by interband excitation. When this is so, the sample is in the *extrinsic region.*

Two different types of extrinsic regions may be distinguished. The first occurs when the donor concentration greatly exceeds the acceptor concentration, that is, when $N_d \gg N_a$. In this case, the concentration of electrons may be evaluated quite readily. Since the donor's ionization energy (the binding energy discussed in Section 6.5) is quite small, all the donors are essentially ionized, their electrons going into the CB. Therefore, to a good approximation,

$$n = N_d. \tag{6.23}$$

The hole concentration is small under this condition. To calculate this concentration, we make the following useful observation. Returning to Section 6.4, we note that Eq. (6.8) is still valid, even in the case of a doped sample. Only when we used (6.14) to evaluate $E_F$ was the discussion restricted to an intrinsic sample. Similarly, Eq. (6.13) is also valid whether the sample is pure or doped. If we multiply these two equations, we find that

$$np = 4\left(\frac{k_B T}{2\pi\hbar^2}\right)^3 (m_e m_h)^{3/2}\, e^{-E_g/k_B T}. \tag{6.24}$$

Note that the troublesome Fermi energy has disappeared from the right side. Thus the product $np$ is independent of $E_F$, and hence of the amount and type of doping; the product $np$ depends only on the temperature. We also see from comparison with (6.21) that the right side is equal to $n_i^2$ [which is reasonable, since Eq. (6.24) is also valid in the intrinsic region, in which case the left side is equal to $n_i^2$]. We may thus write

$$np = n_i^2. \tag{6.25}$$

This equation means that, if there is no change in temperature, the product $np$ is a constant, independent of the doping. If the electron concentration is increased, by varying the doping, the hole concentration decreases, and vice versa.

When the doping is primarily of the donor type, $n \simeq N_d$, as shown by (6.23). According to (6.25), the concentration of holes is

$$p = \frac{n_i^2}{N_d}. \tag{6.26}$$

Since we are in the extrinsic region, $n_i \ll N_d$, and hence $p \ll N_d = n$. Thus the concentration of electrons is much larger than that of holes.

A semiconductor in which $n \gg p$ is called an *n-type semiconductor* ($n$ for negative); this terminology dates back to the early days of semiconductors. Such a sample is characterized, as we have seen, by a great concentration of electrons (donors). (For a strongly $n$-type sample, $n \gg p$, while for a weakly $n$-type sample, $n \gtrsim p$.)

The other type of extrinsic region occurs when $N_a \gg N_d$, that is, the doping is primarily by acceptors. Using an argument similar to the above, one then has

$$p \simeq N_a, \tag{6.27}$$

i.e., all the acceptors are ionized. The electron concentration, which is small, is given by

$$n = \frac{n_i^2}{N_a}. \tag{6.28}$$

Such a material is called a *p-type semiconductor*. It is characterized by a preponderance of holes (acceptors).

In discussing ionization of donors (and acceptors), we assumed that the temperature is sufficiently high so that all of these are ionized. This is certainly true at room temperature. But if the temperature is progressively lowered, a point is reached at which the thermal energy becomes too small to cause electron excitation. In that case, the electrons fall from the CB into the donor level, and the conductivity of the sample diminishes dramatically. This is referred to as *freeze-out*, in that the electrons are now "frozen" at their impurity sites. We can estimate the temperature at which freeze-out takes place from the equation $E_d \simeq k_B T$, which gives a temperature of about 100°K.

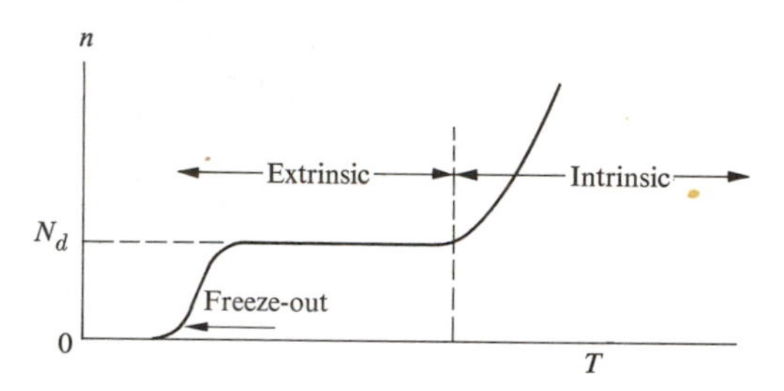


**Fig. 6.13** Variation of electron concentration $n$ with temperature in an $n$-type semiconductor.

The variation of the electron concentration with temperature in an *n-type* sample is indicated schematically in Fig. 6.13, in which the various regions—freeze-out, extrinsic, intrinsic—are clearly marked.

## 6.7 ELECTRICAL CONDUCTIVITY; MOBILITY

Electrical conductivity is, of course, the quantity of primary interest in semiconductors. Both electrons and holes contribute to electric current. But in order to simplify the discussion, let us begin with a sample which contains only one type of carrier: electrons. In other words, the sample is strongly $n$-type.

When an electric field is exerted, electrons drift opposite to the field and carry a net electric current. Since the electrons are represented by an effective mass $m_e$, it follows from Chapter 5 that they may be treated according to the free-electron model of Chapter 4. The electrical conductivity (Section 4.4) is therefore given by

$$\sigma_e = \frac{ne^2 \tau_e}{m_e^*}, \tag{6.29}$$

where $\tau_e$ is the lifetime of the electron. To obtain an order-of-magnitude value for $\sigma_e$, we substitute $n = 10^{15}$ cm$^{-3}$ $= 10^{21}$ $m^{-3}$, $\tau_e = 10^{-12}$ s, and $m_e = 0.1 m_0$. This leads to $\sigma \simeq 1$ (ohm-m)$^{-1}$, which is a typical figure in semiconductors. Although this is many orders of magnitude smaller than the value in a typical metal, where $\sigma \simeq 10^7$ (ohm-m)$^{-1}$, the conductivity in a semiconductor is still sufficiently large for practical applications.

The reason $\sigma_e$ is so small in semiconductors lies, of course, in the smallness of the electron concentration $n$. Although this is usually about $10^{28}$ m$^{-3}$ in metals, it is only about $10^{21}$ m$^{-3}$ in semiconductors. The ratio of these figures substantially accounts for the relative values of the conductivities.

Semiconductor physicists often use another transport coefficient: *mobility*. This is defined as follows: The electron drift velocity in the field may be written as

$$v_e = -\frac{e\tau_e}{m_e}\mathscr{E}, \tag{6.30}$$

according to (4.7). (The negative sign is due to the negative charge on the electron.) The electron mobility $\mu_e$ is defined as the ratio $v_e/\mathscr{E}$, that is, the velocity per unit field strength. Therefore

$$\mu_e = \frac{e\tau_e}{m_e}. \tag{6.31}$$

(The sign is usually disregarded in the definition of mobility.) As defined, the mobility is a measure of the rapidity, or swiftness, of the motion of the electron in the field. The longer the lifetime of the electron and the smaller its mass, the higher the mobility. Table 6.3 provides mobility values for many common substances.

**Table 6.3**

Mobilities for Various Semiconductors
(Room Temperature)

| Crystal | $\mu$, cm$^2$/volt-s | |
|---|---|---|
| | *Electron* | *Hole* |
| C | 1800 | 1600 |
| Si | 1350 | 475 |
| Ge | 3900 | 1900 |
| GaAs | 8500 | 400 |
| GaP | 110 | 75 |
| GaSb | 4000 | 1400 |
| InAs | 33000 | 460 |
| InP | 4600 | 150 |
| InSb | 80000 | 750 |
| CdS | 340 | 18 |
| CdSe | 600 | |
| CdTe | 300 | 65 |
| ZnS | 120 | 5 |
| ZnSe | 530 | 16 |
| ZnTe | 530 | 900 |

We can now express electrical conductivity in terms of mobility. Referring to (6.29) and (6.31), we can write

$$\sigma_e = ne\mu_e, \tag{6.32}$$

indicating that $\sigma_e$ is proportional to $\mu_e$. A typical value for $\mu_e$ may be obtained by substituting $\sigma_e = 1$ (ohm-m)$^{-1}$ and $n = 10^{21}$ m$^{-3}$. This yields

$$\mu_e \simeq 10^{-2}\,\text{m}^2/\text{V-s} = 100\,\text{cm}^2/\text{V-s},$$

in agreement with the figures quoted in Table 6.3.

What we have said about electrons in a strongly $n$-type substance can be carried over to a discussion of holes in a strongly $p$-type substance. The conductivity of the holes is given by

$$\sigma_h = \frac{pe^2\tau_h}{m_h} = pe\mu_h, \tag{6.33}$$

where $\mu_h$ is the hole mobility. The values of $\mu_h$ for semiconductors are also quoted in Table 6.3.

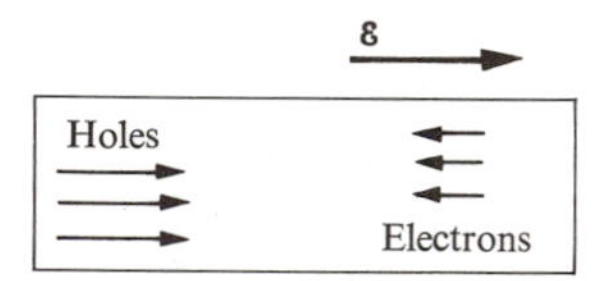


**Fig. 6.14** The drift of electrons and holes in the presence of an electric field.

Let us now treat the general case, in which both electrons and holes are present. When a field is applied, electrons stream opposite to the field and holes stream in the same direction as the field, as Fig. 6.14 shows. The currents of the two carriers are additive, however, and consequently the conductivities are also. Therefore

$$\sigma = \sigma_e + \sigma_h,$$

i.e., both electrons and holes contribute to the currents. In terms of the mobilities, one may write

$$\sigma = ne\mu_e + pe\mu_h. \tag{6.34}$$

The carriers' concentrations $n$ and $p$ need not be equal if the sample is doped, as discussed in the previous section. And one or the other of the carriers may dominate, depending on whether the semiconductor is $n$- or $p$-type. When the substance is in the intrinsic region, however, $n = p$, and Eq. (6.34) becomes

$$\sigma = ne(\mu_e + \mu_h), \tag{6.35}$$

where $n = n_i$, the intrinsic concentration. Even now the two carriers do not contribute equally to the current. The carrier with the greater mobility—usually the electron—contributes the larger share.

### Dependence on temperature

Conductivity depends on temperature, and this dependence is often pronounced. Consider a semiconductor in the intrinsic region. Its conductivity is expressed by (6.35). But in this situation the concentration $n$ increases exponentially with temperature, as may be recalled from (6.16). If we combine this with (6.35), we may write the conductivity in the form

$$\sigma = f(T)e^{-E_g/2k_BT}, \tag{6.36}$$

where $f(T)$ is a function which depends only weakly on the temperature, i.e., as a polynomial. (The function depends on the mobilities and effective masses of the particles.) Thus conductivity increases exponentially with temperature because of the exponential factor in (6.36). Such behavior is amply confirmed by the curve in Fig. 6.15.

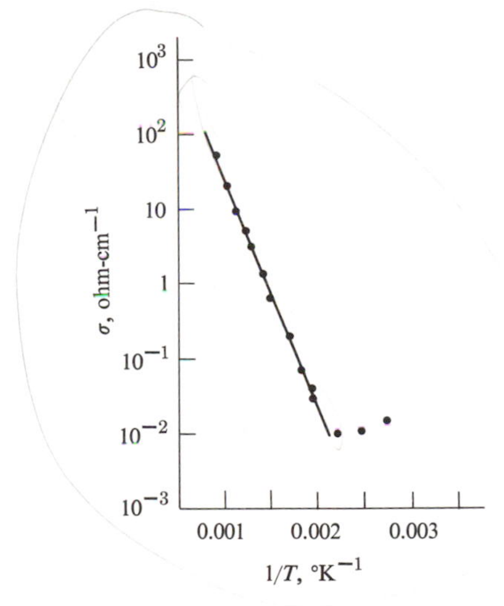


**Fig. 6.15** Conductivity of Si $\sigma$ versus $1/T$ in the intrinsic range. [After Morin and Morita, *Phys. Rev.* **96**, 28, 1954]

The result (6.36) is used to determine the energy gaps in semiconductors. If one takes the logarithms of both sides of the equation, one has

$$\log \sigma = \log f(T) - \frac{E_g}{2k_B}\frac{1}{T}.$$

A plot of log $\sigma$ versus $1/T$ should therefore yield a straight line whose slope, $-E_g/2k_B$, determines the gap. [The weak temperature dependence of $f(T)$

is neglected.] In the early days of semiconductors this was the standard procedure for finding the energy gap. Nowadays, however, the gap is often measured by optical methods (see Section 6.12).

When the substance is not in the intrinsic region, its conductivity is given by the general expression (6.34). In that case the temperature dependence of $\sigma$ on $T$ is not usually as strong as indicated above. To see the reason for this, suppose that the substance is extrinsic and strongly $n$-type. The conductivity is

$$\sigma_e = ne\mu_e.$$

But the electron concentration $n$ is now a constant equal to $N_d$, the donor (hole) concentration, as pointed out in Section 6.6. And any temperature dependence present must be due to the mobility of electrons or holes.

**Mobility versus temperature: scattering mechanisms**

Mobility of electrons (or holes) varies with temperature. This is best seen by referring to (6.31) that is,

$$\mu_e = \frac{e\tau_e}{m_e^*}, \tag{6.31}$$

where (for the sake of discreteness) we have taken the electrons only. Since the lifetime of the electron, or its collision time, varies with temperature (recall Section 4.5), its mobility also varies with temperature. In general, both lifetime and mobility diminish as the temperature rises. (The effective mass of an electron is independent of temperature.)

But the temperature dependence of $\tau_e$ in a semiconductor is quite different from that in a metal. To see this, we write

$$\tau_e = \frac{l_e}{v_r}, \tag{6.37}$$

where $l_e$ is the mean free path of the electron and $v_r$ is its random velocity (Section 4.4). Now electrons at the bottom of the conduction band in a semiconductor obey the classical statistics of Section 6.4, and not the highly degenerate Fermi–Dirac statistics prevailing in metals. These electrons thus have many different speeds, depending on their location in the band. The higher they are in the band, the greater their speed. Thus, in fact, according to (6.37), there is no unique lifetime for the electrons. Different electrons have different lifetimes, and fast-moving elections have shorter lifetimes than slower-moving ones. (The mean free path $l_e$ is the same for all electrons.)

One then defines an *average* lifetime $\bar{\tau}_e$, in which the averaging is over all the electrons. Therefore

$$\bar{\tau}_e = \frac{l_e}{\bar{v}_r}, \tag{6.38}$$

and the mobility is now given by

$$\mu_e = \frac{e\bar{\tau}_e}{m_e}. \tag{6.39}$$

The relation between conductivity and mobility remains intact, as in (6.32). Substituting (6.38) into (6.39) one finds that

$$\mu_e = \frac{el_e}{m_e \bar{v}_r}.$$

We can evaluate the average speed of the electrons by the usual procedure used in the kinetic theory of gases. We recall from basic physics that

$$\tfrac{1}{2} m_e \bar{v}_r^2 = \tfrac{3}{2} k_B T.$$

Thus

$$\mu_e = \frac{el_e}{m_e^{1/2}(3k_B T)^{1/2}}. \tag{6.40}$$

So we see that using the statistical distribution of the electron introduces a factor of $T^{-1/2}$ dependence in the mobility.

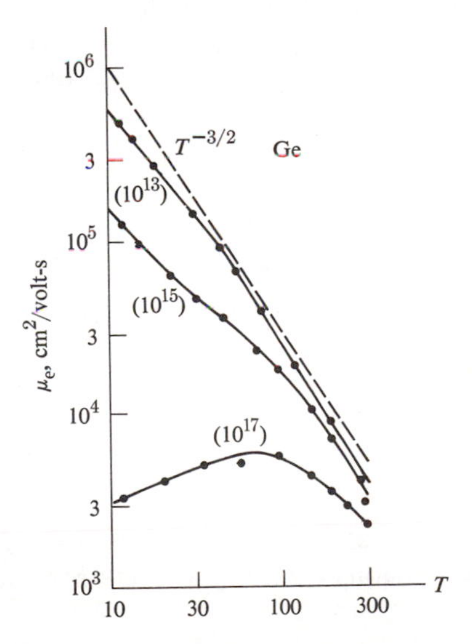


**Fig. 6.16** Electron mobility $\mu_e$ versus $T$ in Ge. The dashed curve represents the pure phonon scattering; numbers in parentheses refer to donor concentrations. [After Debye and Conwell]

The mean free path $l_e$ also depends on the temperature, and in much the same way as it does in metals. We recall from Section 4.5 that $l_e$ is determined by the various collision mechanisms acting on the electrons. (These mechanisms are the collisions of electrons with phonons, the thermally caused lattice vibrations, and collisions with impurities.) At high temperatures, at which collision with phonons is the dominant factor, $l_e$ is inversely proportional to temperature, that is, $l_e \sim T^{-1}$. In that case, mobility varies as $\mu_e \sim T^{-3/2}$. Figure 6.16 shows this for Ge.

Another important scattering mechanism in semiconductors is that of *ionized* impurities. We recall that when a substance is doped, the donors (or acceptors) lose their electrons (or holes) to the conduction band. The impurities are thus ionized, and are quite effective in scattering the electrons (holes), (much as a free ion would scatter an electron passing in the neighborhood). At high temperatures this scattering is masked by the much stronger phonon mechanism, but at low temperatures this latter mechanism becomes weak and the ionized-impurity scattering gradually takes over.

## 6.8 MAGNETIC FIELD EFFECTS: CYCLOTRON RESONANCE AND HALL EFFECT

The effects of a magnetic field on the electronic properties of solids were discussed in Section 4.10 and again in Section 5.19. In Section 4.10, the electrons were treated according to the free-electron model; while in Section 5.19, the band model was employed. Here we shall apply the previously derived results to semiconducting solids, because, as we have said, magnetic effects are very useful in studies of the physical properties of solids.

### Cyclotron resonance

It will be recalled that a charged particle in a magnetic field executes a (circular) cyclotron motion of frequency $\omega_c = eB/m^*$, where $B$ is the magnetic field. Let us apply this result to a semiconductor containing both electrons and holes. When a magnetic field is applied, the electrons execute a cyclotron motion with a frequency

$$\omega_{ce} = \frac{eB}{m_e^*} \tag{6.41}$$

(Fig. 6.17). The sense of the rotation is *counterclockwise*, a fact that you can readily confirm. The holes simultaneously execute a cyclotron motion of frequency

$$\omega_{ch} = \frac{eB}{m_h^*}, \tag{6.42}$$

but the sense of their rotation is clockwise, i.e., opposite to that of the electron. This is, of course, a consequence of the positive charge of the hole.

There are thus two distinct cyclotron frequencies in the system: one corresponding to the electrons, the other to the holes. Cyclotron resonance is achieved by sending an ac signal into the semiconductor slab, where the signal is propagated in the same direction as the magnetic field. When the frequency of the signal $\omega$ is equal to $\omega_{ce}$ or $\omega_{ch}$, power is absorbed by the electrons or by the holes, respectively.

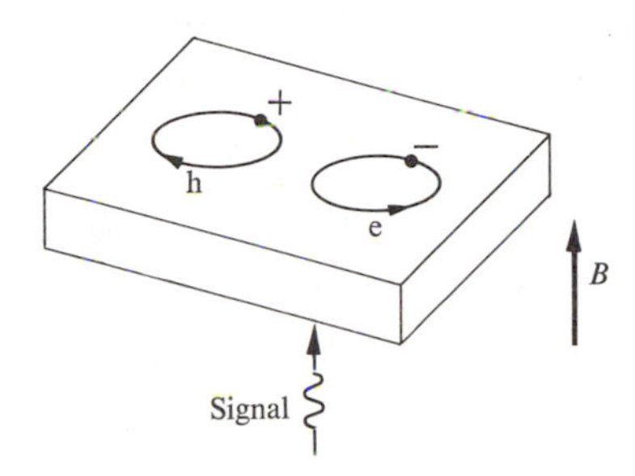


**Fig. 6.17** Cyclotron motions of electrons (e) and holes (h) in a magnetic field $B$.

A useful result of this technique is that one can determine the effective mass of the carriers. By measuring the cyclotron frequency and using (6.41) or (6.42), one may determine the effective masses of the electrons and holes. This is a standard procedure. In fact, the masses quoted in Table 6.1 were determined in this manner.

The technique of cyclotron resonance is also capable of distinguishing between electrons and holes. Suppose that the incident wave is plane-polarized. One can then think of it as being resolved into two circularly polarized waves, one in the clockwise and the other in the counterclockwise direction. The amplitudes of these waves are equal. These waves pass through the sample, and let us suppose that $\omega = \omega_{ce}$, that is, there is an electron resonance. Now, since the electrons orbit in the counterclockwise direction, they absorb energy only from the counterclockwise circular wave, leaving the other wave unaffected. Thus the transmitted wave is no longer plane-polarized, but rather partially polarized in the clockwise direction, and its polarization gives a clear indication that the absorption was by electrons.

In the case of hole resonance, the absorption affects the clockwise wave, and hence the transmitted wave would be polarized in a direction opposite to that of the electrons.

Cyclotron resonance experiments are performed at low temperatures, and on relatively pure samples. In order that the absorption frequency be clearly discernible, it is necessary that the product $\omega_c \tau \gg 1$, where $\tau$ is the collision time. This is equivalent to saying that the particle must execute several circular orbits in a single collision time. When one lowers the temperature to the neighborhood of 4°K, and uses a relatively pure sample, one lengthens the collision time $\tau$, and one makes the quantity $\omega_c \tau$ larger.

In most cyclotron resonance work, the frequency $\omega_c$ falls in the microwave range. Recently, however, it has become possible to make more accurate determinations of cyclotron frequencies by using signals from infrared lasers. Such frequencies are known very accurately. Also, since $\omega_c$ is in the infrared region (which requires a very strong magnetic field, for example, 50 kG), and is so much larger than typical microwave frequencies, the quantity $\omega_c \tau$ is very large, and the cyclotron line is clearly discernible.

**The Hall effect**

We discussed the Hall effect for a single carrier in Section 4.10, where we found that the Hall constant for electrons is

$$R_e = -\frac{1}{ne}. \tag{6.43}$$

Similarly, the Hall constant for holes is

$$R_h = \frac{1}{pe}, \tag{6.44}$$

where the positive sign is due to the positive charge of the hole. Now let us derive the appropriate expression when both types of carriers are present.

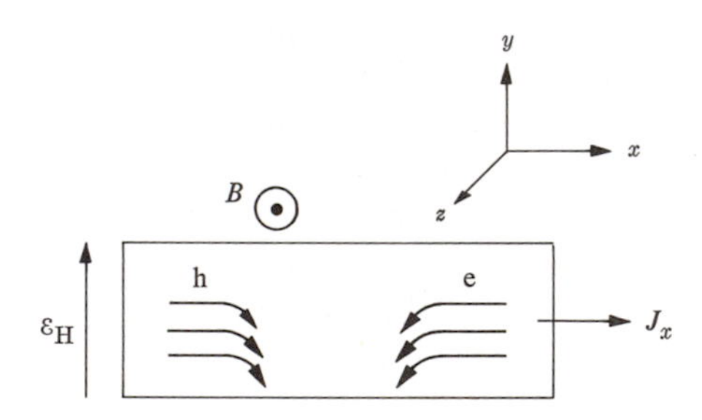


**Fig. 6.18** The Hall effect in a two-carrier semiconductor. The symbols e and h refer to the electrons and holes, respectively.

Figure 6.18 shows the situation. An electric field $\mathcal{E}_x$ is applied in the $x$-direction, and simultaneously a magnetic field $B_z$ is applied in the $z$-direction (normal to the paper). Because of $\mathcal{E}_x$, the carriers drift—electrons to the left, holes to the right. Because of this drift, the magnetic field exerts Lorentz forces on the carriers, which result in their deflections. (The deflections of the electrons and holes are in opposite senses because of their opposite charges.) Both electrons and holes are deflected toward the lower surface of the sample, and therefore tend to cancel each other at the lower surface. But this cancellation is incomplete, as will be shown shortly. Thus there is a net charge which accumulates on the lower surface. An equal and opposite charge accumulates on the upper surface, since the sample as a whole is electrically neutral. Because of these surface charges, an electric field is produced in the $y$-direction. This is the *Hall field*, $\mathscr{E}_{\mathrm{H}}$.

We may calculate the Hall field in the following manner. The Lorentz force acting on an electron is

$$F_{\mathrm{Le}} = -e(\mathbf{v}_{\mathrm{e}} \times \mathbf{B}) = +ev_{\mathrm{e}}B_z,$$

where $v_{\mathrm{e}}$ is the drift velocity of the electrons. The force $F_{\mathrm{Le}}$ is in the $y$-direction. (Since $v_{\mathrm{e}}$ is negative, the force $F_{\mathrm{Le}}$ is actually downward, i.e., in the negative $y$-direction.) This force is equivalent to a Lorentz field

$$\mathscr{E}_{\mathrm{Le}} = -v_{\mathrm{e}}B_z \tag{6.45}$$

acting on the electron. (The minus sign arises because the previous equation has been divided by $-e$, the electron charge.) Since $J_{\mathrm{e}} = -nev_{\mathrm{e}}$, the above equation may also be written as

$$\mathscr{E}_{\mathrm{Le}} = \frac{J_{\mathrm{e}}B_z}{ne}, \tag{6.46}$$

where $J_{\mathrm{e}}$ is that part of the current $J_x$ carried by the electrons.

Following the same procedure, we can establish the fact that the holes experience a Lorentz field in the $y$-direction, given by

$$\mathscr{E}_{\mathrm{Lh}} = -\frac{J_{\mathrm{h}}B_z}{pe}. \tag{6.47}$$

[The carrier charge in (6.46) is simply reversed.]

The problem as a whole is now viewed as follows: The carriers flow in the $x$-direction, but they also experience several electric fields in the $y$-direction. These fields are: $\mathscr{E}_{\mathrm{Le}}$ (felt by the electrons), $\mathscr{E}_{\mathrm{Lh}}$ (felt by the holes), and the Hall field $\mathscr{E}_{\mathrm{H}}$ (felt by both carriers). The total current density in the $y$-direction is therefore

$$J_y = ne\mu_{\mathrm{e}}\mathscr{E}_{\mathrm{Le}} + pe\mu_{\mathrm{h}}\mathscr{E}_{\mathrm{Lh}} + (ne\mu_{\mathrm{e}} + pe\mu_{\mathrm{h}})\mathscr{E}_{\mathrm{H}}. \tag{6.48}$$

But this current vanishes, because the particles are not allowed to flow in the $y$-direction as a result of the presence of the surfaces of the sample. We therefore set $J_y = 0$, and the resulting equation then serves to determine the Hall field $\mathscr{E}_{\mathrm{H}}$. We recall that the Hall constant $R$ is defined as $R = \mathscr{E}_{\mathrm{H}}/J_xB$. By substituting (6.46), (6.47) into (6.48), and noting that $J_{\mathrm{e}} = [n\mu_{\mathrm{e}}/(n\mu_{\mathrm{e}} + p\mu_{\mathrm{h}})]\,J_x$ and $J_{\mathrm{h}} = J_x - J_{\mathrm{e}}$, we find that

$$R = \frac{p\mu_{\mathrm{h}}^2 - n\mu_{\mathrm{e}}^2}{e(n\mu_{\mathrm{e}} + p\mu_{\mathrm{h}})^2}, \tag{6.49}$$

which is the result we have been seeking. It is clear that this expression reduces to the special forms (6.43) and (6.44) for the cases of $n$-type and $p$-type samples, respectively.

According to (6.49), the Hall constant may be either negative, positive, or zero, depending on the carriers' concentrations and mobilities. Thus, by measuring $R$, one obtains information on the relation between these quantities in that particular sample. [Note that the electron and hole terms in (6.49) have opposite signs, and thus tend to cancel each other's contribution, as indicated earlier.]

The primary use of the Hall effect in semiconductors is to enable one to determine carrier concentration. It is clear from (6.43) and (6.44) that, in the case of a single carrier, the concentration is readily obtainable once the Hall constant is measured.

The Hall constant may also be used in determining the mobility. By combining (6.32) and (6.43), one finds that

$$\mu_e = \sigma_e R_e \tag{6.50}$$

for an $n$-type material. A similar relation holds true for the holes in a $p$-type material. Thus the mobilities of electrons and holes can be determined from measuring both the electrical conductivity and Hall constant in extrinsic samples. The product $\sigma R$ is usually referred to as the *Hall mobility*, and denoted by $\mu_H$.

## 6.9 BAND STRUCTURE OF REAL SEMICONDUCTORS

In treating semiconductor properties up to this point, we have assumed the simplest possible band structure, namely, a conduction band of a standard form, centered at the origin, $\mathbf{k} = 0$, and a valence band of a standard inverted form, also centered at the origin. Such a simple structure goes a long way toward elucidating many observed phenomena, but it does not represent the actual band structures of many common semiconductors. Only when one uses the actual band structure is it possible to obtain a quantitative agreement between experiments and theoretical analysis.

A material whose band structure comes close to the ideal structure is GaAs (Fig. 6.19). Consider first the conduction band. It has a minimum at the origin $\mathbf{k} = 0$, and the region close to the origin is well represented by a quadratic energy dependence, $E = \hbar^2k^2/2m_e$, where $m_e = 0.072m_0$. Since the electrons are most likely to populate this region, one can represent this band by a single effective mass.

Note, however, that as $k$ increases, the energy $E(k)$ is no longer quadratic in $\mathbf{k}$, and those states may no longer by represented by a single, unique effective mass. Note in particular that the next-higher energy minimum occurs along the [100] direction. The dependence of energy on $k$ in the neighborhood of this *secondary* minimum is quadratic, and hence an effective mass may be defined locally, but its value is much greater than that of the primary minimum (at the center). (The actual value is $0.36m_0$.)

There are actually other secondary minima equivalent to the one just described. One of these occurs along the [$\bar{1}$00] direction. This follows from the inverse symmetry of $E(\mathbf{k})$ in **k**-space, as discussed in Section 5.4. There are similarly two minima along the $k_y$-axis, and two more along the $k_z$-axis. These follow from the fact that, inasmuch as the crystal has cubic symmetry, the energy band must also have a rotational cubic symmetry [property (iii), Section 5.4]. Thus the band along the $k_y$- and $k_z$-axes must have the same form as along the $k_x$-axis. There are therefore six equivalent secondary minima, or *valleys*, in all along the ⟨100⟩ directions.

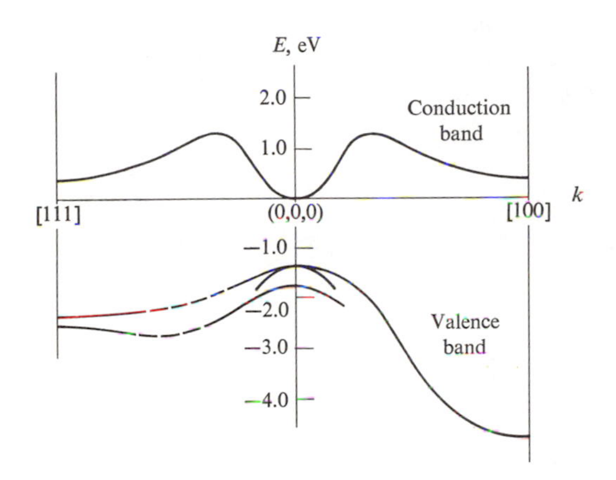


**Fig. 6.19** Band structure of GaAs plotted along the [100] and [111] directions.

It is true that these secondary valleys do not play any role under most circumstances, since the electrons usually occupy only the central, or primary, valley. In such situations, these secondary valleys may be disregarded altogether. There are cases, however, in which an appreciable number of electrons transfer from the central to the secondary valleys, and in those situations these valleys have to be taken into account. Such is the case in the Gunn effect, to be discussed in Section 6.11.

(There are also other secondary valleys in the ⟨111⟩ directions, as shown in Fig. 6.19. These are higher than the ⟨100⟩ valleys, and hence are even less likely to be populated by electrons.)

The valence band is also illustrated in Fig. 6.19. Here it is composed of three closely spaced subbands. Because the curvatures of the bands are different, so are the effective masses of the corresponding holes. One speaks of *light holes* and *heavy holes*.†

---

† The splitting of the valence band is due to the spin–orbit interaction. This interaction is caused by the action of the magnetic field of the nucleus (as seen in the electron's frame of reference) with the spin of the electron. The larger the $Z$ of the atom, the greater the interaction and splitting.

Other III–V semiconductors have band structures quite similar to that of GaAs.

Figure 6.20 shows the band structure of Si, whose conduction band has its lowest (primary) minimum along the [100] direction, at about 0.85 the distance from the center to the edge of the zone. Because of the cubic symmetry, there are actually six equivalent primary valleys located along the ⟨100⟩ directions. These are illustrated in Fig. 6.20(b). The energy surfaces at these valleys are composed of prolate ellipsoidal surfaces of revolution, whose axes of symmetry are along the ⟨100⟩ directions. The longitudinal mass is $m_l = 0.97m_0$, while the two identical transverse masses are $m_t = 0.19m_0$ (see Table 6.1). The mass anisotropy ratio $m_l/m_t \simeq 5$.

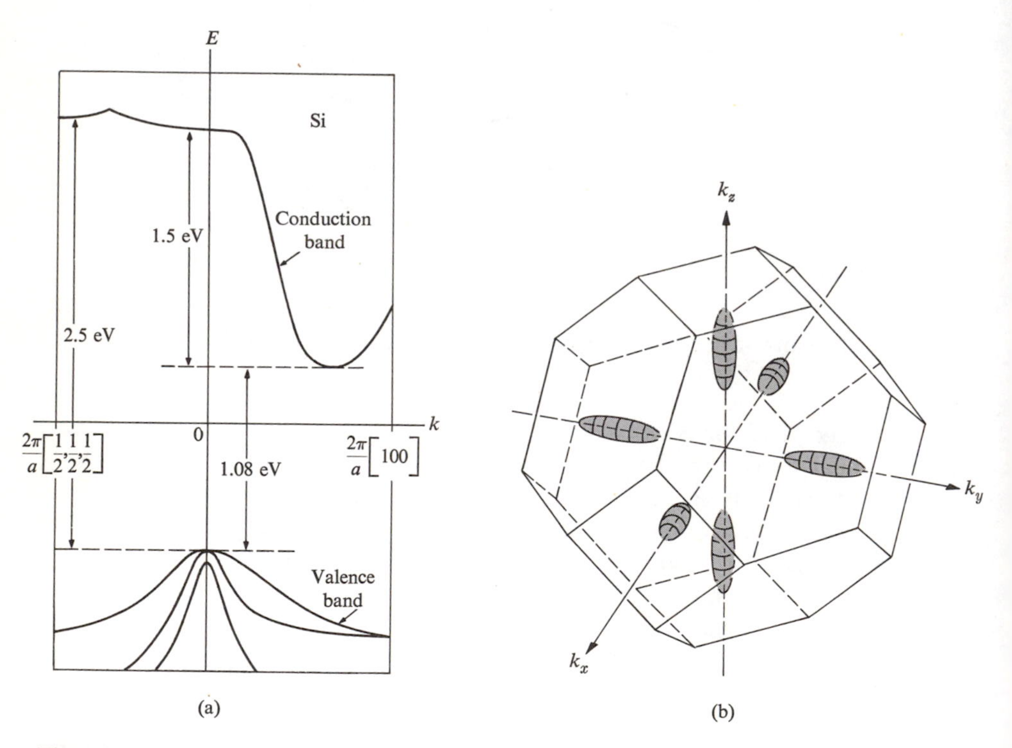


**Fig. 6.20** (a) Band structure of Si plotted along the [100] and [111] directions. (b) Ellipsoidal energy surfaces corresponding to primary valleys along the ⟨100⟩ directions.

The valence band in silicon is represented by three different holes (Fig. 6.20b). One of the holes is heavy ($m_h = 0.5m_0$), and the other two are light.

The energy gap in Si, from the top of the valence band to the bottom of the conduction band, is equal to 1.08 eV. (Note that the bottom of the conduction band does not lie directly above the top of the valence band in **k**-space, but this is irrelevant to the definition of the energy gap.)

Figure 6.21 shows the band structure in Ge. Note in part (a) that the conduction band has its minimum along the [111] direction at the zone edge. (There are actually eight minima, as follows from the cubic symmetry.) These valleys, which are more clearly shown in Fig. 6.21(b), are composed of eight half-prolate ellipsoids of revolution, or four full ellipsoids. (Each two symmetrically placed halves form one full ellipsoid, if we use the periodic-zone scheme of Section 5.6.) The longitudinal and transverse masses are, respectively, $m_l = 1.6m_0$, and $m_t = 0.082m_0$. The mass anisotropy ratio $m_l/m_t \simeq 20$, which is considerable.

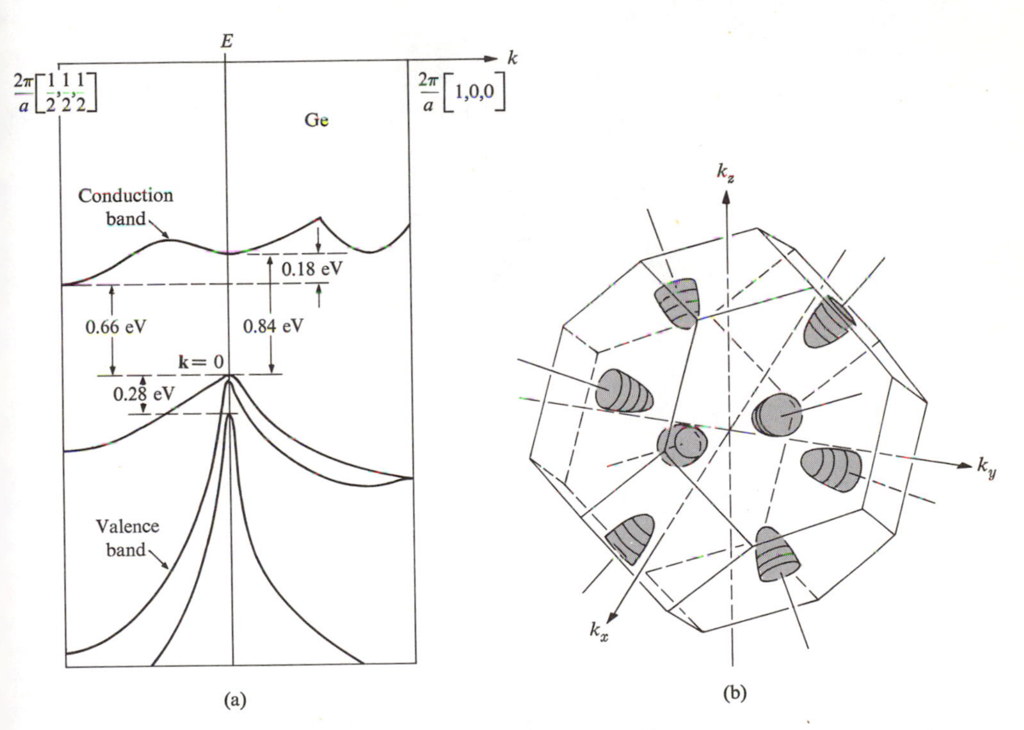


**Fig. 6.21** (a) Band structure of Ge plotted along the [100] and [111] directions. (b) Ellipsoidal energy surface corresponding to primary valleys along the ⟨111⟩ directions.

The valence band of Ge is similar to that of Si, and is represented by one heavy and two light holes. The energy gap is 0.84 eV. (Note also that the conduction band of Ge has secondary valleys along the ⟨100⟩ directions.)

How does one obtain such detailed information regarding the band structure? The shapes of the energy surfaces are determined by techniques of cyclotron resonance. Suppose that the energy surface is an ellipsoid of revolution, and a magnetic field **B** is applied at angle $\theta$ relative to the axis of symmetry (Fig. 6.22). Electrons execute cyclotron motion in planes along the energy surface and normal to the field. It was shown (Problem 4.19) that the cyclotron frequency for this case is

$$\omega_c = eB\left[\frac{\cos^2\theta}{m_t^2} + \frac{\sin^2\theta}{m_t m_l}\right]^{1/2}. \tag{6.51}$$

The cyclotron frequency depends not only on the longitudinal and transverse masses, but also on the angle $\theta$. By measuring the cyclotron frequency at various angles, one can determine the effective masses $m_l$ and $m_t$. (Only two measurements are actually needed, in principle.)

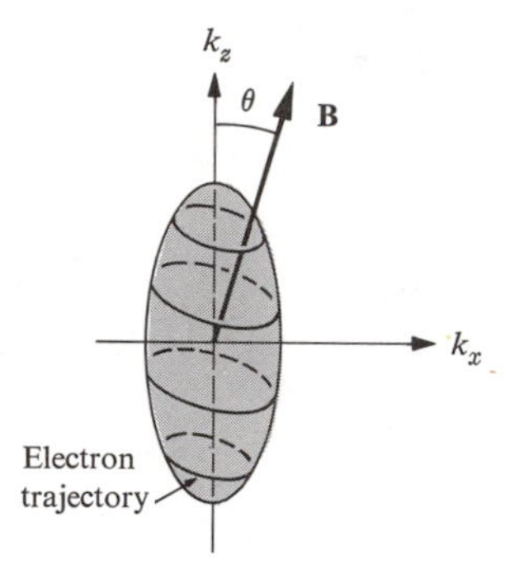


**Fig. 6.22** Cyclotron motion for an electron of ellipsoidal energy surface.

The cyclotron resonance spectrum of Ge is shown in Fig. 6.23. The field **B** is applied along the (110) plane at a 60° angle from the [100] direction. Note that there are only three electron lines, rather than the expected four. This is so

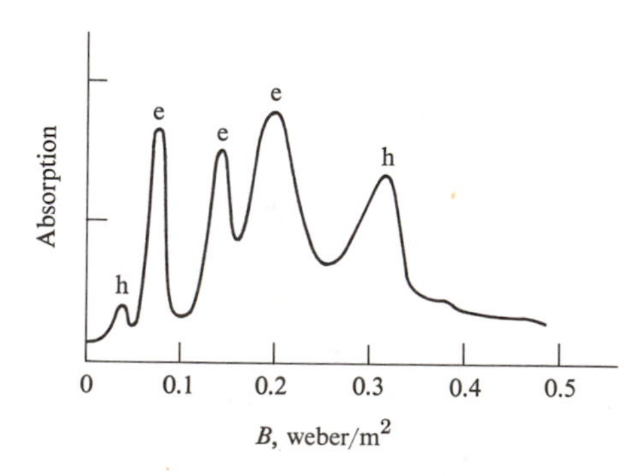


**Fig. 6.23** Cyclotron resonance in Ge at 24 GHz and 4°K. The magnetic field is in the (110) plane at 60° from the [100] axis. [After Dresselhaus, Kip, and Kittel, *Phys. Rev.* **98,** 376, 1955]

because two of the ellipsoids make the same angle with the field, at the chosen field orientation, and hence have the same cyclotron frequency (which two ellipsoids?). Similarly two hole lines (rather than three) appear because the two lighter holes have the same mass, and hence the same frequency. This is indicated by the fact that the line corresponding to the light holes—the one at higher frequencies—is more intense than that of the heavy hole. The reason is, of course, that the two lighter holes absorb more strongly than a heavy one.

Judicious use of cyclotron resonance therefore yields a wealth of information concerning band structure.

## 6.10 HIGH ELECTRIC FIELD AND HOT ELECTRONS

Semiconductors exhibit linear ohmic behavior—that is, $J \sim \mathscr{E}$—in the low electric fields commonly encountered. In the high fields present in some devices, however, considerable deviation from Ohm's law is observed, as shown in Fig. 6.24 for $n$-type germanium. The deviation becomes significant at some field $\mathscr{E}_1$, and for $\mathscr{E}_1 < \mathscr{E}$ the current lies below its expected ohmic value. Above a certain higher field $\mathscr{E}_2$, the current actually saturates at a constant value until, at an extremely high field, usually in the 100 kV/cm range, the sample undergoes an electrical breakdown.

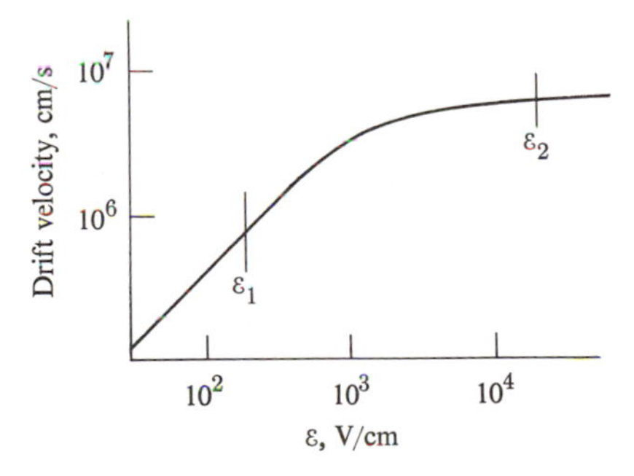


**Fig. 6.24** Drift velocity versus electric field in $n$-type Ge. The current density $J = nev$ is proportional to the velocity.

We shall now present a theory which gives the physical basis underlying this non-ohmic behavior at high fields. Consider the average electron energy $\bar{E} = \frac{3}{2}k_B T$. At high fields (taking an $n$-type sample for concreteness), the electron receives considerable energy from the field because of the acceleration of the electron between collisions, and also loses energy to the lattice (energy which appears as Joule heat). In the steady state the rates of gain and loss of energy must be equal. That is,

$$\frac{d\bar{E}}{dt} = \left(\frac{d\bar{E}}{dt}\right)_{\mathscr{E}} + \left(\frac{d\bar{E}}{dt}\right)_L = (-e\mathscr{E}v) - \frac{\bar{E}(T_e) - \bar{E}(T)}{\tau_E} = 0, \qquad (6.52)$$

where $v$ is the electron drift velocity and $\tau_E$ the *energy relaxation time*. We have allowed for the possibility that the electron temperature $T_e$ may be higher that that of the lattice, $T$, leading to the concept of *hot* electrons. By substituting $\bar{E}(T_e) = \frac{3}{2}k_B T_e$, $\bar{E}(T_L) = \frac{3}{2}k_B T$, $v = \mu_e \mathscr{E}$, and solving the above equation for the electron's temperature, we find that

$$T_e = T + \frac{2}{3}\frac{e\tau_e\mu_e}{k_B}\mathscr{E}^2. \tag{6.53}$$

For $\tau_E = 10^{-11}$ s, $\mu_e = 10^3$ cm$^2$/V-s, and $\mathscr{E} = 10^3$ V/cm, we find that $\Delta T = T_e - T \simeq 100°$K. That is, the electrons are hotter than the lattice by 100°K. The heating would be much greater at higher fields and/or mobility.

We recall from Eqs. (6.31) and (6.37) that $\mu_e = el_e/m_e v_r$, where $v_r$ is the random velocity of the electron, and since $v_r \sim T^{1/2}$, it follows that $\mu_e \sim T^{-1/2}$. We may thus write

$$\mu_e = \mu_{e,0}\left(\frac{T}{T_e}\right)^{1/2}, \tag{6.54}$$

where $\mu_{e,0}$ is the familiar low-field mobility. Equations (6.53) and (6.54) are two equations in $T_e$ and $\mu_e$, and can be employed in solving for these unknowns. In the range in which the field is not too high, one finds

$$\mu_e = \mu_{e,0}\left(1 - \frac{e\tau_e\mu_e}{3k_B T}\mathscr{E}^2\right), \tag{6.55}$$

which explains the initial decrease in mobility just above the field $\mathscr{E}_1$ in Fig. 6.24. The situation in the intermediate field range is complicated, and will not be discussed here.

One can explain the current saturation at high fields by assuming that the electrons dissipate their energy by emitting optical phonons in the lattice. Since these phonons have much greater energy than their acoustic counterparts, they represent the most efficient means for the electrons to rid themselves of the energy gained from the field, thus achieving a steady-state condition.

## 6.11 THE GUNN EFFECT

The *Gunn effect* is named after J. B. Gunn, who made the discovery in 1963, while measuring the currents of hot electrons in GaAs and other III–V compounds. When he was measuring the current $J$ versus the field $\mathscr{E}$ in $n$-type GaAs, he observed an unexpected phenomenon: As $\mathscr{E}$ is increased from zero, the current increases gradually and essentially linearly (Fig. 6.25a) until a field $\mathscr{E}_0$ is reached. As the field is increased beyond $\mathscr{E}_0$, the current suddenly becomes *oscillatory* (versus $t$, not versus $\mathscr{E}$). These oscillations are essentially coherent, provided the sample is sufficiently thin. The field $\mathscr{E}_0$ necessary for the onset of the Gunn oscillation is called the *threshold field*. Typical values for GaAs, as found by Gunn, are $\mathscr{E}_0 \simeq 3$ kV/cm, thickness of the sample $L = 2.5 \times 10^{-3}$ cm, and frequency of the oscillations $\nu \simeq 5$ GHz.

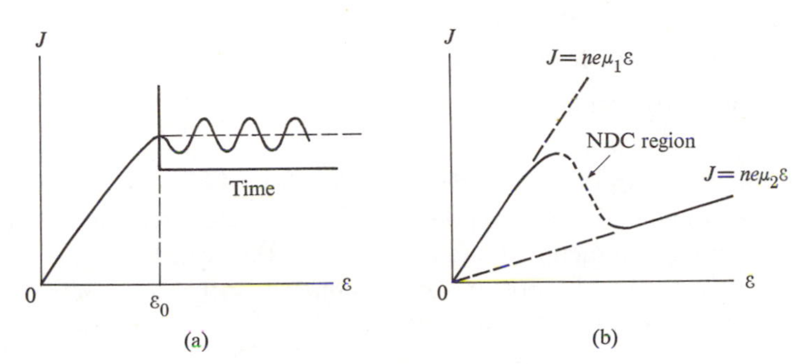


**Fig. 6.25** (a) A graphic summary of the Gunn effect. (b) The current $J$ versus $\mathscr{E}$ in GaAs, showing the NDC region (dashed curve).

Gunn then performed several related experiments to determine the character of the oscillations with a view to discovering the physical mechanism responsible. Figure 6.25(b) shows $J$ versus $\mathscr{E}$ for GaAs, as calculated theoretically. Note an interesting fact: There is a certain field range in which $J$ *decreases* as $\mathscr{E}$ *increases* (the curve corresponding to this range is shown by the dashed line in the figure). This behavior (contrary to the usual one, in which an increase in $\mathscr{E}$ causes an increase in $J$), is described by saying that the sample has a *negative differential conductance* (NDC).† We shall shortly discuss the source of NDC in GaAs, but for the moment let us focus our attention on the NDC itself.

There is a general theorem in electrodynamics which states that an NDC situation is an *unstable* one; in other words, NDC is unlikely to exist in a steady-state situation. A system cannot remain in an unstable state indefiniely, and any fluctuations which may be present will cause it to become increasingly unstable until there is a sudden transition to a stable state. In the case of the Gunn effect, the unstable state of a steady current with an NDC behavior gives way to an entirely different state—one in which the current oscillates coherently in time. The threshold field $\mathscr{E}_0$ for the oscillations is the same as the field in which the NDC region sets in, and this lends further support to the above argument.

In Chapter 7 we shall discuss those aspects of the Gunn effect that concern semiconductor devices. We shall also talk further about instability, and the Gunn mode which arises from it. Here let us simply say that the characteristic oscillation frequency is in the GHz range, and this suggests the Gunn device as a possible

† The word 'differential' is important. The differential conductivity is defined as $\partial J/\partial \mathscr{E}$, compared with the actual conductivity defined by $J/\mathscr{E}$. The negative differential conductance is also often referred to as *negative differential resistance*, and abbreviated NDR.

microwave generator and/or an oscillator. These and other potential applications underlie the great interest in the Gunn effect.

Let us turn now to the physical mechanism responsible for the appearance of the NDC at high field in $n$-type GaAs. For this we must look at the structure of the conduction band of GaAs, shown in Fig. 6.26. There are two types of valleys in the conduction band (see Section 6.9).

a) There is one *central* valley, i.e., the bottom of the valley is located at $\mathbf{k} = 0$, the center of the BZ.

b) In addition, there are six *secondary* valleys whose bottoms are located along the $\langle 100 \rangle$ directions in the BZ. These six valleys are equivalent to each other by symmetry (cubic symmetry of the crystal), but they are not equivalent to the central valley. In GaAs, the bottom of a secondary valley lies above that of the central valley by an amount $\Delta = 0.36$ eV.

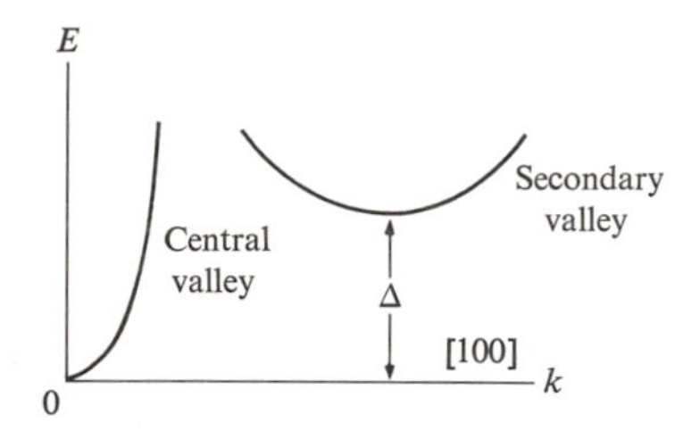


**Fig. 6.26** Conduction band in GaAs, showing central and secondary valleys. (Only half the band is shown.)

The central and secondary valleys have widely different masses and mobilities. If we use the labels 1 and 2 to denote the central and secondary valleys, respectively, then for GaAs $m_1 = 0.072\, m_0$ and $\mu_1 = 5 \times 10^3$ cm$^2$/V-s, while $m_2 = 0.36 m_0$ and $\mu_2 = 100$ cm$^2$/V-s. Note that $m_2$ is considerably larger than $m_1$ ($m_2 = 5m_2$), but, even more important, the mobility $\mu_2$ is very much smaller than $\mu_1$ ($\mu_2 = \mu_1/50$). This means that an electron in the secondary valley drifts much more slowly than an electron in the central valley.

Under normal circumstances, all the electrons reside in the central valley. (Let us suppose for the sake of concreteness that the sample is doped so that the electron concentration $n$ is about $10^{15}$ to $10^{16}$ cm$^{-3}$.) This is so because the bottom of the secondary valley $\Delta$ ($= 0.36$ eV) is so much larger than $k_B T$ at room temperature that only a negligible fraction of the electrons is excited to the secondary valleys. Therefore we may write $n_1 \simeq n$, and the current for a field $\mathscr{E}$ is given by

$$J = n_1 e \mu_1 \mathscr{E} = n e \mu_1 \mathscr{E}. \tag{6.56}$$

Since the secondary valleys are unoccupied, we may ignore them in discussing transport properties (as we did in Section 6.6).

However, when a strong electric field is applied to the system, the situation changes significantly. As we have said (Section 6.10), such a large field causes the electrons to become hot, i.e., to have a higher temperature than the lattice. At sufficiently high fields, the electron temperature $T_e$ may, in fact, become quite high. But at high temperature the secondary valleys become populated. When this happens, the current should be given not by (6.56), but by the more general formula

$$J = J_1 + J_2 = n_1 e\mu_1 \mathscr{E} + n_2 e\mu_2 \mathscr{E}, \tag{6.57}$$

where $J_1$ and $J_2$ are, respectively the currents of electrons in the central valley and in the secondary valleys (all six valleys). We can now see, by examining the two terms in (6.57), how it is possible for an NDC to come about in a material. But we must remember these two facts: The sum $n_1 + n_2$ is equal to $n$, which is a constant independent of $\mathscr{E}$, and $\mu_2 \ll \mu_1$. As $\mathscr{E}$ increases from zero to a value just below $\mathscr{E}_0$, all the electrons are essentially in the central valley and $n_1 \simeq n$. The current is then given by

$$J \simeq ne\mu_1 \mathscr{E}, \tag{6.56}$$

which, when plotted, is a straight line of slope appropriate to the mobility $\mu_1$ (see Fig. 6.25b).† But as $\mathscr{E}$ reaches $\mathscr{E}_0$, and the temperature rises, the concentration $n_1$ decreases abruptly, and $n_2$ increases by the same amount. In this small field increase, $J_1$ decreases rapidly because $n_1$ does. This decrease is partially offset by an increase in $J_2$. However, because $\mu_2$ is so very small, the increase in $J_2$ is very small, and does not even come close to compensating for the decrease in $J_1$. As a result, the total current $J = J_1 + J_2$ actually decreases, leading to an NDC. As $\mathscr{E}$ increases further, more and more of the electrons transfer from the central to the secondary valleys, leading to a further decrease in $J$. The NDC resulting from this intervalley transfer is shown in Fig. 6.26(b). Eventually, at very high field, all the electrons have essentially transferred to the secondary valleys, so we may write

$$J \simeq J_2 = ne\mu_2 \mathscr{E}. \tag{6.58}$$

The current now begins to increase again with $\mathscr{E}$, but with a slope appropriate to $\mu_2$ (see Fig. 6.26b). The interpretation of the Gunn effect in the light of an NDC arising from an intervalley transfer is due to Kroemer.‡

The intervalley transfer and the rapidity with which this occurs is possible because the density of states of the secondary valley is much larger than that of the central valley. According to Section 5.16, $g_2(E) \sim m_2^{*3/2}$, and since there are

---

† We neglect the variation of the mobilities $\mu_1$ and $\mu_2$ with the field, as discussed in Section 6.10, since it is not essential to an understanding of the Gunn effect.

‡ Gunn himself considered this possibility, but rejected it on the grounds that not enough electrons are excited to the secondary valleys at room temperature. He did not take into consideration the fact that the electron temperature rises significantly with the field.

six valleys, it follows that $g_2(E) \sim 6m_2^{*3/2}$; on the other hand, $g_1(E) \sim m_1^{*3/2}$. For GaAs, $g_2(E)/g_1(E) \simeq 60$, so there are many more states available in the secondary valleys than in the central valley for the same energy range.

The Gunn effect has also been observed in InP, $GaAs_xP_{1-x}$, CdTe, ZnSe–InAs, and other semiconducting compounds. All have conduction-band structures similar to that of GaAs, and the intervalley transfer is responsible for Gunn oscillations in every case. Si and Ge have different band structures, and do not show the Gunn effect.†

## 6.12 OPTICAL PROPERTIES: ABSORPTION PROCESSES

In this and the following sections we shall consider the optical properties of semiconductors. These properties span a wide range of phenomena, and aid us greatly in understanding the basic physical properties of semiconductors. These phenomena are also used in the development of optical devices widely used in research and industry.

First we may divide the optical properties into *electronic* and *lattice* properties. Electronic properties, as the name implies, concern processes involving the electronic states of a solid, while lattice properties involve the vibration of the lattice (creation and absorption of phonons). Lattice properties are of considerable current interest, but it is electronic properties which receive most attention in semiconductors, particularly so far as practical applications are concerned. Therefore the discussion here centers almost entirely on electronic properties.

Let us first recollect the important optical lattice properties: Recall Section 3.12, in which we said that ionic crystals exhibit strong absorption and reflection in the infrered region as a result of the interaction of light with optical phonons. Because of the partially ionic character of their bonds, compound semiconductors such as GaAs, GaP, etc., should, and do, exhibit these properties. (Even purely covalent crystals such as Si exhibit infrared lattice absorption, but to explain this would mean consideration of higher-order processes.).

In semiconductors, a number of distinct optical electronic processes take place independently. Let us now look at them one by one.

### The fundamental absorption process

The most important absorption process involves the transition of electrons from the valence to the conduction band (Fig. 6.27). Because of its importance, the process is referred to as *fundamental absorption.*

---

† The Gunn effect has been observed in Ge under uniaxial pressure. The reason is that, under such pressure, the $\langle 111 \rangle$ valleys become inequivalent, leading to sets of inequivalent bands. For certain directions of the field, the effective mobilities may be sufficiently smaller than the lower valley so that Gunn oscillations result at high field. The large anisotropy of the mass of the $\langle 111 \rangle$ valleys plays an important role here.

In fundamental absorption, an electron absorbs a photon (from the incident beam), and jumps from the valence into the conduction band. The photon energy must be equal to the energy gap, or larger. The frequency must therefore be

$$\nu \geqslant (E_g/h). \tag{6.59}$$

The frequency $\nu_0 = E_g/h$ is referred to as the *absorption edge*.

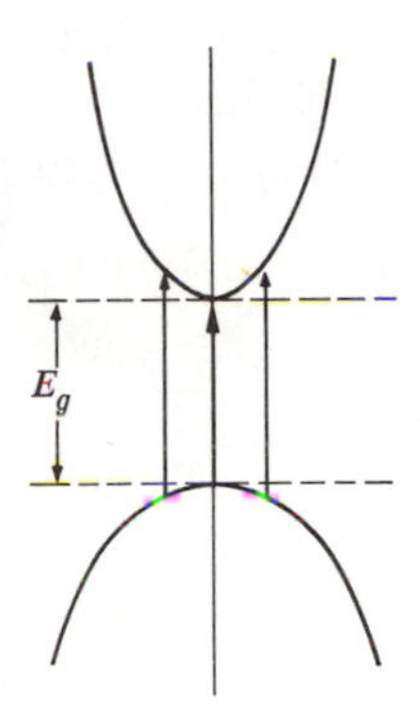


**Fig. 6.27** The fundamental absorption process in semiconductors.

In the transition process (photon absorption), the total energy and momentum of the electron–photon system must be conserved. Therefore

$$E_f = E_i + h\nu \tag{6.60}$$

and

$$\mathbf{k}_f = \mathbf{k}_i + \mathbf{q}, \tag{6.61}$$

where $E_i$ and $E_f$ are the initial and final energies of the electron in the valence and conduction bands, respectively, and $\mathbf{k}_i$, $\mathbf{k}_f$ are the corresponding electron momenta. The vector $\mathbf{q}$ is the wave vector for the absorbed photon. However, recall from Section 3.10 that the wave vector of a photon in the optical region is negligibly small. The momentum condition (6.61) therefore reduces to

$$\mathbf{k}_f = \mathbf{k}_i. \tag{6.62}$$

That is, the momentum of the electron alone is conserved. This *selection rule* means that only vertical transitions in **k**-space are allowed between the valence and conduction bands (Fig. 6.27).

Calculating the absorption coefficient for fundamental absorption requires quantum manipulations. Essentially, these consist of treating the incident radiation as a perturbation which couples the electron state in the valence band to its counterpart in the conduction band, and using the technique of quantum perturbation theory (Section A.6). One then finds that the absorption coefficient has the form (Blatt, 1968)

$$\alpha_d = A(h\nu - E_g)^{1/2}, \tag{6.63}$$

where $A$ is a constant involving the properties of the bands, and $E_g$ is the energy gap. [The meaning of the subscript $d$ will become apparent shortly. Equation (6.63) will also be derived later; see Section 8.9.]

The absorption coefficient increases parabolically with the frequency above the fundamental edge (Fig. 6.28a). (Of course, $\alpha_d = 0$ for $\nu < \nu_0$.) The absorption coefficient for GaAs in Fig. 6.28(b) is consistent with this analysis.

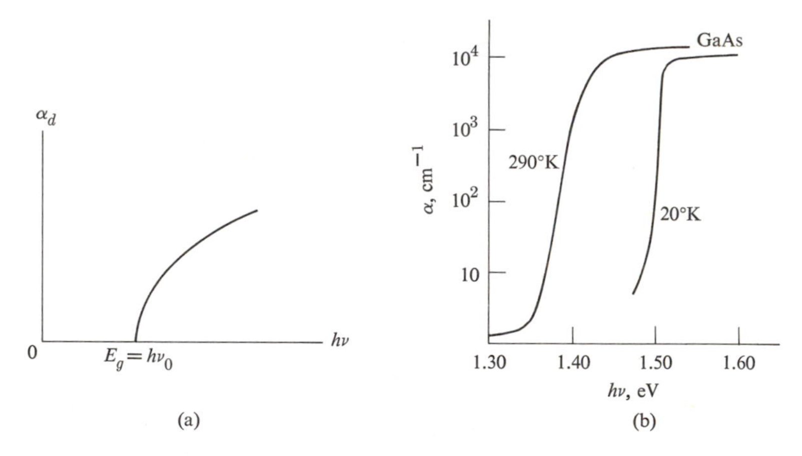


**Fig. 6.28** (a) The absorption coefficient $\alpha_d$ versus $h\nu$ in a semiconductor. (b) The absorption coefficient $\alpha$ versus $h\nu$ in GaAs. [After Hilsum]

A useful application of these results is their use in measuring energy gaps in semiconductors. Thus $E_g$ is directly related to the frequency edge, $E_g = h\nu_0$. This is now the standard procedure for determining the gap, and has all but replaced the earlier method based on conductivity [Section 6.7; also see the discussion after (6.36)], because of its accuracy and convenience. The optical method also reveals many more details about the band structure than the conductivity method.

Note that the absorption coefficient associated with fundamental absorption is large, about $10^4$ cm$^{-1}$. Thus absorption is readily observable even in thin samples. (The sample *must* be thin, in fact, if a transmission is to be observed at all.)

Since the energy gaps in semiconductors are small—frequently 1 eV or less—the fundamental edge usually occurs in the infrared region. Because of this, the study of the infrared region of the spectrum has been greatly expanded by semiconductor research. The development of a large variety of reliable infrared detectors has been one of the many benefits which have accrued from this work (see Section 7.8).

The absorption process occurs in the so-called *direct-gap semiconductors.* Here the bottom of the conduction band lies at $\mathbf{k} = 0$, and hence directly above the top of the valence band. Electrons near the top of the valence band are able to make transitions to states near the bottom of the conduction band, consistent with the selection rule. Examples of such substances are GaAs, InSb, and many other III–V and II–VI compounds.

There are also *indirect-gap* semiconductors, in which the bottom of the conduction band does not lie at the origin (Fig. 6.29a). Recall that both Si and Ge fall in this class: Si has its minimum in the [100] direction, and Ge in the [111] direction. In this case, the electron cannot make a direct transition from the top of the valence band to the bottom of the conduction band because this would violate the momentum selection rule (6.62).

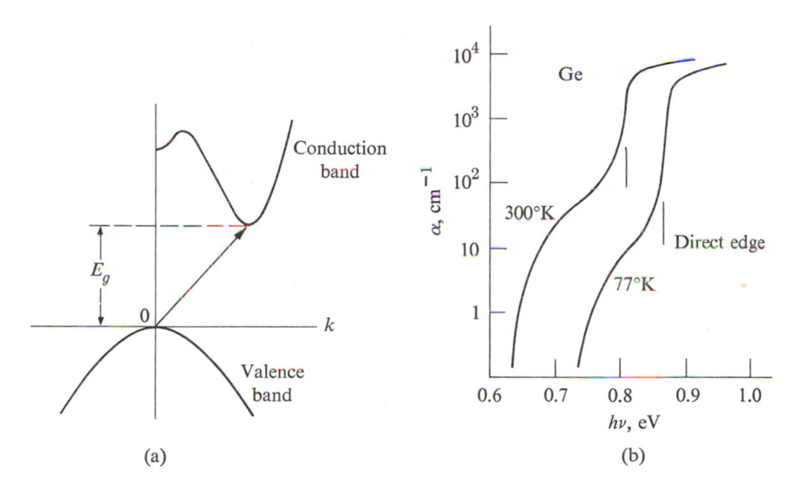


**Fig. 6.29** (a) An indirect-gap semiconductor. (b) The absorption coefficient versus $h\nu$ in Ge. [After Dash and Newman]

Such a transition may still take place, but as a two-step process. The electron absorbs both a photon and a phonon simultaneously. The photon supplies the needed energy, while the phonon supplies the required momentum. (The phonon energy, which is only about 0.05 eV, is very small compared to that of the photon, which is about 1 eV, and hence may be disregarded. The phonon momentum is appreciable, however.)

Calculation of the indirect-gap absorption coefficient, which is more involved than that of direct absorption, shows that the formula, given by Blatt (1968), is

$$\alpha_i = A'(T)\,(h\nu - E_g)^2, \tag{6.64}$$

where $A'(T)$ is a constant containing parameters pertaining to the bands and the temperature (the latter due to the phonon contribution to the process).

Note that $\alpha_i$ increases as the second power of $(h\nu - E_g)$, much faster than the half-power of this energy difference, as in the direct transition. So we may use the optical method to discriminate between direct- and indirect-gap semiconductors, an improvement over the conductivity method. Figure 6.29(b) shows the absorption spectrum for Ge.

### Exciton absorption

In discussing fundamental absorption, we assumed that the excited electron becomes a free particle in the conduction band, and similarly, that the hole left in the valence band is also free. The electron and hole attract each other, however, and may possibly form a *bound* state, in which the two particles revolve around each other. (More accurately, they revolve around their center of mass.) Such a state is referred to as an *exciton*.

The binding energy of the exciton is small, about 0.01 eV, and hence the excitaton level falls very slightly below the edge of the conduction band, as indicated in Fig. 6.30. (The exciton level is in the same neighborhood as the donor level.)

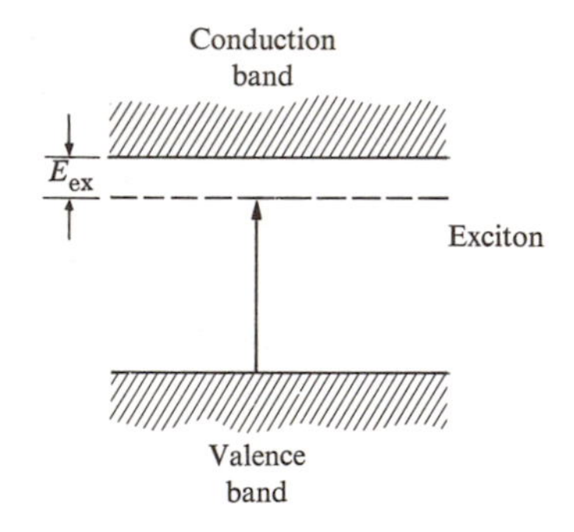


**Fig. 6.30** The exciton level and associated absorption.

The energy of the photon involved in exciton absorption is given by

$$h\nu = E_g - E_{ex}, \tag{6.65}$$

where $E_{ex}$ is the exciton binding energy. The exciton spectrum therefore consists of a sharp line, falling slightly below the fundamental edge. This line is often broadened by interaction of the exciton with impurities or other similar effects, and may well merge with the fundamental absorption band, although often the peak of the exciton line remains clearly discernible. The effect of exciton absorption on the absorption spectrum of Ge is shown in Fig. 6.31.

This illustrates a fact which is often observed: Absorption of an exciton introduces complications into the fundamental absorption spectrum, particularly near the edge, and renders the determination of the energy gap in semiconductors more difficult. However, exciton absorption is important in discussion of optical properties of insulators in the ultraviolet region of the spectrum. For further remarks, refer to Section 8.10.

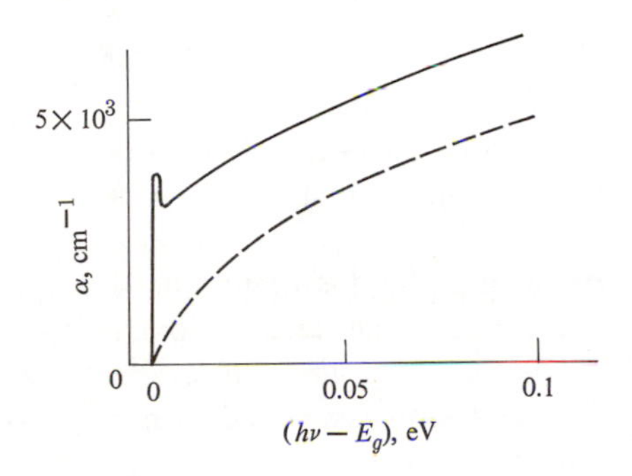


**Fig. 6.31** Excitonic absorption in Ge. Dashed curve represents fundamental absorption (theory); full curve (experiment) includes both fundamental and exciton absorptions. (Measurement at $T = 20°K$.)

### Free-carrier absorption

Free carriers—both electrons and holes—absorb radiation without becoming excited into the other band. In absorbing a photon, the electron (or hole) in this case makes a transition to another state in the same band, as shown in Fig. 6.32. Such a process is usually referred to as an *intraband transition*.

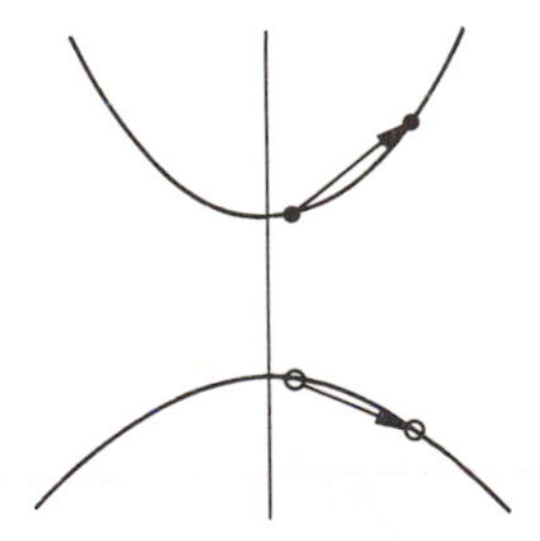

**Fig. 6.32** Free-carrier absorption.

The free-carrier absorption may be treated in precisely the same manner used in Section 4.11 for treating the optical properties of metals. (One now sees, in retrospect, that the optical properties discussed there are associated with intraband transition in the valence band of the metal.) Thus we may simply quote the results. For concreteness, take the substance to be $n$-type, so that only electrons are present. The real and imaginary parts of the dielectric constant are

$$\epsilon_r' = \epsilon_{L,r} - \frac{\sigma_0 \tau}{\epsilon_0(1+\omega^2\tau^2)} = n_O^2 - \kappa^2 \tag{6.66}$$

and

$$\epsilon_r'' = \frac{\sigma_0}{\epsilon_0 \omega(1+\omega^2\tau^2)} = 2n_O\kappa, \tag{6.67}$$

where the symbols have the same meaning as in Section 4.11. (We use $n_O$ rather than $n$ for the index of refraction, to distinguish it from the electron concentration.)

Several different regimes may be distinguished. At low frequency and small conductivity (low concentration), the lattice contribution $\epsilon_{L,r}$ dominates the dielectric polarization in (6.66). Thus the substance acts as a normal dielectric. There is, however, a slight absorption associated with $\epsilon_r''$ of (6.67) which represents the absorption of radiation by free carriers.

In the region of low frequency and high conductivity, the free-carrier term in (6.66) dominates. Thus $\epsilon_r' < 0$, and the substance exhibits total reflection, much as a metal does. This is to be expected, since the electron concentration is very high, approaching (but still much smaller than) the electron concentration in metals.

In the high-frequency (short-wavelength) region, $\omega\tau \gg 1$ (but small conductivity), the material acts like a normal dielectric with $n_O \simeq \epsilon_{L,r}^{1/2}$, and the absorption coefficient is

$$\alpha = \frac{\sigma_0}{\epsilon_0 c n_O \tau^2} \frac{1}{\omega^2}. \tag{6.68}$$

Thus $\alpha \sim \omega^{-2}$ or $\lambda^2$, which is verified experimentally by Fig. 6.33.

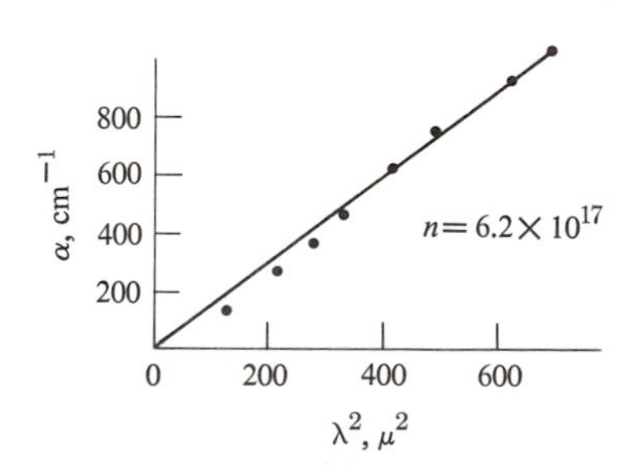


**Fig. 6.33** Free-carrier absorption coefficient versus $\lambda^2$ in $n$-type InSb. [After Moss]

Note that free-carrier absorption takes place even when $h\nu < E_g$, and frequently this absorption dominates the spectrum below the fundamental edge.

For $h\nu > E_g$, of course, both types of absorption—fundamental and free-carrier—occur simultaneously.

### Absorption processes involving impurities

Absorption processes involving impurities often take place in semiconductors. The type and degree of absorption depend on the type of impurity (or impurities) present, and on its concentration.

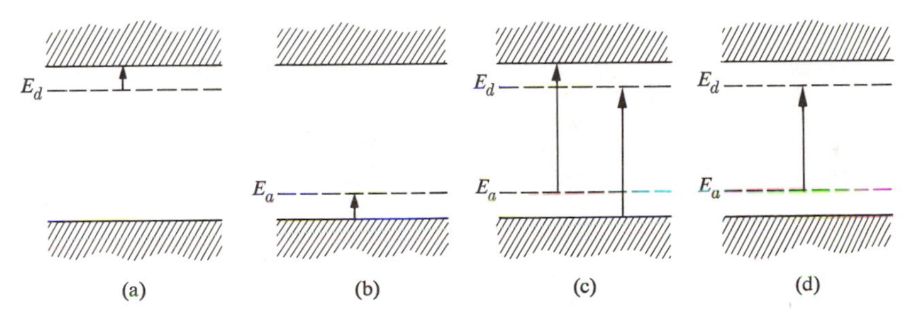


**Fig. 6.34** Various absorption processes involving impurities (see text).

Figure 6.34 depicts the main classes of such processes. Figure 6.34(a) shows the case in which a neutral donor absorbs a photon and the electron makes a transition to a higher level in the impurity itself or in the conduction band. The transitions to higher impurity levels appear as sharp lines in the absorption spectrum. Figure 6.34(b) shows the transition from the valence band to a neutral acceptor, which is analogous to the donor–conduction-band transition above. Figure 6.35 indicates the absorption spectrum associated with the valence-band–acceptor transition in Si.

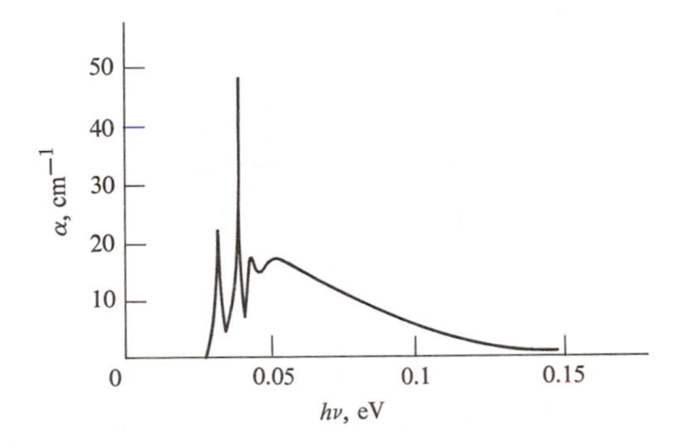


**Fig. 6.35** Absorption coefficient of a boron-doped Si sample versus photon energy $h\nu$. [After Burstein, *et al.*, *Proc. Photoconductivity Conference*, New York: Wiley, 1956]

For shallow impurities, the absorption lines associated with donors and acceptors fall in the far infrared region (since the energy involved is small—only about 0.01 eV). Such processes may serve in principle as a basis for detectors in this rather difficult region of the spectrum. The spectrum may also serve as a diagnostic technique for determining the type of impurity present.

Figure 6.34(c) represents a process in which an electron is excited from the valence band to an ionized donor (it must be ionized; why?), or from an ionized acceptor to the conduction band. Such processes lead to absorption which is close to the fundamental absorption, and are seldom resolved from it.

Figure 6.34(d) illustrates an absorption process involving transition from an ionized acceptor to an ionized donor. The energy of the photon in this case is

$$h\nu = E_g - E_d - E_a. \tag{6.69}$$

This leads to a discrete structure in the absorption curve, but this is often difficult to resolve because of its proximity to the fundamental edge.

Impurities may also affect the absorption spectrum in other, indirect ways. For instance, an exciton is often found to be trapped by an impurity. This may happen as follows: The impurity first traps an electron, and once this happens the impurity—now charged—attracts a hole through the coulomb force. Thus both an electron and a hole are trapped by the impurity. The spectrum of this exciton is different from that of a free exciton because of the interaction with the impurity.

## 6.13 PHOTOCONDUCTIVITY

The phenomenon of *photoconductivity* occurs when an incident light beam impinges upon a semiconductor and causes an increase in its electrical conductivity. This is due to the excitation of electrons across the energy gap, as discussed in Section 6.12, which leads in turn to an increase in the number of free carriers—both electrons and holes—and hence to an increase in conductivity. As we know, excitation can occur only if $\hbar\omega > E_g$. From a practical standpoint, photoconductivity is very important, as it is this mechanism which underlies infrared solid-state detectors.

The concept of photoconductivity is illustrated in Fig. 6.36. A current flows in a semiconductor slab. A light beam is turned on so as to inpinge on the slab in a direction normal to its face. Before the light beam is turned on, the conductivity is given by Eq. (6.34),

$$\sigma_0 = e(n_0\mu_e + p_0\mu_h), \tag{6.70}$$

where $n_0$ and $p_0$ are the concentrations at equilibrium, and $\sigma_0$ is the conductivity in the dark. When the light beam is turned on, the concentrations of the free carriers increase by the amounts $\Delta n$ and $\Delta p$, and the current increases abruptly.

Since electrons and holes are always created in pairs, we have $\Delta n = \Delta p$. The conductivity is now

$$\sigma = \sigma_0 + e\,\Delta n(\mu_e + \mu_h) = \sigma_0 + e\,\Delta\, n\mu_h(1 + b), \tag{6.71}$$

where $b = \mu_e/\mu_h$, the mobility ratio. The relative increase in the conductivity is

$$\frac{\Delta\sigma}{\sigma_0} = \frac{e\Delta n\mu_h(1 + b)}{\sigma_0}. \tag{6.72}$$

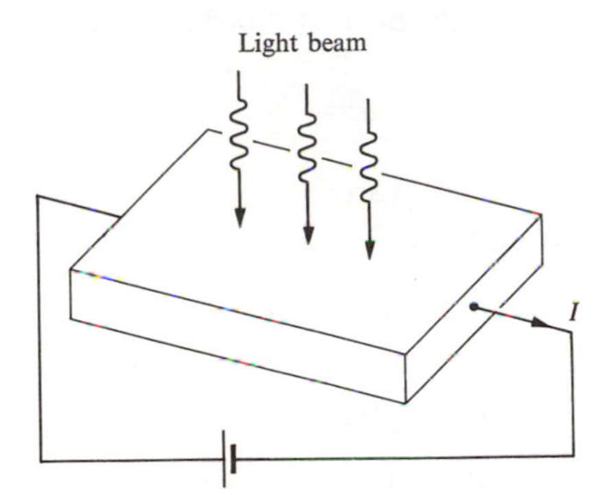


**Fig. 6.36** Basic experimental setup for photoconductivity.

We now need to evaluate $\Delta n$, and it is here that the optical properties of the solid come in. An excess of free carriers is created, so that the situation becomes one of nonequilibrium. There are two factors which lead to the variation of $n$ with time: (a) Free carriers are continually *created* by the incident beam, and (b) excess carriers are also continually *annihilated* by recombining with each other. This recombination is present whenever the concentration of carriers differs from that of equilibrium. The variation of the concentration with time is therefore governed by the following rate equation:

$$\frac{dn}{dt} = g - \frac{n - n_0}{\tau'}, \tag{6.73}$$

where $g$ is the rate of generation of electrons per unit volume due to light absorption, and the second term on the right describes the rate of recombination of electrons; $\tau'$ is called the *recombination time*, which is essentially the lifetime for a free carrier. In the steady state $dn/dt = 0$. That is, the two rates equal each other. Therefore $\Delta\, n = n - n_0$ is given by

$$\Delta n = g\tau'. \tag{6.74}$$

The generation rate can be related to the absorption coefficient and incident intensity as follows. Given that $d$ is the thickness of the slab, then $\alpha\, d$ is the fraction of power absorbed in the slab. [This is the definition of $\alpha$ (see Section 4.11)]. Therefore, if $N(\omega)$ is the number of photons falling on the medium per unit time, it follows that the number of photons absorbed per unit time is $\alpha d N(\omega)$, and hence

$$g = \frac{\alpha d N(\omega)}{V}, \tag{6.75}$$

since each photon absorption leads to an electron–hole creation. The division by the volume $V$ is necessary because we wish to find the rate of generation per unit volume. The number of incident photons $N(\omega)$ is related to the intensity $I(\omega)$ by

$$N(\omega) = \frac{I(\omega)A}{\hbar\omega}, \tag{6.76}$$

where $A$ is the area of the slab, $I(\omega)A$ is the incident power, and $\hbar\omega$ is the photon energy. Combining (6.74) and (6.76), we find that

$$\Delta n = \frac{\alpha I(\omega)}{\hbar\omega}\tau'. \tag{6.77}$$

When we substitute this into (6.72), we obtain

$$\frac{\Delta\sigma}{\sigma_0} = \frac{\alpha I(\omega)\tau'\mu_h(1+b)}{\hbar\omega\sigma_0}. \tag{6.78}$$

Note that $(\Delta\sigma/\sigma_0) \sim I(\omega)$, and also particularly that $(\Delta\sigma/\sigma_0) \sim \alpha(\omega)$, the fundamental absorption coefficient we discussed in Section 6.12. Let us make a numerical estimate. If we take $\tau' \simeq 10^{-4}$ s, $I \simeq 10^{-4}$ watts/cm$^2$, and $\hbar\omega \simeq 0.7$ eV (for Ge), we find $\Delta n \simeq 5 \times 10^{14}$ cm$^{-3}$, which is an appreciable increase in the number of free carriers.

Our treatment of photoconductivity has been idealized in some respects. For example, we have neglected recombination of the free carriers near the surface, an effect known as *surface recombination*. The basic concepts responsible for the phenomenon of photoconductivity have nevertheless been clearly portrayed.

## 6.14 LUMINESCENCE

Section 6.12 presented various processes whereby electrons may be excited by the absorption of radiation. Once electrons have been excited, the distribution of electrons is no longer in equilibrium, and they eventually decay into lower states, emitting radiation in the process. This emission is referred to as *luminescence*. Luminescence is therefore the inverse of absorption. Most of the absorption processes discussed in Section 6.12 may also take place in the opposite direction, leading to several types of luminescence mechanism.

Luminescence—i.e., the electron excitation mechanism—may be accomplished by means other than absorption of radiation. Excitation by an electric current in a *p-n* junction (Section 7.7) results in electroluminescence, while excitation by optical absorption (as in Section 6.12), produces photoluminescence. Cathodoluminescence results when the excitation is achieved by a high-energy electron beam, and thermoluminescence occurs at low temperatures when carriers are first excited by some means, and the electrons are frozen in their trapping states. Then as the solid is heated, thermal agitation assists the electrons to de-excite and release radiation. (The straightforward process whereby the electrons are excited thermally and then release radiation is known as *incandescence.*) There are still other means of excitation, but this list includes the most common methods.

Luminescent emission may take place during the time of excitation, in which case the phenomenon is known as *fluorescence. Phosphorescence* is a luminescence which continues for some time after the excitation has been accomplished.

The physical processes involved in luminescence are the same as those discussed in Section 6.12, except that now they take place in the opposite direction. The most prominent process is that of conduction-band-to-valence-band transition. As an example, we show the photoluminescence spectrum of $n$-type InAs in Fig. 6.37. The broad shape of the spectrum indicates the broad density of states of the bands involved.

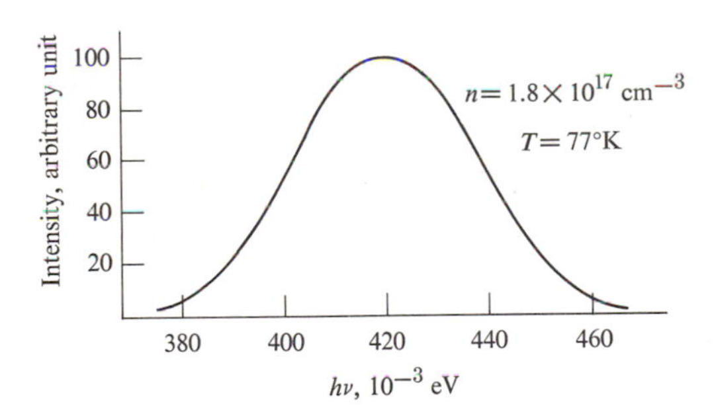


**Fig. 6.37** Photoluminescence spectrum of $n$-type InAs due to electron–hole recombination. [After Mooradian and Fan, *7th Internat. Conf., Physics of Semiconductors,* Pasis, Dunod, 1965]

Excitonic luminescence has also been observed, in addition to luminescence involving transition between impurities and band states, or between band states and impurities. Transition between donors and acceptors is also observed, as well as luminescence due to intraband transition.

The study and characterization of these various processes in different materials form one of the main fields of research on semiconductors today. This research has yielded a variety of technological applications. A very common luminescent device, for example, is the television screen. The images seen are formed by the luminescent coating on the screen, which is excited by an electron beam within the instrument. Another luminescent device is the semiconductor laser, in which the transition from the conduction band to the valence band produces an intense beam of coherent radiation (see Section 7.7).

## 6.15 OTHER OPTICAL EFFECTS

In the last three sections we have described most of the important phenomena associated with the optical properties of semiconductors. There are others, but our limited space does not permit an exhaustive list (Pankove, 1971). We shall mention in passing only some of the other activities which have engaged the attention of physicists in recent years.

Because of the availability of intense beams of radiation—from lasers—nonlinear optical properties have come under increasing investigation. Harmonic generation (multiplication of the incident frequency) has been observed in GaAs, InP, and many other materials. Also frequency mixing (production of various combinations of the frequencies of two incident beams) has been observed in such substances as GaAs, InAs, Si, and Ge.

Other important nonlinear processes are Brillouin and Raman scattering from the phonons in semiconductors. We discussed these nonlinear effects in Section 3.4, in connection with phonons.

## 6.16 SOUND-WAVE AMPLIFICATION (ACOUSTOELECTRIC EFFECT)

The acoustic properties of semiconductors have also been extensively investigated. A primary motive for this research is the design of *acoustic amplifiers*, particularly in the ultrasonic (microwave) region. The mathematical analysis is rather tedious, so we shall indicate here only the physical concepts, and advise anyone interested to consult the bibliography (Wang, 1966).

We have seen how the amplification of sound waves in semiconductors is accomplished (Section 3.11). Free carriers in the sample are set into a drift motion by a high electric field. When the field is sufficiently high so that the drift velocity $v_d$ is greater than the sound velocity, then energy is transferred from the carriers to phonons traveling in the same direction as the carriers, and the sound wave is amplified. For an efficient amplification, the coupling between carriers and phonons, or lattice waves, must be strong. This condition obtains in strongly piezoelectric crystals such as CdS.

The coefficient of acoustic gain as a function of the field has the form shown in Fig. 6.38. The gain coefficient is negative for $v_d < v_s$ (attenuation). That is, in that region, carriers absorb energy from the wave. On the other hand, in the region $v_d > v_s$, the gain is positive.

Acoustic amplification was first observed by Hutson, *et al.*, in 1961. As a medium, they used insulating CdS, so initially there were no carriers. They then introduced carriers by illuminating the sample so that electrons were excited into the conduction band. They sent a shear wave in a direction parallel to the hexagonal axis of the crystal. (The wave was introduced at one end by converting an electromagnetic into an acoustic signal via piezoelectric coupling.) Two frequencies were used, 15 and 45 MHz. When the crystal was in the dark, only attenuation was observed. However, when the crystal was illuminated, amplification was observed above a certain critical field. The experimental result, shown in Fig. 6.39, closely resembles Fig. 6.38.

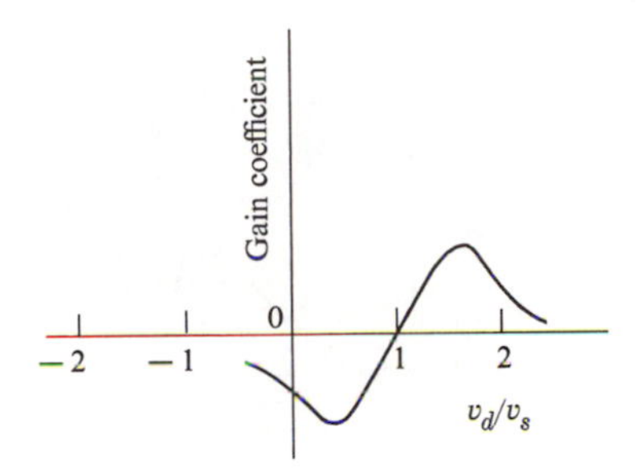


**Fig. 6.38** Acoustic gain coefficient versus $v_d/v_s$.

The crossover point at which the amplification commences corresponds to a field $\mathscr{E} = 700$ V/cm. Since the mobility of electrons in CdS is 285 cm$^2$/V-s, the crossover drift velocity is $2.0 \times 10^5$ cm/s, very close to the velocity of sound in this material.

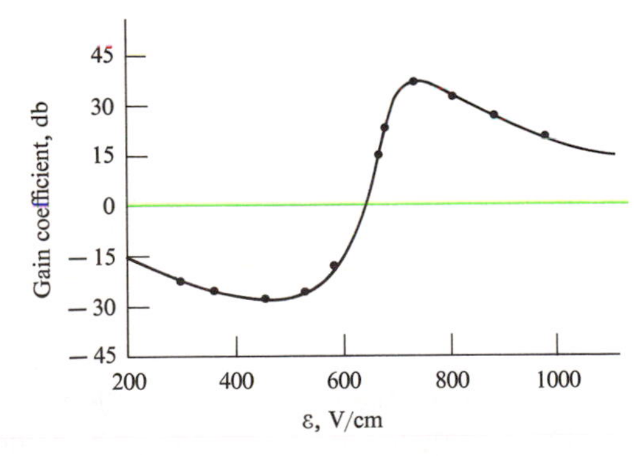


**Fig. 6.39** Gain coefficient (in decibels) versus electric field, at frequency 45 MHz. [Adapted from White, *et al.*, *Phys. Rev. Letters* **7**, 237, 1961]

One can also convert a sound amplifier into an acoustic oscillator by allowing the wave to travel back and forth with the help of good acoustic reflectors at the ends of the sample. Although the wave suffers some attenuation on the return segment of its trip (since the velocity of propagation is opposite to the field), the net gain over the whole trip may be positive. (In fact, for a stable oscillator, this gain must be exactly zero.)

## 6.17 DIFFUSION

Often the concentration of carriers in a semiconductor is *nonuniform* in space. This occurs, for example, in all devices involving *p-n* junctions, such as transistors (see Chapter 7). Whenever there is a nonuniform concentration, the phenomenon of *diffusion* takes place, and it often plays a major role in a given situation. It is because of this that diffusion has received a great deal of attention in semiconductor research.†

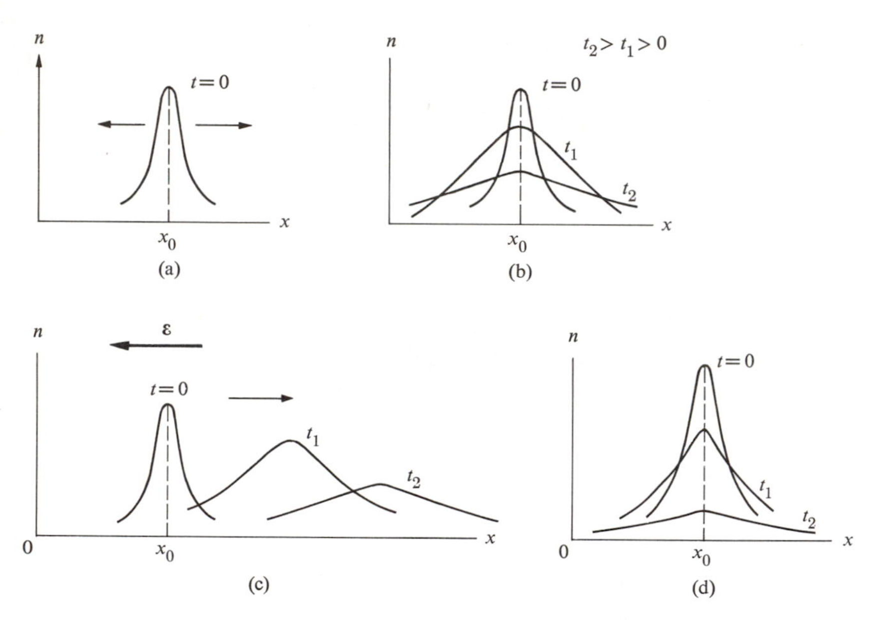


**Fig. 6.40** Diffusion under different circumstances: (a) A concentration pulse, $n$ versus $x$, at $t = 0$. Arrows indicate particle currents. (b) The spread-out of a pulse due to diffusion at three successive instants. (c) An electron pulse diffusing and drifting in an electric field at three successive instants. (d) The disappearance of a pulse due to recombination.

The concept of diffusion can be illustrated as follows. Suppose that a concentration pulse is somehow created in a semiconductor at time $t = 0$, and that the pulse is centered at the position $x = x_0$ (Fig. 6.40a). (A concentration pulse is usually created by injecting carriers into the specimen from an external circuit.) We assume that the concentration of carriers is in a nonequilibrium state (the background is assumed to be uniform, and hence the concentration at equilibrium must be uniform). So therefore a current begins to flow at both edges of the pulse, as shown in Fig. 6.40(a). This flow, in which $n$ changes rapidly with

† Diffusion of atoms in solids, also of great practical importance, will be covered in Section 11.4.

$x$, is called *diffusion*. The effect of the diffusion is eventually to bring the concentration of carriers toward the equilibrium situation, in which the concentration is uniform throughout. The shapes of the pulse at various later instants are illustrated in Fig. 6.40(b). As time progresses, the pulse spreads out in both directions, and the peak *decreases*, although the pulse center remains at $x_0$. We say that the pulse *diffuses*.

If an electric field were also applied to the pulse, at the instant $t = 0$, then the pulse would diffuse as before, but the center of the pulse would also *drift* opposite to the field, as shown in Fig. 6.40(c).

Another, concomitant process is *recombination*. As discussed in Section 6.13, whenever the concentration of carriers is not in an equilibrium state, there is a tendency for the excess carriers to disappear by recombining with carriers of opposite charge, or by being trapped by impurities. The effect of recombination is to bring the concentration of carriers toward equilibrium. Given that the recombination time is $\tau'$, the lifetime of the pulse is essentially equal to $\tau'$, and during the time $t < \tau'$ the pulse diffuses; for $t > \tau'$ the pulse essentially dies out (Fig. 6.40d). Contrast the situation of Fig. 6.40(d) with that of Fig. 6.40(b) in which recombination was neglected.

We have gained a fairly complete physical picture of the diffusion process. Let us consider the above processes in a quantitative manner. The basic law governing diffusion is *Fick's law*, which states that, for a nonuniform concentration, the particle current density $J'$ (that is, the number of particles crossing a unit area per unit time) is given by

$$J' = -D\frac{\partial n}{dx}, \tag{6.79}$$

where $D$ is a constant called the *diffusion coefficient*. This law states that the current is proportional to the *concentration gradient* $\partial n/\partial x$. Thus the more rapidly $n$ varies, the larger the current, which seems plausible.

The negative sign in (6.79) is introduced for convenience, in order to make $D$ a positive quantity. As seen from this equation, and also from Fig. 6.40, $J'$ is opposite to $\partial n/\partial x$. Thus, if $n$ increases to the right, $J'$ is to the left, and vice versa.

Equation (6.79) is valid whether the particles are neutral or charged. In semiconductors, the carriers—electrons and holes—are charged, and hence the particle current $J'$ also carries an electrical current. To obtain the electrical current, one multiplies $J$ by the charge of the carrier. Thus the currents for electrons and holes are given respectively by

$$J_{\mathrm{e}} = eD_{\mathrm{e}}\frac{\partial n}{\partial x} \tag{6.80a}$$

and

$$J_{\mathrm{h}} = -eD_{\mathrm{h}}\frac{\partial p}{\partial x}. \tag{6.80b}$$

One can derive Fick's law by using statistical mechanics (the details are left as an exercise). Statistical mechanics not only enables us to derive this law, but also provides the *Einstein relation*,

$$D = \frac{\mu k_B T}{e}, \tag{6.81}$$

between the diffusion coefficient and the mobility of the carrier. The relation is valid for both electrons or holes, and is a useful formula in that it relates the new quantity $D$ to the mobility, which should be quite familiar to us (Section 6.6). A relation such as (6.81) is expected, since $D$ is, in fact, just another transport coefficient like $\mu$.

Let us now derive the diffusion equation, first for one type and then for two types of carriers.

### The diffusion equation for one type of carrier

Let us assume that we have only one type of carrier, which we shall take to be holes, to avoid confusion arising from the sign of the charge. In the presence of both a concentration gradient and an electric field, the total current is given by

$$J = -eD\frac{\partial p}{\partial x} + pe\mu\mathscr{E}. \tag{6.82}$$

[We have omitted the subscripts on $D$ and $\mu$ (referring to the hole) for simplicity.] The first term on the right is the *diffusion current*, and the second the *drift current*.

We now want to examine how the concentration $p(x,t)$ varies with time at an arbitrary position $x$. Note that the concentration $p$ is a function of both $x$ and $t$. We can see that $p(x)$ varies with time, because of the flow of holes as given by (6.73). This variation is given by the *continuity equation*,† which we write as

$$\left(\frac{\partial p}{\partial t}\right)_{\text{Flow}} = -\frac{1}{e}\frac{\partial}{\partial x}J = D\frac{\partial^2 p}{\partial x^2} - \mu\frac{\partial}{\partial x}(p\mathscr{E}), \tag{6.83}$$

where we have substituted for $J$ from (6.82). In addition to varying with time because of the flow of holes, $p$ varies with time because of recombination. This variation can be written as

$$\left(\frac{\partial p}{\partial t}\right)_{\text{Recomb}} = -\frac{p - p_0}{\tau'}, \tag{6.84}$$

---

† The continuity equation is well known in both electrodynamics and fluid mechanics. Its form in three dimensions is

$$\frac{\partial \rho}{\partial t} + \nabla \cdot \mathbf{J} = 0,$$

where $\rho$ is the density and $\mathbf{J}$ the current (see any textbook on electromagnetism).

where $p_0$ is the equilibrium concentration and $\tau'$ the recombination time of the holes [see Eq. (6.73)]. The total rate of variation is given by

$$\frac{\partial p}{\partial t} = \left(\frac{\partial p}{\partial t}\right)_{\text{Flow}} + \left(\frac{\partial p}{\partial t}\right)_{\text{Recomb}},$$

which, when combined with (6.83) and (6.84), yields the partial differential equation

$$\frac{\partial p}{\partial t} = D\frac{\partial^2 p}{\partial x^2} - \mu\frac{\partial}{\partial x}(p\mathscr{E}) - \frac{p - p_0}{\tau'}. \tag{6.85}$$

This is the diffusion equation† which governs the space-time behavior of the carrier concentration $p$. If we could solve this equation for any specific initial conditions, we would know the concentration at every point $x$ at any instant $t$. We shall not be able to do this in general, however; but we shall solve the equation for a few particular situations, and this will bring out its physical contents.

i) *Stationary solution for* $\mathscr{E} = 0$. We obtain the equation appropriate to this situation from (6.85) by setting $\partial p/\partial t = 0$ and $\mathscr{E} = 0$. The result is

$$D\frac{\partial^2 p}{\partial x^2} - \frac{p - p_0}{\tau'} = 0. \tag{6.86}$$

This is a standard differential equation, whose solution is readily found to be

$$p_1 \equiv p - p_0 = Ae^{-x/(D\tau')^{1/2}}, \tag{6.87}$$

where $A$ is a constant to be determined from boundary conditions. The excess concentration $p_1$ decays exponentially with $x$, and essentially vanishes for $x > (D\tau')^{1/2}$. This distance is known as the *diffusion length*. and is denoted by $L_D$,

$$L_D = (D\tau')^{1/2}. \tag{6.88}$$

The solution (6.87) is appropriate to the physical arrangement whereby we have a semi-infinite specimen, $0 < x < \infty$ (Fig. 6.41), with excess carriers injected at the left face at a constant rate. As the carriers are injected they diffuse to the right, but because of recombination they essentially live only a time $\tau'$, and hence travel a distance $L_D$. One can thus define an effective diffusion velocity $v_D$ as

$$v_D = \frac{L_D}{\tau'} = \left(\frac{D}{\tau'}\right)^{1/2}. \tag{6.89}$$

† The standard diffusion equation is

$$\frac{\partial p}{\partial t} = D\frac{\partial^2 p}{\partial x^2},$$

which is obtained from (6.85) by setting $\mathscr{E} = 0$, and neglecting recombination.

The diffusion current is

$$J_{\text{Diff}} = ep_1 v_D = ep_1 \left(\frac{D}{\tau'}\right)^{1/2}, \tag{6.90}$$

which is the same result that one would obtain by substituting (6.87) into the first term on the right of (6.82).

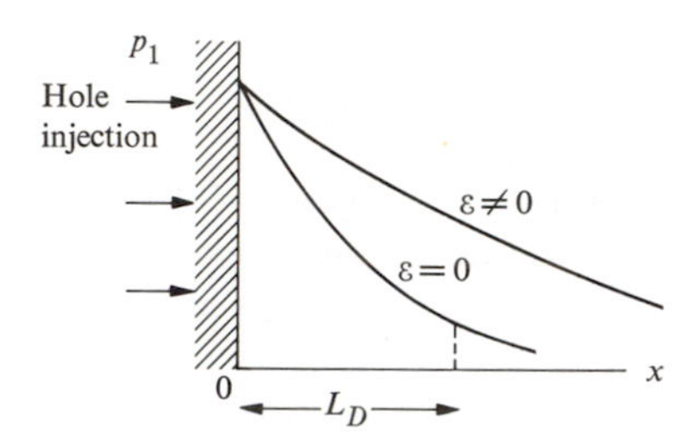


**Fig. 6.41** Steady-state solution for a hole stream injected from the left at $x = 0$, with or without an electric field.

ii) *Stationary solution with a uniform field* $\mathscr{E} \neq 0$. The appropriate equation for the excess concentration $p_1 = p - p_0$, as obtained from (6.85), is now

$$\frac{\partial^2 p_1}{\partial x^2} - \frac{\mu \mathscr{E}}{D} \frac{\partial p_1}{\partial x} - \frac{p_1}{L_D^2} = 0. \tag{6.91}$$

This is also a standard differential equation, whose solution is

$$p_1 = A e^{-\gamma x / L_D}, \tag{6.92}$$

where

$$\gamma = \sqrt{1 + s^2} - s \qquad \text{and} \qquad s = \mu \mathscr{E} L_D / 2D. \tag{6.93}$$

The solution has the same form as (6.87) in the absence of the field, the difference being that the effective diffusion length is now $L_D/\gamma$, where $\gamma$ depends on the field (Fig. 6.41). Since $\gamma < 1$ [from (6.93)], the effective diffusion length is now larger than before. This is expected, since the particles are now "dragged" further by the field as they diffuse. When $\mathscr{E}$ becomes large, $s$ also becomes large, while $\gamma$ becomes small; this leads to a large value for the diffusion length $L_D/\gamma$.

The physical arrangement for the present case is the same as for (i), except that now a uniform field $\mathscr{E}$ is applied to the semi-infinite specimen.

### The diffusion equation for two types of carrier

When there are two types of carrier, we have to deal with two diffusion equations similar to (6.85). However, the two equations are *not* independent. They are coupled together, and hence must be treated simultaneously. The coupling arises because the electric field $\mathscr{E}$, which appears in both equations, is the total field inside the specimen, not the external field $\mathscr{E}_0$. We may, in fact, write

$$\mathscr{E} = \mathscr{E}_0 + \mathscr{E}', \tag{6.94}$$

where $\mathscr{E}'$ is that part of the field which is due to the diffusing electrons and holes. The field $\mathscr{E}'$ is a consequence of the fact that the electrons and holes are electrically charged. Hence these charges create their own field, which again acts on the charges. The effect of this field is to pull the electrons and holes together so that they move together, i.e., to couple the electron's and hole's diffusion equations.

The mathematical treatment for the diffusion of two carriers is fairly complicated, and we shall not attempt it here (see McKelvey, 1966). We shall take only one case, which is simple, but also of practical importance. Suppose that we are dealing with a strongly extrinsic sample, say $n$-type; thus $n_0 \gg p_0$. And suppose that a pulse of holes is injected into the sample. Because of the internal field, an electron pulse is generated which moves with the hole pulse. The electrons and holes are called respectively the *majority* and *minority* carriers in this $n$-type specimen.

The hole pulse moves essentially as if there are no electrons at all, that is, Eq. (6.85) is satisfied, with parameters appropriate to the holes. The motion of the electron pulse is much more complicated, because, since $n_0$ is large, the effect of the pulse on the electron concentration is very small. Thus the neutralizing background into which the hole pulse moves is unaffected by this pulse; hence it moves as an independent hole pulse. The neutralizing background into which the electron pulse moves (the holes) is drastically affected by the pulse.

In summary, it is easier to study the motion of the minority carriers than that of the majority carriers in a two-carrier semiconductor. This point will be important in our discussion of transistors in Chapter 7.

## SUMMARY

Most semiconductors crystallize in the diamond or zincblende structures. In either case, the atom is bonded covalently to its nearest neighbors. Each atom is surrounded by four neighboring atoms, forming a regular tetrahedron.

The band structure of the simplest semiconductor consists of a parabolic, isotropic *conduction band*, below which is an inverted, parabolic, isotropic *valence band.* The two bands are separated by an *energy gap*, which plays an important role in semiconductor phenomena.

When foreign impurities are introduced into a semiconductor, additional localized electronic states are usually created in the energy gap. These states are often very close to the bottom of the conduction band or the top of the valence band. In that case, the impurities—donors and/or acceptors—are readily ionized, supplying free carriers (electrons and holes).

### Carrier concentration

Free carriers are created by thermal excitation of electrons across the energy gap, or by the ionization of donors and acceptors. In an intrinsic (i.e., pure) sample, only thermal excitation takes place, and the numbers of electrons and holes are equal. Their concentration is

$$n = p = n_i$$
$$= 2(k_B T/2\pi\hbar^2)^{3/2}(m_e m_h)^{3/4}\, e^{-E_g/2k_B T}.$$

This concentration rises very rapidly with temperature because of the exponential factor.

In an extrinsic semiconductor, in which cross-gap ionization is negligible compared with the ionization from impurities, the carrier concentrations are given approximately by

$$n = N_d \qquad \text{or} \qquad p = N_a,$$

for $n$- and $p$-type samples, respectively.

### Conductivity and mobility

Conductivity in semiconductors is usually written as a product

$$\sigma = ne\mu,$$

a relation which defines the mobility $\mu$. In intrinsic substances, conductivity increases rapidly with temperature because of the corresponding exponential increase in the carrier concentration $n$. Although mobility also depends on temperature, this dependence is very weak compared with its dependence on the exponential factor.

### Magnetic field effects

These effects, particularly *cyclotron resonance* and the *Hall effect*, are standard techniques for measuring semiconductor parameters. Cyclotron resonance is used primarily to determine effective masses, and the Hall effect to measure carrier concentration.

### Hot electrons and the Gunn effect

When a high electric field is applied to a semiconductor, the carriers (electrons and holes) absorb appreciable energy from the field, and their temperature rises above that of the lattice; i.e., they become "hot." The effect of this is a decrease in their mobility.

In certain semiconductors of appropriate band structure, such as GaAs, the heating of carriers results in a transfer of electrons to high-energy valleys of very low mobility. In such a case the application of the electric field may produce a region of *negative differential conductance.* Because such a situation is inherently unstable, the sample "breaks up" into coherent electrical oscillations, which is the *Gunn effect.*

### Optical absorption, photoconductivity, and luminescence

*Optical absorption* takes place in a semiconductor when an electron in the valence band absorbs a photon from the optical beam, and transfers to the conduction band. Such a *fundamental absorption* takes place only if the photon energy is greater than the energy gap, that is, $E_g < h\nu$. The shape of the absorption curve is a function of the band structure, and hence the absorption curve is often employed in the study of the lattice.

Optical absorption may also take place due to excitonic excitation, or ionization of donors and/or acceptors.

In *photoconductivity*, the conductivity of a semiconductor is raised by shining a light beam on the sample. The optical beam causes additional carriers to be excited across the energy gap, which causes a rise in conductivity.

*Luminescence* is the inverse of optical absorption: Electrons are first excited, in some manner, to higher states, and are subsequently allowed to fall to lower states, emitting photons in the process. It is these photons which give rise to luminescence.

### Diffusion

When the carrier concentration is spatially nonuniform, this nonuniformity causes a current of particles. The direction of this current is such that it tends to remove the nonuniformity, and leads to a uniform distribution of carriers. The basic relation is *Fick's law*,

$$J' = -D\frac{\partial n}{\partial x},$$

where $J'$ is the particle current density. By employing statistical mechanics, one can show that the diffusion coefficient $D$ is related to the carrier mobility by the Einstein relation

$$D = \mu k_B T/e.$$

A dynamical study of diffusion can be made by combining Fick's law with the continuity equation. One can then solve the appropriate differential equation—known as the diffusion equation—in a manner consistent with the initial and boundary conditions of the problems.

## REFERENCES

### General

F. J. Blatt, 1968, *Physics of Electronic Conduction in Solids*, New York: McGraw-Hill

A. F. Gibson, editor, *Progress in Semiconductors*, Volumes I–IX, 1956–1964, New York; Academic Press

J. P. McKelvey, 1966, *Solid State and Semiconductor Physics*, New York: Harper and Row

J. L. Moll, 1964, *Physics of Semiconductors*, New York: McGraw-Hill

R. A. Smith, 1959, *Semiconductors*, Cambridge: Cambridge University Press

E. Spenke, 1958, *Electronics Semiconductors*, New York: McGraw-Hill

S. Wang, 1966, *Solid State Electronics*, New York: McGraw-Hill

R. K. Willardson and A. C. Beer, editors, *Semiconductors and Semimetals*, Volumes I–IV, 1966–1968, New York: Academic Press

H. F. Wolf, 1971, *Semiconductors*, New York: Wiley-Interscience

**Band structure**

D. L. Greenway and G. Harbeke, 1968, *Optical Properties of Semiconductors*, New York: Pergamon Press

D. L. Long, 1968, *Energy Bands in Semiconductors*, New York: John Wiley

R. K. Willardson and A. C. Beer (see General References),Volume I

**Semiconductor statistics**

J. S. Blakemore, 1958, *Semiconductor Statistics*, New York: Pergamon Press

J. P. McKelvey (see General References)

J. L. Moll (see General References)

**Transport phenomena**

F. J. Blatt (see General References)

E. M. Conwell, 1967, *High Field Transport in Semiconductors*, New York: Academic Press

R. A. Smith (see General References)

A. C. Smith, J. F. Janak and R. B. Adler, 1968, *Electronic Conduction in Solids*, New York: McGraw-Hill

S. Wang (see General References)

**Optical properties and photoconductivity**

R. H. Bube, 1960, *Photoconductivity of Solids*, New York: John Wiley

M. Cordona, 1968, "Electronic Optical Properties of Solids," in *Solid State, Nuclear and Particle Physics*, edited by I. Saavedra, New York: Benjamin

D. L. Greenway and G. Harbeke (see Band Structure References)

T. S. Moss, 1959, *Optical Properties of Semiconductors*, New York: Academic Press

J. I. Pankove, 1971, *Optical Properties of Semiconductors*, Englewood Cliffs, N.J.: Prentice-Hall

J. Tauc, 1962, *Photo- and Thermoelectric Effects in Semiconductors*, New York: Pergamon Press

S. Wang (see General References)

R. K. Willardson and A. C. Beer, editors (see General References), Volume III

**Particular classes of materials**

V. I. Fistul, 1969, *Heavily Doped Semiconductors*, New York: Plenum Press

O. Madelung, 1964, *Physics of the III–V Compounds*, New York: John Wiley

B. Ray, 1969, *II–V Compounds*, New York: Pergamon Press
R. K. Willardson and A. C. Beer, editors (see General References)

## QUESTIONS

1. In discussing the tetrahedral bond in the Group IV semiconductors (and other substances), we described the so-called *bond model*, in which each electron is localized along the covalent bond line joining the two atoms. Explain how this may be reconciled with the (delocalized) band model, in which the electron is described by a Bloch function whose probability is distributed throughout the crystal.
2. Do the bond orbitals of the above bonds correspond to the conduction band or the valence band? Why?
3. Describe the bond model associated with the electrons in the conduction band of the group IV semiconductors; i.e., state the spatial region(s) in which these electrons reside.
4. What does the breaking of a bond correspond to in the band model?
5. Give one (or more) experimental reason affirming that the electrons associated with the tetrahedral bond are delocalized.
6. The pre-exponential factor in Eq. (6.8), i.e., the factor preceding $e^{-E_g/k_BT}$, is frequently referred to as "the effective density of states of the conduction band." How do you justify this designation?
7. A cyclotron resonance experiment in $n$-type Ge exhibits only one electron line. In which direction is the magnetic field?
8. Is it possible for a cyclotron resonance experiment in Si to show only one electron line?
9. Does the fact that a sample exhibits intrinsic behavior necessarily imply that the sample is pure?
10. An experimenter measuring the Hall effect in a semiconductor specimen finds to his surprise that the Hall constant in his sample is vanishingly small even at room temperature. He asks you to help him interpret this result. What is the likely explanation?
11. In the expression for the electron temperature (6.53), the first power of the field $\mathscr{E}$ is missing. Can you explain this by symmetry considerations? If the general expression for $T_e$ at an arbitrary field, which would be more complicated than Eq. (6.53), were to be expanded in powers of $\mathscr{E}$, would you expect the terms $\mathscr{E}$, $\mathscr{E}^3$, $\mathscr{E}^5$, etc., to appear? Why? Does your argument apply equally well to such materials as Ge and GaAs?
12. In discussing hot electrons, one finds that the temperature of the electron is greater than that of the lattice. Can you conceive of a situation in which the temperature of the electrons might be lower than that of the lattice?
13. Suppose that, in working with a given semiconductor, you use an incident optical beam which is very strong. Is it possible for a fundamental absorption to take place even at a frequency $\nu < E_g/h$?
14. In an intrinsic semiconductor, is the Einstein relation valid for electrons and holes individually?

## PROBLEMS

1. Derive (6.13) for hole concentration.
2. a) Compute the concentration of electrons and holes in an intrinsic sample of Si at room temperature. You may take $m_e = 0.7\ m_0$ and $m_h = m_0$.
   b) Determine the position of the Fermi energy level under these conditions.
3. Given that the pre-exponential factors in (6.8) and (6.13) are $1.1 \times 10^{19}$ and $0.51 \times 10^{19}\ \text{cm}^{-3}$, respectively, in Ge at room temperature, calculate:
   a) The effective masses $m_e$ and $m_h$ for the electron and the hole.
   b) The carrier concentration at room temperature.
   c) The carrier concentration at 77°K, assuming the gap to be independent of temperature.
4. Gallium arsenide has a dielectric constant equal to 10.4.
   a) Determine the donor and acceptor ionization energies.
   b) Calculate the Bohr radii for bound electrons and holes.
   c) Calculate the temperature at which freeze-out begins to take place in an $n$-type sample.
5. A silicon sample is doped by arsenic donors of concentration $1.0 \times 10^{23}\ m^{-3}$. The sample is maintained at room temperature.
   a) Calculate the intrinsic electron concentration, and show that it is negligible compared to the electron concentration supplied by the donors.
   b) Assuming that all the impurities are ionized, determine the position of the Fermi level.
   c) Describe the effect on the Fermi level if acceptors are introduced in the above sample at a concentration of $6.0 \times 10^{21}\ \text{m}^{-3}$.
6. Given these data for Si: $\mu_e = 1350\ \text{cm}^2/\text{volt-s}$, $\mu_h = 475\ \text{cm}^2/\text{volt-s}$, and $E_g = 1.1$ eV, calculate the following.
   a) The lifetimes for the electron and for the hole.
   b) The intrinsic conductivity $\sigma$ at room temperature.
   c) The temperature dependence of $\sigma$, assuming that electron collision is dominated by phonon scattering, and plot log $\sigma$ versus $1/T$.
7. Repeat Problem 6 for Ge, using Tables 6.1 and 6.2.
8. A sample of extrinsic semiconductor is in the shape of a slab whose length is 5 cm, width 0.5 cm, and thickness 1 mm. When this slab is placed in a magnetic field of 0.6 Wb/m$^2$ normal to the slab, a Hall voltage of 8 mV develops at a current of 10 mA. Calculate: (a) the mobility of the carrier, (b) the carrier density.
9. A sample of $n$-type GaAs whose carrier concentration is $10^{16}\ \text{cm}^{-3}$ has the same dimensions, is in the same field, and carries the same current as in Problem 8. Calculate: (a) the Hall constant in this sample, (b) the Hall voltage developed across the slab.
10. When we derived the Hall constant in Section 4.10, we assumed that the carrier mass is isotropic; the mobility of the carrier is therefore also isotropic. However, we have seen that carriers in some semiconductors have ellipsoidal masses.
    a) Show that when current in an $n$-type Si sample flows in the [100] direction, the Hall constant is given by

$$R = -\frac{3}{ne}\frac{\mu_l^2 + 2\mu_t^2}{(\mu_l + 2\mu_t)^2},$$

where $\mu_l = e\tau/m_l$ and $\mu_t = e\tau/m_t$ are the longitudinal and transverse mobility, respectively.

b) Recalling that $m_l/m_t \simeq 5$ in Si, evaluate the Hall constant for $n = 10^{16}\ \text{cm}^{-3}$.

c) What is the value of $R$, given that the current flows in the [010] direction (with the orientation of the magnetic field appropriately rearranged)? [*Hint*: Note that the populations of the six valleys are equal to each other.]

11. a) Show that the density of states corresponding to an ellipsoidal energy surface is

$$g(E) = \frac{1}{2\pi^2}\left(\frac{2}{\hbar^2}\right)^{3/2} (m_t^2\, m_l)^{1/2}\, E^{1/2},$$

where $m_t$ and $m_l$ are the transverse and longitudinal masses, respectively. (The energy surface is taken to be an ellipsoid of revolution.)

b) If we make the replacement $m_t^2\, m_l = m_d^3$ in the above expression, then $g(E)$ would have the standard form for a spherical mass, (6.6), with $m_d$ substituted for $m_e$. For this reason, the mass $m_d$ is usually called the *density-of-states effective mass*. Taking into account the many-valley nature of the conduction band in Ge, find $m_d$ for this substance (expressing the results in units of $m_0$).

12. When a carrier has an ellipsoidal mass, e.g., the electrons in Si, the mobility is also anisotropic. The longitudinal and transverse mobilities $\mu_l$ and $\mu_t$ are in inverse ratio to the masses, i.e., $\mu_l/\mu_t = m_t/m_l$, as follows from (6.31). (The collision time is isotropic.) In tables such as Table 6.3, the so-called mobility $\mu = (\mu_l + 2\mu_t)/3$ is usually quoted. (This average is for an ellipsoid of revolution.)

a) Calculate $\mu_l$ and $\mu_t$ for silicon.

b) An electric field is applied in the [100] direction, and the field is so high that it heats the electrons (they become hot). But the valleys are heated at different rates because of the difference in carrier mobility in the longitudinal and transverse directions. Indicate which valleys become hotter than others.

c) Calculate the electric field at which the temperature of the hot valleys becomes 1000°K. (The lattice is at room temperature.) Take the energy relaxation time to be $2 \times 10^{-12}$ s. (Assume the mobility to be independent of the field.)

d) Suppose that the valleys are in quasi-equilibrium with each other; electrons then transfer from the hot to the cold valleys, and the valleys' populations are no longer equal. Find the fraction of the total electrons still remaining in the hot valleys at the field calculated in Problem 12(c).

e) Discuss the non-ohmic behavior resulting from this "intervalley transfer." Plot $J$ versus $\mathscr{E}$ up to a field three times the field calculated in Problem 12(c).

13. Estimate the value of the field for which an appreciable transfer of electrons takes place from the central to the secondary valleys in GaAs. [*Hint*: The energy absorbed by an electron in an interval of one lifetime must be of the order of the energy difference between valleys.]

14. a) Calculate the threshold photon energy for direct fundamental absorption of radiation in GaAs at room temperature.

b) Determine the corresponding wavelength.

c) At what wavelength is the absorption coefficient equal to $1000\ \text{cm}^{-1}$?

15. Suppose that you are a solid-state physicist, and a materials engineer asks you: Why should silicon exhibit metallic luster when viewed in visible light, yet be transparent when viewed in infrared light? What is your answer?

16. a) Determine the longest wavelength of light absorbed in ionizing an As donor in Si.
    b) Using data from Table 6.2, repeat Problem 16(a) for a Ga acceptor in Si.
17. A slab of intrinsic GaAs, 3 cm long, 2 cm wide, and 0.3 cm thick is illuminated by a monochromatic light beam, at which frequency the absorption coefficient is 500 $cm^{-1}$. The intensity of the beam is $5 \times 10^{-4}$ W $cm^{-2}$, and the sample is at room temperature.
    a) Calculate the photon flux incident on the slab.
    b) At what depth does the intensity decrease to 5% of its value at the surface?
    c) Calculate the number of electron–hole pairs created per second in the slab. (Assume that the beam entering is totally absorbed through fundamental transition.)
    d) Calculate the increase in the conductivity $\Delta\sigma$ due to the illumination. Take the recombination time to be $2 \times 10^{-4}$ s. [*Data*: The dielectric constant of GaAs is 10.4].
18. Establish the Einstein relation (6.81) between the mobility and diffusion coefficient. Consider a sample in the shape of a rod along which a voltage is applied, but no current may flow because the circuit is open. The sample has now both an electric field and a concentration gradient. Assume Maxwell–Boltzmann statistics for the carriers.
19. It is found experimentally that the mobility in Ge depends on the temperature as $T^{-1.66}$. The mobility of this substance at room temperature is 3900 $cm^2$/volt-s. Calculate the diffusion coefficient at room temperature (300°K) and at the temperature of liquid nitrogen (77°K).
20. Suppose that the concentration of electrons in *n*-type Ge at room temperature decreases linearly from $5 \times 10^{16}$ $cm^{-3}$ to zero over an interval of 2 mm.
    a) Calculate the diffusion current.
    b) What is the value of the electric field required to produce a drift current equal to the diffusion current of part (a)? Use the average value of the concentration in determining the drift current.
    c) Draw a diagram to show the direction of the field.

# CHAPTER 7 SEMICONDUCTORS II: DEVICES



---

*... The morality of art consists*
*in the perfect use of an*
*imperfect medium*

Oscar Wilde

## 7.1 INTRODUCTION

In Chapter 6 we dealt with the physical principles governing the behavior of semiconductors. Here we shall take up the applications of these principles to practical electronic devices. The successful development of these devices, particularly the transistor, stimulated great interest in semiconducting substances, and in solid-state physics and materials science in general. Semiconductor research, particularly since the early 1950's, has also enormously expanded our understanding of the basic structure of matter.

We shall begin with the basic properties of the *p-n* junction, and explain its rectification property. We shall then show how the joining of two such junctions, resulting in the junction transistor, acts as an amplifier. In terms of practical applications, this transistor is the most important of the solid-state devices. We shall then discuss microwave devices operating on the principle of negative differential conductivity (particularly the tunnel and Gunn diodes); the semiconductor laser; and other semiconductor devices. The chapter closes with a section on integrated circuits, an area of increasing importance in solid-state devices.

## 7.2 THE *p-n* JUNCTION: THE RECTIFIER

The *p-n junction* is a specimen made of a single-crystal semiconductor in which there are two adjacent regions, an *n*-type and a *p*-type (Fig. 7.1a). The *n* region is

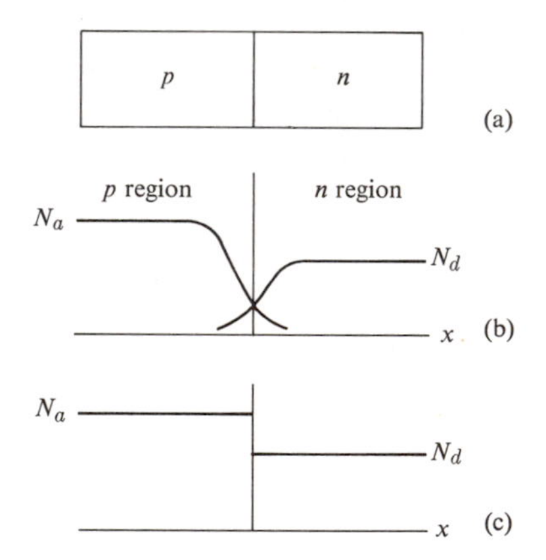


**Fig. 7.1** (a) A *p-n* junction. (b) A graded junction. (c) An abrupt junction.

doped with donor impurities, the *p* region with acceptor impurities. The variations of donor and acceptor concentrations, $N_d$ and $N_a$, across the junction and in its neighborhood, are somewhat as shown in Fig. 7.1(b). Such a junction, in which the impurity concentrations vary gradually, is called a *graded* junction. An abrupt junction, in which the impurities change discontinuously, is shown in Fig. 7.1(c). The donor concentration is a constant, $N_d$, in the *n* region, and zero in the *p* region. The acceptor concentration behaves similarly. To simplify the discussion, we shall consider here only the abrupt junction because we can then illustrate the physical phenomena without inessential mathematical complications. Note, however, that one cannot manufacture a strictly abrupt junction because impurities tend, to some extent, to diffuse across the junction.

When we speak of a junction, we mean the region in which the *p* and *n* regions meet. N.B.: The whole junction is one single piece of crystal, made of the same semiconductor material, in which the two sides are doped differently, rather than two different pieces joined together.

The most interesting electrical property of the junction is that a *potential difference* develops across it even when the junction is at equilibrium. This is called a *contact potential*, and usually has a value in the range 0.1–1.0 volt. To explain how the contact potential arises, let us consider a *p-n* junction at the initial stage, when the junction has just been prepared. At that instant there are only electrons in the *n* region and holes in the *p* region, and the concentrations of free carriers are constant on both sides of the junction. The electric field is zero everywhere, because the charges of the free carriers are balanced by the charges of the ionized impurities, e.g., electrons and ionized donors in the *n* region. The situation is not in equilibrium, however, despite the absence of an electrical field. The electrons begin to diffuse across the junction from the *n* to the *p* region, and similarly the holes diffuse from the *p* to the *n* region. The free carriers diffuse because there is initially a large concentration gradient at the junction for both electrons and holes and, as we saw in Section 6.16, such a gradient leads to a diffusion current in the direction of decreasing concentration. Therefore, in Fig. 7.1(a), electrons diffuse to the left and holes to the right.

The diffusion currents do not continue indefinitely, however, because as electrons diffuse to the *p* region, that region acquires a net negative charge. This is enhanced further by holes leaving the *p* region and diffusing to the *n* side. The net effect of the diffusion current is that the *p* region becomes negatively charged relative to the *n* region. As a result, there develops a potential difference, which is the abovementioned contact potential.

Figure 7.2 shows the positions of the conduction and valence bands near the junction.† Because of charge transfer, the energies on the *p* side have been raised relative to the *n* side. (The *p* side has acquired a net negative charge.) Given that the contact potential is $\phi$, the energy difference between the two sides is $e\phi$. (According to the usual convention in electromagnetism theory, the potential of the *n* side is higher than that of the *p* side by the amount $\phi$. In solid-state physics, however, one plots the energy of an electron, and since the electron has a negative charge $-e$, it follows that the *p* side has a higher electron energy, as shown in Fig. 7.2.)

The presence of the contact potential impedes the flow of the diffusion currents. There is now actually an electric field equal to $-d\phi/dx$ acting at the junction to oppose further diffusion of electrons and holes. Referring to Fig. 7.2, we can see

---

† In a real junction the sharp corners shown in the figure are rounded off, but this point is unimportant for the present discussion.

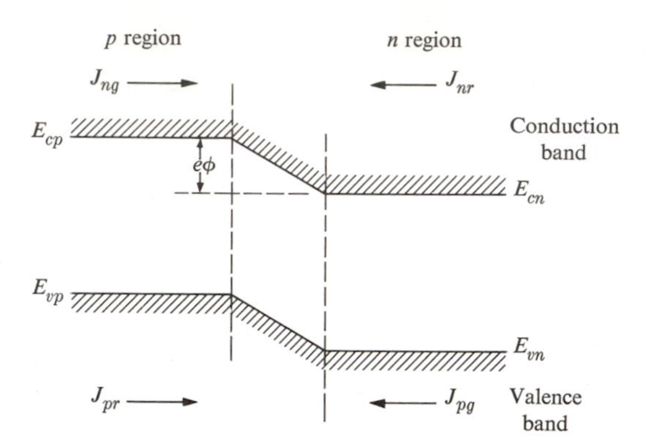


**Fig. 7.2** The *p-n* junction from the point of view of the energy band. Shown are the contact potential $\phi$ and the various fluxes associated with the junction.

that, of the electrons in the conduction band on the $n$ side, only those of kinetic energy larger than the barrier $\phi$ are able to diffuse to the $p$ side. As the charge transfer continues, the potential continues to increase, and hence the diffusion flux continues to decrease until it becomes balanced by an electron flux flowing from the $p$ to the $n$ side. It is called the *generation flux*. Its source lies in the following phenomenon: On the $p$ side, electrons and holes are continually created by thermal generation; the rate of generation depends on the temperature. Simultaneously, these electrons and holes recombine with each other. However, at any one temperature, there is a certain number of electrons and a certain number of holes, the relative concentration of which depends on the concentration of impurities (as discussed in Section 7.6). The electrons in the $p$ region give rise to an electron flux flowing to the $n$ region, because some of them are likely to wander into the junction region itself. Once there, they are quickly swept away to the $n$ side by the strong electric field inside the junction. Thus, looking at electrons alone, there are two fluxes flowing across the junction: (1) A current from the $n$ to the $p$ side due to the large electron concentration on the $n$ side, known as the *recombination flux*† $J_{nr}$ (due to the fact that electrons flowing into the $p$ region eventually combine with holes there). (2) The *generation flux* $J_{ng}$, which flows from the $p$ to the $n$ side, and is due to the generation and subsequent sweeping of electrons by the junction field. Equilibrium is achieved when the two fluxes are equal,

$$J_{nr0} = J_{ng0}. \tag{7.1}$$

By the same token, holes in the valence band also flow across the junction, and

† In this chapter we must differentiate between the particle current and the electric current associated with it. Thus the particle current associated with diffusion, that is, $-D\frac{\partial n}{\partial x}$, will be called the *diffusion flux* and denoted by $J$. The electric current associated with it will be called the *current* and denoted by $I$. Thus for electrons and holes we have, respectively, $I_n = -eJ_n$ and $I_p = +eJ_p$.

there is a hole recombination flux $J_{pr}$ flowing from the $p$ to the $n$ side, and a hole generation flux flowing in the opposite direction. At equilibrium, these fluxes must also be equal,

$$J_{pr0} = J_{pg0}. \tag{7.2}$$

Thus the equilibrium situation is a dynamic one. Fluxes are flowing continually across the junction, but there is no further charge transfer, since the fluxes cancel each other for both types of carrier separately.

We can now explain the rectification property of a $p$-$n$ junction. Suppose that an external voltage $V_0$ is applied to the junction in such a way that the $p$ region is positive, as shown in Fig. 7.3(a) (the $p$ region is connected to the positive electrode

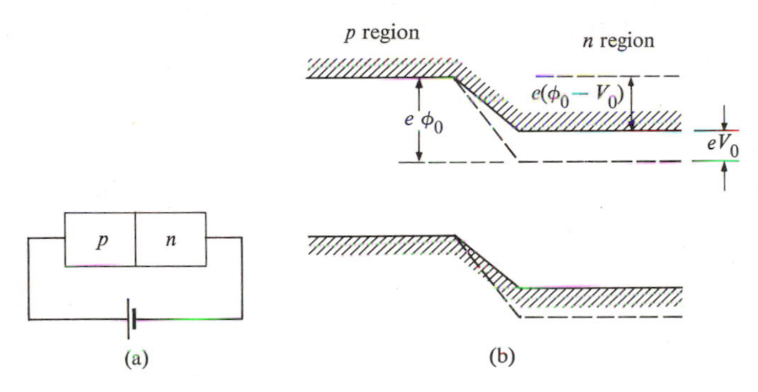


**Fig. 7.3** (a) A forward-biased electrical connection of a $p$-$n$ junction. (b) The effect of a forward bias on the energy-band diagram of a junction. Dashed lines indicate the position of band edges without any bias (at equilibrium).

of the battery). This method of connecting the junction is called the *forward bias*; the effect of this forward bias is shown in Fig. 7.3(b). The $n$ region has been raised by an amount $eV_0$. Let us now see what effect this has on the fluxes discussed above (noting that the present situation is one of nonequilibrium). Starting with the electron currents, we first see that the generation flux is unaffected by $V_0$. That is,

$$J_{ng} = J_{ng0}, \tag{7.3}$$

because there is still a field in the junction strong enough to sweep the electrons coming from the $p$ region, provided that $V_0 < \phi_0$, which is the situation encountered in practice. On the other hand, the recombination flux $J_{nr}$ has been affected considerably. Since the electrons on the $n$ side see a potential hill whose height has been decreased by an amount $eV_0$, the recombination flux is now increased by a factor $e^{eV_0/k_BT}$, assuming that the electrons obey Maxwell–Boltzmann statistics. Thus we have

$$J_{nr} = J_{nr0}\, e^{eV_0/k_BT}. \tag{7.4}$$

There is therefore a net flow of electrons from the $n$ to the $p$ side. The actual electrical current flows from left to right, as the electron has a negative charge $-e$, and has the value

$$I_n = e(J_{nr} - J_{ng}) = e\, J_{ng0}(e^{eV_0/k_BT} - 1), \tag{7.5}$$

where we used Eqs. (7.1) through (7.4).

A forward bias leads to a hole current of a form similar to (7.5). As we see from Fig. 7.3(b), such a bias leads to an increase in the hole recombination current, because the potential hill has also decreased by the amount $eV_0$. (In connection with the energy diagram, we may visualize the hole as an air bubble which has a tendency to float.) This leads to an increase of this current by a factor $e^{eV_0/k_BT}$, while the generation flux remains the same as before. Therefore the electrical current carried by the holes is

$$I_p = e(J_{pr} - J_{pg}) = eJ_{pg0}(e^{eV_0/k_BT} - 1). \tag{7.6}$$

The total electrical current $I$ is the sum of the currents carried by the electrons and the holes. Since both $I_n$ and $I_p$ are in the same direction (from the positive to the negative electrode of the battery), we have, on the basis of (7.5) and (7.6),

$$I = I_n + I_p = e(J_{ng0} + J_{pg0})\,(e^{eV_0/k_BT} - 1),$$

which is of the form

$$I = I_0\,(e^{eV_0/k_BT} - 1), \tag{7.7}$$

where

$$I_0 = e(J_{ng0} + J_{pg0}). \tag{7.8}$$

(Note that $I_0$ is independent of the bias $V_0$.) Figure 7.4 plots $I$ versus $V_0$; we see that $I$ rises sharply with $V_0$. The dependence of $I$ on $V_0$ is essentially exponential, as can be seen from (7.7) by noting that usually $eV_0 \gg k_BT$ (at room temperature $k_BT/e \simeq 0.025$ volt). Therefore, to a very good approximation,

$$I \simeq I_0\, e^{eV_0/k_BT}. \tag{7.9}$$

We have derived the $I$–$V_0$ relation (7.7) for a forward bias. Let us now derive the corresponding relation for the *reverse bias*, which is the case in which the *p-n* junction is so biased that the $p$ side is connected to the negative electrode of the battery, as in Fig. 7.5(a). The effect of such a bias on the energy-level diagram is shown in Fig. 7.5(b), in which we see that the height of the potential is now *increased* by the amount $eV_0$. Here again there are recombination and generation currents for both electrons and holes. In attempting to find the influence of $V_0$ we can follow the same procedure used in the case of forward bias above. The conclusion now is that the generation fluxes are again unaffected by $V_0$, because the junction field is still strong enough to accomplish the sweeping. On the other hand,

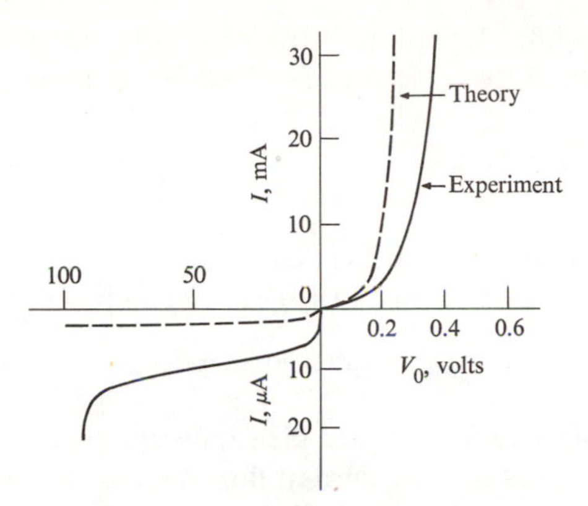


**Fig. 7.4** Current versus voltage ($I$–$V_0$ characteristics) for a junction, illustrating the rectification property. The first quadrant in the $I$–$V_0$ plane refers to the condition of forward bias, while the third quadrant refers to reverse bias. Note the change of scale between these two quadrants.

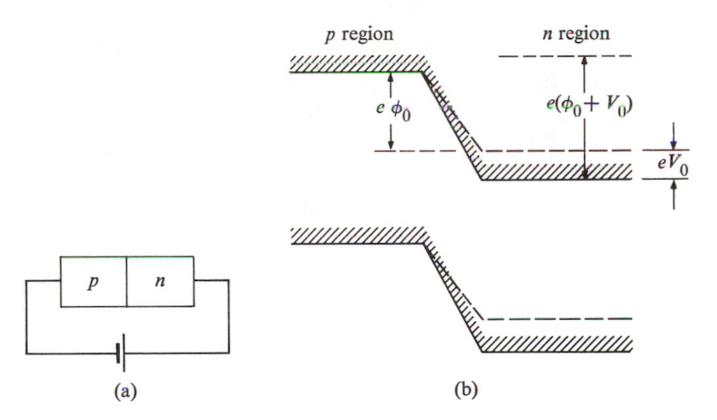


**Fig. 7.5** A reverse-bias connection (a) and its effect on the edges of the energy bands.

since the height of the potential barrier is increased, the generation flux for both electrons and holes decreases by the factor $e^{-eV_0/k_BT}$. In the case of electrons, for example, there are now fewer of them with enough kinetic energy to go over the potential barrier $e(\phi_0 + V_0)$. The total current from the $n$ to the $p$ side (positive to negative electrode) is

$$I = I_n + I_p = e(J_{ng0} - J_{nr}) + e(J_{pg0} - J_{pr})$$

$$= e\, J_{ng0}(1 - e^{-eV_0/k_BT}) + e\, J_{pg0}(1 - e^{-eV_0/k_BT}),$$

or

$$I = I_0(1 - e^{-eV_0/k_BT}), \tag{7.10}$$

where $I_0$ is given by (7.8). Equation (7.10) is the $I$–$V_0$ relation for a reserve bias. Since in usual circumstances $eV_0 \gg k_BT$, it follows that the exponential term in (7.8) is so small that it can be neglected. Therefore, for a reverse bias, we have the simple relation

$$I = I_0. \tag{7.11}$$

That is, the current is a constant independent of $V_0$.

We note that both Eqs. (7.7) and (7.10) can be combined into a single equation,

$$I = I_0(e^{eV_0/k_BT} - 1), \tag{7.12}$$

if $V_0$ is taken to be positive for forward bias and negative for reverse bias. Also, a positive value for $I$ implies that the current flows across the junction from the $p$ to the $n$ side, while a negative value of $I$ indicates a current in the opposite direction. A complete plot of $I$ versus $V_0$, using (7.12), and including negative bias, is shown in Fig. 7.4. Obviously the current for a forward bias is very much larger than that for a reverse bias (for the same $|V_0|$). This means that the junction acts as a rectifier, allowing a current to flow much more readily from $p$ to $n$ than vice versa. The quantity $I_0$, which is the magnitude of the reverse-bias current, is called the *saturation current*. A typical value of reverse-current density is $10^{-5}$ A-cm$^{-2}$, or a current of about 10 $\mu$A/cm$^2$. The forward current depends greatly on the voltage, but a typical value is 100 mA for a bias of 0.2 V.

We have made one implicit assumption: When a bias voltage $V_0$ is applied, all of it appears across the junction region itself, and none is expended across the remainder of the $p$ and $n$ regions. The justification for this is that the junction has a much higher resistance than the remainder of the specimen because, as we shall see in Section 7.3, the junction region is depleted of free carriers. Since the resistance is mainly at the junction, and since the current is usually not very large, taking the voltage across the junction to be the same as the external voltage is a good approximation.

Note also that if the reverse voltage is made very large, finally an electric breakdown occurs, at which point the reverse current suddenly increases very rapidly. The problem of breakdown itself is an interesting one. Two mechanisms may be considered: (1) *Avalanche breakdown*, in which some of the electrons accelerated by the large reverse voltage acquire enough energy to excite electron–hole pairs, which if sufficiently energetic, go on to excite additional electron–hole pairs, and so forth. (2) *Zener breakdown*, which is based on the observation that at very high reverse voltage the thickness (not the height) of the potential barrier between the two sides of the junction becomes so small that quantum tunneling becomes possible. At that point, the current does increase rapidly. (Tunneling in the context of a *p-n* junction is discussed in Section 7.5 on the tunnel diode.) In the lower voltage range ($\simeq$ 4 V), the Zener mechanism dominates, while for large voltage ($\simeq$ 8 V), avalanche breakdown is the dominant mechanism. In the intermediate region, both mechanisms operate simultaneously.

We wish now to derive an explicit expression for the saturation current $I_0$ of (7.8). Consider the case of the forward bias only, and look more closely at the conduction process. The current is carried by both electrons and holes. Let us first discuss the hole current. In the forward bias, the potential barrier is reduced by $eV_0$, and additional holes are injected from the $p$ side, where they are *majority* carriers, to the $n$ side, where they are *minority* carriers (Fig. 7.6).

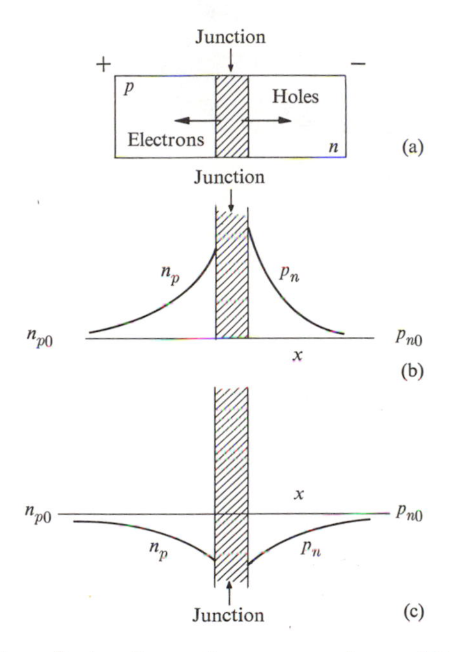


**Fig. 7.6** (a) The injection of minority carriers across a forward-bias junction. (b) Spatial variations of minority-carrier concentrations in the forward condition, showing the effects of minority-carrier injections. (c) Spatial variations of minority carriers in the reverse condition, showing the effects of minority-carrier depletion near the junction.

Once on the $n$ side, these holes diffuse freely, as there is no electric field. But, because of recombination, the excess concentration of holes damps out to its equilibrium value $p_{n0}$ within a length $L_p$. Thus we may write for the excess-hole concentration in the $n$ region

$$p_{n1}(x) = p_n(x) - p_{n0} = (p_{n1})_{x=0}\, e^{-x/L_p}, \qquad x > 0, \tag{7.13}$$

where $(p_{n1})_{x=0}$ is the value of the concentration of excess holes immediately to the right of the junction. The hole concentration decays exponentially in the $n$ region (Fig. 7.6b). We can now understand how the hole current arises: It is a purely diffusive current arising from the concentration gradient of the holes in the $n$ region.

The ultimate source of this current is of course the continuous injection of holes from the $p$ to the $n$ region. Thus we see that, in the case of the forward bias, the current is due to the injection and subsequent diffusion of *minority* carriers.

To calculate the hole current, we need to know $(p_{n1})_{x=0}$, or equivalently $(p_n)_{x=0}$. The reason $(p_n)_{x=0}$ is different from the equilibrium value $p_{n0}$ is that the potential barrier has been reduced by the amount $eV_0$. We therefore expect, from Boltzmann statistics, that

$$(p_n)_{x=0} = p_{n0}\, e^{eV_0/k_BT}. \tag{7.14}$$

By comparing this with the value of (7.13) at $x = 0$, we find that

$$(p_{n1})_{x=0} = p_{n0}\,(e^{eV_0/k_BT} - 1). \tag{7.15}$$

Substituting this into (7.13), we find that

$$p_{n1} = p_{n0}\,(e^{eV_0/k_BT} - 1)\, e^{-x/L_p}, \qquad x > 0, \tag{7.16}$$

for the concentration of excess holes in the $p$ region. Using Fick's law (6.80), we find for the hole diffusion flux

$$J_{pn} = -D_p \frac{\partial p_n}{\partial x} = -D_p \frac{\partial p_{n1}}{\partial x}.$$

If we evaluate this current at $x = 0$, using (7.16), and multiply by $e$ to convert it to an electrical current, we find

$$(I_{pn})_{x=0} = \frac{e\, D_p\, p_{n0}}{L_p}\,(e^{eV_0/k_BT} - 1). \tag{7.17}$$

We have found the hole current at a specific point—immediately to the right of the junction; however, a current of this value, associated with holes, flows at every region of the crystal to the right of the junction.

We can find the electron current in a similar manner by arguing that the forward bias injects electrons from the $n$ to the $p$ region (again injection of minority carriers) which diffuse into the field-free $p$ region, carrying an electron diffusion current. The spatial distribution of the excess electrons is given by an equation similar to (7.16), with suitable modifications, and has the shape shown in Fig. 7.6. The electron current immediately to the left of the junction has a form analogous to (7.17). Again this gives the electron current at every region of the crystal.

The total current $I$ is given by $I_n + I_p$. Therefore, using (7.17) and its analog for the electrons, we have

$$I = e\left(\frac{D_p\, p_{n0}}{L_p} + \frac{D_n\, n_{p0}}{L_n}\right)(e^{eV_0/k_BT} - 1). \tag{7.18}$$

We see that this is of the same form as (7.7). By noting (7.8), we conclude that the saturation current is given by

$$I_0 = e\,(J_{ng0} + J_{pg0}) = e\left(\frac{D_n n_{p0}}{L_n} + \frac{D_p p_{n0}}{L_p}\right). \tag{7.19}$$

We have thus evaluated the saturation current, or the generation current, in terms of the properties of the materials involved, $D_n$, $L_n$, $D_p$, $L_p$, and in terms of the equilibrium concentrations $n_{p0}$ and $p_{n0}$ of the minority carriers in the two regions of the junction.

Equation (7.18) has some implications regarding the choice of material to be used as a rectifier. Thus if the rectifier is to be used under conditions of high forward current, we must make the reverse current $I_0$ small, Let us rewrite (7.19) in terms of the majority concentrations $n_{n0}$ and $p_{p0}$ by using the relation

$$n_{n0}\,p_{n0} = p_{p0}\,n_{p0} = n_i^2(T), \tag{7.20}$$

where $n_i^2(T)$ is the intrinsic concentration, which is $\sim e^{-E_g/2k_BT}$ (see Section 6.4). Thus we may write Eq. (7.19) as

$$I_0 = e\,n_i^2(T)\left(\frac{D_n}{L_n p_{p0}} + \frac{D_p}{L_p n_{n0}}\right). \tag{7.21}$$

We now see that $I_0$ depends strongly on temperature, and although this dependence arises from the dependence of the various quantities in (7.21) on $T$, by far the strongest influence arises from the dependence of $n_i(T)$ on $T$. Since $n_i(T) \sim e^{-E_g/2k_BT}$, one may reduce $I_0$ by choosing a material with a large gap. This is the primary reason for the preference of silicon over germanium for rectifiers operating under conditions of high current and high temperature.

To return to the hole current in the $n$ region [Eq. (7.17)]: It is true that the hole concentration decreases as the holes diffuse to the right, and consequently the diffusion current carried by these holes also decreases. However, since the holes' recombination, just to the right of the junction, depletes the electrons there, other electrons flow into this region from the rest of the circuit to maintain charge neutrality. These replacement electrons ultimately come from the far right side of the $n$ region, where the semiconductor is in contact with the metallic wire completing the electric circuit. These replacement electrons carry their own electric current, also in the $n$ region (which is to the right). When the current is added to the local hole diffusion current, there results a constant current whose value is given by (7.17). Thus as we move from the junction to the right, in the $n$ region, a larger and larger fraction of the current is carried by replacement electrons. This same argument can also be used in the discussion of the electron current in the $p$ region.

Consider the so-called *injection efficiency* $\eta$. As we stated above, in a forward bias, the current is carried by injection of minority carriers, both electrons and holes. What fraction of this current is carried by electrons, and what fraction by holes?

These fractions—called the electron and hole injection efficiencies—are denoted by $\eta_n$ and $\eta_p$, respectively. By inspecting (7.19) and noting (7.20) or (7.21), we readily see that

$$\eta_n = \frac{J_{ng0}}{J_{ng0} + J_{pg0}} = \frac{D_n/L_n p_{p0}}{D_n/L_n p_{p0} + D_p/L_p n_{n0}}$$

and

$$\eta_p = 1 - \eta_n. \tag{7.22}$$

From this we see that if the $D$'s and $L$'s for electrons and holes are comparable, then

$$\eta_n \simeq \frac{n_{n0}}{n_{n0} + p_{p0}} \quad \text{and} \quad \eta_p \simeq \frac{p_{p0}}{n_{n0} + p_{p0}}. \tag{7.23}$$

That is, most of the current is carried by those carriers which are majority carriers in the heavily doped region. In a symmetric junction, where $n_{n0} = p_{p0}$, it follows from (7.23) that the current is carried equally by electrons and holes.

We have not discussed the effect of reverse bias on the carrier concentrations near the junction. We recall that the effect of reverse bias is to increase the height of the potential barrier by $e|V_0|$. Consider the effect of this on the holes near the junction. The generation current, from the $n$ to the $p$ region, remains unaffected, but the recombination current, from the $p$ to the $n$ region, decreases. Therefore more holes flow from the $n$ to the $p$ region, and as a result the concentration of holes in the $n$ region plummets below its equilibrium value near the junction (Fig. 7.6c). Similarly, the concentration of electrons in the $p$ region is reduced below its equilibrium value. Thus the overall effect of a reverse bias in the steady state is to extract minority carriers from the region near the junction.

## 7.3 THE *p-n* JUNCTION: THE JUNCTION ITSELF

In Section 7.2 we derived the rectification properties of a *p-n* junction by using statistical arguments concerning the distribution of free carriers near the junction. We did not need to consider the properties of the junction itself—e.g., the contact potential and the width of the junction—because these quantities were not essential to our discussion of the main topic, the current. A fuller understanding of a *p-n* junction, however, requires some knowledge of the properties of the junction. Let us look at these properties both at equilibrium and in the presence of a bias voltage. Incidentally, our findings in this section do not change those of Section 7.2; rather they shed light on some of the steps we took there.

Consider first the equilibrium case. Because of the large concentration of carriers present at the junction when it was originally formed, the majority carriers diffuse to the opposite side. This emigration of carriers from both sides of the junction leaves layers which are depleted of free carriers on both sides, as seen in

Fig. 7.7(a). On the $n$ side of the junction there is a layer of thickness $w_n$ which is *depleted* of electrons; however, since ionized donors are still present, the layer has a net positive charge. There is another depletion layer on the $p$ side of the junction, of thickness $w_p$, which is negatively charged. We conclude therefore that the immediate neighborhood of the junction is made up of a charged double layer (or a dipole layer). This area of the junction is called the *depletion*, or *space-charge*, region. In this region there is a strong electric field as a result of the charged double layer (the field is directed to the left).

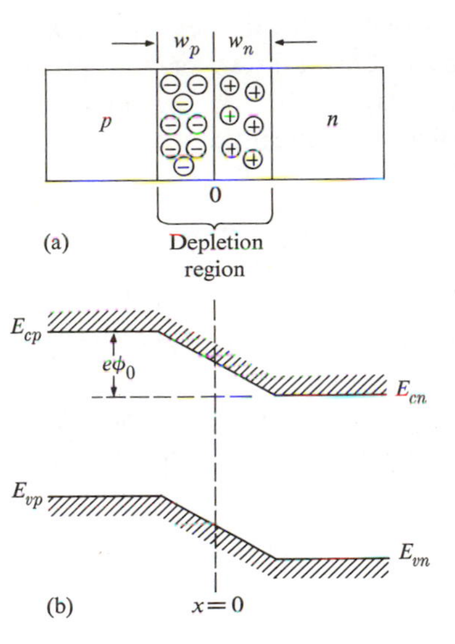


**Fig. 7.7** (a) The depletion region (double layer) at the junction. (b) The positions of band edges at the junction; the contact potential $\phi_0$.

Outside the depletion region, the carrier concentrations are unaffected by the junction, and hence are uniform, so the field is zero because there is charge neutrality. Figure 7.7(b) shows the effect of the junction on the energy-level diagram, as well as the potential barrier $e\phi_0$, as discussed in Section 7.2. (The equilibrium contact potential, denoted by $\phi$ in Section 7.2, will henceforth be designated by $\phi_0$.)

Let us calculate the contact potential $\phi_0$. As seen from Fig. 7.7(b),

$$e\phi_0 = E_{cp} - E_{cn}, \tag{7.24}$$

where $E_{cp}$ and $E_{cn}$ are the energies of the edges of the conduction bands in the $p$ and $n$ regions, respectively. These energies can be related to the equilibrium concentration as follows,

$$n_{p0} = U_c e^{-(E_{cp} - E_F)/k_B T}, \qquad n_{n0} = U_c e^{-(E_{cn} - E_F)/k_B T}, \tag{7.25}$$

where $U_c = 2(m_e k_B T / 2\pi\hbar^2)^{3/2}$, as we see by referring to (6.8). Here $E_F$ is the Fermi energy, which is the same throughout the junction, since we are discussing an equilibrium situation. By finding the ratio $n_{n0}/n_{p0}$ from (7.25) and using (7.24), we establish that

$$\frac{n_{n0}}{n_{p0}} = e^{e\phi_0/k_B T}. \tag{7.26}$$

This gives $\phi_0$ in terms of the equilibrium electron concentrations on both sides of the junction. It is more convenient, however, to express $\phi_0$ in terms of the majority carriers on both sides, that is, $n_{n0}$ and $p_{p0}$. To eliminate $n_{p0}$ from (7.26) in favor of $p_{p0}$, we use (7.20), involving the intrinsic concentration $n_i$. Combining these two relations, one finds

$$\phi_0 = \frac{k_B T}{e} \log\left(\frac{n_{n0}\, p_{p0}}{n_i^2}\right). \tag{7.27}$$

We recall from Section 6.6, however, that usually $n_{n0} \simeq N_d$ and $p_{p0} \simeq N_a$, where $N_d$ and $N_a$ are the concentrations of donors and acceptors, respectively. This means that essentially all the impurities are ionized, which is true, except at fairly low temperature, for example, $< 50°$K. Therefore the contact potential is given approximately by

$$\phi_0 \simeq \frac{k_B T}{e} \log\left(\frac{N_d N_a}{n_i^2}\right), \tag{7.28}$$

a potential which depends on the properties of the semiconductor, the doping, and the temperature. To get an idea of the magnitudes involved, recall that $k_B T/e \simeq 0.025$ volt at room temperature. This gives $\phi_0 = 0.3$ volt for germanium with dopings $N_d = N_a = 10^{16}\ \text{cm}^{-3}$.

Finding the contact potential was a relatively easy matter. One has to work harder in order to find other quantities, such as the width of the junction and the electric field inside it. To obtain these, one usually needs to solve a Poisson's equation which leads to a nonlinear differential equation.

For example: Suppose that we have a plane junction, perpendicular to the $x$-axis. In this case, the Poisson's equation for the potential $\phi$ reduces to

$$\frac{d^2\phi}{dx^2} = -\frac{\rho(x)}{\epsilon}, \tag{7.29}$$

where $\rho(x)$ is the charge density and $\epsilon$ the dielectric constant of the medium. It is through $\rho(x)$ that the properties of the semiconductor and impurities enter. In the most general case, we may write at an arbitrary point $x$,

$$\rho(x) = e\,[p(x) + N_d(x) - n(x) - N_a(x)], \tag{7.30}$$

where $N_d(x)$ and $N_a(x)$ are the concentrations of *ionized* donors and acceptors, and $p(x)$ and $n(x)$ are the carrier concentrations, all at point $x$. If we were to pursue this general discussion, we would have to compute the quantities $p(x)$, $n(x)$, etc., which turn out to be functions of the local $\phi(x)$, and when we substituted all these into (7.30) and then into (7.29), we would find a nonlinear differential equation.

Let us instead simplify the discussion by assuming that the junction is abrupt, and that there are no carriers at all in the depletion region, i.e., complete depletion. These assumptions are realizable in practice. In the depletion region, Eq. (7.29) now becomes (recall Fig. 7.7)

$$\begin{aligned} \frac{d^2\phi_n}{dx^2} &= -\frac{eN_d}{\epsilon}, \qquad 0 < x < w_n, \\ \frac{d^2\phi_p}{dx^2} &= \frac{eN_a}{\epsilon}. \qquad -w_p < x < 0. \end{aligned} \tag{7.31}$$

Here $N_d$ and $N_a$ are the concentrations of ionized impurities on both sides of the junction; they are independent of $x$. We want to solve Eqs. (7.31) subject to the following boundary conditions: (i) The electric field is zero outside the depletion region (recall that $\mathscr{E} = -d\phi/dx$). (ii) The electric field is continuous at the point $x = 0$, the center of the junction. (iii) The potential is continuous at $x = 0$ (and is chosen to be zero, since the potential has an arbitrary additive constant). (iv) The potential difference between the far ends of the depletion layers, $x = w_n$ and $x = -w_p$, is equal to $\phi_0$, which we calculated above. This means that our solution is restricted to the equilibrium case. Solving (7.31) subject to the above boundary conditions is a straightforward matter, and the details are left as an exercise. The results are

$$\begin{aligned} w_n N_d &= w_p N_a & \text{(a)} \\ w_n &= [2\epsilon\phi_0 N_a/N_d(N_d + N_a)e]^{1/2} & \text{(b)} \\ w_p &= [2\epsilon\phi_0 N_d/N_a(N_d + N_a)e]^{1/2} & \text{(c)} \\ w = w_n + w_p &= \left[\frac{2\epsilon\phi_0}{e}\left(\frac{1}{N_d} + \frac{1}{N_a}\right)\right]^{1/2} & \text{(d)} \\ \mathscr{E}_0 &= 2\phi_0/w, & \text{(e)} \end{aligned} \tag{7.32}$$

where $\mathscr{E}_0$ is the magnitude of the electric field at $x = 0$, the center of the junction. From (a) we see that the total charges of the two depletion layers are equal in magnitude. From (b) (or c) we see that $w_n \sim N_d^{-1/2}$ (the dependence of $\phi_0$ on $N_d$ is weak, since it is logarithmic), and hence the heavier the doping, the thinner the layer. Equation (b) states again that heavier doping leads to a thinner junction. Finally, Eq. (e) states that $\mathscr{E}_0$ is also equal to twice the average field $(\phi_0/w)$ over the *whole* junction.

Figure 7.8 shows a plot of the variation of the field across the junction (the negative sign means that the field is pointing to the left in Fig. 7.7). To show the numerical values involved: For Ge, when $N_d = N_a = 10^{16}\ \text{cm}^{-3}$, $\phi_0 = 0.3$ V, $w_n = w_p = 0.16\mu$, $w = 0.33\mu$, and $\mathscr{E}_0 = 21$ kV/cm. The junction width is of the order of a micron, and the field present is considerable.†

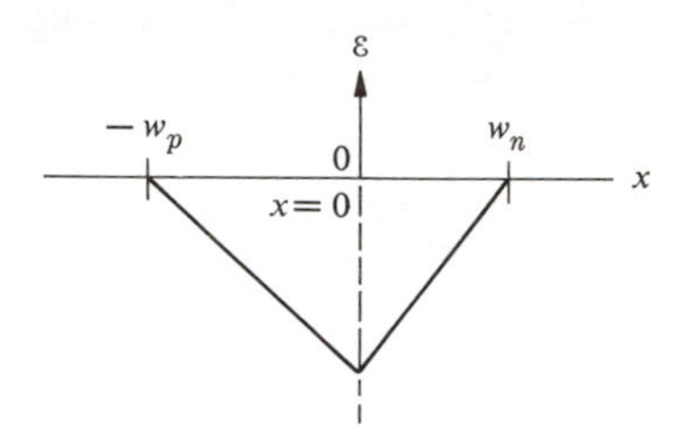


**Fig. 7.8** Spatial variation of the internal electric field in the neighborhood of the junction.

Now let's extend the above results to the nonequilibrium case, in which a certain bias voltage is applied across the junction. We can obtain the nonequilibrium results from those at equilibrium, (7.32), by making the following observations. The claim is that when an external bias $V_0$ is applied across the junction, almost the whole of this voltage actually appears across the depletion region only, the voltage drop across the remainder of the junction being essentially negligible. That is,

$$\text{Voltage across depletion region} = \phi_0 - V_0, \tag{7.33}$$

where $V_0$ is positive or negative according to whether the bias is positive or negative, respectively. The justification for (7.33) is that, since the depletion region essentially has no carriers, it has a very high resistance. The rest of the junction has an abundance of carriers, and hence a small resistance. Because the two resistances are connected in series, almost all the voltage appears across the high resistance region, i.e., the junction. However, the approximation (7.33) is more valid for

† The equilibrium contact potential $\phi_0$ cannot be used to drive an electric current in an external circuit. If the junction is connected to a metallic wire which completes the circuit, additional contact potentials develop at the semiconductor–metal interface between the wire and the two sides of the junction, and the effect of these additional contact potentials is to cancel the original potential $\phi_0$ entirely (see the question section).

reverse than for forward bias, since in forward bias a large current flows, and hence some voltage drop occurs outside the depletion region, even though its resistance is quite small. Equation (7.33) is usually a good approximation for both directions.

To solve for the width of the depletion region and the field in the junction in the presence of a bias, we solve the appropriate Poisson's equation, subject to certain boundary conditions. The procedure is exactly the same as in the equilibrium case, except that in boundary condition (iv) we replace $\phi_0$ by $\phi_0 - V_0$, in accordance with (7.33). We therefore obtain results such as (7.32), except that $\phi_0$ is replaced everywhere by $\phi_0 - V_0$. That is, $\phi_0 - V_0$ for forward bias and $\phi_0 + |V_0|$ for reverse bias. Note in particular that $w \sim (\phi_0 - V_0)^{1/2}$ and $\mathscr{E}_0 \sim (\phi_0 - V_0)^{1/2}$. Thus for forward bias the depletion region has contracted and the field has decreased from their equilibrium values. Note also that if in the latter case $|V_0| \gg \phi_0$, which is readily realizable, then $w$, $\mathscr{E}_0 \sim |V_0|^{1/2}$, that is, both the width and field increase as the square root of $|V_0|$.

## 7.4 THE JUNCTION TRANSISTOR

Of all the semiconductor devices, the most useful is the transistor.† It has revolutionized the communications industry and made computer technology possible. Of the many available transistors, the junction transistor is the one most commonly used. Therefore we shall discuss only this transistor in this chapter, and touch on other transistors briefly later.

Figure 7.9 illustrates the basic concepts involved in the operation of a junction transistor, a *p-n-p* transistor. Only the *p-n-p* transistor will be discussed here; the operation of an *n-p-n* junction can then be deduced from the obvious symmetry of the situation.

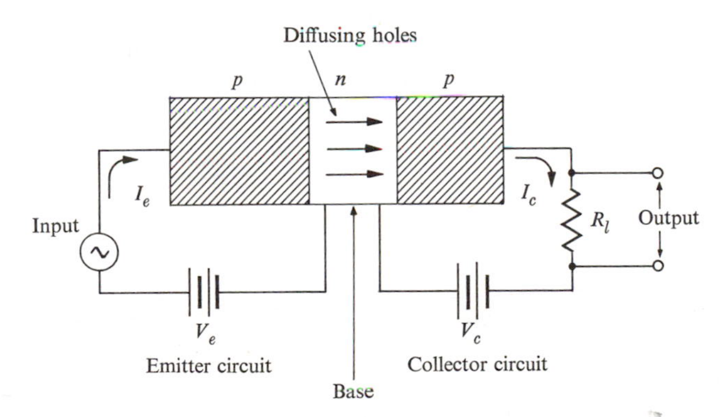


**Fig. 7.9** The basic construction and operation of the junction transistor.

† The name transistor was originally used as an abbreviation for "transferred resistor," referring to the fact that the operation of a transistor involves the transfer of a current from one circuit to another.

A piece of single-substance crystal is so doped that the end regions are *p*-type, while the middle region is *n*-type. In other words, we have two *p-n* junctions joined together back-to-back, with a common *n* region. The junction to the left is forward biased, the junction to the right is reverse biased. The forward-biased junction and its circuit are the *emitter* electrode, the reverse-biased junction and its circuit are the *collector* electrode (the reason for this terminology will become evident shortly). The *n* region in the middle is called the *base*.

We can see the basic idea for the transistor acting as an amplifier by looking at Fig. 7.9, and thinking about our previous discussion of the *p-n* junction. The forward-biased circuit on the left injects (or emits) holes across the junction and into the base. Thereafter the holes diffuse into the base until they are collected by the reverse-biased junction to the right. leading to a current flowing in the collector circuit. A voltage signal applied to the emitter circuit leads to the injection of a hole pulse across the emitter junction which, after diffusing through the base and being received by the collector, appears as a current pulse which can be picked up across a load resistor in the collector circuit. The reason for the amplification is that the currents flowing in both circuits can be made essentially equal to each other, regardless of the resistance of the load $R_l$. Thus the output voltage $V_l$ across $R_l$ can be made much larger than that of the input signal, and the same applies to the input and output powers. Let us now go through the appropriate mathematical analysis.

We denote the voltage and current in the emitter circuit by $V_e$ and $I_e$; they are related by (7.12). That is,

$$I_e = I_{e0}\, e^{eV_e/k_B T}, \tag{7.34}$$

where $I_{e0}$ is the saturation current in the emitter. [We neglected the term unity in (7.12) in comparison with the exponential.] As we saw in Section 7.2, a forward bias emits holes into the *n* region, the base. The holes diffuse through the base and are collected by the collector junction, but some of these holes may decay on the way. Suppose that a fraction $\alpha$ of these holes survive; we can write for the current in the collector,

$$I_c = I_{c0} + \alpha I_e, \tag{7.35}$$

where the first term is the saturation current of the collector (reverse bias and $e\,|V_c| \ll k_B T$), and the second term is due to the surviving holes. Since $I_{c0}$ is very small, we may neglect it and write

$$I_c \simeq \alpha I_e. \tag{7.36}$$

The voltage drop across the load resistor is

$$V_l = R_l I_c = \alpha R_l I_e. \tag{7.37}$$

The above equations can be used in a straightforward manner to evaluate the gain in voltage and also in power.

Suppose that an input signal in the emitter circuit leads to a current increment $dI_e$. We can calculate the voltage gain $dV_l/dV_e$ from (7.34) and (7.37),

$$\frac{dV_l}{dV_e} = \frac{\alpha R_l I_e}{k_B T/e}. \tag{7.38}$$

We can also calculate the power gain $dP_l/dP_e$. By writing $P_l = V_l I_c$ and $P_e = V_e I_e$, and carrying out the necessary straightforward differentiations, we find that

$$\frac{dP_l}{dP_e} = \frac{2\alpha^2 I_e R_l}{(k_B T/e)\,(1 + \log(I_c/I_{e0}))}. \tag{7.39}$$

The above gain equations give the *small-signal* dc gains of the transistor. If we take $I_e = 10$ mA, $I_{e0} = 10\ \mu$A, and $k_B T/e \simeq 0.025$ V at $T = 300°$K, $\alpha \simeq 1$, and $R_l = 2 \times 10^3 \Omega$, we find the voltage and power gains to be about 800 and 200, respectively, which are quite appreciable.

The current gain of the device $dI_c/dI_e$ is equal to $\alpha$ from (7.36); that is, it is equal to the fraction of holes which survive between the emitter and the collector. Clearly it is desirable to make $\alpha$ as large as possible in order to maximize the voltage and power gains. Of course $\alpha$ cannot be larger than unity because some of the holes —on account of recombination— do decay while diffusing through the base.

There is actually another reason why $\alpha$ is less than unity: The current at the emitter junction is not wholly carried by holes injected into the base; a part of this current is carried by electrons injected from the base into the $p$ region to the left [see (7.22) and the related discussion]. These electrons eventually move into the external parts of the circuit, and hence do not contribute to the amplification process.

Including both hole-injection-efficiency and the hole-recombination factors, we write

$$\alpha = \eta_p f, \tag{7.40}$$

where $\eta_p$ is the hole efficiency (Eq. 7.23) and $f$ is a parameter called the *base transport factor*. To maximize $\alpha$, one increases $f$ by reducing the width of the base so that the two junctions are quite close to each other. One also increases $\eta_p$ by doping the $p$ region more heavily than the $n$ region [see (7.23)]. By proper design, including minimizing surface recombination, we can make $\alpha$ very close to unity; for example, 0.99.

There is one fundamental limitation on the operation of a junction transistor: the restriction to low frequency. Since the operation is inherently dependent on the diffusion of holes in the base, complications arise at high frequencies due to a "secondary" diffusion process between the peaks and troughs of the signal, as shown in Fig. 7.10. These effects have a tendency to "wash out" the signal increase at higher frequencies.

We shall not go through the details here, but the result is that essentially there is an upper cut-off frequency beyond which the transistor cannot function properly.

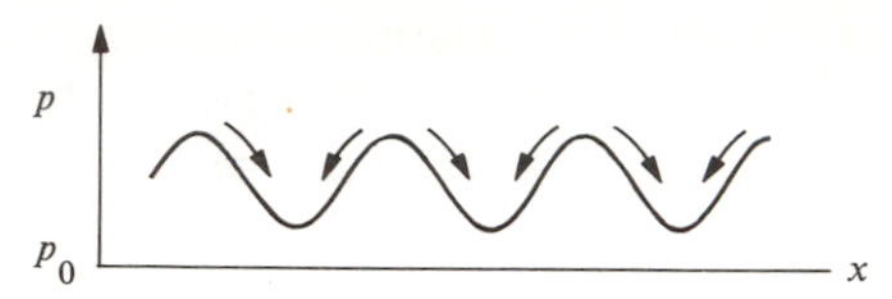


**Fig. 7.10** The additional dynamical diffusion arising at high frequencies.

This frequency, which depends on the diffusion properties of the holes as well as the thickness of the base, is given by

$$\omega_0 \simeq 2\frac{D_p^{1/2}}{L_b}\left[\frac{D_p\tau_p}{L_b^2} - 1\right]^{1/2}, \tag{7.41}$$

where $\tau_p$ is the hole recombination time in the base and $L_b$ is the base width. For example, in Ge, $\nu_0 = 0.56$ MHz for $L_b = 5 \times 10^{-3}$ cm, while for $L_b = 5 \times 10^{-4}$ cm, $\nu_0 = 56$ MHz. The higher the desired cut-off frequency, the smaller the base width must be. However, there are technological limitations on how thin the base can be made, which makes the junction transistor a low-frequency device. The search for devices of higher frequency range—e.g., in the microwave region—has led to other types of transistors, particularly to types such as the Gunn oscillator, which will be discussed later in the chapter.

## 7.5 THE TUNNEL DIODE

The *tunnel diode*, invented by Esaki in 1958, is a device that can function efficiently either as an amplifier or an oscillator. It may operate well even in the microwave range. The principle on which it operates can be quite readily understood. First we form a *p-n* junction in which both the *n* and *p* regions are heavily doped, e.g., about $10^{19}$ cm$^{-3}$. Under such heavy doping conditions, the contact potential $\phi_0$ is large; the space-charge (depletion) region is very narrow, and the field in this region is extremely high, about 900 kV/cm [see Eq. (7.32)], as shown in Fig. 7.11(a).

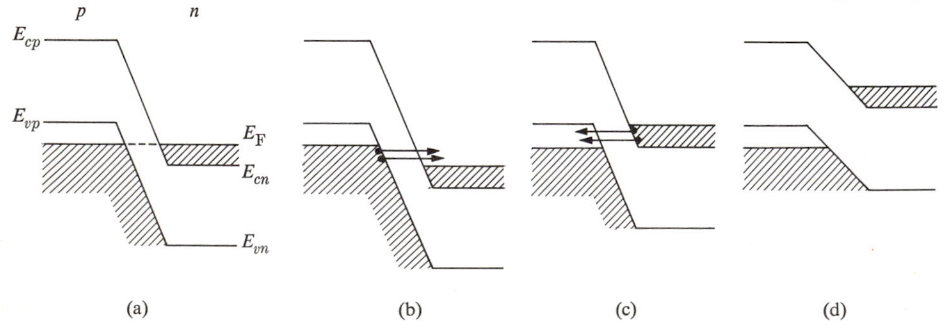


**Fig. 7.11** The principle of the tunnel diode: (a) Situation at equilibrium. (b) A large tunneling current for reverse bias. (c) Some tunneling current for small forward bias. (d) Zero tunneling current for larger forward bias.

The impurity levels have also "broadened" into impurity bands, which overlap the conduction and valence bands. In fact, the concentration of carriers is so large that the electrons and holes obey the degenerate Fermi–Dirac statistics characteristic of metals, and the Fermi level $E_F$ lies in the bands themselves. This is shown in Fig. 7.11(a), in which it is also indicated that $E_F$ is the same throughout the junction at equilibrium. Let us now see what type of $I - V_0$ characteristics such a diode possesses.

Consider the reverse-bias case first (Fig. 7.11b). Since the conduction band in the $n$ region has been lowered, by $e|V_0|$, electrons in the valence band in the $p$ region can *tunnel* through the potential barrier and end up in the conduction band in the $n$ region. An electrical current flows in the process. The tunneling process is entirely quantum mechanical in nature, and depends on the fact that the wave function in quantum mechanics does penetrate a potential barrier [see Eisberg 1961]. Energy is converted in the tunneling, and the tunneling current is appreciable only if the potential barrier is quite thin, a condition prevailing in Fig. 7.11(b). [Note that tunneling is inhibited in Fig. 7.11(a) by the exclusion principle, because the final states are already occupied.] The effect of a reverse bias is to introduce empty states in the $n$ region at energy levels parallel to those of the electrons in the $p$ region. The concentration of electrons in the valence band is large, and the field in the space-charge region is also large. Therefore a very large tunneling current flows. Essentially the device can support no reverse bias, and may be considered as exhibiting a Zener breakdown (Section 7.2), even at very small voltage.

The interesting features of the tunnel diode appear only when we consider the forward-bias situation. Figure 7.11(c) shows the effect of a small forward bias. The current now flows because the electrons on the $n$ side are able to tunnel into empty states on top of the valence band on the $p$ side. As $V_0$ increases initially, the current increases, as more electrons are able to tunnel. However, beyond a certain bias, the number of available empty states begins to decrease, and the bands begin to "uncross." The current then begins to decrease, essentially reaching a zero value (Fig. 7.11d). As $V_0$ continues to increase, the current begins to increase, because minority carriers begin to diffuse across the junction. As the barrier decreases in height, some electrons and holes begin to flow over it.

Figure 7.12 illustrates $I$–$V_0$ characteristics of the tunnel diode. The interesting feature of the curve is the presence of an NDC (negative differential conductivity) region, in which an increase in the voltage actually leads to a decrease in the current. A tunnel diode in the NDC region can be used either as an amplifier or an oscillator in an electronic circuit.†

---

† The physical basis for amplification is as follows: When a signal is applied to a circuit element of an NDC character, the current produced is opposite to the field. Hence energy absorbed from the element and the signal field is amplified. To design an oscillator, one connects the NDC element to a resistor whose resistance is equal and opposite to that of the NDC element. The total resistance of the circuit is then equal to zero, and an oscillation, once started, continues without decay.

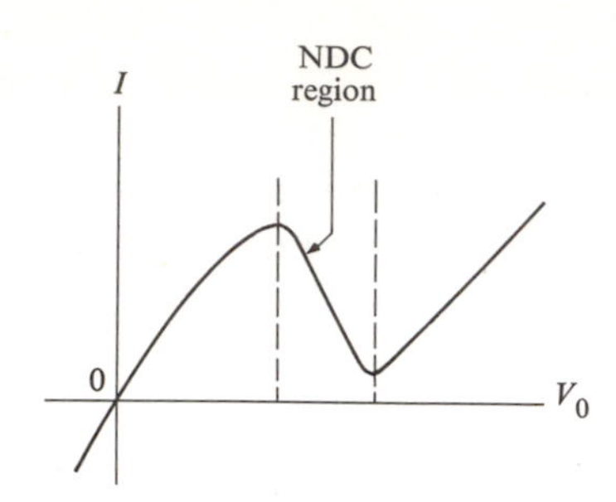


**Fig. 7.12** The $I$–$V_0$ characteristics for a tunnel diode.

The efficiency of the tunnel diode depends on the ratio of the peak to the valley currents in Fig. 7.12. (The valley current does not appear to go to zero exactly, probably because of the presence of some density-of-states tails, or some trapping states.) This ratio can be made as large as 15 by using heavy doping densities, and this may lead to really large peak currents for, say, dopings of the order of $10^{20}$ cm$^{-3}$. Tunnel diodes have been made of silicon and gallium arsenide, to name a few materials.

Finally, note that, because the tunneling process occurs almost instantaneously, the tunnel diode can operate even at fairly high frequencies, for example, 10 GHz.

## 7.6 THE GUNN DIODE†

The Gunn diode operates on the principle of negative difference conductance (NDC). It may be used as an amplifier, oscillator, or other related device. The NDC property of the Gunn diode is a result of the transfer of electrons from the low-energy high-mobility valley to the high-energy low-mobility valley at strong electric-field values in GaAs and other semiconductors of similar band structure (see Section 6.11). An important characteristic of the Gunn diode is that it is a *bulk* device, i.e., a device whose function depends on the microscopic properties of the homogeneous material itself rather than on the surface properties of a *p-n* junction. This sets the Gunn diode apart from most of the devices discussed earlier. The bulk characteristic endows the Gunn diode with an advantage over junction devices in that the number of carriers participating in a bulk-effect device can be made much larger than in a junction device. (The name Gunn diode, incidentally, is a misnomer, as there is no diode involved; the device is symmetric and can be used equally well in either direction.)

A Gunn diode in the NDC region can operate in several different modes, depending on the properties of the sample as well as on the external circuit to

† A detailed discussion of the Gunn diode and other hot electron devices is given in J. E. Carroll, 1970, *Hot Electron Microwave Generators*, London: Edward Arnold Ltd.

which the sample is connected. (Review Section 6.11 regarding the role of the electric field in producing an NDC.)

Let us begin with the *Gunn mode*, which was the first to be discovered. In this mode the sample acts as a microwave generator whose frequency, typically in the GHz range, is essentially given by

$$\nu_0 = \frac{v_d}{L}, \tag{7.42}$$

where $L$ is the length of the sample and $v_d$ the average drift velocity of the electrons. This relation establishes the Gunn mode as a *transit-time* effect, since the period of the signal is equal to the time of transit of the electron from one end of the sample to the other. What sort of a thing is propagating in the sample which leads to the periodic signal observed?

Suppose that the sample is biased so that there is a uniform field $\mathscr{E}$ ($= V/L$) inside the sample. and that the field is large enough so that the sample is in the NDC region (Fig. 7.13a). That is, $\mathscr{E} > \mathscr{E}_{\text{th}}$, where $\mathscr{E}_{\text{th}}$ is the threshold field. We want to show that this condition is an unstable one, and not likely to be observed (but see below). Figure 7.13(b) shows a thin layer of the sample, in which there is a small excess of electrons; that is, $n > n_0$, where $n_0$ is the equilibrium uniform concentration throughout the sample. This will now be called an *accumulation* layer. Its initial existence may be due to thermal fluctuations of the electrons, or, more likely, to some slight inhomogeneity in the doping. Under normal conditions, the accumulation layer would quickly damp out, and the carrier concentration

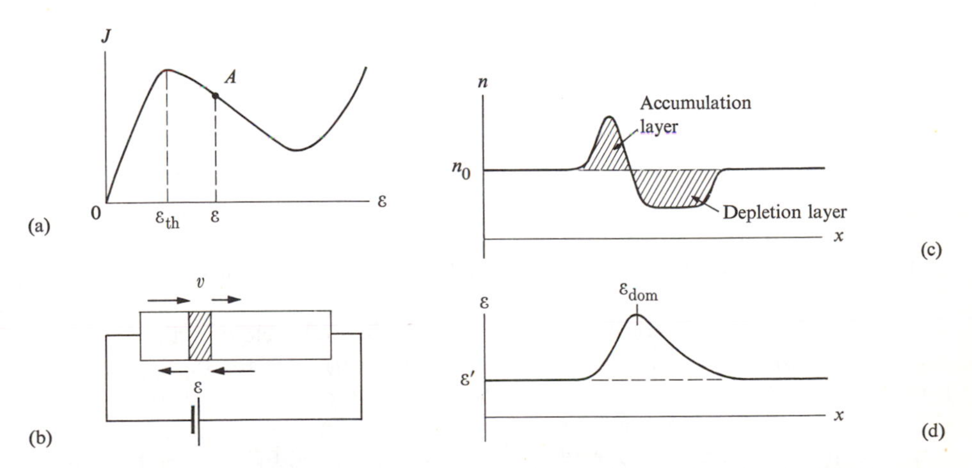


**Fig. 7.13** (a) $J$ versus $\mathscr{E}$, showing an NDC region. (b) Instability in the NDC region. The trailing edge of the accumulation layer moves faster than the leading edge, which leads to further growth of the domain. (c) Concentration $n$ versus distance $x$, showing the double layer associated with the domain. (d) Electric field $\mathscr{E}$ versus $x$, showing the high-field domain.

would remain essentially constant throughout the sample. In the unusual condition of NDC, however, the accumulation layer grows instead.

Note first that the uniform field $\mathscr{E}$ is directed to the left in Fig. 7.13(b), and that the electrons drift to the right. Note also that the accumulation layer itself is drifting, due to the drift of electrons both inside and outside the layer. Because of the net charge inside the layer, the field in the neighborhood is no longer uniform. The field at the leading, front edge of the layer is slightly larger than that at the trailing edge. (This can be deduced from simple reasoning, using Coulomb's law, or more formally from Poisson's equations.) Figure 7.13(a) shows that a larger field in the NDC region means a smaller velocity. Therefore electrons at the leading edge of the layer move slowly, while those at the trailing edge move fast, both contributing to growth in the layer. Thus, as the layer drifts from the cathode to the anode, it grows simultaneously. The growth is eventually checked by nonlinear effects (which need not be considered here), after which the accumulation layer achieves a stable shape which drifts down the length of the sample with a constant drift velocity.

The actual situation is even more interesting than described above. The drifting object turns out to be not a single layer, but a double layer (Fig. 7.13c). The trailing portion is composed of a narrow accumulation layer ($n > n_0$), while the leading portion is composed of a somewhat broad depletion layer ($n < n_0$). The presence of such an electric dipole layer modifies the field distribution as shown in Fig. 7.13(d), and a very large field is produced at the dipole layer. Looking at the field distribution throughout the sample, we see that the distribution has split into two parts: a strong-field region or *domain* at the layer, and a low-field region throughout the remainder of the sample. Viewing the situation in terms of the high-field domain, we can summarize the Gunn mode as being one in which a domain begins to grow at the cathode, continues to grow as it drifts towards the anode, matures and drifts, and eventually reaches the anode, where it collapses and disappears. The cycle is then repeated again.

Figure 7.14 shows a computer-generated picture of the growth and drift of the domain. Every time the domain disappears, a current pulse (an increase in the current) is generated in the external circuit, and this is what Gunn observed originally. The shape of the field domain along the sample has also been measured experimentally.

It is interesting to determine the properties of the high-field domain, such as its speed, the internal field, etc. However, the exact formulas for these factors are rather complicated, and the answers can be obtained only by numerical solution of the equations. We shall therefore be content with an approximate treatment. Suppose that the average field inside the domain is denoted by $\mathscr{E}_{\text{dom}}$ and the outside field by $\mathscr{E}'$ (Fig. 7.13d). We determine these two fields as follows.

Using the $J$-versus-$\mathscr{E}$ curve, we draw a horizontal line such that the two shaded regions have equal areas (Fig. 7.15). Then we determine $\mathscr{E}'$ and $\mathscr{E}_{\text{dom}}$ as indicated in Fig. 7.15. This method for determining the fields is called the *equal-areas rule*.

Note that this rule also determines the domain velocity $v_{\text{dom}}$ ($J_{\text{dom}} = nev_{\text{dom}}$). We see that the original NDC unstable state has given way to a state in which the field distribution has split into two regions: one with a low field $\mathscr{E}'$ and the other with a high field $\mathscr{E}_{\text{dom}}$. The drift velocities for these two fields are equal, and hence the

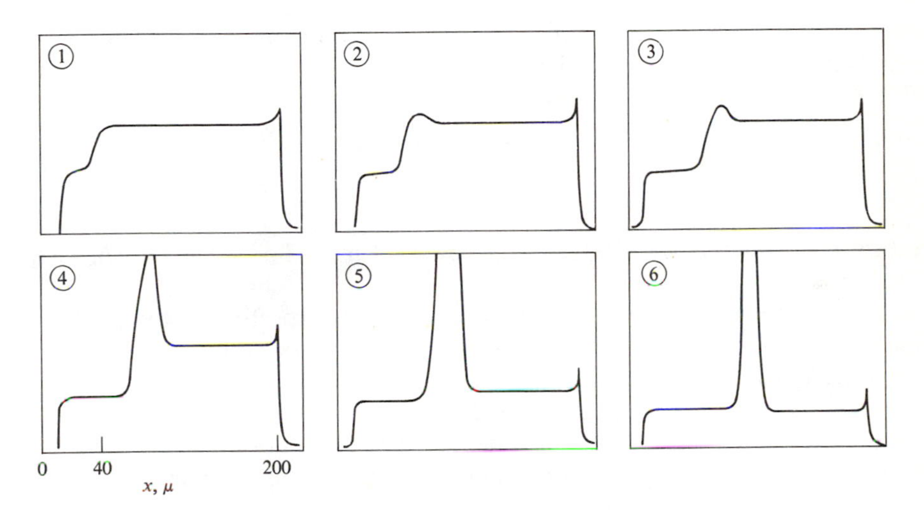


**Fig. 7.14** Sequence showing development in time of the high-field domain in a 200μ sample of GaAs at room temperature (after McCumber and Chynoweth).

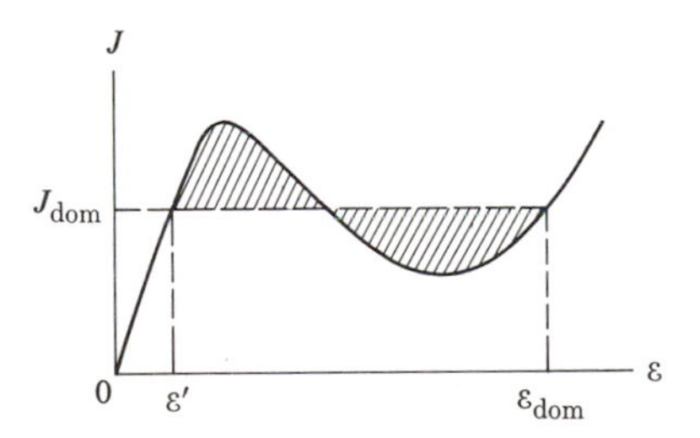


**Fig. 7.15** The "equal-areas" rule.

domain shape is stable. Note that now the differential conductivity is positive in both regions. In terms of the conduction band, this means that the high-field domain is populated essentially only with low-mobility electrons, while the remainder of the specimen is populated with high-mobility electrons. Because of the $N$-shape of the $J$–$\mathscr{E}$ curve, the two sets of electrons have the same speed, even though their mobilities are widely different. The field in the domain can be as large as 100 kV/cm, while $\mathscr{E}'$ can be less than 1 kV/cm.

The width of the domain may be determined as follows: Given that the external voltage is $V$ and the width of the domain is $w$, then

$$V = \mathscr{E}_{\text{dom}}\, w + \mathscr{E}'(L - w), \tag{7.43}$$

where the first term on the right is the drop across the domain, and the second term is the drop across the remainder of the sample. Solving for $w$, assuming that $\mathscr{E}' \ll \mathscr{E}_{\text{dom}}$, we find that

$$w \simeq \frac{V - \mathscr{E}'L}{\mathscr{E}_{\text{dom}}}. \tag{7.44}$$

Substituting $V = 6$ V, $L = 200\ \mu$, $\mathscr{E}_{\text{dom}} = 10^2$ kV/cm, and $\mathscr{E}' = 0.1$ kV/cm, we find that $w \simeq 40\ \mu$.

In order for the oscillations to have a satisfactory spectral purity, the sample must be quite thin, i.e., of the order of 200 $\mu$. Otherwise several domains may form along the length of the sample at any one time, and this contributes to the noise. (The domain usually starts at the cathode, but this need not be so if the sample is long, as the domain may then "nucleate" at some region of high resistivity along the sample; after forming, the domain drifts toward the anode and collapses.)

As we have stated above, Gunn oscillations usually occur in the microwave range. The power of the ac signal comes ultimately from the source of the dc field. CW (continuous wave) devices with up to 100 GHz and 100 mW output and 5% efficiency have been built, and pulsed devices of 200-W peak output at 1.5 GHz and 5% efficiency have also been reported. Table 7.1 outlines the performance of the Gunn diode.

**Table 7.1**

Maximum CW and Pulsed Powers and Performance Data for Gunn Diodes

| | Active-layer thickness, $\mu$ | Threshold voltage, V | Maximum output power (CW) or peak power (pulsed) | $\nu$, GHz | Efficiency, % | $n_0$, $\text{cm}^{-3}$ |
|---|---|---|---|---|---|---|
| CW | | | | | | |
| | 40 | 13.0 | 56 mW | 2.5 | 5.2 | $8 \times 10^{18}$ |
| | 25 | 7.5 | 65 | 5 | 2.3 | $3 \times 10^{15}$ |
| | 12 | 4.2 | 110 | 11 | 3.0 | |
| | 5 | 2.0 | 20 | 50 | 3.3 | $6 \times 10^{15}$ |
| Pulsed | | | | | | |
| | 100 | 27 | 205 W | 1.5 | 6 | $3 \times 10^{14}$ |
| | 5 | 2.0 | 0.4 | 50 | 9 | $6 \times 10^{15}$ |

We have thus far discussed only the Gunn mode, but other modes of operation are also possible. The Gunn domain requires especially favorable conditions, particularly enough time for the growth process. If these conditions are not satisfied, the domains are unable to grow, and no Gunn oscillations are observed. The diode can then be used as a different circuit element. A crucial factor in discriminating between the various modes is the quantity $n_0L$, the *concentration-length product*. Thus, in GaAs, we must have $n_0L \gtrsim 10^{12}\ \text{cm}^{-2}$ in order for the specimen to operate in the Gunn mode.

Since the quantity $n_0L$ is so important, let us look into its origin. It is well known that when excess charge density $\Delta\rho$ is placed in a medium, the excess density decays in time as

$$\Delta\rho(t) = \Delta\rho(0)\, e^{-t/\tau_D},$$

where $\tau_D$ is called the *dielectric relaxation time*, which is related to the properties of the medium by

$$\tau_D = \frac{\epsilon}{\sigma}, \tag{7.45}$$

where $\epsilon$ and $\sigma$ are, respectively, the dielectric constant and the conductivity of the medium [$\Delta\rho(0)$ is the initial charge density]. In the NDC region, the effective $\sigma$ which enters (7.45) is actually negative, and hence an excess charge would grow in time, as discussed above, according to

$$\Delta\rho(t) = (\Delta\rho)_{t=0}\, e^{t|\sigma|/\epsilon}. \tag{7.46}$$

If we now apply this idea to the charge associated with the traveling domain, then, in order for the domain to mature during its transit time, the exponent in (7.46) must exceed unity. That is

$$\frac{t|\sigma|}{\epsilon} = \left(\frac{L}{v_d}\right)\frac{|\sigma|}{\epsilon} > 1.$$

When we substitute $|\sigma| = n_0\, e\bar{\mu}$, the inequality becomes

$$n_0L > \epsilon v_d/e\bar{\mu}, \tag{7.47}$$

where $\bar{\mu}$ is the average mobility [$\bar{\mu} = (n_1\,\mu_1 + n_2\,\mu_2)/n$]. When we substitute the numerical values appropriate to GaAs, we find that $n_0L \gtrsim 10^{12}\ \text{cm}^{-2}$.

Another important mode for the Gunn diode is the LSA (*limited space charge accumulation*) mode, discovered by Copeland in 1966. In this mode, the domain is inhibited from forming, or is quenched, and hence the field remains uniform throughout the sample. The sample is also in the NDC region, unlike the case of the Gunn mode, in which the NDC property actually disappears. In the LSA mode, the domain is quenched by the application of a high-frequency bias to the sample; see Fig. 7.16.

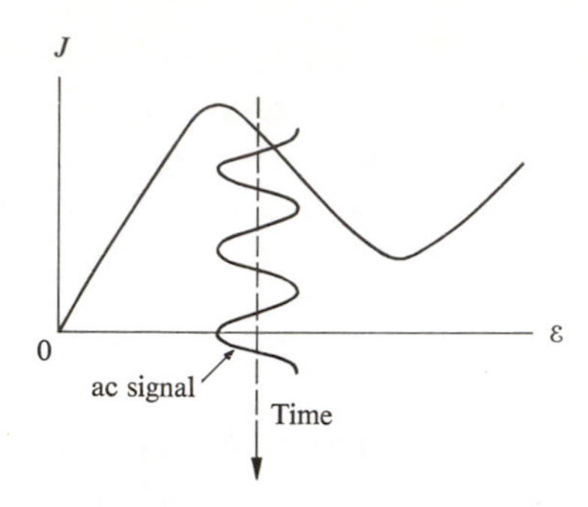


**Fig. 7.16** The LSA mode.

During part of the cycle of the ac bias (there is also a driving dc bias), the resultant field is lowered below the threshold $\mathscr{E}_{\text{th}}$. The domain which had been growing during the earlier part of the cycle suddenly collapses (unless it has already reached the cathode). Therefore a necessary condition of the LSA mode is that essentially

$$\nu > \nu_0, \qquad (7.48)$$

where $\nu$ is the circuit frequency and $\nu_0$ the *intrinsic frequency* of (7.42). Substituting from (7.42), we find that $\nu L > v_d \simeq 10^7$ cm/s in GaAs. Note that, if $\nu L > 10^7$ cm/s, the diode would operate in the LSA mode regardless of the value of $n_0L$. When the diode is in this mode it essentially retains its NDC property, and can be used as an amplifier.

We see then that the mode of operation depends on both the $n_0L$ and $\nu L$ properties. In addition to the Gunn and LSA modes, there are also other modes, depending on the values of these products. For further details, refer to the literature cited at the end of the chapter.

## 7.7 THE SEMICONDUCTOR LASER

One of the most interesting semiconductor devices is the semiconductor laser, invented in 1963. It is, in fact, the only semiconductor device which can be used for amplification in the infrared and optical ranges. All other devices are restricted to the microwave range or below.

It is assumed that the reader is acquainted with the basic concepts of the laser in general,† so that we need to review this subject only briefly here. Consider a gas composed of atoms, and focus on two of the excited atomic energy levels: levels 1

† See for example, B. A. Lengyel (1971), *Lasers*, second edition, Wiley, New York.

and 2 shown in Fig. 7.17(a). The populations of these levels $n_1$ and $n_2$ per unit volume at equilibrium are related by

$$\frac{n_2}{n_1} = e^{-\Delta E/k_B T}, \tag{7.49}$$

where $\Delta E = E_2 - E_1$ is the difference in energy between the two levels. We see that $n_2 < n_1$.

When an external signal of frequency $\omega \simeq \Delta E/\hbar$ is passed through the gas, it is strongly absorbed. Photons from the signal are absorbed by electrons in the lower level 1, which then make quantum transistions to the upper level 2. Given that a sample has length $L$ (the length of the tube containing the gas), then the intensity of the signal at the end of the sample is given by

$$I = I_0 e^{-\alpha L}, \tag{7.50}$$

where $\alpha$ is the absorption coefficient and $I_0$ is the initial intensity.

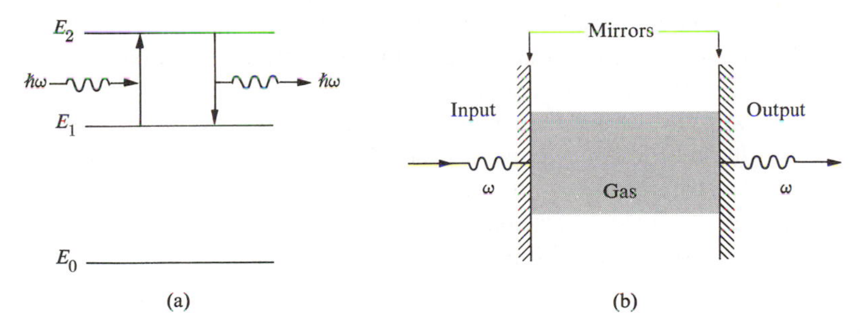


**Fig. 7.17** Basic principles of laser operation: (a) The two active levels $E_1$ and $E_2$, and the absorption and emission processes. (b) A laser cavity.

There is actually another process which occurs simultaneously. Not only does the signal help cause transitions from level 1 to level 2, but it also *stimulates* transitions from upper level 2 to lower level 1 (provided that $\omega \simeq \Delta E/\hbar$). In this process a photon of frequency $\omega$ is *emitted* for every quantum transition. Since this emission process is aided by the signal, it is called *stimulated emission*. It reinforces the signal. It is difficult to explain stimulated emission classically, but quantum mechanically it appears on the same footing as the absorption process, and in fact the two processes have equal probabilities of occurring. We can now understand why the signal undergoes absorption in passing through the gas: The two processes—absorption and stimulated emission—act competitively on the signal. Absorption weakens it and stimulated emission enhances it. Although they have equal probabilities, the net result is absorption because the lower level has a larger population, $n_2 > n_1$. Thus we may write

$$\alpha = B(n_1 - n_2), \tag{7.51}$$

where $B$ is a proportionality constant, indicating that the population is proportional to the population difference. Because $n_1 > n_2$, it follows that $\alpha > 0$.

The basic premise of the laser device lies in an attempt to amplify the signal by making $n_2 > n_1$. That is, by *inverting* the populations from their equilibrium values. We can readily see that if $n_2 > n_1$, then $\alpha < 0$; and, using (7.50), it follows that

$$I = I_0 \, e^{|\alpha| L} > I_0.$$

That is, the signal is *amplified*, because there are more emitting atoms than absorbing ones.

In a laser operation, the sample is placed in an *optical cavity* composed of two parallel mirrors [a Fabry–Perot interferometer (Fig. 7.17b)], and the beam is extracted through one of the mirrors. The laser beam has many advantages over conventional sources: (i) extremely high spectral purity; (ii) high directionality (the ability of the beam to travel a long distance with very small divergence), both properties being due to the fact that the photons of stimulated emission are emitted in phase with the signal; (iii) very high intensity.

A necessary condition for *lasing* is that $|\alpha| \, 2L = 1$.† Usually the signal is also absorbed partially by some mechanisms other than the atomic transitions described above—e.g., other foreign ions—and we must include that part also. If we represent all these losses by an absorption coefficient $\alpha'$, and denote the $|\alpha|$ due to amplification by $g$ (for gain), the condition for lasing becomes

$$(g - \alpha') = \frac{1}{2L} \quad \text{or} \quad g = \frac{1}{2L} + \alpha'. \tag{7.52}$$

Population inversion can be achieved by pumping the gas with a strong beam of frequency $\omega_p = (E_2 - E_0)/\hbar$ to excite more electrons to level 2, and hence give it a larger population than level 1.

The operation of the semiconductor laser is basically the same as that of the gas laser. The necessary modifications appropriate to solids must be made, however. First we recall from Section 6.12 that a light beam passing through a semiconductor undergoes strong absorption near the band edge, that is, that $\hbar\omega \gtrsim E_g$. The absorption is due to interband transition between the valence and conduction bands. It follows from this and from our discussions above of the laser that amplification should also be possible here if the population of the valence and conduction bands near the band edges could be inverted.

---

† This condition is derived for the situation where the laser operates as a cw oscillator. The intensity must return to its original value after the beam travels a full round trip inside the cavity, i.e., a distance $2L$.

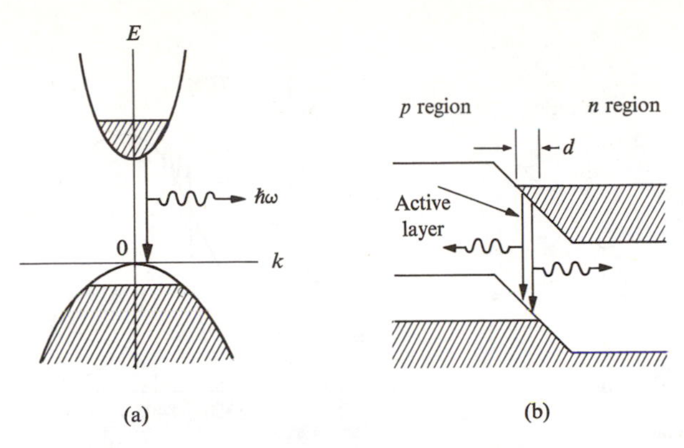


**Fig. 7.18** (a) A population-inversion arrangement in a semiconductor, suitable for laser action. (b) Achievement of population inversion in a heavily doped junction.

Figure 7.18 illustrates the idea. Suppose that the material is so heavily doped with $n$- and $p$-type impurities that the free carriers have essentially degenerate distributions, with Fermi energies $E_{\mathrm{F}c}$ and $E_{\mathrm{F}v}$ in the two bands. (The distribution is not an equilibrium one, since it decays rapidly, and hence two different *quasi-Fermi* levels are possible.) A distribution such as this leads to amplification because electrons, stimulated by the signal, make transitions from the conduction band to the empty states (holes) at the top of the valence band, emitting photons of frequency $\simeq E_g/\hbar$ in the process. Strong amplification is expected here, since we found the interband transition to be quite strong. The necessary condition for amplification is

$$(E_{\mathrm{F}c} - E_{\mathrm{F}v}) \geqslant \hbar\omega. \tag{7.53}$$

The question now is how to create population inversion. This was first accomplished by using a highly doped $p$-$n$ junction. Figure 7.18(b) shows that, when such a junction is forward biased, there is a certain region in space in which the population inversion of Fig. 7.18(a) is accomplished. Figure 7.18(b) shows a steady-state situation. Electrons are continually injected from the right and recombine with holes in the active region. These holes in turn have been injected from the left. Because of the method of excitation, this device is known as an *injection laser*.

The active region is parallel to the junction face (Fig. 7.19a), and the laser beam is extracted from the side of the junction. The optical cavity is formed by the faces of the crystal itself, which are usually taken along the cleavage plane in GaAs, for example, and are then polished.

There is a threshold requirement for the operation of a junction laser: The population inversion must be strong enough for the gain made by downward transitions to be larger than other absorption effects, e.g., by free carriers or other draining effects, such as the partial transmission incurred every time the beam hits the faces of the crystal. Thus there is a threshold carrier concentration $n$, which can be related to a measurable quantity, i.e., the junction current. By using the

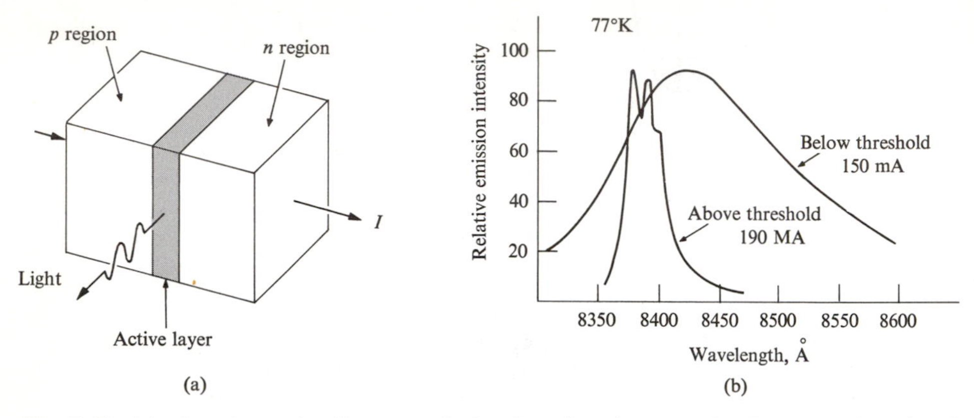


**Fig. 7.19** (a) A schematic diagram of the junction laser. (b) Spectra of emitted radiation from a GaAs junction laser below and above the threshold condition (after Quist, *et al.*).

continuity equation in combination with the diffusion equation, we can relate $n$ and $I$ by

$$\frac{I}{d} \simeq \frac{en}{\tau'}, \tag{7.54}$$

where $d$ is the width of the active region and $\tau'$ the carrier recombination time. Figure 7.19(b) shows the emission from GaAs below and above the threshold current. The line narrowing above the threshold indicates the onset of laser action. A typical value for the density of the threshold current is $100\ \text{A/cm}^2$ at $T \simeq 0°\text{K}$. Threshold current depends on temperature. At intermediate and high $T$ it varies as $T^3$. This temperature dependence is due to several factors, one of which is that some of the empty states at the top of the valence band become occupied at $T > 0°\text{K}$, even before the stimulated emission process takes place, thus inhibiting some of the transition (the exclusion principle).

The first laser action in a semiconductor was observed in a GaAs junction. The laser line in GaAs lies in the deep red (infrared) at $\lambda = 8400$ Å at 77°K, and corresponds quite closely to the energy gap in GaAs.† As the temperature increases,

† The junction laser described can operate continuously only at low temperatures (<77°K) because the large threshold current density ($5 \times 10^4 \text{A/cm}^2$) produces far more Joule's heat than can be transferred away. Recently a diode laser was developed which operates continuously at room temperature, by requiring a much lower threshold density ($100\ \text{A/cm}^2$). In this device the *p-n* junction is joined to other suitable crystals on both sides of the junction, hence the name *heterojunction laser*; the effect of these new substances is to increase the efficiency of the laser by: (a) confining the carriers to the active region; and (b) confining the light to the active region. See M. B. Parish and I. Hayashi, *Scientific American*, **225**, July 1972. For a more technical treatment, see Milnes (1971).

the wavelength of the laser increases, in agreement with the decrease of the energy gap with temperature. The frequency can be increased and brought into the optical region by alloying the material with phosphor, i.e., by using $Ga(As)_{1-x}P_x$ which has a wider gap.

Since 1963, many other semiconductor compounds have been found to lase—for example, InSb, InP, etc. (See Table 7.2), extending in frequency from the far infrared into the ultraviolet. Table 7.2 shows that other means (besides current injection) of creating population inversion are also possible. In the *electron-beam method*, an energetic electron beam is impinged on the medium, exciting many electron–hole pairs, which then recombine radiatively, emitting photons. In the *optical method*, a laser beam from one semiconductor may be used to invert the population in another material.

**Table 7.2**
Some Semiconductor Laser Materials

| Material | Photon energy, eV | Wavelength, $\mu$ | Method of excitation or pumping |
|---|---|---|---|
| ZnS | 3.82 | 0.32 | Electron beam |
| CdS | 2.50 | 0.49 | Electron beam, optical |
| CdSe | 1.82 | 0.68 | Electron beam |
| CdTe | 1.58 | 0.78 | Electron beam |
| $Ga(As_{1-x}P_x)$ | 1.41–1.95 | 0.88–0.63 | *p-n* junction |
| GaAs | 1.47 | 0.84 | *p-n* junction, electron beam, optical, avalanche |
| InP | 1.37 | 0.90 | *p-n* junction |
| GaSb | 0.82 | 1.5 | *p-n* junction, electron beam |
| InSb | 0.23 | 5.2 | *p-n* junction, electron beam, optical |
| PbS | 0.29 | 4.26 | *p-n* junction, electron beam |
| PbTe | 0.19 | 6.5 | *p-n* junction, electron beam, optical |

No laser action has been observed in silicon or germanium. This is not surprising, since these materials are indirect-gap semiconductors (Section 6.12). In them, electrons and holes cannot recombine directly, since this would violate the law of conservation of momentum. All the materials listed in Table 7.2 are direct-gap compounds.

A semiconductor laser has many advantages over a gas laser. Its small size, simplicity, and high efficiency—in addition to the fact that it can be mass-manufactured and readily connected to electronic circuits—are among the most obvious advantages. The semiconductor can also be tuned continuously by changing the energy gap by pressure, for example. Disadvantages include its relatively poor monochromaticity, due to the fact that the transitions are between bands and not between sharp atomic levels. There are also other effects which occur in the solid which contribute to line broadening. The small size of the semiconductor laser makes the quality of the beam collimation rather poor. Notwithstanding these limitations, the semiconductor laser is an important device which would be exceedingly useful in the quest for new developments in optical electronics.

Another semiconductor laser which has been discovered quite recently and which promises to be a very useful device is the so-called *spin-flip Raman* (SFR) laser. Consider the motion of electrons in the conduction band of, say, InSb, in the presence of a strong magnetic field. The effect of the field is (1) it makes the electrons move into cyclotron orbits (Section 7.4) and (2) it orients the spin magnetic moment of the electron in a direction either parallel to or opposite to the field. The difference in energy between the two levels is $\Delta E = g\mu_B B$, where $g$ is the *Landé factor*, $\mu_B$ the *Bohr magneton*, and $B$ the *magnetic field*. Orientation of the dipole moment in a magnetic field is discussed at length in Chapter 9 on magnetism, particularly in Section 9.6.

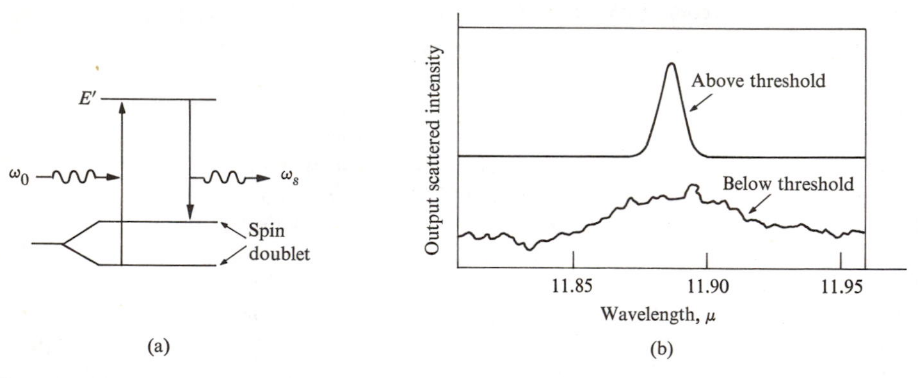


**Fig. 7.20** (a) The spin-flip Raman scattering process responsible for the SFR laser. (b) Spectra of scattered radiation from an InSb sample below and above threshold (after Patel and Shaw).

Figure 7.20(a) illustrates the operation of an SFR laser. A light beam of frequency $\omega_0$ falls on the system. An electron in the lower of the two spin levels absorbs a photon of energy $\hbar\omega_0$ and makes a transition to a higher level $E$, after which it makes a downward transition to the upper level of the same spin doublet. In the downward process it emits a photon of energy $\hbar\omega_s$ The net effect of the whole process is that a photon $\hbar\omega_0$ is absorbed, a photon $\hbar\omega_s$ is emitted, and the electron flips its spin. From energy conservation, we know that the emitted frequency is given by

$$\omega_s = \omega_0 - g\mu_B B. \tag{7.55}$$

Laser action according to the above scheme was reported by Patel and Shaw† in InSb and other materials (Fig. 7.20b), obtaining an efficiency of about 1% in a pulsed operation. The great advantage of the SFR is its *tuneability*. By varying $B$, a relatively simple operation, one can change $\omega_s$ continuously, and attain a continuous tuneability. Using a $Q$-switched $CO_2$ laser at $10.6\mu$ as the pump, Patel and Shaw obtained a continuous tuning in the range $10.9$–$13.0\mu$ by varying $B$ in the range 15–100 kG in InSb. In further development of the SFR laser, an efficiency as high as 50% at a threshold power of 50 mW in a CW operation was reported (Bruek and Mooradian). Note from (7.55) that the range of tuneability increases linearly with $g$. Therefore materials with large $g$-factors, such as InSb, are desirable.

## 7.8 THE FIELD-EFFECT TRANSISTOR, THE SEMICONDUCTOR LAMP, AND OTHER DEVICES

There are many important semiconductor devices in addition to the few major ones covered thus far. In fact their number is extremely large, making semiconductors by far the most versatile substances in electronics technology today. Our discussion here is necessarily brief, but the reader will find ample coverage in the many references appended to this chapter.

### Field-effect and drift transistors

Although the transistor junction is the backbone of the semiconductor industry, there are many other useful ones, many of which are particularly suited to specific purposes. A most important member of this group is the *field-effect transistor*, or FET, first proposed in its modern form by Shockley in 1952. This device, shown in Fig. 7.21, is a small piece of semiconductor—e.g. silicon—which consists of

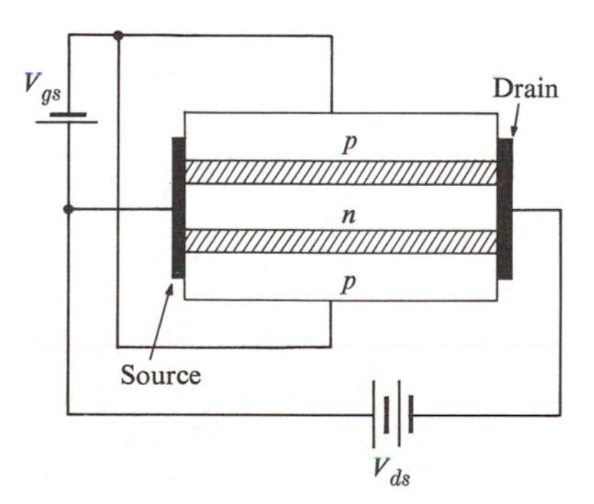


**Fig. 7.21** The FET. The cross-hatched regions represent depletion layers and the solid region the ohmic contacts.

† C. K. N. Patel and E. D. Shaw, *Phys. Rev. Letters* **24**, 451 (1970); also *Phys. Rev.* 3B, 1279 (1971).

three layers of $p$, $n$, and $p$ formed by proper doping of the sample (an $n$-$p$-$n$ device is also possible). A battery is connected across the middle $n$ layer, causing a current to flow parallel to its surface. This current is then modulated by a transverse electric field established by another battery. By superposing a signal on this field, one can amplify the signal at the load resistor in the current circuit.

The physical principle underlying the operation of the FET is related to the presence of the two depletion layers separating the $n$ and $p$ regions (cross-hatched in the figure). Recall from Section 7.3 that a depletion layer forms at any junction due to the diffusion of free carriers across the junction. The effect of this on the situation in Fig. 7.21 is to reduce the width of the conducting layer (the $n$ type), known as the *channel*, and hence reduce the conductance of the device, because the depletion layer (having no free carriers) does not contribute to the conduction process. When we denote the geometrical width of the $n$ layer by $w$, the width of the channel in the absence of the transverse electric field is

$$w_c = w - 2w_0, \tag{7.56}$$

where $w_0$ is the width of each of the depletion layers in the $n$ region and the factor 2 accounts for the presence of two of these layers, associated with each of the junctions. The width $w_0$ is given by (7.32b). That is,

$$w_0 = \left(\frac{2\,\epsilon\phi_0}{e\,N_d}\right)^{1/2}, \tag{7.57}$$

where $\phi_0$ is the equilibrium junction voltage, and we have assumed that $N_d \ll N_a$.

Let us now consider the effect of the transverse field. This field is established by a battery which is connected so that the $p$-$n$ junctions are reverse-biased (Fig.7.21). The result of this bias is to increase the width of the depletion layer, thus decreasing the width of the channel still further, and raising the resistivity. This field therefore acts as a *gate* which controls the flow of electrons between the negative electrode of the current circuit (the *source*) and the positive electrode (the *drain*). The greater the transverse reverse-bias potential, the narrower the channel, and consequently the smaller the current. In fact, at a sufficiently large potential, the depletion layers may move so far into the channel that the channel vanishes, and further flow of the current is blocked. The necessary voltage for this *pinch-off* to take place can be found by replacing $\phi_0$ by $\phi_0 + V_p$ in (7.57) and setting $w_c$ in (7.56) equal to zero. The result is

$$V_p = \frac{e\,N_d\,w_i^2}{8\epsilon}, \tag{7.58}$$

where we have assumed that $V_p \gg \phi_0$, so that the equilibrium junction voltage may be dropped.

Figure 7.22 shows the electrical characteristics—the drain current $I_d$ versus the drain-source voltage $V_{ds}$—of the FET. Note that the current increases in an essentially ohmic manner at first, and then begins to round off, eventually saturating at high voltage values. To understand why, one must consider the IR voltage drop along the length of the channel. At high currents this voltage is appreciable, and its effect is to make the region near the drain far more positive than that near the source. This therefore sets up an internal gate voltage of its own, which causes the depletion layer to bulge into and narrow the channel, particularly near the drain, as shown in Fig. 7.23.

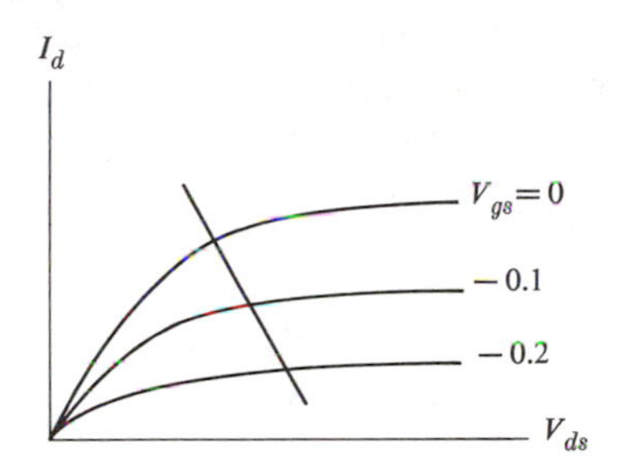


**Fig. 7.22** Electrical characteristics of an FET. Curves represent characteristics at different gate voltages (expressed in arbitrary units); straight line represents operation of the amplifier.

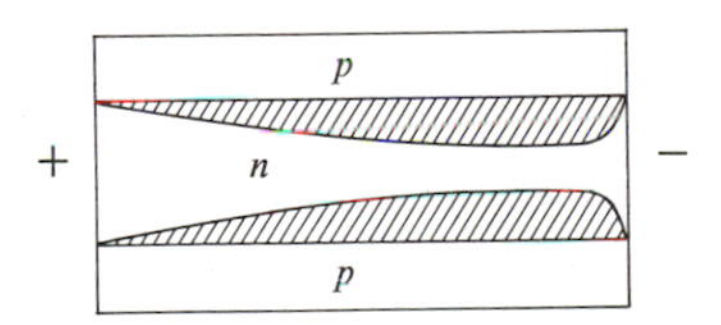


**Fig. 7.23** The bulging of the depletion layer near the drain.

This internal voltage, which is present even in the absence of external voltage, limits the current, and at sufficiently high value causes the current to saturate by closing off the channel at the drain. (The current does not vanish entirely due to this internal pinch-off, because then the IR voltage drop would also vanish, and no channel narrowing would occur.)

The FET amplifier operates in the pinch-off region. The incident signal, superimposed on the gate voltage, causes the gate voltage to vary (straight line in Fig. 7.22), and the output signal is then picked up at the load resistor. Theoretical analysis shows that the current in this region is given by

$$I_d = A\left(\frac{V_{gs}}{V_p} - 1\right)^2, \tag{7.59}$$

where $V_{gs}$ is the source–gate bias voltage. The mutual conductance parameter $g_m = \partial I_d / \partial V_{gs}$ is thus given by

$$g_m = \frac{2A}{V_p}\left(\frac{V_{gs}}{V_p} - 1\right) \sim (V_{gs} - V_p) \tag{7.60a}$$

or

$$g_m \sim I_d^{1/2}. \tag{7.60b}$$

To make the device sensitive to the signal and the operating voltage as low as possible, the doping in the *n* channel is made small compared with that in the *p* region. Consequently even a small change in $V_{gs}$ produces a large change in the width of the depletion layer, and a correspondingly appreciable change in the current circuit. (Note that no transverse current flows in the FET, because the junctions are reverse-biased; see the electrical connections in Fig. 7.21.)

The primary advantage the FET has over the junction transistor is that the amplification in the FET is accomplished by the flow of majority carriers. The junction transistor, on the other hand, operates by the flow of minority carriers, and consequently is often quite sensitive to small disturbances in these carriers, e.g., changes in temperature or exposure to atomic radiation.

The device discussed above is sometimes referred to as the junction FET, to distinguish it from a similar device known as the MOS—or *MOS field-effect*—transistor, which operates on the same principle as the FET, except that its outer *p* layers are replaced by two insulating thin films, for example, $SiO_2$, deposited on the surface of the channel. A third thin film, this time of metal, is deposited on top of the insulator, and the gate voltage is connected to this last layer. The three layers consist of a metal, insulator (or oxide), and a semiconductor; hence the name MOS. Although the gate is electrically insulated from the channel, the modulation in this MOSFET device takes place via the transverse field transmitted through the insulator.

Another transistor particularly suited to high-frequency operations is the *drift* transistor, which has the same design as the ordinary junction transistor, except that the base element is not uniformly doped. Instead the doping is *graded*, so that it is greatest near the emitter, and decreases almost exponentially to a small value near the emitter junction. The effect of this nonuniform doping is that, in the absence of electrical connections, carriers in the base diffuse toward the collector, and in the process an electrical field is set up to balance this flow, in a manner similar to that discussed in Section 6.17. When the transistor is connected for operation, the minority carriers injected into the base from the emitter find an already existing field, whose polarity is such that it sweeps the carriers quickly toward the collector. In the ordinary transistor, the flow of the minority carriers in the base region is governed by diffusion. In the drift transistor, the flow is governed by an electric field (hence the name *drift*), and by proper doping one can make the field so large as to significantly reduce the transit time, or equivalently, the effective width of the base. A narrow base raises the operating frequency limit of the transistor, as will be recalled from Section 7.4 [see Eq. (7.41)]. Transistors operating at frequencies higher that a gigahertz have been manufactured by this technique.

**Microwave devices**

There are several devices in the microwave region besides the Gunn diode. For example, a junction can be used as a *varactor* (variable-reactance element). The junction has a capacitance, associated with the space-charge region, which depends, in a nonlinear fashion, on the applied voltage (Section 7.2). Thus the varactor can be used as a switching or modulation device for harmonic generation and frequency conversion, since the controlling voltage has a much lower frequency than the signal. Another microwave device is the IMPATT (impact and transit time) diode, which employs the avalanche and transit-time properties of the junction to produce negative differential conductivity (NDC) at microwave frequencies. The device can thus be used an an amplifier or oscillator.

Semiconductors can also be used as ultrasonic generators—performing the same function as a transducer—and as ultrasonic amplifiers. For example, the high-field domain in a Gunn diode is large enough to cause the ions of the lattice to oscillate, and this can be used to generate sound waves in the microwave range.

**Photodetectors and related devices**

Another major use of semiconductors is in radiation detectors. Most of these devices are based on the phenomenon of photoconductivity (Section 6.13). The radiation to be detected is allowed to fall on a semiconductor sample, where it excites electrons from the valence into the conduction band, thus creating electron–hole pairs. These carriers are detected by the current they carry across the biased sample. This current, being proportional to the intensity of the radiation, serves as an electrical measure of the incident radiation. A typical photoconductor is CdS, which has long been in use as a light meter in cameras. Its energy gap is 2.4 eV, and it is sensitive to radiation of wavelength $\lambda \leqslant 0.5\ \mu$, that is, of photon energy at least equal to the energy gap.

A photoconductive detector has several advantages over thermal detectors such as the thermopile or bolometer. It has a much shorter response time. Also a bolometer measures the integrated intensity, over all wavelengths, while a photodetector responds only to those photons whose energies are greater than the energy gap, and thus makes possible a certain degree of spectral analysis. A properly designed photodetector should have a small, dark conductivity ($\sigma_0$ in Section 6.13), so that the relative change in $\sigma$ upon illumination is appreciable and readily detectable. This requires a large energy gap so that only a small amount of cross-gap excitation takes place at room temperature. In addition, a relatively pure substance must be used, so that carriers may not be thermally excited from the impurity levels. Moreover, a good photodetector should have a large *gain*, or number of electrons recorded per absorbed photon. This number is often greater than unity, because if the electron–hole recombination time is long, and if the holes are trapped on some crystal defect, then after the original electron has drifted out of the sample into the external circuit, another electron is recalled from the cathode to electrically balance the trapped hole. This newly arrived electron is then also swept across the sample, and recorded as another photoelectron. Therefore, if the recombination time is much greater than the transit time, several electrons per photon are recorded and the gain is high. This illustrates once more the influential role played by traps and other impurities in the operation of semiconductor devices.

Although we have talked explicitly only about photodetectors in the visible optical region, the same type of substances can also be employed as infrared (IR) detectors. This area of research has received particular attention in recent years because radiation in this region, though invisible, is emitted in great quantities by bodies at room temperature. IR detectors are also increasingly helpful in connection with research in far-infrared spectroscopy.

A useful infrared photodetector must meet several requirements. First, the gap must be narrow enough for cross-gap excitation to take place even with the low-energy infrared photons. To minimize dark conductivity, the sample must be fairly pure. An IR photodetector is often cooled well below room temperature to quench excitation either across the gap or from impurities.

Lead salts—the chalcogenides PbS, PbSe, and PbTe—have been widely used as IR photodetectors, up to a wavelength of about 5 $\mu$. Also InSb and InAs, which can be produced with high purity, have been used for this purpose. The former, with an energy gap of 0.18 eV, is useful up to a wavelength of 7.3 $\mu$, and the latter, with an energy gap of 0.35 eV, up to a wavelength of 3.5 $\mu$,

One can extend the detection capability further into the IR region by employing substances which have smaller gaps. A smaller gap can be achieved nowadays in a variety of ways; one is to alloy a semimetal with a semiconductor. Figure 7.24 shows, for instance, the "tuning" of the gap as a result of alloying PbTe with SnTe, the gap varying continuously between zero and 0.33 eV.

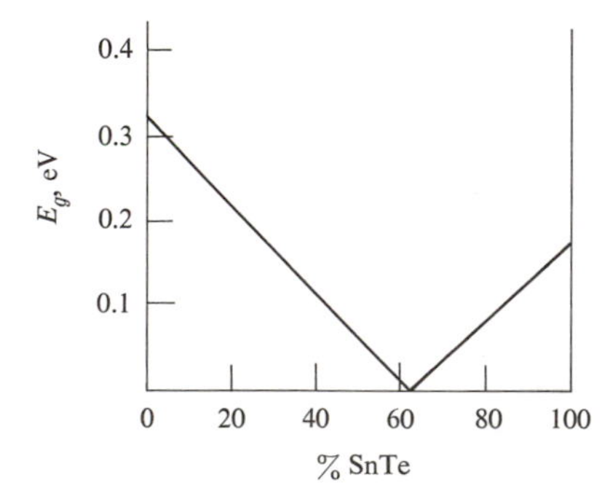


**Fig. 7.24** Energy gap versus composition in PbTe–SnTe at room temperature. (After Dimmock, *et al.*)

In the extreme long-wavelength IR region, it becomes more convenient to detect radiation not through cross-gap excitation but by excitation of the carriers from the impurities, these carriers being then detected by their photoconductive currents. These are descirbed as *extrinsic* photoconductors, as distinct from *intrinsic* photoconductors. Germanium, which can be readily prepared with the proper purity, is normally employed for this purpose, particularly since the energy levels of many types of impurities in germanium have been studied extensively. Recall (Section 6.5) that the impurity levels from the bands are usually about 0.01 eV, a fact which has been confirmed by experimental studies, such as optical measurements (Section 6.14). The use in Ge of impurities from the III or V column in the periodic table produces reasonably good extrinsic detectors covering the range 10 to 100 $\mu$.

The photodetecting devices discussed thus far are all of the bulk type, i.e., utilizing homogeneous samples. But junction photodetectors have also been developed. An important group is based on the *photovoltaic* effect, in which light falls on a *p-n* junction, and at sufficiently high frequency electron–hole pairs are created by cross-gap excitation. The carriers created in the depletion layer of the junction—i.e., the region toward which the light is primarily directed—are swept quickly by the high electric field in the junction. The electrons are swept toward the *n* region, the holes towards the *p* region. This flow of charge produces a momentary current in the circuit which, under open-circuit conditions, creates a voltage across the terminals, and this voltage serves as a measure of the intensity of the incident radiation; i.e., the photocell serves as a photodetector.

When the circuit is closed a continuous current flows. We can now establish the current–voltage relationship by referring to the junction equations of Section 7.2. The total current in the circuit is now

$$I = I_0 \left(e^{eV/k_B T} - 1\right) - I_s, \tag{7.61}$$

where $I_s$, the short-circuit current, is the current due to the carriers created specifically by the radiation and swept by the junction field. This current may be written as $I_s = e\eta P$, where $\eta$ is the quantum efficiency and $P$ the photon flux. The first term on the right of (7.61) is the familiar junction current of Eq. (7.7), due to the injection of minority carriers, as discussed in Section 7.2, and is present whether the junction is illuminated or not. The illumination-induced current is taken to be negative, $-I_s$, because it is opposite to the current of a forward-biased junction (why?). Equation (7.61) indicates that the open-circuit voltage $V_{oc}$, that is, $I = 0$, is

$$V_{oc} = \frac{k_B T}{e} \log\left(\frac{I_s + I_0}{I_0}\right). \tag{7.62}$$

This photovoltaic device, the *photodiode*, can also operate on a battery, thus converting radiation into electrical energy. The maximum power output can be close to $0.75\, V_{oc}/I_s$.

A potentially useful photovoltaic battery is the *solar cell*, which converts solar radiation that strikes the earth into useful electrical energy. Such a battery has indeed been built and operated, but its efficiency is not as great as one would wish. The problem is complicated by the fact that solar radiation covers a wide range of wavelengths. The greatest spectral intensity falls near $\lambda = 0.5\ \mu$, but a good deal of the incident energy falls well below and well above this wavelength. A diode with a wide band gap, say 2 eV, would be effective in converting the solar energy at the peak wavelength, but would lose all the infrared energy. Conversely, a narrow-gap diode, say 0.5 eV, would absorb the incident photons, but much of the energy in the near-infrared, visible, and ultraviolet regions would be lost, because of the absorbed energy only the fraction necessary for band–band excitation is recovered as useful electrical energy. Thus one must choose a band gap which will strike a happy medium between these conflicting requirements. The maximum theoretical efficiency for GaAs—the most suitable of the known semiconductors—is 24%; GaAs solar cells of 11% efficiency have been built. Silicon cells of 14% efficiency have also been built, compared to the theoretical limit of 20% for silicon.

**The semiconductor lamp**

A semiconducting *p-n* junction, properly biased, can also serve as a light emitter, i.e., a semiconductor lamp. A *light-emitting diode* (LED) has several advantages over conventional lamps. It has much greater emission power per unit volume, greater ruggedness, and can be built at considerably less cost. The LED can also be modulated much more rapidly than the conventional lamp. The basic design of an LED consists of a forward-biased *p-n* junction, as in the semiconductor laser (see Fig. 7.18). However, the LED operates below the lasing threshold, and the light is emitted as incoherent radiation. As explained in Section 7.7, electrons are injected from the *n* region across the depletion layer and into the *p* region, where they recombine with the majority carriers (the holes), thereby emitting radiation from the diode. Simultaneously holes are injected from the *p* to the *n* region, where they recombine with electrons, again contributing to the emitted radiation.

Several factors influence the operation of the LED as an electroluminescent device, chief of which is the energy gap, which in fact determines the wavelength of radiation. The best LED substance, GaAs, which has an energy gap of 1.35 eV, emits radiation at $\lambda = 0.9\ \mu$, in the infrared, while a GaP diode, of $E_g = 2.25$ eV, emits in the visible region (the deep red).

Another important factor is the so-called internal quantum efficiency of the LED. An electron–hole pair may recombine *radiatively*, emitting a photon, or *nonradiatively*, through several intermediate steps, in which energy is eventually lost in the form of heat. In an efficient device, the radiative process must be dominant, so that most of the energy is converted into light. The *quantum efficiency* is the ratio of radiative to nonradiative transitions. Other losses, such as loss of joule heat, must also be minimized.

The band structure of the substance has an important bearing on the quantum efficiency. Recall (Section 6.12) that radiative recombination and excitation are strongest in direct-gap semiconductors, such as GaAs, in which both conduction-band and valence-band extrema are at the origin, $k = 0$. Recombination in indirect-gap semiconductors, such as Ge and Si, takes place only by the intervention of phonons, and this reduces the transition probability (although a recombination and radiation does take place). This explains why GaAs is so much preferable to, say, Si, for semiconductor lamps. Thus the overall efficiency of a device made of GaP (an indirect-gap substance) is only about 1%. (Overall efficiency is to be distinguished from quantum efficiency. Overall efficiency takes into account other factors; see below.)

As we have said, GaAs is the best-known substance for semiconductor lamps, but it has the disadvantage that its emitted radiation falls in the infrared rather than the visible region, because of its small gap. This gap can be widened by alloying the GaAs with some other semiconductor such as GaP or AlAs, but since these are indirect-gap substances their overall ratio must be rather small or the material will turn into an indirect-gap diode. Researchers have been able to build a lamp of red light, $\lambda = 0.62\ \mu$, from an alloy of the semiconductors GaAs and AlAs.

A major factor in the consideration of the overall efficiency—particularly in GaAs—is the total internal reflection of the radiation at the surface as it attempts to leave the solid. GaAs has a relatively high index of refraction, $n = 3.4$, and the radiation making an angle of incidence $\theta > \sin^{-1}(1/n)$ with the surface is reflected back into the diode, and eventually lost in the form of heat. Considering these various factors, the overall efficiency—i.e., the ratio of the radiation energy to the electrical energy of the injection current—of a GaAs lamp is slightly less than 10%.

Radiation is usually extracted from a window opened in the metallic contact at the $n$ side of the junction. Semiconductor lamps are used in many ways; for example, they are often used in display devices.

**Solid-state counters**

Semiconductors are invaluable as particle detectors in high-energy physics. When a high-energy particle—such as an electron, proton, $\alpha$-particle, $\gamma$-particle, or x-ray—falls on a semiconductor crystal such as silicon or germanium, it causes a large number of electron–hole pairs to be created. These can be readily detected by photoconductive devices, such as light photodetectors. The number of such pairs can be estimated by determining the ratio of the energy of the incident particle (usually several MeV) to the energy of creation of the electron–hole pair, of the order of 3 eV. The pulse of the photoconductive current is proportional to the energy of the stopped particle, and one can conveniently use a semiconductor counter to determine this energy.

## 7.9 INTEGRATED CIRCUITS AND MICROELECTRONICS

One of the advantages of a transistor over a thermionic tube amplifier, it will be recalled, is its considerably smaller size. Extremely small circuit elements—such as resistors, capacitors, and transistors—can now be made by employing special techniques, as will be discussed below. Moreover, these microelements can be deposited on the same surface, the *substrate*. Metallic leads between them can also be deposited on the same surface, leading to a very sophisticated *integrated circuit* (IC), capable of performing highly complex functions. The testament to this field of microminiaturization can be readily appreciated when one considers the intricacy and sophistication of such complicated devices as television sets, computers, and electronic desk calculators, whose development without the integrated circuit would have been prohibitively difficult.

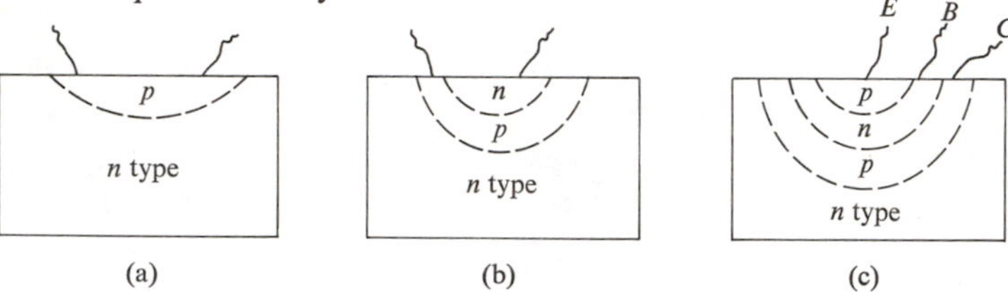


**Fig. 7.25** (a) A microresistor, (b) a microcapacitor, (c), a microtransistor. Wavy lines represent electrical leads, and letters in (c) refer to emitter, base, and collector.

Figure 7.25 shows three circuit elements manufactured from semiconductor materials. Figure 7.25(a) is a resistor, whose resistance is determined by the doping of the $p$ layer. Figure 7.25(b) is a capacitor, made up of a $p$-$n$ junction. We can appreciate the capacitor property of such a junction by referring to the treatment in Section 7.3. Recall that a depletion bilayer is formed; the portion of this layer in the $n$ region is positively charged, while that in the $p$ region is negatively charged. The total charges of the two parts are equal in magnitude. It is this positive–negative charge distribution which is responsible for the capacity of the junction. The total charge in each of the two layers per unit area of the junction is $|Q| = eN_d w_n$. If we use (7.32), we find that

$$|Q| = \left[\frac{2\epsilon e \phi_0 N_d N_a}{N_d + N_a}\right]^{1/2}.$$

When a bias voltage $V_0$ is applied across the junction, then $\phi_0$ should be replaced by $\phi_0 - V_0$ in the above equation ($V_0$ is positive for forward bias). Thus

$$|Q| = \left[\frac{2\epsilon e(\phi_0 - V_0) N_d N_a}{N_d + N_a}\right]^{1/2}. \tag{7.63}$$

And if the voltage $V_0$ is changed by a small increment, the charge also changes, and the differential capacitance is given by

$$C = \left|\frac{dQ}{dV_0}\right| = \frac{1}{2}\left[\frac{2\epsilon e}{(\phi_0 - V_0)} \frac{N_d N_a}{N_d + N_a}\right]^{1/2}. \tag{7.64}$$

The capacitance depends, as expected, on the dopings; it is large when the dopings are large, and vice versa. Note also that for a large reverse-bias voltage $C \sim (-V_0)^{-1/2}$, which is confirmed experimentally in abrupt junctions.

Figure 7.25(c) is a FET, and requires no further elaboration, as it was covered in Section 7.8.

Recall that, in an integrated circuit, the elements are deposited on the same Si substrate or chip, as illustrated in Fig. 7.26 for a resistor and a junction transistor. Obviously, for proper operation, the two elements must be electrically insulated from each other; i.e., no conduction through the *n*-type substrate should take place. This is ensured by biasing this substrate with a high positive potential, so that the resistor–substrate junction and the transistor–substrate junction are both reverse-biased, and consequently only a negligible current can flow between the two elements.

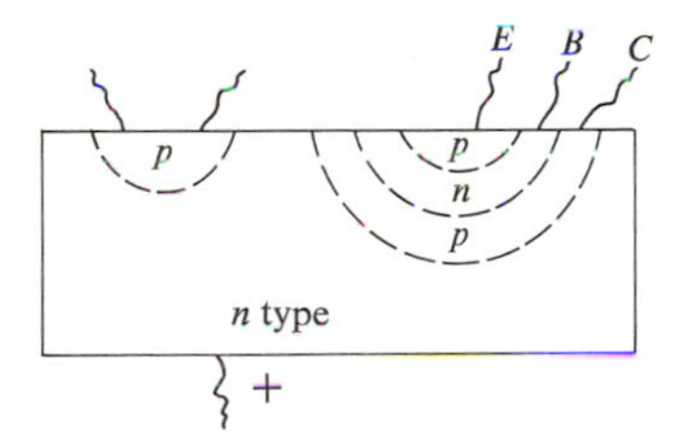


**Fig. 7.26** A microresistor and microtransistor on the same substrate. The positive potential at the bottom is used for mutual isolation of the elements.

And finally some discussion is in order regarding the manner in which the various dopings are deposited on the substrate. The traditional method of manufacturing a *p-n* junction was to start with, say, a rod-shaped piece of *n*-type germanium, and alloy one side of it with a trivalent metal. One then heated the crystal to a temperature of several hundred degrees until the metal melted and dissolved into the Ge, then cooled the whole sample again to room temperature, allowing the Ge to recrystallize with sufficient acceptors to form a *p* region on one side of the sample.

But this technique is too crude to be used in integrated circuits, since the widths of the layers involved here are usually very small and require highly controlled and accurate techniques. There are now several such methods available.

1) *Controlled diffusion.* A chip of the substrate material is placed in a chamber, and a steady concentration of the desired impurities is maintained in a gaseous phase surrounding it. As the whole system is raised to a high temperature, the impurities diffuses into the chip. The depth of penetration depends on the temperature, the duration of the process, and the nature of the impurities (Section 11.4) By controlling these variables one can obtain a precise depth. [The resistor in Fig. 7.20(a) can be made in this manner.]

If a different and new layer is required, as in the transistor, the new layer is formed by diffusing it on top of the old layer, but to a lesser depth. This technique has the disadvantage, however, that the new layer still has the old impurities embedded in it, since diffusing the new impurities on top does not remove the old ones. If one repeats the process with several layers, then one obtains a concentration of different impurities, resulting in a greater and greater conductivity. To avoid this, a new technique, the epitaxial growth method, has been developed.

2) *Epitaxial growth.* Layers of the desired impurities (Si or Ge) are deposited on the chip by placing it in a chamber within a gaseous reaction system, and the Si or Ge layers are precipitated directly from the system. The precipitation takes place so slowly and gradually that the crystalline continuity between the chip and the new layers is maintained.

3) *Ion implantation.* A third technique coming increasingly into use is the ion implantation method, in which the desired impurities are shot toward the surface of the semi-conductor, after being accelerated in a static accelerator. The depth of penetration depends on the accelerating potential. By varying this depth, one can prepare a wide range of impuritiy profiles. Potentials used for this purpose are of the order of a few kV, and typical depths are about 100 Å.†

By employing these techniques, one can make circuit elements extremely small. For example, a silicon chip of area about 2 mm$^2$ contains more than 300 elements. This trend toward microminiaturization is clearly the wave of the future. In recent years the field has been developing very rapidly; already it amounts to about 30% of the total dollar market.

The integrated circuit's advantages over the conventional circuit are as follows. (i) A drastic reduction in volume, particularly important in sophisticated devices such as computers, (ii) greater reliability, (iii) considerable reduction in cost. The main disadvantage of an IC is that once a part of the circuit—even a single element—is damaged, the entire circuit is rendered useless, and must be replaced.

The impact of the IC concept on future engineering education may be illustrated by this quotation from Beeforth (1970): "Until the advent of integrated circuits, it was necessary for electronic engineers to be familiar with basic circuit design. In the future, this will no longer be so important, as a wide range of basic circuits becomes readily available in the integrated form. The engineer will be free to deal with overall systems, without having the actual circuitry involved; 'the architect no longer needs to worry about how the individual bricks are made.' "

---

† A readable discussion of the ion-implantation method, including tips on experimental techniques and many illustrations, can be found in F. F. Morehead, Jr., and B. L. Crowder, *Scientific American*, **228**, April 1973. An important point brought out is the fact that, after exposure, the sample is annealed in order to eliminate the large number of displaced host atoms and vacancies created by the collision of the incident beam with the host atoms. The mechanism involved is discussed in Section 11.3.

## SUMMARY

### The *p-n* junction and rectification

When a *p-n* junction is formed, free carriers—both electrons and holes—diffuse across the junctions. Electrons flow from the $n$ to the $p$ side of the junction, while holes flow in the opposite direction. Because of the charge flow, the $p$ side acquires a *negative contact potential* $-\phi_0$ relative to the $n$ side. The value of $\phi_0$ is

$$\phi_0 = \frac{k_B T}{e} \log\left(\frac{N_d N_a}{n_i^2}\right),$$

where $N_d$ and $N_a$ are the concentrations of the donors and acceptors, respectively, on the two sides of the junction.

The junction acts as a rectifier. The current–voltage relationship has the form

$$I = I_0\,(e^{eV_0/k_B T} - 1),$$

where $V_0$ is the bias voltage. When this voltage is in the forward direction, $V_0 > 0$, $e^{eV_0/k_B T} \gg 1$, and hence

$$I \simeq I_0(e^{eV_0/k_B T}).$$

The current is large, and increases rapidly with the voltage. But for a reverse bias, $V_0 < 0$, $e^{eV_0/k_B T} \ll 1$, and

$$I = -I_0.$$

The current is now small, and independent of the voltage.

### The junction transistor

A junction transistor is a structure comprised of two junctions connected back to back. One, called the *emitter*, is forward biased; the other, called the collector, is reverse biased. The emitter injects minority carriers into the base. The carriers diffuse through the base, and are received at the collector.

When an electric signal is applied at the emitter, a corresponding carrier pulse passes through the base and the collector, and the amplified signal is picked up at a load resistor inserted into the collector circuit. The voltage gain is

$$\frac{d V_l}{d V_e} = \frac{\alpha R_l I_e}{k_B T/e}.$$

The gain may be increased by raising the value of the parameter $\alpha$, which is accomplished by increasing the injection efficiency and reducing the thickness of the base layer.

### The tunnel diode

When the dopings in a *p-n* junction are very high, the width of the junction becomes very small, and tunneling of the carriers across the energy gap becomes possible.

If the junction is biased in the forward direction, a region of negative differential conductivity (NDC) results. This NDC is utilized in the design of electric amplifiers and oscillators in the microwave range.

### The Gunn diode

When a high electric field is impressed on a thin GaAs sample, or samples of similar band structure, again a region of NDC is produced. This property may also be used in the design of microwave amplifiers and oscillators.

Two different modes of oscillation are possible. In the Gunn mode, the sample separates into two regions: a region of very high electric field—the Gunn domain—and a region of low field (the rest of the sample). The oscillation frequency of the Gunn mode is equal to the transit-time frequency,

$$\nu_0 = \frac{v_d}{L},$$

where $v_d$ is the electron drift velocity and $L$ is the length of the sample.

In the other mode, the LSA mode, one prevents the domain from forming by impressing a signal whose frequency is larger than $\nu_0$. In this mode, the diode acts as an element of true NDC character.

### The semiconductor laser

Laser action in semiconductors is achieved by inverting the electron populations of the valence and conduction bands. This inversion was first accomplished in a highly doped *p-n* junction, although other means have also proved possible. The frequency of the emitted coherent radiation is close to the fundamental edge of the semiconductor, that is, $E_g/h$. Laser action is possible only in direct-gap semiconductors because of the requirement of conservation of momentum.

### Other semiconductor devices

There are many other semiconductor devices, chief among which are the light or infrared detector, in which the intensity of the radiation is determined by measuring the photoconductive current in the semiconductor specimen, and the field-effect transistor (FET). Both have certain advantages over the junction transistor.

## REFERENCES

### General references

W. R. Beam, 1965, *Electronics of Solids*, New York: McGraw-Hill

T. H. Beeforth and H. J. Goldsmid, 1970, *Physics of Solid State Devices*, London: Pion Ltd.

D. F. Dunster, 1969, *Semiconductors for Engineers*, London: Business Books Ltd.

J. K. Jonscher, 1965, *Principles of Semiconductor Device Operation*, London: Bell and Sons

J. H. Leck, 1967, *Theory of Semiconductor Junction Devices*, New York: Pergamon Press
J. P. McKelvey, 1966, *Solid State and Semiconductor Physics*, New York: Harper and Row
J. L. Moll, 1964, *Physics of Semiconductors*, New York: McGraw-Hill
M. J. Morand, 1964, *Introduction to Semiconductor Devices*, Reading, Mass.: Addison-Wesley
A. Nussbaum, 1964, *Semiconductor Device Physics*, Englewood Cliffs, N.J.: Prentice-Hall
J. F. Pierce, 1967, *Semiconductor Junction Devices*, Merrill Books, Inc.
E. Spenke, 1958, *Electronic Semiconductors*, New York: McGraw-Hill
S. M. Sze, 1969, *Physics of Semiconductor Devices*, New York: John Wiley
L. V. Valdes, 1961, *Physical Theory of the Transistor*, New York: McGraw-Hill
A. van der Ziel, 1968, *Solid State Physical Electronics*, second edition, Englewood Cliffs, N.J.: Prentice-Hall

**Microwave devices; the Gunn diode**

H. A. Watson, editor, 1964, *Microwave Semiconductor Devices and their Circuit Applications*, New York: McGraw-Hill

See also books by Dunster, van der Ziel, and Sze listed above under General References

**Semiconductor lasers**

A. G. Milnes and D. L. Feucht, 1972, *Heterojunctions and Metal–Semiconductor Junctions*, New York: Academic Press
J. I. Pankove, 1971, *Optical Processes in Semiconductors*, Englewood Cliffs, N.J.: Prentice-Hall
A. E. Siegman, 1971, *An Introduction to Lasers and Masers*, New York: McGraw-Hill
A. Yariv, 1967, *Quantum Electronics*, New York: John Wiley

**Integrated circuits**

R. M. Burger and R. P. Donovan, editors, 1968, *Fundamentals of Silicon Integrated Device Technology*, Volume 2, Englewood Cliffs, N.J.: Prentice-Hall
K. J. Dean, 1967, *Integrated Electronics*, London: Chapman and Hall
J. Eimbinder, 1968, *Linear Integrated Circuits: Theory and Applications*, New York: John Wiley
D. K. Lynn and C. S. Meyer, 1967, *Analysis and Design of Integrated Circuits*, New York: McGraw-Hill
S. Schartz, editor, 1967, *Integrated Circuit Technology*, New York: McGraw-Hill
R. M. Warner and J. N. Fordenwalt, editors, 1965, *Integrated Circuits: Design, Principles and Fabrication*, New York: McGraw-Hill

## QUESTIONS

1. Show qualitatively the position of the Fermi level in a *p-n* junction at equilibrium. Use a figure similar to Fig. 7.2.
2. In the derivation of the rectification equation in Section 7.2 the approximation was made that the whole bias voltage appeared across the junction. Does this approximation hold better for forward or reverse bias? Explain.

3. Describe a metal–semiconductor junction at equilibrium (McKelvey, 1966).
4. Suppose that a *p-n* function at equilibrium is short-circuited with a metallic wire. Could the contact potential of the junction drive an electric current in the circuit? Explain. Draw the appropriate energy-band diagram for the whole circuit.
5. Show qualitatively the position of the Fermi level(s) in a biased *p-n* junction.
6. When the holes in a *p-n-p* transistor diffuse through the base, a certain fraction of them recombine with electrons and disappear. Does the fact that Si is an indirect-gap semiconductor improve or hamper the operation of a silicon transistor?
7. Suppose that the difference in energy between the bottoms of the central and secondary valleys in GaAs is gradually reduced until it vanishes. Do you expect the Gunn effect to be observed throughout this range? (Assume that the masses and mobilities of the various valleys remain unchanged.)
8. The wavelength of the coherent radiation emitted from a GaAs laser decreases from 9000 Å to 7000 Å as the substance is alloyed with phosphorus, producing the compound GaAsP. Explain why.

## PROBLEMS

1. Establish Eq. (7.6) for the hole current in a forward-biased *p-n* junction.
2. The saturation current for a *p-n* junction at room temperature is $2 \times 10^{-6}$ amp. Plot the current versus voltage in the voltage range − 5 to 1 volt. Find the differential resistance at a reverse bias of 1 volt and forward bias of 0.25 volt, and compare the two values thus obtained.
3. Derive Eqs. (7.32) by solving the Poisson's equation (7.31), subject to the appropriate boundary conditions.
4. a) Determine the contact potential for a *p-n* junction of germanium at room temperature, given that the donor concentration is $10^{18}$ cm$^{-3}$ and the acceptor concentration is $5 \times 10^{16}$ cm$^{-3}$. Assume the impurities to be completely ionized.
   b) Calculate the widths of the depletion layer of the junction.
   c) Calculate the electric field at the center of the junction.
   d) The depletion double layer also acts as a capacitor, with the depletion regions on the opposite sides of the junction having equal and opposite charges. Evaluate the capacitance per unit area of the junction.
5. Repeat Problem 4 for silicon, whose dielectric constant is $12\,\epsilon_0$.
6. Using the rectifier equation, determine the differential resistance of a 1 mm$^2$ *p-n* junction of Ge (Problem 4) under a condition of forward bias at 0.25 volt. Take the recombination times $\tau_e = \tau_h = 10^{-6}$ s. Compare the answer with the resistance of an intrinsic sample of the same length as the depletion layer of the junction.
7. Draw the energy-band diagram for the *p-n-p* transistor at equilibrium. Plot the hole concentration versus the position along the length of the structure.
8. Repeat Problem 7 with the appropriate biases applied to the transistor.
9. Derive Eq. (7.38) for the voltage gain in a junction transistor.
10. Derive Eq. (7.39) for the power gain in a junction transistor.
11. Describe the operation of an *n-p-n* transistor, and derive expressions for the voltage and power gains in such a structure.
12. Read the description of the operation of the field-effect transistor given in Sze (1969). Summarize the physical processes involved and the characteristics of this device.

13. Estimate the dopings required for the operation of a GaAs tunnel diode. Take $n_d = n_a$, and assume that tunneling becomes appreciable when the horizontal distance of the energy gap becomes 75 Å. You may employ the results developed in Section 7.3.
14. a) Using the continuity equation and Poisson's equation, show that an excess localized charge in a semiconductor decays in time according to the equation $\Delta\rho(t) = \Delta\rho(0)\, e^{-\tau_D}$, where $\tau_D = \epsilon/\sigma$ is the dielectric relaxation time and $\Delta\rho(0)$ is the initial excess density.
    b) Calculate $\tau_D$ for GaAs at low field for a carrier concentration of $10^{21}$ m$^{-3}$.
15. Draw a Cartesian coordinate system in which the abscissa represents the product $n_0 L$ and the ordinate the product $vL$. Mark the various regions in this plane corresponding to the Gunn mode and the LSA mode in GaAs.
16. Look up the derivation of (7.61) for the threshold current in an injection laser (Sze, 1969).
17. The lasing operation in a semiconductor laser may be influenced by several factors, such as temperature, pressure, magnetic field, etc. These effects are summarized in Chapter 10 of Pankove (1971). Read this chapter and give a brief summary.
18. Various procedures for population inversion in semiconductor lasers have been employed in addition to the injection technique in a *p-n* junction. Read the review of these procedures given in Pankove (1971), and give a brief summary of the results, including diagrams of experimental setups.

# CHAPTER 8 DIELECTRIC AND OPTICAL PROPERTIES OF SOLIDS



---

*When life is true to the poles of nature, the streams of truth will roll through us in songs.*

Ralph Waldo Emerson

## 8.1 INTRODUCTION

In this chapter we shall discuss dielectric and optical properties of solids and other phases of matter. These properties span an enormous range of frequencies, from the static to the ultraviolet region, and provide valuable information on the physical properties as well as the structure of matter.

After an elementary review, we shall relate the dielectric constant to the polarization properties of the molecules constituting a given substance. Then we shall consider the various sources of molecular polarization: dipolar, ionic, and electronic contributions. Finally we shall consider two important properties: piezoelectricity and ferroelectricity. Both are related to ionic polarizability.

## 8.2 REVIEW OF BASIC FORMULAS

Let us review some of the basic formulas which will be useful in the following sections. A concept most important in this chapter is that of the *electric dipole* and its *moment*. Think of an electric dipole as an entity composed of two opposite charges of equal magnitudes, $q$ and $-q$, as in Fig. 8.1. The moment of this dipole is defined as

$$\mathbf{p} = q\mathbf{d}, \tag{8.1}$$

where $\mathbf{d}$ is the vector distance from the negative to the positive charge.† The electric moment is therefore equal to one of the charges times the distance between them.

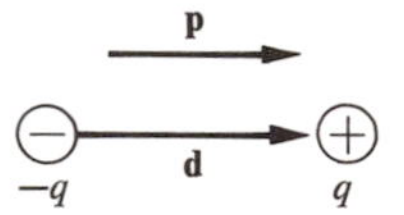


**Fig. 8.1** An electric dipole.

An electric dipole produces an electric field, which may be calculated by applying Coulomb's law to find the fields of the two charges separately, and then adding the results. The field for the dipole is given by

$$\mathscr{E} = \frac{1}{4\pi\epsilon_0}\frac{3(\mathbf{p}\cdot\mathbf{r})\mathbf{r} - r^2\mathbf{p}}{r^5}, \tag{8.2}$$

which gives the field in terms of $\mathbf{r}$, the vector joining the dipole to the field point, and the moment $\mathbf{p}$. In deriving this expression, we have assumed that $r \gg d$, that is, expression (8.2) is valid only at points far from the dipole itself. In atoms and molecules this condition is well satisfied, since $d$, being of the order of an atomic diameter, is very small indeed.

---

† Using the symbol $\mathbf{p}$ to denote the dipole moment should not lead to confusion with linear momentum, denoted by the same symbol, since linear momentum does not enter into this chapter.

When a dipole is placed in an external electric field, it interacts with the field. The field exerts a torque on the dipole which is given by

$$\boldsymbol{\tau} = \mathbf{p} \times \boldsymbol{\mathscr{E}}, \tag{8.3}$$

where $\mathscr{E}$ is the applied field (Fig. 8.2). The magnitude of the torque is $\tau = p\mathscr{E} \sin \theta$, where $\theta$ is the angle between the directions of the field and the moment, and the direction of $\boldsymbol{\tau}$ is such that it tends to bring the dipole into alignment with the field. This tendency toward alignment is a very important property, and one which we shall encounter repreatedly in subsequent discussions.

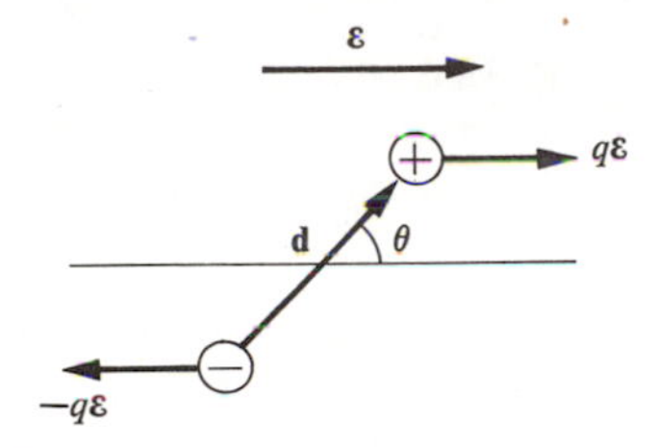


**Fig. 8.2** The torque exerted on one dipole by an electric field. Vectors $q\mathscr{E}$ and $-q\mathscr{E}$ represent the two forces exerted by the field on the point charges of the dipole.

Another, and equivalent, way of expressing the interaction of the dipole with the field is in terms of the potential energy. This is given by

$$V = -\mathbf{p} \cdot \boldsymbol{\mathscr{E}} = -p\mathscr{E} \cos \theta , \tag{8.4}$$

which is the potential energy of the dipole. We can see that the energy depends on $\theta$, the angle of orientation, and varies between $-p\mathscr{E}$, when the dipole is aligned with the field, and $p\mathscr{E}$, when the dipole is opposite to the field. Because the energy is least when the dipole is parallel to the field, it follows that this is the most favored orientation, i.e., the dipole tends to align itself with the field. This is, of course, the same conclusion reached above on the basis of torque consideration.

In discussing dielectric materials, we usually talk about the *polarization* **P** of the material, which is defined as the dipole moment per unit volume. If the number of molecules per unit volume is $N$, and if each has a moment **p**, it follows that the polarization is given by†

$$\mathbf{P} = N\mathbf{p}, \tag{8.5}$$

where we have assumed that all the molecular moments lie in the same direction.

† In this chapter, the symbol $N$ (not $n$) stands for the concentration, i.e. the number of entities (molecules, atoms, etc.) per unit volume.

When a medium is polarized, its electromagnetic properties change; this is expressed through the well-known equation

$$\mathbf{D} = \epsilon_0 \mathscr{E} + \mathbf{P}, \tag{8.6}$$

where $\mathbf{D}$ is the electric displacement vector and $\mathscr{E}$ the electric field in the medium.

It is also well known that the displacement vector $\mathbf{D}$ depends only on the *external* sources producing the external field, and is completely unaffected by the polarization of the medium.† It follows that the external field $\mathscr{E}_0$, that is, the field outside the dielectric, satisfies the relation

$$\mathbf{D} = \epsilon_0 \mathscr{E}_0. \tag{8.7}$$

When we compare this with (8.6), we find that

$$\mathscr{E} = \mathscr{E}_0 - \frac{1}{\epsilon_0}\mathbf{P}, \tag{8.8}$$

showing that the effect of the polarization is to modify the field inside the medium. In general, this results in a reduction of the field.

Equation (8.6) is usually rewritten in the form

$$\mathbf{D} = \epsilon \mathscr{E} = \epsilon_0 \epsilon_r \mathscr{E}, \tag{8.9}$$

where the *relative dielectric constant*

$$\epsilon_r = \epsilon/\epsilon_0 \tag{8.10}$$

expresses the properties of the medium. All the dielectric and optical characteristics of the substance are contained in this constant, and indeed much of this chapter is concerned with evaluating it under a variety of circumstances. Thus it follows that we can gain much information about a medium by measuring its dielectric constant. From this point on, we shall refer to the relative dielectric constant $\epsilon_r$ as simply the dielectric constant, since we rarely need to use the actual dielectric constant $\epsilon = \epsilon_0 \epsilon_r$.

Figure 8.3 shows a simple procedure for measuring dielectric constant. The plates of a capacitor are connected to a battery which charges the plates. When there is no dielectric inside the capacitor, the electric field produced by the charges is $\mathscr{E}_0$, which can be determined by measuring the potential difference $V_0$ across the capacitor, and using the relation

$$\mathscr{E}_0 = V_0/L, \tag{8.11}$$

where $L$ is the distance between the plates. This relation should be familiar to the reader from his study of elementary physics. If a dielectric slab is now inserted between the plates, the field $\mathscr{E}_0$ induces the polarization of the

† See, for example, J. B. Marion (1965), *Classical Electromagnetic Radiation*, New York: Academic Press.

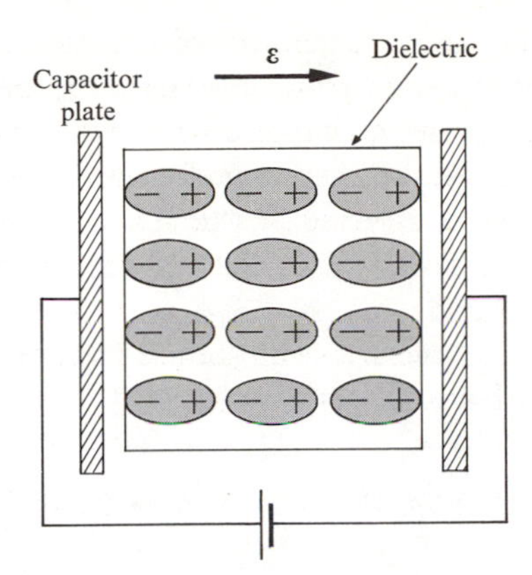


**Fig. 8.3** Simple experimental setup for measuring dielectric constant. Note polarization of molecules in the solid.

medium—i.e., the lining up of the dipole moments along the field—which, in turn, modifies the field to a new value $\mathscr{E}$. This new field can be determined by measuring the new potential difference $V$ by a voltmeter, and using the relation

$$\mathscr{E} = V/L. \tag{8.12}$$

The dielectric constant is given in terms of the fields $\mathscr{E}_0$ and $\mathscr{E}$ by the relation

$$\epsilon_r = \mathscr{E}_0/\mathscr{E}, \tag{8.13}$$

as can be seen by comparing (8.9) with (8.10). It follows, therefore, that

$$\epsilon_r = V_0/V, \tag{8.14}$$

where we used (8.11) and (8.12). We can thus obtain the dielectric constant by measuring the potential differences across the capacitor, with and without the presence of the dielectric, and taking their ratio.

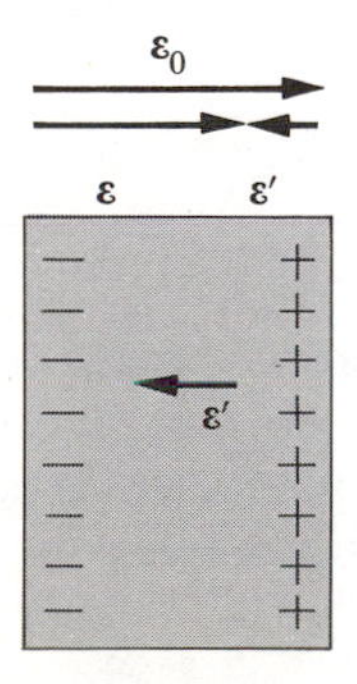


**Fig. 8.4** The field $\mathscr{E}'$ due to polarization charges at the surfaces opposes external field $\mathscr{E}_0$. Resultant internal field is $\mathscr{E}$.

Figure 8.4 shows why the polarization of the medium reduces the electric field. The effect of the polarization produces net polarization charges situated at the faces of the dielectric, a positive charge on the right and a negative on the left. (The dipolar charges inside the medium cancel each other.) These charges create their own electric field which is directed to the left, and thus opposes the external field $\mathscr{E}_0$. When we add this polarization field to the external field $\mathscr{E}_0$, to obtain the resultant field $\mathscr{E}$, we find that $\mathscr{E} < \mathscr{E}_0$, as previously stated. When we combine this result with (8.12), we arrive at the useful conclusion that the dielectric constant of a medium is always larger than unity.†

## 8.3 THE DIELECTRIC CONSTANT AND POLARIZABILITY; THE LOCAL FIELD

Since the polarization of a medium—i.e., the alignment of the molecular moment—is produced by the field, it is plausible to assume that the molecular moment is proportional to the field. Thus we write

$$\mathbf{p} = \alpha \mathscr{E}, \tag{8.15}$$

where the constant $\alpha$ is called the *polarizability* of the molecule. The expression (8.15) is expected to hold good, except in circumstances in which the field becomes very large, in which case other terms must be added to (8.15) to form what is, in effect, a Taylor-series expansion of $\mathbf{p}$ in terms of $\mathscr{E}$. Equation (8.15) may be regarded as the first term in this expansion. (Higher-order terms lead to nonlinear effects.)

The polarization $\mathbf{P}$ can now be written as

$$\mathbf{P} = N\alpha\mathscr{E}, \tag{8.16}$$

which, when substituted into (8.6), yields

$$\mathbf{D} = \epsilon_0\mathscr{E} + N\alpha\mathscr{E} = \epsilon_0\left(1 + \frac{N\alpha}{\epsilon_0}\right)\mathscr{E}. \tag{8.17}$$

Comparing this result with (8.9), one finds

$$\epsilon_r = 1 + (N\alpha/\epsilon_0), \tag{8.18}$$

giving the dielectric constant in terms of the polarizability. This is a useful result in that it expresses the *macroscopic* quantity, $\epsilon_r$, in terms of the *microscopic* quantity, $\alpha$, thus forming a link between the two descriptions of dielectric materials.

The *electric susceptibility* $\chi$ of a medium is defined by the relation

$$\mathbf{P} = \epsilon_0\chi\mathscr{E}, \tag{8.19}$$

† This is not necessarily true at high frequencies (see Section 8.9).

which relates the polarization to the field. By comparing this equation with (8.16), we find that the susceptibility and polarizability are interrelated by

$$\chi = \frac{N\alpha}{\epsilon_0}, \tag{8.20}$$

and hence Eq. (8.18) may be written simply as

$$\epsilon_r = 1 + \chi. \tag{8.21}$$

Thus the departure of the dielectric constant from unity, the value for vacuum, is equal to the electric susceptibility.† (If several gaseous species are present, than the factor $N\alpha$ in (8.20) should be replaced by $\sum_i N_i \alpha_i$.)

Equation (8.18) may also be written in terms of the density of the medium by noting that $N = \rho N_A / M$, where $\rho$ is the density, $M$ the molar mass, and $N_A$ Avogadro's number. Thus

$$\epsilon_r = 1 + (\rho N_A / \epsilon_0 M)\alpha. \tag{8.22}$$

This expression, indicating that $\epsilon_r$ increases linearly with density, holds in gases, in which density can be conveniently varied over a wide range. This fact lends support to the argument used in the derivation of (8.19), and in particular to (8.15).

Experiments do show, however, that Eqs. (8.18) or (8.22) do not hold well in liquids or solids, i.e., in condensed physical systems. This point is important to us here, as our primary interest lies in describing solid substances, and we must therefore seek a better expression for the dielectric constant than (8.18). The root of the difficulty lies in (8.15). It is implied here that the field acting on and polarizing the molecules is equal to the field $\mathscr{E}$, but a closer examination reveals that this is not necessarily so. If it develops that the polarizing field is indeed different from $\mathscr{E}$, relation (8.15) should then be replaced by

$$\mathbf{p} = \alpha \mathscr{E}_{\text{loc}}, \tag{8.23}$$

where $\mathscr{E}_{\text{loc}}$ is, by definition, the polarizing field—also called the *local field.*

To evaluate $\mathscr{E}_{\text{loc}}$ we must calculate the total field acting on a certain typical dipole, this field being due to the external field as well as all *other* dipoles in the system. This was done by Lorentz as follows: The dipole is imagined to be surrounded by a spherical cavity whose radius $R$ is sufficiently large that the matrix lying outside it may be treated as a *continuous medium* as far as the dipole is

† Actual dielectric media are anisotropic, i.e., the value of $\epsilon_r$, or $\chi$, depends on the direction of the field. Thus the parameters $\epsilon_r$ and $\chi$ are tensor quantities of the second rank. In order to concentrate on the physical principles, we shall, however, ignore the anisotropy and regard the dielectric as an isotropic medium, in which case the dielectric constant is represented by a scalar, i.e., a single number.

concerned (Fig. 8.5). The interaction of our dipole with the other dipoles lying inside the cavity is, however, to be treated microscopically, which is necessary since the discrete nature of the medium very close to the dipoles should be taken into account. The local field, acting on the central dipole, is thus given by the sum

$$\mathscr{E}_{\text{loc}} = \mathscr{E}_0 + \mathscr{E}_1 + \mathscr{E}_2 + \mathscr{E}_3, \tag{8.24}$$

where $\mathscr{E}_0$ is the external field, $\mathscr{E}_1$ the field due to the polarization charges lying at the external surfaces of the sample, $\mathscr{E}_2$ the field due to the polarization charges lying on the surface of the Lorentz sphere, and $\mathscr{E}_3$ the field due to other dipoles lying within the sphere. Note that the part of the medium between the sphere and the external surface does not contribute anything since, in effect, the volume polarization charges compensate each other, resulting in a zero net charge in this region.

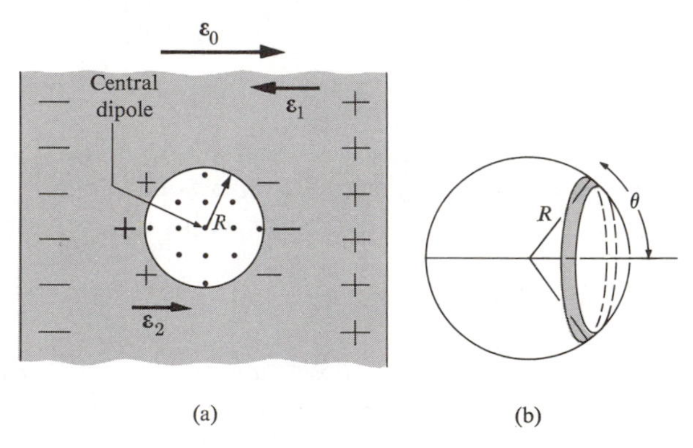


**Fig. 8.5** (a) The procedure for computing the local field. (b) The procedure for calculating $\mathscr{E}_2$, the field due to the polarization charge on the surface of the Lorentz sphere.

Let us now evaluate the various fields which appeared above.

$\mathscr{E}_1$: This field, due to the polarization charges on the external surface, is also known as the *depolarization field*, since it is obviously opposed to the external field. The value of this field depends on the geometrical shape of the external surface, and for the simple case of an infinite slab is given by

$$\mathscr{E}_1 = -\frac{1}{\epsilon_0}\mathbf{P}, \tag{8.25}$$

which you may confirm by using Gauss' law. The depolarization fields for other geometrical shapes can be found in the references (Kittel, 1971), as well as in the problems.

$\mathscr{E}_2$: The polarization charges on the surface of the Lorentz cavity may be considered as forming a continuous distribution (recall that the cavity is large) whose density is $-P \cos \theta$. The field due to the charge at a point located at the center of the sphere is. according to Coulomb's law, given by

$$\mathscr{E}_2 = \int_0^\pi \left( -\frac{P \cos \theta}{4\pi\epsilon_0 R^2} \right) \cos \theta \, (2\pi R^2 \sin \theta \, d\theta), \tag{8.26}$$

where the additional factor $\cos \theta$ is included because we are, in effect, evaluating only the component of the field along the direction of **P** (other components vanish by symmetry), and the factor $2\pi R^2 \sin \theta \, d\theta$ is the surface element along the sphere (see Fig. 8.5b). Integration of (8.26) leads to the simple result

$$\mathscr{E}_2 = \frac{1}{3\epsilon_0} \mathbf{P}, \tag{8.27}$$

a field in the same direction as the external field.

$\mathscr{E}_3$: This field, which is due to other dipoles in the cavity, may be evaluated by summing the fields of the individual dipoles using (8.2). The result obtained depends on the crystal structure of the solid under consideration, but for the case of a cubic structure it may readily be shown that the sum vanishes. That is,

$$\mathscr{E}_3 = 0, \tag{8.28}$$

as the reader will be asked to show in the problem section. In other structures the dipolar field $\mathscr{E}_3$ may not vanish, and it should then be included in the rest of the discussion.

If the various fields are now substituted into (8.24), one finds that

$$\mathscr{E}_{\text{loc}} = \mathscr{E}_0 - \frac{2}{3\epsilon_0} \mathbf{P}, \tag{8.29}$$

which gives the polarizing field in terms of the external field and the polarization.

We may compare the value of $\mathscr{E}_{\text{loc}}$ obtained above with that of $\mathscr{E}$ in (8.8). We discover that

$$\mathscr{E}_{\text{loc}} = \mathscr{E} + \frac{1}{3\epsilon_0} \mathbf{P}, \tag{8.30}$$

which shows that $\mathscr{E}_{\text{loc}}$ is indeed different from $\mathscr{E}$, as we have suspected. The former field is, in fact, larger than the latter, so the molecules are more effectively polarized than our earlier discussions have indicated. Equation (8.30) is known as the *Lorentz relation*.

The difference between $\mathscr{E}$, which is known as the *Maxwell field*, and the Lorentz field $\mathscr{E}_{\text{loc}}$ may be explained as follows. The field $\mathscr{E}$ is a macroscopic quantity, and as such is an average field, the average being taken over a large number of molecules (Fig. 8.6). It is this field which enters into the Maxwell

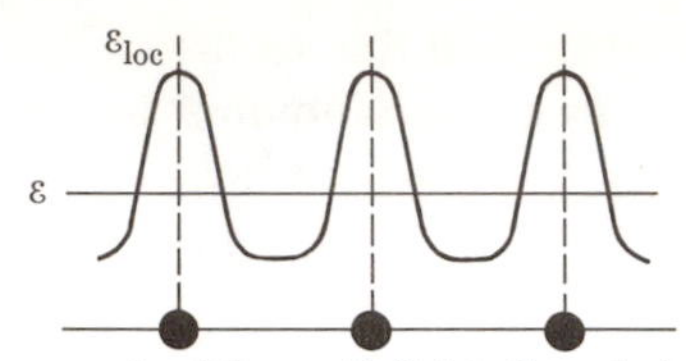


**Fig. 8.6** The difference between the Maxwell field $\mathscr{E}$ and the local field $\mathscr{E}_{\text{loc}}$. Solid circles represent molecules.

equations, which, you will recall, are used for the macroscopic description of dielectric media. In the present situation the field $\mathscr{E}$ is a constant throughout the medium.

On the other hand, the Lorentz field $\mathscr{E}_{\text{loc}}$ is a *microscopic* field which fluctuates rapidly within the medium. As the figure indicates, this field is quite large at the molecular sites themselves, and hence the molecules are more effectively polarized than they would be in the average field $\mathscr{E}$.

Let us now evaluate the dielectric constant. The polarization, according to (8.23) and (8.16), is given by

$$\mathbf{P} = N\alpha\mathscr{E}_{\text{loc}}, \tag{8.31}$$

which, when used in conjunction with (8.30), yields

$$\mathbf{P} = \left( \frac{N\alpha}{1 - \dfrac{N\alpha}{3\epsilon_0}} \right) \mathscr{E}. \tag{8.32}$$

This relation between $\mathbf{P}$ and $\mathscr{E}$ supersedes the earlier one, (8.16), and we note the fact that the denominator being less than unity contributes to the enhancement of the polarization; the enhancement is due to the local field correction. When the result (8.32) is substituted into (8.16) and (8.17), one finds the following expression for the dielectric constant

$$\epsilon_r = \frac{1 + \dfrac{2}{3\epsilon_0} N\alpha}{1 - \dfrac{N\alpha}{3\epsilon_0}}, \tag{8.33}$$

which is the relation we have been seeking. It is the generalization of (8.18) when the local field correction is taken into account.

In gases, in which the molecular concentration $N$ is small, the expression (8.33) reduces to the earlier (8.18) without the field correction. This can be seen by noting that $(N\alpha/3\epsilon_0) \ll 1$ in the denominator of (8.33), since $N$ is small, so that one may expand this denominator in powers of $(N\alpha/3\epsilon_0)$, which in first order reduces precisely to (8.18). This is expected, of course, because for small $N$ the polarization $\mathbf{P}$ is also small, which, according to (8.27), means that the local field becomes more or less the same as the average field. In liquids and solids, however, the polarization is no longer small, and Eq. (8.33) has a wider range of applicability.

Equation (8.30) is also frequently rewritten in the form

$$\frac{\epsilon_r - 1}{\epsilon_r + 2} = \frac{N\alpha}{3\epsilon_0}, \tag{8.34}$$

which is referred to as the *Clausius–Mosotti relation*. We can also write this equation as

$$\frac{M}{\rho}\left(\frac{\epsilon_r - 1}{\epsilon_r + 2}\right) = \frac{N_A\alpha}{3\epsilon_0}, \tag{8.35}$$

which shows that the polarizability $\alpha$ may be determined from the measurable quantities $M$, $\rho$, and $\epsilon_r$. The expression on the right (and on the left) is known as the *molar polarizability.*

## 8.4 SOURCES OF POLARIZABILITY

Let us now examine more closely the physical process which gives rise to polarizability. Basically, polarizability is a consequence of the fact that the molecules, which are the building blocks of all substances, are composed of both positive charges (nuclei) and negative charges (electrons). When a field acts on a molecule, the positive charges are displaced along the field, while the negative charges are displaced in a direction opposite to that of the field. The effect is therefore to pull the opposite charges apart, i.e., to polarize the molecule.

There are different types of polarization processes, depending on the structure of the molecules which constitute the solid. If the molecule has a *permanent moment*, i.e., a moment even in the absence of an electric field, we speak of a *dipolar* molecule, and a *dipolar* substance.

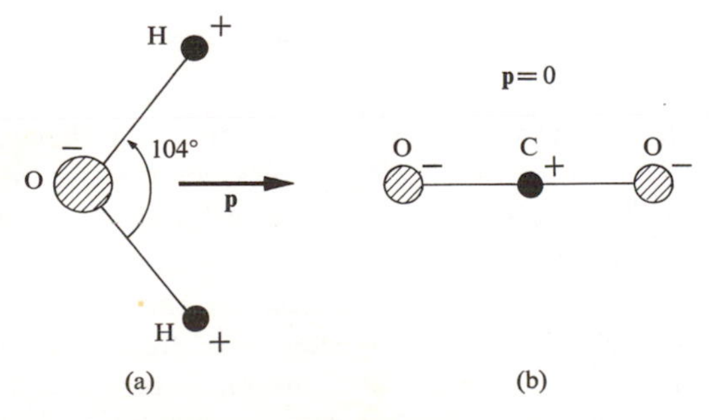


**Fig. 8.7** (a) The water molecule and its permanent moment. $p = 1.9$ debye units (1 debye $= 10^{-29}$ coul·m). (b) $CO_2$ molecule.

An example of a dipolar molecule is the $H_2O$ molecule in Fig. 8.7(a). The dipole moments of the two OH bonds add vectorially to give a nonvanishing net dipole moment. Some molecules are nondipolar, possessing no permanent moments; a common example is the $CO_2$ molecule in Fig. 8.7(b). The moments of the two CO bands cancel each other because of the rectilinear shape of the molecule, resulting in a zero net dipole moment.

The water molecule has a permanent moment because the two OH bands do not lie along the same straight line, as they do in the $CO_2$ molecule. The moment thus depends on the geometrical arrangement of the charges, and by measuring the moment one can therefore gain information concerning the structure of the molecule.

Despite the fact that the individual molecules in a dipolar substance have permanent moments, the net polarization vanishes in the absence of an external field because the molecular moments are randomly oriented, resulting in a complete cancellation of the polarization. When a field is applied to the substance, however, the molecular dipoles tend to align with the field, as stated in Section 8.2, and this results in a net nonvanishing polarization. This leads to the so-called *dipolar polarizability* which will be evaluated in Section 8.5.

If the molecule contains ionic bonds, then the field tends to stretch the lengths of these bonds. This occurs in NaCl, for instance, because the field tends to displace the positive ion $Na^+$ to the right (see Fig. 8.8), and the negative ion $Cl^-$ to the left, resulting in a stretching in the length of the bond. The effect of this change in length is to produce a net dipole moment in the unit cell where previously there was none. Since the polarization here is due to the relative displacements of oppositely charged ions, we speak of *ionic polarizability*.

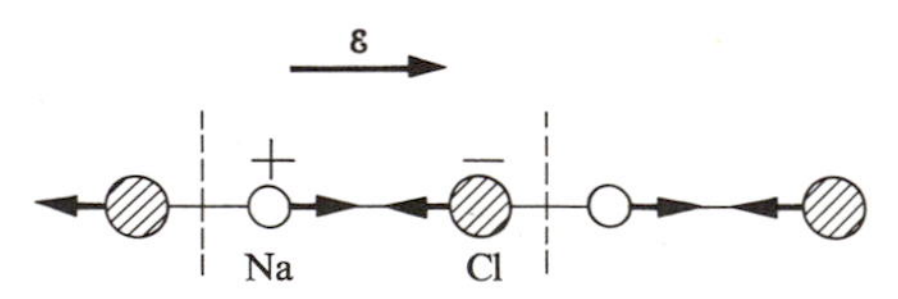


**Fig. 8.8** Ionic polarization in NaCl. The field displaces the ions $Na^+$ and $Cl^-$ in opposite directions, changing the length of the bond.

Ionic polarizability exists whenever the substance is either ionic, as in NaCl, or dipolar, as in $H_2O$, because in each of these classes there are ionic bonds present. But in substances in which such bonds are missing—such as Si and Ge—ionic polarizability is absent.

The third type of polarizability arises because the individual ions or atoms in a molecule are themselves polarized by the field. In the case of NaCl, each of the $Na^+$ and $Cl^-$ ions are polarized. Thus the $Na^+$ ion is polarized because the electrons in its various shells are displaced to the left relative to the nucleus, as shown in Fig. 8.9. We are clearly speaking here of *electronic polarizability*.

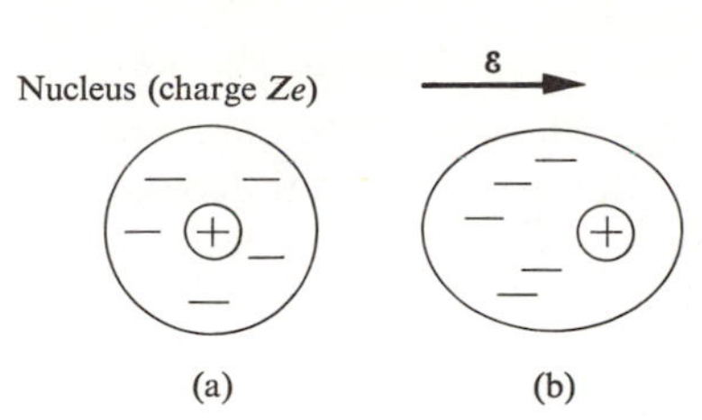


**Fig. 8.9** Electronic polarization. (a) Unpolarized atom. (b) Atom polarized as a result of the field.

Electronic polarizability arises even in the case of a neutral atom, again because of the relative displacement of the orbital electrons.

In general, therefore, we may write for the total polarizability

$$\alpha = \alpha_e + \alpha_i + \alpha_d, \tag{8.36}$$

which is the sum of the various contributions; $\alpha_e$, $\alpha_i$, and $\alpha_d$ are the electronic, ionic, and dipolar polarizabilities, respectively. The electronic contribution is present in any type of substance, but the presence of the other two terms depends on the material under consideration. Thus the term $\alpha_i$ is present in ionic substances, while in a dipolar substance all three contributions are present. In covalent crystals such as Si and Ge, which are nonionic and nondipolar, the polarizability is entirely electronic in nature.

The relative magnitudes of the various contributions in (8.36) are such that in nondipolar, ionic substances the electronic part is often of the same order as the ionic. In dipolar substances, however, the greatest contribution comes from the dipolar part. This is the case for water, as we shall see.

The various polarizabilities may be segregated from each other because each contribution has its own characteristic features which distinguish it from the others, as we shall see in the remainder of this chapter. Dipolar polarizability, for instance, exhibits strong dependence on temperature, while the other two contributions are essentially temperature independent.

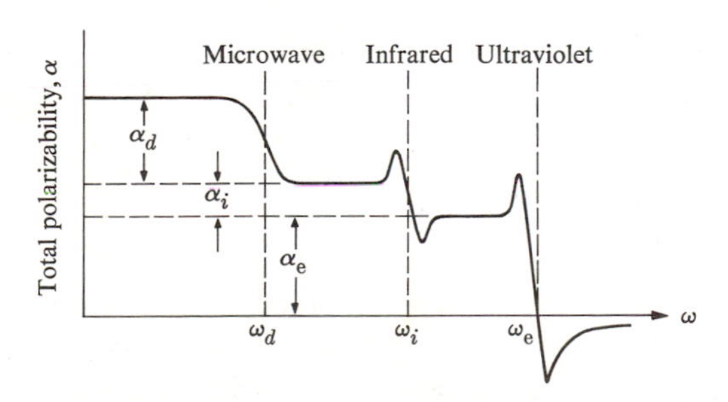


**Fig. 8.10** Total polarizability $\alpha$ versus frequency $\omega$ for a dipolar substance.

Another important distinction between the various polarizabilities emerges when one examines the behavior of the ac polarizability that is induced by an alternating field. Figure 8.10 shows a typical dependence of this polarizability on frequency over a wide range, extending from the static all the way up to the ultraviolet region. It can be seen that in the range $\omega = 0$ to $\omega = \omega_d$, where $\omega_d$ ($d$ for dipolar) is some frequency usually in the microwave region, the polarizability is essentially constant. In the neighborhood of $\omega_d$, however, the polarizability decreases by a substantial amount. This amount corresponds precisely, in fact, to the dipolar contribution $\alpha_d$. The reason for the disappearance of $\alpha_d$ in the frequency range $\omega > \omega_d$ is that the field now oscillates too rapidly for the dipole to follow, and so the dipoles remain essentially stationary.

The polarizability remains similarly unchanged in the frequency range $\omega_d$ to $\omega_i$, and then plummets at the higher frequency. The frequency $\omega_i$ lies in the infrared region, and corresponds to the frequency of the transverse optical phonon in the crystal $\omega_t$ (Section 3.12). For the frequency range $\omega > \omega_t$, the ions with their heavy masses are no longer able to follow the very rapidly oscillating field, and consequently the ionic polarizibility $\alpha_i$ vanishes, as shown in Fig. 8.10.

Thus in the frequency range above the infrared, only the electronic polarizability remains effective, because the electrons, being very light, are still able to follow the field even at the high frequency. This range includes both the visible and ultraviolet regions. At still higher frequencies (above the electronic frequency $\omega_e$), however, the electronic contribution vanishes because even the electrons are too heavy to follow the field with its very rapid oscillations.

We see, therefore, that the dielectric constant of a dipolar substance may decrease substantially as the frequency is increased from the static to the optical region. For example, the dielectric constant of water is 81 at zero frequency, while it is only 1.8 at optical frequencies.

The frequencies $\omega_d$ and $\omega_i$, characterizing the dipolar and ionic polarizabilities, respectively, depend on the substance considered, and vary from one substance to another. However, their orders of magnitude remain in the regions indicated above, i.e., in the microwave and infrared, respectively. The various polarizabilities may thus be determined by measuring the substance at various appropriate frequencies.

Let us now evaluate the various polarizabilities, and show how measuring them may give us information about the internal microscopic structure of a given substance.

## 8.5 DIPOLAR POLARIZABILITY

We can obtain the expression for dipolar polarizability (also called *orientational polarizability*) by applying the basic formulas of Section 8.2 and some elementary statistical mechanics. Imagine that an electric field is applied to a dipolar system in which the dipoles are able to rotate freely, as in a gas or liquid. Before the field was applied, the dipoles were oriented randomly, resulting in a vanishing average polarization, but the presence of the field tends to align the dipoles, resulting in a net polarization in the direction of the field. It is this polarization that we wish to calculate.

Suppose the field is along the $x$-direction. The potential energy of the dipole is given, according to (8.4), by

$$V = -\mathbf{p}\cdot\mathscr{E} = -p\mathscr{E}\cos\theta, \tag{8.37}$$

where $\theta$ is the angle made by the dipole with the $x$-axis (Fig. 8.11). The dipole is no longer oriented randomly. The probability of finding it along the $\theta$-direction is given by the distribution function

$$f = e^{-V/kT} = e^{p\mathscr{E}\cos\theta/kT}. \tag{8.38}$$

This expression is simply the Boltzmann factor, well known from statistical mechanics, with the potential energy being the orientational energy of (8.37). This distribution function, shown in Fig. 8.11(b), indicates that the dipole is more likely to lie along the field $\theta \simeq 0$ than in other directions, in agreement with the picture developed previously.

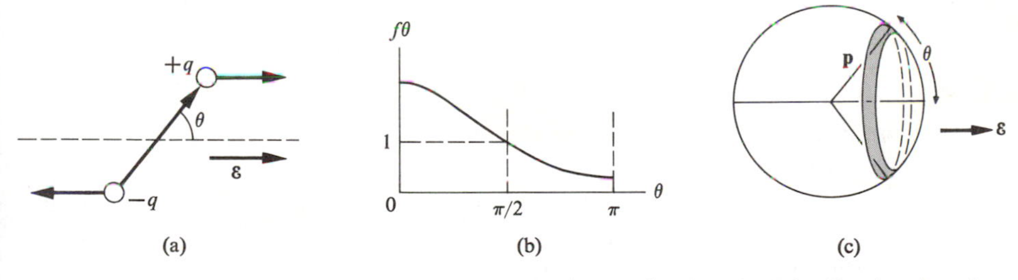


**Fig. 8.11** (a) Aligning torque applied by the field to a dipole. (b) Distribution function $f(\theta)$ versus angle of orientation. (c) The integration over the solid angle defining the orientation of the dipole. Shaded area represents the element of the spherical shell specifying the orientation of the dipole.

The average value of $p_x$, the $x$-component of the dipole moment, is given by the expression

$$\bar{p}_x = \frac{\int p_x f(\theta)\, d\Omega}{\int f(\theta)\, d\Omega}, \tag{8.39}$$

where the integration is over the solid angle, whose element is $d\Omega$. By carrying out the integration over the whole solid angle range (Fig. 8.11c), we take into account all the possible orientations of the dipole. The function $f(\theta)$ is the distribution function of (8.38) with its dependence on $\theta$ indicated, and the denominator in (8.39) is included for a proper normalization of this distribution function. In evaluating expression (8.39), we use the formulas $p_x = p\cos\theta$, $d\Omega = 2\pi\sin\theta\, d\theta$

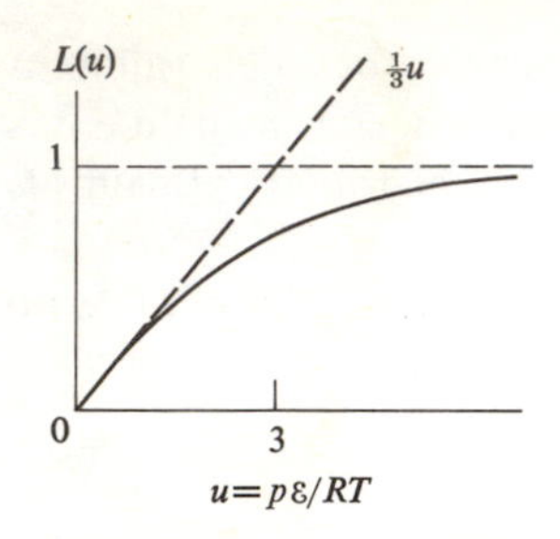


**Fig. 8.12** The Langevin function $L(u)$ versus $u$.

(where the factor $2\pi$ arises from the integration over the azimuthal angle $\phi$), $f(\theta)$ taken from (8.38), and the limits on the integrals $\theta = 0$ and $\theta = \pi$. Thus

$$\bar{p}_x = \int_0^\pi p \cos\theta \, e^{p\mathscr{E}\cos\theta/kT}\, 2\pi \sin\theta \, d\theta \bigg/ \int_0^\pi e^{p\mathscr{E}\cos\theta/kT}\, 2\pi \sin\theta \, d\theta,$$

which, when evaluated, yields†

$$\bar{p}_x = p\, L(u), \tag{8.40}$$

where

$$L(u) = \operatorname{Coth}(u) - \frac{1}{u} \qquad \text{and} \qquad u = \frac{p\mathscr{E}}{kT}. \tag{8.41}$$

The function $L(u)$, known as the *Langevin function*, is plotted in Fig. 8.12. Near the origin the function increases *linearly*, and one may show that $L(u) \simeq \frac{1}{3}u$. As $u$ increases, the function continues to increase, monotonically, eventually saturating at the value unity as $u \to \infty$. The dipole moment $\bar{p}_x$, as a function of $p\mathscr{E}/kT$, has the same shape as Fig. 8.12, except for a change of the vertical scale by a constant $p$. Thus, for small values of the field, $\bar{p}_x$ increases linearly, while at very high field, $\bar{p}_x$ saturates at the maximum value $p$. This shows that at very high field the dipole points exactly along the field, which is a plausible result.

In most experimental situations, the ratio $u = p\mathscr{E}/kT$ is very small. For example, if we take $p \simeq 10^{-29}$ coul · m, $\mathscr{E} = 10^5\ V/\text{m}$, and $T = 300°\text{K}$, we find $u \simeq 10^{-4}$, which is very small indeed compared with unity. Thus we may use the low-field approximation

$$\bar{p}_x = \frac{p^2}{3kT}\mathscr{E}. \tag{8.42}$$

† The evaluation of $\bar{p}_x$ is facilitated by noting the following point: If the integral in the denominator is denoted by $Z$, then it may be readily verified that the integral in the numerator is $[\partial/\partial(p\mathscr{E}/kT)]Z$. That is, the derivative of $Z$ with respect to the quantity $p\mathscr{E}/kT$. Thus $\bar{p}_x = [\partial/\partial(p\mathscr{E}/kT)]Z/Z = [\partial/\partial(p\mathscr{E}/kT)] \log Z$. Therefore $\bar{p}_x$ may be evaluated by finding $Z$, taking its logarithm, and carrying out the indicated integration. The actual value one finds for $Z$ is $4\pi \sinh(p\mathscr{E}/kT)/(p\mathscr{E}/kT)$.

That is, the net dipole moment is directly proportional to the field, and inversely proportional to the temperature.

The result (8.42) may also be obtained from the following physical argument. As we know, the effect of a field is to align the dipoles, whereas the effect of temperature is to oppose this and to randomize the direction of the dipoles. Therefore one may write

$$\bar{p}_x = p\,\frac{\text{orientational energy}}{\text{thermal energy}}.$$

If we substitute the values orientational energy $= p\mathscr{E}$ and thermal energy $\simeq kT$, we obtain

$$\bar{p}_x = p\,\frac{p\mathscr{E}}{kT} = \frac{p^2\mathscr{E}}{kT},$$

which is the same as (8.42), except for the numerical factor $\frac{1}{3}$, which is of the order of unity. We see therefore that at low field orientational energy is much less than thermal energy, and consequently the net dipole moment $\bar{p}_x$ is only a small fraction of its maximum value $p$. On the other hand, at high field, orientational energy dominates thermal energy, and consequently the net moment $\bar{p}_x$ is very close to its maximum value, that is, $\bar{p}_x \simeq p$.

Dipolar polarizability, on the basis of (8.42), is given by

$$\alpha_d = \frac{p^2}{3kT}. \tag{8.43}$$

When this is substituted into the Clausius–Mosotti relation (8.35), one finds that

$$\frac{M}{\rho}\left(\frac{\epsilon_r - 1}{\epsilon_r + 2}\right) = \frac{N_{\mathrm{A}}}{3\epsilon_0}\left(\alpha_{ei} + \frac{p^2}{3kT}\right), \tag{8.44}$$

where $\alpha_{ei}$ is the combined polarizability due to both electronic and ionic contributions. This polarizability is essentially temperature independent, as we shall see in later sections.

If we plot the molar polarizability $(M/\rho)\,[(\epsilon_r - 1)/(\epsilon_r + 2)]$ versus the inverse temperature, $1/T$, we should obtain a straight line the slope of which is proportional to $p^2$, and its intercept should be proportional to $\alpha_{ei}$. This graph therefore leads to the determination of both the molecular dipole moment and the nondipolar polarizability, both of which are very useful quantities.

Such a plot is shown in Fig. 8.13 for several gaseous substances. We can see that the linear behavior predicted by (8.44) is borne out experimentally.

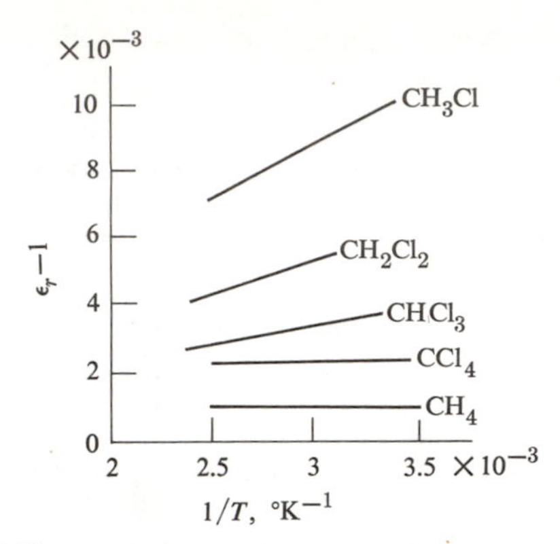


**Fig. 8.13** Total susceptibility $\chi = \epsilon_r - 1$ versus $1/T$ for several gaseous substances. (Note that denominator on left side of Eq. (8.44) is $\epsilon_r + 2 \simeq 3$ for these gaseous materials.)

The graph indicates that the molecules $CH_3Cl$, $CH_2Cl_2$, and $CHCl_3$ are all dipolar, while the molecules $CCl_4$ and $CH_4$, whose graphs are horizontal, are nonpolar (no permanent moment). Indeed it is easy to understand why the methane molecule $CH_4$ is nondipolar. Its structure, as shown in Fig. 8.14, is such that the hydrogen atoms are located at the corners of a regular tetrahedron, with the carbon atom at the center. There are four bonds joining the carbon to each of the hydrogen atoms, and although each of these bonds has an electric moment, the total dipole moment of the molecule vanishes because of the symmetric arrangement of the bonds. Note, however, that when one of the hydrogen atoms is replaced by a chlorine atom, the resulting $CH_3Cl$ molecule, no longer symmetrical, acquires a permanent moment, in agreement with Fig. 8.13.

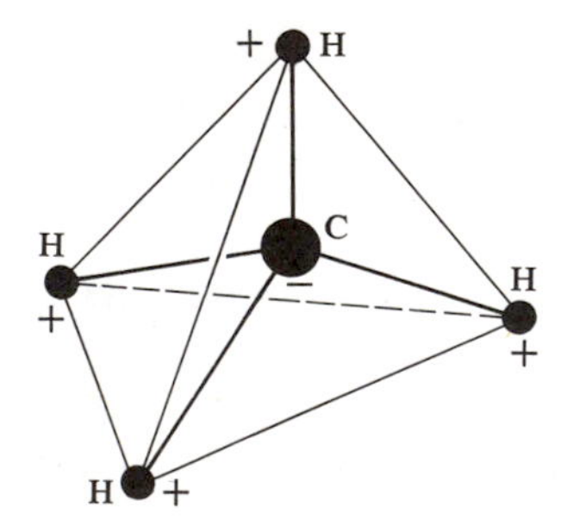


**Fig. 8.14** Geometrical structure of the methane molecule ($CH_4$).

Table 8.1 gives dipole moments for various molecules, measured in the manner indicated above. The moments are expressed in terms of the *Debye unit*, which is equal to $10^{-29}$ coul · m. This convenient microscopic unit corresponds to a dipole of charge $q = 10^{-19}$ coul ($= e/1.6$) and length $10^{-10}$ m ($= 1$ Å). Since the distances encountered in molecules are of the order of angstroms, and the charges of the order of $e$, the moments encountered are of the order of the Debye unit.

**Table 8.1**

Permanent Dipole Moments of Some Dipolar Molecules

| Substance | Dipole moment, debyes | Substance | Dipole moment, debyes |
|---|---|---|---|
| HF | 1.91 | $NH_3$ | 1.5 |
| HCl | 1.1 | $CH_3Cl$ | 1.97 |
| HBr | 0.8 | $CH_2Cl$ | 1.59 |
| HI | 0.38 | $CHCl_3$ | 0.95 |
| NO | 0.1 | $H_2O$ | 1.9 |
| CO | 0.1 | $H_2S$ | 1.10 |
| NaI | 4.9 | $SO_2$ | 1.6 |
| KCl | 6.3 | | |

## 8.6 DIPOLAR DISPERSION

Let us now discuss ac dipolar polarizability. When an electric field oscillates, the dipoles in the system tend to follow the field, flipping back and forth as the field reverses its direction during each cycle. However, a dipole experiences some friction due to its collision with other molecules in the system. This means that some energy is absorbed from the field, and we speak of *dielectric loss*. This energy appears eventually in the form of heat, which raises the temperature of the substance. Therefore studying the ac polarizability and the dielectric loss gives information on the interaction between the molecules in the medium.

The equation we shall use to describe the motion of the dipolar polarization is

$$\frac{dp_d(t)}{dt} = \frac{1}{\tau}\left[p_{ds}(t) - p_d(t)\right], \tag{8.45}$$

where $p_d(t)$ is the actual dipolar moment at the instant $t$, while $p_{ds}(t)$ is the saturated (or equilibrium) value of the moment, which would be the value approached by $p_d(t)$ if the field were to retain its instantaneous value for a long time. We have assumed that the rate of increase of $p_d(t)$ is proportional to the departure of this moment from its equilibrium value, and the quantity $\tau$ is called the *relaxation time*, also referred to as the *collision time*.

Let us illustrate the meaning of (8.45) in a very simple situation. Suppose that a static field is applied at the instant $t = 0$. In that case, $p_{ds}(t) = \alpha_d \mathscr{E} = p_0$ ($p_0$ is the permanent moment of the molecule), because this is the value reached

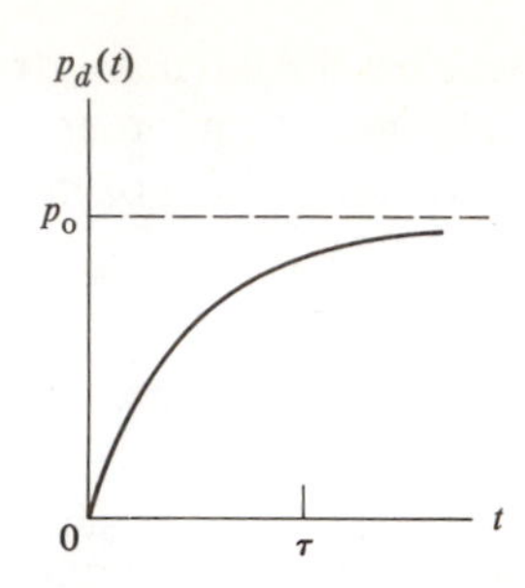


**Fig. 8.15** Instantaneous dipole moment $p_d(t)$ versus time $t$ in a static electric field.

by the moment long after the application of the field, where $\alpha_d$ is the static polarizability calculated in Section 8.5. Equation (8.45) now reduces to

$$\frac{dp_d}{dt} + \frac{p_d(t)}{\tau} = \frac{p_0}{\tau}, \tag{8.46}$$

which, as a first-order linear differential equation, can be readily solved, yielding

$$p_d(t) = p_0(1 - e^{-t/\tau}). \tag{8.47}$$

Thus the moment rises toward its equilibrium value in an exponential fashion, (Fig. 8.15), much like the direct-current rise in an R–L electrical circuit (of time constant $\tau$) when the battery has just been connected.

Suppose, on the other hand, that the medium has been placed in a static field for a sufficiently long interval for the moment to have achieved its equilibrium value $p_0$, and let this field be suddenly removed at $t = 0$. In applying (8.45), we now take $p_{ds} = 0$, since this is the equilibrium value, and the equation now leads to the solution

$$p_d(t) = p_0 e^{-t/\tau}, \tag{8.48}$$

showing that the moment *relaxes* to its equilibrium value of zero polarization exponentially, where the rate of relaxation is determined by the relaxation time $\tau$. The situation is the same as that of the current decay in an R–L circuit, of time constant $\tau$, when the switch has just been opened.

Let us now apply (8.45) to the case of an ac field

$$\mathscr{E}(t) = A\, e^{-i\omega t}. \tag{8.49}$$

The equilibrium moment is given by

$$p_{ds}(t) = \alpha_d(0)\mathscr{E}(t) = \alpha_d(0) A\, e^{-i\omega t}, \tag{8.50}$$

where $\alpha_d(0)$ is the static dipolar polarizability discussed in Section 8.5. Clearly the expression (8.50) is the value which would be reached by $p_d(t)$ if the field were to remain equal to $\mathscr{E}(t)$ at all subsequent times (that is, for $t' > t$). Equation (8.45) now reduces to

$$\frac{dp_d(t)}{dt} + \frac{p_d(t)}{\tau} = \frac{\alpha_d(0)}{\tau}\mathscr{E}(t). \tag{8.51}$$

Since the driving term on the right is varying harmonically in time, as indicated by (8.49), we try a solution of the form

$$p_d(t) = \alpha_d(\omega)\mathscr{E}(t) = \alpha_d(\omega)A\, e^{-i\omega t}, \tag{8.52}$$

where $\alpha_d(\omega)$ is, by definition, the ac polarizability. When this is substituted into (8.51), one readily arrives at

$$\alpha_d(\omega) = \frac{\alpha_d(0)}{1 - i\omega\tau}. \tag{8.53}$$

It can be seen that the ac polarizability is now a complex quantity, indicating that the polarization is no longer in phase with the field. This gives rise to energy absorption, as we shall see shortly.

To derive the corresponding expression for the dielectric constant $\epsilon_r(\omega)$, we write

$$\epsilon_r(\omega) = 1 + \chi_e(\omega) + \chi_d(\omega),$$

where $\chi_e(\omega)$ and $\chi_d(\omega)$ are the electronic and dipolar susceptibilities, respectively. We have assumed for simplicity that the ionic contribution is sufficiently small to be negligible, and we have also ignored the local field correction, i.e., we have used (8.18). Now in the frequency region in which dipolar dispersion is significant—i.e., the microwave region—the electronic susceptibility is constant because the electrons, being so light, can respond to the field essentially instantaneously. We may therefore write the above equation as

$$\epsilon_r(\omega) = n^2 + \chi_d(\omega), \tag{8.54}$$

where $n^2 = 1 + \chi_e$ is the optical dielectric constant and $n$ is the index of refraction.

The dipolar contribution $\chi_d(\omega) = \epsilon_r(\omega) - n^2$ does not follow the field instantaneously. There is a phase lag, as implied by the complex polarizability of (8.53). Since $\chi_d$ is proportional to $\alpha_d$ (see 8.20), it follows that $\chi_d(\omega)$ has the same complex form as $\alpha_d(\omega)$ in (8.53), and one may then write (8.54) in the form

$$\epsilon_r(\omega) = n^2 + \frac{\epsilon_r(0) - n^2}{1 - i\omega\tau}, \tag{8.55}$$

where the numerator on the right gives the static value of the dipolar susceptibility, that is, $\chi_d(0) = \epsilon_r(0) - n^2$. Equation (8.55) is the expression we have been seeking for the dielectric constant. This quantity is clearly frequency dependent, signifying that the medium exhibits *dispersion*.

This dielectric constant, being a complex quantity, can be written as

$$\epsilon_r(\omega) = \epsilon_r'(\omega) + i\epsilon_r''(\omega), \tag{8.56}$$

yielding for the real and imaginary parts

$$\epsilon_r'(\omega) = n^2 + \frac{\epsilon_r(0) - n^2}{1 + \omega^2\tau^2}, \tag{8.57a}$$

and

$$\epsilon_r''(\omega) = \frac{\epsilon_r(0) - n^2}{1 + \omega^2\tau^2}\,\omega\tau, \tag{8.57b}$$

which are known as *Debye's equations*.

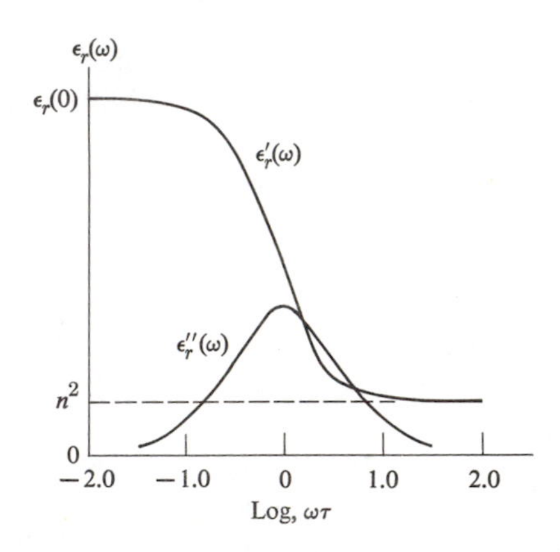


**Fig. 8.16** Real and imaginary parts $\epsilon_r'(\omega)$ and $\epsilon_r''(\omega)$ versus log $(\omega\tau)$ for a dipolar substance.

Figure 8.16 plots the components of the dielectric constant versus log $\omega\tau$. Note that the real part $\epsilon_r'(\omega)$ is a constant, equal to $\epsilon_r(0)$ for all frequencies at which $\omega \ll 1/\tau$ (the quantity $1/\tau$ is often called the *collision frequency*), a frequency range which usually covers all frequencies up to the microwave region. As the frequency increases to such an extent that $\omega \gtrsim 1/\tau$, the real part $\epsilon_r'(\omega)$ decreases, and eventually reaches the value $n^2$, the high-frequency dielectric constant. This confirms the statements made in Section 8.5.

Figure 8.16 also shows that the imaginary part, $\epsilon_r''(\omega)$, achieves its maximum, equal to $(\epsilon_r(0) - n^2)/2$, at the frequency $\omega = 1/\tau$, and decreases as the frequency departs from this value in either direction. The curve decreases to half its maximum value when

$$\omega\tau = (1 + \omega^2\tau^2)/4,$$

which gives the frequencies $\omega = 0.27/\tau$ and $\omega = 3.73/\tau$, the two values corresponding respectively to the low and high frequencies of the $\epsilon_r''(\omega)$ curve. The function $\epsilon_r''(\omega)$ is appreciable over a frequency range of more than one order of magnitude, the range being centered around the collision frequency $1/\tau$.

The rate of energy loss in the system may be calculated as follows: The polarization current density is

$$J = \frac{dP}{dt}, \tag{8.58}$$

and therefore the rate of joule heating per unit volume is given by

$$Q = J\mathscr{E}. \tag{8.59}$$

The polarization vector is given in terms of the dielectric constant by the relation

$$\begin{aligned} P(t) &= \epsilon_0 \left[\epsilon_r(\omega) - 1\right] \mathscr{E}(t) \\ &= \epsilon_0 \left[(\epsilon_r'(\omega) - 1) + i\epsilon_r''(\omega)\right] \mathscr{E}(t), \end{aligned} \tag{8.60}$$

which can also be written as

$$P(t) = \epsilon_0\, \epsilon_r^*(\omega)\, e^{i\phi}\, \mathscr{E}(t), \tag{8.61}$$

where $\epsilon_r^*(\omega) = [(\epsilon_r(\omega) - 1)^2 + \epsilon_r''^2(\omega)]^{1/2}$ and $\phi$ is an angle given by

$$\tan\phi = \frac{\epsilon_r''(\omega)}{\epsilon_r'(\omega) - 1}. \tag{8.62}$$

It is evident from (8.61) that the polarization lags behind the field by an angle $\phi$ (recall that $\mathscr{E}(t) \sim e^{-i\omega t}$).

The density of the polarization current is now given according to (8.58) and (8.61) by

$$\begin{aligned} J &= -\, i\omega\epsilon_0\epsilon_r^*(\omega)\, e^{i\phi}\, \mathscr{E}(t) \\ &= \omega\epsilon_0\epsilon_r^*(\omega)\, e^{i(\phi - \pi/2)}\, \mathscr{E}(t), \end{aligned} \tag{8.63}$$

which precedes the field by a phase angle $\phi' = (-\,\phi + \pi/2)$. [Draw the figure.] If we now substitute this value into (8.59) and determine the time average, we obtain

$$\begin{aligned} Q &= \tfrac{1}{2}\,|J|\,|\mathscr{E}|\cos\phi' \\ &= \tfrac{1}{2}\,\epsilon_0\omega\epsilon_r^*(\omega)\sin\phi\,|\mathscr{E}|^2 \\ &= \tfrac{1}{2}\,\epsilon_0\omega\epsilon_r''(\omega)\,|\mathscr{E}|^2, \end{aligned} \tag{8.64}$$

where we have used (8.62) in the last equation. Note that the loss rate is proportional to $\omega\,\epsilon_r''(\omega)$, that is, essentially to $\epsilon_r''(\omega)$. Thus the loss rate is greatest near the collision frequency.

Measuring the dielectric constant enables us to determine the relaxation time, as we have just seen. This time depends on the interaction between the dipolar molecule and the fluid in which it rotates. Debye has shown that, when we treat the surrounding medium as a viscous fluid, the relaxation time for a spherical molecule is given by

$$\tau = \frac{4\pi\eta R^3}{kT}, \tag{8.65}$$

where $\eta$ is the viscosity of the fluid and $R$ the radius of the molecule. For water at room temperature, $\eta \simeq 0.01$ poise, $R \simeq 2\text{Å}$, leading to $\tau \simeq 2.5 \times 10^{-11}$s, in approximate agreement with experiment.

The time $\tau$ increases as the temperature is lowered both because of $T$ in the denominator and because viscosity increases as temperature decreases. For example, the relaxation time in ice at $-20°$C is of the order of $10^{-7}$s, which is five orders of magnitude greater than the value at room temperature. Table 8.2 lists relaxation times for a few simple liquids at room temperature.

**Table 8.2**

Relaxation Times at 20°C

| Substance | $\tau$ |
|---|---|
| Water | $9.5 \times 10^{-11}$ |
| Alcohol | 13 |
| Chloroform | 7.5 |
| Acetone | 0.33 |
| Chlorobenzene | 0.12 |
| Toluene | 0.75 |
| *t*-butyl chloride | 0.48 |

The relaxation times in solids are much longer than in liquids, because the dipoles in solids are more rigidly constrained against rotation, as we shall see in Section 8.7.

## 8.7 DIPOLAR POLARIZATION IN SOLIDS

We derived the result (8.43) for dipolar polarizability on the basis of a model in which the molecular dipole moment may rotate continuously and freely, except for occasional collisions with the surrounding medium. Such a model is applicable in gases and liquids, but not in solids, because in solids the molecular moment does not rotate freely. It is constrained to a few *discrete* orientations determined by the interaction of this dipole with neighboring ones. A dipole may hop back and forth between these various discrete orientations in a manner which depends on the temperature and the electric field, but it is not *a priori* obvious that the resulting polarizability would be governed by an expression similar to (8.43). What is the actual behavior of the dipolar polarizability in a solid?

The answer depends on the particular solid and on the range of temperature. In some solids, dipolar moments seem indeed to be frozen in their orientations, and are unaffected by the field. In these solids, the dipolar polarizability vanishes altogether. In other solids, however, applying a field results in transitions between the orientations in such a manner as to result in a net polarization. One then often finds that the polarizability shows essentially the same behavior as (8.43).

Consider, for instance, the case of hydrogen sulfide ($H_2S$). The melting point of this substance is $T_m = 188°K$, yet, as Fig. 8.17 demonstrates, the dielectric constant continues to rise as the temperature is lowered, just as it does in the liquid state. The rise continues until a temperature $T_0 = 103°K$ is reached, at which the dielectric constant drops appreciably, from 20 to 3. Below this it remains constant. Although for the low-temperature range $T < T_0$ the dipoles indeed seem to be frozen, in the intermediate range $T_0 < T < T_m$ the dipoles are able to polarize, even though the substance is in the solid state. It is this ability to polarize that we now wish to explore.

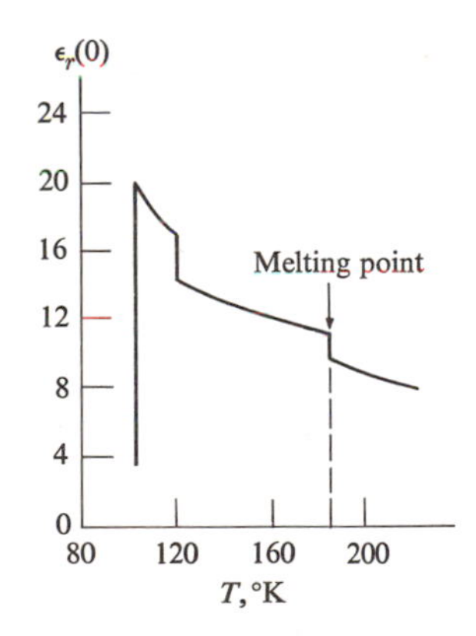


**Fig. 8.17** Static dielectric constant $\epsilon_r(0)$ for $H_2S$ versus temperature. [After Smyth and Walls]

Consider the following model which, despite its oversimplifications, illustrates the basic concepts involved. We assume that each dipole of the lattice has only two possible orientations, either to the right or to the left. The potential curve is shown in Fig. 8.18, in which the potential energy is plotted versus the orientation angle of a dipole. The bottom of the potential wells correspond to the two allowed orientations. Intermediate orientations are forbidden because of the high potential energy involved.

In the absence of an external field, the dipole is equally likely to point in the left or right direction, and as a result the net polarization is zero in this equilibrium situation. When a field is applied to the right, however, the well to the right is lowered by an amount $+\, p\mathscr{E}$, as shown by the dashed line in the figure, since it corresponds to a dipole orientation parallel to the field, that is, $\theta = 0$ in (8.5). At the same time, the well to the left is raised by an amount $p\mathscr{E}$, corresponding to $\theta = \pi$. The two wells are no longer equivalent, and since the left well is now higher, it is populated to a lesser extent than the right well. Hence the net polarization.

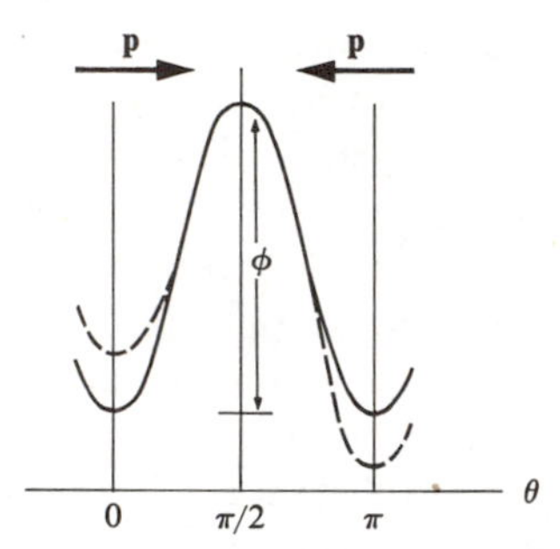


**Fig. 8.18** Potential of a dipole in a solid versus orientation angle $\theta$. The height of the barrier $\phi$ is called the activation energy. Solid curve represents the situation in absence of field; dashed curve the situation in presence of field.

When we denote the probability of the leftward orientation by $w$, it follows that

$$\frac{w}{1 - w} = e^{-2p\mathscr{E}/kT}, \tag{8.66}$$

where the term on the right is the Boltzmann factor, corresponding to a potential difference $2p\mathscr{E}$ (note that $1 - w$ is the probability of the rightward orientation). Solving for $w$, we find that

$$w = \frac{e^{-2p\mathscr{E}/kT}}{1 + e^{-2p\mathscr{E}/kT}},$$

which, in the condition $p\mathscr{E} \ll kT$ which usually prevails, reduces to

$$w \simeq \tfrac{1}{2} e^{-2p\mathscr{E}/kT}. \tag{8.67}$$

The net moment along the field direction, the $x$-direction, is

$$\bar{p}_x = p(1 - w) - pw = p(1 - 2w), \tag{8.68}$$

which, by use of (8.67), leads to

$$\bar{p}_x = p(1 - e^{-2p\mathscr{E}/kT}). \tag{8.69}$$

If one expands the exponential in powers of the field, retaining terms up to the first power only, which is justified insofar as $p\mathscr{E} \ll kT$, one finds

$$p_x = \frac{2p^2\mathscr{E}}{kT}, \tag{8.70}$$

leading to a dipolar polarizability

$$\alpha_d = \frac{2p^2}{kT}. \tag{8.71}$$

This, except for a numerical factor, is of the same form as the result (8.43) obtained on the basis of the model of continuous rotation.

The two-orientation model explains, in principle at least, the decrease in dipolar polarizability with temperature in $H_2S$ (Fig. 8.17). At low temperatures the field is able to orient all the dipoles to point to the right, but as the temperature increases the dipole can flip its orientation more readily (the necessary energy is supplied by thermal excitation), and the polarizability diminishes.

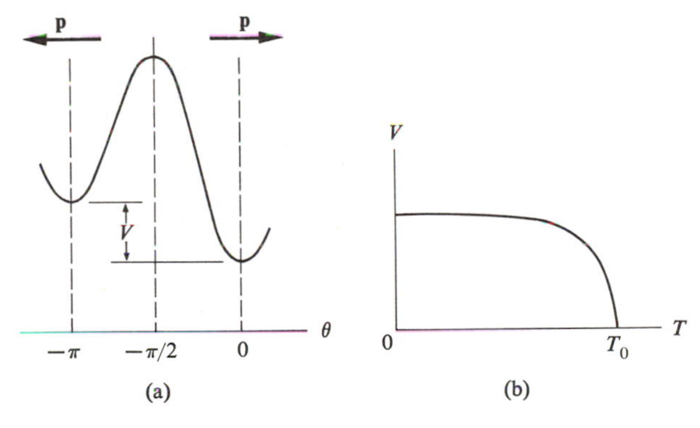


**Fig. 8.19** (a) Potential energy versus orientation angle $\theta$ in an asymmetric potential barrier. (b) Variation of potential $V$ with temperature.

The model we have used to describe the solid does not, however, explain the apparent freezing of the molecular dipoles for $T < T_0 = 103°\text{K}$ in Fig. 8.17, but this can be rectified by a slight change in the model. Suppose that the potential curve versus the orientation is as shown in Fig. 8.19(a). Here again the dipole has only two possible orientations, but the rightward orientation is favored because it is lower than the leftward by a potential $V$. If $V \gg kT$, then all the dipoles point to the right, in the absence of the field. Even when the field is applied, the dipoles remain frozen in their original orientation, unaffected by the field (unless the field is very strong).

To explain the behavior of $H_2S$, the potential must depend on the temperature in a manner somewhat like that shown in Fig. 8.19(b). The potential is large and constant at low temperature, but it vanishes as $T$ approaches and passes $T_0$.† In this manner, polarization is inhibited below the transition temperature $T_0$, but it is allowed for the range $T > T_0$.

The model we used in connection with Fig. 8.18 may also be used to study dielectric dispersion in solids. Thus the *jumping frequency* $\nu$ may be written

$$\nu = \nu_D e^{-\phi/kT}, \tag{8.72}$$

where $\nu_D$ is of the order of the Debye frequency, $\nu_D \simeq 10^{13}$ Hz, and $\phi$ is the activation energy‡ (see Fig. 8.18). The relaxation time (the jumping period) is therefore

$$\tau \simeq \frac{1}{\nu_D} e^{\phi/kT}, \tag{8.73}$$

which is to be used in conjunction with the dispersion equations (8.57) to describe dispersion in solids.

## 8.8 IONIC POLARIZABILITY

We turn now to ionic polarizability. We discussed this subject in Section 3.12 in connection with the optical properties of lattice vibrations, and therefore we shall be content here with quoting the results of that section, and with a brief discussion of their relation to our present purpose. We found there that the frequency-dependent dielectric constant is given by

$$\epsilon_r(\omega) = \epsilon_r(\infty) + \frac{\epsilon_r(0) - \epsilon_r(\infty)}{1 - (\omega^2/\omega_t^2)}, \tag{8.74}$$

where $\omega_t$ is the frequency of the optical phonon and $\epsilon_r(0)$, $\epsilon_r(\infty)$ are, respectively, the static dielectric constant and the dielectric constant at high frequency ($\omega \gg \omega_t$).

In (8.74) the first term on the right, $\epsilon_r(\infty)$, contains only the electronic polarizability, which is constant in the infrared region, where this expression is useful. The second term on the right is the ac polarizability, the quantity $[\epsilon_r(0) - \epsilon_r(\infty)]$ being the static ionic susceptibility, and the frequency dependence shown was derived in Section 3.12 from the equations of motion of the ions. We ignored the local field correction in (8.74), since in calculating the

† The dependence of the potential on temperature, shown in this figure, is not as arbitrary (or strange) as it may seem at first. Actually this potential is a "cooperative" interaction, due to all the dipoles in the substance. As the temperature rises more and more dipoles are able to flip over, and there are fewer and fewer dipoles in the original orientation which produces the restraining potential.

‡ The exponential increase of $\gamma$ with temperature, given in (8.72), is due to the fact that the dipole is able to flip only if the ion (or ions) involved has sufficient energy to go over the potential barrier $\phi$ in Fig. 8.18.

dielectric constant we have simply added the electronic and the ionic susceptibilities.

Equation (8.74) may also be rewritten in another form by recalling that $\epsilon_r(\infty) = n^2$, where $n$ is the optical index of refraction, and the result is

$$\epsilon_r(\omega) = n^2 + \frac{\epsilon_r(0) - n^2}{1 - (\omega^2/\omega_t^2)}. \tag{8.75}$$

The dielectric constant $\epsilon_r(\omega)$ is plotted versus $\omega$ in Fig. 8.20. For $\omega \ll \omega_t$, $\epsilon_r(\omega) = \epsilon_r(0)$, the static dielectric constant, which is expected, since at low frequency the ions are able to respond to the ac field essentially instantaneously. However, in the range $\omega \gg \omega_t$, $\epsilon_r(\omega) \simeq n^2$; the ionic contribution has vanished because the field now oscillates too rapidly for the massive ions to follow.

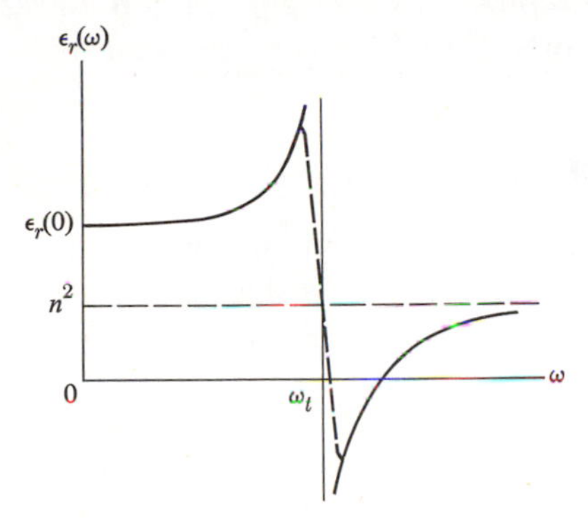


**Fig. 8.20** Dielectric constant $\epsilon_r(\omega)$ versus $\omega$, showing dispersion in infrared region due to optical phonons in an ionic crystal. Dashed curve indicates removal of divergence due to collisions of ions.

The optical dielectric constant $n^2$ may therefore be appreciably smaller than the static dielectric constant $\epsilon_r(0)$, due to the absence of the ionic contribution. In NaCl, for example, $n^2 = 2.25$, while $\epsilon_r(0) = 5.62$. Table 8.3 illustrates this point further for several alkali halide crystals.

**Table 8.3**

Static and Optical Dielectric Constants for Some Ionic Crystals

| Substance | $\epsilon_r(0)$ | $\epsilon_r(\infty) = n^2$ |
|---|---|---|
| LiF | 9.27 | 1.90 |
| LiC | 11.0 | 2.7 |
| NaCl | 5.62 | 2.32 |
| KCl | 4.64 | 2.17 |
| RbCl | 5.10 | 2.18 |

We note from Fig. 8.20 that the substance exhibits great dispersion near the optical phonon frequency $\omega_t$. This leads to strong optical absorption and reflection in the infrared region, as discussed in Section 3.12.

We also observe from Fig. 8.20 that the dielectric constant diverges at $\omega = \omega_t$. This divergence is attributable to the ionic susceptibility, and is expected since, as the signal frequency becomes equal to the natural frequency of the system $\omega_t$, a *resonance* condition is satisfied, and the response of the system becomes infinitely large. In practice such a divergence is not observed, of course, because of collisions experienced by the ions. These collisions arise from several mechanisms which cause scattering of the optical phonons in the crystal, e.g., anharmonic interaction, scattering by defects, etc., as discussed in Section 3.9. The effect of collision is to round off the dielectric constant, as indicated by the dashed line in Fig. 8.20, so that even though this constant is still quite large near the resonance frequency, the troublesome divergence has been removed.

## 8.9 ELECTRONIC POLARIZABILITY

Now that we have discussed dipolar and ionic polarizabilities, let us look at electronic polarizability and dispersion. We shall give a classical treatment first as a preliminary to the quantum discussion to follow.

### Classical treatment

To find the static polarizability, we assume that the electrons form a uniform, negatively charged sphere surrounding the atom. It can be shown through the laws of electrostatics that when a field $\mathscr{E}$ is applied to this atom, the nucleus is displaced from the center of the sphere by a distance

$$x = \left(\frac{4\pi\epsilon_0 R^3}{Ze}\right)\mathscr{E}, \tag{8.76}$$

where $R$ is the radius of the sphere (the atomic radius), and $Ze$ the nuclear charge (see the problem section). The atom is thus polarized, and the dipole moment, $p = Zex$, yields the electronic polarizability

$$\alpha_e = 4\pi\epsilon_0 R^3. \tag{8.77}$$

If we substitute the typical value $R = 10^{-10}$ m, we find that $\alpha_e \simeq 10^{-41}$ farad $\cdot$ m$^2$, in an order of magnitude which has actual polarizabilities given in Table 8.4.

To find the ac polarizability, we assume that the electrons in the atom experience an elastic restoring force corresponding to a resonant frequency $\omega_0$.†

† Although an electron interacts with a bare nucleus according to the Coulomb law, the classical screening of the nucleus by other electrons results in a harmonic like force between the electron and the nucleus.

**Table 8.4**

Electronic Polarizabilities for Some Inert Gases and Closed-Shell Alkali and Halogenic Ions (in units of $10^{-40}$ farad m$^2$).

| Inert gases | | Alkali cores | | Halogenic closed-shell | |
|---|---|---|---|---|---|
| He | 0.18 | $Li^+$ | 0.018 | $F^-$ | 0.76 |
| Ne | 0.35 | $Na^+$ | 0.20 | $Cl^-$ | 2.65 |
| Ar | 1.74 | $K^+$ | 0.86 | $Br^-$ | 3.67 |
| Kr | 2.2 | $Rb^+$ | 1.34 | $I^-$ | 5.5 |
| Xe | 3.6 | $Cs^+$ | 2.20 | | |

When the ac field is polarized in the $x$-direction, the appropriate equation of motion for the electron is

$$m\frac{d^2x}{dt^2} + m\omega_0^2 x = -e\mathscr{E}. \tag{8.78}$$

Assuming an ac field $\mathscr{E} = \mathscr{E}_0 e^{-i\omega t}$, one can readily solve for $x$ and the polarization. The polarizability is found to be

$$\alpha_e(\omega) = \frac{e^2/m}{\omega_0^2 - \omega^2}. \tag{8.79}$$

If there are $Z$ electrons per atom and $N$ atoms per unit volume, the resulting electric susceptibility is

$$\chi_e(\omega) = \frac{NZe^2/\epsilon_0 m}{\omega_0^2 - \omega^2}, \tag{8.80}$$

and the index of refraction is given by

$$n^2(\omega) = 1 + \frac{NZe^2/\epsilon_0 m}{\omega_0^2 - \omega^2}. \tag{8.81}$$

Figure 8.21 plots the function $n^2(\omega)$ versus $\omega$, and shows strong dispersion at the resonance frequency $\omega_0$. Such behavior is typical of all resonant systems, and reflects the strong interaction between the driving field and the system when the frequency-matching condition is satisfied, that is, when $\omega \simeq \omega_0$. The annoying divergence at $\omega = \omega_0$ can be removed by including a collision term in Eq. (8.78), as we did in Section 4.11. [Indeed, the results thus obtained should be the same as those in Section 4.11, if we set $\omega_0 = 0$, that is, if we treat the electrons as free particles.] Note that at high frequencies, that is, $\omega_0 \ll \omega$, $n^2(\omega) \to 1$, as for a vacuum, because at such high frequencies the electrons cannot follow the rapid oscillations of the field.

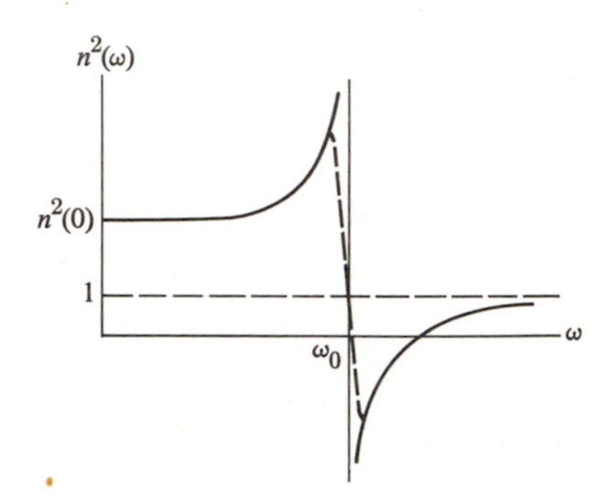


**Fig. 8.21** Square of index of refraction $n^2(\omega)$ versus frequency, illustrating dispersion in ultraviolet region due to motion of electrons.

### Quantum theory

The motion of an electron in an atom is governed by quantum laws, and hence an accurate treatment of electronic polarizability necessitates the use of quantum mechanics (a brief review of the subject is given in the Appendix). Suppose that the energy spectrum of an atom consists of two levels only, the ground state $E_0$ and the excited level $E_1$. It can then be shown (Van Vleck, 1932), that the electronic polarizability is given by

$$\alpha_e(\omega) = \frac{e^2}{m} \frac{f_{10}}{\omega_{10}^2 - \omega^2}, \tag{8.82}$$

where $\omega_{10} = (E_1 - E_0)/\hbar$, the Einstein frequency for the two levels, and $f_{10}$ is a quantity expressing the coupling between the two wave functions $\psi_0$ and $\psi_1$ by the incident electric field; $f_{10}$ is referred to as the *oscillator strength*, and is usually of the order of unity. Note that the quantum result (8.82) is quite similar to the classical expression (8.79). The static polarizability, $\alpha_e(0) = (e^2 f_{10}/m\omega_{10}^2)$ from (8.82), can also be similarly related to $\alpha_e$ of (8.77).

In an atom containing many excited levels, expression (8.82) is generalized to

$$\alpha_e(\omega) = \frac{e^2}{m} \sum_{j \neq 0} \frac{f_{j0}}{\omega_{j0}^2 - \omega^2}, \tag{8.83}$$

where $\omega_{j0} = (E_j - E_0)/\hbar$, and $j$ refers to the $j^{\text{th}}$ excited level. The system now has a number of resonance frequencies, and strong dispersion appears near each of them.

We can now see why $\alpha_e(\omega)$ is independent of temperature. Since $E_j - E_0$ is typically of the order of a few electron volts, the thermal energy $kT$ is too small to excite the electrons to the higher levels; thus in the absence of the field the electrons all lie in the ground level, which is the level to be used as the initial state in (8.83).

**Interband transition in solids**

The expression (8.83) for $\alpha_e(\omega)$ is applicable to a single, isolated atom. It is thus useful in the dielectric treatment of gases, since a gas may be considered as an aggregate of independent atoms. However, the result (8.83) is not applicable to a solid, since a solid's energy spectrum consists of continuous bands rather than discrete levels, and the electron states are represented by delocalized Bloch functions (Section 5.2) rather than localized atomic orbitals.

The quantum treatment which led to (8.83) can also be modified to yield the appropriate expression for the case of a solid. It is convenient to begin the discussion with $\epsilon_r''$, the imaginary component of the dielectric constant, which represents the absorption of the EM wave by the system, as discussed in Section 4.11. It can be shown (Greenway, 1968) that $\epsilon_r''$ is given by

$$\epsilon_r''(\omega) = \frac{A}{\omega^2} \int ds \frac{f_{cv}(\mathbf{k})}{\nabla[E_c(\mathbf{k}) - E_v(\mathbf{k})]}, \tag{8.84}$$

where $E_v(\mathbf{k})$ and $E_c(\mathbf{k})$ are the energies of the valence and conduction bands, respectively, and $\mathbf{k}$ is the wave vector of the electron which absorbs the photon and transfers from the valence to the conduction band. The integral in (8.84) is over a surface contour in the Brillouin zone which conserves the energy

$$E_c(\mathbf{k}) - E_v(\mathbf{k}) = \hbar\omega. \tag{8.85}$$

[The momentum conservation is guaranteed because $\mathbf{k}$ has the same value in both bands, as shown in Eq. (8.84). The photon's momentum is negligibly small (Section 3.4).] The quantity $f_{cv}(\mathbf{k})$ is the band-to-band oscillator strength, as in Eq. (8.82).

Figure 8.22 illustrates the application of (8.84) to a direct-gap semiconductor. The integration region consists of a sphere surrounding the origin, part of which is shown in the figure. It can be shown (see the problem section) that $E_c(\mathbf{k}) - E_v(\mathbf{k}) = E_g t\hbar^2 k^2/2\mu$, where $E_g$ is the energy gap and $\mu = m_e m_h/(m_e + m_h)$ is the electron–hole reduced mass. Substituting this into (8.84), and carrying out the integration, one finds

$$\epsilon_r''(\omega) = \frac{B}{\omega^2} (\hbar\omega - E_g)^{1/2}, \tag{8.86}$$

where $B = \pi(2\mu/h^2)^{3/2} f_{cv} A$. This expression is valid for $E_g < \hbar\omega$ [$\epsilon_r''(\omega) = 0$ for $\hbar\omega < E_g$, as discussed in Section 6.12], and shows that $\epsilon_r''(\omega)$ increases parabolically with $\omega$ near the absorption edge; that is, $\epsilon_r''(\omega) \sim (\hbar\omega - E_g)^{1/2}$, as noted in Section 6.12.

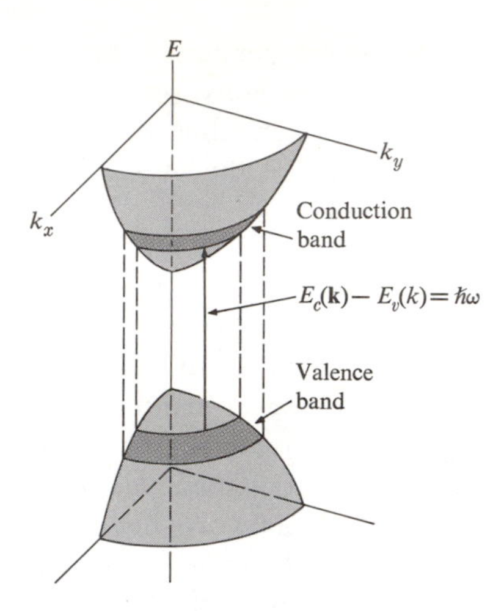


**Fig. 8.22** The various states in **k**-space involved in the absorption process at light frequency $\omega$.

When expression (8.84) is applied to bands of more complicated shapes, the integration may become exceedingly complex. In general, the integration contour is multiply connected, and consists of several distinct "pockets" in the Brillouin zone, each of which satisfies Eq. (8.85). But note also that the largest contribution comes from those points in the zone at which $E_c(\mathbf{k})$ and $E_v(\mathbf{k})$ have the same slope, because such points, known as the *critical points*, produce singularities in the integrand of Eq. (8.84).

Figure 8.23 shows $\epsilon_r''(\omega)$ for Ge, and correlates the various "shoulders" in the curve with the critical points responsible for the high absorption values. One can see that studies of optical absorption can be highly useful in the determination of band structure, and particularly in delineating the various critical points in the zone.

The real component of the dielectric constant $\epsilon_r'(\omega)$ describes the polarization aspects of the electronic system (Section 4.11). Although $\epsilon_r''$ and $\epsilon_r'$ describe physically distinct phenomena, they are, in fact, mathematically related by an important theorem known as the *Kramers–Kronig relation* (Brown, 1967). In particular, the static dielectric constant may be written as

$$\epsilon_r(0) = \epsilon_r'(0) = 1 + \frac{2}{\pi} P \int_0^\infty \frac{\epsilon_r''(\omega)}{\omega}\, d\omega, \tag{8.87}$$

where $P$ implies that the principal part of the integral is to be taken. Thus we may evaluate $\epsilon_r(0)$ by substituting $\epsilon_r''(\omega)$ from (8.84) and carrying out the frequency integration which illustrates that, like $\epsilon_r''(\omega)$, $\epsilon_r(0)$ is also directly dependent on the band structure of the solid. Note in particular that a significant correlation between $\epsilon_r(0)$ and the energy gap of the solid exists; since $\epsilon_r''(\omega) = 0$ for $\hbar\omega < E_g$,

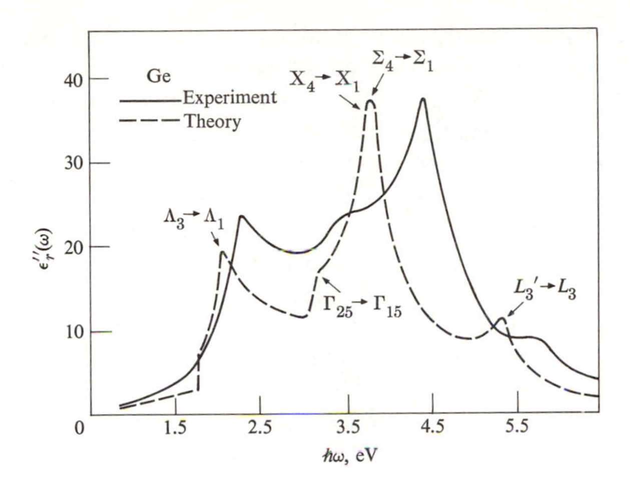


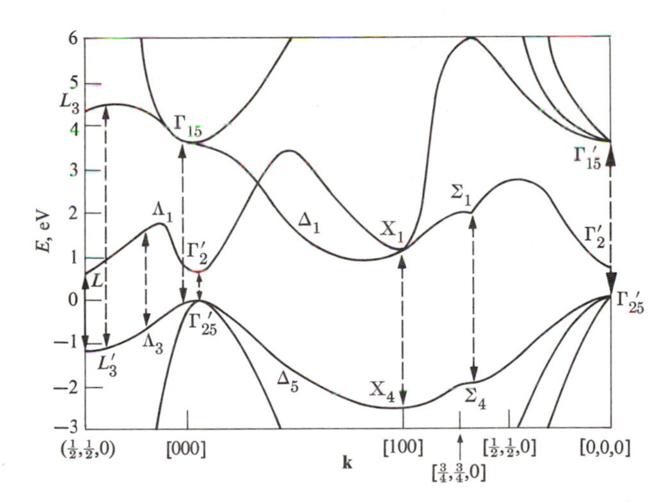


**Fig. 8.23** (a) Imaginary dielectric constant $\epsilon_r''(\omega)$ versus photon energy $\hbar\omega$ for Ge. (b) The band structure of Ge. Dashed arrows indicate various critical points. [After Phillips, 1966]

we may write Eq. (8.87) as

$$\epsilon_r(0) = 1 + \frac{2}{\pi} \int_{\omega_0}^{\infty} \frac{\epsilon_r''(\omega)}{\omega} \, d\omega, \tag{8.88}$$

where $\omega_0 = E_g/\hbar$ is the frequency at the absorption edge. Clearly, the smaller the gap, the smaller $\omega_0$, and the greater $\epsilon_r(0)$, because of the factor $\omega^{-1}$ in the integrand. This explains why $\epsilon_r(0) = 16$ in Ge, whose $E_g \simeq 1$ eV, while $\epsilon_r(0) = 5.6$ in NaCl, whose $E_g = 7$ eV.

Interband electronic polarizability and its associated dielectric constant are responsible for the optical properties of solids, particularly insulators and semiconductors, in the visible and ultraviolet ranges, because only such polarizability is effective at high frequency ranges. Also of importance in insulators and semiconductors is exciton absorption (Section 6.14).

As pointed out, the critical points assume a particularly significant role in the interpretation of interband-transition spectroscopic data. Since these points usually occur at symmetry points or along symmetry direction in the BZ, a knowledge of the interband energy difference $E_c(\mathbf{k}) - E_v(\mathbf{k})$ and the symmetry character (i.e., the location in the zone) of these points are highly useful in elucidating the band structure of the solid. Although the energy difference may be determined from the curve of $\epsilon_r''(\omega)$ versus $\omega$ (for example, Fig. 8.23), the accuracy is limited due to the background absorption associated with the noncritical regions of the zone. A special technique, known as *modulation spectroscopy*, has been developed in recent years to overcome this difficulty. The technique consists basically of devising an experimental procedure for extracting the first (or higher) derivative, $d\,\epsilon_r''(\omega)/d\omega$, as a function of $\omega$. The reader can readily see that one can locate the critical points more readily on the derivative curve than on the original curve. Experimentally, this is achieved by superposing on the solid, in addition to the signal, an external time-dependent perturbation varying with a modulation frequency $\omega_m$, and measuring the relative change in the dielectric function $\Delta\epsilon_r''/\epsilon_r''$ induced by the perturbation. Many different types of perturbations have been used, e.g., temperature and hydrostatic pressure. The symmetry character of the critical point is determined by applying a vector perturbation, like an electric field, or a tensor perturbation as a uniaxial pressure. For a brief review, see J. E. Fischer and D. E. Aspner, *Comments on Solid State Physics*, IV, 131; IV, 159. For a thorough treatment, see M. Cardona, 1968, *Modulation Spectroscopy*, New York, Academic Press.

## 8.10 PIEZOELECTRICITY

In this and the following sections we turn to certain phenomena associated with ionic polarization. The term *piezoelectricity* refers to the fact that, when a crystal is strained, an electric field is produced within the substance. As a result of this field, a potential difference develops across the sample, and by measuring this potential one may determine the field. The inverse effect—that an applied field produces strain—has also been observed. (It was discovered in about 1880.)

The piezoelectric effect is very small. A field of 1000 V/cm in quartz produces a strain of only $10^{-7}$. That is, a rod 1 cm long changes its length by 10Å. Conversely, even small strains can produce enormous electric fields.

The piezoelectric effect is often used to convert electrical energy into mechanical energy, and vice versa; i.e., the substance is used as a *transducer*. For instance, an electric signal applied to the end of a quartz rod generates a mechanical strain, which consequently leads to the propagation of a mechanical wave—a sound wave—down the rod. (One can reconvert the mechanical energy into electrical energy at the other end of the rod, if desired, by picking up the electric field produced there.) Quartz is the most familiar piezoelectric substance, and the one most frequently used in transducers.

The microscopic origin of piezoelectricity lies in the displacement of ionic charges within the crystal. In the absence of strain, the distribution of the charges at their lattice sites is symmetric, so the internal electric field is zero. But when the crystal is strained, the charges are displaced. If the charge distribution is no longer symmetric, then a net polarization, and a concomitant electric field, develops. It is this field which operates in the piezoelectric effect.

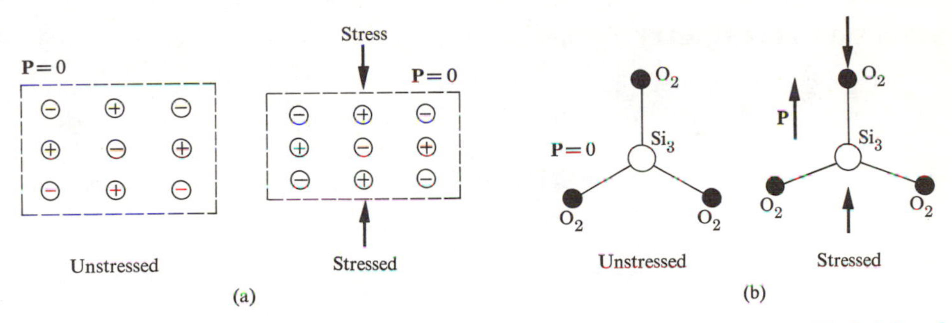


**Fig. 8.24** Crystal with center of inversion exhibits no piezoelectric effect. (b) Origin of piezoelectric effect in quartz: crystal lacks a center of inversion.

It follows that a substance can be piezoelectric only if the unit cell *lacks a center of inversion*. Figure 8.24(a) shows this, and demonstrates that if a center of inversion *is* present, it persists even after distortion, and consequently the polarization remains zero. However, when there is no center of inversion, as in Fig. 8.24(b), distortion produces a polarization. We can now understand, for example, why no regular cubic lattice can exhibit piezoelectricity.

**Table 8.5**

Some Piezoelectric Crystals (in Decreasing Value of Piezoelectric Coefficient)

| Crystal | Chemical formula | Relative strength |
|---|---|---|
| Rochelle salt | $NaKC_4H_4O_6 \cdot 4H_2O$ | Very strong |
| ADP | $NH_4H_2PO_4$ | Strong |
| KDP | $KH_2PO_4$ | Moderate |
| α-Quartz | $SiO_2$ | Weak |

Of the 32 crystal classes, 20 are noncentrosymmetric, and these are candidates for piezoelectric materials. The lack of inversion center, however, is not sufficient to guarantee piezoelectricity, and only relatively few substances, some of which are listed in Table 8.5, exhibit this phenomenon.

Another common application of piezoelectrics, in addition to their use in transducers, is in delay lines. When an electric signal is converted into a mechanical wave, it travels through a quartz rod at the velocity of sound, which, since it is much less than the velocity of light, leads to considerable delay of the signal .[Also piezoelectrics and related electro-optic crystals are now widely used in the fields of laser technology and modern optics. For instance, the cavity length of a laser may be varied continuously in a controlled manner by the application of a voltage to a piezoelectric crystal situated at one end of the cavity.]

## 8.11 FERROELECTRICITY

We have often commented that ionic susceptibility is not sensitive to variations in temperature. Although this is true for most substances, there is a class of materials which exhibits a marked departure from this rule: the *ferroelectric* materials. In these substances, the static dielectric constant changes with temperature according to the relation

$$\epsilon_r = B + \frac{C}{T - T_C}, \qquad T > T_C, \tag{8.89}$$

where $B$ and $C$ are constants independent of temperature. This relation is known as the *Curie–Weiss law*, and the parameters $C$ and $T_C$ are called the *Curie constant* and *Curie temperature*, respectively.

This behavior is valid in the temperature range $T > T_C$. In the range $T < T_C$, the material becomes *spontaneously* polarized, i.e., an electric polarization develops in it without the help of an external field. (This phenomenon is analogous to the spontaneous magnetization which takes place in ferromagnetic materials.)

A *phase transition* occurs at the temperature $T_C$. Above the transition temperature, the substance is in the *paraelectric phase*, in which the elementary dipoles of the various unit cells in the crystal are oriented randomly. The dielectric constant is given by (8.89), whose form is illustrated in Fig. 8.25a.

Below the transition temperature, the elementary dipoles interact with each

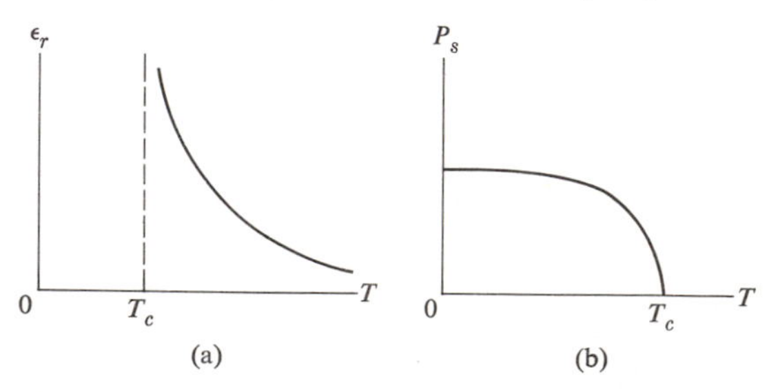


**Fig. 8.25** (a) Dielectric constant $\epsilon_r$ versus $T$ in a ferroelectric substance. (b) Spontaneous polarization $P_s$ versus $T$ in a ferroelectric substance.

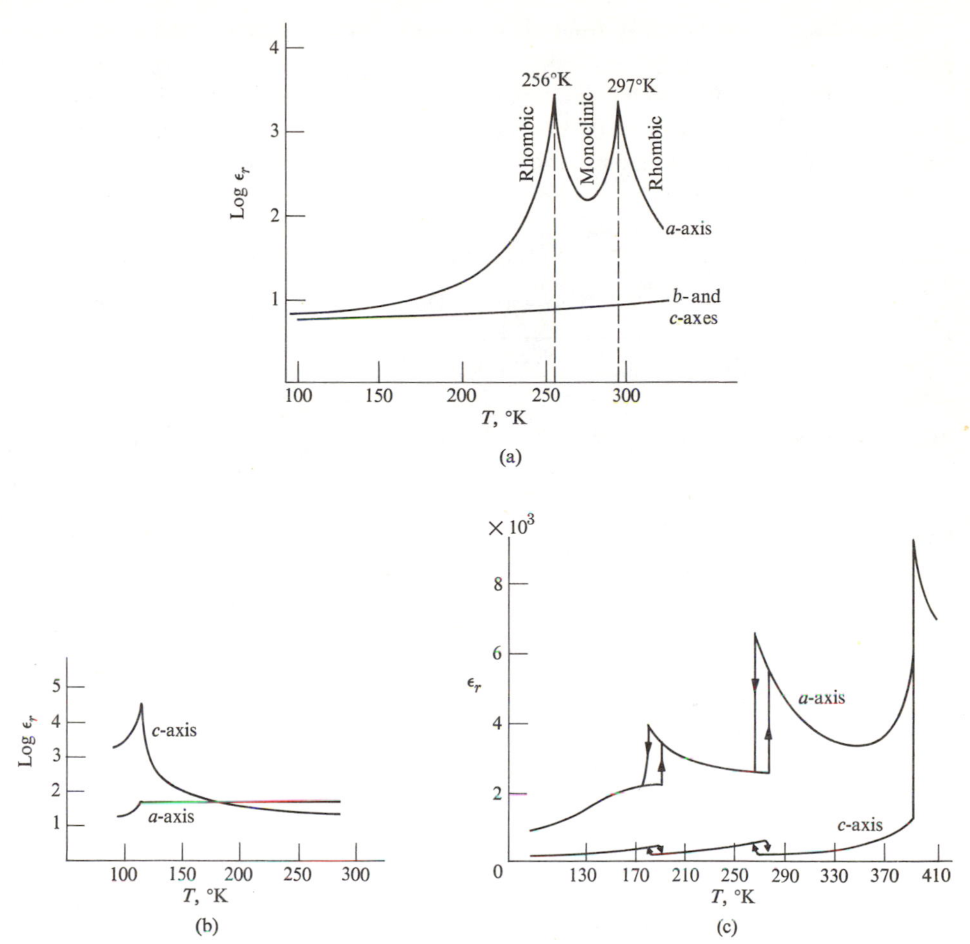


**Fig. 8.26** (a) Log $\epsilon_r$ versus $T$ for Rochelle salt along $a$-, $b$- and $c$-axes. (After Halblützel) (b) Log $\epsilon_r$ versus $T$ for KDP along $a$- and $c$-axes. (After Busch) (c) $\epsilon_r$ versus $T$ for $BaTiO_3$. (After Merz)

other, and this gives rise to an internal field, which lines up the dipoles. The direction of this field and the associated polarization lie in a certain favorable orientation in the crystal. Figure 8.25(b) shows the variation of the spontaneous polarization $P_s$ with temperature for $T < T_C$. This polarization increases gradually as the temperature is lowered.

The second term in (8.89) is usually much larger than the first. Thus, although typically $B \simeq 5$, $\epsilon_r \simeq 1000$ or even larger near the transition temperature. We may therefore ignore $B$, and write to a good approximation

$$\epsilon_r = \frac{C}{T - T_C}. \tag{8.90}$$

There are three major ferroelectric groups: The Rochelle salt group, the KDP (potassium dihydrogen phosphate) group, and the perovskites group, headed by barium titanate. Table 8.6 gives data on these substances, and Fig. 8.26 presents the variation of temperature of the dielectric constants. Note in particular the enormous value of the dielectric constant in barium titanate, for which $\epsilon_r \simeq 10^5$ near the transition temperature.

**Table 8.6**

Ferroelectric Data

| Crystal | Chemical formula | $T_C$(°K) | $C$, °K | $P_s$, coul/m² | |
|---|---|---|---|---|---|
| Rochelle-salt group | $NaK(C_4H_4O_6) \cdot 4H_2O$ | 297 (upper)<br>255 (lower) | 178 | $267 \times 10^{-5}$ | [at 278°K] |
| | $LiNH_4(C_4H_4O_6 \cdot H_2O$ | 106 | | 220 | [95] |
| KDP group | $KH_2PO_4$ | 123 | 3100 | 5330 | [96] |
| | $KD_2PO_4$ | 213 | | 9000 | — |
| | $RbH_2PO_4$ | 147 | | 5600 | [90] |
| | $CsH_2AsO_4$ | 143 | | — | — |
| Perovskites | $BaTiO_3$ | 393 | $1.7 \times 10^5$ | 26,000 | [296] |
| | $SrTiO_3$ | 32 | | 3000 | [4] |
| | $WO_3$ | 223 | | — | — |

**The microscopic model**

Let us now inquire into the microscopic source of ferroelectricity. The most obvious explanation is to assume a dipolar substance and use the Lorentz local field correction obtained in (8.30). This leads to a dielectric constant

$$\epsilon_r = \frac{1 + \frac{2}{3}\chi}{1 - \frac{1}{3}\chi}. \tag{8.91}$$

If we set $\chi \simeq \chi_d$, thus neglecting electronic and ionic contributions, which is appropriate in dipolar substances, and substitute for $\chi_d$ from (8.43), we find that

$$\epsilon_r = \frac{T + C}{T - T_C}, \tag{8.92}$$

where $C = 2Np^2/9\epsilon_0 k$ and $T_C = Np^2/9\epsilon_0 k$. If we ignore the term $T$ in the numerator, it is evident that Eq. (8.92) has the same form as (8.90). In particular, the dipolar model predicts that $\epsilon_r$ diverges as $T$ approaches $T_C$ from above, and consequently one expects the system to become unstable and make a transition to a new phase, the "ferroelectric phase." The divergence at $T_C$ is referred to as the *polarization catastrophe*.

Despite the fact that the dipolar model seems to lead naturally to ferroelectricity, the model is inadequate to account for observations. If we apply this model to water, for instance, for which $N \simeq \frac{1}{3} \times 10^{29}\ \mathrm{m}^{-3}$ and $p = 0.62$ debye, it predicts that water would become ferroelectric at $T_C = 1100°\mathrm{K}$. In fact, however, water never becomes ferroelectric, not even below its freezing point.

Another fact which underscores the failure of the model is its prediction that any dipolar substance should become ferroelectric at a sufficiently low temperature. Instead, however, all known ferroelectrics are *nondipolar* in nature. We must therefore look elsewhere for the explanation of ferroelectricity.

Ferroelectricity is associated with ionic polarizability. To see this, let us consider an ionic substance. The ac dielectric constant is given by

$$\epsilon_r(\omega) = n^2 + \frac{A}{\omega_t^2 - \omega^2}, \tag{8.93}$$

where we have used (8.74), and denoted $\chi_i(0)\omega_t^2$ by the constant $A$. The static dielectric constant, according to (8.93), is given by

$$\epsilon_r(0) = n^2 + \frac{A}{\omega_t^2}. \tag{8.94}$$

This expression shows that $\epsilon_r(0)$ increases as $\omega_t$ decreases, and indeed $\epsilon_r(0)$ diverges as $\omega_t \to 0$.

But why should $\omega_t$ decrease? We shall now show that the inclusion of the local field does indeed lead to a reduction in the value of this frequency. According to Eqs. (3.83 and 3.84), the transverse motion for the unit cell is governed by the equation

$$\mu \frac{d^2u}{dt^2} + 2\beta u = 0, \tag{8.95}$$

where $\mu$ is the reduced mass of the unit cell, $u$ the relative displacement between the ions, and $\beta$ the force constant between the ions† (Section 3.6). This expression leads to a mode of oscillation with a frequency

$$\omega_t^2 = \frac{2\alpha}{\mu}, \tag{8.96}$$

which is the frequency of the long-wavelength optical phonon.

† The force constant is denoted here by $\beta$ rather than $\alpha$, as in Chapter 3, in order to avoid any confusion with polarizability.

The equation of motion (8.95), however, requires modification if we consider the local field correction, because there is a polarization $P = Ne^*u$ associated with the displacement $u$, and hence a Lorentz electric field

$$\mathscr{E} = \frac{P}{3\epsilon_0} = \frac{Ne^*u}{3\epsilon_0}, \tag{8.97}$$

where $e^*$ is the effective charge on the ion. Because of this field there is now an electric force acting on the unit cell given by $2e^*\mathscr{E}$, which modifies the equation of motion (8.95) to

$$\mu \frac{d^2u}{dt^2} + 2\beta u = 2e^*\mathscr{E}.$$

If one substitutes for $\mathscr{E}$ from (8.97), and rearranges the equation, one finds that

$$\mu \frac{d^2u}{dt^2} + \left(2\beta - \frac{2Ne^{*2}}{3\epsilon_0}\right) u = 0,$$

which is the equation for a harmonic oscillator of frequency $\omega_t^*$ given by

$$\omega_t^{*2} = \frac{2\beta}{\mu} - \frac{2Ne^*}{3\epsilon_0\mu}$$

or

$$\omega_t^{*2} = \omega_t^2 - \frac{2Ne^*}{3\epsilon_0\mu}, \tag{8.98}$$

where we have used Eq. (8.96).

The frequency $\omega_t^*$ is less than $\omega_t$, the frequency obtained by neglecting the local field. It is easy to see the reason for this reduction: When the lattice is displaced, a local field is created in the same direction as $u$. The effect of this field is to reduce the restoring force, and consequently the oscillator frequency. The origin of the force constant $\beta$ lies in the short-range elastic forces between the ions, while the local field is due to the familiar long-range Coulomb forces between these ions.

The expression (8.94) for the dielectric constant should now be replaced by

$$\epsilon_r(0) = n^2 + \frac{A}{\omega_t^{*2}}. \tag{8.99}$$

The effect of the local field is to increase the dielectric constant. If the second term on the right of (8.98) is large enough to cancel the first term, then $\omega_t^* \to 0$, and the dielectric constant becomes infinite. What happens, in fact, is that the system feels the *instability* and makes an adjustment to avoid the divergence, i.e., undergoes a transition to the ferroelectric phase. It is thus expected that the system would also undergo a simultaneous transition into a more stable crystal structure. This is indeed found to be the case in all ferroelectric transitions.

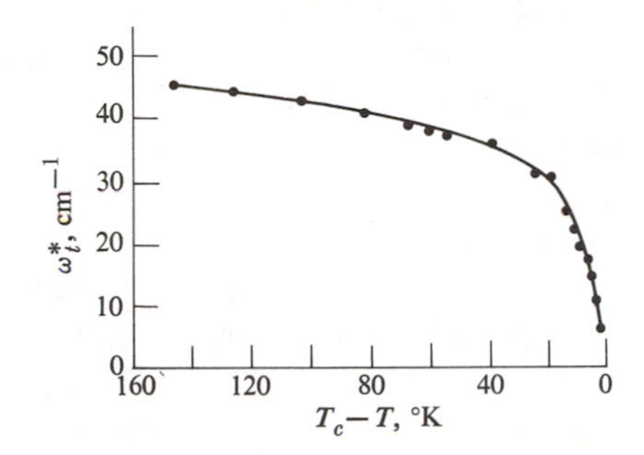


**Fig. 8.27** Transverse frequency $\omega_t^*$ versus ($T_C - T$) in antimony sulphoiodide (SbSI). (After Perry and Agrawal, *Solid State Comm.* **8**, 225, 1970)

Figure 8.27 illustrates the observed decrease in phonon frequency as the temperature approaches the Curie temperature. Note that the frequency here is about 10 cm$^{-1}$, or $\nu = 3 \times 10^9$ Hz, considerably smaller than a typical optical phonon frequency of $10^{13}$ Hz.†

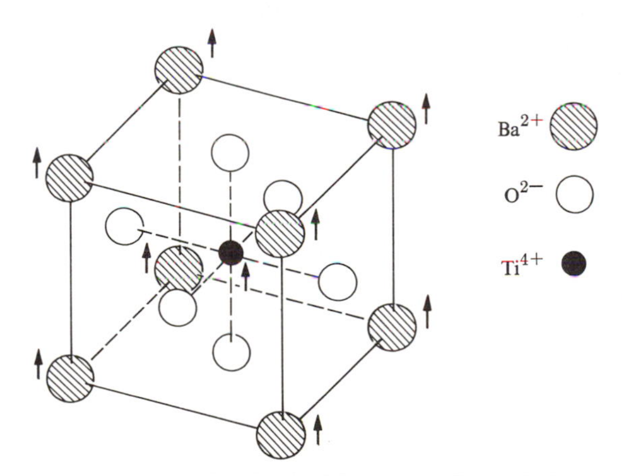


**Fig. 8.28** Structure of $BaTiO_3$ in cubic phase (above $T_C$).

As a concrete example of ferroelectric structure transformation, Fig. 8.28 shows the appropriate structure for $BaTiO_3$. Above the Curie temperature the structure is cubic, but as the temperature is lowered to $T_C$, the $Ba^{2+}$ and $Ti^{4+}$ ions are displaced as shown, producing a slightly compressed cubic structure. Although the displacement is small—only about 0.15 Å—it is enough to give the

† The mode whose frequency vanishes at the Curie temperature is called the *soft* mode.

observed polarization. It is this relative displacement of the internal structure which gives the model its name: the *displacive model.*

We have shown that the Lorentz calculation of the local field is misleading when applied to dipolar substances; yet we have used the same procedure for evaluating this field when it is associated with the ionic polarization. There is no contradiction here, because Onsager showed, many years ago, that while the Lorentz procedure is valid in evaluating the field associated with electronic and ionic polarizabilities, the procedure is inapplicable when one is dealing with orientational polarizability. Onsager demonstrated that the actual local field associated with the dipolar polarizability is much smaller than that provided by the Lorentz procedure, and it is this overestimation which leads to the erroneous conclusions concerning ferroelectricity. You can find a detailed discussion of this point in Frölich (1958).

Ferroelectricity, like piezoelectricity, can occur only in noncentrosymmetric crystals. The requirements of ferroelectricity are, however, more stringent, requiring the existence of a favorable axis of polarity. Only 10 crystal classes have sufficiently low degrees of symmetry to permit the occurrence of ferroelectricity.

### Ferroelectric domains

A substance which is in its ferroelectric phase undergoes spontaneous polarization, but the *direction* of the polarization is not the same throughout the sample. The material is divided into a number of small *domains*, in each of which the polarization is constant. But the polarization in the different domains are different, so that the net total polarization of the whole sample vanishes in the equilibrium situation (Fig. 8.29).

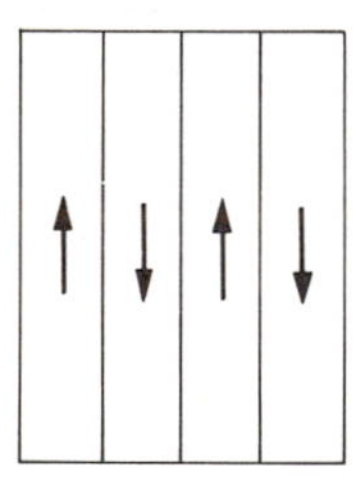

**Fig. 8.29** Domain structure in an unpolarized ferroelectric sample.

When an external field is applied, the domains whose polarization is parallel to the field grow, while the domains of opposite polarization shrink. These growing and shrinking processes continue as the field increases until, at a sufficiently high field, the whole of the sample is polarized parallel to the field.

We shall discuss the concept of domains, and the associated hysteresis loop, in detail in connection with ferromagnetic materials (Section 9.11).

## SUMMARY

### The dielectric constant and molecular polarizability

The dielectric constant $\epsilon$ is defined by the equation

$$\mathbf{D} = \epsilon\mathcal{E},$$

where $\mathbf{D}$ is the electric displacement and $\mathcal{E}$ the average field inside the dielectric. In terms of the polarization $\mathbf{P}$, the displacement vector $\mathbf{D}$ is

$$D = \epsilon_0\mathcal{E} + \mathbf{P}.$$

The polarization $\mathbf{P}$ arises as a result of the polarization of the molecules, and is given by

$$\mathbf{P} = N\mathbf{p},$$

where $N$ is the concentration of molecules and $\mathbf{p}$ the electric moment of each of these molecules. The electric moment is proportional to the field, and is given by

$$\mathbf{p} = \alpha\mathcal{E},$$

where $\alpha$ is the *molecular polarizability*. Substituting this into the above equations, we may express the relative dielectric constant in terms of the polarizability,

$$\epsilon_r = 1 + (N\alpha/\epsilon_0).$$

This result, which ignores the local-field correction, holds well in gases. In liquids and solids, however, the local-field correction is appreciable, and must be included. We then find the local field to be

$$\mathcal{E}_{\text{loc}} = \mathcal{E} + (\tfrac{1}{3}\epsilon_0)\,\mathbf{P},$$

which leads to the Clausius–Mosotti relation,

$$\frac{\epsilon_r - 1}{\epsilon_r + 2} = \frac{N\alpha}{3\epsilon_0}.$$

### Dipolar polarizability

Molecular polarizability is, in general, the additive result of dipolar, ionic, and electronic contributions. Statistical treatment of dipolar polarization gives the following expression for dipolar polarizability,

$$\alpha_d = p^2/3kT,$$

which decreases as the inverse of the temperature. The dielectric constant is

$$\epsilon_r = 1 + N\alpha_{ei}/\epsilon_0 + Np^2/3\epsilon_0 kT.$$

By plotting $\epsilon_r$ versus $1/T$, one may determine both the permanent moment $p$ and the electronic-ionic polarizability $\alpha_{ei}$. This information sheds light on the geometrical structure of the molecules.

The ac dipolar polarizability may be calculated by assuming that the dipole does not follow the field instantaneously, but with a certain relaxation time $\tau$. One then finds the frequency-dependent dielectric constant

$$\epsilon_r(\omega) = n^2 + \frac{\epsilon_r(0) - n^2}{1 - i\omega\tau},$$

where $n$ is the optical index of refraction and $\epsilon_r(0)$ the static dielectric constant. As the frequency $\omega$ increases from the range $\omega \ll 1/\tau$ to the range $1/\tau \ll \omega$, the dipolar contribution decreases from the value $[\epsilon_r(0) - n^2]$ to 0, because at high frequencies the dipoles no longer follow the field. The imaginary part of the dielectric constant is related to the energy absorbed by the dielectric from the field.

**Ionic polarizability**

Ionic crystals exhibit dispersion in the infrared region, as a result of the strong interaction of the electromagnetic wave with the optical phonons of the substance. The dielectric constant is

$$\epsilon_r(\omega) = n^2 + \frac{\epsilon_r(0) - n^2}{1 - (\omega^2/\omega_t^2)},$$

where $\omega_t$ is the optical phonon frequency. As $\omega$ varies from the range $\omega \ll \omega_t$ to the range $\omega_t \ll \omega$, the ionic contribution decreases from $[\epsilon_r(0) - n^2]$ to 0, because the ions no longer follow the field at high frequencies.

**Electronic polarizability**

A simplified classical treatment of static electronic polarizability yields

$$\alpha_e = 4\pi\epsilon_0 R^3,$$

where $R$ is the atomic radius. The classical ac electronic polarizability, obtained by treating the electron as a classical particle bound to the remainder of the atom by a harmonic force, is

$$\alpha_e(\omega) = \frac{e^2/m}{\omega_0^2 - \omega^2},$$

where $\omega_0$ is the natural oscillation frequency of the bound electron. This yields an optical dielectric constant $n^2$ given by

$$n^2 = 1 + \frac{NZe^2/\epsilon_0 m}{\omega_0^2 - \omega^2}.$$

Quantum treatment leads to a similar result.

In solids, dielectric and optical properties are related directly to the structure of the energy band of the substance.

### Piezoelectricity

In noncentrosymmetric ionic crystals, the mechanical straining of a substance produces an internal electric field, and vice versa. This property is widely utilized in transducers, i.e., devices which convert electrical into mechanical energy, and vice versa.

### Ferroelectricity

A ferroelectric substance is one which exhibits spontaneous polarization below a certain temperature. Above this *Curie temperature* $T_C$ the dielectric constant is given by the *Curie–Weiss* law,

$$\epsilon_r = B + \frac{C}{T - T_C}.$$

The ferroelectric property can be explained by the *displacive model*: As the temperature approaches $T_C$ from above, one of the optical phonon modes becomes so soft—due to the local-field correction—that $\epsilon_r \to \infty$, causing a structural phase transition and a concomitant spontaneous polarization.

## REFERENCES

### General references

J. Birks, editor, 1959–1961, etc., *Progress in Dielectrics* (series), New York: John Wiley; Academic Press

C. J. F. Böttcher, 1952, *Theory of Electric Polarization*, Amsterdam: Elsevier

F. C. Brown, 1967, *The Physics of Solids*, New York: Benjamin

W. F. Brown, Jr., 1956, *Encyclopedia of Physics*, Volume 17, New York: Springer-Verlag

H. Fröhlich, 1958, *Theory of Dielectrics*, second edition, Oxford: Oxford University Press

J. C. Slater, 1967, *Insulators, Semiconductors and Metals*, New York: McGraw-Hill

C. P. Smyth, 1955, *Dielectric Behavior and Structure*, New York: McGraw-Hill

J. H. Van Vleck, 1932, *Theory of Electric and Magnetic Susceptibilities*, Oxford: Oxford University Press

A. R. Von Hipple, 1954, *Dielectrics and Waves*, New York: John Wiley

### Dipolar polarizability

P. Debye, 1945, *Polar Molecules*, New York: Dover

See also books cited under General References by Böttcher, Brinks, Fröhlich, and Smyth.

### Electronic interband transitions, excitons

M. Cordona, 968, "Electronic Optical Properties of Solids," in *Solid State Physics, Nuclear Physics and Particle Physics*, I. Saavedra, editor, New York: Benjamin

D. L. Greenway and G. Harbeke, 1968, *Optical Properties and Band Structure of Semiconductors,* New York: Pergamon Press

R. S. Knox, 1963, "Theory of Excitons," in *Solid State Physics,* Supplement 5, New York: Academic Press

H. R. Phillips and H. Ehrenreich, 1967, "Ultraviolet Optical Properties," in *Semiconductors and Semimetals*, R. K. Willardson and A. C. Beer, editors, New York: Academic Press

*Note:* Other relevant review articles in this volume are by M. Cordona, E. J. Johnson, J. O. Dimmock, H. Y. Fan, B. O. Seraphin, and H. E. Bennet.

J. C. Phillips, 1966, "The Fundamental Optical Spectra of Solids," in *Solid State Physics*, Volume 18, New York: Academic Press

F. Wooten, 1972, *Optical Properties of Solids*, New York: Academic Press

### Piezoelectricity and ferroelectricity

J. C. Burfoot, 1967, *Ferroelectrics*, New York: Van Nostrand

W. G. Cady, 1947, *Piezoelectricity*, New York: McGraw-Hill

W. Cochran, 1960, "Crystal Stability and the theory of Electricity," in *Advances in Physics*, Volume 9, p. 387; *ibid.*, Volume 10, p. 401 (1961)

P. W. Forsbergh, 1956, "Piezoelectricity, Electrostriction and Ferroelectricity," in *Encyclopedia of Physics*, Volume 17

## QUESTIONS

1. Let $A$ and $B$ refer to two different atoms. Using symmetry arguments, determine whether the following types of molecules are dipolar or not: $AA$, $AB$, $ABA$ (rectilinear arrangement), $ABA$ (triangular arrangement), $AB_3$ (planar arrangement with $A$ at center of triangle), $AB_4$ (tetrahedral arrangement). Give one example of each type.
2. The static dielectric constant of water is 81, and its index of refraction 1.33. What is the percentage contribution of ionic polarizability?
3. For a typical atom, estimate the field required to displace the nucleus by a distance equal to 1% of the radius. [Refer to Eq. (8.79).]
4. Explain physically why ionic polarizability is rather insensitive to temperature. Do you expect a slight change in temperature to lead to an increase or a decrease in the polarizability as $T$ rises? Explain.
5. Referring to Table 6.4, one notes that the polarizabilities of the alkali ions are consistently lower than those of the halide ions. Give a physical, i.e., qualitative, explanation of this fact.
6. In the classical treatment of electronic ac polarizability, the restoring force on the electron is assumed to have a harmonic form. How do you justify this in view of the fact that the force due to the nucleus has a coulomb form which is very different from the harmonic form? Give an expression for the natural frequency $\omega_0$ in terms of the properties of the atom.
7. If one sets $\omega_0$ equal to zero in (8.85), one obtains the same electron dielectric constant found in Section 4.11. Explain why.

8. Suppose that a light beam passing through a semiconductor is absorbed either by electrons excited from the valence band to the conduction band (fundamental absorption), or by excitons. Describe an experimental electrical procedure for testing which of these two mechanisms is the operative one.

## PROBLEMS

1. Using Coulomb's law, derive the expression (8.2) for the field of an electric dipole. Assume that $d \ll r$.
2. a) Derive Eq. (8.3), that is, show that the torque exerted on a dipole $\mathbf{p}$ by a uniform field $\mathscr{E}$ is given by

$$\boldsymbol{\tau} = \mathbf{p} \times \mathscr{E}.$$

   b) Derive Eq. (8.4), that is, show that the potential energy of a dipole in a field is given by

$$V = -p\mathscr{E}\cos\theta,$$

   where $\theta$ is the angle between the dipole and the field.
3. The dipole moment for a general distribution of charges is defined as the sum

$$\mathbf{p} = \sum_i q_i \mathbf{r}_i,$$

   where $q_i$ and $\mathbf{r}_i$ are the charge and position, respectively, of the $i^{\text{th}}$ charge, and the summation is over all the charges present. The choice of the origin of coordinates is arbitrary.
   a) Show that the above reduces to expression (8.1) for the special case of two equal and opposite charges. (Take an arbitrary origin.)
   b) Prove that if the charge system has an overall electrical neutrality, then the dipole moment is independent of the choice of origin.
4. Determine the dipole moment for the following charge distributions: 1.5 $\mu$coul each at the points (0,3), (0,5), where the coordinate numbers are given in centimeters.
5. A parallel-plate capacitor of area $4 \times 5$ cm$^2$ is filled with mica ($\epsilon_r = 6$). The distance between the plates is 1 cm, and the capacitor is connected to a 100-V battery. Calculate:
   a) The capacitance of this capacitor
   b) The free charge on the plates
   c) The surface charge density due to the polarization charges
   d) The field inside the mica. (What would the field be if the mica sheet were withdrawn?)
6. Prove that when a molecule is polarized by a field $\mathscr{E}$, a potential energy is stored in this molecule. The value of this energy is $\frac{1}{2}\alpha\mathscr{E}^2$, where $\alpha$ is the molecular polarizability. What is the value of this energy for an Ar atom in a field of $10^3$ volt/m? The polarizability of this atom is $1.74 \times 10^{-40}$ farad-m$^2$.
7. a) Show that the surface charge density of the polarization charges on the outer surface of a dielectric is given by

$$\sigma_p = \mathbf{P}\cdot\hat{\mathbf{n}},$$

   where $\hat{\mathbf{n}}$ is a unit vector normal to the surface.

b) Prove Eq. (8.25). That is, show that the depolarization field in an infinite slab, in which the field is normal to the slab, is given by

$$\mathscr{E}_1 = -\frac{1}{\epsilon_0}\mathbf{P}.$$

c) The depolarization field $\mathscr{E}_1$ depends on the geometrical shape of the specimen. When the shape is such that the polarization inside is uniform, the depolarization factor $L$ is defined such that

$$\mathscr{E}_1 = -\frac{L}{\epsilon_0}\mathbf{P}.$$

Show that the depolarization factor for an infinite slab with field normal to the slab is 1, while for a slab in which the field is parallel to the face, $L = 0$. Also show that $L = \frac{1}{3}$ for a sphere, and $L = 0$ or $\frac{1}{2}$ for a cylinder, depending on whether the field is parallel or normal to the axis of the cylinder, respectively. Put these results in tabular form.

8. a) Prove Eq. (8.28), showing that the field $\mathscr{E}_3$ due to the dipoles inside a spherical cavity vanishes in a cubic crystal.
   b) Suppose that the Lorentz cavity is chosen to have a cubic shape. Calculate the field $\mathscr{E}_2$ due to the charges on the surface of this cavity.
   c) Does this new choice of cavity modify the value of the local field? Explain. Use your answer to evaluate the field $\mathscr{E}_3$ due to the dipoles inside the cavity. (You may take the crystal to be cubic.)
9. The field $\mathscr{E}_3$ of Eq. (8.24) due to the dipole inside a cavity depends on the symmetry of the crystal, and in general does not vanish in a noncubic crystal. Assuming that this field has the form

$$\mathscr{E}_3 = (b/\epsilon_0)\,\mathbf{P},$$

where $b$ is a constant, calculate the dielectric constant $\epsilon_r$ in such a substance.

10. Show that Eq. (8.33) reduces to (8.18) in gaseous substances, i.e., substances in which $N\alpha/\epsilon_0$ is very small.
11. Establish Eq. (8.40) by carrying out the necessary integration.
12. a) Expand the Langevin function $L(u)$ of (8.41) in powers of $u$ up to and including the third power in $u$, and show that

$$L(u) = u/3 - u^3/45 + \cdots, \quad u \ll 1.$$

   b) Calculate the field required to produce polarization in water equal to 10% of the saturation value at room temperature.
13. a) Using Fig. 8.13 and Table 8.1, calculate the molecular concentration of $CHCl_3$, $CH_2Cl_2$, and $CH_3Cl$ at which the measurements reported in the figure were made.
   b) Calculate the electronic-ionic polarizability $\alpha_{ei}$ in each of these substances.
14. The molar polarizability of water increases from $4 \times 10^{-5}$ to $6.8 \times 10^{-5}$ m$^3$ as the temperature decreases from 500°K to 300°K. Calculate the permanent moment of the water molecule.
15. Calculate the real and imaginary parts of the dielectric constant $\epsilon_r'(\omega)$ and $\epsilon_r''(\omega)$ for water at room temperature. Plot these quantities versus $\omega$ up to the frequency $10^{12}$ Hz. (Use semilogarithmic graph paper.)

16. We expressed the absorption in dipolar substances in terms of the imaginary dielectric constant, $\epsilon_r''(\omega)$. It is also frequently expressed in terms of the so-called *loss angle* $\delta$, which is defined as

$$\tan\delta = \frac{\epsilon_r''}{\epsilon_r'},$$

where the quantity $\tan\delta$ is called the *loss tangent*.

a) Show that the electric displacement vector is

$$D = \epsilon_0[\epsilon_r'^2 + \epsilon_r''^2]^{1/2}\, e^{i\delta}\,\mathscr{E}.$$

b) Calculate the loss tangent as a function of the frequency, and plot the result versus $\omega\tau$.

c) Show that the power absorbed by a dielectric (per unit volume) is

$$Q = \tfrac{1}{2}\,\epsilon_0\epsilon_r'\omega\tan\delta\,\mathscr{E}^2.$$

Express the loss angle $\tan\delta$ in terms of the ratio of the dissipated energy to the energy stored in the dielectric.

d) Calculate the loss tangent in water at room temperature at frequency 10 GHz. Also calculate the energy dissipated per unit volume, given that the field strength is 5 volts/m.

17. Assuming that the jumping period $\tau$ decreases exponentially with temperature as in (8.73), explain how the real and imaginary parts of the dielectric constant $\epsilon_r'$ and $\epsilon_r''$ vary with temperature. Plot the results versus $1/T$. (Assume that all quantities other than $\tau$ are independent of temperature.) Does the loss tangent increase or decrease with temperature? Explain.

18. In deriving the result (8.74) for the dielectric constant involving ionic polarizability, it was assumed that the ions experience no collision or loss during their motion. Postulate the existence of a collision mechanism whose time is $\tau_i$, and reevaluate the (complex) dielectric constant. Plot the real and imaginary parts $\epsilon_r'(\omega)$, $\epsilon_r''(\omega)$ versus $\omega$, and compare with Fig. 8.20.

19. The crystal NaCl has a static dielectric constant $\epsilon_r(0) = 5.6$ and an optical index of refraction $n = 1.5$.

a) What is the reason for the difference between $\epsilon_r(0)$ and $n^2$?

b) Calculate the percentage contribution of the ionic polarizability.

c) Use the optical phonon for NaCl quoted in Table 3.3, and plot the dielectric constant versus the frequency, in the frequency range $0.1\,\omega_t$ to $10\,\omega_t$.

20. Using the data in the previous problem and Table 8.4, calculate the nearest distance between Na and Cl atoms. Calculate the lattice constant of NaCl, and compare the result with the value quoted in Table 1.2. (Sodium chloride has an fcc structure.)

21. Calculate the static polarizability for the hydrogen atom, assuming that the charge on the electron is distributed uniformly throughout a sphere of a Bohr radius. Also calculate the natural electron frequency $\omega_0$.

22. Show that expression (8.80) leads to a static susceptibility equal to that given by (8.77). Use elementary electrostatic arguments to find $\omega_0$ in terms of atomic characteristics.

23. Modify expression (8.80) for the electronic polarizability to include the presence of a collision mechanism of time $\tau$. Evaluate the high-frequency dielectric constant, both real and imaginary parts.
24. Carry out the steps leading to the expression (8.86) for $\epsilon_r''(\omega)$ due to interband transition in solids.
25. The Kramers–Kronig relations, which lead to (8.88), are derived in Brown (1966). Read the discussion there and present your own summary.
26. a) An acoustic oscillator is made of a quartz rod. Explain why the resonant frequency of this oscillator is given by

$$\nu = \frac{v_s}{2l},$$

where $l$ is the length of the rod and $v_s$ the velocity of sound in the specimen.

b) Show that this frequency is also given by the expression

$$\nu = \frac{1}{2l}\sqrt{\frac{Y}{\rho}},$$

where $Y$ is Young's modulus and $\rho$ the mass density of the rod.

c) Taking $Y = 8.0 \times 10^{11}$ dyne/cm$^2$ and $\rho = 2.6$ g/cm$^3$ for quartz, calculate the length of a 5-kHz-oscillator.

d) Calculate the potential difference across the rod for a strain of $2 \times 10^{-8}$. The piezoelectric coefficient $P/S = 0.17$ coul/m$^2$.

27. Many applications of piezoelectric crystals are discussed in Mason (1950). Make a summary of these.
28. In evaluating the local field correction in (8.97), we neglected the electronic contribution. Reevaluate the correction including this contribution, and calculate the new optical phonon frequency $\omega_t^*$ and the dielectric constant.
29. A dielectric has a very small electrical conductivity. However, if a very strong electric field is applied, the conductivity suddenly increases as the field reaches a certain high value. This phenomenon, known as *dielectric breakdown*, is due to the fact that a strong field ionizes the electrons from their atoms, and as these electrons are accelerated they ionize other atoms, etc. Read the discussion of dielectric breakdown presented in N. F. Mott and R. W. Gurney (1953), *Electronic Processes in Ionic Crystals*, second edition, Oxford University Press, and write your own review of this phenomenon.
30. The discussion of dielectric and optical properties in the text was limited to the *linear* region, i.e., the field is sufficiently small that polarization is a linear function of the field. Nonlinear effects become important at high fields, which are now conveniently available from laser sources. Read the discussion of such effects given in A. Yariv (1971), *Introduction to Optical Electronics*, Holt, Rinehart, and Winston, and write a brief summary.

# CHAPTER 9 MAGNETISM AND MAGNETIC RESONANCES



---

*Where order in variety we see,*
*and where, though all things differ, all agree.*

Alexander Pope

## 9.1 INTRODUCTION

The magnetic properties of matter have fascinated physicists, chemists, and engineers for many years. In recent times these properties have been the subject of especial interest because of the information they yield about the constitution of matter, and the interactions involved therein. This information, being truly interdisciplinary, is of interest not only to physicists, but also to other scientists and engineers. Magnetic materials have wide-ranging technical applications, from transformer cores in electrical machinery, to magnetic tapes in computers.

After a brief elementary review, we shall discuss the magnetic behavior of free independent atoms, and then the magnetic properties of conduction electrons in ordinary metals. Next we shall talk about ferromagnetic insulators and metals, with special attention to the internal magnetic field responsible for ferromagnetism. We shall also cover the practical subject of ferromagnetic domains and their role in the magnetization process. Then we shall move on to the various types of magnetic resonances. This subject is very important because it yields information on the dynamical aspects of magnetic moments, unlike the previous studies which gave only information of a static nature. The chapter ends with a discussion of spin waves, the excitation modes in a spin system.

## 9.2 REVIEW OF BASIC FORMULAS

In this section we shall review some of the basic formulas which will be used in this chapter; most of these should be familiar to the reader from his study of elementary physics. Let us begin with the fundamental concept of the *magnetic dipole moment*. Consider two magnetic charges, $q_m$ and $-q_m$, which are equal and opposite. They form a *magnetic dipole* whose moment is given by

$$\boldsymbol{\mu}_m = q_m \mathbf{d}, \tag{9.1}$$

where $\mathbf{d}$ is the vector joining the negative to the positive charge, as shown in Fig. 9.1. Note the similarity between this definition and that used in connection with the moment of an electric dipole, (8.1). This similarity will appear frequently in our discussion.

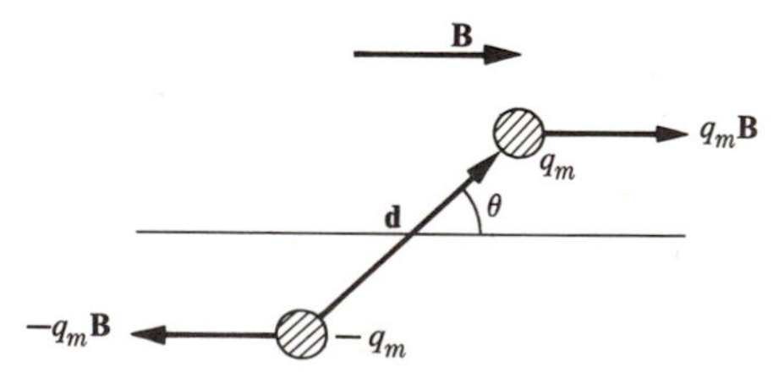


**Fig. 9.1** Magnetic dipole and torque exerted on it by a magnetic field.

When our magnetic dipole is placed in a magnetic field whose induction is $\mathbf{B}$, then, because a charge $q_m$ experiences a force

$$\mathbf{F} = q_m \mathbf{B}, \tag{9.2}$$

the dipole itself (composed of two opposite charges) experiences a couple whose torque is

$$\boldsymbol{\tau} = \boldsymbol{\mu}_m \times \mathbf{B}. \tag{9.3}$$

The effect of this torque is to turn the dipole and align it with the field, in a manner similar to the way an electric field aligns an electric dipole. Because of the torque, the dipole has an orientation potential energy given by

$$V = -\boldsymbol{\mu}_m \cdot \mathbf{B} = -\mu_m B \cos\theta, \tag{9.4}$$

where $\theta$ is the angle between the field and the dipole directions. The minimum energy, $-\mu_m B$, occurs at $\theta = 0$, where the dipole lies along the field. The maximum energy is achieved at $\theta = \pi$, where the dipole is oriented opposite to the field.

We have defined the magnetic dipole in terms of magnetic charges, but such charges do not, in fact, exist. All the known magnetic properties of matter are attributable to the rotation of electric charges. We recall from elementary physics that an electric current loop acts like a magnetic dipole of moment

$$\mu_m = IA, \tag{9.5}$$

where $I$ is the current and $A$ the area of the loop. The direction of $\boldsymbol{\mu}_m$, which is a vector, is normal to the plane of the loop, and such that the current flows counterclockwise relative to an observer standing along $\boldsymbol{\mu}_m$ (Fig. 9.2).

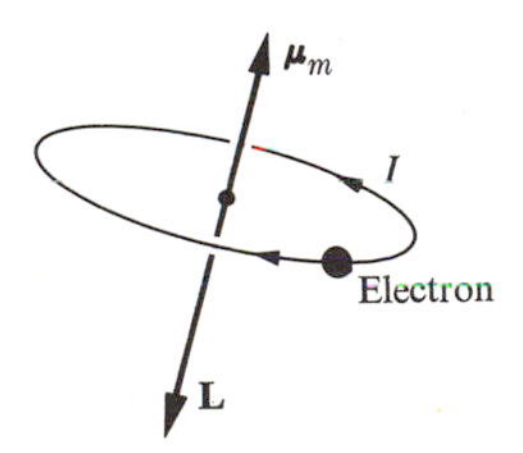


**Fig. 9.2** Magnetic dipole moment $\boldsymbol{\mu}_m$ associated with a current loop; $I$ represents electric current. Vector **L** is angular momentum of electron producing the current.

The current loops in an atom are composed of rotating electrons. In this case we can establish a simple relation between $\boldsymbol{\mu}_m$ and the angular momentum **L** of the electron. Noting that $I = e(\omega/2\pi)$, $A = \pi r^2$, and $L = mr^2\omega$, where $\omega$ is the angular velocity, we may show with the help of (9.5) that

$$\boldsymbol{\mu}_m = \left(-\frac{e}{2m}\right)\mathbf{L}. \tag{9.6}$$

The negative sign indicates that $\boldsymbol{\mu}_m$ is opposite to **L**. The coefficient $-e/2m$

relating $\mathbf{\mu}_m$ to $\mathbf{L}$ is called the *gyromagnetic ratio*. The usefulness of (9.6) derives from the fact that it expresses $\mathbf{\mu}_m$ in terms of $\mathbf{L}$, which is a familiar quantity in quantum mechanics (see Appendix).

In addition to its orbital rotation, the electron also rotates about its own axis, a motion referred to as *spin*. Thus there is a magnetic moment associated with the spin, and this moment may be related to the spin angular momentum $\mathbf{S}$. The relation is

$$\mathbf{\mu}_m = \left( -\frac{e}{m} \right) \mathbf{S}, \tag{9.7}$$

which shows that the spin gyromagnetic ratio $(-e/m)$ is twice the value obtained for the orbital motion in (9.6). The classical derivation of this ratio does not apply to the spin motion because this motion is entirely quantum in nature.

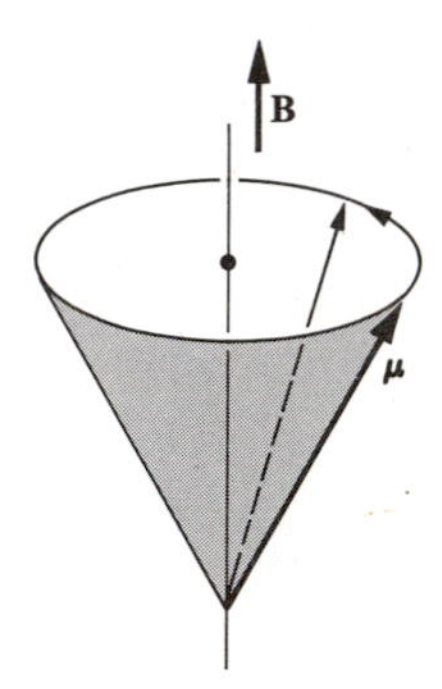


**Fig. 9.3** Precession of a dipole moment $\mathbf{\mu}$ in a magnetic induction $\mathbf{B}$.

Let us now think about the dynamics of a classical dipole in a magnetic field. The equation of motion is

$$\frac{d\mathbf{L}}{dt} = \boldsymbol{\tau}, \tag{9.8}$$

where $\boldsymbol{\tau}$ is the torque. If we substitute for $\boldsymbol{\tau}$ from (9.3), and for $\mathbf{L}$ from (9.6), we obtain

$$\frac{d\mathbf{\mu}}{dt} = -\left(\frac{e}{2m}\right) \mathbf{\mu} \times \mathbf{B}. \tag{9.9}$$

(The subscript on $\mathbf{\mu}$ will be deleted henceforth, for brevity, and since this leads to no confusion.) This relation represents a *precessional* motion (see Fig. 9.3), of frequency

$$\omega_{\mathrm{L}} = \frac{eB}{2m}, \tag{9.10}$$

known as the *Larmor frequency*. The dipole simply precesses around the direction of the field, always maintaining the same angle. For an electron at $B = 0.1\ \text{W/m}^2$, $\omega_L \simeq 10$ GHz.

This statement concerning the Larmor precession seems to cast some doubt on our earlier assumption that the dipole tends to align itself with the field. But in the precession, the dipole merely *rotates* around **B** without ever getting closer to the direction of the field. The point is well taken. In a pure Larmor precession no alignment takes place. In practical situations, however, this precession is usually accompanied by numerous collisions, during which the dipole loses energy. As it does so, it gradually approaches the direction of **B**, until eventually it lines up exactly with the field. This process of gradual magnetization is referred to as *relaxation*. We shall discuss it in more detail in Section 9.12.

The potential energy of the dipole, Eq. (9.4), can also be written as

$$E = +\frac{e}{2m} L_z B, \tag{9.11}$$

where we have used (9.6). Here **B** is taken to be in the $z$-direction, and $L_z$ is the $z$-component of the angular momentum. We recall that, according to quantum mechanics (Section A.4), the component $L_z$ is quantized by $L_z = m_l \hbar$, where $m_l$ is an integer which takes the values $-l, -l+1, \cdots, l-1, l$, and where $l$ is the orbital quantum number for the angular momentum of the electron. Thus Eq. (9.11) may also be written as

$$E = \left(\frac{e\hbar}{2m}\right) B m_l.$$

The ratio $\mu_B \equiv (e\hbar/2m)$ is called the *Bohr magneton*, and has the numerical value $9.3 \times 10^{-24}\ \text{J m}^2/\text{W}$. We may therefore write

$$E = \mu_B B m_l. \tag{9.12}$$

As $m_l$ takes its various allowed values, the energy also takes its appropriate values in the presence of the magnetic field. For $l = 1$, $m_l$ takes the values $-1$, 0, and 1, and the corresponding three energy levels are illustrated in Fig. 9.4.

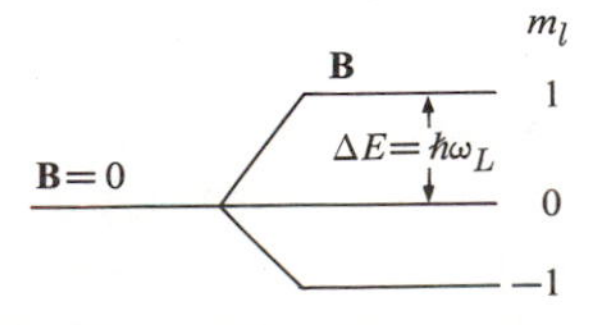


**Fig. 9.4** Splitting of an atomic level by a magnetic field (Zeeman effect) for $l = 1$.

The splitting of an atomic level by the magnetic field in the manner just described is known as *Zeeman splitting*. The interval between any two adjacent levels is the same, and is given by

$$\Delta E = \mu_B B. \tag{9.13}$$

Note that the lowest level, $m_l = -1$, corresponds to the orientation in which **L** is the opposite to **B**, and hence $\boldsymbol{\mu}$ is parallel to **B**, in agreement with the classical picture. Similarly, the highest level, $m_l = 1$, corresponds to the orientation in which $\boldsymbol{\mu}$ is opposite to the field.

In the case of the spin, the Zeeman energy (9.12) is given by

$$E = 2\mu_B B m_s, \tag{9.14}$$

where the factor 2 arises from the fact that the spin gyromagnetic ratio is twice the classical value. Since the spin quantum number $s = \frac{1}{2}$, the allowed values are $m_s = -\frac{1}{2}$ and $+\frac{1}{2}$. The corresponding Zeeman splitting, composed of only two lines, is shown in Fig. 9.5. The difference in energy between the two levels is

$$\Delta E = 2\mu_B B, \tag{9.15}$$

which is double the orbital separation.

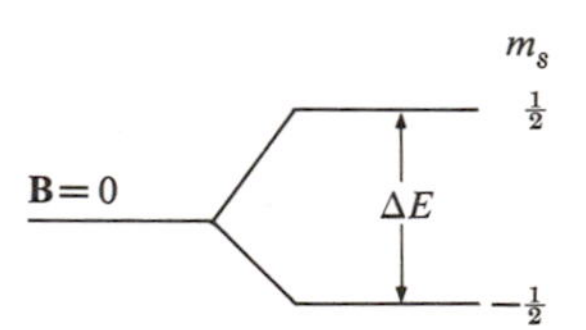


**Fig. 9.5** Zeeman effect for the spin case.

It is useful to establish a correspondence between the classical and quantum descriptions of the dipole motion in a magnetic field. Note that for the orbital case, the energy splitting $\Delta E = \mu_B B$ is also equal to $\hbar\omega_L$, where $\omega_L$ is the Larmor frequency given by (9.10). Thus the quantum frequency for the transition between the Zeeman levels is equal to the classical Larmor frequency. The same correspondence applies to spin, where both the Larmor frequency and the Zeeman spacing are twice as large.

In this section the orbital and spin angular momenta have been treated separately, but when an electron has both types of momenta, as frequently happens, there is an interaction between them, and the two should be treated simultaneously. This situation and its bearing on magnetism will be discussed in Section 9.6.

## 9.3 MAGNETIC SUSCEPTIBILITY

A magnetic field can be described by either of two vectors: The *magnetic induction* $\mathbf{B}$ or the *magnetic field intensity* $\mathscr{H}$, which are related in vacuum by

$$\mathbf{B} = \mu_0 \mathscr{H}, \tag{9.16}$$

where $\mu_0 = 4\pi \times 10^{-7}$ Hz/m is the permeability of free space.

When a material medium is placed in a magnetic field, the medium is magnetized. This magnetization is described by the magnetization vector $\mathbf{M}$, the dipole moment per unit volume. The magnetic induction inside the medium is then given by the relation

$$\mathbf{B} = \mu_0 \mathscr{H} + \mu_0 \mathbf{M}, \tag{9.17}$$

which you should recall from basic physics. The induction is composed of two parts: The part $\mu_0 \mathscr{H}$ generated by the external sources, and the part $\mu_0 \mathbf{M}$, due to the magnetization of the medium.

Since the magnetization is induced by the field, we may assume that $\mathbf{M}$ is proportional to $\mathscr{H}$. That is,

$$\mathbf{M} = \chi \mathscr{H}, \tag{9.18}$$

the proportionality constant $\chi$ being known as the *magnetic susceptibility* of the medium.† When this expression for $\mathbf{M}$ is inserted into (9.17), it leads to

$$\mathbf{B} = \mu_0 (1 + \chi) \mathscr{H}. \tag{9.19}$$

Thus the vectors $\mathbf{B}$ and $\mathscr{H}$ are proportional to each other,

$$\mathbf{B} = \mu \mathscr{H}, \tag{9.20}$$

and the proportionality constant

$$\mu = \mu_0 (1 + \chi) \tag{9.21}$$

is known as the *permeability of the medium*. It is often more convenient to use the relative permeability $\mu_r$, which is defined as $\mu_r = \mu/\mu_0$. Therefore

$$\mu_r = 1 + \chi, \tag{9.22}$$

a relation connecting the permeability and susceptibility of the medium.‡

---

† The magnetic susceptibility $\chi$ bears no physical relationship to the electric susceptibility of Section 8.2, although the same symbol is used for both. No confusion should arise, however, since electric susceptibility will not appear in this chapter.

‡ Our discussion assumes that the medium is magnetically isotropic. But real crystals are anisotropic, and the susceptibility and permeability are represented by second-rank tensors. In order to avoid mathematical complications, however, we shall ignore anisotropic effects in our treatment.

Our approach here clearly parallels that used in the electric case (Section 8.3), and relation (9.22) is the analog of (8.21). Note, however, that in writing (9.18) we assumed that **M** is proportional to $\mathscr{H}$, the external field, and in doing so we in effect ignored such things as demagnetization field, local field correction, etc., which we felt obliged to include in the electric case. The neglect of these factors is justifiable in the magnetic case because $M$ is very small compared to $\mathscr{H}$ (typically $\chi = M/\mathscr{H} \simeq 10^{-5}$), unlike the electric case, in which $\chi \simeq 1$. But when we deal with ferromagnetic materials, where $M$ is quite large, this omission is no longer tenable, and the above effects must be included, as we shall see in Section 9.11.

## 9.4 CLASSIFICATION OF MATERIALS

Materials may be grouped into three magnetic classes, depending on the sign and magnitude of the susceptibility. Those materials for which $\chi$ is positive—that is, for which **M** is parallel to $\mathscr{H}$—are known as *paramagnetic*, whereas those for which $\chi$ is negative—that is, for which **M** is opposite to $\mathscr{H}$—are *diamagnetic*. Table 9.1 gives susceptibilities for some representative substances, and emphasizes once more the smallness of the magnitude of $\chi$.

**Table 9.1**

Magnetic Susceptibilities (per $cm^{-3}$)

| Material | $\chi$ |
|---|---|
| *Paramagnetic* | |
| Al | $+ 2.2 \times 10^{-5}$ |
| Mn | $+ 98$ |
| W | $+ 36$ |
| *Diamagnetic* | |
| Cu | $- 1.0 \times 10^{-5}$ |
| Au | $- 3.6$ |
| Hg | $- 3.2$ |
| Water | $- 9.0$ |
| H | $- 0.2 \times 10^{-8}$ |

1) *Paramagnetic materials.* The best-known examples of paramagnetic materials are the ions of transition and rare-earth ions. The fact that these ions have incomplete atomic shells is what is responsible for their paramagnetic behavior.

2) *Diamagnetic materials.* Ionic and covalent crystals are diamagnetic. These substances have atoms or ions with complete shells, and their diamagnetic behavior is due to the fact that a magnetic field acts to distort the orbital motion.

3) *Ferromagnetic materials.* The magnetic susceptibility of ferromagnetic materials

may be very large ($10^5$ $cm^{-3}$) and a ferromagnetic substance becomes spontaneously magnetized below a certain temperature. Examples are the ferromagnetic metals Fe, Co, and Ni, and their alloys. We shall discuss them, and enumerate some of the interesting phenomena they display, later in this chapter.

## 9.5 LANGEVIN DIAMAGNETISM

Let us now establish the fact that the effect of a magnetic field on the orbital motion of an electron is such as to produce a diamagnetic susceptibility. Consider an electron rotating about the nucleus in a circular orbit, and let a magnetic field be applied perpendicular to the plane of the paper, as shown in Fig. 9.6(a). Before this field is applied, we have, according to Newton's second law,

$$F_0 = m\omega_0^2 r, \tag{9.23}$$

where $F_0$ is the attractive coulomb force between the nucleus and the electron, and $\omega_0$ is the angular velocity. The magnetic moment of the electron is

$$\mu_0 = IA = \frac{e}{2}\omega_0 r^2, \tag{9.24}$$

where $r$ is the radius of the electron's orbit. This moment is parallel to the field for the geometry and sense of rotation shown in the figure.

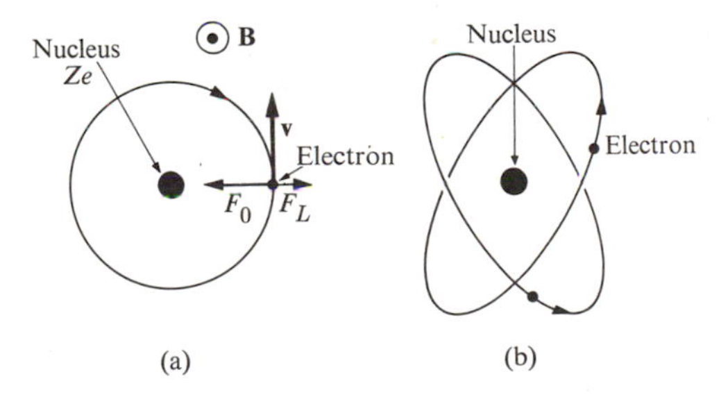


**Fig. 9.6** Atomic origin of diamagnetism. (a) The Lorentz force $F_L$ opposes the Coulomb force $F_0$; **v** is the electron velocity. (b) Three-dimensional nature of electron orbits.

When the field is applied, an additional force starts to act on the electron: the *Lorentz force* $-e(\mathbf{v} \times \mathbf{B})$. For the geometry of Fig. 9.6, the effect is to produce a radially outward force given by $eBr\omega$, and Eq. (9.23) should therefore be amended to

$$F_0 - eBr\omega = m\omega^2 r. \tag{9.25}$$

Thus the angular frequency is now different from $\omega_0$, and its value may be determined from this relation. The solution of this quadratic equation in $\omega$, in the limit of the small field, is given by

$$\omega = \omega_0 - \frac{eB}{2m}, \tag{9.26}$$

which shows that the rotation of the electron has been slowed down. This reduction in frequency produces a corresponding change in the magnetic moment which, according to (9.24), is

$$\Delta\mu = -\left(\frac{e^2r^2}{4m}\right)B. \tag{9.27}$$

Since the moment parallel to the field has been reduced, the induced moment is opposite to the field, i.e., the response of the electron is diamagnetic.

It can be readily appreciated that if we initially chose an electron which was rotating counterclockwise, the initial moment would be opposite to the $z$-axis, i.e., negative. The effect of the field would then be to speed up the electron, resulting in an even more negative moment. That is, the induced moment would again be negative—diamagnetic—and given by (9.27). Thus the diamagnetic response of an orbiting electron holds good in general, and in fact may be shown to follow directly from the familiar Lenz's law.

When applied to an atom, Eq. (9.27) requires some modification, because the electron orbits around a spherical surface rather than in a circle (see Fig. 9.6b). However, only the cross section normal to the field is effective in the diamagnetic response, and hence on the average we should replace $r^2$ in (9.27) by $\frac{2}{3}r^2$, the new $r$ being the radius of the sphere, which leads to

$$\Delta\mu = -\left(\frac{e^2r^2}{6m}\right)B. \tag{9.28}$$

We can now readily evaluate the magnetic susceptibility. Given that the atom has $Z$ electrons and that there are $N$ atoms per unit volume, the susceptibility $\chi = M/\mathscr{H} = \mu_0 NZ\Delta\mu/B$, or

$$\chi = -\frac{\mu_0 e^2}{6m}(NZ\overline{r^2}), \tag{9.29}$$

where $\overline{r^2}$ is the average square radius of the electron. The averaging is done over all the occupied orbitals in the atom. This expression yields values which are of the same order of magnitude as those obtained by measurements. Thus for $N = 10^{29}\,\text{m}^{-3}$, $Z = 10$, $\overline{r^2} = 10^{-20}\ \text{m}^2$, and appropriate values for the constants in (9.29), we find $\chi \simeq 10^{-5}$, in agreement with the values listed in Table 9.2.

**Table 9.2**

Diamagnetic Susceptibilities per Gram-Ion for Rare-Gas Atoms and for Ions of Filled Shell Structure

| Element | $\chi$ | Element | $\chi$ |
|---|---|---|---|
| He | $-1.9 \times 10^{-6}$ | $Li^+$ | $0.7 \times 10^{-6}$ |
| Ne | $-7.6$ | $Na^+$ | 6.1 |
| Ar | $-19$ | $K^+$ | 14.6 |
| Kr | $-29$ | $Mg^{2+}$ | 4.3 |
| Xe | $-44$ | $Ca^{2+}$ | 10.7 |
| $F^-$ | $-9.4$ | $Sr^{2+}$ | 18.0 |
| $Cl^-$ | $-24.2$ | | |
| $Br^-$ | $-34.5$ | | |

Diamagnetic susceptibility is observed most clearly in those solids in which the atomic shells are completely filled. Examples of these are provided by rare-gas crystals, and also ionic crystals, whose magnetic susceptibilities are given in Table 9.2. In the case of covalent crystals, which are also diamagnetic, Eq. (9.29) can be applied only to the core electrons. The electrons forming the bond have orbitals which are far from circular, and hence the derivation leading to (9.29) does not apply here. The susceptibility of a covalent crystal may be written as

$$\chi = \chi_i + \lambda, \tag{9.30}$$

where $\chi_i$ includes the effect of the core electrons (the ions), and $\lambda$ the effect of the bonding electrons. One can determine the value of $\lambda$ for a specific bond empirically in a given compound, and use this value in other compounds in which it occurs.

When some of the atomic shells in a solid are incompletely filled, the substance then has a paramagnetic contribution in addition to the diamagnetic contribution. The *net susceptibility* is the difference between the two contributions, but since the paramagnetic one is usually larger, it masks the diamagnetic contribution.

## 9.6 PARAMAGNETISM

An atom whose shells are not completely filled has a *permanent magnetic moment*, which (as we shall see) arises from the combination of the orbital and spin motions of its electrons. For the time being we shall accept this moment as a given quantity, and discuss the effect of a magnetic field on such a moment—first classically, and then quantum mechanically.

## Classical theory

The potential energy of a magnetic dipole in a magnetic field is given by

$$V = -\boldsymbol{\mu}\cdot\mathbf{B},$$

according to (9.4). The energy is least when the moment is parallel to the field, and thus the moment tends to line up with the field. The effect of temperature is to randomize the direction of the dipole. The result of these two competing processes is that some magnetization is produced. We can solve the problem analytically the same way we solved the problem of dipolar electrical polarization (Section 8.5), which leads to

$$\bar{\mu}_z = \mu L(v), \tag{9.31}$$

where $\bar{\mu}_z$ is the average of $\mu_z$, the component of the moment along the direction of the field (taken in the $z$-direction) and $L(v)$ is the Langevin function,†

$$L(v) = \text{Coth}\, v - \frac{1}{v} \quad \text{and} \quad v = \frac{\mu B}{kT}. \tag{9.32}$$

Figure 9.7 shows a plot of $\bar{\mu}_z$ versus $\mu B/kT$. We see that, at low field, $\bar{\mu}_z$ is proportional to the magnetizing field $B$, but as $B$ increases in value, $\bar{\mu}_z$ begins to saturate, eventually reaching the maximum value $\mu$. It achieves this maximum value when the dipole lies exactly parallel to the field.

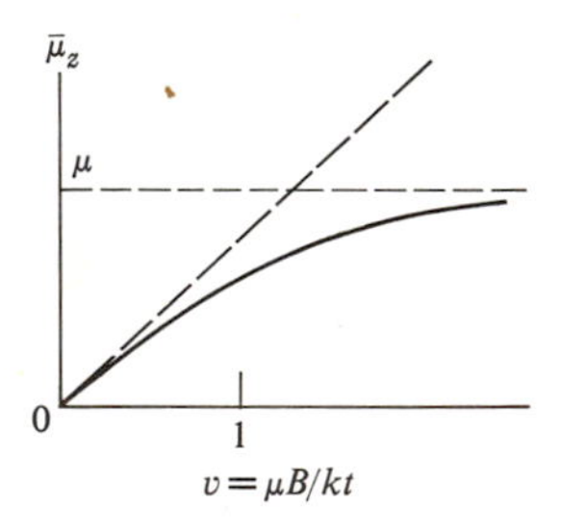


**Fig. 9.7** Average dipole moment component $\bar{\mu}_z$ versus $v = \mu\text{B}/kT$. Dashed line represents low-field approximation.

In most practical situations, the ratio $\mu B/kT$ is very small compared to unity. Thus for $\mu = \mu_\text{B}$, $B = 0.1\ \text{W/m}^2$, and $T = 100°\text{K}$, this ratio is about 0.001. Therefore we may approximate the function $L(v) \simeq \frac{1}{3}v$, which leads to

$$\bar{\mu}_z = \frac{\mu^2 B}{3kT}. \tag{9.33}$$

† Langevin derived the formula (9.31) for dipolar magnetization before its electrical analog. His treatment was adapted to the electrical case by Debye.

The magnetization is given by

$$M = N\bar{\mu}_z = \frac{N\mu^2 B}{3kT},$$

where $N$ is the atomic concentration, and the susceptibility is given by

$$\chi = \frac{N\mu_0\mu^2}{3kT}, \tag{9.34}$$

which, you will note, is of the same form as the electric susceptibility of (8.43). Equation (9.34) is referred to as the *Curie law*. If one substitutes $N = 10^{26}$ m$^{-3}$ and $T = 100°$K, one finds that $\chi \simeq 10^{-5}$, in agreement with observation (see Table 9.1).

The susceptibility given by Eq. (9.34) is also referred to as the *Langevin paramagnetic susceptibility*. Note in particular that $\chi$ is inversely proportional to temperature. This is in marked contrast to the diamagnetic susceptibility, which is essentially temperature independent.

**Quantum theory**

We can express the magnetic moment $\mu$ of the atom in terms of the total angular momentum **J** as

$$\boldsymbol{\mu} = g\left(-\frac{e}{2m}\right)\mathbf{J}, \tag{9.35}$$

where $g$ is a constant known as the *Landé factor*. Its value depends on the relative orientations of the orbital and spin angular momenta. Expression (9.35) is the same as the classical expression (9.6), except for the factor $g$.

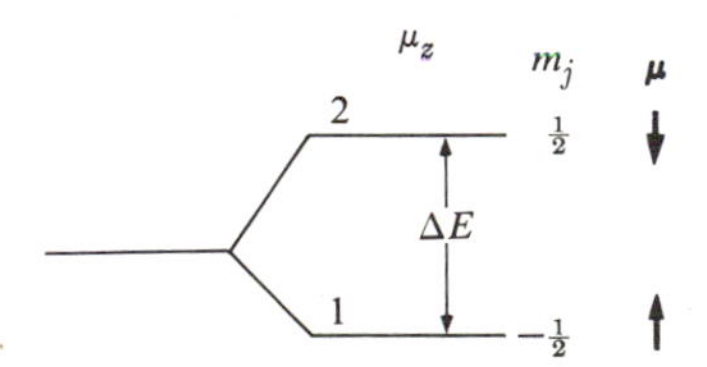


**Fig. 9.8** Quantum description of paramagnetic susceptibility. Arrows under $\boldsymbol{\mu}$ indicate the orientation of the moment for the various levels.

When a magnetic field is applied to the atom, a Zeeman splitting results (Section 9.2). The Zeeman energy is

$$E = -\boldsymbol{\mu}\cdot\mathbf{B} = g\mu_B B m_j. \tag{9.36}$$

If the angular momentum quantum number $j$ is $\frac{1}{2}$, the component $m_j$ can take the values $m_j = -\frac{1}{2}$ or $+\frac{1}{2}$, resulting in a double splitting, as shown in Fig. 9.8. The difference in energy between the levels is

$$\Delta E = g\mu_B B. \tag{9.37}$$

Note that the lower level, $m_j = -\frac{1}{2}$, corresponds to the moment parallel to the field, while the upper level corresponds to the moment opposite to the field.

The magnetization $M$ is given by

$$M = g\mu_B(N_1 - N_2), \tag{9.38}$$

where $g\mu_B$ is the $z$-component of the moment when it is fully aligned with the field, and $N_1$, $N_2$ are the concentrations of atoms in the lower and upper levels, respectively. These two concentrations are related by

$$\frac{N_2}{N_1} = e^{-\Delta E/kT}, \tag{9.39}$$

where the term in the exponent on the right is the familiar Boltzmann factor. Since these concentrations also satisfy the relation $N_1 + N_2 = N$, where $N$ is the total concentration, we may use these two equations to solve for $N_1$ and $N_2$. When we do this, and substitute the results into (9.38), we obtain

$$M = Ng\mu_B \frac{e^x - e^{-x}}{e^x + e^x} = Ng\mu_B \tanh(x), \tag{9.40}$$

where $x = g\mu_B B/kT$.

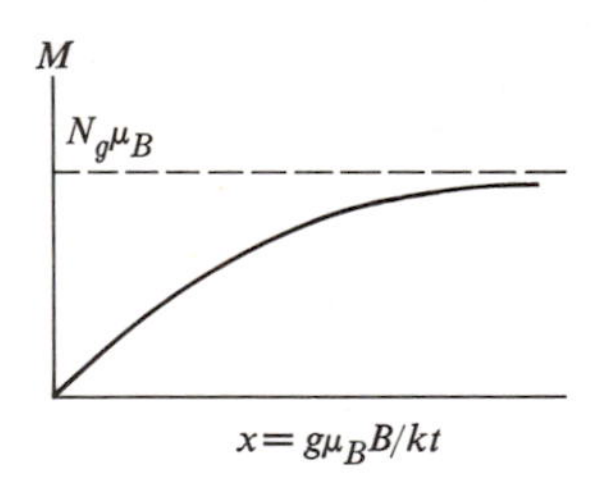


**Fig. 9.9** Magnetization $M$ versus $x = g\mu_B B/kT$ for a system with $j = \frac{1}{2}$.

The magnetization is plotted versus the field in Fig. 9.9. At low field, $M$ is proportional to $B$, but at higher fields $M$ begins to saturate, eventually reaching the maximum value $Ng\mu_B$ when all the dipoles are in exact alignment with the field. Qualitatively, this is the same conclusion reached earlier on the basis of the classical treatment.

Let us take a closer look at the physical process of magnetization in the quantum treatment. For $j = \frac{1}{2}$, the dipole can take only two orientations, one parallel to the field, corresponding to the lower level of Fig. 9.8, and the other opposite to the field, corresponding to the upper level. As the magnetic field is raised, the spacing between the levels increases and the dipoles drop from the higher to the lower level, leading to magnetization.

For a weak field, the ratio $x \ll 1$ and $\tanh x \simeq x$, which, when substituted into (9.40), leads to the susceptibility

$$\chi = \frac{\mu_0 N (g\mu_B)^2}{kT}. \tag{9.41}$$

This is the same as the classical result, provided we assume that the effective moment of the atom is given by $\mu_{eff} = \sqrt{3}\, g\mu_B$.

Our quantum derivation was based on the simplest type of Zeeman splitting, i.e., one involving only two levels. If $j$ were larger than $\frac{1}{2}$, then, in general, the number of levels would be $(2j + 1)$, which leads to the susceptibility

$$\chi = \frac{\mu_0 N \mu_{eff}^2}{3kT}, \tag{9.42}$$

where $\mu_{eff} = p\mu_B$, and

$$\mu_{eff} = p\mu_B \qquad \text{and} \qquad p = g[j(j+1)]^{1/2}. \tag{9.43}$$

The number $p$ is called the *effective number* of Bohr magnetons for the atom. (We shall consider the derivation of (9.42) in the problem section.) We can see, therefore, that quantum-mechanical treatment leads to the same conclusions as classical treatment.

**The atomic origin of magnetism**

We can explain the atomic origin of magnetism by considering orbital and spin motions, and the interaction between them. The total orbital angular momentum of the atom is defined as

$$\mathbf{L} = \sum_i \mathbf{L}_i,$$

where the sum is over all the electrons. Since the sum over a complete shell is zero, however, we must carry out the summation only over the incomplete shell. Similarly, the total spin angular momentum is

$$\mathbf{S} = \sum_i \mathbf{S}_i,$$

where again the sum is over the incomplete shell. The total angular momentum of the atom $\mathbf{J}$ is given by

$$\mathbf{J} = \mathbf{L} + \mathbf{S}. \tag{9.44}$$

The angular momenta **L** and **S** interact with each other via the *spin–orbit interaction*. [This interaction exists above and beyond the Coulomb interaction, between the electrons and the nucleus, and the interaction between the electrons themselves.] Because of this interaction, the vectors **L** and **S** are no longer constants. However, the total angular momentum **J** remains constant.† Thus the vectors **L** and **S** precess around **J**, as indicated in Fig. 9.10.

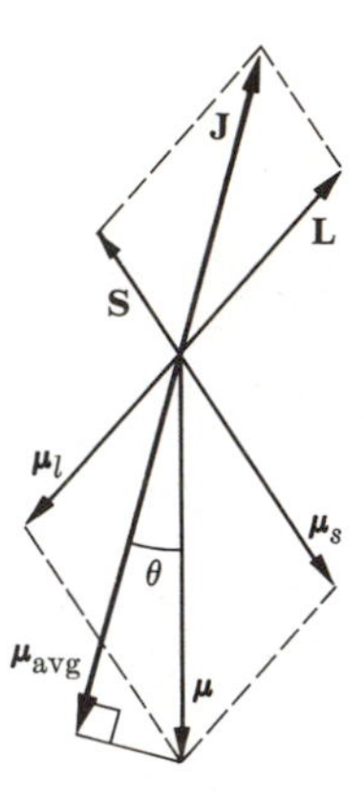


**Fig. 9.10** The spin-orbit interaction, including derivation of the Landé factor.

The dipole moments, $\boldsymbol{\mu}_L = -(e/2m)\mathbf{L}$ and $\boldsymbol{\mu}_S = -(e/m)\mathbf{S}$, corresponding to the orbital and spin momenta, also precess around **J**. However, note that the total moment $\boldsymbol{\mu} = \boldsymbol{\mu}_L + \boldsymbol{\mu}_S$ is not collinear with $J$, as in the pure moments $\boldsymbol{\mu}_L$ and $\boldsymbol{\mu}_S$, but is tilted toward the spin because of its larger gyromagnetic ratio. The vector $\boldsymbol{\mu}$ makes an angle $\theta$ with **J**, and this moment also precesses around **J**. Since the precession frequency is usually quite high, only the component of $\boldsymbol{\mu}$ along **J** is observed; the other component averages out to zero. One can show that

$$\mu_{\text{avg}} = \mu \cos\theta = g\left(-\frac{e}{2m}\right)J,$$

where

$$g = 1 + \frac{j(j+1) + s(s+1) - l(l+1)}{2j(j+1)}. \tag{9.45}$$

This is the $g$-factor, which we have used previously. For a pure orbital motion, $s = 0$, $j = l$, and $g = 1$, while for a pure spin motion $l = 0$, $j = s$, and $g = 2$. These values agree with the discussion in Section 9.2. Note now that the $\mu$ we used earlier in this section is equal to $\mu_{\text{avg}}$, where the subscript was dropped.

† In effect, we are saying that each of the vectors **L** and **S** applies a torque on the other which causes it to precess. There is no torque on the total momentum **J**, however, and hence it does not precess; i.e., it remains constant.

How may we determine $l$, $j$, and $s$ for an atom if all we know is the number of atoms in the incomplete shell and the angular momentum of this shell? We do this by following *Hund's rules*.

i) The spin number $s$ takes its maximum value allowed by the exclusion principle.

ii) Then $l$ also takes its maximum value allowed by the same principle, consistent with (i).

iii) If the shell is less than half full, $j = |l - s|$, and if the shell is more than half full, $j = l + s$.

Let us apply these rules to the carbon atom. There are only two electrons in the 2p subshell ($l = 1$) which can accommodate a maximum of six electrons. To determine the angular momentum, we can make the two spins parallel to each other without violating the exclusion principle, resulting in $s = 2 \times s = 2 \times \frac{1}{2} = 1$. The maximum $l$ consistent with the exclusion principle is $l = 1$ (why?). Since the shell is less than half full, $j = |l - s| = 0$. Thus in the ground state the carbon atom has zero magnetic moment, and exhibits no paramagnetism. In most cases involving incomplete shells, however, $j$ is other than zero, and the atom then shows paramagnetism.

**Rare-earth ions**

Experiments on rare-earth ions in crystals show that they obey the Curie law, with an effective number of magnetons in agreement with the theory of spin–orbit interaction. Table 9.3 confirms this. In these ions, therefore, the angular momenta **L** and **S** are strongly coupled, and the moment of the ion can respond freely to the external field.

**Table 9.3**

Effective Number of Magnetons for Rare-Earth Ions

| Ion | Ground state | Theory $p = g\sqrt{j(j+1)}$ | Experiment (Eq. 9.43) $p$ |
|---|---|---|---|
| $La^{3+}$ | ${}^1S_0$ | 0 | Diamagnetic |
| $Pr^{3+}$ | ${}^3H_4$ | 3.58 | 3.6 |
| $Nd^{3+}$ | ${}^5I_{9/2}$ | 3.62 | 3.6 |
| $Dy^{3+}$ | ${}^6H_{15/2}$ | 10.6 | 10.6 |

This result is not surprising. In these ions—from La to Lu in the periodic table—the 4f shell is incompletely filled. The outer 5p shell is completely filled, while the 5d and 6s shells which are still further out are stripped of their electrons to form the ionic crystal (Fig. 9.11). Thus the only incomplete shell is the 4f shell, and this is the one in which the magnetic behavior occurs. Since electrons in this shell lie deep within the ion, screened by the outer 5p and 5d shells, they are not appreciably affected by other ions in the crystal. Magnetically their behavior is much like that of a free ion. Typical values for the spin–orbit and the crystal-field interactions in these materials are $10^3$ cm$^{-1}$ and $10^2$ cm$^{-1}$, respectively.†

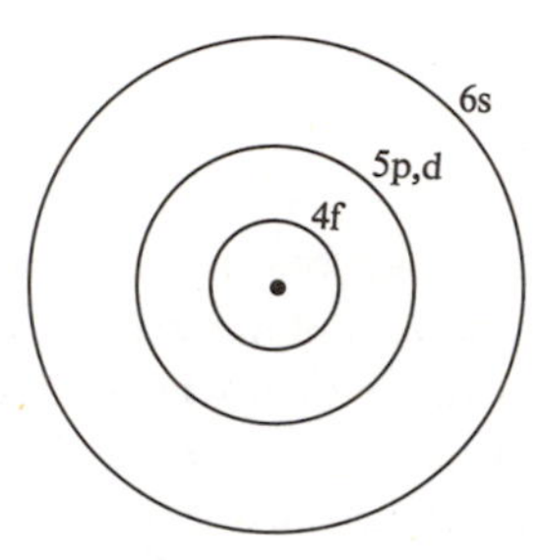


**Fig. 9.11** Various shells in rare-earth ions. The incomplete 4f shell is screened from other atoms by the fifth shell. (The sixth shell is usually ionized.)

### Iron-group ions

Table 9.4 shows that iron-group (ferric or ferrous) ions behave magnetically as if $j \simeq s$; that is, only the spin moment can contribute to magnetization. We can see this by means of the following argument. The magnetic properties of this group of elements are due to the electron in the incomplete 3d shell. Since

**Table 9.4**

Iron–Group Ions

| Ion | Ground state | Theory $p = g\sqrt{j(j+1)}$ | Theory $p = 2\sqrt{s(s+1)}$ | Experiment (Eq. 9.43) $p$ |
|---|---|---|---|---|
| $K^+$, $Ca^{3+}$ | $^1S_0$ | 0 | 0 | Diamagnetic |
| $Ti^{3+}$, $V^{4+}$ | $^2D_{3/2}$ | 1.55 | 1.73 | 1.7 |
| $V^{3+}$ | $^3F_2$ | 1.63 | 2.83 | 2.8 |
| $V^{2+}$, $Cr^{3+}$, $Mn^{4+}$ | $^4F_{3/2}$ | 0.77 | 3.87 | 3.8 |
| $Mn^{2+}$, $Fe^{3+}$ | $^6S_{5/2}$ | 5.92 | 5.92 | 5.9 |
| $Fe^{2+}$ | $^5D_4$ | 6.70 | 4.90 | 5.4 |
| $Ca^{2+}$ | $^2D_{5/2}$ | 3.55 | 1.73 | 1.9 |

† Another reason why the free-ion treatment applies to the rare-earth ions is that the spin–orbit interaction is strong in these substances, because this interaction is proportional to $Z$, the atomic number of the element concerned, and all the rare-earth ions have large $Z$'s.

electrons in this outermost shell interact strongly with neighboring ions, the orbital motion is essentially destroyed, or *quenched*, leaving only the spin moment to contribute to the magnetization. In other words, in these ions, the strength of the crystal field is much greater than the strength of the spin–orbit interaction, just the reverse of the situation in rare-earth ions. Typical strengths of the crystal field and spin–orbit interactions in the iron group are $10^4$ cm$^{-1}$ and $10^2$ cm$^{-1}$, respectively.

## 9.7 MAGNETISM IN METALS

Most metals are paramagnetic. There are also a few important metals which exhibit ferromagnetism. In this section we shall treat only paramagnetic metals; we shall discuss ferromagnetic metals in Section 9.10. The conduction electrons in the metal make two contributions: A paramagnetic one due to their spins, and a diamagnetic one due to their orbital motions, induced by the magnetic field. The *net electronic susceptibility* is the difference between these contributions.

### Spin paramagnetism

Spin paramagnetism arises from the fact that each conduction electron carries a spin magnetic moment which tends to align with the field. In calculating the susceptibility, one may be inclined to use result (9.41), with $j = s = \frac{1}{2}$, which gives

$$\chi = \frac{\mu_0 N \mu_B^2}{kT}, \tag{9.46}$$

where we have also set $g = 2$, since we are dealing with a pure spin motion. This shows that $\chi \sim 1/T$.

Experiments show, however, that spin susceptibilities in metals are essentially independent of temperature. The observed values are also considerably smaller than predicted by (9.46). These facts clearly cast strong doubts on the applicability of (9.41) to the conduction electrons.

The source of the difficulty lies in the fact that Eq. (9.41) was derived on the basis of localized electrons obeying the Boltzmann distribution. The conduction electrons, on the other hand, are delocalized, and satisfy the Fermi–Dirac distribution (see Section 4.6).

The proper treatment, taking this into account, is illustrated in Fig. 9.12. In the absence of the field, half the electrons have spins pointing in the positive $z$-direction, and the other half in the negative direction (Fig. 9.12a), resulting in a vanishing net magnetization. When a field is applied along the $z$-direction, the energy of the spins parallel to $B$ is lowered by the amount $\mu_B B$, while the energy of spins opposite to $B$ is raised by the same amount (Fig. 9.12b). The situation which ensues is energetically unstable, and hence some electrons near the Fermi level begin to transfer from the opposite-spin half to the parallel-spin one, leading to a net magnetization. Note that only relatively few electrons near the Fermi level are able to flip their spins and align with the field. The other electrons, lying deep within the Fermi distribution, are prevented from doing so by the exclusion principle (see the similar discussion in Section 4.6).

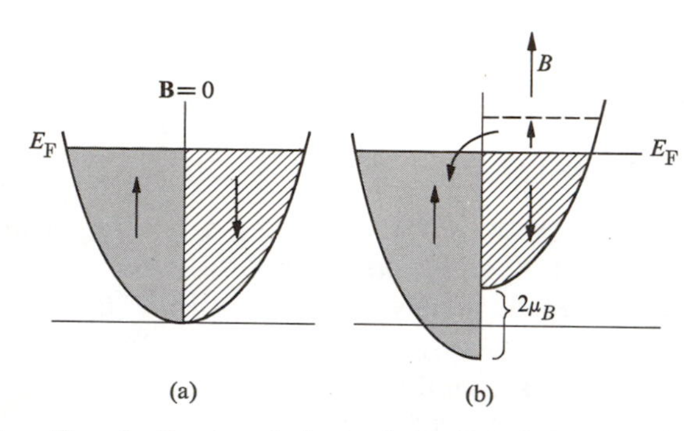


**Fig. 9.12** (a) When **B** = 0, the two halves of the Fermi–Dirac distribution are equal, and thus $M = 0$; (b) When a field **B** is applied, spins in the antiparallel half flip into the parallel half, resulting in a net parallel magnetization.

We can now derive a good estimate of the magnetic susceptibility. The electrons participating in the spin flip occupy an energy interval of thickness about equal to $\mu_B B$ (Fig. 9.12). Thus their concentration is given by $N_{eff} = \frac{1}{2}\, g(E_F)\mu_B B$, where $g(E_F)$ is the density of states at the Fermi energy level [the factor $\frac{1}{2}$ is inserted because $g(E_F)$ as defined in Section 6.11 includes both spin directions, while in the present circumstances only one spin direction is involved in the flipping]. Since each spin flip increases the magnetization by $2\mu_B$ (from $-\mu_B$ to $+\mu_B$), it follows that the net magnetization is given by

$$M \simeq N_{eff}\, 2\mu_B = \tfrac{1}{2}\, g(E_F)\mu_B 2\mu_B = \mu_B^2 g(E_F)B,$$

leading to a paramagnetic susceptibility

$$\chi_p \simeq \mu_0 \mu_B^2 g(E_F). \tag{9.47}$$

The susceptibility is thus determined by the density of states at the Fermi level; and the quantity $g(E_F)$, which is so important in transport phenomena (Section 6.18), plays a major role here also. One can thus obtain information on $g(E_F)$ by measuring $\chi_p$.

According to (9.47), $\chi_p$ is essentially independent of temperature. This is seen from the fact that temperature has only a small effect on the Fermi–Dirac distribution of the electrons, and consequently the derivation leading to (9.47) remains valid.

If we apply (9.47) to a band of standard type, we have $g(E_F) = 3N/2E_F$ [see Eqs. (5.63) and (4.34)]. The equation then leads to

$$\chi_p \simeq \tfrac{3}{2}\chi\frac{T}{T_F}, \tag{9.48}$$

where $\chi$ is the classical susceptibility of (9.46) and $T_F$ the Fermi temperature ($E_F = kT_F$). Since $T_F$ is very large, often 30,000°K or higher, we can see that $\chi_p$ is smaller than $\chi$—by a factor of about $10^2$—again in agreement with experiment.

In transition metals, the paramagnetic susceptibility is exceptionally large, because $g(E_F)$ is large, by virtue of the narrow and high 3d band.

### Diamagnetism

Conduction electrons also exhibit diamagnetism on account of the cyclotron motion they execute in the presence of the magnetic field, as shown in Fig. 9.13. Each electron loop is equivalent to a dipole moment whose direction is opposite to that of the applied field.

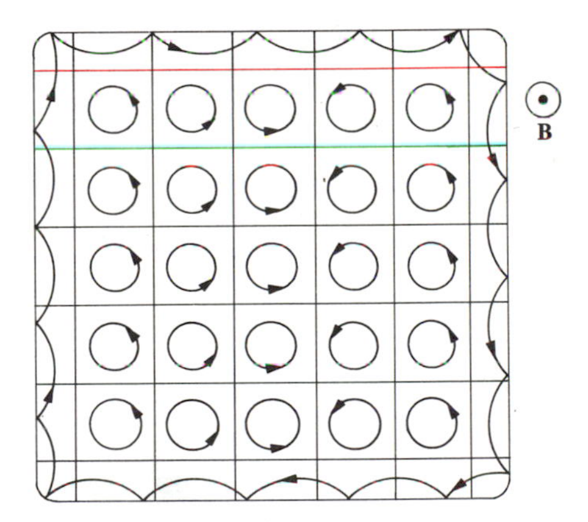


**Fig. 9.13** Diamagnetic effect of cyclotron motion in metals. Electrons at the boundaries tend to cancel the effect of the bulk electrons.

Classical treatment shows that the total diamagnetic contribution of all electrons is zero. The effect of the closed electron loops is canceled by the effect of the electron traveling in the opposite direction along the boundaries.†

† Note that the area of the loop at the boundary is equal to the total area of all the inner loops. Since the magnetic moment of a current loop is proportional to the area of the loop, it follows that the moment of the surface loop just cancels that of the inner loop. The fact that the total magnetic susceptibility for any charge and current is zero, according to classical electrodynamics, is known as the *Van Leeuwen theorem*. For a proof, see Van Vleck (1932).

Quantum treatment, which is too complicated to present here (see Martin, 1967), shows that there is a nonvanishing diamagnetic contribution which is equal to one-third of the spin paramagnetic susceptibility given by (9.48). The net response is therefore paramagnetic.

**Table 9.5**

Susceptibilities of Some Monovalent and Divalent Metals × $10^6$
(Room Temperature)

| | Experimental | | | Theoretical |
|---|---|---|---|---|
| Element | $\chi_{\text{total}}$ (expt) | $\chi_{\text{core}}$ | $\chi_{\text{electron}} = \chi_{\text{total}} - \chi_{\text{core}}$ | $\chi_{\text{electron}} = \chi_{\text{spin}} + \chi_{\text{orbit}}$ |
| K | 0.47 | − 0.31 | 0.76 | 0.35 |
| Rb | 0.33 | − 0.46 | 0.79 | 0.33 |
| Cu | − 0.76 | − 0.20 | 1.24 | 0.65 |
| Ag | − 2.1 | − 3.0 | 0.9 | 0.60 |
| Au | − 2.9 | − 4.3 | 1.4 | 0.60 |
| Mg | 0.95 | − 0.22 | 1.2 | 0.65 |
| Ca | 1.7 | − 0.43 | 2.1 | 0.5 |

In comparing theoretical results with experiment, one must also include the diamagnetic effect of the ion cores, which can be treated according to Section 9.5. Table 9.5 gives the results for some metallic elements.

## 9.8 FERROMAGNETISM IN INSULATORS

*Ferromagnetism* is the phenomenon of spontaneous magnetization. The best-known examples of ferromagnets are the transition metals Fe, Co, and Ni, but other elements and alloys involving transition or rare-earth elements also show ferromagnetism. Thus the rare-earth metals Gd, Dy, and the insulating transition metal oxide $CrO_2$ all become ferromagnetic under suitable circumstances.

Ferromagnetism involves the alignment of an appreciable fraction of the molecular magnetic moments in some favorable direction in the crystal. The fact that the phenomenon is restricted to transition and rare-earth elements indicates that it is related to the unfilled 3d and 4f shells in these substances.

Ferromagnetism appears only below a certain temperature, which is known as the *ferromagnetic transition temperature* or simply as the *Curie temperature*. This temperature depends on the substance, but its order of magnitude is about 1000°K, as seen from Table 9.6. Thus the ferromagnetic range often includes the whole of the usual temperature region.

Above the Curie temperature, the moments are oriented randomly, resulting in a zero net magnetization. In this region the substance is paramagnetic, and its susceptibility is given by

$$\chi = \frac{C}{T - T_f}, \tag{9.49}$$

which is known as the *Curie–Weiss law*. The constant $C$ is called the *Curie constant* and $T_f$ the *Curie temperature*. Expression (9.49) is of the same form as (9.34), the Langevin susceptibility, except that the origin of temperature is shifted from 0 to $T_f$. Figure 9.14 illustrates the applicability of the Curie–Weiss law to Ni; notable deviation appears only near the Curie point.

**Table 9.6**

Curie Temperature and Saturation Magnetizations for Ferromagnetic Substances†
($n_B$ is the number of magnetons per unit at 0°K)

| Substance | $T_f$, °K | $M_s$, at 0°K | $n_B$, at 0°K |
|---|---|---|---|
| Fe | 1043 | $1.74 \times 10^6$ amp·m$^{-1}$ | 2.22 |
| Co | 1403 | 1.45 | 1.72 |
| Ni | 631 | 0.5 | 0.54 |
| Gd | 289 | 2.01 | 7.10 |
| Dy | 105 | 2.92 | 10.1 |
| $CrO_2$ | 515 | — | — |
| $MnOFe_2O_3$ | 410 | — | 2.03 |
| $FeOFe_2O_3$ | 480 | — | 5.0 |
| $Y_3Fe_5O_{12}$(YIG) | 130 | 0.2 | 5.0 |

† Temperatures listed are actual ferromagnetic transition temperatures, which are slightly lower than those values for the Curie law in the paramagnetic region. The law does not hold well very near the transition point.

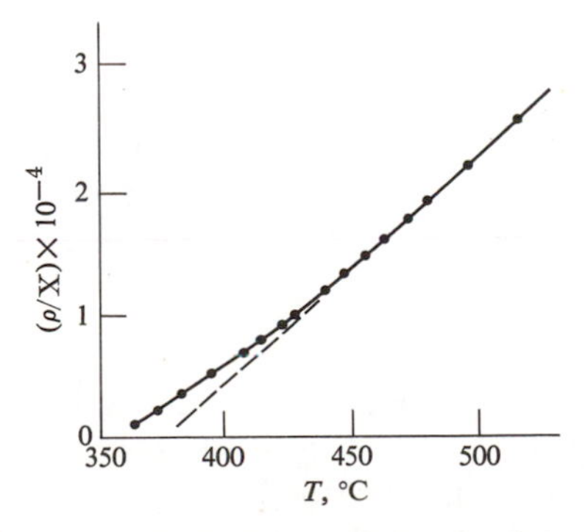


**Fig. 9.14** The reciprocal of susceptibility per gram of Ni (a ferromagnetic substance) near the Curie point (358°C). The quantity $p$ is the mass density. Dashed line represents extrapolation from high-temperature region.

Note that Eq. (9.49) predicts a divergence in $\chi$ as the temperature is lowered toward $T_f$. This is an indication of the oncoming transition to the ferromagnetic phase.

In the temperature range $T < T_f$, spontaneous magnetization is referred to as *saturation magnetization*. This magnetization increases as the temperature is lowered (Fig. 9.15), reaching its maximum at $T = 0°\text{K}$. Thus, as the temperature is reduced, more and more dipoles line up in the magnetization direction. Table 9.6 gives values of saturation magnetization in various materials.

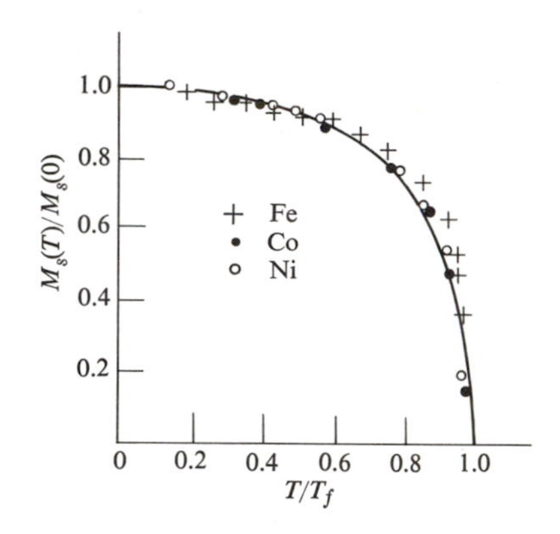


**Fig. 9.15** Ratio of saturation magnetization at temperature $T$ to that at 0°K, $M_S(T)/M_S(0)$ versus $T/T_f$ for Fe, Co, and Ni. Solid curve is obtained from Weiss theory, Eq. (9.55), for $j = \frac{1}{2}$.

Ferromagnetism appears in both metals and insulators. It is simpler to treat the latter materials, however, and therefore we shall limit our discussion here to these materials. We shall talk about ferromagnetic materials in Section 9.10.

### The molecular field theory

In the ferromagnetic region the moments are magnetized spontaneously, which implies the presence of an internal field to produce this magnetization. We shall follow Weiss and assume that this field is proportional to the magnetization

$$\mathscr{H}_\text{W} = \lambda M, \tag{9.50}$$

where $\lambda$ is the *Weiss constant*. For agreement with experiments, $\lambda$ turns out to be very large—about $10^4$. The origin of this enormous field $\mathscr{H}_\text{W}$ will be discussed later in the section, but for the moment we shall take it as a phenomenologically given field which acts to align the molecules. Ultimately, of course, it must arise as a result of the interaction between the molecules, and is referred to as the *molecular field*.

We can use (9.40) to evaluate the magnetization produced by the field $\mathscr{H}_W$. Assuming for simplicity that $j = \frac{1}{2}$, for which (9.40) is applicable, we write

$$M = Ng\mu_B \tanh\,(\mu_0 g\mu_B \lambda M/kT), \tag{9.51}$$

in which the field is entirely due to the internal field of (9.50). This is a transcendental equation in $M$, which we shall solve in the following graphical fashion. We denote the argument of the hyperbolic function by $x$. That is,

$$M = \frac{kT}{\mu_0 g\mu_B \lambda}\, x. \tag{9.52}$$

Equation (9.51) now takes the form

$$M = Ng\mu_B \tanh x. \tag{9.53}$$

These two equations are solved simultaneously by plotting them on an $M$-*versus*-$x$ graph, and finding the points of intersection. Figure 9.16 shows the hyperbolic curve corresponding to (9.53), and the straight line corresponding to (9.52); this line is plotted for several temperatures. For temperatures below a certain critical value, the two curves intersect at a point such as $A$, which represents finite spontaneous magnetization. Thus the molecular field (9.50) does indeed lead to ferromagnetism. The other intersection point, at the origin, represents a nonmagnetized state, but this state is energetically unstable.

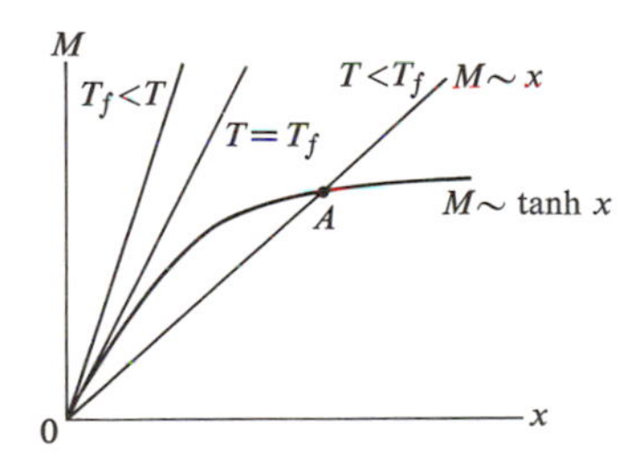


**Fig. 9.16** The curves $M \sim x$, a straight line, and $M \sim \tanh x$ versus $x$. The intersection point $A$ represents spontaneous magnetization, i.e., a ferromagnetic state.

The critical temperature is the temperature at which the straight line (9.54) becomes tangential to the hyperbolic curve at the origin. Making the approximation $\tanh x \simeq x$, valid for small $x$, and equating $M$ in the two equations (9.52) and (9.53) yields the result

$$\lambda = \frac{kT_f}{\mu_0 N(g\mu_B)^2}, \tag{9.54}$$

which relates the Weiss constant $\lambda$ to the Curie temperature $T_f$, and since the latter is a measurable quantity, we have here a method for determining $\lambda$. If one sets

$T_f = 10^{3}$°K, $N = 10^{29}$ m$^{-3}$, and appropriate values for the other constants, one finds $\lambda \simeq 10^4$, as we have previously stated.

It is evident from Fig. 9.16 that the maximum magnetization is $M(0) = Ng\mu_B$, which is achieved as $T \to 0$°K. Equation (9.51) may also be written as

$$\frac{M}{M(0)} = \tanh\left(\frac{T}{T_f}\right), \tag{9.55}$$

where we used (9.54). Thus if we plot the reduced magnetization $M/M(0)$ versus the reduced temperature $T/T_f$, we obtain a universal curve applicable to all magnetic substances of the same value of $j$. This is confirmed by Fig. 9.15.

The molecular field also leads to the Curie–Weiss law in the paramagnetic region $T > T_f$. The total field is now

$$\mathscr{H}_{\text{tot}} = \mathscr{H} + \mathscr{H}_{\text{W}},$$

where $\mathscr{H}$ is the applied field and $\mathscr{H}_{\text{W}}$ the molecular field. When we use (9.40), assuming that the total field is small, we have

$$M = M(0)\frac{\mu_0 g\mu_B}{kT}(\mathscr{H} + \lambda M),$$

which may be written, with the help of (9.54), as

$$M = \left(\frac{T_f}{\lambda}\right)\frac{1}{T - T_f}\mathscr{H}.$$

The susceptibility is given by

$$\chi = \frac{C}{T - T_f},$$

where $C = T_f/\lambda = \mu_0 N(g\mu_B)^2/k$, which is of the form of the Curie–Weiss law.

**The physical origin of the molecular field**

The presence of the molecular field indicates that neighboring moments interact with each other, and that the interaction is spin-dependent. The interaction energy between two moments may be written as

$$V_{\text{ex}} = -J'\,\mathbf{s}_1 \cdot \mathbf{s}_2, \tag{9.56}$$

where $\mathbf{s}_1$ and $\mathbf{s}_2$ are the two spins,† and $J'$ is called the *exchange constant*. The energy $V_{\text{ex}}$ is referred to as the *exchange energy*.

† The vectors $\mathbf{s}_1$ and $\mathbf{s}_2$ are related to the actual angular momenta by the relations $\mathbf{S}_1 = \mathbf{s}_1\hbar$, and $\mathbf{S}_2 = \mathbf{s}_2\hbar$. Thus $\mathbf{s}$ is a dimensionless vector in the same direction as $\mathbf{S}$ and has the length $[s(s+1)]^{\frac{1}{2}}$ where $s$ is the angular momentum quantum number. The constant $J'$ has the dimension of energy. The definition of dimensionless spin vectors is made here for convenience.

In order for the above interaction to lead to ferromagnetism, the constant $J'$ must be positive, because the parallel-spin state—that is, $\mathbf{s}_1 = \mathbf{s}_2$—has an energy $-J's^2$, while the antiparallel-spin state, $\mathbf{s}_1 = -\mathbf{s}_2$, has an energy $J's^2$. Consequently the former is lower than the latter only if $J' > 0$.

The exchange constant $J'$ is related to the Weiss constant $\lambda$. If we assume that the dipole experiences exchange interaction only with its nearest neighbors (the constant $J'$ decreases very rapidly with the distance between the dipoles), the total exchange for the dipole is $-zJ's^2$, where $z$ is the number of nearest neighbors. This is equivalent to a molecular magnetic field $\mathscr{H}_W$ given by

$$zJ's^2 = (gs\mu_B)(\mu_0 \mathscr{H}_W), \tag{9.57}$$

where $gs\mu_B$ is the value of the magnetic moment. The maximum value of $\mathscr{H}_W$ is equal to $\lambda M(0) = \lambda N g s u_B$, according to (9.50), which, when inserted in (9.57), yields

$$J' = \frac{\mu_0 N (g\mu_B)^2}{z} \lambda. \tag{9.58}$$

As expected, $J'$ is proportional to $\lambda$, both being measures of the strength of the molecular field, and consequently also proportional to the Curie temperature. Substitution of the appropriate values for the various constants yields a value $J' \simeq 0.1$ eV, which is a typical value for the exchange energy between two neighboring moments in a ferromagnetic crystal.

We now turn to the origin of the interaction energy (9.56). The most natural suggestion is the so-called dipole–dipole interaction, which gives an energy of the order

$$V_{12} \simeq \mu_0 \frac{\mu_B^2}{r^3},$$

where $r$ is the distance between the dipoles. If one substitutes a typical value for $r$, however, one finds that $V_{12} \simeq 10^{-4}$ eV, which is about three orders of magnitude smaller than the observed value. Thus the dipole–dipole interaction cannot account for ferromagnetism, and we must look for another, much stronger, type of interaction.

The correct approach to the problem was made first by Heisenberg. The requirement of the Pauli exclusion principle introduces forces which are *spin-dependent*, because the statement of the principle includes the spin. These so-called *exchange forces* are strong because they are of the same order as the Coulomb force.† Consider, for example, the hydrogen molecule. There are two

† The reason for using the word "exchange" in connection with these forces is that they follow from a quantum principle which states that electrons cannot be distinguished from each other. Thus if any two electrons are permuted or exchanged, the observable properties of the system do not change. This principle is essentially equivalent to the Pauli exclusion principle.

electrons moving in the Coulomb field of two nuclei, and there are two possible arrangements for the spins of the electrons: either parallel or antiparallel. If they are parallel, the exclusion principle requires the electrons to remain far apart. If they are antiparallel, the electrons may come closer together and their wave functions overlap considerably. These two arrangements have different energies because, when the electrons are close together, the energy rises as a result of the large Coulomb repulsion. This factor alone favors the parallel-spin state, but there are other factors which compensate and favor the antiparallel-spin state. Which state actually exists depends on which of these factors prevails. In the hydrogen molecule, the ground state corresponds to the antiparallel arrangement, i.e., the nonmagnetic state. In ferromagnetic substances, however, the opposite situation prevails, and the parallel arrangement has the lower energy.

The point is that the exclusion principle gives rise to a spin-dependent force between the moments, whose strength is essentially given by the Coulomb interaction,

$$V_{12} \simeq \frac{e^2}{4\pi\epsilon_0 r},$$

which is far stronger than the dipole–dipole interaction. You can show that this gives the correct order of magnitude for the interaction.

Slater suggested a criterion for the occurrence of ferromagnetism. The critical factor is the ratio $r/2r_a$, where $r$ is the interatomic distance and $r_a$ the atomic radius. Figure 9.17 is a plot of $J$ versus the above ratio for various transition metals. It is only when the ratio exceeds 1.5 that $J'$ becomes positive and the material shows ferromagnetism. The substances Fe, Ni, and Co satisfy the criterion, but Cr and Mn fail, and these latter are not, in fact, ferromagnetic.

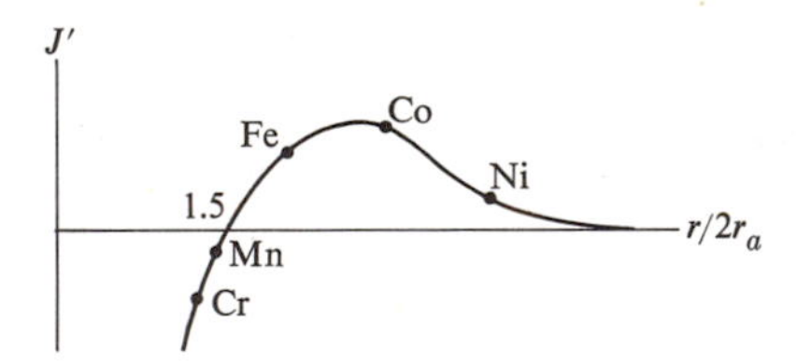


**Fig. 9.17** Exchange constant $J'$ versus interatomic distance for transition elements.

Slater's criterion underscores the importance of the 3d shell in the origin of ferromagnetism. The fact that the radius of this shell is small plays a crucial role in the appearance of the phenomenon. A similar comment applies to the 4f shell in the rare-earth ferromagnets.

## 9.9 ANTIFERROMAGNETISM AND FERRIMAGNETISM

The only type of magnetic order which has been considered thus far is ferromagnetism, in which, in the fully magnetized state, all the dipoles are aligned in exactly the same direction (Fig. 9.18a). There are, however, substances which show different types of magnetic order. Figure 9.18(b) illustrates an *antiferromagnetic* arrangement, in which the dipoles have equal moments, but adjacent dipoles point in opposite directions. Thus the moments balance each other, resulting in a zero net magnetization. Another type of arrangement commonly encountered is the *ferrimagnetic* pattern shown in Fig. 9.18(c). Neighboring dipoles point in opposite directions, but since in this case the moments are unequal, they do not balance each other completely, and there is a finite net magnetization. Other more complicated arrangements, some of which are variations on the ones already mentioned, have been observed, but the three major classes of Fig. 9.18 will suffice for our purposes here. Let us now briefly discuss the antiferromagnetic and ferrimagnetic arrangements.

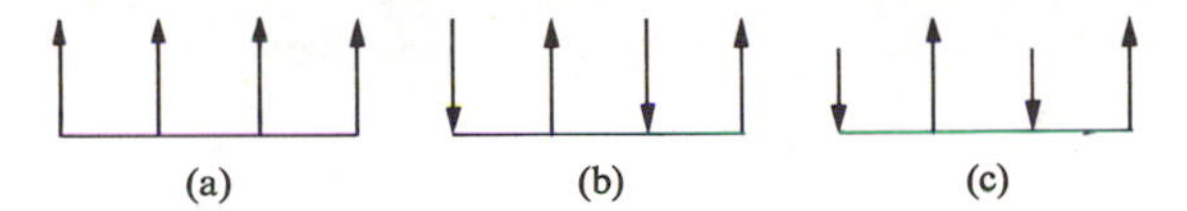


**Fig. 9.18** Magnetic arrangements: (a) ferromagnetic, (b) antiferromagnetic, (c) ferrimagnetic.

### Antiferromagnetism

Antiferromagnetism is exhibited by many compounds involving transition metals. The crystal $MnF_2$ shown in Fig. 9.19 is an ionic crystal in which electrons have been transferred from the manganese to the fluorine atoms (chemical notation $Mn^{2+}F_2^-$). The manganese ions are magnetic because of their incomplete 3d shell, and are distributed over an fcc structure. The substance is antiferromagnetic because the ions at the corners all point in one direction, while the ions at the cube center all point in the opposite direction.

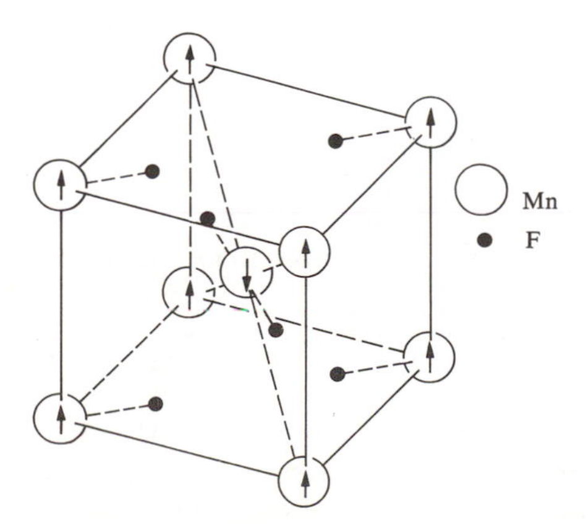


**Fig. 9.19** Spin structure of $MnF_2$.

As in ferromagnetism, antiferromagnetism also disappears at a certain point as the temperature is raised. The transition point is called the *Néel temperature* $T_N$. Above this point the substance is paramagnetic, and the susceptibility is well represented by the formula

$$\chi = \frac{C}{T + T'_N}, \tag{9.59}$$

where $C$ and $T'_N$ are constants depending on the substance. This behavior is shown in Fig. 9.20 for $MnF_2$, whose Neel temperature is $T \;\; = 72°K$. Note that the susceptibility does not diverge at the transition point, unlike the ferromagnetic case.

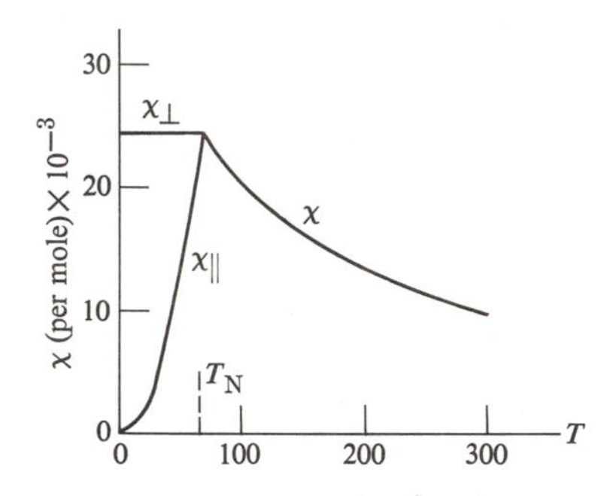


**Fig. 9.20** Susceptibility $\chi$ versus $T$ for $MnF_2$, whose $T_N = 78°K$. (The quantities $\chi_{\|}$ and $\chi_{\perp}$ below $T_N$ refer to susceptibilities for the field parallel to and perpendicular to the spontaneous spin direction, respectively. [After Bizette and Tsai, *Compt. rend.* (Paris), **238**, 1575 (1954).]

The temperatures $T_N$ and $T'_N$ are listed in Table 9.7 for some substances. One can relate these temperatures to parameters characterizing the magnetic interactions in the material. This is done by generalizing the molecular-field theory of ferromagnetism to the present situation by introducing two Weiss constants, $\lambda_1$ and $\lambda_2$, where $\lambda_1$ describes the interaction of the dipole with other equivalent dipoles, and $\lambda_2$ the interaction with the dipoles of the opposite orientation (nearest neighbors). One may then establish that

$$T_N = \frac{C}{2}(\lambda_1 - \lambda_2) \qquad \text{and} \qquad T'_N = \frac{C}{2}(\lambda_1 - \lambda_2). \tag{9.60}$$

We may well ask: Since the net magnetization $M = 0$ for an antiferromagnetic phase, how can this be distinguished from a nonmagnetic state when there is no magnetic order at all? An obvious answer can be given on the basis of the behavior of susceptibility as a function of temperature. A paramagnetic substance obeys the Curie law $\chi \sim 1/T$ at all temperatures, while an antiferromagnetic substance exhibits the behavior shown in Fig. 9.20. One can also ascertain the magnetic order in the antiferromagnetic phase by means of neutron diffraction. Below the Néel temperature, the dipoles form what amounts to two interpenetrating magnetic lattices of opposite spins, which give rise to Bragg reflection of the neutron beam.

**Table 9.7**

Antiferromagnetic Data

| Substance | $T_N$, °K | $T'_N$, °K |
|---|---|---|
| MnO | 116 | 610 |
| FeO | 198 | 570 |
| CoO | 291 | 330 |
| NiO | 525 | ~ 2000 |
| MnS | 160 | 528 |
| MnTe | 307 | 690 |
| $MnF_2$ | 67 | 82 |
| $Cr_2O_3$ | 307 | 485 |

**Ferrimagnetism**

Ferrimagnetic substances, often referred to as *ferrites*, are ionic oxide crystals whose chemical composition is of the form $XFe_2O_4$, where X signifies a divalent metal. These often crystallize in the *spinel* structure, shown in Fig. 9.21 (spinel is actually the compound $MgAl_2O_4$).

The most familiar example of this group is magnetite (lodestone), whose chemical formula is $Fe_3O_4$. More explicitly, the chemical composition is $(Fe^{2+}O^{2-})$ $(Fe_2^{3+}O_3^{2-})$, showing that there are two types of iron ions: ferrous (doubly charged), and ferric (triply charged). The compound crystallizes in the spinel structure of Fig. 9.21, with the ferrous ions replacing Mg and the ferric ions replacing aluminium. The unit cell contains 56 ions, 24 of which are iron ions and the remainder oxygen. The magnetic moments are located on the iron ions.

If we study the unit cell closely, we find that the Fe ions are located in either of two different coordinate environments: A tetrahedral one, in which the Fe ion is surrounded by 4 oxygen ions, and an octahedral one, in which it is surrounded by 6 oxygen ions. Of the 16 ferric ions in the unit cell, 8 are in one type of position and 8 are in the other. Furthermore, the tetrahedral structure has moments oriented opposite to those of the octahedral one, resulting in a complete cancellation of the contribution of the ferric ions. The net moment therefore arises entirely from the 8 ferrous ions which occupy octahedral sites. Each of these ions has six 3d electrons, whose spin orientations are $\uparrow\uparrow\uparrow\uparrow\uparrow\downarrow$. Hence each ion carries a moment equal to 4 Bohr magnetons. Since the length of the edge of the cubic cell, as given by x-ray analysis, is 8.37 Å, it follows that the saturation magnetization is $M_s = 4\mu_B/a^3 = 0.56 \times 10^6$ A/m.

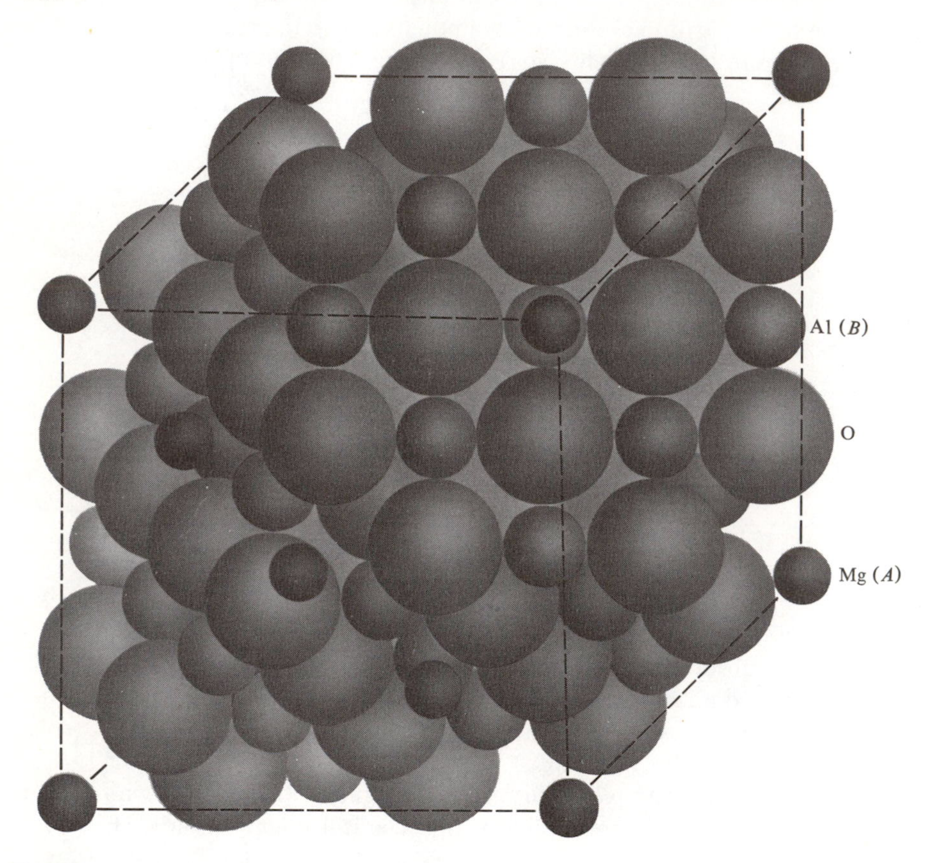


**Fig. 9.21** The spinel structure of $MgAl_2O_4$. The $A$ and $B$ sites are occupied by Mg and Al atoms, respectively. (After Azaroff)

Other metallic ions may be substituted for the ferrous ions in $Fe_3O_4$, resulting in other ferrimagnetic compounds. Examples of these are Ni, Mn, Mg, Zn, etc.

In modern applications, ferrites are the most useful of all magnetic materials, because, in addition to their magnetic properties, they are also good electrical insulators, unlike the ferromagnetic metals. Thus losses due to free electrons are eliminated.

## 9.10 FERROMAGNETISM IN METALS

The model we have used in discussing ferromagnetism in insulators cannot be applied directly to metals. This model assumes that the electrons are localized around the lattice sites, while in metals the electrons are delocalized, extending over the whole crystal. The scheme used to describe the magnetic properties of such electrons is called the *itinerant-electron model*, and was first developed by Stoner.

The failure of the localized model to account for ferromagnetism in metals can be illustrated by the following. If this model were applicable, then the magnetic moment per atom would be $s\mu_B$, where $s$ is an integer or half integer. By contrast, this number is found to be 2.22, 1.72, and 0.54 for Fe, Co, and Ni, respectively.

We shall now proceed with the itinerant model. The electrons of interest occupy the 3d band (this band overlaps the 4s band, but the latter does not contribute to ferromagnetism and hence is ignored in the present discussion).

Figure 9.22(a) shows this band divided into two subbands, representing the two possible orientations, up and down. In the nonmagnetic state shown in Fig. 9.22(a), the two subbands are equally populated, resulting in a zero magnetization.

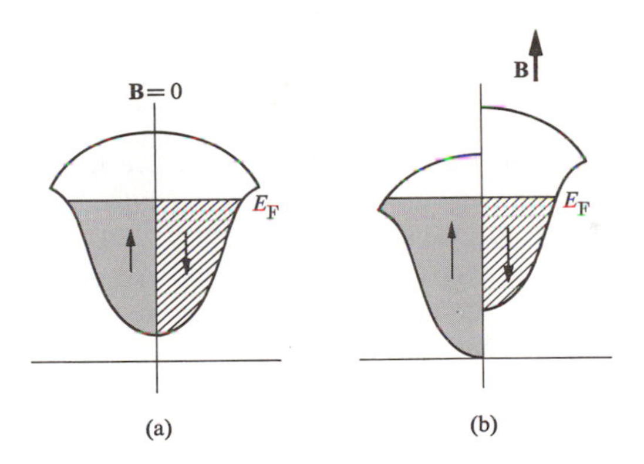


**Fig. 9.22** Magnetization process in the itinerant model.

Let us now assume that there is an exchange interaction. This tends to align the moments in the up direction. Thus, in order to lower their energies, the electrons transfer from the down to the up direction. But when this happens, a net magnetization develops, and the energies of the two subbands are no longer equal. The down-subband is displaced upward relative to the up-subband, as shown in Fig. 9.22(b). The resulting magnetization is the saturation magnetization observed in ferromagnetism. The amount of this magnetization depends on the relative displacement of the subbands, which, in turn, is determined by the strength of the exchange interaction and the shape of the band.

Let us express these ideas quantitatively. When an electron flips its moment, it loses an amount of exchange energy $\frac{1}{2}BM = \frac{1}{2}(\mu\mathscr{H}_W)M = \frac{1}{2}\mu_0\lambda M^2$, where $\mathscr{H}_W$ is the molecular field (the factor $\frac{1}{2}$ arises because we are calculating the self energy). For a flip of one electron, $M = 2\mu_B$, because the electron has reversed its moment from $-\mu_B$ to $+\mu_B$. Thus the loss of energy is $\frac{1}{2}\mu_0\lambda(2\mu_B)^2 = 2\mu_0\lambda\mu_B^2$. It would seem at first that the system could achieve the lowest energy when all the down electrons flipped their moments, so that the system was completely magnetized in the up direction.

This is not the case, however, because, as Fig. 9.22(b) shows, the transferred electrons gain in kinetic energy; they are now farther from the bottom of the band. Therefore, in order for the electron to make the transfer the loss in exchange energy must exceed the gain in kinetic energy. We calculated the loss in exchange energy above, and we can estimate the gain in kinetic energy as follows. Suppose that $n$ electrons near the Fermi level are transferred from the down- to the up-subband. The new energy range $\Delta E$ occupied above $E_F$ in the up-subband is given by $n = \frac{1}{2}g(E_F)\Delta E$, where $g(E_F)$ is the density of states at the Fermi level. [The factor $\frac{1}{2}$ is included because $g(E_F)$ was defined to include both spin directions, while here we are considering only the up-subband.] For a transfer of one electron, $n = 1$, and hence the kinetic energy gain is $\Delta E = 2/g(E_F)$. Therefore the condition for ferromagnetism may be expressed as

$$2\mu_0\lambda\mu_B^2 > \frac{2}{g(E_F)}. \tag{9.61}$$

For this to be satisfied, the exchange constant must be large, which requires an atomic shell of small radius (see Fig. 9.17). Also $g(E_F)$ must be large, which requires a narrow band. These requirements are consistent because the smaller the radius of the shell, the less the overlap of the wave functions, and hence the narrower the band. These requirements are satisfied by the 3d band in Fe, Co, and Ni, and also by the 4f band in Gd and Dy.

The fact that a large $g(E_F)$ enhances ferromagnetism is evident from the following consideration. When $g(E_F)$ is large, the band can accommodate a large number of electrons in a small energy range, and thus the gain in kinetic energy occasioned by the electron flipping its moment is small. But when $g(E_F)$ is small, the band is essentially flat, like the 4s band, and the gain in kinetic energy is quite large. This rules out ferromagnetism in such a band.

Figure 9.23 illustrates the band picture of the ferromagnetic state in Ni.

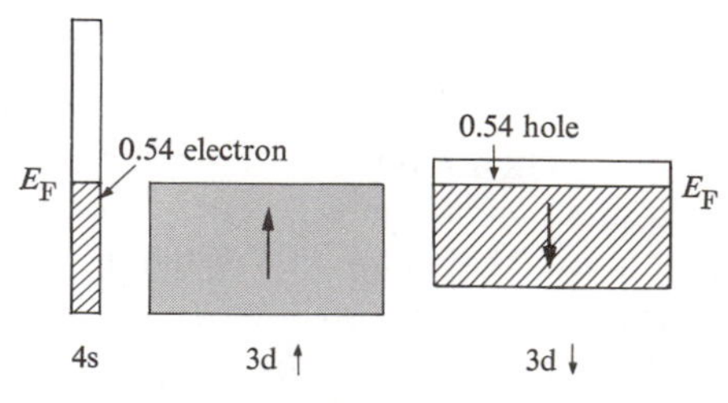


**Fig. 9.23** Occupation of the 3d and 4s bands in nickel; 0.54 electron per atom, on the average, is transferred from the 3d↓ to the 4s band.

Our presentation of the itinerant model is naturally a simplified one, and condition (9.61) should be viewed only as a semi-quantitative guide. The basic difficulty in constructing such a model is that the band concept, despite its usefulness, begins to break down somewhat when applied to narrow bands. In these bands the electrons tend to have a measure of localization around atomic sites, which means that the electron–electron correlation also becomes important. Yet such correlation is entirely ignored in the usual band model. This point is relevant to ferromagnetism because both the 3d and the 4f are narrow bands. Although much work is now being applied to it—and much progress has been made—this problem remains essentially unsolved.

## 9.11 FERROMAGNETIC DOMAINS

Ferromagnetic materials in their natural state are usually found to be demagnetized even below the Curie temperature. To explain this, Weiss postulated that the substance is divided into a large number of small *domains*, in which each domain is magnetized, but the directions of magnetization in the various domains are such that they tend to cancel each other, leading to a vanishing net magnetization. Though Weiss originally formulated this postulate on theoretical grounds, it has since been confirmed experimentally. One can observe the domain structure by carefully polishing the surface of the ferromagnetic substance, and spreading over it a fine powder of ferromagnetic particles. The particles collect along the domain boundaries. Figure 9.24 shows the powder pattern for a silicon-iron crystal. (Domains may also be observed by the use of a polarizing microscope; see the question section at the end of this chapter.)

The formation of the domain, and its shape, depend on the competition among a number of energy terms present in the magnetic crystal. Suppose that the whole crystal is in a state of uniform magnetization, as in Fig. 9.25(a). This state has the lowest possible exchange energy, since all adjacent spins are parallel to each other. However, it also has a large amount of magnetostatic energy. Because of the magnetization, there is a positive magnetic charge on the lower surface. These charges produce a magnetic field opposite to **M**, which is called the *demagnetization field* $\mathbf{B}_d$. Because **M** is opposite to $\mathbf{B}_d$, there is a positive magnetostatic energy whose density is given, according to (9.4), by

$$E_m = \tfrac{1}{2} M \mathbf{B}_d. \tag{9.62}$$

The value of $\mathbf{B}_d$ depends on the shape of the surface, and is usually written as $\mathbf{B}_d = -\mu_0 D\mathbf{M}$, where $D$ is the *demagnetization factor*.† This factor, which is large for a flat sample and small for an elongated sample, is equal to unity for a sample in the shape of a thin, flat disc normal to the field. The magnetostatic energy is of the order of $10^6$ J/m$^3$.

† The demagnetization factor is the same as the depolarization factor for a sample of the same geometrical shape (see Problem 8.7).

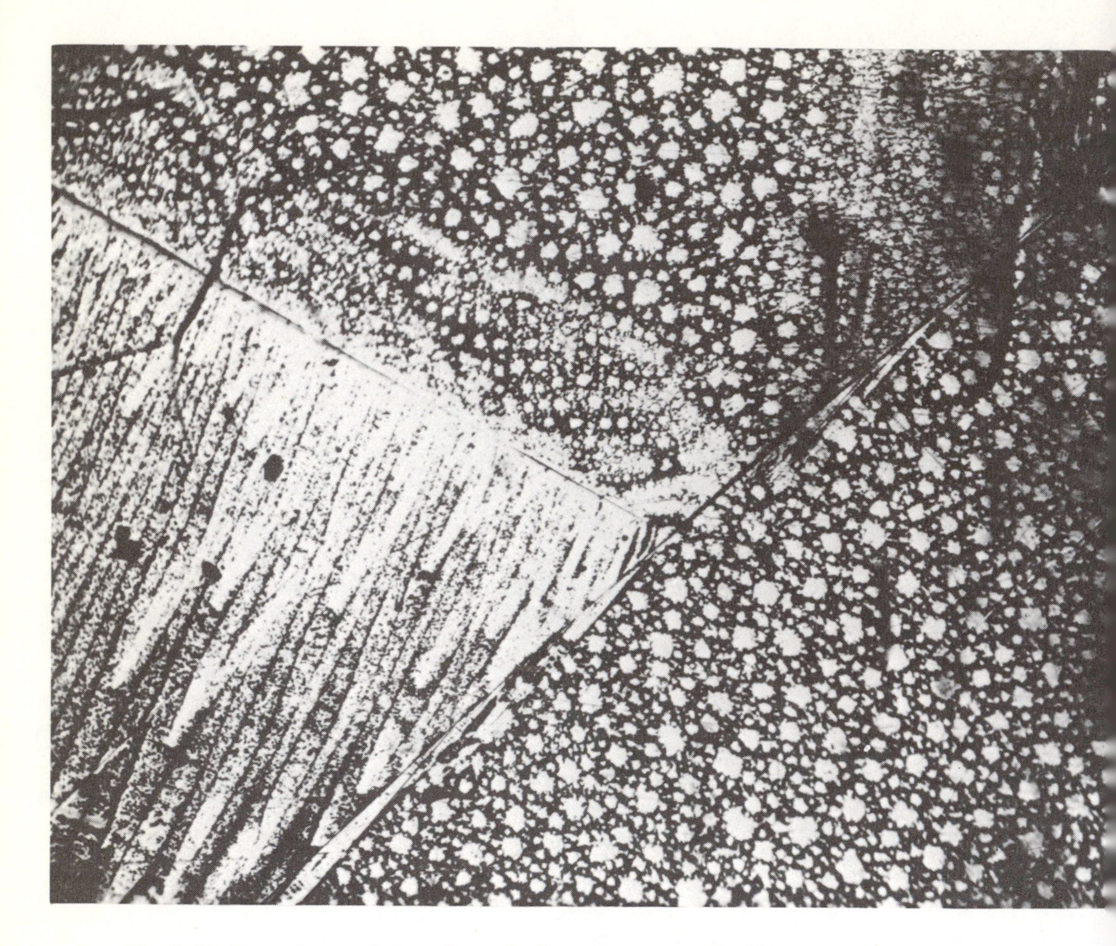

**Fig. 9.24** Domains and domain walls in a ferromagnetic Si-Fe crystal. (From Walter J. Moore, *Seven Solid States*, New York: W. A. Benjamin, 1967.)

In order to reduce the magnetostatic energy, the sample divides into domains. Thus, a division into two opposite domains, as in Fig. 9.25(b), causes the sample's magnetostatic energy to be reduced by about one-half, because the demagnetizing field inside the sample is reduced significantly. Much of this field is now confined to the end regions of the specimen. (Note that the crystal structure is unaffected by the domains.) Further reduction in energy can be achieved if the sample divides into still smaller domains, and it may seem at first that the divisions can continue indefinitely.

There are other factors, however, which should be considered. It requires some energy to create the "wall" separating two domains, because the direction of spin changes in that region. We recall from (9.56) that the exchange energy between two neighboring moments is

$$E_{ex} = -J'\mathbf{s}_1 \cdot \mathbf{s}_2 = -J's^2 \cos\theta. \tag{9.63}$$

If the wall is infinitely thin, then $\theta = \pi$, for the two moments on opposite sides of the wall are antiparallel, and $E_{ex} = J's^2$. When we estimate this for a unit area, we find that its value is appreciable. Furthermore, the more domains present, the larger the total area of the domains and the greater the total exchange energy. This fact therefore opposes the magnetostatic energy by acting to limit the number of domains.

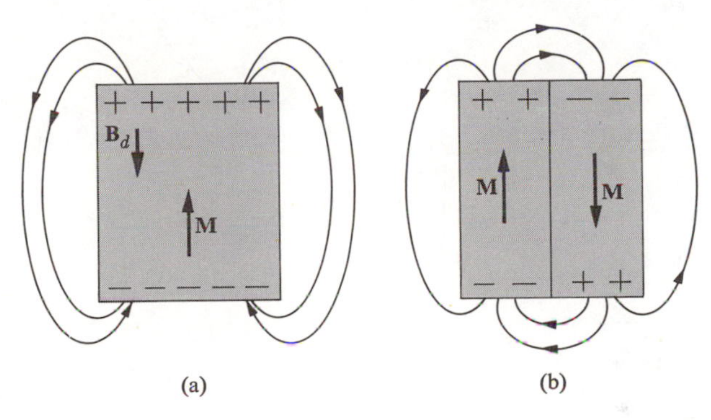


**Fig. 9.25** (a) A ferromagnet in a state of uniform magnetization; **B***d* represents demagnetization field due to surface magnetic charges. Note the field lines. (b) A ferromagnet divided into two ferromagnetic domains. Note that field lines are now confined primarily to end regions.

The wall described is known as a *Bloch wall*. Its thickness is not infinitely small, but it has a finite value, i.e., the spin orientation changes gradually in the transition region (Fig. 9.26). In this manner the spin reversal is accomplished over a number of steps, and hence the spin rotation between two neighboring moments is rather small. This leads to a reduction in the exchange energy associated with the wall. For iron, the wall is about 1000 Å thick, and its energy about $10^{-3}$ J/m$^2$.

On the subject of the Bloch wall, we may also mention another factor which plays a role in determining its thickness. Experiments on ferromagnetic materials show that it is easier to magnetize a substance in one direction than in another. Figure 9.27 shows that iron is more easily magnetized in the [100] direction than in the [111] direction. The more favorable direction is referred to as the *easy direction*, while the least favorable is known as the *hard direction*. Since it requires a larger field to magnetize the substance in the hard direction, the magnetization requires a larger energy. The difference in energy between the easy and hard directions is called the *magnetic anisotropy energy*. The effect of this energy on the wall is to reduce its thickness, because the thicker the wall, the more dipoles point in the hard direction. Thus, although exchange energy favors a thick wall, anisotropic energy favors a thin wall, and a balance is struck by minimizing the sum of these two energy terms.

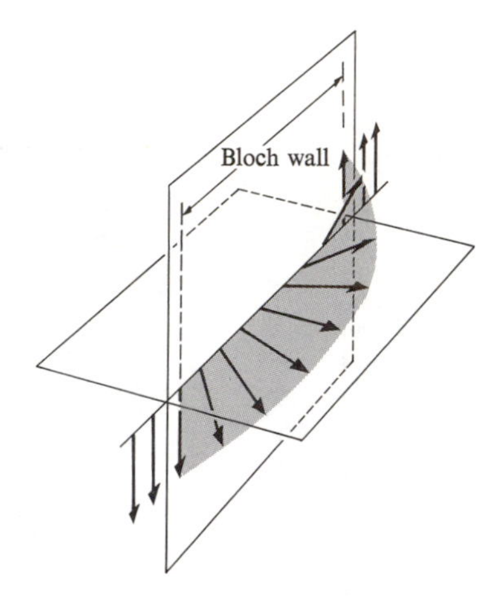


**Fig. 9.26** Successive rotation of spin direction inside Bloch wall.

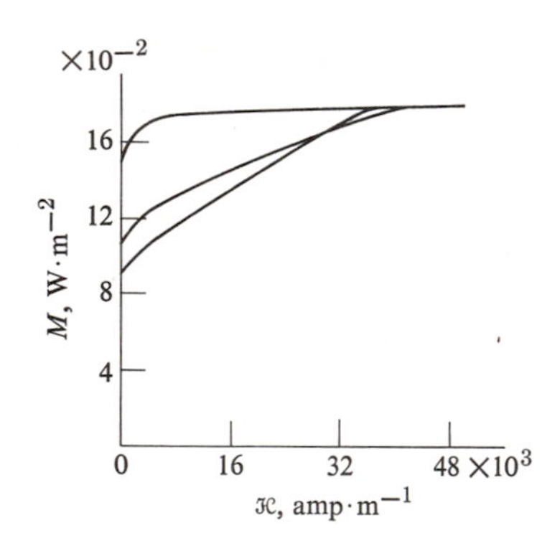


**Fig. 9.27** Magnetization curve for single-crystal iron.

Closer examination of the domain structure reveals the presence of small transverse domains near the end of the sample (Fig. 9.28). These are called *closure domains*, and for good reason, as they have the effect of closing the "magnetic loop" between two adjacent domains, resulting in a further decrease in magnetostatic energy. These closure domains are small, however, and the reason lies in yet another energy term, the *magnetostriction energy*. These regions, whose magnetization is not along the easy axis, undergo an elastic deformation because of the magnetization, an effect known as *magnetostriction*. The magnitude of this energy is about 50 J/m$^3$. Thus an additional elastic energy is required for these domains, and the larger these are, the greater is this energy. Again a balance is struck between this term and the reduction in the magnetostatic energy.

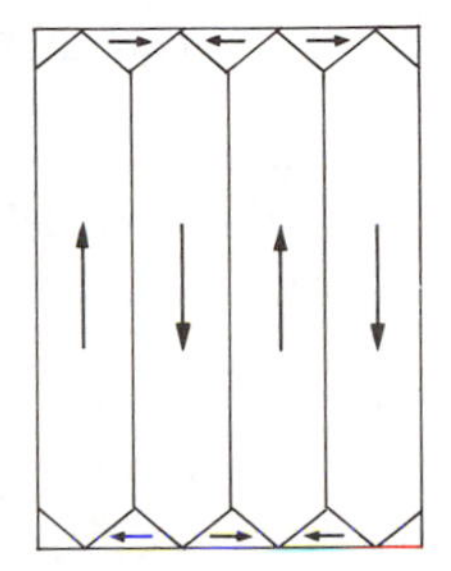

**Fig. 9.28** Closure domains, at end regions of sample.

### The magnetization process

As we have stated previously, a ferromagnetic sample is usually in the demagnetized state. In order to magnetize it, one applies an external field. Figure 9.29 illustrates the progress of the magnetization process as the external field increases. Starting at the origin, the magnetization $M$ increases slowly at first, but more rapidly as the field is increased, and eventually $M$ saturates at the point $A$.

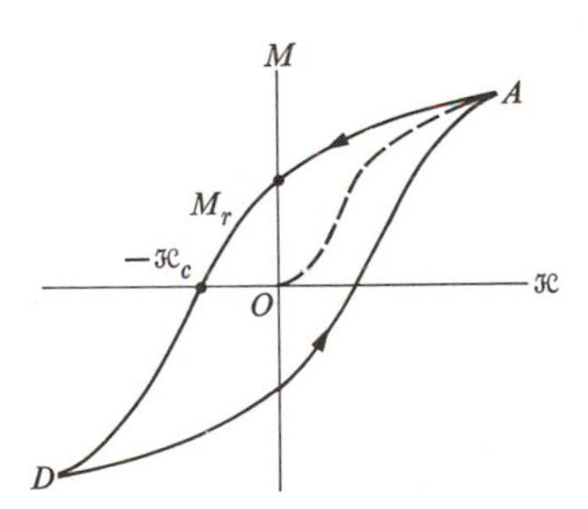


**Fig. 9.29** Hysteresis loop in a ferromagnet.

If the field is now reduced, the new curve does not retrace the original curve $OA$; rather it follows the line $AD$ shown in the figure. Even when the field is reduced to zero, some magnetization $M_r$, known as *remanent magnetization*, still survives. To destroy the magnetization completely, a negative field $-\mathcal{H}_c$ is required, which is called the *coercive force*. The sample clearly exhibits *hysteresis*, and if the field $\mathcal{H}$ alternates periodically, the magnetization traces the solid curve in Fig. 9.29, which is the *hysteresis loop*.

Hysteresis implies the existence of energy losses in the system. These losses are proportional to the area of the loop. One may demonstrate this by noting that as $M$ increases by the amount $dM$, the energy absorbed by the system (per unit volume) is $\mu_0 \mathscr{H} dM$. When this is integrated over the closed loop, it yields the total loss

$$E = \mu_0 \oint H dM,$$

which, aside from the factor $\mu_0$, is indeed the area of the loop.

The relative mobility $\mu_r$, as we recall, is defined as $\mu = 1 + (M/\mathscr{H})$—see (9.17). But in this region, in which the magnetization curve departs appreciably from linearity, as in Fig. 9.29, it is more useful to define the differential permeability as

$$\mu_r = 1 + \frac{dM}{d\mathscr{H}},$$

which is, of course, related to the slope of the magnetization curve. In ferromagnetic materials, this quantity can be very large—as much as $10^5$.

How is magnetization accomplished? Starting from the demagnetized state, and as the field is raised, the domains whose magnetization is parallel to the field are energetically more favored than the others, and hence they grow at the expense of the less-favored domains. For a small field this growth is reversible, and if the field is removed the sample returns to the original demagnetized state. But for large field the growth becomes irreversible, and some magnetization is retained even if the field is removed altogether. When a very large field is applied, not only is the maximum growth accomplished, but even the last few remaining unfavorable domains rotate so as to align with the field.†

But just how does the growth process take place, and why is it reversible in some circumstances and irreversible in others? The answer is not simple, and not as yet fully understood. However, broadly speaking we can say that the growth of a favorable domain is accomplished by the outward motion of its Bloch walls. The higher the field, the greater the motion. For a small field, the walls move back once the field is removed, but for a large field they cannot quite return to their

---

† A type of domain known as a *magnetic bubble* has been discovered recently. It is of great potential importance to computer technology. In thin films of certain orthoferrites, for example, $Y_3Fe_5O_{12}$, as a magnetic field is applied normal to the film, the size of the domains of magnetization opposite to the field decreases until at higher fields they shrink into very small (few $\mu$'s) cylinders which are the bubbles mentioned above. The bubbles are stable, mobile, and repel each other. They can also be moved and manipulated by the application of a suitable magnetic field in the plane of the film.

In computer design, the bubbles may be used as digital bits. It is also necessary that their density (number/cm$^2$) be high, as well as their mobility in the magnetic film. Their advantage over electromechanical storage devices is that the latter's inherent difficulties, such as wear, head crash, dirt, etc., are eliminated. Also the new device would have greater lifetime, e.g. 40 years. See G. S. Almasi, *Proc. IEEE* **61**, 438 (1972).

original positions, particularly if the sample contains appreciable amounts of impurities and other crystalline imperfections. These tend to prevent a complete return by "pinning down" the walls in their final positions. Experiments show that the more imperfect the sample, the greater the remanent moment $M_r$.

**Table 9.8**

Data for Permanent (Hard) and Soft Magnetic Materials (After Hutchison and Baird, 1963, *Engineering Solids*, New York: Wiley)

| *Permanent materials made from powder* | | |
|---|---|---|
| | $B_r = \mu_0 M_r$, $Wb \cdot m^{-2}$ | $\mathscr{H}_c$, $amp \cdot m^{-1}$ |
| Cobalt ferrite | 0.4 | $4 \times 10^4$ |
| Fe-Co | 0.92 | 8 |
| Fe-Co ferrite | 0.60 | 13 |
| *Permanent materials made from alloys* | | |
| Alnico 11 | 0.73 | $4.7 \times 10^4$ |
| Alnico 5 | 1.27 | 5.4 |
| Carbon steel | 1.0 | 0.4 |
| Cobalt steel | 1.0 | 2 |

| *Soft materials* | | | |
|---|---|---|---|
| | $\mu_r$(max) | $B_s$, $Wb \cdot m^{-2}$† | $\mathscr{H}_c$, $amp \cdot m^{-1}$ |
| Fe (commercial) | 6,000 | 2.16 | 90 |
| Fe (pure) | 350,000 | 2.16 | 0.9 |
| Fe (4% Si) | 6,500 | 2.01 | 40 |
| Supermalloy | $10^6$ | 0.80 | 0.34 |

† $B_s$ refers to the saturation value $B_s = \mu_0 M_s$.

Generally speaking, magnetic materials are employed in two main types of application: (a) permanent magnets or (b) transformer cores. In permanent magnets, one requires a large remanent magnetization and large coercive force, resulting in *magnetically hard* materials, which are often impure, strained, and contain grain boundaries. Transformer cores utilize *magnetically soft* materials, which have low values of $\mathscr{H}_c$ and high permeability. These should be highly purified, carefully annealed, and properly oriented for magnetization in the easy direction, so as to leave the Bloch walls free to move without hindrance. Table 9.8 gives data for an assortment of hard and soft substances.

## 9.12 PARAMAGNETIC RESONANCE; THE MASER

So far, in discussing magnetic effects, we have concerned ourselves only with static situations: A static field is applied and the induced magnetization is observed after sufficient time has elapsed for the system to have reached its final equilibrium state. Although much information can be gleaned from these measurements, as we have seen, a great deal more can be attained by using alternating magnetic fields. We can then obtain accurate information on the magnetic state of the dipoles, the interaction between dipoles, and also the interaction between dipoles and lattice.

In this section we shall deal with paramagnetic systems only, systems in which the interaction between dipoles is weak. (Ferromagnetic systems will be considered in Section 9.14.) We shall find that, with appropriate field arrangements, the system may exhibit *paramagnetic resonance* corresponding to the case in which the external frequency is equal to the Larmor frequency of the system. From studying the position and shape of the resonance line, one can obtain the above information.

### Resonance

Let us begin with the mathematical description. The magnetization vector **M** represents the magnetic state of the system. When a magnetic field is applied, the vector **M** moves according to the equation

$$\frac{d\mathbf{M}}{dt} = -\gamma\mathbf{M} \times \mathbf{B}, \tag{9.64}$$

where we have used (9.9), and $\gamma$ is the *gyromagnetic ratio* $(ge/2m)$.† Our concern now is with the type of motion executed by **M** as a function of time. When **B** is a constant field, **M** simply precesses around **B** with the Larmor frequency

$$\omega_L = \gamma B, \tag{9.65}$$

as we recall from the discussion in Section 9.2. But if the field is variable, then the motion is more complicated.

We suppose that the field **B** is composed of two parts, a large static component $B_0$ in the $z$-direction, and a small alternating transverse component **b** in the $xy$ plane. That is,

$$\mathbf{B} = \hat{\mathbf{k}}B_0 + \mathbf{b}, \tag{9.66}$$

where $\hat{\mathbf{k}}$ is a unit vector in the $z$-direction (Fig. 9.30). Because **b** is so small, we

† We obtain Eq. (9.64) from (9.9) by multiplying (9.9) by the factor $N$, the concentration of dipoles.

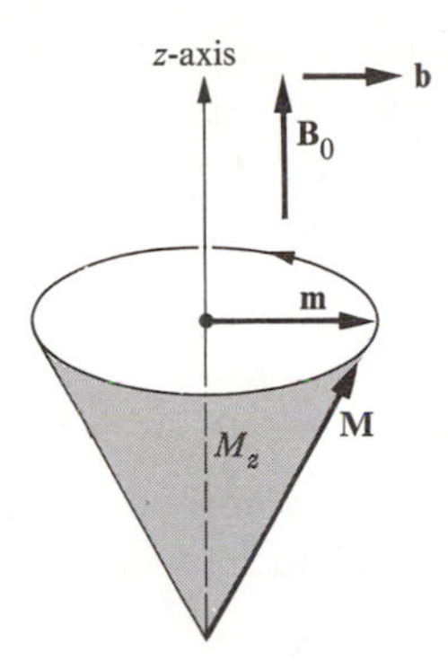


**Fig. 9.30** Arrangement of magnetic fields, both $\mathbf{B}_0$ and $\mathbf{b}$, and precession of magnetization vector in paramagnetic resonance.

may neglect it in the zero$^{\text{th}}$ order and visualize the vector **M** as precessing around the $z$-axis with a Larmor frequency

$$\omega_0 = \gamma B_0. \tag{9.67}$$

The presence of **b**, does, however, affect this motion, and we can study this by returning to the equation of motion (9.64). For the sake of simplicity, the calculations will be carried only to the first order in **b**. For convenience, we shall split the magnetization **M** as follows:

$$\mathbf{M} = \hat{\mathbf{k}} M_z + \mathbf{m}, \tag{9.68}$$

where $M_z$ is the "longitudinal" component, parallel to $B_0$, and **m** the "transverse" component in the $xy$ plane. It is assumed that the transverse component is much smaller than the longitudinal one. If we substitute from (9.66) and (9.68) into (9.64), we find that

$$\frac{dm_x}{dt} = -\gamma(m_y B_0 - M_z b_y) \tag{9.69a}$$

$$\frac{dm_y}{dt} = -\gamma(M_z b_x - m_x B_0) \tag{9.69b}$$

$$\frac{dM_z}{dt} = -\gamma(m_x b_y - m_y b_x) = 0, \tag{9.69c}$$

which are three equations in the unknowns $m_x$, $m_y$, and $M_z$ that should be solved simultaneously.

In (9.69c), the quantity $dM_z/dt$ has been set equal to zero because it is of second order, e.g., the term $m_x b_y$ is a product of two small quantities. Thus, to first order, the projection $M_z$ is a constant independent of time, meaning that the vector **M** simply precesses around the $z$-axis.

The complete solution depends on the form of the small transverse field **b**. We shall assume that the system is subjected to a plane-polarized alternating signal of frequency $\omega$. That is,

$$\mathbf{b} = \mathbf{b}_0 e^{i\omega t}, \tag{9.70}$$

where we have employed the usual complex notation.† Because of this, the transverse magnetization is expected to have a similar form also, and hence we attempt the solution

$$\mathbf{m} = \mathbf{m}_0 e^{i\omega t}, \tag{9.71}$$

where $\mathbf{m}_0$ is the vector amplitude of the magnetization. When we substitute from (9.71) into the two equations (9.69a) and (9.69b), we are led to two simultaneous equations in $m_x$ and $m_y$ only, whose solution is (see the problem section)

$$m_x = \frac{\gamma M_z}{\omega_0^2 - \omega^2}(\omega_0 b_x + i\omega b_y) \tag{9.72a}$$

$$m_y = \frac{\gamma M_z}{\omega_0^2 - \omega^2}(-i\omega b_x + \omega_0 b_y). \tag{9.72b}$$

These two equations, giving the magnetization in terms of the applied field, can be used to determine the susceptibility. One readily finds that

$$\chi_{xx} = \chi_{yy} = \frac{\mu_0 \gamma \omega_0 M_z}{\omega_0^2 - \omega^2} \tag{9.73a}$$

and

$$\chi_{yx} = -\chi_{xy} = -i\frac{\mu_0 \gamma \omega M_z}{\omega_0^2 - \omega^2}. \tag{9.73b}$$

There are several interesting features in these results: First, the susceptibility $\chi$ is a tensor with nonvanishing off-diagonal components. Thus the magnetization **m** is *not* in the same direction as **b**, but **m** lags behind **b**, as shown in Fig. 9.31(a). If we follow the curve traced by the magnetization vector **m** as a function of time, we obtain an ellipse whose major axis lies in the direction of the applied field, as

† In previous discussions of oscillatory phenomena, we have taken the time factor as $e^{-i\omega t}$ rather than $e^{i\omega t}$. The mathematical difference between the two cases is trivial, however, and one can modify all the subsequent results in this section by simply reversing the sign of $\omega$ everywhere.

in Fig. 9.31(b) [this may be demonstrated simply by setting $b_y = 0$ in (9.72), and noting the phase relation and amplitudes of $m_x$ and $m_y$]. Thus the total vector $\mathbf{M}$ precesses around $\mathbf{B}_0$, tracing an elliptical cone, with a frequency $\omega$.

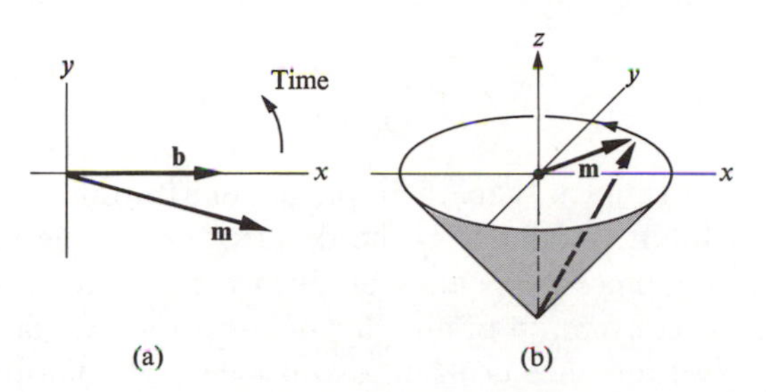


**Fig. 9.31** (a) Phase difference between transverse field **b** and transverse magnetization. (b) Elliptical curve traced by transverse magnetization **m**.

Second, and more important, the results (9.73) show that the susceptibility becomes infinite when $\omega = \omega_0$. This is hardly surprising, because, as we noted in (9.65), $\omega_0$ is the natural frequency for the system, and when $\omega = \omega_0$ the applied field is synchronous with the precessional motion, leading to a very large increase in the magnetization. This is the condition for *electron paramagnetic resonance*, often abbreviated EPR.

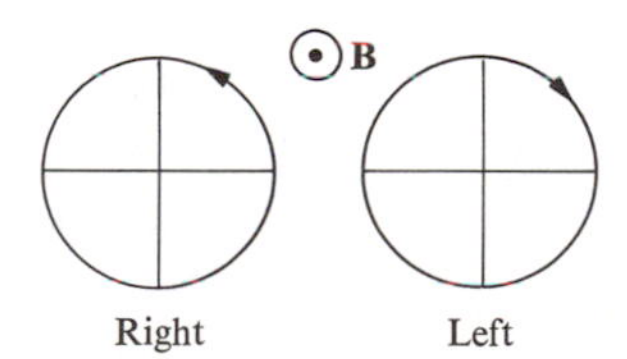


**Fig. 9.32** Right and left circular polarization.

We can simplify the analysis significantly by using circular polarization instead of the plane polarization used above. Thus, for a right-handed polarization, $b_x = b_0 \cos \omega t$ and $b_y = b_0 \sin \omega t$. That is, $\mathbf{b} = \mathbf{b}_0 e^{i\omega t}$, and hence

$$b_y = -ib_x, \tag{9.74a}$$

according to the complex notation (see also Fig. 9.32). Analogously, the following relation holds true for a left-handed polarization:

$$b_y = ib_x. \tag{9.74b}$$

If one substitutes (9.74a) into (9.69), one finds for the right-handed susceptibility

$$\chi_R = \frac{\mu_0 \gamma M_z}{\omega_0 - \omega}. \tag{9.75a}$$

Similarly, one finds the susceptibility for the left-handed polarization to be

$$\chi_L = \frac{\mu_0 \gamma M_z}{\omega_0 + \omega}. \tag{9.75b}$$

These expressions are simpler than their plane-polarization counterparts (9.73). Also note that $\chi_R$ exhibits resonance at the frequency $\omega_0$ while $\chi_L$ does not. This is because the Larmor precession takes place in the counterclockwise direction. Thus by adopting the convention of circular polarization, we have discovered not only the frequency of resonance $\omega_0$, but also its sense of rotation. For this type of precessional motion, circular polarization offers the natural choice. A plane-polarized motion can, of course, always be analyzed into its two circular components, and then treated by the use of (9.75).

**Relaxation**

Our description of the precessional motion of the dipole is still incomplete in one respect: We have not introduced a coupling mechanism to account for the interaction between dipoles and their environment. That such a mechanism exists should be evident from the following considerations. When a static field is applied to a system of dipoles, these dipoles eventually turn around and align themselves predominantly with the field. But in doing so, they lose some magnetic energy. Since total energy is always conserved, this loss of dipole energy must be dissipated, which can happen only if the dipoles are coupled to their environment in some manner. Let us now take this coupling into account.

Instead of (9.64), we shall use the following expressions as our equations of motion:

$$\frac{dm_{x,y}}{dt} = -\gamma(\mathbf{M} \times \mathbf{B})_{x,y} - \frac{m_{x,y}}{\tau_2}. \tag{9.76}$$

$$\frac{dM_z}{dt} = -\gamma(\mathbf{M} \times \mathbf{B})_z - \frac{M_z - M_0}{\tau_1}. \tag{9.77}$$

These are known as the *Bloch equations*. The first describes the motion of either $m_x$ or $m_y$, where an obvious notation is used. The second describes the motion of $M_z$. The quantities $\tau_1$ and $\tau_2$ are time constants whose meaning will be elucidated shortly. Let us think about the justification for—and significance of—these important equations.

Consider Eq. (9.77). It is the same as (9.64), except for the new second term on the right side. In this term, $M_z$ is the instantaneous $z$-component of the magnetization, while $M_0$ is the equilibrium magnetization in that direction. The term expresses the fact that, starting at some arbitrary value, $M_z$ will approach $M_0$ at a rate proportional to the departure from equilibrium, and the proportionality constant is $1/\tau_1$. Such a form, characteristic of relaxation phenomena (see, for example, Section 9.6), represents a *magnetic relaxation* whose time is $\tau_1$. If we solve (9.77) for a static field $\mathbf{B}_0$ in the $z$-direction, we find that $M_z$ *spirals* toward its equilibrium value $M_0$, as shown in Fig. 9.33, and the time taken to complete the magnetization process is approximately $\tau_1$.

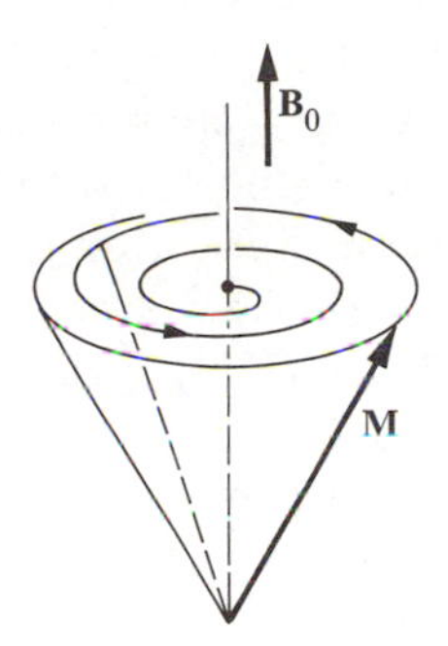


**Fig. 9.33** Magnetic relaxation.

The time $\tau_1$ is known as the *longitudinal time*, or, more descriptively, as the *spin–lattice relaxation time*. The designation indicates that an exchange of energy is involved: the magnetization loses some energy, and this is transferred to the lattice. The details of the interaction are complicated, but broadly speaking the vibration of the lattice atoms surrounding the dipole creates an oscillating field which acts on the dipole and absorbs energy from it. Thus the higher the temperature, the greater the interaction, and the shorter the time $\tau_1$. It is usually found that $\tau_1 \sim 1/T$; a typical value at nitrogen temperature is $\tau_1 \simeq 10^{-6}$ s.

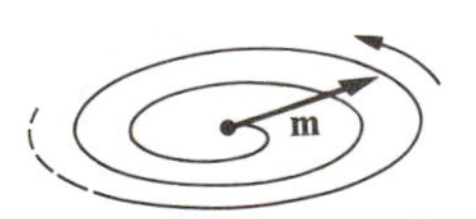


**Fig. 9.34** Relaxation of transverse magnetization $\mathbf{m}$ when transverse field $\mathbf{b} = 0$.

We now return to (9.76), which describes the transverse motion. We should now be able to understand it more easily in light of the above discussion. Again the new term is the second one on the right side. To understand its effect clearly, we may also consider the case in which $\mathbf{B} = \mathbf{B}_0$, a static field in the $z$-direction. Figure 9.34 indicates what happens here: Starting with its initial value, the vector $\mathbf{m}$ approaches its final equilibrium value $\mathbf{m} = 0$ after a time approximately equal to $\tau_2$. Thus $\tau_2$ is a relaxation time for the transverse motion.

The time $\tau_2$ is known as the *transverse relaxation time*, or, more descriptively, as the *spin–spin relaxation time*. It arises because neighboring dipoles, coupled via the familiar magnetic dipole–dipole interaction, attempt to break up any initial coherence between the directions of the individual transverse moments. The time $\tau_2$ is usually very short, often of the order of $10^{-10}$ s, and is independent of temperature. It does, however, depend strongly on the concentration of the magnetic atoms; the larger the concentration the closer the dipoles, which leads to a strong interaction and consequently a shorter relaxation time $\tau_2$.

How does this great disparity between $\tau_1$ and $\tau_2$ affect our picture of the magnetization process, and why does such a disparity exist in the first place? Let us begin with the first question. Suppose that there are only three dipoles, which were originally in complete alignment with each other at the instant $t = 0$ (Fig. 9.35a). A static field $\mathbf{B}_0$ in the $z$-direction is applied, after which we observe the subsequent precessional motion of the dipoles. Since $\tau_2 \ll \tau_1$, the first thing to take place is that $\mathbf{m} \to 0$ (Fig. 9.35b). The phases between the individual moments have been quickly reshuffled to yield a vanishing transverse magnetization. For this reason, the time $\tau_2$ is sometimes referred to as the *dephasing time*. After the dephasing, the moments begin to spiral toward the direction of the field, resulting in an increased magnetization in that direction (Fig. 9.35c) after a time $\tau_1$.

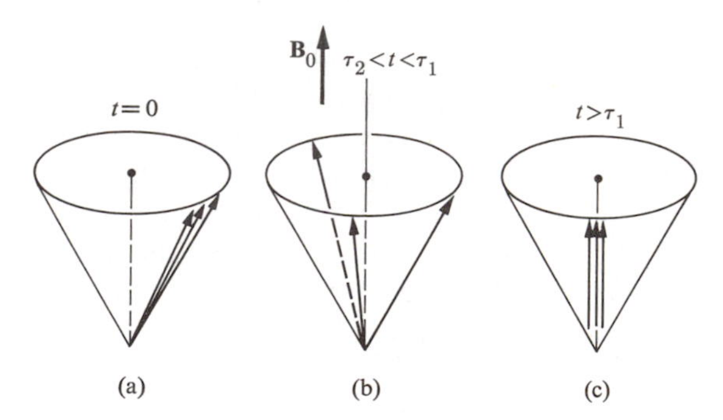


**Fig. 9.35** (a) Initial orientation of the three spins. (b) Situation after transverse relaxation. (c) Situation after longitudinal relaxation.

The essential reason for the disparity in the magnitudes or $\tau_1$ and $\tau_2$ is that the longitudinal relaxation process involves a dissipation of energy, whereas the transverse does not. When the moments try to tilt toward $\mathbf{B}_0$, they do so only if they can release some of their energy, and the faster they can do this the shorter is $\tau_1$. However, conditions for exchanging large amounts of energy are rather stringent in magnetic interactions. It is this difficulty in disposing of their energy quickly that makes the moments magnetize slowly. Note that transverse relaxation requires no tilt toward $\mathbf{B}_0$, and hence no energy exchange.

Now let us solve Eqs. (9.76) and (9.77), using a method similar to that employed in solving (9.69). We assume a steady-state situation, that is, we set $dM_z/dt = 0$ in (9.77) and $\mathbf{m} = \mathbf{m}_0\, e^{i\omega t}$ in (9.76), presuming, of course, that the ac signal is circularly polarized in the right-hand direction. We then find (see the problem section) for the susceptibility $\chi = \chi' + i\chi''$,

$$\chi'(\omega) = \gamma\mu_0 M_0 \frac{(\omega_0 - \omega)\tau_2^2}{1 + (\omega_0 - \omega)^2\tau_2^2 + \tau_1\tau_2(\gamma b_0)^2}, \tag{9.78a}$$

$$\chi''(\omega) = \gamma\mu_0 M_0 \frac{\tau_2}{1 + (\omega_0 - \omega)^2\tau_2^2 + \tau_1\tau_2(\gamma b_0)^2}. \tag{9.78b}$$

The susceptibility is complex because the signal is now partially absorbed. Figure 9.36 is a sketch of $\chi'$ and $\chi''$ versus the frequency, and shows a typical resonance behavior at $\omega = \omega_0$, which is of course anticipated. The troublesome divergence at $\omega_0$ has been removed by the inclusion of relaxation mechanisms.

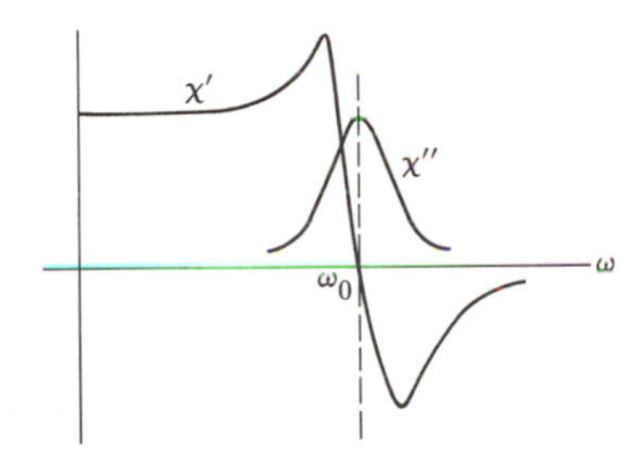


**Fig. 9.36** Real and imaginary susceptibilities, $\chi'$ and $\chi''$, as functions of frequency $\omega$, in EPR.

The information to be gained from these curves is as follows.

a) *The g-factor.* This quantity can be determined by measuring the resonance frequency $\omega_0$, because

$$\omega_0 = \gamma B_0 = g\frac{e}{2m}B_0.$$

In a crystal, this factor is not given by (9.45), but is strongly influenced by the crystal field.

b) *The time* $\tau_2$. This is determined most conveniently from the linewidth of $\chi''$. It can be seen from (9.78b) that

$$\tau_2^{-2} = (\omega' - \omega_0)^2, \tag{9.79}$$

where $\omega'$ is the half-width frequency, i.e., the frequency at which $\chi''$ reduces to half its resonance value.

c) *The time* $\tau_1$. This is again determined from $\chi''$. Near resonance, this susceptibility has the approximate expression

$$\chi'' = \gamma\mu_0 M_0 \frac{\tau_2}{1 + \tau_1 \tau_2(\gamma b_0)^2}, \tag{9.80}$$

which shows that $\chi''$ decreases as the signal strength, represented by $b_0$, increases. This phenomenon, referred to as *saturation*, can be used to determine $\tau_1$, because $\chi''$ decreases to half its value when $\tau_1\tau_2(\gamma b_0)^2 = 1$, and $\tau_2$ and $b_0$ can be determined independently.

The quality of the resonance—i.e., its sharpness—is enhanced by a long $\tau_2$; otherwise the line would be too broad to be detected. This is usually accomplished by diluting the magnetic ions in the host crystal. Similarly the time $\tau_1$ must not be too short, or else the resonance will be masked again. This often happens at room temperature, and to offset this, experiments are usually conducted at liquid-nitrogen temperatures, or even lower.

Figure 9.37 is a diagram of an assembly used to observe paramagnetic resonance. The resonance frequency $\omega_0$, for a field of a few kilogauss, lies in the microwave range, i.e., about $10^{10}$ Hz. The real part of the susceptibility $\chi'$ is measured by the change in the inductance of the coil due to the presence of the sample, while the imaginary part $\chi''$ is determined from the absorption in the sample. One can show that the power absorbed per unit volume is given by

$$P = \frac{1}{\mu_0} \omega\chi''(\omega) b_0^2. \tag{9.81}$$

An important practical comment which should be interjected here is that in resonance measurements one usually varies the static field $B_0$ rather than the frequency $\omega$, because one can vary $B_0$ far more conveniently than $\omega$. The measurement is made at a fixed frequency $\omega$ which is swept by the field.

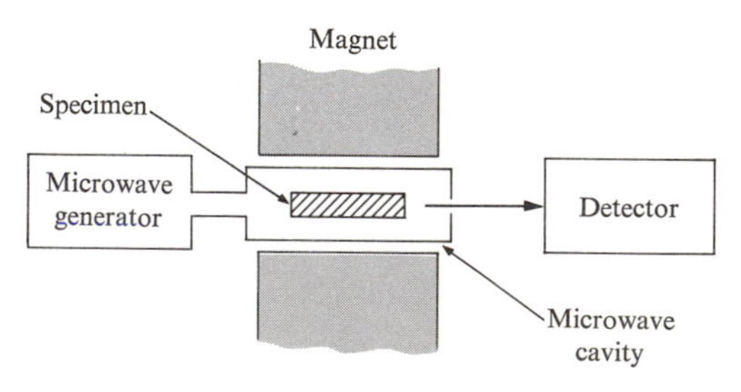


**Fig. 9.37** Schematic diagram for experimental arrangement in EPR measurements.

The technique of paramagnetic resonance is widely used in physics, chemistry, biology, and other fields, and is one of the principal tools employed in the analysis of matter. We shall indicate here its primary uses in physics, and in Chapters 13 and 14 we shall deal with its applications in chemistry and biology.

Paramagnetic resonance has been used intensively in the study of the magnetic properties of 3d, 4d, and 5d ions, as well as 4f and 5f ions in salts. Table 9.9 gives data on these materials. Note the wide range of $g$-values, engendered by the crystal field. Most of these have been calculated theoretically, and the agreement with experiment is generally good.

**Table 9.9**

Data on Transition Metals, Rare-Earth and Actinide Ions, Obtained from EPR Measurements (After Morrish, 1965)

| Ion | Configuration | Effective spin | Range of $g$-values |
|---|---|---|---|
| $Cr^{3+}$, $V^{2+}$ | $3d^3$ | $\frac{3}{2}$ | 1.992 |
| $Fe^{3+}$ | $3d^5$ | $\frac{1}{2}$ | 0.7–2.6 |
| $Mo^{5+}$ | $4d^1$ | $\frac{1}{2}$ | 1.95 |
| $Ru^{3+}$ | $4d^5$ | $\frac{1}{2}$ | 1.0–3.24 |
| $Ce^{3+}$ | $4f^1$ | $\frac{1}{2}$ | 0.95–2.18 |
| $Nd^{3+}$ | $4f^3$ | $\frac{1}{2}$ | 2.0–3.6 |
| $Er^{3+}$ | $4f^{11}$ | $\frac{1}{2}$ | 1.47–8.9 |
| $(NpO_2)^{2+}$ | $5f^1$ | $\frac{1}{2}$ | 0.20–3.40 |
| $(PuO_2)^{2+}$ | $5f^2$ | $\frac{1}{2}$ | 0–0.59 |

The EPR technique is also used in studying paramagnetic molecules, notably in $O_2$, NO, and $NO_2$. Unlike most other molecules, the spins in these molecules are unbalanced, resulting in a net magnetic moment, and consequently they exhibit EPR.

### The maser

Much of the research on EPR was originally sparked by the observation of maser action in ruby in the mid-1950's.† We can see the principle of this action and its relation to EPR by looking at Fig. 9.38. The two levels $E_1$ and $E_2$ are a Zeeman doublet formed by the application of a static magnetic field $B_0$. The energy difference $\Delta E = E_2 - E_1$ is given by

$$\Delta E = \hbar\omega_0 = \hbar\gamma B_0 = g\mu_B B_0. \tag{9.82}$$

Suppose that the system is at equilibrium in the presence of the field $B_0$. The populations $N_1$ and $N_2$ of the two levels are related by

$$\frac{N_2}{N_1} = e^{-\Delta E/kT}, \tag{9.83}$$

† The word *maser* is an acronym for Microwave Amplification by Stimulated Emission of Radiation.

which shows that $N_2 < N_1$. That is, the upper level is less densely populated, a conclusion we may have anticipated. At room temperature and usual field, $kT \gg \Delta E$, and the two levels are essentially equally populated, the thermal energy being so large that the atoms can easily transfer from one level to the other.

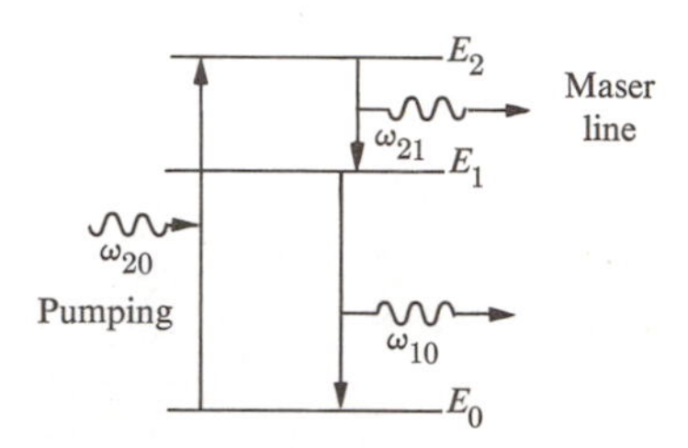


**Fig. 9.38** Principles of maser emission: Electrons are pumped from level $E_0$ to level $E_2$, they then flip their spins, going into a lower level $E_1$ and emitting coherent radiation of frequency $\omega_{21}$.

At low temperature, however, $kT \ll \Delta E$, and most of the atoms fall to the lower level, i.e., their moments are parallel to the field. If under these circumstances a signal of frequency $\omega$ passes through the system, then the signal may be absorbed. This occurs when an atom absorbs a photon, and transfers from the lower to the upper level. This can happen only if $\Delta E = \hbar\omega$, according to Bohr's rule (Section A.5). Comparing this with (9.82), we find that

$$\omega = \omega_0, \tag{9.84}$$

which is, of course, the same criterion for resonance obtained previously on classical grounds. Thus we see that the signal is absorbed at low temperature, the absorbed energy being used in exciting the spin system.

Let us suppose that the population of the levels are arranged so that $N_2 > N_1$, which means that the upper level is the more densely populated. [This condition of *population inversion* cannot be achieved at equilibrium, of course, but it can be realized by other means.] In that case, and when the resonance condition $\omega = \omega_0$ is satisfied, the signal is *amplified*, because more atoms are stimulated to transfer downward (emitting photons) than upward, with a net enhancement of the signal. This amplification is what is responsible for the maser action.

Figure 9.38 illustrates how the condition of population inversion may be accomplished, in the simplest possible case. Atoms are transferred from the ground level $E_0$ to the upper level $E_2$ of the Zeeman doublet by "pumping" the system with an external radiation of frequency $\omega_{20} = (E_2 - E_1)/\hbar$. These atoms now make spontaneous transitions to $E_1$ and $E_0$, but if the spin–lattice relaxation time $\tau_1$ for transitions from $E_2$ to $E_1$ is long, then it is possible—at low temperature—for more atoms to exist in $E_2$ than in $E_1$. The system is then inverted, and if a signal with a frequency $\omega_{21} = (E_2 - E_1)/\hbar$ passes through the system, it may be amplified.

It is evident that maser action is simply the converse of EPR. The best-known maser, made of ruby, involves the spin states of the chromium impurities in this material. More details can be found in Yariv (1966), and in other references on quantum electronics.

## 9.13 NUCLEAR MAGNETIC RESONANCE

*Nuclear magnetic resonance* (NMR) is the nuclear analog of electron paramagnetic resonance. Instead of the electrons spinning, the nuclei have the resonant moments. A nucleus is composed of protons and neutrons packed into an extremely small volume (radius about $10^{-13}$ cm). These *nucleons* have spins, and the spin of the nucleus as a whole is the vector sum of the spins of the individual nucleons. The spin quantum number for the nucleus is denoted by $I$, whose value, in analogy with the electron, must be either an integer or a half-integer. The angular momentum is therefore given by

$$\mathbf{I} = \mathbf{I}\hbar, \tag{9.85}$$

where $\mathbf{I}$ specifies both the magnitude and direction of the angular momentum vector. [Actually the magnitude of the angular momentum vector is $[I(I+1)]^{\frac{1}{2}}\hbar$.]

Associated with the nuclear spin motion there is also a nuclear magnetic moment, which we shall denote by $\boldsymbol{\mu}_n$. In analogy with the electron case, this moment is related to the angular momentum, and we may write

$$\boldsymbol{\mu}_{\mathrm{n}} = g_{\mathrm{n}}\left(\frac{e}{2M_{\mathrm{p}}}\right)\mathbf{I} = g\mu_{\mathrm{Bn}}\mathbf{I}, \tag{9.86}$$

where $g_{\mathrm{n}}$ is the nuclear $g$-factor, $M_{\mathrm{p}}$ the proton mass, and $\mu_{\mathrm{Bn}} = e\hbar/2M_{\mathrm{p}}$, the *nuclear Bohr magneton*. Table 9.10 provides values of $I$ and $g_{\mathrm{n}}$ for a number of nuclei. The values of $g_{\mathrm{n}}$ are of the order of unity.

The nuclear moment differs from the electronic moment in two respects. First, the nuclear moment is of the order of $\mu_{\mathrm{Bn}}$ which—because the proton mass $M_{\mathrm{p}}$ is 1839 times larger than the electron mass—is about one-2000th the size of the electron moment. Nuclear moments are therefore much smaller than electron moments, as a result of the enormous difference in mass. Second, the value of $g_{\mathrm{n}}$ may be either positive or negative, depending on the nucleus. Thus the moment vector $\mu_{\mathrm{n}}$ may be either parallel or antiparallel to $\mathbf{I}^*$, unlike the case of the electron, in which the two vectors are always antiparallel.

When a field $\mathbf{B}_0$ is applied to a system of nuclei, their moments precess around $\mathbf{B}_0$ with the Larmor frequency,

$$\omega_0 = \gamma_{\mathrm{n}}B_0 = (g_{\mathrm{n}}e/2M_{\mathrm{p}})B_0, \tag{9.87}$$

**Table 9.10**

Nuclear Magnetic Moments and Spins ($\mu_{Bn} = 5.05 \times 10^{-27}$ amp m$^2$)

| Isotope | $\mu_n$ (in units of $\mu_{Bn}$) | $I$ |
|---|---|---|
| $n^1$ | $-1.913$ | $\frac{1}{2}$ |
| $p^1$ | 2.793 | $\frac{1}{2}$ |
| $H^2$ | 0.8574 | 1 |
| $He^3$ | $-2.127$ | $\frac{1}{2}$ |
| $Li^7$ | 3.256 | $\frac{3}{2}$ |
| $C^{13}$ | 0.7022 | $\frac{1}{2}$ |
| $N^{14}$ | 0.4036 | 1 |

as follows from the equation of motion (9.9) and from (9.86). If an alternating signal of frequency $\omega$, whose field is normal to $\mathbf{B}_0$, then impinges on the system, nuclear resonance takes place. It is accompanied by strong absorption at $\omega = \omega_0$, that is,

$$\omega = \omega_0 = (g_n e/2M_p)B_0 \tag{9.88}$$

or

$$\nu = 0.213\, g_n B_0 \text{ MHz},$$

where $B_0$ is in W/m$^2$. For a typical value of the field—for example, $B_0 = 0.5$ W/m$^2$—the resonance frequency lies in the rf region, that is, $10^6 - 10^7$ Hz. This is a very convenient range, as lumped-parameter circuits may be used here, and great accuracy attained.

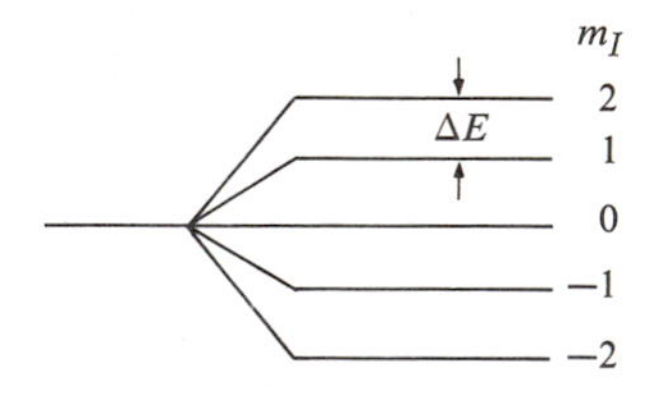


**Fig. 9.39** Zeeman splitting of a nuclear level for $I = 2$.

Quantum mechanically, NMR may be viewed as a transition between nuclear Zeeman sublevels. If $I = 2$, for example, the level splits into $(2I + 1) = 5$ sublevels as a result of the application of the magnetic field $B_0$ (Fig. 9.39). The spacing between the equidistant sublevels is given by

$$\Delta E = g_n \mu_{Bn} B_0. \tag{9.89}$$

When an external ac signal is applied, a transition between these sublevels is induced when

$$\Delta E = \hbar\omega, \tag{9.90}$$

and this results in the absorption of the signal. Note that Eq. (9.90) is the same as the resonance condition (9.87), showing that significant absorption takes place only at resonance, a conclusion we could have anticipated.

Equation (9.90) is based on the transition between adjacent Zeeman sublevels only. Transitions between nonadjacent levels are forbidden by the selection rule $\Delta m_I = \pm 1$, where $m_I$ is the quantum number for the $z$-component of the angular momentum.

We have not as yet discussed nuclear magnetic relaxation, but we can do this by using the same method we used in the case of EPR. Again we have two relaxation times $\tau_1$ and $\tau_2$, characterizing the interactions of the nuclear moment with its environment. These times can be determined from the height and width of the NMR line, much as in the case of the electron. The details need not be repeated here.

Resonance is commonly achieved by varying the field $B_0$ rather than the frequency $\omega$ until the resonance condition is satisfied. Thus the experiment is performed at a constant frequency; most common NMR spectrometers are designed for the frequencies 60, 100, or 220 MHz, with the field to be adjusted for the various nuclei.

In order to determine a nuclear property accurately—for example, $\mu_n$—one must determine both the frequency and the field to the same degree of accuracy [see (9.87)]. The frequency can be determined to one part in $10^6$ or better, but unavoidable inhomogeneities in the field introduce an accuracy limit of about one part in $10^4$. This is the accuracy of NMR measurements!

The NMR technique, like the EPR technique, is a tool that is widely used in physics, chemistry, biology, etc., because it yields information about the microscopic constitution of matter. Let us now look at some of its uses in physics; we shall get around to the chemical and biological applications of NMR in Chapters 12 and 13.

i) *Nuclear data.* One obtains data on the nucleus from NMR because, as indicated by (9.87), these measurements give $g_n$, or equivalently, the nuclear moment $\mu_n$. The spin $I$ is not determined, but other types of resonance measurements can yield this also. Therefore NMR is a highly useful technique in nuclear physics.

ii) *Environmental effects.* In our discussion of NMR, we have treated the nucleus as if it were isolated. In solid-state phenomena, however, the nucleus is not isolated, but is surrounded by its electrons, as well as by nearby atoms and molecules, which form its natural environment. The interaction of the nucleus with its environment causes the shape of the resonance line to be altered in a manner characteristic of the environment. Thus NMR can be used to study the microscopic environment. It is for this reason that NMR is so valuable in physics, as well as chemistry and other fields.

iii) *Resonance in solids.* The transverse relaxation time $\tau_2$, which may be determined from the linewidth, is due to spin–spin interaction. When one resonant nuclear moment interacts with another nearby, there is a magnetic field acting on the first moment given by

$$B = \pm \frac{\mu_0}{4\pi} \frac{\mu_n(3\cos^2\theta - 1)}{r^3}, \tag{9.91}$$

where $r$ is the distance between the moments and $\theta$ the angle between $\mathbf{B}_0$ and the line joining the moments (Fig. 9.40). The plus sign refers to the case in which the moments are parallel, and the minus sign the case in which they are antiparallel. If there are several other moments simultaneously acting on the central one, we can find the local field, of course, by adding all the individual contributions.

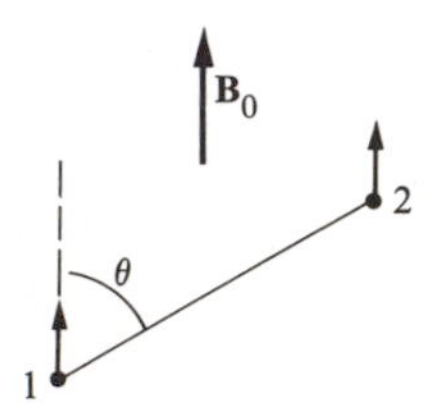


**Fig. 9.40** Spin–spin interaction between two dipoles, 1 and 2.

The total field experienced by the resonant moment is found by adding (9.90) to the applied field $B_0$. Since the field $B$ can take many different values, depending on the relative spin orientations of the neighboring moments, we see that the total field may take a number of values which are close to $B_0$, which lead, in effect, to the splitting and broadening of the resonance line. From the shape of this line, we can go a long way toward identifying the environment surrounding the nuclear moment.

iv) *Resonance in liquids and gases.* Scientists also use NMR techniques to examine liquids and gases. The linewidth in liquids and gases is much narrower than in solids. For the reason, recall Eq. (9.22), which describes spin–spin interaction. This interaction is also present in liquids, except that in a liquid the nuclei rotate and move rapidly. Therefore the local environment changes rapidly, and the local field averages out to a very small value, resulting in a small linewidth. This phenomenon is known as *motional narrowing.*

v) *Cooling by adiabatic demagnetization.* There are some experiments which necessitate the use of very low temperatures—well below 1°K. These temperatures cannot be attained by the usual method of direct cooling using liquid helium. So the standard technique is adiabatic demagnetization, a method which calls into play the phenomena of paramagnetic or nuclear spins.

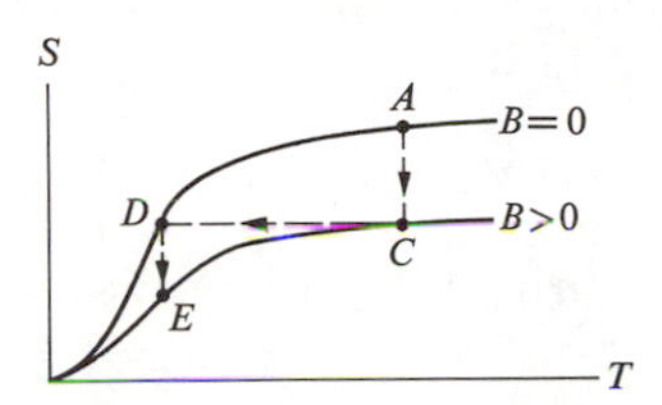


**Fig. 9.41** Entropy $S$ versus temperature $T$ for a spin system, with and without a magnetic field. Adiabatic cooling (dashed line) is accomplished by repeated application and removal of the magnetic field.

Consider a paramagnetic salt which has been cooled to a temperature of about 1°K. The sample is placed in a magnetic field of 1 $W/m^2$, and the spins align with the field. As they do so, these spins lose energy, which is dissipated into the lattice. Consequently the lattice warms up to some extent, though the temperature is held down by keeping the sample in thermal contact with the liquid-helium bath. When the heat due to magnetization has been removed completely, the sample is thermally isolated, and the magnetic field is reduced to zero. In the absence of the field, the spins tend to randomize (demagnetize adiabatically), which requires some energy. This energy is absorbed from the lattice, which consequently cools even further, resulting in a lower temperature. These steps can be repeated many times (Fig. 9.41), until the desired temperature is reached. Temperatures as low as 0.01°K have been obtained using paramagnetic demagnetization, and these have been reduced even further by the use of nuclear demagnetization.

## 9.14 FERROMAGNETIC RESONANCE; SPIN WAVES

When an external field is applied to a ferromagnetic material, the magnetization vector **M** begins to precess around the field. We therefore have the possibility of *ferromagnetic resonance* (FMR), which is analogous to the spin resonance discussed in Section 9.12. As a matter of fact, the two types of resonance were observed almost simultaneously in 1946, at about the time of the first reported observation of NMR.

FMR is particularly important in physics and engineering, for several reasons. (1) It is a powerful tool for studying ferromagnetic substances and their fascinating phenomena. (2) It is the basis of many useful microwave devices.

Let us begin our analytic discussion with the Bloch equation,

$$\frac{d\mathbf{M}}{dt} = -\gamma\mathbf{M} \times \mathbf{B}, \tag{9.92}$$

which is the analog of (9.64). Here $\gamma = (ge/2m)$ is the gyromagnetic ratio. Unlike the case of EPR, however, the field **B** is generally quite different from the applied field because of the many other contributions usually present in a ferromagnetic substance. Given that $\mathscr{H}_0$ is the applied external field, the most general expression for the internal field **B** is

$$\mathbf{B} = \mu_0(\mathscr{H}_0 - D\mathbf{M} + \mathscr{H}_a + \mathscr{H}_e), \tag{9.93}$$

where $-D\mathbf{M}$ is the demagnetizing field, $\mathscr{H}_a$ the field due to the magnetic anisotropy, and $\mathscr{H}_e$ the field due to exchange energy. The various terms of (9.93) are the magnetic analogs of the electrical terms in (8.24), although the nature of the interactions is very different in the two cases.

Obviously the complicated nature of the field in (9.93) means that the resonance frequency depends in a complex manner on the various interactions in the crystal, and for this reason the resonance can yield information concerning these interactions. To keep the discussion simple, we shall illustrate the situation by taking the simplest possible circumstances: We shall neglect $\mathscr{H}_a$ and $\mathscr{H}_e$, and retain only the demagnetizing field $-D\mathbf{M}$ in addition to the external field $\mathscr{H}_0$. The demagnetization factor $D$ depends on the shape and orientation of the sample. For a flat disc normal to $\mathbf{B}_0$, $D = 1$, and hence

$$\mathbf{B} = \mu_0(\mathscr{H}_0 - \mathbf{M}_0), \tag{9.94}$$

where $\mathbf{M}_0$ is the saturation magnetization. Since $\mathbf{M}_0$ lies in the same direction as $\mathscr{H}_0$, we may also write

$$B = \mu_0(\mathscr{H}_0 - M_0). \tag{9.95}$$

If we substitute (9.94) into (9.92), we are led to conclude that **M** precesses around $\mathscr{H}_0$ with a Larmor frequency

$$\omega_0 = \mu_0\gamma(\mathscr{H}_0 - M_0), \tag{9.96}$$

which is the *ferromagnetic resonance frequency*. Note that the demagnetizing field has acted to reduce the resonance frequency. The frequency $\omega_0$ lies in the microwave region for fields in the range $B_0 \simeq 1\ \mathrm{W/m^2}$, that is, in the same region as the EPR, which is not surprising, since $\gamma$ is essentially the same in both cases. The FMR in supermalloy is shown in Fig. 9.42.

Resonances have also been reported in antiferromagnetic and ferrimagnetic substances. The above classical treatment can be readily extended to deal with these materials.

A major area of application related to FMR involves *Faraday rotation*: Suppose that a plane-polarized signal, in the $xy$ plane, passes through a block of a ferromagnetic material with a field $B_0$, and that the signal travels along the $z$-axis, as in Fig. 9.43. We can split the signal into two circular components, one right-

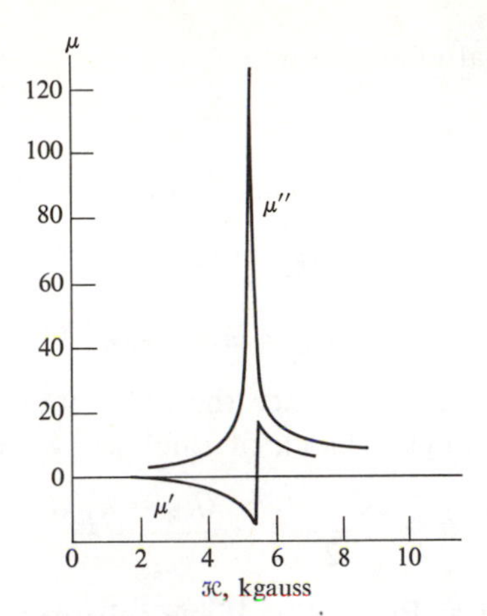


**Fig. 9.42** Real and imaginary permeability in supermalloy, showing FMR. [After Bloembergen, *Phys. Rev.* **78,** 572 (1950)]

handed and the other left-handed. The two components travel with different velocities, and consequently when the wave emerges from the other side of the block, its plane of polarization has rotated by a certain *Faraday angle*.

Let us evaluate the Faraday angle. Because of the gyromagnetic motion of the magnetization $M$, the medium presents a permeability

$$\mu_R = \mu_0 \left(1 + \frac{\omega_M}{\omega_0 - \omega}\right) \tag{9.97a}$$

or

$$\mu_L = M_0 \left(1 + \frac{\omega_M}{\omega_0 + \omega}\right) \tag{9.97b}$$

depending on whether the wave is right-handed or left-handed, respectively. We have used (9.75), where $\omega_M = \gamma\mu_0 M_0$.

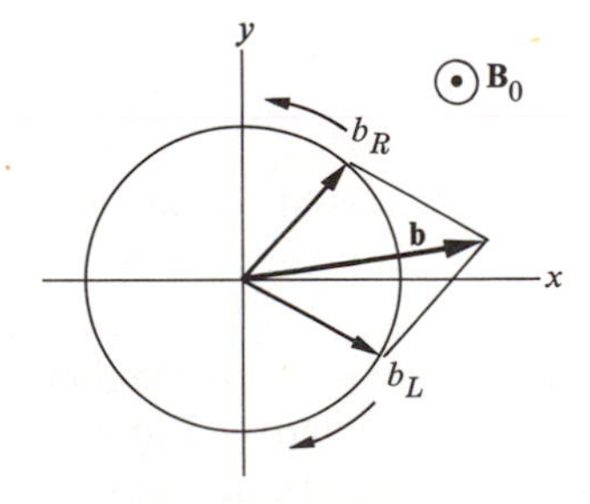


**Fig. 9.43** A plane polarized wave **k** is resolved into circularly polarized waves in a Faraday rotation.

For an incident signal of the form

$$\mathbf{b} = \mathbf{b}_0 e^{-i(\omega t - kz)},$$

the two circular components have wave vectors

$$k_R = \omega\sqrt{\mu_R \epsilon} \tag{9.98a}$$

and

$$k_L = \omega\sqrt{\mu_L \epsilon}, \tag{9.98b}$$

where $\epsilon$ is the dielectric constant. Since the phase angle changes by the amount $kd$ as the wave passes through a block of thickness $d$, it follows that

$$\theta = \frac{\theta_R - \theta_L}{2} = \frac{(k_R - k_L)d}{2}, \tag{9.99}$$

as we can see by looking at Fig. 9.43. If we substitute from (9.97) into (9.98), and then into (9.99), we find for the Faraday rotation, per unit length,

$$\frac{\theta}{d} = -\sqrt{\frac{\epsilon}{\epsilon_0}}\frac{\omega_M}{2c}, \tag{9.100}$$

where $c$ is the velocity of light. For manganese zinc ferrite, where $\omega_M = 2.6 \times 10^{10}\ \text{s}^{-1}$ and $\epsilon/\epsilon_0 = 23$, we find $\theta/d = 118$ degrees/cm. (It has been assumed that $\omega_0, \omega_m \ll \omega$.)

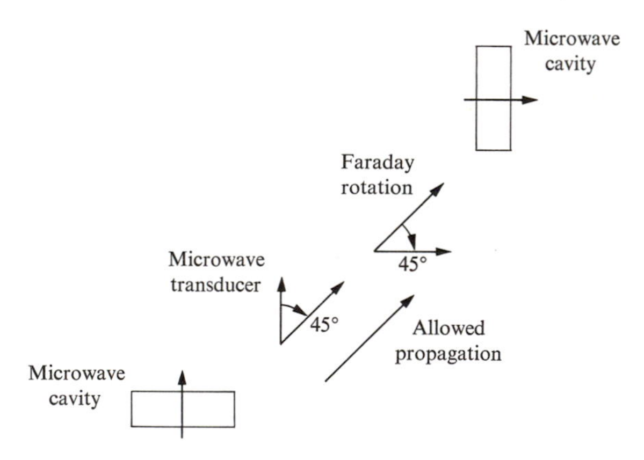


**Fig. 9.44** Principle of the isolator: Vectors at the cavities indicate direction of resonant modes in each cavity.

Faraday rotation is the basis of many ferromagnetic microwave devices. The simplest of these, the *isolator*, is illustrated in Fig. 9.44. Two microwave cavities are arranged so that their polarizations are at right angles to each other. A wave traveling forward has its plane rotated by 90° so that it can pass through the second cavity. However, a wave traveling backward retains the same polarization, because Faraday rotation is nonreciprocal, and hence is rejected by the first cavity. The arrangement therefore allows waves to travel only in one direction, and hence can be used to isolate waves according to their direction of propagation. Isolators used in combination can be used to design other devices, such as gyrators, circulators, etc. More details can be found in Wang (1966).†

## Spin waves

There is another interesting dynamical aspect of the spin motion in a ferromagnet: *spin waves*. The spins precess around the vector $\mathbf{M}_0$ in such a manner that the orientations of the various spins along the line are correlated, as depicted in Fig. 9.45(a). Such a wave propagates through the lattice with a certain velocity, which will be calculated below.

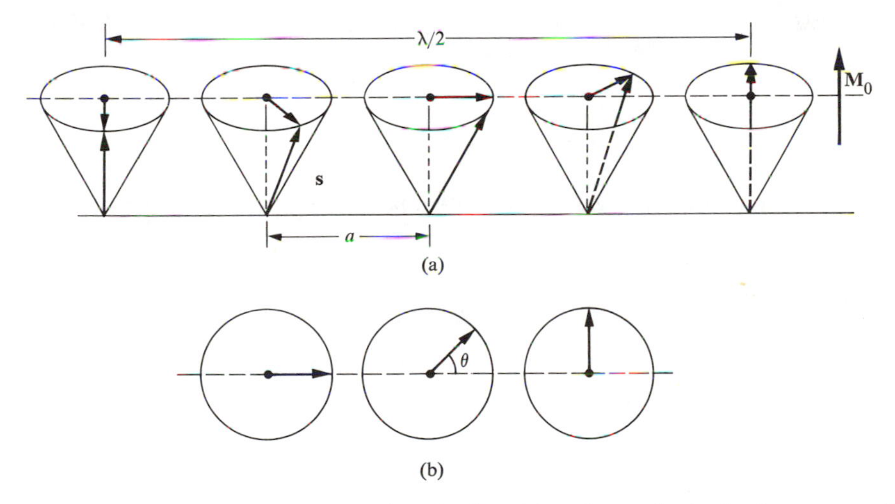


**Fig. 9.45** (a) Spin wave motion. (Wavelength $\lambda = 16a$, where $a$ is the lattice constant.) (b) Angle $\theta$ between adjacent spins. (Planar view from top of spin system.)

The restoring force responsible for the oscillation is the exchange force between the spins (Section 9.8). The lowest energy of the system occurs when all spins are parallel to each other in the direction of $\mathbf{M}_0$. When one of the spins is tilted or disturbed, however, it begins to precess—due to the field of the other spins—and because of the exchange interaction the disturbance propagates as a wave through the system.

Spin waves are analogous to lattice waves (Section 3.6). In lattice waves, atoms oscillate around their equilibrium positions, and their displacements are correlated

† The Faraday rotation and related effects are coming into use also in modern magneto-optic technology. See, for example, R. F. Pearson, *Contemp. Physics*, **14**, 201, 1973.

through elastic forces. In spin waves, the spins precess around the equilibrium magnetization and their precessions are correlated through exchange forces. We shall note many points of similarity between the two types of waves.

But first let us calculate the dispersion relation for spin waves. For this, we shall use a procedure used earlier in connection with lattice waves. Recall from Section 9.8 that the exchange energy between two spins is given by $E = -J's^2 \cos\theta$, where $\theta$ is the angle between the spins. Referring to Fig. 9.45(b), we note that each spin is influenced by two other spins on its opposite sides (nearest-neighbor interaction), and hence the exchange energy is

$$E = -2J's^2 \cos\theta.$$

The relative energy—i.e., the increase in energy above the ground state—is therefore

$$\begin{aligned}\Delta E &= -2J's^2\cos\theta - (-2J's^2)\\ &= +2J's^2(1-\cos\theta)\\ &= 4J's^2\sin^2\left(\frac{\theta}{2}\right).\end{aligned}$$

The frequency of oscillation is, according to quantum mechanics, given by $\omega = \Delta E/\hbar$, which yields

$$\omega = \left(\frac{4J's^2}{\hbar}\right)\sin^2\left(\frac{\theta}{2}\right).$$

The angle $\theta$ is the phase difference between two adjacent spins. Therefore $\theta = (a/\lambda)2\pi = aq$, where $a$ is the lattice constant and $q$ the wave vector of the wave. When we substitute this into the above equation, we find that

$$\omega = \left(\frac{4J's^2}{\hbar}\right)\sin^2\left(\frac{aq}{2}\right), \tag{9.101}$$

which is the desired dispersion relation.

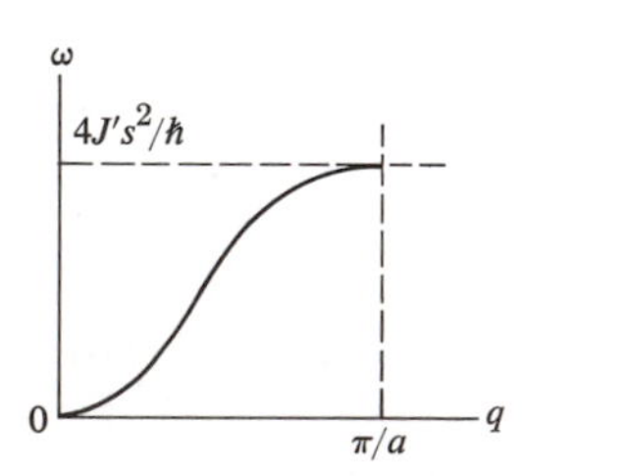


**Fig. 9.46** Dispersion curve for a spin wave.

The dispersion curve for the spin wave is shown in Fig. 9.46. The spin "lattice" can support waves whose frequencies range between zero and the maximum value

$\omega_m = 4J's^2/\hbar$. There is therefore an upper cutoff frequency, above which the wave is scattered very strongly, and no propagation takes place. This again is reminiscent of lattice waves. The upper frequency lies in the microwave range, as one can see by substituting for $J'$ the approximate value derived in Section 9.8, that is, $J' \simeq 0.1$ eV.

Note that for small $q$, $\omega$ is also very small. This is because the spins at long wavelength are still almost parallel to each other, and hence the restoring exchange force is small. For large $q$, however, the spins become appreciably *un*parallel, and hence a large restoring exchange force arises. The maximum value of the restoring force occurs when the lattice constant is equal to $\lambda/2$, that is, at $q = \pi/a$, in agreement with Fig. 9.46. This occurs at the boundary of the Brillouin zone.

Our remarks concerning the symmetry of the lattice-wave curve in the Brillouin zone (see Section 3.6) apply here also, and hence they will not be repeated.

Note that in the long-wavelength limit,

$$\omega \simeq Aq^2. \tag{9.102}$$

That is, $\omega$ is proportional to $q^2$. This differs from the case of lattice waves, in which $\omega \sim q$. Because of the form (9.102), the phase and group velocities of the spin wave are unequal, even in the long-wavelength region.

The dispersion curve can be determined by neutron diffraction. Since the neutron carries a magnetic moment, it is coupled to the field of the spin wave, and this results in the diffraction of the neutron beam. The equation for the conservation of momentum is

$$\mathbf{k}' = \mathbf{k} + \mathbf{q}, \tag{9.103}$$

where $\mathbf{k}$ and $\mathbf{k}'$ are the initial and final wave vectors of the neutron and $\mathbf{q}$ is the wave vector of the spin wave.

Spin waves are quantized in much the same way as lattice waves. The unit of quantization (or magnetic excitation) is called the *magnon*. A magnon of wave vector $\mathbf{q}$ carries an energy

$$E = \hbar\omega(\mathbf{q}), \tag{9.104}$$

a momentum

$$\mathbf{p} = \hbar\mathbf{q}, \tag{9.105}$$

and a magnetic moment

$$m = g\mu_B, \tag{9.106}$$

this moment being oriented opposite to $\mathbf{M}_0$. As far as magnetization is concerned, each magnon excitation is equivalent to a reversal of one spin.

In addition to evaluating the magnetization associated with thermal magnon excitation, we can evaluate the specific heat. The number of magnons in the mode $\mathbf{q}$, at thermal equilibrium, is given by the Bose–Einstein function,

$$\bar{n}(q) = \frac{1}{e^{\hbar\omega(\mathbf{q})/kT} - 1}. \tag{9.107}$$

Thus the energy is

$$\bar{E} = \int_0^{\omega_m} \hbar\omega\bar{n}g(\omega)d\omega, \tag{9.108}$$

where $g(\omega)$ is the magnon density of states. We can calculate this density of states from the dispersion relation, as in the phonon case (Section 3.7). When we do this, and when we use the long-wavelength approximation (9.102) (the Debye approximation!), we find that $\bar{E} \sim T^{5/2}$, and hence

$$C_v \sim T^{3/2}. \tag{9.109}$$

This is in agreement with experiment at low temperature.

The magnetization $M(T)$ can similarly be written as

$$M(T) = M(0) - g\mu_B \int_0^{\omega_m} \bar{n}g(\omega)d\omega.$$

In the same approximation used above, we find that the decrease in magnetization as a result of thermal excitation is

$$\Delta M = M(0) - M(T) \sim T^{3/2}, \tag{9.110}$$

which is known as *Bloch's law*, in honor of the man who first postulated the existence of spin waves. This result is also in agreement with experiment.

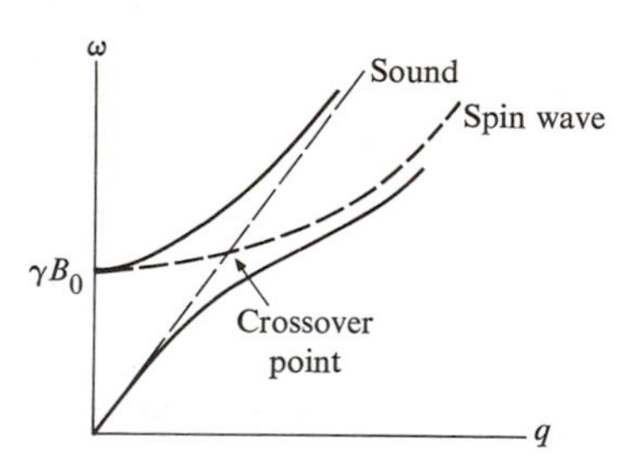


**Fig. 9.47** Interaction between acoustic and spin-wave modes. Dashed lines indicate free modes, while solid curves represent coupled modes.

Now let us take a look at the interaction between spin and elastic waves. When an external field is applied to a ferromagnetic sample, the dispersion relation (9.102) for the spin wave is modified to

$$\omega = \gamma B_0 + Aq^2, \tag{9.111}$$

where the first term on the right is due to the external field, and the second to the exchange interaction.

Figure 9.47 plots the dispersion curves for both spin and elastic waves. In the region of crossover, the two modes couple strongly and repel each other, in much the same fashion (Section 3.12) as electromagnetic and lattice waves. Note also how the modes change character as $q$ increases. In YIG (yttrium iron garnet) the crossover frequency is $\omega = 0.46$ GHz for $B_0 = 0.3\ \mathrm{W/m^2}$.

This coupling has been employed to generate acoustic waves by converting magnetic energy into acoustical energy. This suggests possible interesting applications in acoustical amplifiers.

## SUMMARY

The magnetic induction **B** and the magnetic field $\mathscr{H}$ in a material medium are related by the equation

$$\mathbf{B} = \mu_0 \mathscr{H} + \mu_0 \mathbf{M},$$

where **M** is the magnetization vector of the medium. The magnetization **M** is proportional to the field,

$$\mathbf{M} = \chi \mathscr{H},$$

and the constant $\chi$ is the magnetic susceptibility. Substituting this equation into the previous relation, one may write this as

$$\mathbf{B} = \mu \mathscr{H},$$

where the permeability $\mu$ is defined as

$$\mu = \mu_0(1 + \chi).$$

The relative permeability $\mu_r = \mu/\mu_0$ is thus

$$\mu_r = 1 + \chi.$$

There are two basic types of contributions to the susceptibility: A *diamagnetic* contribution resulting from the deformation of the orbits of the electrons by the magnetizing field, and a *paramagnetic* contribution due to the alignment of the magnetic moments of the electrons (if such moments are present) with the field.

### Langevin diamagnetism

When one treats the orbits of electrons around atoms as circular current loops, one finds that a magnetic field produces a diamagnetic susceptibility

$$\chi = -\frac{\mu_0 e^2}{6m}(NZ\overline{r^2}).$$

**Langevin paramagnetism**

Given that the typical atom has a net moment $\mu$, one can then show that the alignment of the moments with the field leads to a classical paramagnetic susceptibility

$$\chi = \frac{N\mu_0\mu^2}{3kT}.$$

The quantum treatment yields the same result, provided that $\mu = g[j(j+1)]^{1/2}\mu_B$. This formula holds well in transition and rare-earth ions.

**Magnetism in metals**

In metals, the conduction electrons make a spin paramagnetic contribution

$$\chi_p = \mu_0\mu_B^2 g(E_F),$$

which is independent of temperature. The conduction electrons also have a diamagnetic effect due to their cyclotron motion. In a metal of simple band structure, the magnitude of the electrons' cyclotron contribution is equal to one-third of their spin contribution. The ion cores also introduce a diamagnetic effect, which must be added to the previous two for comparision with experiments.

**Ferromagnetism**

A ferromagnetic substance is one which exhibits spontaneous magnetization below its Curie temperature. Above this temperature, the substance is paramagnetic, and obeys the *Curie–Weiss law*,

$$\chi = \frac{C}{T - T_f},$$

where $T_f$ is the Curie temperature.

The ferromagnetic phase appears because of an internal magnetic field $\mathscr{H} = \lambda M$. This field, in turn, has its genesis in an exchange interaction between the magnetic dipoles of the substance.

Other magnetic structures besides the ferromagnetic are observed; examples are the antiferromagnetic and ferrimagnetic substances. These also owe their existence to exchange interactions between magnetic moments.

Ferromagnetism is also observed in some metals, but the theoretical treatment there becomes difficult because the 3d electrons are only partially localized. Ferromagnetism is favored in those metals of narrow but dense energy bands, and large exchange constants.

**Magnetic resonance**

When a magnetic field $B_0$ is applied to a substance, the dipole moments of the atoms precess around the field with frequency

$$\omega_0 = \gamma B_0,$$

where $\gamma$ is the gyromagnetic ratio of the dipole. When a signal of frequency $\omega = \omega_0$ passes through the system, power is absorbed from the signal by the dipoles. This is the phenomenon of *electron paramagnetic resonance* (EPR). One can determine relaxation effects by introducing appropriate relaxation times.

*Nuclear magnetic resonance* (NMR) is the nuclear analog of EPR.

*Ferromagnetic resonance* (FMR) is the same as EPR, except that the internal magnetic field, which is now strong, has to be explicitly taken into account. This field is a function of several types of interaction; these can therefore be studied by the FMR technique.

### Spin waves

Spin waves are collective excitations in spin systems. The interaction responsible for these modes is the exchange interaction between the moments of the system. Spin waves carry both energy and momentum. As the temperature is raised, energy is absorbed by the excitations of the spin waves, and the magnetization also decreases.

## REFERENCES

R. Kubo and T. Nagamiya, editors, 1969, *Solid State Physics*, New York: McGraw-Hill
D. H. Martin, 1967, *Magnetism in Solids*, Cambridge, Mass.: M.I.T. Press
A. H. Morrish, 1965, *Physical Principles of Magnetism*, New York: John Wiley
J. H. van Vleck, 1932, *The Theory of Electric and Magnetic Susceptibility*, Oxford: Oxford University Press
R. M. White, 1970, *Quantum Theory of Magnetism*, New York: McGraw-Hill

### Ferromagnetism, antiferromagnetism, and ferrimagnetism

R. M. Bozorth, 1951, *Ferromagnetism*, New York: Van Nostrand
S. Chikazumi, 1964, *Physics of Magnetism*, New York: John Wiley
B. Lax and K. J. Button, 1962, *Microwave Ferrites and Ferrimagnetism*, New York: McGraw-Hill
D. H. Martin, *op. cit.*
A. H. Morrish, *op. cit.*
G. T. Rado and H. Suhl, editors, 1963, *Magnetism*, New York: Academic Press
J. S. Smart, 1965, *Effective Field Theories of Magnetism*, Philadelphia: W. B. Saunders
J. S. Smart and H. P. J. Wijn, 1959, *Ferrites*, New York: John Wiley
J. Smit, editor, 1971, *Magnetic Properties of Materials*, New York: McGraw-Hill

### Magnetic resonances

A. Abragam, 1961, *Nuclear Magnetism*, Oxford: Oxford University Press
W. Low, 1960, "Paramagnetic Resonance in Solids," *Solid State Physics*, Supplement 2, New York: Academic Press
B. Lax and K. J. Button, *op. cit.*
A. H. Morrish, *op. cit.*

G. E. Pake, 1956, "Nuclear Magnetic Resonance," in *Solid State Physics*, Volume 2, New York: W. A. Benjamin
G. E. Pake, 1962, *Paramagnetic Resonance*, New York: W. A. Benjamin
R. T. Schumacher, 1970, *Introduction to Nuclear Magnetic Resonance*, New York: W. A. Benjamin
C. P. Slichter, 1963, *Principles of Magnetic Resonance*, New York: Harper and Row
M. Sparks, 1964, *Ferromagnetic Relaxation Theory*, New York: McGraw-Hill
A. Yariv, 1966, *Quantum Electronics*, New York: John Wiley

## QUESTIONS

1. The text stated that the diamagnetic response associated with the orbital motion of atomic electrons can be predicted on the basis of Lenz's law. Prove this statement.
2. Do you expect the constant $\lambda$ in (9.30) describing the susceptibility of the covalent bond to be positive or negative? Why?
3. Given that the total angular momentum quantum number $j$ for an atom is $j = \frac{1}{2}$, does this necessarily mean that the angular momentum is pure spin, and hence $g = 2$? Illustrate your answer with an example.
4. You may have realized, after reading Section 9.6, that the formula for paramagnetic susceptibility is valid only if one considers the ground state of the atom. But other excited atomic levels are also present. Explain the following.

   a) Why is it usually permissible to disregard these higher levels when calculating the susceptibility?

   b) How you would modify Eq. (9.42), or the original formula from which it is derived, if the temperature were high enough for some of the excited levels to be appreciably populated?

5. Given that the precession frequency due to spin–orbit interaction is 10 GHz, estimate the effective magnetic field experienced by the spin moment as a result of this interaction.
6. Referring to Questions 4 and 5, estimate the temperature above which the simple formula (9.42) breaks down for the strength of spin–orbit interaction given in Question 5.
7. Give a sufficient condition for the existence of paramagnetic susceptibility in terms of the number of electrons in the atom (or ion).
8. The spin paramagnetic susceptibility of conduction electrons is given in (9.47). What is its value for a full band? Is the answer surprising? Explain.
9. Neither Mn nor Cr are ferromagnetic by themselves, yet some of their alloys (with other elements) are. Explain how this may be possible. Refer to Fig. 9.17.
10. Solid-state theorists often conjecture that any spin system would eventually become ferromagnetic at sufficiently low temperature. Can you justify this conjecture in light of the discussion in Section 9.8? Given that the dipole–dipole electrostatic interaction is the one responsible for such a ferromagnetic transition, estimate the Curie temperature. (How would you account for the fact that only relatively few spin systems are observed in the ferromagnetic phase, even at very low temperatures?)

11. Can the domain structure in a ferromagnetic substance be detected by x-ray diffraction? By neutron diffraction?
12. Equation (9.80) shows that $\chi''$ decreases as the strength of the signal is increased, a phenomenon known as *saturation*. Explain the physical original of this phenomenon. *Hint*: Think of $\chi''$ as it relates to the rate of absorption. Also note that the quantum picture of the EPR is more helpful in explaining this phenomenon than the classical picture.
13. Explain why the condition $\tau_1\tau_2(\gamma b_0)^2 \ll 1$ is necessary for the observation of EPR. Refer to Eqs. (9.78).
14. Prove Eq. (9.81).
15. The condition of population inversion in a maser is often stated by ascribing a negative absolute temperature (!) to the system. Explain why this is meaningful; refer to Eq. (9.83). Calculate the temperature of the system, given that $\Delta E = 1$ GHz and $N_2/N_1 = 2$.
16. Is the nuclear factor $g_n$ positive or negative for the nucleus illustrated in Fig. 9.39?
17. The neutron has a magnetic moment (Table 9.10), and yet this particle is electrically neutral. Does the existence of this moment puzzle you? Explain. Also discuss how such a moment may be possible if one endows the neutron with a submicroscopic structure.
18. What is the precise physical meaning of the word adiabatic in connection with the technique of cooling by adiabatic demagnetization? Why are nuclear rather than electron spins used at very low temperatures?
19. The NMR technique is most useful in organic chemistry, due to the proton resonance of hydrogen. What are the two other commonest elements in this field of chemistry, and why are they not usually useful in NMR?
20. Another standard technique for observing ferromagnetic substances is by using a polarizing microscope. If a thin section is cut off the substance, and the plane of the section is normal to the easy-axis direction, then, when one adjusts the polarizing filter on the microscope, half the domains appear bright and the other half dark. Explain why.
21. Using the fact that the specific heat of the spin system is $C \sim T^{3/2}$ at low temperature, give a physical derivation for the dependence of the magnon density of states $g(\omega)$ on $\omega$ in the long-wavelength region. Compare your result with the answer given in Problem 25.

## PROBLEMS

1. Prove the validity of Eqs. (9.3) and (9.4).
2. Establish the result (9.6).
3. a) Prove the Larmor theorem, i.e., that a classical dipole $\boldsymbol{\mu}$ in a magnetic field $\mathbf{B}$ precesses around the field with a frequency equal to the Larmor frequency $\omega_L = eB/2m$.

   b) Evaluate the Larmor frequency, in hertz, for the orbital moment of the electron in a field $B = 1\ \text{W/m}^2$.

   c) What is the precession frequency for a spin dipole moment in the same field?

4. The diamagnetic susceptibility due to the ion cores in metallic copper is $-0.20 \times 10^{-6}$. Knowing that the density of Cu is 8.93 g/cm$^3$ and that its atomic weight is 63.5, calculate the average radius of the Cu ion.
5. a) The susceptibility of Ge is $-0.8 \times 10^{-5}$. Taking the radius of the ion core to be 0.44 Å, estimate the percentage of the contribution of the covalent bond to the susceptibility. Germanium has a density of 5.38 g/cm$^3$ and an atomic weight of 72.6.

   b) Given that the applied field is $\mathscr{H} = 5 \times 10^4$ amp $\cdot$ m$^{-1}$, calculate the magnetization in Ge; also the magnetic induction.
6. A system of spins ($j = s = \frac{1}{2}$) is placed in a magnetic field $\mathscr{H} = 5 \times 10^4$ amp $\cdot$ m. Calculate the following.

   a) The fraction of spins parallel to the field at room temperature ($T = 300$°K).

   b) The average component of the dipole moment along the field at this temperature.

   c) Calculate the field for which $\bar{\mu}_z = \frac{1}{2}\mu_B$.

   d) Repeat parts (a) and (b) at the very low temperature of 1°K.
7. Establish the result (9.42) for an arbitrary value of $j$. (This result is derived under the condition $\mu_B B \ll kT$.) Estimate the field below which the result is valid at room temperature.
8. Prove that the average dipole moment of an atom, including the effect of the spin–orbit interaction, is given by $\mu_{avg} = g(-e/2m)J$, where the Lande factor $g$ is given by (9.45).
9. Verify the theoretical values of $p$ given in the third column in Table 9.3.
10. Repeat Problem 9 for the third and fourth columns in Table 9.4.
11. a) The spin susceptibility of conduction electrons at $T = 0$°K is given in Eq. (9.47). Express this result in terms of the electron concentration for an energy band of standard form.

    b) Calculate the spin susceptibility for K, whose density is 0.87 g/cm$^3$ and whose atomic weight is 39.1.

    c) Calculate the diamagnetic susceptibility of the conduction electrons in K.

    d) Using the above results and Table 9.5, calculate the average radius of the K ion in the metallic state.
12. Iron has a bcc structure with a lattice constant $a = 2.86$ Å.

    a) Using the value of the saturation magnetization in Table 9.6, show that the dipole moment of an Fe atom is equal to 2.22 $\mu_B$. The density of Fe is 7.92 g/cm$^3$, and its atomic weight is 55.6. (You may assume, for the present purpose, that the 3d electrons are completely localized.)

    b) Calculate the Weiss exchange constant $\lambda$ and the molecular field in iron.

    c) Evaluate the Curie constant in iron.

    d) Estimate the exchange energy for a dipole interacting with its nearest neighbors.
13. Repeat Problem 12 for Co (hcp, $a = 2.51$, $c = 4.1$ Å), and Ni (hcp, $a = 2.66$, $c = 4.29$ Å). The densities of Co and Ni are 8.67 and 9.04 g/cm$^3$, respectively.
14. a) Applying the Weiss model, with two exchange constants $\lambda_1$ and $\lambda_2$, to an anti-

ferromagnetic substance, derive the Néel formula for the susceptibility at high temperature [Eq. (9.59)].

b) Evaluate the exchange constants $\lambda_1$ and $\lambda_2$ for $MnF_2$.

c) Explain why $\lambda_2 > \lambda_1$.

15. Carry out the steps leading to Eq. (9.72).
16. a) Discuss the splitting of a $Cr^{3+}$ ion in a static magnetic field.

    b) Calculate the field for which the electron resonance for this ion occurs at 10 GHz.
17. Solve the Bloch equations (9.66) and (9.67) in the presence of a static field $\mathbf{B}_0$ but in the absence of the signal, and show that the magnetization spirals toward its equilibrium value as described in Fig. 9.33. Take the initial angle between magnetization and the field to be $10^0$, the longitudinal and transverse time to be $10^{-6}$ and $5 \times 10^{-7}$ s, respectively, and plot the longitudinal and transverse components of the magnetization versus time, in the interval $0 < t < 5 \times 10^{-6}$ s.
18. Carry out the steps leading to Eq. (9.78).
19. Nuclear magnetic resonance in water is due to the protons of hydrogen.

    a) Find the field necessary to produce NMR at 60 MHz.

    b) Find the maximum power absorbed per unit volume, given that the strength of the signal is such that $\tau_1 \tau_2 \gamma^2 b_0^2 = 1$ and $\tau_1 = \tau_2 = 3$ s.
20. Carry out the steps leading to (9.100).
21. Many microwave magnetic devices are discussed in Lax and Button (1962). Make a brief study of these devices, and present a review report.
22. The text said that spin waves are modes which describe the collective excitations of a spin system. It also pointed out the close analogy between spin waves (magnons) and lattice waves (phonons). What is the spin mode of excitation analogous to the Einstein mode in the lattice? That is, what are the localized spin excitation modes? Assuming that these are the only modes of excitation possible (which is incorrect), calculate the magnetization and spin specific heat for the system as functions of the temperature.
23. Discuss why spin waves are more favorable as modes of excitation than local spin modes, particularly at low temperatures.
24. Determine the expressions for the phase and group velocities of spin waves. Calculate the group velocity in iron at wavelength $\lambda = 1$ cm. (Use results of Problem 12.)
25. Show that the magnon density of states $g(\omega)$ in the long-wavelength limit is given by $g(\omega) = (\frac{1}{4}\pi^2)(\hbar/J's^2a^2)^{3/2} \times \omega^{1/2}$.
26. Many ferromagnetic, ferrimagnetic, and antiferromagnetic substances, such as the oxides and chalcogenides of the 3d transition metals, exhibit a small amount of electrical conductivity, i.e., they are semiconductors. Although we have not discussed this subject here, it is a lively area of research today and is reviewed in depth in J. P. Suchet, 1971, *Crystal Chemistry and Semiconduction*, New York: Academic Press. Study the highlights of this book and write a review report.

# CHAPTER 10 SUPERCONDUCTIVITY



---

*Take her up tenderly,*
*Lift her with care;*
*Fashion'd so slenderly*
*Young, and so fair!*

Thomas Hood

## 10.1 INTRODUCTION

One of the most interesting properties of solids at low temperatures is that, in many metals and alloys, electrical resistivity vanishes entirely below a certain temperature—the temperature depending on the substance. This zero resistivity (or infinite conductivity) is known as *superconductivity*. Scientists' efforts to explain this fascinating phenomenon have contributed greatly to our understanding of solids in general, and particularly solids at low temperatures. Superconductivity has been applied in the design of superconducting magnets, computer switches, and many other technical devices. In addition, engineers are proposing the use of superconductivity in the field of transportation and in the transmission of power without pollution.

We begin our discussion with the electrical properties of superconductors. We shall then consider the Meissner effect—the fact that a superconductor expels a magnetic flux—and we shall further explore the effects of a magnetic field by showing that it may destroy the superconducting property.

Thermodynamic considerations suggest the presence of a gap in the energy spectrum of the electrons at the Fermi surface.

We next discuss the London theory of superconductor electrodynamics, which describes the distributions of fields and currents in superconductors. Although semi-phenomenological in nature, this theory is extremely useful in giving a simple account of many phenomena associated with superconductivity. This is followed by consideration of the microscopic theory of superconductivity—due to Bardeen, Cooper, and Schrieffer—commonly referred to as the BCS theory. This theory shows that the physical origin of superconductivity lies in the interaction between conduction electrons and the ions of the lattice.† We shall also discuss tunneling phenomena involving superconductors, including the Josephson effect. These phenomena are among the best-known illustrations of the basic concepts of quantum mechanics.

The chapter closes with a look at miscellaneous subjects related to superconductivity.

## 10.2 ZERO RESISTANCE

Superconductivity was first observed in 1911 by the Dutch physicist H. K. Onnes in the course of his experiments on the electrical conductivities of metals at low temperatures. He observed that as purified mercury is cooled, its resistivity vanishes abruptly at 4.2°K (Fig. 10.1). Above this temperature the resistivity is small, but finite, while the resistivity below this point is so small that it is essentially zero. The temperature at which the transition takes place is called the *critical temperature*.

Onnes surmised correctly that he was dealing with a new state of matter below the critical temperature, and coined the term *superconducting state*. Above the

---

† For this great scientific achievement, Bardeen, Cooper, and Schrieffer were awarded the 1972 Nobel prize in physics. Bardeen, who had already received a Nobel prize for his work on the transistor, thus has the unprecedented distinction of getting two Nobel prizes in the same field.

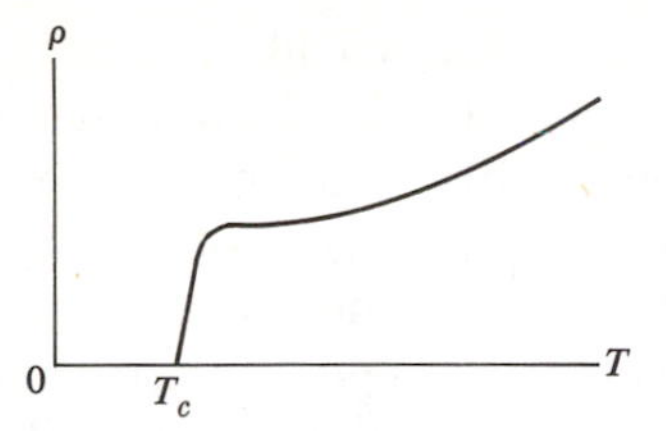


**Fig. 10.1** Resistivity $\rho$ versus temperature $T$ for a superconductor. The resistivity vanishes for $T \leqslant T_c$.

critical temperature $T_c$, the substance is in the familiar *normal state*, but below $T_c$ it enters an entirely different superconducting state. This transition may be likened to other familiar phase transitions, such as that of vapor–liquid at the vaporization point, or the ferromagnetic transition at the Curie point.

Onnes found that the superconducting transition is *reversible*: When he heated the superconducting sample it recovered its normal resistivity at the temperature $T_c$. This confirmed his supposition that here was a new state of matter, one which depends on the state variables, such as temperature, rather than on the history of the sample.

We can gain some insight into the nature of superconductivity using the free-electron model of Chapter 4. It was shown in Section 4.4 that the resistivity of a metal may be written as

$$\rho = \frac{m}{ne^2\tau},$$

where $\tau$ is the collision time, and pointed out that $\rho$ decreases as the temperature is lowered, because, as $T$ decreases, the lattice vibrations begin to "freeze," and hence the scattering of the electrons diminishes. This results in a longer $\tau$ and hence a smaller $\rho$, as indicated by the above equation. If $\tau$ becomes infinite at sufficiently low temperatures, then the resistivity vanishes entirely, which is what is observed in superconductivity. We shall see in Section 10.5 that, as the temperature is lowered below $T_c$, a fraction of the electrons become superconducting, in the sense that they have infinite collision times. These electrons undergo no scattering whatsoever, even though the substance may contain some impurities and defects. It is these electrons which are responsible for superconductivity.

One usually measures the resistivity of a superconductor by causing a current to flow in a ring-shaped sample (one can start the current by induction after removing a magnetic flux linking the ring), and observing the current as a function of time. If the sample is in the normal state, the current damps out quickly because of the resistance of the ring. But if the ring has zero resistance, the current, once set up, flows indefinitely without any decrease in value. Physicists have made experiments to test this, and found that even after several years of operation the current in the ring remained constant, as far as they could tell. For instance, they found that the upper limit for the resistivity of a superconducting lead ring was about $10^{-25}$ ohm-m. The fact that this is about $1/10^{+17}$ as large as the value at room temperature does indeed justify taking $\rho = 0$ for the superconducting state.

The superconducting transition is not always sharp. But if the specimen is made up of a metallic element, which is pure and structurally perfect, the transition is usually sharp. Pure Ga, prepared under these conditions, has a transition range of less than $10^{-5}$°K. By contrast, a metallic alloy which is strained may have a broad transition range of 0.1°K or more. This is illustrated by Fig. 10.2.

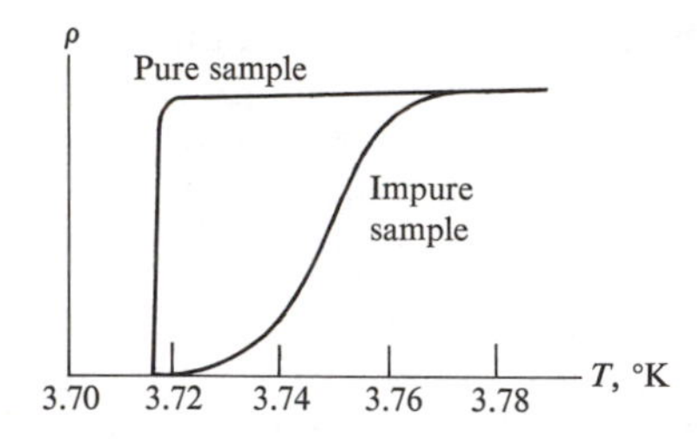


**Fig. 10.2** Effect of impurities on superconducting transition in tin.

### Occurrence of superconductivity

Superconductivity is not a rare phenomenon. It is exhibited by an appreciable number of elements (27 as of now), and many alloys. Table 10.1 lists most of the superconducting elements, and the better-known superconducting alloys, together with their critical temperatures.

Note that the critical temperature varies widely—from 0.01°K for W to 20.8°K for NbAlGe. It would be useful to have superconductors with much higher critical temperatures, particularly approaching room temperature, but efforts to achieve this have met with failure. The highest known critical temperature is close to 20°K, and this has remained the case for a number of years, although physicists still hope that someday they will find materials that have higher critical temperatures.†

Since superconductivity appears only in some substances, and not in all, and since $T_c$ varies widely, it is useful to have criteria which indicate the expected value of $T_c$ and the likelihood of observing superconductivity in a particular substance. The rules given below are due to B. Matthias, who, on the basis of these rules, discovered thousands of new superconductors.

†The latest record critical temperature is 23.2°K and occurs in $Nb_3Ge$. This discovery, made during the Fall of 1973, is expecially significant because the new temperature lies above the boiling temperature of hydrogen, and it is possible therefore to begin moving superconductivity technology from one based on liquid helium to a more practical one using liquid hydrogen.

**Table 10.1**

Transition Temperatures of Superconductors

| Element | $T_c$, °K | Compound | $T_c$, °K |
|---|---|---|---|
| Al | 1.2 | $Nb_3Al_{0.8}Ge_{0.2}$ | 20.1 |
| Cd | 0.5 | $Nb_3Sn$ | 18.1 |
| Ga | 1.1 | $Nb_3Al$ | 17.5 |
| In | 3.4 | $Nb_3Au$ | 11.5 |
| Ir | 0.1 | $Nb_3N$ | 16.0 |
| La ($\alpha$) | 4.8 | $M_0N$ | 12.0 |
| La ($\beta$) | 4.9 | $V_3Ga$ | 16.5 |
| Pb | 7.2 | | |
| Hg ($\alpha$) | 4.2 | | |
| Hg ($\beta$) | 4.0 | | |
| Mo | 0.9 | | |
| Nb | 9.3 | | |
| Os | 0.7 | | |
| Rh | 1.7 | | |
| Ru | 0.5 | | |
| Ta | 4.5 | | |
| Tc | 8.2 | | |
| Tl | 2.4 | | |
| Th | 1.4 | | |
| Sn | 3.7 | | |
| Ti | 0.4 | | |
| W | 0.01 | | |
| U($\alpha$) | 0.6 | | |
| U ($\beta$) | 1.8 | | |
| V | 5.3 | | |
| Zn | 0.9 | | |
| Zr | 0.8 | | |

a) Superconductivity occurs only in substances in which the valence number per atom is between 2 and 8. In general, superconducting elements lie in the inner columns of the periodic table. The phenomenon has not yet been observed in the alkali or noble metals.

b) The valence numbers 3, 4.7, 6.4 (nearly odd) are particularly favorable, i.e., they result in higher critical temperatures, while the numbers 2, 4 and 5.6 (nearly even) are particularly unfavorable. This is illustrated in Fig. 10.3, in which the valence number is varied continuously in the alloy ZrNbMoRe.

c) A small atomic volume, accompanied by a small atomic mass, favors superconductivity.

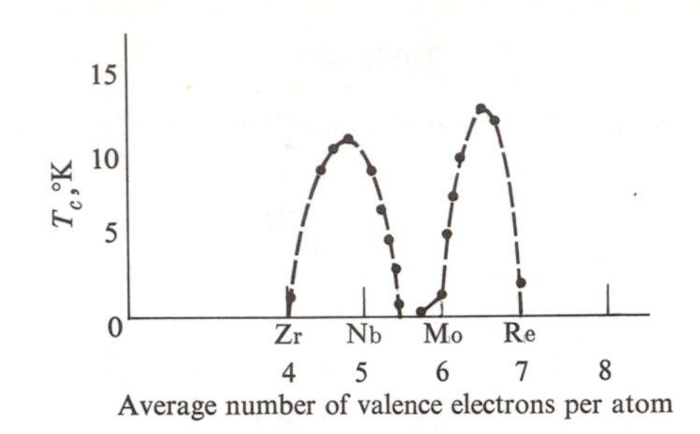


**Fig. 10.3** Variation of critical temperature with valence number for alloys of elements in the second transition series of the periodic table.

Although these rules were prescribed by Matthias on empirical grounds, some of them may be related, albeit loosely, to the BCS theory.

## 10.3 PERFECT DIAMAGNETISM, OR THE MEISSNER EFFECT

In 1933, two German physicists, Meissner and Ochsenfeld, observed that a superconductor expels magnetic flux completely, a phenomenon known as the *Meissner effect.* In a series of experiments on superconducting cylinders, they demonstrated that, as the temperature is lowered to $T_c$, the flux is suddenly and completely expelled as the specimen becomes superconducting, as shown in Fig. 10.4. (The flux expulsion continues for all $T < T_c$.) They established this by carefully measuring the magnetic field in the neighborhood of the specimen. Furthermore, they demonstrated that the effect is *reversible*: When the temperature is raised from below $T_c$, the flux suddenly penetrates the specimen after it reaches $T_c$, and the substance is in the normal state.

The magnetic induction inside the substance is given by

$$\mathbf{B} = \mu_0(\mathscr{H} + M) = \mu_0(1 + \chi)\mathscr{H}, \tag{10.1}$$

where $\mathscr{H}$ is the external intensity of the magnetic field, $M$ the magnetization in the

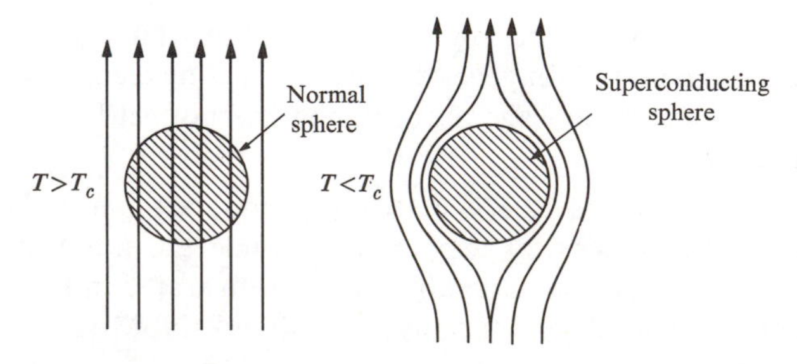


**Fig. 10.4** The Meissner effect: The magnetic flux is expelled from a superconductor, that is, for $T < T_c$.

medium, and $\chi$ its magnetic susceptibility. Since $B = 0$ in the superconducting state, it follows that

$$M = -\mathscr{H}, \tag{10.2}$$

meaning that the magnetization is equal to and opposed to $\mathscr{H}$. The medium is therefore diamagnetic, and the susceptibility is

$$\chi = -1. \tag{10.3}$$

Such a condition—in which the magnetization cancels the external intensity exactly—is referred to as *perfect diamagnetism.* Figure 10.5 illustrates the magnetization in a superconductor.

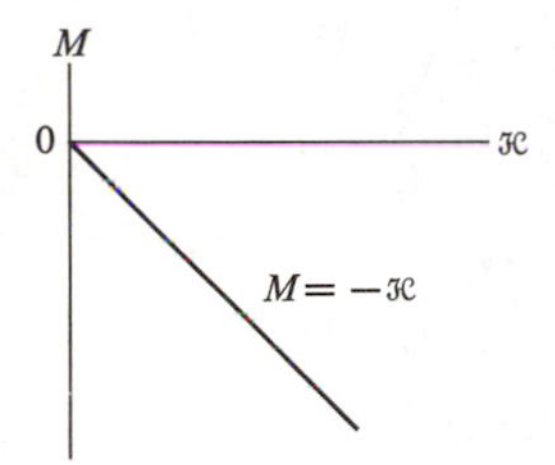


**Fig. 10.5** Magnetization curve for a superconductor.

Compare this behavior with that of a normal metal. The metal is also diamagnetic—if the spin susceptibility is ignored—but in that case $\chi \simeq -10^{-5}$, which is much smaller than that given by (10.3). It follows that some new mechanism operates in superconductors in order to give such an enormous diamagnetism.

The Meissner effect is a powerful means of shedding light on the superconducting state, and it has been speculated that, had the effect been discovered before 1933, the full understanding of superconductivity would have come much earlier. The Meissner effect is particularly interesting because it contradicts classical laws, as we shall see shortly.

## 10.4 THE CRITICAL FIELD

Shortly after Onnes first observed superconductivity, it was found that superconductivity can be destroyed by the application of a magnetic field. If a strong enough magnetic field, called the *critical field,* is applied to a superconducting specimen, it becomes normal and recovers its normal resistivity even at $T < T_c$.

The critical field depends on the temperature. For a given substance, the field decreases as the temperature rises from $T = 0°\text{K}$ to $T = T_c$. It has been found empirically that the variation is represented by

$$\mathscr{H}_c(T) = \mathscr{H}_c(0)\left[1 - \left(\frac{T}{T_c}\right)^2\right], \tag{10.4}$$

which holds approximately for many substances, as shown in Fig. 10.6. Thus the field has its maximum value, $\mathscr{H}_c(0)$, at $T = 0°K$, and vanishes at $T = T_c$. This result is expected, of course, because at $T = T_c$ the specimen is already normal, and no field is necessary to accomplish the transition. The critical field is typically of the order of several hundred gauss. Table 10.2 gives the critical fields for some superconductors.

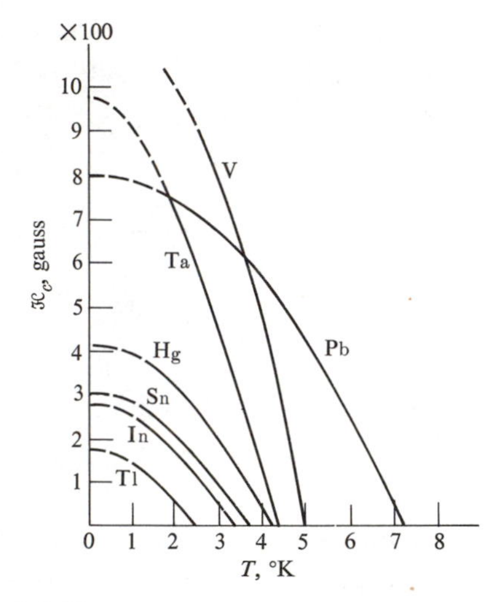


**Fig. 10.6** Critical field versus temperature for several superconductors.

**Table 10.2**

Critical Fields of Some Superconductors

| Element | $\mathscr{H}_0$ | |
|---|---|---|
| | A/m | gauss |
| Al | $0.79 \times 10^4$ | 99 |
| Cd | 0.24 | 30 |
| Ga | 0.41 | 51 |
| Pb | 6.4 | 803 |
| Hg($\alpha$) | 3.3 | 413 |
| Hg($\beta$) | 2.7 | 340 |
| Ta | 6.6 | 830 |
| Sn | 2.4 | 306 |

The critical field need not be external. A current flowing in a superconducting ring creates its own magnetic field, and if the current is large enough so that its own field reaches the critical value, then superconductivity is also destroyed.

This places a limitation on the strength of the current which may flow in a superconductor, and this is, in fact, the primary limitation in the manufacture of high-field superconducting magnets.

## 10.5 THERMODYNAMICS OF THE SUPERCONDUCTING TRANSITION

The purpose of the discussion of thermodynamics in this section is to unify the various observations described thus far. Even though the discussion will not lead to conclusive statements about the microscopic forces involved (thermodynamics is essentially a macroscopic science), it will provide clues as to the nature of the transition.

Figure 10.7 illustrates the variation of specific heat with temperature for a superconductor. The peaking of $C_v$ just below $T_c$ indicates an appreciable increase in *entropy*—or disorder—as $T$ increases toward $T_c$, and transition to the normal state becomes imminent. Thus the superconducting state has a greater degree of order than the normal state.

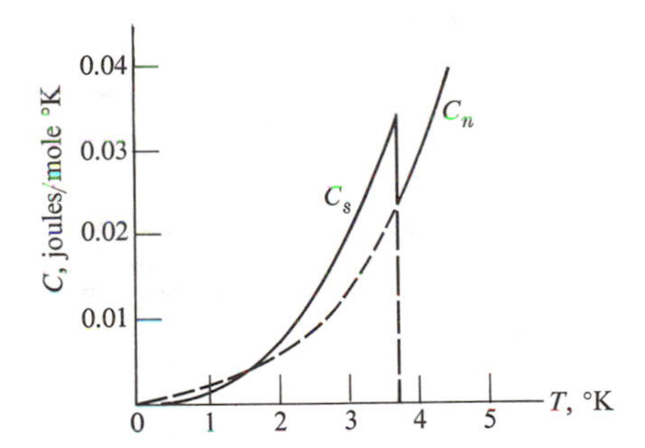


**Fig. 10.7** Molar specific heat of tin versus temperature. The dashed curve is an extrapolation which represents what the specific heat would have been if the normal state had persisted for $T < T_c$.

Experiments at very low temperatures indicate that the specific heat of the electrons in that region decreases exponentially†

$$C_v = ae^{-b(T/T_c)}. \tag{10.5}$$

This exponential behavior implies the presence of an *energy gap* in the energy spectrum of the electrons. This gap, which lies just at the Fermi level (Fig. 10.8), prevents the electrons from being readily excitable. It also leads to a very small specific heat. The width of the gap $\Delta$ must be of the order of $kT_c$, because when the

† To obtain the total specific heat of a superconductor, one must add to this the specific heat of the lattice. The lattice contribution $\sim T^3$ at low temperatures, as we recall from Chapter 3.

substance is raised to $T_c$, it becomes normal and its electrons are then readily excited. Thus

$$\Delta \simeq kT_c. \tag{10.6}$$

Substituting $T_c = 5°K$, a typical value, one finds that $\Delta \simeq 10^{-4}$ eV. This energy gap is very small compared with the gaps we have encountered previously, and it is for this reason that superconductivity appears only at very low temperatures.

We have noted that the superconducting state has a higher degree of order than the normal state. One may, in fact, view the superconducting transition as similar to the condensation of a vapor into the more ordered liquid state. Similarly, one expects a reduction in energy as a result of the transition. Let us now calculate the "condensation" (or latent) energy associated with the superconducting transition.

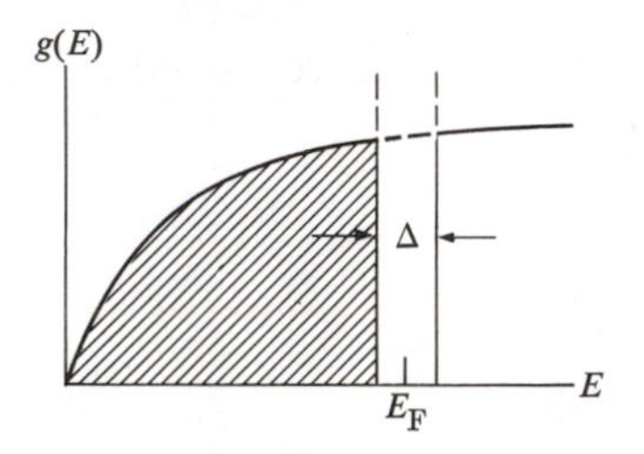


**Fig. 10.8** The density of states $g(E)$ versus $E$ for a superconductor, illustrating the superconducting gap at the Fermi energy level. The gap is greatly magnified for purposes of illustration, the actual value of $\Delta/E_F$ being about 0.0001. The screened area represents the region occupied at $T = 0°K$.

Figure 10.9 plots the critical field $\mathcal{H}_c$ versus $T$. The curve divides the $\mathcal{H}_c$–$T$ plane into two regions: the normal and the superconducting. Suppose that the specimen is at temperature $T_1 < T_c$. When the specimen starts at point $A$ and follows the vertical path $AN$—that is, gradually increasing the field—it becomes normal at the point $N$. Thus the "condensation" energy is

$$\Delta E = E_N - E_A. \tag{10.7}$$

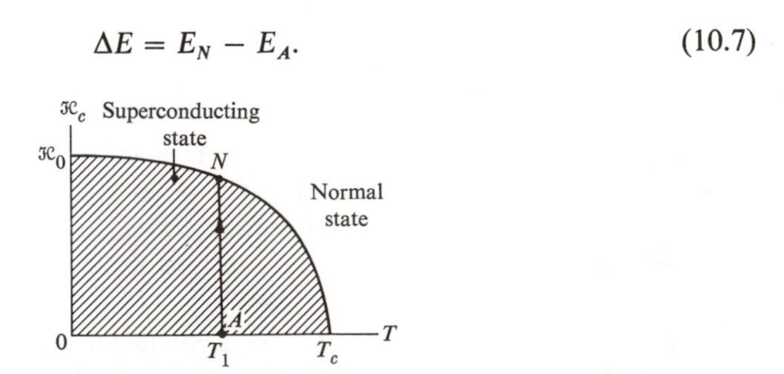


**Fig. 10.9** Calculation of the superconducting "condensation" energy.

This energy can be readily calculated. Since the specimen acts as a perfect diamagnet along the path $AN$, $\Delta E$ is equal to the demagnetization energy,

$$\Delta E = -\int_0^{\mathscr{H}_c} B\,dM = -\mu_0 \int_0^{\mathscr{H}_c} \mathscr{H}(-\,d\mathscr{H}) = \tfrac{1}{2}\mu_0 \mathscr{H}_c^2 \tag{10.8}$$

per unit volume. This is the amount of energy needed to convert a system from the superconducting into the normal state, and, conversely, it is the amount lost by the system when it makes the transition from the normal to the superconducting state. Since a system always seeks to be in a state which has the lowest possible energy, it follows that the superconducting state is the more stable one for $T < T_c$ (Fig. 10.10).

The maximum amount of condensation energy is

$$\Delta E = \tfrac{1}{2}\mu_0 \mathscr{H}_c^2(0), \tag{10.9}$$

and occurs, of course, at $T = 0°\mathrm{K}$. If one substitutes a typical value of $\mathscr{H}_c(0) = 500\ \mathrm{G}$, one finds $\Delta E = 10^3\ \mathrm{J/m^3}$.

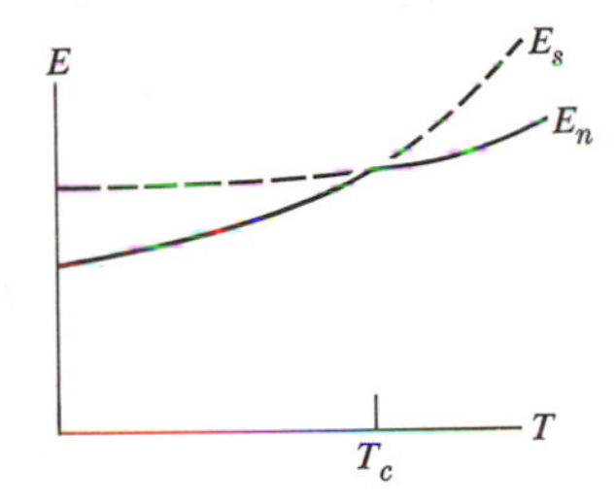


**Fig. 10.10** The energies $E_S$ and $E_N$ of the superconducting and normal states, versus temperature.

A useful relation can now be established between the critical field and the critical temperature, We calculated the condensation energy in terms of the field, but it may also be estimated in terms of $T_c$. To do this, we must realize that only a fraction of the electrons—those lying within a shell $kT_c$ of the Fermi surface—are affected by the superconducting transition. This is because those electrons lying deep inside the Fermi sphere require much greater energy for excitation, in the neighborhood of 5 eV per electron, while we have seen that, in superconductivity, energies of the order of only $10^{-4}$ eV are involved. Thus we may estimate that the concentration of *effective* electrons is

$$n_{\mathrm{eff}} \simeq n \frac{kT_c}{E_{\mathrm{F}}}, \tag{10.10}$$

where $n$ is the total concentration of conduction electrons. Each of the effective electrons acquires an additional energy of about $kT_c$ in order to be excited across the gap. Therefore

$$\Delta E \simeq n_{\text{eff}}\, kT_c = n\frac{(kT_c)^2}{E_{\text{F}}}, \tag{10.11}$$

which is the same as the energy calculated in (10.9). Equating these energies, one finds that

$$\mathscr{H}_c(0) \simeq \left(\frac{2nk^2}{\mu_0 E_{\text{F}}}\right)^{1/2} T_c. \tag{10.12}$$

That is, the critical field is proportional to the critical temperature. Thus the higher the transition temperature, the greater the field required to destroy superconductivity. You may readily verify the validity of (10.12) by comparing the figures in Tables 10.1 and 10.2.

Equation (10.12) may be used to estimate $\mathscr{H}_c(0)$ if $T_c$ is given, and vice versa. Thus if one substitutes $T_c = 5°\text{K}$, $E_{\text{F}} = 5\,\text{eV}$, and $n = 10^{29}\,\text{m}^{-3}$, one finds that $B_c(0) = 0.01\ \text{W/m}^2$ ($= 100\,\text{G}$), which is in excellent agreement with observed values.

**The two-fluid model**

In 1934, in order to explain thermodynamic properties of superconductors in fuller detail, Gorter and Casimir introduced the *two-fluid model of superconductivity*. According to this model, the conduction electrons in a superconducting substance fall into two classes: *superelectrons* and *normal electrons*. The normal electrons behave in the usual fashion discussed in Chapter 4, that is, as charged particles flowing in a viscous medium. But the superelectrons have several novel properties which endow the superconductor with its distinctive features. These electrons experience no scattering, have zero entropy (perfect order), and a long *coherence length* (*about* $10^4$ Å), or spatial extension over which the superelectron is spread.

The number of superelectrons depends on the temperature. To obtain agreement with experiment, Gorter and Casimir found that the concentration of these electrons is given by the formula

$$n_s = n\left[1 - \left(\frac{T}{T_c}\right)^4\right], \tag{10.13}$$

which is plotted in Fig. 10.11. Thus, at $T = 0°\text{K}$, all the electrons are superelectrons, but as $T$ increases, the superelectrons decrease in number, and eventually they all become normal electrons at $T = T_c$.

The two-fluid model explains the zero-resistance property of the superconductor. For $T < T_c$, some superelectrons are present, and since these have infinite conductivity—recall that they experience no scattering—they essentially short-

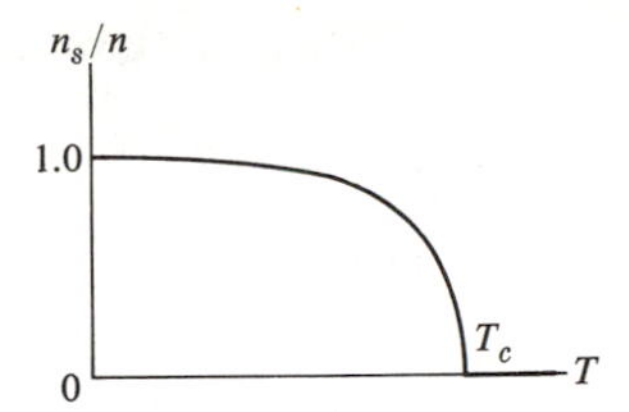


**Fig. 10.11** The fraction of superelectrons $n_s/n$ versus temperature.

circuit the normal electrons, resulting in infinite conductivity for the sample as a whole.

This model may be readily related to the concept of the energy gap discussed above. All the electrons below the gap are essentially frozen in their state of motion by virtue of the gap (see Fig. 10.8); hence these are the superelectrons. Those above the gap are normal electrons. The gap decreases as the temperature increases, and vanishes entirely at $T = T_c$, as shown in Fig. 10.12. Thus, as $T \to T_c$ and the gap vanishes, all the electrons become normal.†

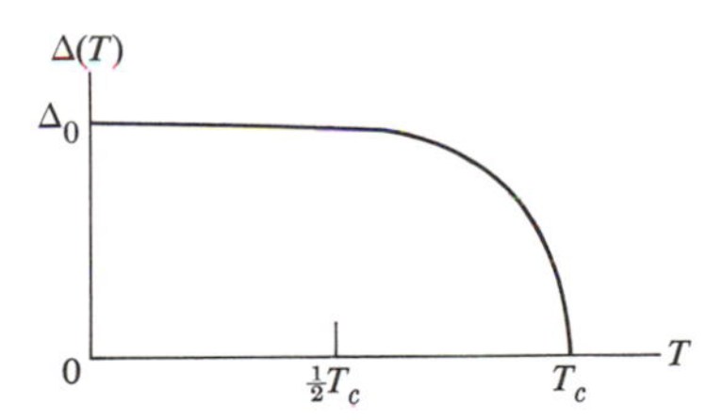


**Fig. 10.12** Decrease of the superconducting gap $\Delta(T)$ with temperature.

## 10.6 ELECTRODYNAMICS OF SUPERCONDUCTORS

The most interesting superconducting phenomena, and the most useful in practice, are the electrodynamic properties. The theory underlying these phenomena was put forward by the London brothers (F. and H. London) in 1935, and was elaborated and expanded by F. London in his book published in 1950. The theory is semi-phenomenological in nature, in that it uses an additional equation which could not at that time be derived from first principles, but it is nevertheless extremely useful because it explains known observations with minimum mathematical effort.

---

† The decrease of the gap with temperature and the vanishing of the gap at $T = T_c$ is expected, since the superconducting transition is a collective effect. (See a similar remark made in connection with dipolar polarization in solids, Section 8.7).

Let us use the two-fluid model. The equation of motion for a superelectron in the presence of an electric field is

$$m \frac{d\mathbf{v}_s}{dt} = -e\mathscr{E}, \tag{10.14}$$

which follows, since the only force acting on the electron is the force due to the electric field. The collision force is absent because this type of electron undergoes no collision. The density of the supercurrent $J_s$ is thus given by

$$\mathbf{J}_s = n_s(-e)\,\mathbf{v}_s, \tag{10.15}$$

which, when combined with (10.14), yields

$$\dot{\mathbf{J}}_s = \frac{n_s e^2}{m} \mathscr{E}, \tag{10.16}$$

where the dot over $\mathbf{J}_s$ denotes time differentiation. In the steady state, the current in a superconductor is constant. Therefore it follows from (10.16) that $\dot{\mathbf{J}}_s = 0$, or

$$\mathscr{E} = 0. \tag{10.17}$$

This important conclusion asserts that, in the steady state, *the electric field inside a superconductor vanishes.* In other words, the voltage drop across a superconductor is zero.

Equation (10.17) leads immediately to another important result. When this relation is combined with the Maxwell equation,

$$\dot{\mathbf{B}} = -\nabla \times \mathscr{E}, \tag{10.18}$$

one finds that

$$\dot{\mathbf{B}} = 0. \tag{10.19}$$

This affirms that in the steady state the magnetic field is constant.

But Eq. (10.19) is at variance with the Meissner effect. This equation states that $\mathbf{B}$ is constant regardless of the temperature, whereas we recall from Section 10.3 that when $T$ is raised toward $T_c$, the flux suddenly penetrates the sample as the transition point is reached. Thus the above formalism requires some modification.

To proceed with this modification, let us substitute for $\mathscr{E}$ from (10.16) into (10.18), which yields

$$\dot{\mathbf{B}} = -\frac{m}{n_s e^2} \nabla \times \dot{\mathbf{J}}_s. \tag{10.20}$$

This equation is invalid, as has just been seen, because it predicts that $\dot{\mathbf{B}} = 0$. To rectify this, London postulated the relation

$$\mathbf{B} = -\frac{m}{n_s e^2} \nabla \times \mathbf{J}_s, \tag{10.21}$$

which has the same form as (10.20), except that the time differentiations have been eliminated. We shall see presently that relation (10.21), known as the *London equation*, leads to results that are in agreement with experiment.

Equation (10.21) is a relation between **B** and $\mathbf{J}_s$. These quantities are also related by the Maxwell equation

$$\nabla \times \mathbf{B} = \mu_0 \mathbf{J}_s. \tag{10.22}$$

If we eliminate $\mathbf{J}_s$ between (10.21) and (10.22) [we can take the curl of (10.22), substitute for $\nabla \times \mathbf{J}_s$ from (10.21), and then use the identity of $\nabla \times \nabla \times \mathbf{B} = \nabla(\nabla \cdot \mathbf{B}) - \nabla^2 \mathbf{B} = -\nabla^2 \mathbf{B}$, where use is made of $\nabla \cdot \mathbf{B} = 0$], we find that

$$\nabla^2 \mathbf{B} = \frac{\mu_0 n_s e^2}{m} \mathbf{B}. \tag{10.23}$$

Let us apply this field equation to a situation of simple geometry. The specimen is semi-infinite, with its surface lying in the $yz$ plane (Fig. 10.13), and the field is applied in the $y$-direction. Since quantities vary only in the $x$-direction, Eq. (10.23) reduces to

$$\frac{\partial^2}{\partial x^2} B_y = \frac{\mu_0 n_s e^2}{m} B_y. \tag{10.24}$$

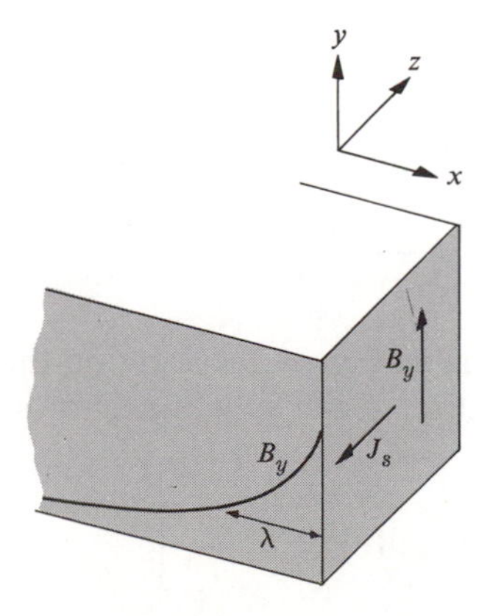


**Fig. 10.13** Solution of the London equation. The magnetic field decays exponentially within the superconductor.

The solution to this simple differential equation is

$$B_y(x) = B_y(0)\, e^{-x/\lambda}, \tag{10.25}$$

where

$$\lambda = (m/\mu_0\, n_s\, e^2)^{1/2}. \tag{10.26}$$

Equation (10.25) shows that the field decreases exponentially as one proceeds from the surface into the superconductor. Thus the field vanishes inside the bulk of the medium, in accord with the Meissner effect. This lends support to the London equation (10.21). As a matter of fact, this agreement was the primary motivation for postulating the London equation in the first place.

Note, however, that Eq. (10.25) predicts that the field penetrates the sample to some extent, the distance of penetration being roughly equal to $\lambda$. Thus the flux is not expelled entirely from the superconductor, as was once thought, but there is a small region near the surface in which there is an appreciable field. The parameter $\lambda$ is known as the *London penetration depth.*

This prediction was later verified experimentally, and was a great triumph for the London theory. If one substitutes appropriate values for the parameters in (10.26), one finds that $\lambda \simeq 500$ Å, which is close to the experimentally observed values, as shown in Table 10.3.

**Table 10.3**

Penetration Depths (Measured Values)

| Element | $\lambda(0)$, Å |
|---|---|
| Al | 500 |
| Cd | 1300 |
| Hg | 380–450 |
| In | 640 |
| Nb | 470 |
| Pb | 390 |
| Sn | 510 |

Another impressive confirmation of the London theory is its prediction of the variation of $\lambda$ with temperature. If one substitutes for $n_s$ from (10.13) into (10.26), one obtains

$$\lambda = \lambda(0)\left[1 - \frac{T^4}{T_c^4}\right]^{-1/2}, \tag{10.27}$$

where

$$\lambda(0) = (m/\mu_0 n e^2)^{1/2} \tag{10.28}$$

is the penetration depth at $T = 0$. According to (10.27), $\lambda$ increases as $T$ increases from 0°K, and becomes infinite at $T = T_c$, as shown in Fig. 10.14. This latter conclusion is expected, because at $T = T_c$ the substance becomes normal, and the field penetrates the whole specimen, i.e., the specimen has an infinite depth of penetration. The temperature dependence predicted by (10.27) is well confirmed by experiment.

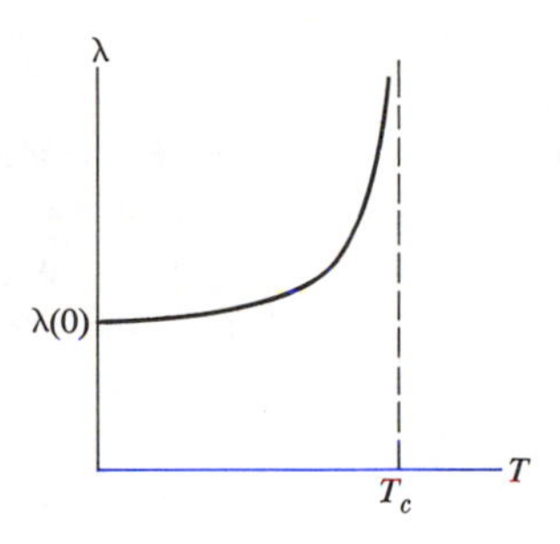


**Fig. 10.14** Increase of the penetration depth $\lambda$ with temperature, according to the London theory.

A third conclusion from the London theory is the existence of an electric current flowing near the surface. If one substitutes for **B** from (10.25) into (10.21) and solves for the current, one finds

$$J_z(x) = -\left(\frac{n_s e^2}{\mu_0 m}\right)^{1/2} B_y(x) = -J_s(0)\, e^{-(x/\lambda)}, \tag{10.29}$$

which is a current flowing in the negative $z$-direction. Since this current decays exponentially as one moves into the superconductor, it is essentially a *surface current*. Therefore the Meissner effect is accompanied by a surface current, and it is this current which acts to shield the inner superconductor from the external magnetic field, resulting in a perfectly diamagnetic medium. (In other words, the magnetic field due to the surface current completely cancels the external field inside the medium.)

So we get a very interesting picture: The current in a current-carrying superconductor (the *supercurrent*) is restricted to the region very close to the surface. If, for example, the specimen is in the shape of a cylinder, the current flows only along the surface of the cylinder, leaving the whole inner region free of any current. This is very different from a normal conductor, in which the current is uniform throughout the sample.

## 10.7 THEORY OF SUPERCONDUCTIVITY

The modern theory of superconductivity was promulgated by Bardeen, Cooper, and Schrieffer in their classic paper in 1957.† The BCS theory has now gained universal acceptance because it has proved capable of explaining all observed phenomena relating to superconductivity. Starting from first principles and employing a completely quantum treatment, their theory explains the various observable effects such as zero resistance, the Meissner effect, etc. Because their theory is so steeped in quantum mechanics, one cannot discuss it meaningfully without using advanced quantum concepts and mathematical techniques. Therefore, in the interest of simplicity, let us instead give a brief, qualitative, conceptual exposition of the BCS theory.

Consider a metal in which the conduction electrons lie inside the Fermi sphere. Suppose that two electrons lie just inside the Fermi surface (Fig. 10.15), and repel each other because of coulomb interaction. But this coulomb force is reduced substantially on account of the screening due to the presence of other electrons in the Fermi sphere (recall the discussion of the Fermi hole, Section 4.3). After the screening is taken into account, the interaction between the two electrons disappears almost entirely, although a small repulsive residue persists.

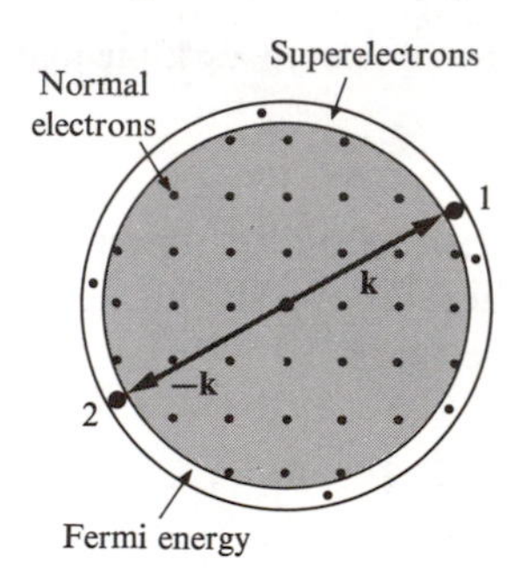


**Fig. 10.15** Interaction between two electrons, 1 and 2, near the Fermi surface in a metal.

However, something new may occur. Suppose that, for some reason, the two electrons attract each other. Cooper showed that the two electrons would then form a *bound* state (provided they were very close to the Fermi surface). This is very important, because, in a bound state, electrons are paired to form a single system, and their motions are correlated. The pairing can be broken only if an amount of energy equal to the binding energy is applied to the system.

Our two electrons are called a *Cooper pair*. The binding energy is strongest when the electrons forming the pair have opposite moments and opposite spins, that is, $\mathbf{k}\uparrow$, $-\mathbf{k}\downarrow$. It follows, therefore, that if there is any attraction between them, then all the electrons in the neighborhood of the Fermi surface condense into a system of Cooper pairs. These pairs are, in fact, the superelectrons discussed in Sections 10.5

† *Phys. Rev.* **106**, 162 (1957). A similar theory was published shortly afterward by N. Bogolubov, *Nuovo Cimento* **7**: 6, 794 (1958).

and 10.6, and the binding energy corresponds to the energy gap introduced in Section 10.5.

We have been talking about the consequences of electron–electron attraction, but how does this attraction come about in the first place? In superconductive materials, it results from the electron–lattice interaction (Fig. 10.16).

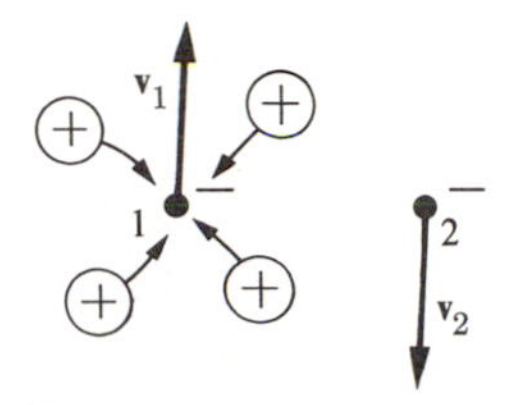


**Fig. 10.16** The screening of electron 1 by the positive ions of the lattice. Solid circles represent the two electrons considered.

Suppose that the two electrons, 1 and 2, pass each other. Because electron 1 is negatively charged, it attracts positive ions toward itself (electron–lattice interaction). Thus electron 2 does not "see" the bare electron 1. Electron 1 is *screened* by ions. The screening may greatly reduce the effective charge of this electron; in fact, the ions may overrespond and produce a net positive charge. If this happens, then electron 2 will be attracted toward 1. This leads to a net attractive interaction, as required for the formation of the Cooper pair.

The ions' overresponse may be understood qualitatively. Since electron 1 is near the Fermi surface, its speed is great. At the same time the ions, because of their heavy masses, respond rather slowly. By the time they have felt and completely responded to electron 1, electron 1 has left its initial region, at least partially, thus stimulating the overcompensation. One can also reason that this process is most effective when electron 1 and electron 2 move in opposite directions (why?).

(In technical literature, one says that each electron is surrounded by a "phonon cloud," and that the two electrons establish an attractive interaction by exchanging phonons; for example, electron 1 emits phonons which are very quickly absorbed by electron 2, as in Fig. 10.17.† Since the phonon is involved twice—once in emission and once in absorption—the attraction between electrons is a second-order process.)

As a result of this binding between electron 1 and electron 2, an energy gap appears in the spectrum of the electron. This gap straddles the Fermi energy level,

† Imagine a situation in which one person throws massive balls to another person, who receives them. We can readily see that such a process leads to a *repulsive* force between the persons; the first person recoils backward when he throws the ball; the second person recoils by the same amount when he receives the ball. However, if the two persons were to exchange helium-filled balloons in air, the result would be an attractive force between them.

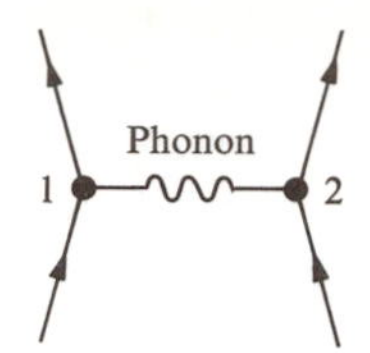


**Fig. 10.17** The phonon exchange responsible for the attractive interaction between electrons 1 and 2.

as shown in Fig. 10.18, in which are plotted the density of states for a superconductor.

The states in the energy range $(E_F - \frac{1}{2}\Delta_0, E_F + \frac{1}{2}\Delta_0)$ are now forbidden. These states have been "pulled" both down and up, resulting in a peaking of the density of states just below and just above the gap. Far from the Fermi energy, the density of states for the superconductor is the same as in the normal metal.

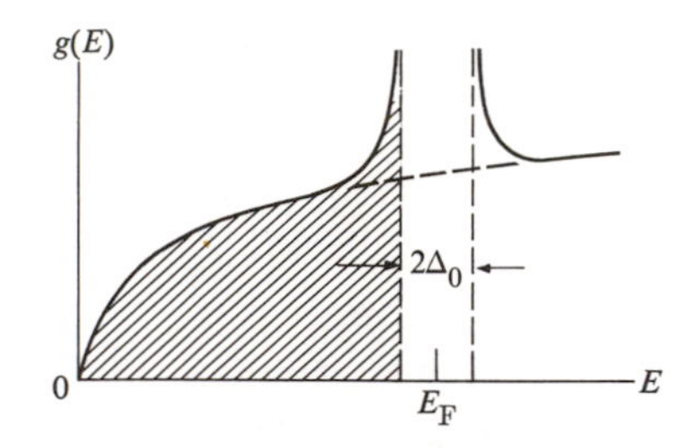


**Fig. 10.18** The density of states $g(E)$ versus $E$ for a superconductor, illustrating the energy gap. The cross-hatched region is fully occupied at $T = 0$°K.

The theory shows that the gap at zero temperature is given by

$$\Delta_0 = 4\hbar\omega_D\, e^{-(2/g(E_F)V')}, \tag{10.30}$$

where $\omega_D$ is the Debye frequency, $g(E_F)$ is the density of states for the normal metal at the Fermi level, and $V'$ is the strength of the electron–lattice interaction. The Debye frequency appears in (10.30) because of the exchange of a phonon between the electrons of the Cooper pair. Several interesting results follow at once from this expression.

a) Roughly speaking, $\Delta_0 \simeq \hbar\omega_D$, the latter being the energy of a typical phonon. This also yields the correct order of magnitude, since $\hbar\omega_D \simeq 10^{-27} \times 10^{+13} = 10^{-14}$ erg $\simeq 10^{-2}$ eV. When the exponential factor of (10.30) is included, it reduces $\Delta_0$ to about $10^{-4}$ eV, in agreement with observation.

b) Since $\omega_D \sim M^{-1/2}$, where $M$ is the mass of the vibrating ion [Eq. (3.39)], it follows that $\Delta_0 \sim M^{-1/2}$. Thus the gap—and hence the critical temperature $T_c$—

decrease as $M$ increases. This is observed experimentally by varying the isotope ratio of the metal, and is referred to as the *isotope effect*. This effect was observed long before the advent of the BCS theory, and was the strongest clue to the fact that the lattice is somehow involved in the process of superconductivity.

c) The gap increases, and so does $T_c$, as the electron–lattice interaction $V'$ increases. In other words, a strong $V'$ favors superconductivity. This is plausible because the ions are then attracted more strongly to the electron, increasing the chance of overcompensation occurring. This result is curious, on the other hand, because $V'$ is also responsible for the resistivity in the normal state, in which the larger the $V'$, the higher the resistivity. We therefore reach the seemingly paradoxical conclusion that poor normal conductors make good superconductors, while good normal conductors are poor superconductors. This is, however, in agreement with experiment: The former group includes, for instance, Pb and Nb, and the latter group includes the alkali and noble metals, which exhibit no superconductivity whatsoever, even at the lowest attainable temperatures.

The BCS theory shows that the critical temperature is given by

$$\Delta_0 = 3.52\, kT_c. \tag{10.31}$$

This relation can be tested experimentally, because $\Delta_0$ and $T_c$ can be measured independently; observations confirm the validity of the relation to a good approximation. Typically it is found that $\Delta_0 = 4\,kT_c$ (Table 10.4).

**Table 10.4**

Ratio of Measured Gap to $kT_c$, $\Delta_0/kT_c$

| Element | $\Delta_0/k_B T_c$ |
|---|---|
| In | 4.1 |
| Sn | 3.6 |
| Hg | 4.6 |
| V | 3.4 |
| Pb | 4.1 |

The energy gap decreases with temperature, as mentioned in connection with Fig. 10.11. This behavior is also derived from the BCS theory, and the agreement with experiment is good.†

The energy gap can be determined experimentally by any of several means.

† According to the BCS theory, the variation of the gap with temperature is given by $\Delta(T)/\Delta_0 = \tanh[T_c\Delta(T)/T\Delta_0]$.

One involves infrared absorption. If a beam of infrared light shines on a superconductor at very low temperature, absorption of infrared radiation takes place only when the frequency of the radiation is sufficiently high that a Cooper pair is excited across the gap.† That is,

$$\hbar\omega \geqslant 2\Delta_0. \tag{10.32}$$

The gap can therefore be determined from the frequency at which the absorption commences. Since $\Delta_0 \simeq 10^{-4}$ eV, the corresponding frequency lies in the infrared region. (N.B.: The BCS theory explains the zero-resistance property as follows: Once set in a drift motion, a Cooper pair may be scattered only if the collision mechanism imparts an energy to the pair which is at least equal to $2\Delta_0$. But at low temperatures this amount of energy cannot be supplied by the phonons, because only very low-energy phonons are excited. Thus the Cooper pair continues its drift motion indefinitely.)

## 10.8 TUNNELING AND THE JOSEPHSON EFFECT

When a thin junction involving a superconductor is prepared, tunneling may take place across the junction, and the tunneling current may be used to study the physical properties of the superconductor. Figure 10.19(a) shows such a junction, in which two pieces of metal, in the superconducting and normal state, respectively, are joined by a thin insulating film, of thickness of about 50 Å. The film acts as a potential barrier as far as the flow of electrons across the junction is concerned, but because the film is thin, it does not completely inhibit the flow. According to quantum mechanics, electrons are still able to tunnel under a thin potential barrier.

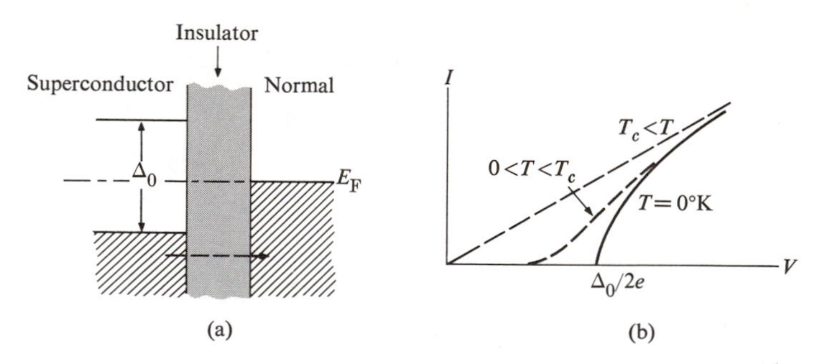


**Fig. 10.19** Tunneling in a normal–superconductor junction. (a) Tunneling (dashed arrow) is inhibited by the exclusion principle. Note that the Fermi level $E_F$ is the same throughout the system, and passes through the midgap of the superconductor. (b) Current $I$ versus voltage $V$ in the junction.

---

† The minimum energy required to excite a Cooper pair is $2\Delta_0$, twice the gap, and not $\Delta_0$. It is not possible to excite only one member of an electron pair, because the pair form an indivisible whole unit. If the pair is broken up for any reason, then we have two single normal electrons, i.e., both electrons have been excited across the gap simultaneously.

If a small voltage $V$ is applied across the junction (taking the field to be directed to the left), the energy band of the left side is raised by the amount $eV$, but the electrons are still unable to flow to the right because the states lying horizontally across are already occupied. But if the voltage is increased further so that the energy band of the superconductor is raised by $\Delta_0/2$, then the corresponding horizontal states on the right are now empty, and the current proceeds to flow. This results in the current–voltage characteristic shown in Fig. 10.19(b). The voltage at which the current begins to flow is such that

$$\Delta_0/2 = eV, \tag{10.33}$$

and from this the superconducting gap may be determined.

The above tunneling is referred to as *normal*, or *single-electron tunneling*, because single electrons tunnel to the right. Another type of tunneling, one which involves Cooper pairs, has received a great deal of attention recently, and is responsible for the *Josephson effect*.† The underlying principle is that if the insulating film is very thin—i.e., about 10 Å —then the pairs would not tunnel readily across the junction, but also their (quantum) wave functions on both sides would be highly correlated. In fact, the effect of the film is merely to introduce a phase difference $\phi_0$ between the two parts of the wave function on opposite sides of the junction, as shown in Fig.10.20. The current density across the junction is given in terms of this phase by the relation (see Feynman, 1963)

$$J = J_1 \sin \phi_0, \tag{10.34}$$

where $J_1$ is a measure of the probability of transition across the junction.

In the absence of any potential difference across the junction, the phase $\phi_0$ adjusts its value to that of the actual current, so that Eq. (10.34) is satisfied.

Let us now suppose that a static potential $V_0$ is applied across the junction. We recall from quantum mechanics that the phase of the wave function in quantum

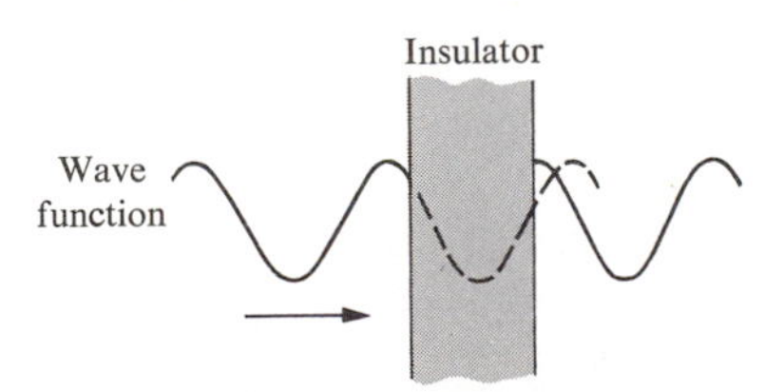


**Fig. 10.20** Wave function of an electron at the junction of two superconductors; note the phase shift in the wave function.

†For predicting this effect bearing his name, Brian D. Josephson received the 1973 Nobel Prize in physics. Also sharing the prize were Ivor Giaver and Leo Esaki for their work on normal tunneling in superconductors (above) and the tunnel diode (see 7.5), respectively.

mechanics is given by

$$\Delta\phi = \frac{Et}{\hbar}, \tag{10.35}$$

where $E$ is the total energy of the system. Let us apply this to calculate the additional phase difference experienced by the Cooper pair as it tunnels across the junction. In this case $E = (2e)V_0$, in which the factor 2 is introduced because the system here involves a pair of electrons. Therefore

$$\Delta\phi = \frac{2eV_0 t}{\hbar}, \tag{10.36}$$

which now alters (10.34) to the new form

$$\begin{aligned} J &= J_1 \sin(\phi_0 + \Delta\phi) \\ &= J_1 \sin\left(\phi_0 + \frac{2eV_0}{\hbar} t\right), \end{aligned} \tag{10.37}$$

which represents an alternating current. This result is interesting because a static potential is seen to lead to an ac current, and the frequency

$$\omega = \frac{2eV_0}{\hbar} \tag{10.38}$$

is readily tuned by varying $V_0$. Numerically

$$\nu = 484\ V_0\ \mathrm{GHz},$$

for $V_0$ is in millivolts. Since $V_0$ is usually of the order of several millivolts, the Josephson frequency falls in the microwave range. The tunneling current (10.37) was observed very soon after it was first predicted by Josephson in 1962. One method of observation involves measuring the emission of microwave radiation from the junction. Agreement between theory and experiment is very good.

The Josephson effect has many applications. An important one is its recent use in the redetermination of the fundamental physical constants. We can see this from the fact that the frequency in (10.38) includes the ratio $2e/h$, containing both the charge on the electron and Planck's constant. It has been possible to determine the ratio to an accuracy of 6 ppm.

## 10.9 MISCELLANEOUS TOPICS

Now let us take a look at some other important topics related to superconductivity.

### The intermediate state

In discussing the critical field (Section 10.4), we said that the flux was expelled by a superconductor until a field $\mathscr{H}_c$ was reached, whereupon the whole specimen made a discontinuous transition into the normal state. Actually, however, the transition is discontinuous only for specimens with simple geometries and particular field orientations: for example, a cylinder whose axis is oriented parallel to the field.

Consider, on the other hand, the case in which the axis of the cylinder is normal to the field. Figure 10.21 shows the distribution of the field in the neighborhood of the cylinder. The field is stronger at the points $AA'$ than at the points $DD'$ because of the "crowding" of the field lines at points $AA'$. It can be shown, in fact, that the $AA'$ field is twice as strong as the $DD'$ one. Thus as the intensity of the field is raised, it reaches its critical value at the points $AA'$ before it does at $DD'$, and the sides of the cylinder thus turn into a normal state at the field $\mathscr{H} = \frac{1}{2}\mathscr{H}_c$. As the intensity of the field is raised further, the specimen divides into alternate normal and superconducting laminae parallel to the field, as shown in Fig. 10.21, and the specimen is said to be in the *intermediate state*. And when the intensity of the field is raised still further, the normal regions grow until, at $\mathscr{H} = \mathscr{H}_c$, the whole specimen becomes completely normal.

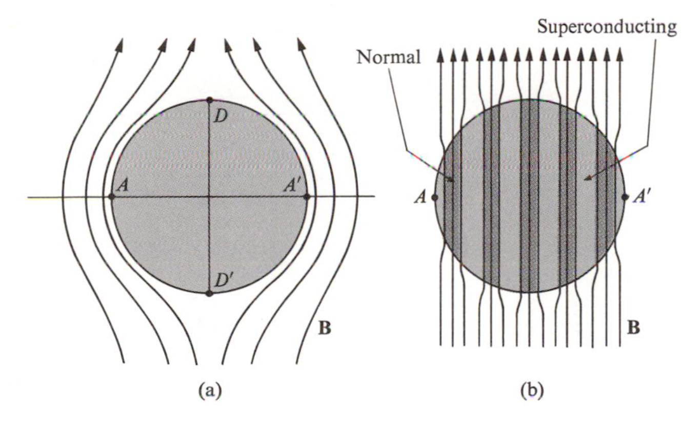


**Fig. 10.21** Intermediate state in a cylinder whose axis is normal to the field: (a) The situation for $\mathscr{H} < H_c/2$. (b) The situation for $\frac{1}{2}\mathscr{H}_c < \mathscr{H} < \mathscr{H}_c$, showing the intermediate state.

Because of the division into thin laminae, the field distribution is "straightened out," which leads to a reduction in the demagnetization energy of the superconducting regions, i.e., essentially the whole flux passes through the normal region. The number of laminae, however, is kept finite by virtue of the fact that there is a *surface energy* associated with the wall between the superconducting and normal regions.

### Critical field in a small specimen

The critical field depends also on the size of the specimen, if the specimen is small. We shall demonstrate this by estimating the field for a thin film. The field distribution for such a sample is illustrated in Fig. 10.22. We recall from Section 10.5 that the condensation energy per unit volume is proportional to $\mathscr{H}_c^2$. However, since the width of the demagnetized region is essentially $(d - 2\lambda)$ rather than $d$, where $d$ is the thickness of the film, because of the flux penetration, it follows that

$$(d - 2\lambda)\,\mathscr{H}_c'^2 \simeq d\,\mathscr{H}_c^2,$$

where $\mathscr{H}_c'$ is the critical field for the film, while $\mathscr{H}_c$ is the field for a bulk sample (where the effects of field penetration may be ignored). Therefore

$$\mathscr{H}_c' \simeq \left(\frac{1}{1 - (2\lambda/d)}\right)^{1/2} \mathscr{H}_c. \tag{10.39}$$

The field $\mathscr{H}_c'$ is larger than $\mathscr{H}_c$, and if $d$ is small the increase may be by as much as a factor of 10. This property finds applications in some switching devices employing thin superconducting films.

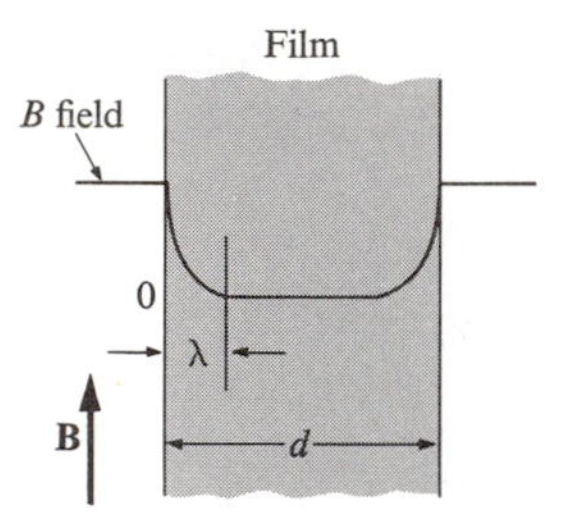


**Fig. 10.22** Field distribution in a superconducting thin film.

### Type II superconductors

There is an important class of superconductors which does not behave in quite the manner described so far. The Meissner effect begins to break down in these substances, at least partially, well before the critical field is reached, even when the field distribution is uniform. This class is referred to as type II superconductors, in contrast to the substances we have hitherto described, which are called type I superconductors. Figure 10.23(a) shows the magnetic induction $B$ versus the intensity $\mathscr{H}$ for a type II specimen. The Meissner effect is satisfied up to a field $\mathscr{H}_{c_1}$, after which the flux partially penetrates the specimen, and the substance becomes completely normal at the still higher field $\mathscr{H}_c$, which is the critical field. Type II materials are *hard* superconductors because they usually have high critical fields.

In the field interval $\mathscr{H}_{c_1}$ to $\mathscr{H}_c$, the substance is said to be in the *mixed state*. A close examination of the structure of the specimen in this state reveals the presence of small circular regions in the normal state, which are surrounded by a large super-

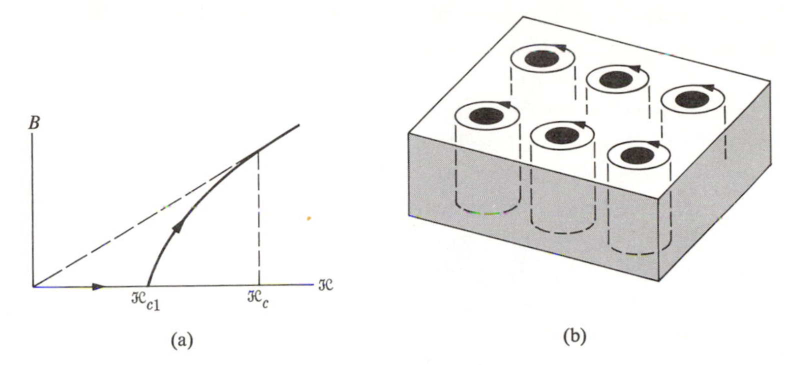


**Fig. 10.23** (a) Induction $B$ versus $\mathcal{H}$ for a type II superconductor. Dashed line represents a type I superconductor, and dashed line a normal metal. (b) The mixed state, showing normal cores surrounded by circulating supercurrents.

conducting region forming the remainder of the specimen (Fig. 10.23b). The small normal regions are referred to as *vortices* or *fluxoids*. The vortex structure of the mixed state is too fine to be seen by the naked eye, but its existence has been experimentally verified.

The reason for the appearance of the vortices is that the coherence length $\xi$ in type II superconductors is very short; specifically, $\xi < \lambda$, where $\lambda$ is the penetration depth. It can be shown (Rose-Innes, 1969) that, if this condition is satisfied, the surface energy is negative, which means that the substance tends to reduce its energy by forming normal-superconducting surfaces by creating vortices well below the critical field.

Materials with high critical temperatures tend to fall in the type II category, and the reason is qualitatively as follows. The coherence length represents the extension of the wave function of the superelectron. Using the position-momentum uncertainty relation, we write

$$\xi \simeq \frac{\hbar}{\Delta p}, \tag{10.40}$$

where $\Delta p$ is the uncertainty in momentum. But a superelectron lies within an energy interval $\approx kT_c$ from the Fermi surface, and hence the uncertainty of its energy is

$$\Delta E \simeq kT_c.$$

Since $E = p^2/2m$, it follows that $\Delta E = p\Delta p/m$, or $\Delta p \sim \Delta E = kT_c$, which, when substituted into (10.40) yields

$$\xi \sim \frac{1}{T_c}. \tag{10.41}$$

Thus $\xi$ is inversely proportional to $T_c$, or $\mathcal{H}_c$, and the greater the $T_c$ the shorter the coherence length.

Transition metals and alloys usually fall in the type II class. The coherence length in these substances is shortened by the relatively large amount of scattering present.

Superconductivity, in a sense, has had a rather unfortunate history. Most of the substances studied up to the late 1940's were actually type II materials to which, as we now know, the simple London theory does not quite apply. Yet workers in the field tried to apply the theory to these substances, resulting in only partial success and much frustration. It was only in the 1950's that the situation was completely clarified, and the theory of superconductivity reached its golden age.†

## SUMMARY

### Zero resistance

When the temperature of a would-be superconductor is lowered below the critical temperature $T_c$, the substance enters a new state of matter, the *superconducting state*, in which its resistivity vanishes entirely. The critical temperature depends on the substance, the observed values ranging from about 0.01°K to about 20°K.

### The Meissner effect

A superconductor expels a magnetic flux completely, so that the magnetic induction inside the substance is zero, that is, $\mathbf{B} = 0$.

### Critical field

If a sufficiently large magnetic field is applied to a superconductor, the substance reverts to the normal state. This *critical field* decreases with temperature as

$$\mathscr{H}_c = \mathscr{H}_c(0)\left(1 - \frac{T^2}{T_c^2}\right),$$

and vanishes at $T = T_c$.

† Recently Heeger, *et al.*, have reported what may turn out to be a very significant development. They claim to have observed the onset of superconducting-like transition in a material at the high temperature of 60°K. The material involved is an organic salt (ATTF) (TCNQ). Above 60°K the substance is in a one-dimensional metallic state, and as the temperature is lowered toward 60°K, the conductivity increases very rapidly in a manner analogous to the usual superconducting transition. Unfortunately just then the lattice itself becomes unstable and the crystal deforms into a new structure, and the substance becomes a semiconductor instead of a superconductor. Efforts are currently underway to stabilize the hoped-for superconducting state by preventing the lattice transformation. See Heeger, *et al.*, in *Solid State Communications* (March 1973).

### Thermodynamical aspects

The specific heat of a superconductor decreases exponentially with temperature, as $e^{-b(T/T_c)}$, which implies the existence of an energy gap in the energy spectrum of the superconductor. The gap is of the order of $kT_c$; more accurately it is close to 3.5 $kT_c$. The existence of this gap, later derived by the BCS theory, is the most basic feature of the superconducting state.

Many properties of superconductors can be explained by the two-fluid model, in which the electrons are divided into two classes: normal electrons and superelectrons. The unusual properties of superconductors are due to the superelectrons, which experience no collision and also have zero entropy (perfect order).

### Electrodynamics

In order to explain the Meissner effect, the Londons postulated the field equation $\nabla \times \mathbf{B} = -(m/n_s e^2)\nabla \times \mathbf{J}_s$ in a superconductor, where $n_s$ and $\mathbf{J}_s$ are the concentration and current density, respectively, of the superelectrons. When this equation is combined with Maxwell equations, it yields the solution $\mathbf{B} = 0$ inside a superconductor (Meissner effect).

Two other effects also predicted by the London equation: (a) A penetration of the superconductor by magnetic flux for a small distance $\lambda$ (the penetration depth); and (b), a supercurrent flowing along the surface of the superconductor.

### The BCS theory

According to the BCS theory, superelectrons exist as *Cooper pairs*. Each pair forms a bound state, the attractive interaction necessary for such a state being created by phonons exchanged between the pair.

### Tunneling

When a metal and a superconductor, or two superconductors, are separated by a thin insulating film, electrons can tunnel across the film. The current–voltage characteristics of the junction may be used in the determination of the superconducting gap.

If the film between the superconductors is very thin, Cooper pairs themselves may tunnel across the junction, leading to the Josephson effect. A static voltage across the junction produces an ac current of frequency $\nu = 2\,eV_0/h$.

## REFERENCES

P. G. deGennes, 1966, *Superconductivity of Metals and Alloys*, New York: W. A. Benjamin

R. Feynman, 1963, *Lectures in Physics*, Volume III, Reading, Mass.: Addison-Wesley

C. G. Kuper, 1968, *An Introduction to the Theory of Superconductivity*, Oxford: Oxford University Press

E. A. Lynton, 1969, *Superconductivity*, third edition, London: Methuen

V. L. Newhouse, 1964, *Applied Superconductivity*, New York: John Wiley
B. W. Petley, 1971, *An Introduction to the Josephson Effect*, London: Mills and Boon, Ltd.
G. Rickayzen, 1965, *Theory of Superconductivity*, New York: John Wiley
A. C. Rose-Innes and E. H. Rhoderick, 1969, *Introduction to Superconductivity*, New York: Pergamon Press
B. Serin, 1956, *Encyclopedia of Physics*, Volume 15, New York:
D. Shoenberg, 1960, *Superconductivity*, second edition, Oxford: Oxford University Press
J. E. C. Williams, 1970, *Superconductivity and Applications*, London: Pion Ltd.

## QUESTIONS

1. What is the expected composition of a ZrNb alloy which has the highest $T_c$? Answer the same question for a NbSn alloy.
2. It was stated, following Eq. (10.12) that the critical field $\mathscr{H}_c(0)$ is essentially proportional to the critical temperature $T_c$. (This will also be confirmed by your plot in Problem 3.) Yet the electron concentration $n$ also appears in (10.12), and this concentration differs from one superconductor to another. Why does the linear relationship still hold, nonetheless?
3. Discuss at least two different experimental methods for determining the critical temperature of a superconductor.
4. Experiments show that even though a superconductor exhibits zero static resistance, its ac resistance is finite, albeit very small. Explain how this is possible. [*Hint*: Use the two-fluid model. An electric circuit representation is also useful.]
5. Derive Eq. (10.29) for the surface current in a superconductor.
6. A footnote in Section 10.5 said that the gap $\Delta(T)$ decreases with temperature because of the collective nature of the superconducting transition. Explain this point more fully, relying on the concept of the Cooper pair.
7. Is the superconductor–normal junction of Fig. 10.19(a) electrically symmetric, or not?
8. A cylinder in the intermediate state is shown in Fig. 10.21(b). Describe one experimental electrical method for distinguishing this state from the superconducting state shown in Fig. 10.21(a).

## PROBLEMS

1. Consider a lead solenoid wound around a doughnut-shaped tube. The total number of turns is 2500, and the diameter of the lead wire is 30 cm. The solenoid is cooled below the critical point, at which an electric current is induced in the coil. Assuming the lead resistivity in the superconducting state to be less than $10^{-25}$ ohm-m, calculate the minimum time interval needed for the current to damp out by 0.01%. (Assume the length of the wire to be sufficiently large for the infinite-length approximation to hold.)
2. a) Figure 10.7 indicates a discontinuity in specific heat at the transition point as the substance becomes superconducting. The size of the discontinuity can be calculated using a thermodynamical argument. Show that the size of the

discontinuity per mole is given by

$$C_s - C_n = V_m T_c \left( \frac{\partial \mathscr{H}_c}{\partial T} \right)^2_{T_c},$$

where $V_m$ is the molar volume.

b) Calculate this difference for tin, and compare your answer with the value given in Figure 10.7. The density and atomic weight of tin are 7.0 g/cm$^2$ and 119, respectively. [*Hint*: In part (a), recall that

$$C = T \frac{\partial S}{\partial T} \quad \text{and} \quad S = - \frac{\partial E}{\partial T},$$

where $S$ is the entropy and $E$ the free energy of the system.]

3. Plot $\mathscr{H}_c(0)$ versus $T_c$ for a few superconductors using data from Tables 10.1 and 10.2, and verify the linear relationship predicted in Eq. (10.12).
4. The superconducting gap $\Delta(T)$ decreases with temperature, as indicated in Fig. 10.12. The BCS theory shows that this decrease is given by $\Delta(T)/\Delta_0 = \tanh (T_c\Delta(T)/T\Delta_0)$, for $T < T_c$. Using this relation and Table 10.4, plot $\Delta(T)$ versus $T$ for tin, in the range $0 < T < T_c$.
5. Section 10.5 said that the exponential behavior of the specific heat (10.5) implies the existence of an energy gap. This can be seen most readily by calculating the specific heat of an intrinsic semiconductor, in which the gap plays a very important role. Carry out this calculation, and establish the exponential behavior indicated above.
6. The London equation (10.21) is equivalent to the condition of perfect diamagnetism of a superconductor. A basic (and controversial) question often arises: Which is the more electrodynamic property of a superconductor, perfect conductivity or perfect diamagnetism? By this we mean: Does one of these two properties imply the other, or are they independent? Answer this question. [*Hint*: Note that the electric field and magnetic induction are related to each other by the Maxwell equations, in particular, $\mathscr{E} = -\partial \mathbf{A}/\partial t$ and $\mathbf{B} = \nabla \times \mathbf{A}$, where $\mathbf{A}$ is the vector potential.]
7. Prove that the magnetic flux linking a superconducting ring is quantized according to $\Phi = n(h/2e)$, where $\Phi$ is the flux and $n$ an integer. This quantization was predicted by F. London (1950), and verified experimentally in 1961. [*Hint*: Use the Wilson–Sommerfeld quantization condition,† and take the path of integration in the interior of the ring. Recall also that the momentum of an electron in a magnetic field is given by $\mathbf{p} = m\mathbf{v} + e\mathbf{A}$.] (The quantization formula given by London was actually erroneous in one respect, because the concept of the Cooper pair was unknown in 1950. What do you expect the original London formula to have been?)
8. Discuss the Josephson tunneling current, given that, in addition to the static bias, an alternating voltage is also impressed across the junction. Enumerate the frequencies of the various modes of excitation.

---

† This condition is $\oint p_i dq_i = nh$, where $n$ is an integer, and $q_i$ and $p_i$ are a coordinate and its conjugate momentum. Thus the Bohr condition for quantizing the angular momentum, $L = nh$, can be obtained from the integral by taking $q_i = \theta$ and $p_i = L$, where $\theta$ is the angle and $L$ the angular momentum.

9. The coherence length $\xi$ of a superelectron, which is the spatial extension of a superelectron (or of a Cooper pair), may be viewed as the quantum uncertainty in the position resulting from the uncertainty in the electron energy. Estimate the value of this coherence length for a typical superconductor.
10. Applications of superconductivity to the design of technical devices, including superconducting magnets, are discussed in Newhouse (1964) and Williams (1970). Study the highlights of these books and write a brief report.

# CHAPTER 11 TOPICS IN METALLURGY AND DEFECTS IN SOLIDS



---

*Truth is never pure, and rarely simple.*

Oscar Wilde

## 11.1 INTRODUCTION

In our discussion of metals, which has occupied a large part of this book, the emphasis has been on electrons, which play the central role in electrical, optical, and magnetic properties. Little has been said about atoms, apart from their somewhat passive role in forming a crystal-lattice background in which the electrons move. There is some justification for this emphasis. Great strides have indeed been made recently in our understanding of the electronic properties of metals, and semiconductors in particular. However, we should not forget the important role of atoms in solids, particularly in determining structural and mechanical properties which are of great concern to the metallurgist, the materials scientist, and the engineer. We shall therefore devote this chapter to atoms, their arrangement, and their motion in metals, alloys and other solids, with special attention to structural and mechanical effects.

We shall first classify types of crystalline imperfections, with emphasis on vacancies. The motion of these vacancies leads us to the subject of atomic diffusion, and thence to metallic alloys, with special attention to the various metallic phases and their stability. We shall then discuss dislocations and their influence on the mechanical strength of metals and alloys. After a brief treatment of ionic conductivity, the chapter closes with a discussion of the photographic process and radiation damage in solids, two topics of great practical value.

## 11.2 TYPES OF IMPERFECTIONS

The concept of a perfect crystal is an extremely useful and appealing one. In fact, it formed the underpinning for most of this book. But we have said repeatedly that real crystals are not perfect. By taking great pains, one can reduce crystal *imperfections*, or *defects*, considerably, but one can never eliminate them entirely. In some situations defects are, in fact, highly desirable, as in the case of donor and acceptor impurities, which are essential to the operation of the transistor.

As the name implies, a defect is a region involving a break, or an irregularity, in the crystal structure. The most important types of defects are: (1) point defects, (2) line defects, and (3) surface defects, depending on the geometrical shape of the defect.

**1.** a) *Point defect.* An irregularity in the crystal structure, localized in the lattice. An example is a foreign atom, or impurity, in the crystal. A crystal usually contains all sorts of impurities which attach themselves to it during the crystallization process, particularly small atoms present in the atmosphere in which the crystal is grown, such as oxygen, hydrogen, and nitrogen. An impurity is *substitutional* if it occupies a lattice site from which a host atom has been expelled, and *interstitial* if it occupies a position between the host atoms. The region surrounding an impurity is strained, the extent of the strain depending on the kind of impurity atom and its location. An appreciable number of substitutional impurities may be present only if the size of the impurity is not far from that of the host atom, otherwise the strain energy required would be prohibitive. Similarly, only small atoms can exist in large numbers as interstitial impurities because the space between host atoms is small, especially in metals, in which the atoms are tightly packed.

b) *Vacancy*. An empty lattice site from which the regular atom has been removed. In metals, as in other solids, vacancies are created by thermal excitation, provided the temperature is sufficiently high, because, as the atoms vibrate around their regular positions, some acquire enough energy to leave the site completely. When the regular atom leaves, the region surrounding a vacancy is distorted because the lattice *relaxes*, as it were, in order to partially fill the void left by the atom. This contributes further to the irregularity of the lattice in the immediate neighborhood of the vacancy.

c) *A regular atom in an interstitial position*. Considerable energy is needed to pull an atom from a regular to an interstitial position. This type of defect is created thermally only at high temperatures, near the melting point of the solid. One can also create this kind of defect by subjecting the solid to an external radiation—e.g., a neutron beam in a reactor—in which collision of incident particles with atoms causes these atoms to be dislodged from their sites into interstitial positions. It is evident that vacancies are also created in this same process.

**Fig. 11.1** Grain boundaries in molybdenum ( × 250). (After O. K. Riegger).

2. *Line defect.* A line defect, also called a *dislocation*, is a linear array of misplaced atoms extending over a considerable distance inside a lattice. As we shall see in Section 11.6, in which we shall consider dislocations in some detail, this type of defect is primarily responsible for the softness and ductility of pure metals.

3. *Surface defect.* In a surface defect, the crystalline irregularity extends in two dimensions. Most solids are not single crystals but *polycrystals*, in which a sample is composed of a large number of single crystal pieces, or *grains*, joined together to form one solid (Fig. 11.1). At each *grain boundary*, the crystal undergoes an abrupt change of orientation; the whole boundary therefore acts as a surface defect. These defects exert much influence on the properties of a polycrystal, particularly on its mechanical strength. Another surface defect, almost too obvious to be noticed, is the surface of the sample itself. This surface has a decisive effect on the properties of samples such as thin films and fibers.

All these types of defects play important roles in metallurgical and chemical processes in solids, and for this reason there has been much research on defects lately, with the result that they are now much better understood. The interested reader will find a great deal of new information in the references at the end of the chapter.

## 11.3 VACANCIES

There are two types of vacancy. In one type the displaced atom migrates in successive steps and eventually settles at the surface; this is a *Schottky defect* (Fig. 11.2). In the second type, called a *Frenkel defect*, the defect includes both atom and vacancy. Because of the additional elastic energy involved in squeezing an atom into an interstitial position, the Frenkel defect requires a large amount of energy, and for this reason is not usually present in metals except under special circumstances. Therefore vacancies usually exist only near free surfaces, grain boundaries, and dislocations, rather than inside a perfect crystal, because only at surfaces, boundaries, or dislocations can they be created without a concomitant formation of interstitials. In other words, these extended defects act as vacancy sources. We shall therefore talk primarily about Schottky defects.

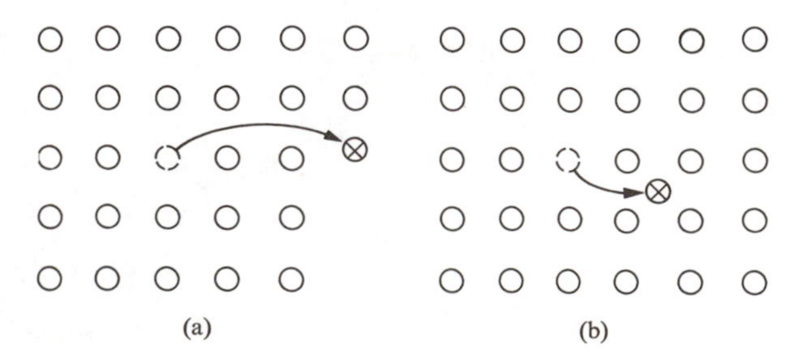


**Fig. 11.2** (a) A Schottky defect. (b) A Frenkel defect.

We can estimate the energy required to create a vacancy—the *formation energy*—in order of magnitude, by visualizing an atom, at its lattice site, "attached" to its neighbors by a number of chemical bonds which hold it in its position. When it is removed to create the vacancy, the bonds are broken, and this requires an energy roughly equal to the number of bonds (the *coordination number*) times the energy per bond. Since the energy per bond is typically about 0.5 eV, we should expect a formation of about 1 eV per vacancy, and this agrees with observations.

We can evaluate the number of vacancies generated by thermal excitation by using well-known results from statistical mechanics. Suppose that the crystal is at equilibrium at temperature $T$. Now, according to Chapter 3, each atom oscillates back and forth around its equilibrium position. Its average energy, at high temperature, is $3kT$. This energy is about 0.08 eV at room temperature, which is much less than the formation energy $E_v$ of the vacancy. One may therefore conclude that no vacancies can be created when the crystal is at room temperature. However, the quantity $3kT$ gives only the *average* energy of the atom, and not the actual energy at every instant during the motion. According to statistical mechanics, the atom, at thermal equilibrium, may have any energy $E$. The probability of its having this energy is given by the Boltzmann factor $e^{-E/kT}$. The fact that the exponential decreases rapidly with energy means that the atom has a very small probability of having a high energy. Now a vacancy is created when the atom has an energy equal to the formation energy, because the atom has then acquired sufficient energy to leave its site. Since the probability of this happening is given by $e^{-E_v/kT}$, it follows that the number of vacancies $N_v$ is given by

$$N_v = N\, e^{-E_v/kT}, \tag{11.1}$$

where $N$ is the total number of atoms in the crystal. At low temperature the number of vacancies is small because $kT \ll E_v$, but this number increases rapidly as the temperature rises. Thus for $N = 10^{29}$ atoms/m$^3$ and $E_v = 1$ eV, the number of vacancies at 300°K is $N_v \simeq 10^{12}$ vacancies/m$^3$, while at 900°K it is about $10^{22}$ vacancies/m$^3$. Therefore raising the temperature by only a factor of three causes the number of vacancies to increase by ten orders of magnitude.

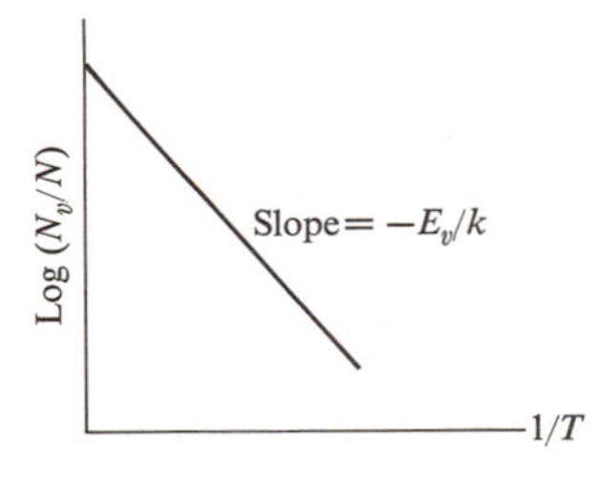


**Fig. 11.3** Log ($N_v/N$) versus $1/T$, where $N_v$ is the number of vacancies.

We see from (11.1) that if we plot log($N_v/N$) versus $1/T$ we obtain a straight line whose slope is $-E_v/k$, as shown in Fig. 11.3, and the slope can therefore be used to determine the formation energy. This procedure is commonly followed, and gives good results. Table 11.1 gives the vacancy formation energies for a few common metals.

**Table 11.1**
Energies of Vacancy Formation (k cal/mole)

| Ag | Au | Cu | Al |
|---|---|---|---|
| 25.1 | 26.5 | 21.6 | 15 |

Several methods can be used to measure the number of vacancies—or rather their concentration—all of which lead to the same result, as they should when properly employed. The basis of all of them is that the presence of vacancies leads to a change in some physical property of the sample, and therefore by measuring the change one may obtain information about the vacancy. For instance, the presence of vacancies causes an expansion in volume proportional to the concentration of the vacancies, and by measuring the change in volume one can deduce this concentration. Another common method is to measure the changes in electrical resistivity, which should be proportional to the concentration of vacancies. Since the vacancies act as scattering centers (see Section 4.5), we should expect that

$$\Delta\rho = A\, e^{-E_v/kT},$$

where $A$ is some positive constant. This type of experiment is usually performed on samples that have been quenched, that is, samples that have been cooled rapidly from a high temperature. The rapid cooling, in effect, freezes the vacancies, which are plentiful at high temperature, because they are not allowed enough time to move about in the crystal and disappear at grain boundaries and dislocations. In this manner one can prepare a sample with a large number of vacancies at a relatively low temperature. Because of this, the temperature in the above equation should refer to the quenching temperature $T_Q$, and not the lower temperature at which the resistivity is measured. By varying $T_Q$ continuously and measuring the corresponding resistivity, one can evaluate the variation of concentration of vacancies with temperature and obtain the formation energy.

When a quenched sample is heated slowly, vacancies are able to move through the crystal again, and tend to move toward surfaces and other extended defects where they may disappear. In this way some of the excess vacancies are annihilated, and the equilibrium distribution is reestablished, at least in part. When the concentration of vacancies is large, some of the vacancies sometimes cluster into groups of two, three, or more. One speaks of such entities as *divacancies*, *trivacancies*, etc. For further information on this and related subjects, refer to the review article by Takamura cited in the bibliography.

## 11.4 DIFFUSION†

When the concentration of atoms in a sample is not uniform, atoms migrate from a region of high concentration to one of low concentration, the process continuing until the distribution of atoms becomes uniform throughout the solid. This flow down the concentration gradient is referred to as *atomic diffusion*, and is of major importance in many metallurgical processes. For instance, the hardening of steel involves the diffusion of atoms of carbon and other elements through iron, which is accomplished by heating the iron in an environment rich in carbon and other required elements. In the manufacture of transistors, the sample has to be doped by impurities, both donors and acceptors, in a controlled manner. This is most commonly accomplished by diffusing the required impurities into a highly purified specimen of a semiconductor in such a way that they have the proper spatial distribution. Since the operation of many solid-state devices depends on very careful distribution of atoms, the control of diffusion is becoming increasingly important.

Let us begin our discussion with a macroscopic treatment involving setting up and solving the appropriate differential equations, followed by a microscopic treatment in terms of the movement of individual atoms. Then we shall connect the two treatments and arrive at a microscopic expression for the diffusion parameter.

The basis for the macroscopic treatment is the *first Fick's law*, which states that the diffusion current (number flux density) $J$ is related to the concentration gradient by

$$J = -D\frac{\partial c}{\partial x}, \tag{11.2}$$

where the parameter $D$, supposedly a constant, is called the *diffusion coefficient*. The minus sign is inserted to make $D$ a positive quantity and ensure that the current flows down the concentration gradient. The expression (11.2) also applies to unidirectional flow in which the concentration varies along the $x$-direction only, but a suitable generalization to a three-dimensional situation can readily be made.

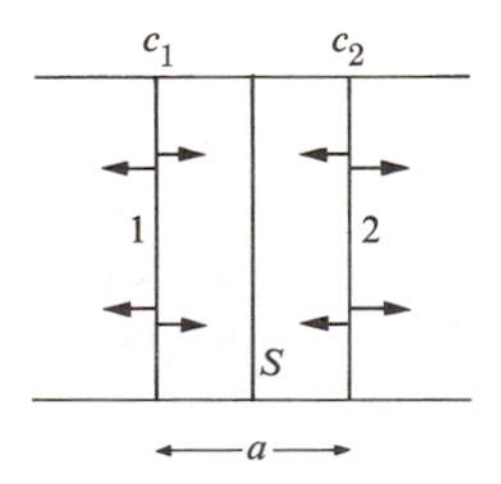


**Fig. 11.4** Jumping motion of atoms in planes 1 and 2 which leads to diffusion.

† Note a partial similarity between the discussion here and the discussion of the diffusion of carriers in semiconductors (Section 6.17).

A justification for Fick's law may be given in terms of the following kinetic model. Consider a sample (Fig. 11.4), and let us calculate the rate of flow across a section $S$ normal to the concentration gradient. The concentrations at two adjacent atomic planes straddling the section are indicated by $c_1$ and $c_2$ $(c_1 > c_2)$. Now atoms on plane 1 jump both to the left and to the right randomly, but only when they jump to the right do they cross section $S$. Similarly, atoms on plane 2 cross $S$ only when they jump to the left. However, since $c_1 > c_2$, there are more atoms crossing to the right than to the left, and consequently there is a net diffusion to the right, i.e., down the concentration gradient. Quantitatively, if the frequency of the jump of the atoms is $\nu$, then the diffusion current to the right is

$$J = \tfrac{1}{2} n_1 \nu - \tfrac{1}{2} n_2 \nu,$$

where the two terms on the right give the diffusion rates for atoms starting from the $c_1$ and $c_2$ planes, respectively, and the factor $\frac{1}{2}$ is inserted because atoms on each plane can jump either to the right or left, but they cross $S$ only half the time. The quantities $n_1$ and $n_2$ refer to the number concentrations in the two planes, and are related to $c_1$ and $c_2$, the fractional concentrations, by the relations $n_1 = c_1 a$ and $n_2 = c_2 a$, where $a$ is the distance between the planes. Substituting into the above equation yields

$$J = \tfrac{1}{2} \nu a (c_1 - c_2).$$

If the concentration does not vary rapidly between adjacent planes, a condition which obtains in almost all practical situations, we may write $c_1 - c_2 \simeq - a\, \partial c/\partial x$, which amounts to treating $c$ as a continuous function. When we substitute this into the above expression for $J$, we find that

$$J = -\tfrac{1}{2} \nu a^2 \frac{\partial c}{\partial x}, \tag{11.3}$$

which is the same expression as Fick's law, with a diffusion constant given by

$$D = \tfrac{1}{2} \nu a^2. \tag{11.4}$$

We have actually imposed more restriction on the motion than necessary, because, since the problem is in fact a three-dimensional one, we have to allow for circumstances in which the atoms in, say, plane $c_1$ may jump parallel to the plane rather than to the right or left. This means that under random jumping the atom crosses $S$ only one-sixth of the time, and Eq. (11.4) should therefore be replaced by

$$D = \tfrac{1}{6} \nu a^2. \tag{11.5}$$

Numerical values for $D$ in metals and in semiconductors near room temperature fall in the broad range of $10^{-20}$ to $10^{-50}$ m/s$^2$. This enormous difference must be accounted for by a difference in $\nu$, the jump frequency, since the interplanar distance $a$ is roughly the same for all these materials. As an estimate of $\nu$, if we substitute the values $D = 10^{-20}$ m/s$^2$ and $a = 3$ Å into (11.5), we find a jump frequency of about 1 s$^{-1}$. On the other hand, a value of $D = 10^{-50}$ m/s$^2$ yields $\nu \simeq 10^{-30}$ s$^{-1}$, that is, an atom makes one jump every $10^{27}$ years! We shall examine the physical meaning of the jump frequency more closely later in this section.

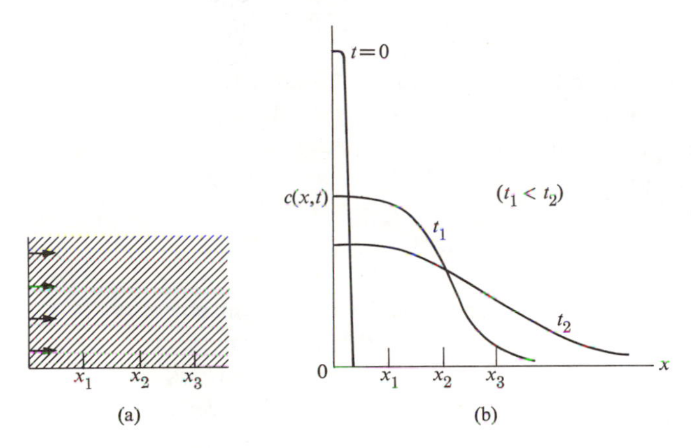


**Fig. 11.5** (a) Diffusion of atoms in a metallic bar. (b) Profile of a diffusion pulse as a function of distance and time.

One can measure the diffusion coefficient by depositing on the sample a thin film of the atoms whose diffusion in a specific metal is sought, and monitoring the concentration of the solute atoms at several depths $x_1$, $x_2$, $x_3$, etc. (Fig. 11.5a), after allowing sufficient time for diffusion to take place. From these measurements one can calculate the coefficient $D$.

There are two methods of measuring the concentration of solute atoms versus depth of diffusion: One is to employ an ordinary chemical analysis at various depths in the sample. The other is much more convenient, and employs a radioactive isotope to tag the diffusing atoms. One then determines the concentration of solute atoms by measuring radioactive intensity as a function of depth. One does this by slicing the sample parallel to the $x$-axis and placing it over a prepared film. The emulsion in the film is sensitized by the radioactivity, and the degree of darkness over various parts of the film is a measure of the concentration of the diffusing isotope.†

† In practice, this autoradiographic technique does not yield sufficiently accurate data. Instead one measures the concentration of solute atoms in the slices by using electronic counters.

The equation which describes this experiment is derived by combining Fick's law with the continuity equation in one dimension,

$$\frac{\partial c}{\partial t} + \frac{\partial J}{\partial x} = 0.$$

When we substitute from (11.2) in the above equation, the result is

$$\frac{\partial c}{\partial t} = \frac{\partial}{\partial x}\left(D\frac{\partial c}{\partial x}\right) = D\frac{\partial^2 c}{\partial x^2}, \tag{11.6}$$

where it is assumed, in the second equation, that $D$ is independent of $x$, or, equivalently, that $D$ is independent of the concentration. The solution of Eq. (11.6)—this equation is known as *Fick's second law*—under the boundary conditions imposed in the experiment is

$$c(x, t) = \frac{A}{t^{1/2}} e^{-x^2/4Dt}, \tag{11.7}$$

where $A$ is a constant determined by the initial concentration of solute atoms at $x = 0$. That is, by the concentration in the initially deposited layer. Figure 11.5(b) gives the profiles of concentration versus depth at various instants.

One can find a simple measure of depth of diffusion of solute atoms as a function of time by evaluating the average root mean square (rms) value of $x$. That is,

$$\bar{x} \equiv \left[\int_0^\infty x^2\, c(x, t)\, dx \Big/ \int_0^\infty c(x, t)\, dx\right]^{1/2}.$$

By substituting from (11.7) into this equation, and by using tables of integrals to evaluate standard integrals, you should arrive at the important result,

$$\bar{x} = \sqrt{2\,Dt.} \tag{11.8}$$

Thus the diffusion front propagates to the right with a travel distance proportional to $t^{1/2}$, a time dependence characteristic of all diffusion processes, with an ever-decreasing speed. Later in this section we shall be able to derive (11.8) from a microscopic model, and so the equation serves as a bridge between the macroscopic and microscopic descriptions.

There are two types of diffusion: *self-diffusion* and *interdiffusion*. In self-diffusion, the diffusing atoms are of the same type as the solvent background, e.g., copper in copper. In interdiffusion, the two types of atom are different. If the concentration of the solute is appreciable—as in the case of an alloy, in which the distinction between solvent and solute tends to disappear—then the two kinds of atom tend to diffuse into each other. Since the diffusion coefficients of the two kinds are usually unequal, the boundary between them moves along the bar of sample material progressively as time passes. This was observed in the Kirkendall experiment, in which two bars, one zinc and the other brass (a copper–zinc alloy), were joined. The boundary moved into the brass region, indicating that zinc diffuses more rapidly than copper. In the case of interdiffusion, the *effective* diffusion coefficient for the combined system is

$$D = C_A D_B + C_B D_A,$$

in which $A$ and $B$ refer to the two different atoms. (The $C$'s refer to fractional concentrations.)

**The microscopic model**

Diffusion occurs when atoms migrate between adjacent lattice sites, and therefore one should be able to describe the process in terms of a microscopic model involving the motion of the individual atoms. However, the atomic motion is random, and this must somehow lead to an organized macroscopic motion down the concentration gradient.

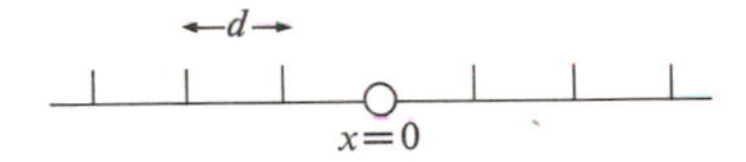


**Fig. 11.6** Diffusion of an atom in a one-dimensional lattice.

The simplest model we can choose (Fig. 11.6) is a one-dimensional lattice and an atom placed originally at a lattice site, say $x = 0$. The atom hops to adjacent sites at a frequency $\nu$, but at every site the direction of hopping is completely random, i.e., it may jump either to the right or to the left with equal probability. Where is the atom likely to be found after it has made $n$ jumps, given that it started at $x = 0$?

The problem is exactly equivalent to a well-known one in statistics, the *random-walk problem*, in which one asks where a person is likely to be after taking $n$ steps if, after each step, he is equally likely in the next step to move either forward or backward. Let $x_n$ denote the position of the atom after $n$ jumps. Then

$$x_n = d_1 + d_2 + \cdots + d_n,$$

where $d_1, d_2, \ldots,$ all have a magnitude equal to $d$, the lattice constant, although they may differ in sign. The average value of $x_n$ is zero, because the average of each step is zero, and the atom is most likely to be found at $x = 0$, the initial point, which is to be expected from the symmetrical nature of the problem. However, there is a finite probability that the atom is to be found at other sites—i.e., that the atom will have made a net displacement—and this probability is measured by the standard mean deviation or, equivalently, the rms value of $x_n$. Denoting this by $\bar{x}$, one has

$$\bar{x} = \sqrt{\overline{x_n^2}}\,.$$

Substituting for $x_n$ from the previous equation and noting that, after squaring, $d_i^2$ is the same for all steps and equal to $d^2$, one arrives at

$$\bar{x} = \sqrt{n\,d^2}. \tag{11.9}$$

The distance $\bar{x}$, which is proportional to $n^{1/2}$, should be compared with $L = nd$, which is the distance the atom would have covered had it jumped in only one direction. Thus

$$\bar{x} = L/n^{1/2},$$

which means that, for large $n$, the situation in which this statistical analysis has any validity, $\bar{x} \lll L$. Although a random motion may lead to a net motion, the motion is greatly impeded by the random character. Thus if the atom is to move a distance of 1 micron, it has to make, with $d = 1$ Å, $10^0$ jumps; for a jump frequency of $1\ \text{s}^{-1}$ this requires about two years!

One can generalize the discussion by allowing the atom to move along a three-dimensional lattice in all directions consistent with the lattice. The result is the same as (11.9), except that $\bar{x}$ should be replaced by the radial distance $\bar{R}$, because the atom now migrates radially outward in all directions. If the macroscopic geometrical arrangement is unidirectional, as in Fig. 11.3, then we would be interested only in $\bar{x}$ and not $\bar{R}$, although the atom in fact diffuses in all directions. In the symmetrical case, in which $\overline{x^2} = \overline{y^2} = \overline{z^2}$, one can readily show that $\overline{x^2} = \frac{1}{3}\overline{R^2}$, and therefore Eq. (11.9) is generalized to

$$\bar{x} = \sqrt{\frac{n\,d^2}{3}}. \tag{11.10}$$

It is now convenient to introduce the jump frequency into the expression. We do this by noting that $n = \nu t$, where $t$ is the time interval. Equation (11.10) can then be written as

$$\bar{x} = \sqrt{\frac{\nu\,d^2}{3}t}. \tag{11.11}$$

This is seen to be of exactly the same form as (11.8), and the two equations become identical if the diffusion coefficient is taken to be

$$D = \frac{\nu\,d^2}{6}, \tag{11.12}$$

which is the same expression as Eq. (11.5), derived earlier on the basis of macroscopic kinetic analysis [note that $d$ in (11.12) is the same as $a$ in (11.5)]. In developing the microscopic analysis here, we have unearthed the statistical basis for the diffusion process.

**Diffusion mechanisms and temperature dependence**

Experiments show that the diffusion coefficient increases sharply with temperature, and that it is well interpreted by the formula

$$D = D_0 \, e^{-Q/kT}, \tag{11.13}$$

where $D_0$ is a constant insensitive to temperature. The energy $Q$ in the exponent, called the *activation energy*, is typically about 2 eV or 46,000 cal/mole. To explain the origin of the temperature dependence of (11.13), we must inspect the microscopic expression for the diffusion coefficient, Eq. (11.12), and pay particular attention to the jump frequency $\nu$. The question of jump frequency requires a close examination of the diffusion mechanism on a microscopic level.

Several different mechanisms may lead to diffusion, or atomic migration, in a lattice. If enough interstitial atoms are present, diffusion can occur by their hopping between interstices down the concentration gradient, as in the case of diffusion of carbon atoms in steel. In the case of substitutional alloys or substitutional impurities diffusion may occur through the consecutive exchange of places among the atoms (Fig. 11.7). A third mechanism involves migration of vacancies (also illustrated in Fig. 11.7), which occurs when many atoms in succession fill the vacancy, and the vacancy consequently migrates in the opposite direction.

Thus if foreign atoms are deposited on the left side of a sample, the vacancies inside it diffuse to the left to compensate for those annihilated by the intruding atoms, and the foreign atoms migrate to the right. Other diffusion mechanisms could be included; in general they all operate simultaneously to produce the net diffusion observed. In metals, however, the vacancy diffusion mechanism is most important, and for this reason it is the one we shall discuss in detail.

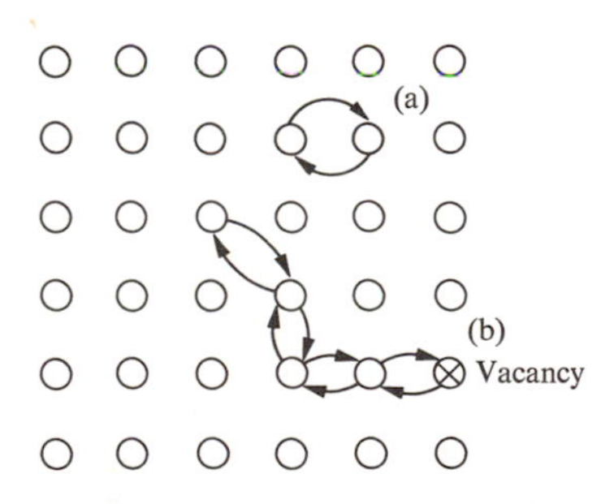


**Fig. 11.7** Process (a), diffusion through interchange of substitutional atoms. Process (b), diffusion by vacancy migration.

Figure 11.8 shows the energy involved in a vacancy migration. The solid circle represents the atom whose migration is under consideration; it must have an energy at least equal to $E_m$ in order to be able to leave its site and exchange places with the vacancy. The origin of this potential barrier lies in the fact that the atom, in moving, pushes other atoms sideways, and consequently the lattice is strained in that region; $E_m$ represents the maximum strain energy incurred. In the present context, $E_m$ is often called the activation energy for the transition, i.e., it is the minimum energy required for the transition to proceed. Its value varies from one metal to another, but is typically about 1 eV.

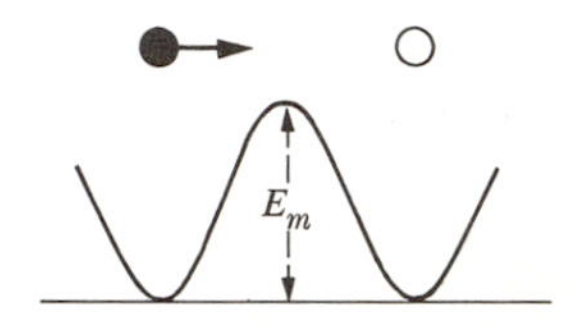


**Fig. 11.8** Energy barrier $E_m$ seen by diffusing atom. Solid circle indicates an atom, open circle a vacancy.

The atom in Fig. 11.8 oscillates around its equilibrium position with a frequency $\nu_0$, as discussed in Chapter 3, but usually its energy is far too small at ordinary temperatures to allow it to jump the potential barrier. However, during a fraction of time equal to the Boltzmann factor $e^{-E_m/kT}$, the atom has energy equal to $E_m$, and is then able to make the transition. At an oscillation frequency $\nu_0$, the atom hits the barrier $\nu_0$ times per second. The probability of escape for each time is $e^{-E_m/kT}$. Thus the jump frequency of the atom is

$$\nu = \nu_0\, e^{-E_m/kT}.$$

This expression has to be generalized to allow for the fact that an atom in a three-dimensional lattice can jump into any of its $z$ neighboring sites, and also that this can take place only if the final site has a vacancy. Since the probability of a vacancy at a site is $e^{-E_v/kT}$, it follows that the jump frequency for the diffusing atom is

$$\nu = z\,\nu_0\, e^{-E_v/kT}\, e^{-E_m/kT} = z\,\nu_0\, e^{-(E_v+E_m)/kT},$$

which may also be written as

$$\nu = z\,\nu_0\, e^{-Q/kT}, \tag{11.14}$$

where $Q = E_v + E_m$. Substituting this into expression (11.12) for the diffusion coefficient yields

$$D = \tfrac{1}{6} z\,\nu_0\, a^2\, e^{-Q/kT} = D_0\, e^{-Q/kT}, \tag{11.15}$$

which is precisely the same form as (11.13), where $D_0 = \frac{1}{2} z\,\nu_0\, a^2$. We see that the activation energy $Q$ is the sum the of vacancy-formation energy plus the atomic-transition energy, and is divided roughly equally between them, each being about 1 eV.

The values of $D_0$ and $Q$ for a substance can be determined by plotting $\log D$ versus $1/T$ which, according to (11.15), should give a straight line whose intercept is $\log D_0$ and slope $-Q/k$; an example is shown in Fig. 11.9. Table 11.2 gives a list of $D_0$ and $Q$ for some substances, as determined by this method.

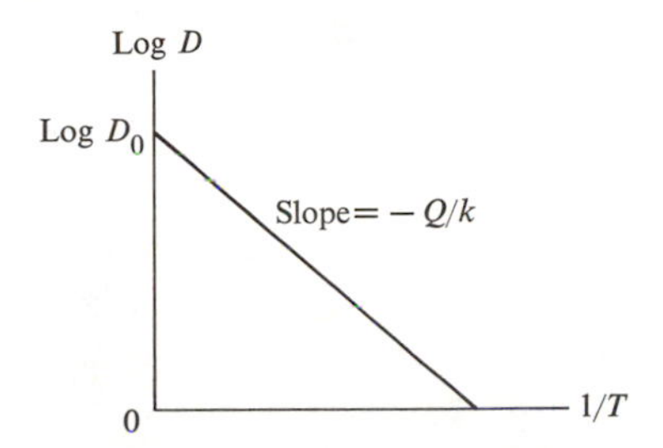


**Fig. 11.9** Variation of diffusion coefficient with temperature.

The vacancy-diffusion method, besides yielding the correct temperature dependence of $D$, also gives a value for $D_0$ in agreement with experiment. If we choose the reasonable values $z = 12$, as for an fcc lattice, $\nu_0 = 10^{13}\ \text{s}^{-1}$, $a = 2$ Å, and $Q = 2$ eV, we find $D_0 = 8 \times 10^{-5}\ \text{m}^2/\text{s}$, which is of the same order of magnitude as the values given in the table.

**Table 11.2**
Diffusion Parameters

| Metals | $D_0 \times 10^4$, m$^2$/s | $Q$, cal/mole |
|---|---|---|
| Fe in Fe ($\gamma$ phase) | 1.4 | $5.65 \times 10^4$ |
| Cu in Cu | 0.69 | 5.02 |
| Pb in Pb | 0.28 | 2.42 |
| Ag in Ag | 0.4 | 4.41 |
| Au in Au | 0.09 | 4.17 |
| Al in Al | 0.18 | 3.01 |
| Zn in Cu | 0.34 | 4.56 |
| C in Fe ($\alpha$ phase) | 0.02 | 2.00 |
| C in Fe ($\gamma$ phase) | 0.15 | 3.4 |

The point that should be especially stressed in connection with the diffusion coefficient is that its increase with temperature is very rapid. Thus for the $D_0$ estimated above, and for an activation energy $Q = 2$ eV, one finds that at $T = 300°$K, $D \simeq 8 \times 10^{-5} \times 10^{-34} \simeq 10^{-38}\ \text{m/s}^2$, while at $T = 1500°$K, $D \simeq 8 \times 10^{-5}\ e^{-16} \simeq 8 \times 10^{-5} \times 10^{-7} \simeq 10^{-11}\ \text{m/s}^2$, an increase of 27 orders of magnitude due to raising the temperature by a factor of five. Therefore diffusion rates can be greatly enhanced by raising the temperature, a fact often used in practice.

Diffusion by interstitial atoms can be treated in a similar manner, and the result is much the same, except that now the activation energy $Q$ is the same as the transition energy $E_m$, which is needed to overcome the elastic-strain energy incurred when the interstitial atom pushes the regular atoms aside.

Diffusion also takes place in liquids. A typical value for the diffusion coefficient is about $10^{-10}$ m/s$^2$ at a temperature just above the freezing point. It is found experimentally that the diffusion distance is well represented by an equation such as (11.11), that is, that the $t^{1/2}$ dependence also obtains in liquids. A theoretical treatment for diffusion in liquids is more complicated than that for solids, because the concepts of crystal lattice, a fixed jumping step, and a well-defined jump frequency do not apply. Instead we use a model in which the atoms move freely between collisions, and apply the concept of mean free path. The result obtained is the same as Eq. (11.11). This method is essentially that used by Einstein in his explanation of Brownian motion.

## 11.5 METALLIC ALLOYS

Metals are rarely used in their pure form in industrial applications because pure metals, among other drawbacks, are too soft and ductile. Thus carbon and other metals are added to iron to harden it to steel, and aluminum is strengthened by adding copper, silicon, and other elements to it. Brass is an alloy of copper and zinc. In all these cases one deals with a metallic alloy in which atoms of one or more elements are dissolved in a metal. An alloy is therefore a *solid solution*. It differs from a chemical compound in that, in a solid solution, the range of concentration of the solute relative to the solvent may vary, while in a chemical compound this concentration is fixed.

In an alloy the solute atoms take up positions either at the interstices or at regular lattice sites. In the first case, the alloy is *interstitial*, e.g., carbon in steel. In the second case, the alloy is *substitutional*, e.g., zinc in copper (brass). Clearly an interstitial alloy can be formed only if the solute atoms are small enough to fit into the interstices without the expenditure of a large amount of energy. In general, interstitial solubility in metals is limited because the atoms in a metal are relatively closely packed. In a substitutional alloy, the solute atoms occupy regular lattice sites in more or less random fashion. As more solute atoms are added they occupy more sites and the crystal simply grows in size. The crystal structure remains unaltered except, perhaps, for a slight change in the lattice constant. This type of alloy is called a *primary solid solution*. In some cases, however, the crystal structure may undergo a change as the concentration of the solute becomes appreciable, in which case we speak of a *secondary solid solution*. In general, when the crystal structure of the solution is different from the crystal structures of the pure metal components, the solution is said to be in an *intermediate phase*.

In this section we are interested primarily in substitutional alloys and their properties. To simplify the discussion, we shall confine ourselves to binary (two-component) alloys, and keep the physical concepts in the foreground.

### Rules for primary solubility

Two metals can form a primary solution only when they are similar. For instance, silver and gold, which are quite similar, form a primary solution over the entire composition range, from pure silver all the way to pure gold. Under less-ideal conditions, two metals form a primary solution only over a limited range. For example, copper can be dissolved in silver only up to about 15% atomic weight before the alloy undergoes a phase change. The conditions favoring primary solubility were studied carefully by Hume–Rothery and coworkers, whose results are summarized by the following four rules.

a) *Atomic size effect.* The solute and solvent atoms should be close in size. The difference in diameter of atoms should not exceed 15%. For silver and gold, the difference is only 0.2%.

b) *Crystal structure effect.* In order for there to be extensive solubility, the structures of the solute and solvent metals should be similar. Both silver and gold, for example, have an fcc structure.

c) *Electronegative valence effect.* The two elements must have similar electrochemical characteristics. By contrast, an electropositive element such as silver and an electronegative element such as bromine would form a chemical compound, not an alloy.

d) *Relative valence effect.* This rule asserts that it is easier to dissolve a metal of higher valence into one of lower valence than the reverse. For instance, aluminum dissolves more readily in copper than copper in aluminum because, apparently, in the former situation it is relatively easy for the excess aluminum electrons to detach themselves from their own atoms and accommodate themselves in the alloy. If copper is dissolved in aluminum, however, there is a deficiency of conduction electrons at the copper sites, and the electrons that tend to neutralize this deficiency have high energy.

Even if all these rules are satisfied, the two metals still may not dissolve into each other appreciably, because—although these rules are necessary—they are not sufficient by themselves. They only state the circumstances most favorable for stability.

### The phase diagram†

The phase diagram is a convenient graphical summary of the melting characteristics of an alloy. The simplest type of phase diagram is illustrated in Fig. 11.10, in which the abscissa represents the concentration of component $B$ and the ordinate represents the temperature. The *solidus* and *liquidus* lines divide the figure into three separate regions: (a) Below the solidus, the alloy is a homogeneous solid solution

† Also called the *equilibrium diagram.*

phase for every composition. (b) Above the liquidus, the alloy is a homogeneous liquid solution phase for every composition. (c) Between solidus and liquidus lines, the alloy is composed of two different phases, a solid and a liquid, coexisting in equilibrium with each other.

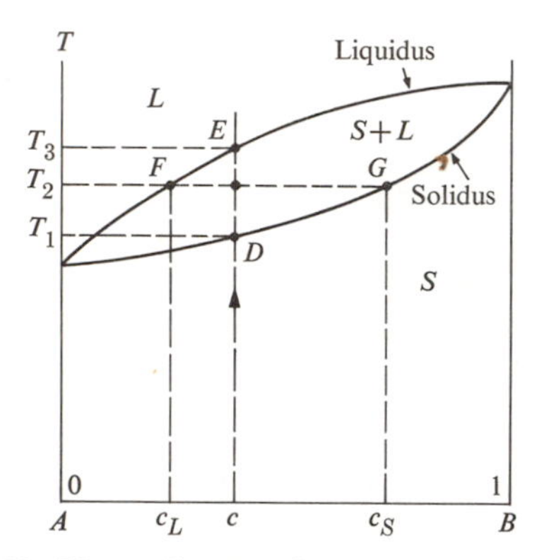


**Fig. 11.10** Phase diagram for a binary alloy *A-B*.

Suppose that an alloy of composition $c$ is gradually heated. The vertical arrow indicates the path followed by the specimen. At the outset, the alloy is a homogeneous solid, and it remains so until it reaches $D$, the intersection with the solidus line. At this point some of the solid begins to melt, and the first few droplets of molten alloy begin to form. As the temperature is raised further, more of the solid is converted into liquid. The system is now a solid and a liquid phase coexisting with each other. However, the phases have different compositions. At temperature $T_2$, for instance, the composition of the liquid is given by the point $F$ on the liquidus line, while that of the solid is given by the point $G$ on the solidus line. These two concentrations are indicated by $c_L$ and $c_S$, respectively. Since $c_S > c_L$, the solid has more element $B$ than the liquid phase does. If the temperature is raised still further, more and more of the solid is converted into liquid, until point $E$ is reached, when the alloy is entirely in the liquid phase. The composition is now equal to $c$, the same as that of the initial solid. Any further heating leaves the alloy in the liquid phase, until evaporation takes place.

The composition of the solid and liquid phases in the region between the solidus and liquidus lines can be evaluated from the phase diagram. Suppose that the temperature is $T_2$; because mass is conserved, we may write

$$c(S + L) = c_S S + c_L L,$$

where $S$ and $L$ are the amounts of solid and liquid. By rearranging, we can write this equation as

$$\frac{L}{S} = \frac{c_S - c}{c - c_L}, \tag{11.16}$$

which is known as the *lever formula*. It expresses $L$ and $S$ in terms of concentrations obtainable from the phase diagram. (Note that $c_S - c$ is the length along the abscissa between $c$ and $c_S$.)

An interesting conclusion follows from the phase diagram: An alloy of a fixed composition does not melt at a fixed temperature. The melting takes place over a range of temperatures, usually a few degrees. This is unlike a pure metal, for which there is a fixed, well-defined melting point. This contrast is also evident in the phase diagram, because the solidus and liquidus lines converge only at the endpoints at which the alloy reduces to one or the other of the two pure metals.

### Thermodynamics: free energy and entropy

To gain a deeper understanding of phase diagrams and related effects, we need the concept of *free energy*. In thermodynamics there are two different free energies: The *Helmholtz energy*, $F = E - TS$, and the *Gibbs energy*, $G = F + pV$, where the symbols on the right sides of these equations have the usual meaning. Most experiments are performed at atmospheric pressure, which is so low that the $pV$ term may be neglected, compared with the other terms, without serious error. Therefore it is sufficient for our purpose to use the free energy,

$$F = E - TS. \tag{11.17}$$

The term $E$ represents the total internal energy, both potential and kinetic, of the system, and $S$ is the total entropy. A well-known principle in thermodynamics—the principle of *minimum free energy*—asserts that if a system is allowed several alternative states, it will choose the one with the *lowest* free energy.

To clarify the meaning of this principle, we shall apply it quantitatively to a solid. The energy $F$ has its lowest possible value when the internal energy $E$ is as low as possible, and at the same time the entropy is as large as possible. Now $E$ (which in a solid is primarily potential) is a negative quantity, and is minimized by placing all atoms at their regular sites, because each atom then rests at the bottom of its potential well. But this arrangement has a very low entropy value, because entropy is proportional to disorder, and the above arrangement has a high degree of order. So the requirements of minimizing $E$ and maximizing $S$ conflict with each other. The actual state adopted by the system is one which balances these two factors, that is, a state in which most of the atoms oscillate around their positions. The maximization of $F$ necessitates the presence of vacancies in the amount given by Eq. (11.1) (see the problems section at the end of this chapter).

The thermodynamic definition of entropy is

$$dS = \frac{dQ}{T} = \frac{C_p\,dT}{T}, \tag{11.18}$$

where $dQ$ is the amount of heat absorbed by the system in a reversible process and

$C_p$ the specific heat at constant pressure. In statistical mechanics, the entropy of a system is defined as

$$S = k \log p, \tag{11.19}$$

where $p$ is the number of *microstates* corresponding to the same *macrostate* of the system. One defines a macrostate, or thermodynamic state, by specifying the average values of a few macroscopic parameters of the system, such as volume, pressure, energy, etc. A macrostate can be verified experimentally. On the other hand, to define a microstate would require specifying the positions and velocities of *all particles* making up the system. In general, there are many different microstates which correspond to the same macrostate, and consequently these are not distinguishable experimentally. This number is the quantity $p$ given in (11.19). Texts on thermodynamics and statistical mechanics show that the two definitions for entropy, (11.18) and (11.19), are entirely equivalent, and therefore they will be used interchangeably here. We shall shortly have the chance to use (11.19) in calculating the entropy of a substitutional alloy.

## Polymorphic transformation

When a metal or alloy is heated, at some temperature it undergoes a transformation to a new crystal structure (or solid phase). This happens most frequently in the transition metals and their alloys. A well-known example is iron which, when heated to 910°C, makes a transition from a bcc ($\alpha$-iron) to an fcc ($\beta$-iron) structure. Other transition metals show similar *polymorphic* transformations.

The phenomenon can be understood in terms of the free-energy principle. Using Eqs. (11.17) and (11.18), one can show (see the problem section) that the free energy at temperature $T$ is given by the expression

$$F = E_0 - TS_0 - \int_0^T \left( \int_0^T \frac{C_p(T')}{T'}\, dT' \right) dT, \tag{11.20}$$

where $E_0$ and $S_0$ are the internal energy and entropy at absolute zero,† respectively.

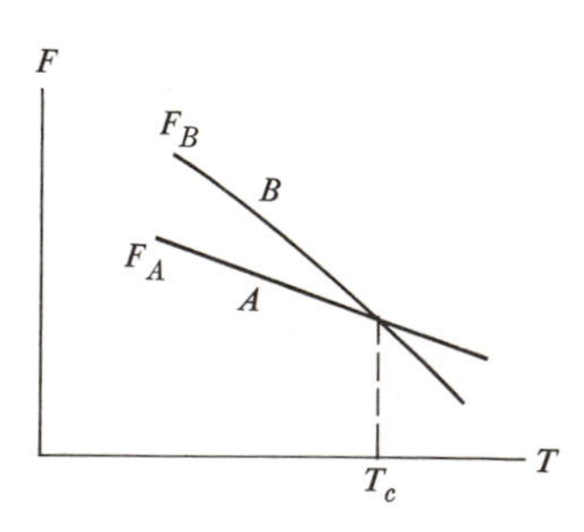


**Fig. 11.11** Polymorphic transformation. Free energy $F$ versus $T$, for a system in two different solid phases, $A$ and $B$.

† In a pure metal the entropy $S_0$ vanishes, according to the third law of thermodynamics.

Let us compare two possible but different structures for the system, $A$ and $B$ (Fig. 11.11). The structure $A$ has a lower $E_0$, and hence is more stable at low temperature than $B$. Since $A$ is more tightly bound, it also has a higher Einstein, or Debye, temperature, and consequently a rather low specific heat $C_p$ (Section 3.4). It follows from (11.20) that as the temperature increases, the free energy $F_A$ decreases at a lower rate than $F_B$, and hence the curves for $F_A$ and $F_B$ versus $T$ will intersect at some temperature $T_c$, as shown in Fig. 11.11. Below $T_c$, $F_A < F_B$, and $A$ is the more stable of the two structures, while above $T_c$, the situation is reversed. Of course, the transformation is observed only if the transition temperature is below the melting point; otherwise the solid would melt before it had a chance to undergo the polymorphic transformation.

Figure 11.11 can also be used to describe the melting transition of a metal, where $A$ and $B$ then refer to the solid and liquid phases, respectively.

**The mixing entropy of a substitutional alloy**

When two metals at the same temperature are mixed to form an alloy, the entropy of the system increases by virtue of the mixing process. This increase is called the *entropy of mixing*, or *entropy of disorder*. We expect the increase intuitively because the system becomes more disordered. We shall now calculate the increase in entropy using the statistical definition (11.19).

Suppose that the alloy has $N$ atoms in all, of which $n$ are $B$ type and the remainder, $N - n$, are $A$ type. To assign numbers for the two types of atom is, in effect, to specify the macrostate, and to calculate $p$ of (11.19) we need to count the number of different macrostates, i.e., the number of atomic arrangements consistent with the macrostate. If all the atoms really were different, there would be $N!$ different arrangements, because the first atom can be placed in any of $N$ different lattice sites, the second atom in any of $(N - 1)$ sites, and so forth. Since all the $A$ atoms are identical, however, any two arrangements differing from each other only by the interchange of two or more of these atoms cannot be counted as different arrangements. The same reservation applies to the identity of the $B$ atoms. Since the $A$ atoms can be interchanged among themselves in $(N - n)!$ different ways, and the $B$ atoms in $n!$ different ways, it follows that the number of distinct arrangements, or microstates, for the $N$ atoms is

$$p = \frac{N!}{n!\,(N - n)!}. \tag{11.21}$$

When we take the logarithm of $p$ and use the *Stirling formula*,

$$\log N! \simeq N \log N - N,$$

[the factorials $n!$, $(N - n)!$ are treated similarly], an approximation valid for the large numbers of atoms encountered in solids, and then substitute into (11.19), we find for the entropy of mixing

$$S = -Nk\,[c \log c + (1 - c) \log (1 - c)], \tag{11.22}$$

where $c = n/N$, the concentration of the $B$ atoms. The concentration of the $A$ atoms is, of course, $(1 - c)$. Note that $S$ is positive, since the logarithmic terms are negative, by virtue of the fact that $c$, $(1 - c) < 1$.

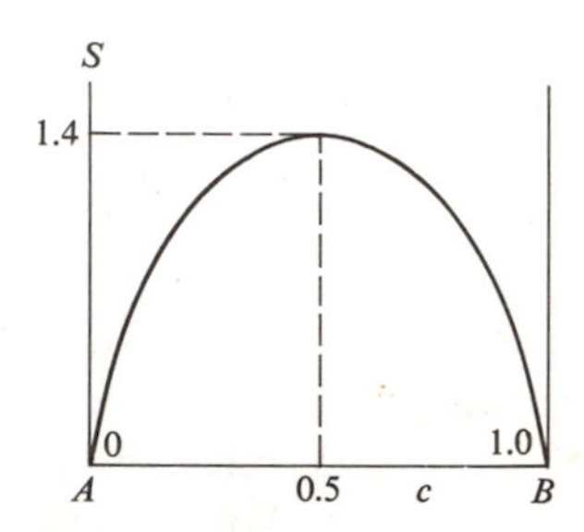


**Fig. 11.12** The mixing entropy of a substitutional alloy $S$ versus concentration $c$. The entropy has a maximum at $c = 0.5$, whose value is 1.4 cal/mole.

The variation of entropy with composition is indicated in Fig. 11.12, where $S$ has a maximum at $c = 0.5$, the point of maximum disorder, and decreases on either side of this point; it reaches zero at the endpoints at which, in each case, a state of complete order prevails. Numerically the maximum entropy is 1.4 cal/mole. It is important to note that near the endpoints $S$ increases very rapidly as the other element is added. This means that there is a strong tendency toward solution at low concentration, regardless of other possibly unfavorable factors.

### Melting and structure

Why does a solid melt when it is heated, and why does this take place at a certain fixed temperature, different for each substance? The answer must be that above this temperature the free energy of the liquid phase becomes lower than that of the solid phase. We may compare the free energies of the two phases at the same temperature by writing

$$\Delta F = E_L - E_S - T(S_L - S_S) = \Delta E - T\,\Delta S. \tag{11.23}$$

The energy difference $\Delta E$ is positive because in the liquid phase many of the atoms occupy interstitial positions, which results in a high energy. The volume is also larger in the liquid than in the solid phase, so that the atoms are pulled away from each other with some expenditure of energy. However, $\Delta S$ is also positive, because the liquid phase, being more disordered than the solid, has a higher entropy. For $T < T_m$, where $T_m$ is the melting temperature, the term $\Delta E$ dominates, that is, $\Delta F > 0$, and consequently no melting takes place, while for $T > T_m$ the entropy term dominates, and the solid melts completely. At $T = T_m$, the energy and entropy terms exactly balance each other, $\Delta E = 0$, and the two phases are in equilibrium with each other.

## Free energy of alloy phases

Free energy can be used to study the stability of an alloy as the composition is varied over a wide range. Suppose that the free energy of the alloy in its homogeneous solid phase is given by the solid $U$-shaped curve of Fig. 11.13(a). We also suppose that in this phase the alloy is a primary solid solution throughout the concentration range. We shall show that the homogeneous-solution phase is more stable than any other structure.

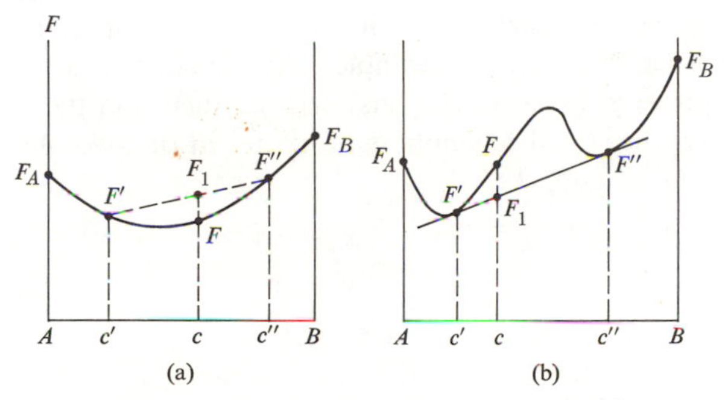


**Fig. 11.13** (a) A free-energy-versus-concentration diagram leading to a stable, homogeneous phase. (b) A free-energy-versus-concentration diagram leading to a stable phase mixture in the concentration range $c' < c < c''$.

Referring to Fig. 11.13, note that at composition $c$ the free energy for the homogeneous-solution phase is $F$. Compare this with another possibility, namely, that the system breaks up into two coexisting solid phases, one of concentration $c'$ and the other of concentration $c''$. A state of this type is called a *phase mixture*. It can be shown (see the problems at the end of this chapter) that the free energy for a phase mixture of components $c'$ and $c''$ varies with concentration along the straight line $F'F''$ as the concentration increases from $c'$ to $c''$. Therefore at concentration $c$ the free energy of the phase mixture is $F_1$. Since the free energy of the homogeneous phase $F$ is less than that of the phase mixture $F_1$, the former is the more stable structure. By choosing different $c'$ and $c''$, one can change the energy $F_1$, but, for the type of free-energy curve of Fig. 11.13(a), one cannot make it less than $F$. Therefore the homogeneous-single phase is the stable structure. Examples of systems with free-energy curves resembling this figure are the Ag-Au and Cu-Ni alloys.

The situation is quite different when the free-energy curve has the W-shape of Fig. 11.13(b). Again the homogeneous-solid phase is represented by the solid curve. The straight line $F'F''$ is the common tangent to this curve, and $c'$ and $c''$ are the concentrations corresponding to the tangential points. There are now three possibilities: If $c < c'$, the lowest free energy is given by the curve $F_AF'$, that is, the system is a primary solid solution rich in $A$. Similarly, if $c > c''$, the free energy is given by $F''F_B$, and the system is a solid solution rich in $B$. However, in the range $c' < c < c''$, the lowest free-energy curve is not given by the solid curve $F'F\,F''$, but rather by the straight line $F'F''$. In physical terms, this means that in this last concentration range the system breaks into a phase mixture whose components have concentrations $c'$ and $c''$, the former being richer in $A$ than the latter. The concentrations $c'$ and $c''$ mark the limits of primary solubility of the elements $A$ and $B$ into each other. When $0 < c < c'$, for example, the whole system is in a single homogeneous phase, in which $A$ and $B$ atoms are distributed randomly on the lattice sites. On the other hand, for the range $c' < c < c''$, the system breaks up into two phases, of concentrations $c'$ and $c''$, coexisting side by side (clusters of $c'$ and $c''$ intermingled with each other) in equilibrium, not unlike a liquid–solid phase mixture of ice in water, for example. As the concentration increases from $c'$ to $c''$, the $c''$ phase grows at the expense of the other, and the transformation is completed as $c$ reaches $c''$. The amounts of matter in the two phases are given by the lever formula

$$\frac{x''}{x'} = \frac{c - c'}{c'' - c},$$

which we can derive the same way we did (11.16).

The justification for the assertion that the free energy in the range $c' < c < c''$ is given by the common tangent straight line $F'F''$ follows from an argument used to establish a similar significance for $F'F''$ in Fig. 11.13(a). Since at every $c$ in this range $F_1 < F'$ (Fig. 11.13b), it follows that the phase-mixture structure is the more stable one in the range $c' < c < c''$.

Most binary metallic alloys exhibit the behavior shown in Fig. 11.13(b), including, for example, the Cu-Ag and Cd-Bi alloys.

**Free energy of a substitutional alloy: microscopic model**

Let us calculate the free energy of a substitutional solid solution, using a simple atomic model, and compare the results with the free-energy curves we have discussed. The free energy for the solution is

$$F = E - TS = E_0 + \int_0^T C_p\,dT - T\int_0^T C_p/T\,dT + NkT[c\log c + (1-c)\log(1-c)], \tag{11.24}$$

where the various terms of energy and entropy mean the following: $E_0$ is the energy at absolute zero, the first integral is the increase in thermal energy, the second integral results from the thermal entropy [see (11.18)], and the last term is the mixing entropy (11.22). We note that if the two types of atoms are not dissimilar, then the integral terms are insensitive to compositional changes, and may be ignored if we are interested only in the shape of the curve $F$ versus $c$.

We can calculate $E_0$ as follows: If we call the energy of an $A$–$A$ bond $V_{AA}$, then the total energy of the $A$–$A$ bonds in the whole crystal is

$$\tfrac{1}{2}\cdot N(1-c)\,Z\,(1-c)\,V_{AA} = \tfrac{1}{2}N\,Z\,(1-c)^2\,V_{AA},$$

where $Z$ is the coordination number of the crystal structure. We can arrive at this expression by noting that $N(1 - c)$ is the total number of $A$ atoms, while $Z(1 - c)$ is the number of $A$ atoms surrounding an $A$ site, on the average, provided the atoms are distributed randomly. The factor $\frac{1}{2}$ is necessary because otherwise each bond would have been doubly counted. The energies of the $B$–$B$ and $A$–$B$ bonds can be similarly calculated, and the result for the total internal energy $E_0$ is therefore

$$E_0 = \tfrac{1}{2}NZ(1 - c)^2\, V_{AA} + \tfrac{1}{2}NZ\, c^2\, V_{BB} + \tfrac{1}{2}NZ\, c(1 - c)\, V_{AB}, \qquad (11.25)$$

where $V_{BB}$ and $V_{AB}$ are the energies for a $B$–$B$ and $A$–$B$ bond, respectively. This equation may be recast in the following useful form:

$$E_0 = \tfrac{1}{2}NZ\left[c\, V_{AA} + (1 - c)\, V_{BB} + 2c(1 - c)\left(V_{AB} - \frac{V_{AA} + V_{BB}}{2}\right)\right]. \qquad (11.26)$$

This expression now has to be inserted in (11.24), and the result plotted versus $c$. You can verify that only curves of the types shown in Fig. 11.13 are obtained. More specifically, the U-shaped curve of Fig. 11.13(a) is obtained when $V_{AB} \leqslant (V_{AA} + V_{BB})/2$, while the type shown in Fig. 11.13(b) is obtained when $V_{AB} > (V_{AA} + V_{BB})/2$. Thus the latter type holds true when the attraction between the different atoms is less strong than the average attraction between similar atoms.† From this point of view, you can see why, in this case, like atoms prefer to segregate into two separate phases, as we discovered previously. There is a range of primary solubility near the endpoints because the mixing entropy there increases very rapidly, forcing a certain amount of solubility, limited though it may be.

**The phase diagram and free energy**

The concept of free energy leads readily to the phase diagram (Fig. 11.10) for a binary alloy. This can be seen from Fig. 11.14, in which we plot the free energy for a

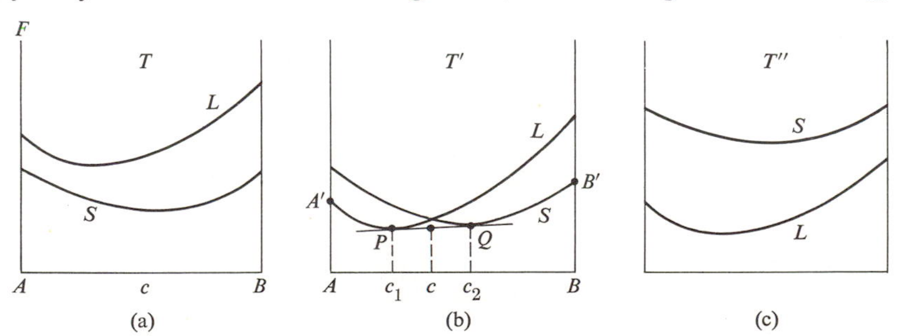


**Fig. 11.14** (a) Free energies of solid and liquid phases of an alloy below its melting range. (b) Free energies of solid and liquid phases within the melting range of the alloy. (c) Free energies of the two phases above melting point.

† Recall that the potential terms $V_{AA}, V_{BB}, V_{AB}$ are all negative, because they represent attractive forces.

completely miscible binary alloy at three progressively higher temperatures $T$, $T'$, and $T''$ ($T < T' < T''$). The diagrams show the energy for both the solid and liquid phases at each temperature. At temperature $T$ the solid's curve lies entirely below that of the liquid curve, and therefore the alloy is in the solid phase. However, at some higher temperature $T'$, the liquid's curve crosses the solid curve (Fig. 11.14b). In this situation, the structure of the system depends on the average composition $c$. Once we know $c$, we can infer the structure by using the rules developed previously for minimizing the free energy. Thus in Fig. 11.14(b) we draw the common tangent $PQ$ which determines the concentrations $c_1$ and $c_2$. Then, if $c < c_1$, the minimum free energy is represented by the curve $A'P$, corresponding to a solid solution rich in $A$. For $c > c_2$, the free energy is given by $QB$, corresponding to a liquid solution rich in $B$. However, in the intermediate-composition range $c_1 < c < c_2$, the free energy is given by the straight line $PQ$, representing a mixture

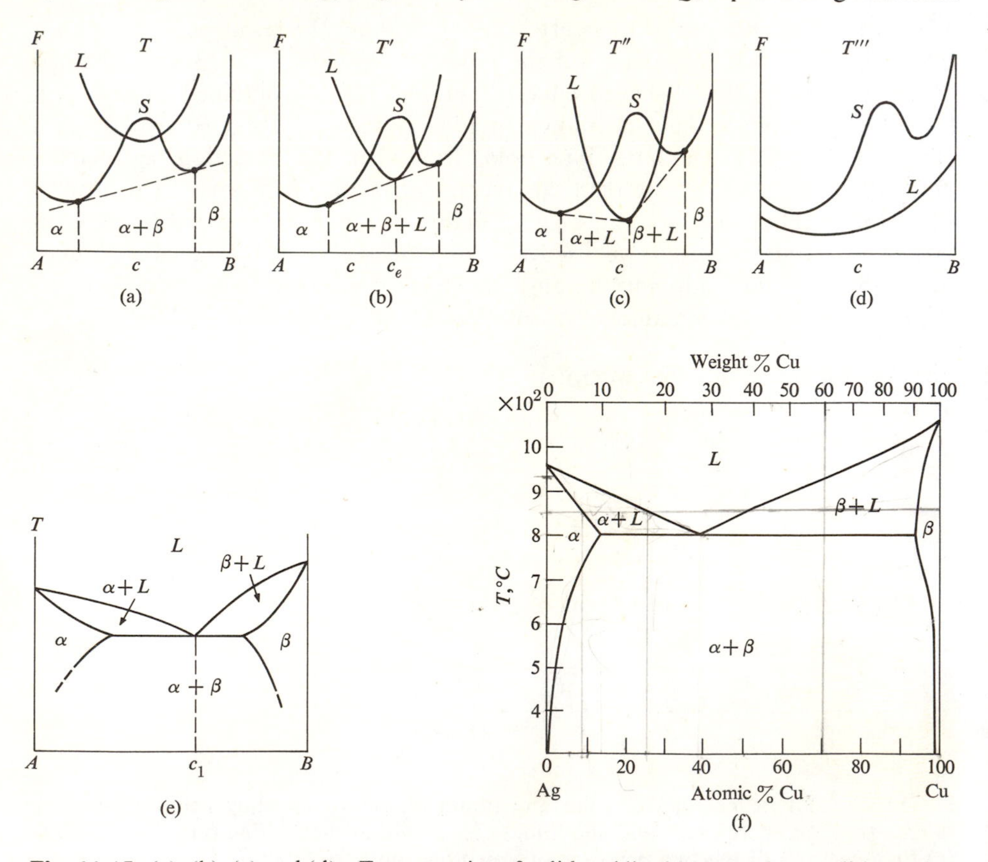


**Fig. 11.15** (a), (b), (c) and (d): Free energies of solid and liquid phases for a solid solution described by Fig. 11.13(b), at increasingly higher temperatures ($T < T' < T'' < T'''$). (e) Phase diagram for the system. (f) Phase diagram of Ag-Cu system. (After Cottrell, 1948)

of solid phase of composition $c_1$ and liquid phase of composition $c_2$. The amount of liquid phase versus solid phase can be found by the appropriate lever formula. The situation at the temperature $T'$ corresponds to that at $T_2$ in the phase diagram in Fig. 11.10, and the concentrations $c_1$, $c_2$, used there are the same as the ones we find here. The reader can see, after a little reflection, that if the free-energy curves for the liquid and solid phases are given at all temperatures at which they cross each other—i.e., near $T'$—then he can, in fact, plot the solidus and liquidus line of Fig. 11.10 and determine the phase diagram for the alloy. We can also see why the melting process of an alloy extends over a range of temperatures. The reason is that the crossing and uncrossing of the solid and liquid curves in Fig. 11.14 is accomplished over a finite range of temperature.

At the temperature $T''$ the alloy is completely melted, because the liquid curve lies entirely below that of the solid.

It is now useful to infer the phase diagram for a system whose free energy, for the solid solution, is given by Fig. 11.13(b). Figure 11.15 plots the free energies for the solid and liquid phases at four different temperatures, $T$, $T'$, $T''$, and $T'''$, near the melting range at which $T < T' < T'' < T'''$. In Fig. 11.15(a) the system is either a primary solution of phase $\alpha$, rich in $A$, or a solution in phase $\beta$, rich in $B$, or a phase mixture of $\alpha$ and $\beta$, depending on the concentration as indicated above. No liquid phase appears because the free energy of the liquid phase is too high. At a higher temperature $T'$, shown in Fig. 11.15(b), a situation obtains in which the tangents of the $\alpha$ and $\beta$ phases also touch the liquid curve, and this gives rise to several possibilities, depending on the concentration. A particularly interesting one occurs when the composition is equal to $c_e$; here the three phases—$\alpha$, $\beta$, and the liquid phase—coexist. Such a composition is called the *eutectic composition*, and the corresponding temperature is called the *eutectic temperature*. At still higher temperatures, the curves appear as shown in Figs. 11.15(c) and (d). The phase diagram resulting from this situation is shown in Fig. 11.15(e). A characteristic feature of such a phase diagram is that elements $A$ and $B$ show only limited solid solubility in each other. They tend to segregate into phase mixtures or turn into a liquid phase. A well-known example of this type of system is the Cu-Ag alloy shown in Fig. 11.15(f).

### Intermediate phases

In our discussion of solid solutions, we have so far assumed that the solution has the same crystal structure throughout the entire composition range. However, some other solid phases may have a low free energy at intermediate compositions. This possibility is illustrated in Fig. 11.16(a) for three different solid phases, $\alpha$, $\beta$, and $\gamma$. Using the rules developed for minimizing free energy, one can determine the possible phase structure at various values of composition.

When the temperature is raised, the positions of the various intermediate phases may change relative to each other. Eventually, when the temperature is sufficiently high, melting starts. The phase diagram for this system can be inferred from observing the evolution of the free-energy diagram with temperature, and using the rules of minimization. The phase diagram of the Mg-Pb alloy of Fig. 11.16(b) is a typical result. This diagram is more complex than Fig. 11.15(e), and in fact may be viewed as a set of two eutectic diagrams joined together. In most practical alloys in which there are several elements and several intermediate phases, the phase diagram is very complex indeed.

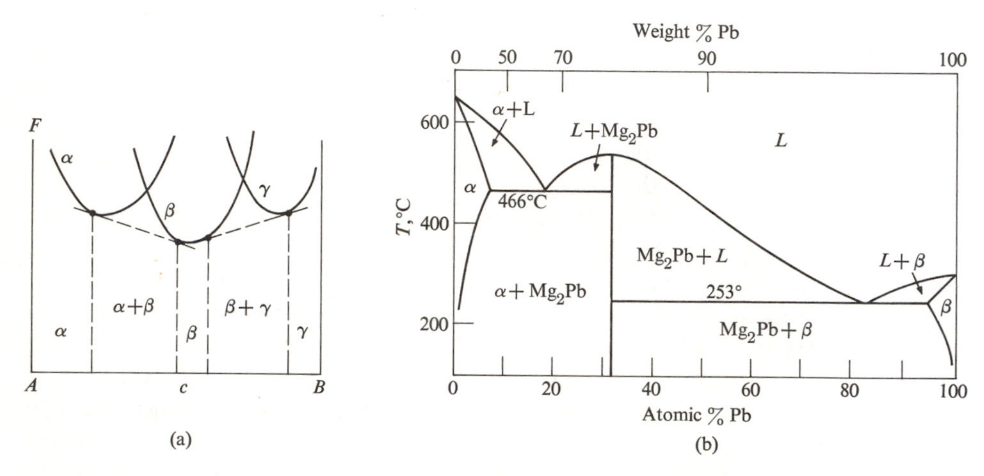


**Fig. 11.16** (a) Intermediate phases of a solid solution. (b) Intermediate phases of the Mg-Pb system. (After Wert, 1970)

### Electron concentration and the zone theory of alloy phases

In our discussion of structural properties of alloys, the conduction electrons have so far played no role. The energy of these electrons should, in fact, be added to the internal energy $E_0$ to arrive at the *total* internal energy, but in an alloy in which the two elements are of the same valence, such as the Ag-Au or Cu-Ag systems, the electron/atom ratio remains unchanged as the composition is varied, and consequently the energy of conduction electrons remains essentially unaffected throughout the composition range. This is why we were justified in omitting this energy term. However, in alloys involving Cu, Ag, or Au with other metals of higher valence, the electron/atom ratio changes with composition, and this leads to interesting effects on the crystal structure. It was first observed by Hume-Rothery that the $\alpha$-phase (fcc) of such alloys as CuZn, CuAl and AgMg becomes unstable when the electron/atom ratio approaches the value 1.4, and a complete transformation to the $\beta$-phase (bcc) takes place when the ratio is near 1.5.

Since the critical factor in this type of transformation is the concentration of electrons, we look for an explanation of the above transition in terms of the band structure, as described in Chapter 6. Figure 11.17 shows the density of states for the free-electron model, as well as for the fcc and bcc structures. As we stressed in

Chapter 5, the density-of-states function deviates from the free-electron value appreciably only when the Fermi surface begins to touch the Brillouin zone, which occurs at the energies $E'$ and $E''$ for the fcc and bcc structures, respectively.

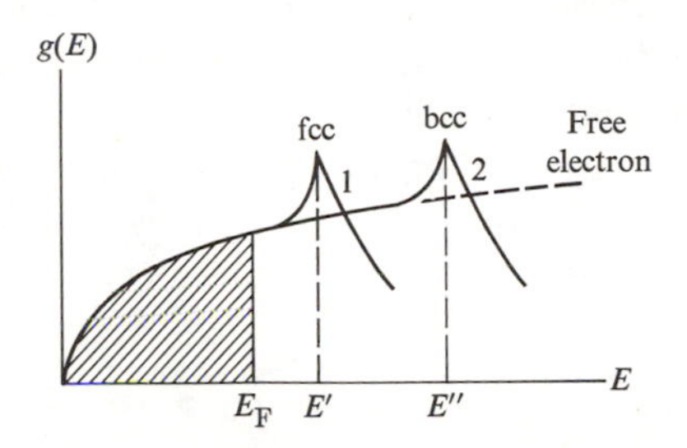


**Fig. 11.17** Density of states $g(E)$ versus energy $E$ for an fcc structure and a bcc structure. Dashed line represents free-electron model; cross-hatched area represents region occupied by electrons.

As the concentration of electrons increases, starting in the fcc structure, $E_F$ also increases, until the energy $E'$ is reached. Beyond this point, any increase in electrons would lead to a rapid increase in $E_F$ [because $g(E)$ of curve 1 decreases rapidly], and hence to a rapid increase in the energy of the electrons. To obviate this increase in energy, the system begins to transform partially into the bcc phase and thus lessens its energy, the transformation being completed at $E''$. We leave it to you to show that, according to the free-electron model (Section 4.7), the points $E'$ and $E''$ occur at electron/atom ratios of 1.36 and 1.48, respectively, in close agreement with the observed values.

## 11.6 DISLOCATIONS AND THE MECHANICAL STRENGTH OF METALS

Metals have distinctive mechanical properties which make them highly useful in industrial applications. They can be stretched into wire, hammered into sheets, molded, and bent. They can also, with proper treatment, be made to withstand a great amount of stress, e.g., the steel beams used in building construction. Also metals, when pure, are very soft and can be readily deformed. For instance, pure iron itself is a rather soft material, and it is only after it has been alloyed with carbon and other metals that it acquires enough strength for industrial uses. Mechanical strength—and its relation to the impurity contents of a given substance—is of great concern to the engineer and metallurgist. It is also of interest to the physicist because, as it turns out, the explanation requires explicit consideration of the movement of individual atoms in the crystal lattice. The concept of *dislocation* lies at the very heart of this discussion.

A dislocation is a linear defect in a crystal, i.e., it involves a large number of atoms arranged along a line. There are basically two different types, an *edge dislocation* and a *screw dislocation*. Figure 11.18 illustrates an edge dislocation: An extra half-plane $AB$ of atoms has been embedded into the upper half of the crystal, as shown in cross section. The half-plane terminates at the point $A$, which, in three dimensions, represents a linear array of atoms normal to the plane of the paper, and this array is the dislocation. Typically it extends over many tens of angstroms. The region in the neighborhood of the dislocation experiences a noticeable distortion relative to the normal crystalline arrangement. The upper region, in which the half-plane is introduced, is compressed because the atoms are squeezed against each other, while the lower region of the crystal is somewhat expanded. Far away from the dislocation the crystal regains its regularity.

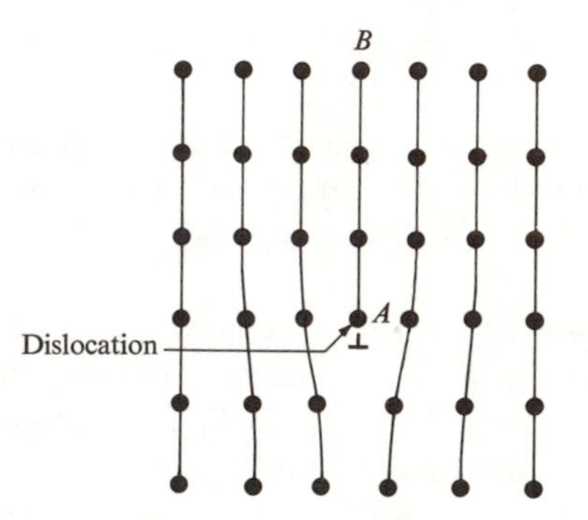


**Fig. 11.18** An edge dislocation. The dislocation is a line of atoms perpendicular to the paper at point $A$.

The formation energy for a dislocation—i.e., the associated elastic-strain energy—is usually about 10 eV per atomic length, which is considerable when compared with thermal energies at, say, room temperature. Consequently dislocations are not created thermally, but by mechanical treatment such as bending or hammering.

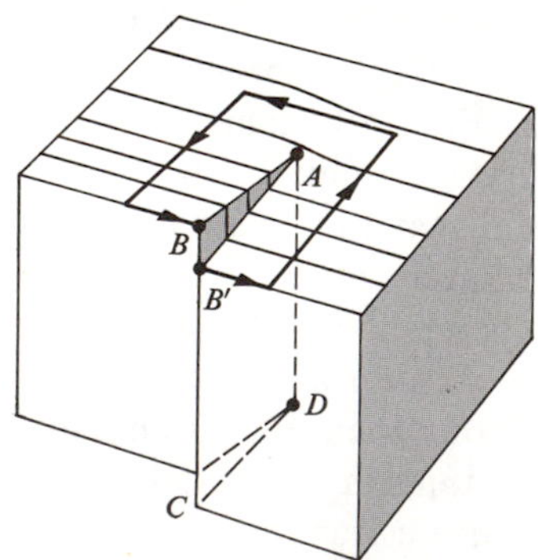


**Fig. 11.19** A screw dislocation. The dislocation is represented by line $AD$. Lines on top represent vertical atomic planes. Shaded area $ABB'$ indicates region of slippage. (Points $B$ and $B'$ were coincident before the dislocation was created.)

A screw dislocation is illustrated in Fig. 11.19. This may be created, one may imagine, as a planar cut $ABCD$ made in the crystal. The left side is then slipped up past the right side. The line $AD$ is the dislocation, and it lies at the end of the step $BAB'$ created by the slip. The reason for referring to this as a screw dislocation is that if one moves in the atomic plane around the dislocation, as indicated by the arrows, one finds that the plane actually spirals. The region of a screw dislocation is also one of considerable strain due to the slippage, but it is a shear-type strain with no attendant change in volume, unlike an edge dislocation, which involves considerable dilatation. The energy of formation of a screw dislocation has about the same value as the energy of formation of an edge dislocation, so these dislocations must also be created by nonthermal methods.

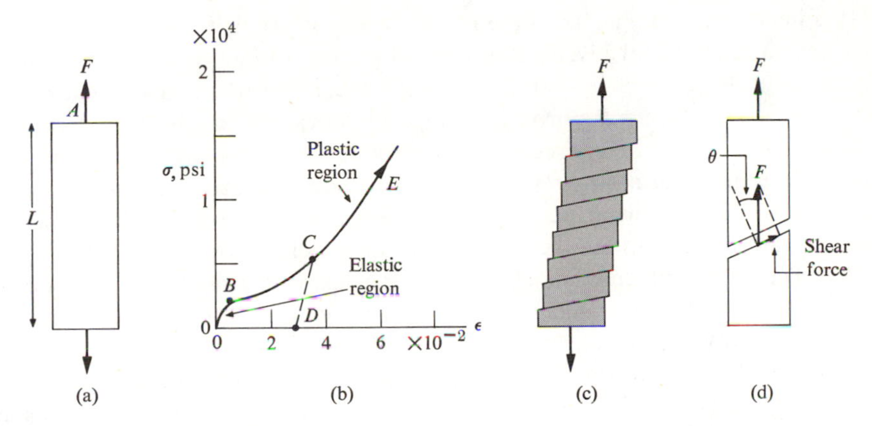


**Fig. 11.20** (a) Application of a stress to a metallic bar. (b) Stress–strain curve for a Cu single crystal (nearly pure) at room temperature. [After H. Birnbaum, quoted in Wert (1970)] (c) Microscopic view of actual strain process, showing slippage of the atomic planes past each other. (d) Calculation of shear force along slip plane.

Let us now try to relate this concept of dislocation to the mechanical strength of metals. A force $F$ is applied to a metallic sample, usually rod-shaped, of length $L$ (Fig. 11.20a), and as the force is gradually increased the elongation is measured. When the elongation is small, the sample returns to its original shape once the load force is removed. This *elastic* property is shared by all solids. However, if the stretching process is continued, a point is reached beyond which the deformation becomes permanent, even when the load is removed. This is called *plastic* deformation. Instead of using force and elongation to discuss this phenomenon, we use stress $\sigma$ and strain $\epsilon$, which are defined, respectively, as the force per unit area and the fractional increase in length,

$$\sigma = \frac{F}{A}, \qquad \epsilon = \frac{\Delta L}{L}.$$

The advantage of using $\sigma$ and $\epsilon$ instead of $F$ and $L$ is that they are independent of the shape of the specimen.† Figure 11.20(b) shows the observed stress–strain curve for a sample of copper, where the elastic and plastic regions are clearly indicated. In the elastic region the strain is proportional to the stress (Hooke's law), and the proportionality constant

$$Y = \frac{\sigma}{\epsilon} \tag{11.27}$$

is known as *Young's modulus*, as you will recall from basic physics. Of course, Hooke's law is not obeyed in the plastic region. The line $DCE$ in Fig. 11.20(b) indicates the stress–strain curve for a sample which had already suffered some plastic deformation.

It is important to understand the phenomenon of plasticity, as it is a common occurrence in pure metals even at very small strain. In fact, pure metals start to deform plastically at much less strain than expected, a fact which gives some clue to their internal structure. Returning to Fig. 11.20(a), one might expect, at first thought, that strain is a consequence of atomic planes being pulled apart by applied force, and that a larger strain (larger atomic separation) requires a larger stress. This is indeed what occurs in the elastic region. In the plastic region, however, various regions of the crystal appear to slip against each other (Fig. 11.20c). The crystalline units undergoing slippage are called *slip bands*, and it is the sliding of these bands past each other that is responsible for plastic elongation. It is now clear why our metallic rod does not recover its original length: because the bands do not slip back to their original positions once the load is removed.

Before we talk about how the slippage takes place microscopically, we may note that it is caused by the shear component of the applied stress. Imagine a plane cut into the sample (Fig. 11.20d). The applied force $F$ can then be decomposed into two forces, one parallel and the other normal to the plane. The parallel force is a shear force, and has a value $F \sin \theta$, where $\theta$ is the angle between $F$ and the normal to the plane. The shear stress $\tau$ in this plane is given by

$$\tau = \frac{F \sin \theta}{A/\cos \theta} = \sigma \sin \theta \cos \theta = \frac{\sigma}{2} \sin 2\theta, \tag{11.28}$$

where we used the fact that the sliced surface has an area of $A/\cos \theta$. The maximum value of $\tau$, which is $\sigma/2$, occurs at $\theta = 45°$. Slippage along a plane occurs when $\tau$ along that plane exceeds a certain critical value. In an isotropic material the slippage should therefore take place in a plane inclined at 45° relative to the applied force. Crystals are not isotropic, however, and certain planes having lower critical stresses than others act as slippage planes; these planes usually have high atomic

† We used $\sigma$ and $\epsilon$ in earlier chapters to denote electrical quantities; now we are using $\sigma$ and $\epsilon$ to denote mechanical quantities. But this should cause no confusion, because here we are discussing mechanical properties only.

concentration. For instance, the (111) planes in an fcc lattice show the least resistance to shear, and are therefore the planes along which slippage takes place. In these "easy-slip" planes, some directions are more favorable than others, and act as easy-slip directions. These directions also have large concentrations of atoms, e.g., the [110] direction in the fcc lattice.

Now that we are convinced that the slip process does occur, the question is just how the slip takes place on a microscopic scale. An obvious model is that one whole plane of atoms slips past a neighboring one—along the slip plane. But such a model cannot be correct, because it would lead to critical stress larger than the observed value by several orders of magnitude.

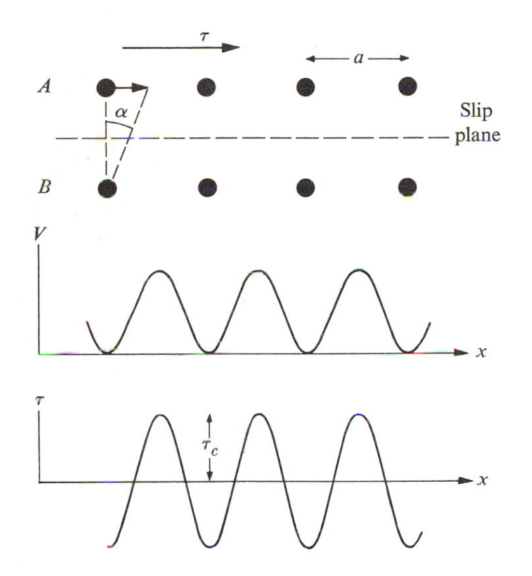


**Fig. 11.21** Rigid model of slip motion. Top: the slippage process. Middle: potential energy versus slip displacement. Bottom: shear stress versus slip displacement.

For example, Fig. 11.21 shows a row of atoms $A$ slipping past another row $B$, and also shows the potential that an atom of $A$ "feels" as it moves to the right. Since the shear stress at any position is proportional to the derivative of the potential, the curve of the shear versus position may have the shape shown in the figure. This can be represented approximately by the sinusoidal expression

$$\tau = \tau_c \sin\left(2\pi \frac{x}{a}\right), \tag{11.29}$$

where $\tau_c$ is the critical stress. When $\tau < \tau_c$ the atoms of $A$ are displaced from equilibrium only slightly, and return to this state as soon as $\tau$ is removed. However, for $\tau > \tau_c$, the atoms "roll" over the potential hill, and therefore never return to their original positions even if the stress is removed. The value of $\tau_c$ can be estimated by comparing the results of (11.29) for small displacement with those of elastic theory, which are known to hold under these conditions of small displacement. For small displacement $x$, Eq. (11.29) can be written approximately as

$$\tau = 2\pi\,\tau_c \frac{x}{a} = 2\pi\,\tau_c \alpha, \tag{11.30}$$

$\alpha \simeq x/a$ being the shear angle, as shown in Fig. 11.21. In the theory of elasticity the ratio $\tau/\alpha$ is the shear modulus, or rigidity. Denoting this by $\mu$, and using (11.30), one arrives at

$$\tau_c = \frac{\mu}{2\pi}, \tag{11.31}$$

relating the critical stress $\tau_c$ to the elastic shear modulus. A typical value for $\mu$ in metals is about $10^{11}$ N/m$^2$, yielding $\tau_c \simeq 10^{10}$ N/m$^2$. Observed values for $\tau_c$ in pure crystals are, however, much smaller than this, typically about $10^6$ N/m$^2$, four orders of magnitude less than the predicted value. In other words, the observed limit of elastic strain is much smaller than the model of Fig. 11.21 suggests. Instead of an $\alpha$-value of about 0.1 radian, or 6°, the observed angle is about $10^{-5}$, or half a millidegree. In metals, this surprising softness, or great tendency toward plastic flow, needs explanation; here the concept of dislocation comes to the rescue.

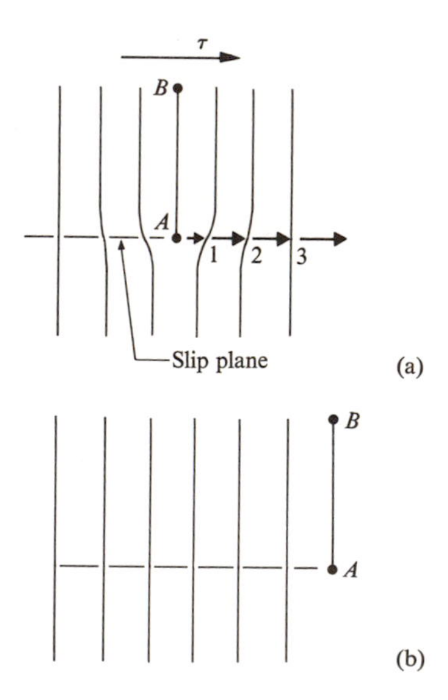


**Fig. 11.22** The real model of slip motion: (a) Arrows indicate successive displacements of an edge dislocation under the influence of an external shear stress. (b) Final shape of crystal after slip motion has taken place.

Figure 11.22 shows an edge dislocation. When a shear stress $\tau$ is applied as shown, one can think of plastic deformation as a result of a consecutive motion of dislocation $A$ to positions 1, 2, 3, etc., until it reaches the surface, as shown in Fig. 11.22(b). The net result is the same as if the upper half of the crystal had been rigidly displaced by one interatomic distance past the lower half. However, the stress required to trigger the motion of the dislocation is much less than that needed in a rigid displacement because the number of atoms involved in the motion, in each step, is much smaller.† Also, since the dislocation region is already strained, the atoms are not in a highly stable equilibrium situation to begin with, so it requires only a small stress to accomplish the motion, and consequently the plastic deformation.

We conclude that the softness of pure metal is due essentially to the presence of dislocations, which lessen the resistance of the metal-to-plastic flow, as the dislocations are relatively free to move.

The concept of dislocations helps us to understand other phenomena of metallic behavior. For example, impure metals and alloys are usually much stronger than pure metals because the impurities and defects are usually attracted to a dislocation, since they can accommodate themselves more easily there. But a dislocation which is "loaded" with impurities is no longer able to move readily because it has to drag all the impurities with it. In effect the dislocation is "pinned down" by the impurities. The crystal therefore shows greater resistance to plastic deformation, a phenomenon known as *impurity hardening*.

Another interesting effect is that a metal becomes harder after being strained (the slope of the curve in Fig. 11.20 increases in the plastic region), which is known as *work hardening*. For instance, if one attempts to bend a wire which has already been bent at one point, the new bend forms far from the original bend. The explanation is that a plastically deformed sample already contains an unusually large number of dislocations created by the original strain; further dislocation motion is impeded because the dislocations interfere with each other, being oriented at different angles.

Many other effects confirm the existence of dislocations. For further details refer to the literature cited at the end of the chapter.

Direct observation of dislocations can be accomplished by chemical etching of the crystal surface, which, after magnification, reveals those dislocations that intersect the surface, as shown in Fig. 11.23. Other dislocations which lie entirely within the crystal are more difficult to observe, but even these can be seen in some transparent substances, such as silver bromide.

We have said that under normal circumstances the mechanical strength of pure metals is determined not by interatomic forces but by structural defects. What is the mechanical strength of a dislocation-free metal, i.e., a structurally perfect metal?

† As a familiar analogy, when you try to smooth a ripple in a rug, it is easier to push the ripple gradually than to cause the rug to slide by pulling at the edge.

**Fig. 11.23** Spiral growth pattern in $CdI_2$. (Photograph by K. A. Jackson)

It is hard to grow such a specimen, especially because dislocations play an important role in the process of crystal growth. As the melt begins to freeze into various crystalline nuclei, the new atoms attaching themselves to the nuclei already formed prefer to settle at a dislocation step ($ABB'$ in Fig. 11.19) because there are more atoms there to attract them than there are on a flat surface. Thus, as a crystalline nucleus grows, a dislocation fills up, and spirals in the process. We can see this type of spiraling in many crystals after the surface has been polished and etched. Notwithstanding the difficulties in growing dislocation-free metals, it has recently become possible to grow such crystals. Whiskers of some metals such as tin have been obtained, either intentionally or accidentally, which were free of any dislocations. Experiments on these samples have revealed them to be exceptionally strong, with a strength closely approaching the theoretical value (11.31), determined by the cohesive energy due to interatomic forces.

## 11.7 IONIC CONDUCTIVITY

Ionic crystals such as alkali and silver halides are excellent insulators because their wide band gaps inhibit electronic excitation and conduction. Nevertheless, the electrical conductivity in these crystals—small as it is—does not vanish entirely, and an appreciable value, for example, 0.1 $(\Omega\ \mathrm{cm})^{-1}$, can be measured at high temperatures. The conduction here takes place not via the transport of electrons or holes, but by the movement of the lattice ions themselves, in a manner similar to the electrolytic current which flows in water containing a solution of, say, NaCl. In fact, metal is deposited at the cathode end of the sample. When one weighs the amount deposited, one finds that Faraday's law is satisfied, as in the familiar electrolytic solutions, confirming that ions are indeed transported in this situation.

Ionic conductivity depends strongly on temperature. In most cases it obeys the simple relation

$$\sigma = \sigma_0 e^{-(E_1/kT)}, \tag{11.32}$$

where $E_1$ is a constant energy parameter and $\sigma_0$ is a slowly varying function of temperature. Thus $\sigma$ increases rapidly with temperature, and reaches appreciable value near the melting point of the solid.

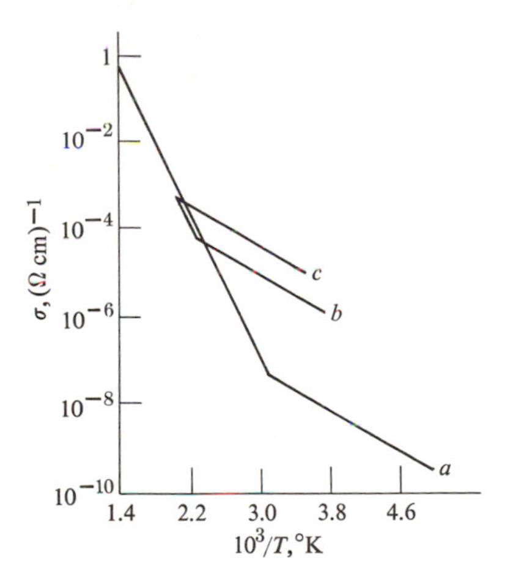


**Fig. 11.24** Ionic conductivity $\sigma$ versus inverse temperature for AgBr. Curve (*a*) is for a high-purity sample; curves (*b*) and (*c*) are for a crystal containing 0.028 and 0.12 mole% of $CdBr_2$. (After Brown)

It is found experimentally that two regions may be distinguished, the high-temperature (or *intrinsic*) region and the low-temperature *extrinsic* region (Fig. 11.24). In both regions the exponential relation (11.32) is obeyed, but in the high-temperature case the energy $E_1$ (the slope in the figure) is much larger, and the factor $\sigma_0$ is several orders of magnitude larger than in the low-temperature region. The factor $\sigma_0$ in the low-temperature region, unlike $\sigma_0$ in the intrinsic region, is also highly dependent on the impurity contents in the crystal. The parameters for NaCl and AgBr in the intrinsic region are, respectively, $\sigma_0 = 3.5 \times 10^6\,(\Omega\text{cm})^{-1}$, $E_1 = 1.86\,\text{eV}$, and $\sigma_0 = 1.8 \times 10^5\,(\Omega\,\text{cm})^{-1}$, $E_1 = 0.79\,\text{eV}$.

The form of (11.32) suggests that the energy $E_1$ is an activation energy for the movement or hopping of the ions. Clearly such an ionic movement is not possible in a perfect crystal. The presence of defects, especially vacancies, is essential to the occurrence of this phenomenon. The activation energy $E_1$ must therefore be related to the formation and activation energies of the vacancy, as discussed in Sections 11.3 and 11.4.

Think of conduction taking place by the ion jumping from one vacancy to another, or, equivalently, by the motion of the vacancy in the opposite direction. Employing this model, we may give a simplified derivation of (11.32) by using the Einstein relation between mobility and the diffusion coefficient, Eq. (6.81), at least in the intrinsic region. Thus ionic conductivity may be written as

$$\sigma = N_v\, e\, \mu_v = N_v\, e \frac{k\,T\,D_v}{e}$$

$$= k\,T\,N_v\,D_v,$$

where $N_v$ and $D_v$ are the concentration and diffusion coefficients of the vacancy, respectively. Substituting for $N_v$ from (11.1), and for $D_v$ from (11.13), one finds

$$\sigma = (k\,T\,N\,D_0)\, e^{-(E_v+Q)/kT}, \tag{11.33}$$

which is the form of (11.32) with $E_1 = E_v + Q$.

For comparison with actual experiments, this argument must be extended to account for the presence and transport of both positive and negative ions in the crystal. The treatment must also take special notice of the type of vacancy, whether of the Schottky or Frenkel type (Section 11.2). In Frenkel vacancies, interstitial ions are also present, and contribute to conduction, adding further to the conductivity. This explains why silver halides, whose defects are primarily of the Frenkel type, generally have higher conductivity than alkali halides, whose defects are primarily of the Schottky type.

The behavior in the extrinsic region is more complicated, and depends on a variety of additional new factors. Thus the conductivity could be appreciable if the sample is quenched from high temperature by rapid cooling, so that the substance may contain a large number of vacancies even at low temperature (the vacancies are essentially "frozen in," as discussed in Section 11.3). This reduces the activation energy, as can readily be seen from the above discussion, since the vacancies are present, and need not be generated thermally.

## 11.8 THE PHOTOGRAPHIC PROCESS

Because modern photographic techniques are such an essential part of present-day technology, this section is devoted to some basic aspects of this subject. The active

**Fig. 11.25** Photomicrograph of AgBr grains in emulsion, showing photolysis of silver after exposure. (After Webb)

component of a negative emulsion plate consists of a large number of silver halide microcrystals suspended in a gelatin solution. When the plate is exposed to light, the silver ions in the crystal, usually AgBr or AgBr-AgI, are chemically reduced by being supplied with the necessary electrons. The resulting neutral silver atoms cluster into dark colloidal specks which are visible under a microscope (Fig. 11.25). It is these specks, several million atoms in size, which form the image of the photographed object. The photographic plate may be considered as a light detector, not unlike a semiconductor detector (Section 7.8). Properly made films can be extremely sensitive, recording radiation of only a few quanta of light.

Viewed chemically, the photographic process is essentially a photolysis of, say, AgBr, into Ag and Br atoms. But an understanding of the process also requires some solid-state concepts. Before discussing the physical processes involved, let us distinguish between the following two situations: First, the level of light exposure is so high that the dark metallic silver specks may be seen; this is the *printout* effect. Second, in ordinary photography, the specks are far too small to be seen except after the film has been developed. In this case we also speak of the *latent image*. Let us take the printout effect first; our treatment closely follows Mott (1948).

We have said that the structure of an emulsion plate consists of a large number of AgBr microcrystals, or grains, suspended in a gelatin solution. These grains have a flat shape, about $0.1\,\mu$ along the surface, and about $0.01\,\mu$ thick. Depending on the quality and type of film, the crystallites may be smaller or larger, and there may even be a distribution of sizes. There are two basic experimental facts which demand explanation: First, the quantum efficiency is nearly equal to unity, except at too great exposures. By this we mean that for every photon absorbed, a silver atom is added to one of the silver specks. Second, these silver atoms are finally localized on a few small specks on the surfaces of the crystallites. The theory must therefore explain how a photon absorbed at one point can give rise to a silver atom located a considerable distance away.

Gurney and Mott in 1938 advanced a very useful and simple theory which explains most of the observed effects. They assumed that when a photon is absorbed, an electron is excited to the conduction band of the AgBr. Being light and mobile, the electron can travel a considerable distance in a very short time, and thus if it encounters an already formed silver speck, it adheres to this speck. (How the speck formed originally will be discussed later in this section.) The speck, now negatively charged, sets up an electric field in the surrounding region. Responding to this field, silver ions in the crystal move toward the speck to neutralize the charge, and this results in additional silver atoms on the speck. The ionic conductivity of the silver ions is thus essential to the photographic process, and this explains the enhancing effect of high temperature. The higher the temperature, the greater the ionic conductivity, and the faster the rate of growth of the silver speck. Illumination is also essential for the continual growth, for so long as new electrons are newly excited and transported to the specks, more thermally generated Ag ions are attracted to neutralize the additional electronic charge.

An important assumption of the Gurney–Mott theory is that holes, created simultaneously with the electrons, are trapped, and thus can't move to neutralize (and thereby inhibit the growth of) the specks. Just how and why the holes are trapped is not clear, and this weakness of the theory has been criticized by some workers. The holes are not trapped indefinitely, however. Presumably they eventually find their way to the surface and recombine with some halogen ions, after which the resulting atoms escape from the surface of the halide crystal.

It is also plausible to expect, on the basis of this model, that the specks do not grow within the body of the grain, which would create a considerable strain, but on the surface, or possibly on cracks within the grain.

The model also suggests a simple treatment for the growth kinetic of the specks, and its relation to illumination intensity and temperature. Suppose that the speck contains $n$ unneutralized electrons. The electric field set up by this charge is $\mathscr{E} = ne/4\pi\epsilon r^2$. If the neutralizing ions flow from the inner side of the speck, the total ionic current is $2\pi r^2 \sigma\mathscr{E} = ne\sigma/2\epsilon$. Thus the number of ions flowing in per second is $p = n\sigma/2\epsilon$. In the steady state, this number $p$ must also be equal to the number of electrons created per second from photon absorption.

The number of electrons on the speck is also governed by another consideration: $n$ must not be so large as to keep further electrons from adhering to the speck. The requirement can be written semi-quantitatively as

$$\frac{n\, e^2}{4\pi\, \epsilon R} < kT,$$

where the left side is the coulomb repulsion energy, and the right the random thermal kinetic energy of the electron. If this inequality is satisfied, the new electron has enough energy to overcome the repulsion, with a certain probability of being captured by the speck.

Substituting this value for $n$, we find that the limiting value for the number of electrons per second which may lead to future silver atoms is

$$p = \frac{2\pi\, kTR\,\sigma}{e^2}. \tag{11.34}$$

If there are $N$ specks per grain, each of radius $R$, the number of photons which leads to silver atoms is

$$p = \frac{2\pi\, kTN\, R\,\sigma}{\sigma}.$$

Figure 11.26 shows the dependence of the rate of growth intensity on illumination $I$ and on temperature. Thus the rate of growth saturates at a value of $I \sim P$, and the saturation value rises with temperature, in agreement with experiment. Taking $N = 10$ and $R = 50\mu$, Mott found that in AgBr ($\epsilon = 12\,\epsilon_0$), $p = 2$ ($\sigma = 10^{-13}$ $\text{cm}^{-1}\,\Omega^{-1}$) at $-100$°C and $p = 2 \times 10^5$ ($\sigma = 10^{-8}\,\text{cm}^{-1}\,\Omega^{-1}$) at 20°C.

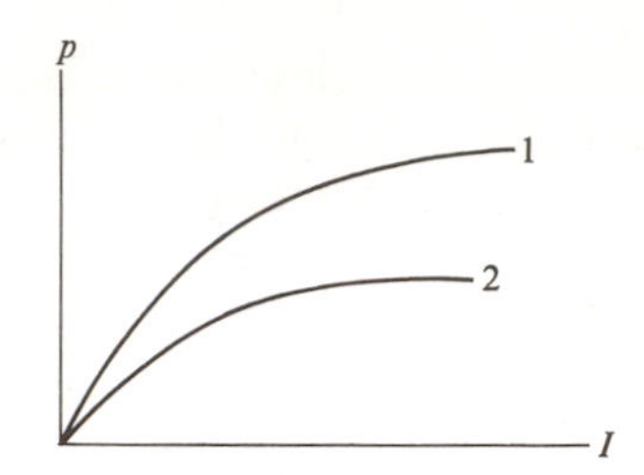


**Fig. 11.26** Rate of growth of silver specks $p$ versus illumination intensity $I$, according to the Gurney–Mott theory. Curve 1 is for room temperature, and curve 2 is for low temperature.

We turn now to the question of low exposure and the latent image. Presumably even here very small silver specks are also formed (though too small to be visible) and serve as nuclei for growth during the development process. These submicroscopic specks have their origin in the trapping of an initial electron at some foreign impurity, known as a "sensitivity speck." The action of the developer is thought to proceed as follows: The AgBr dissolves in the developer, and the Ag ions move through the solution toward the silver speck due to the difference in potential between the bromide and the silver.

## 11.9 RADIATION DAMAGE IN SOLIDS

We conclude this chapter by discussing the effects of radiation on solids. The term radiation here is not restricted to photons, but is quite general, and includes incident neutrons, protons, and almost any particle that may bombard the solid. Our interest here is not so much in detecting radiation (as in Section 7.8), but in the changes in the solid due to radiation. In the case of strong radiation, these changes are sometimes deleterious, significantly affecting, for example, the electrical conductivity and mechanical stability of the substance.

There are at least two reasons for the importance of studies of radiation damage. Practically speaking, materials are necessarily subjected to strong radiation in some circumstances by virtue of the operation concerned. For instance, materials used for moderating nuclear reactors are bombarded by strong fluxes of neutrons and other fission fragments emitted by the active radiative substance. Similarly, transistors used at high altitude (as in space flight) are exposed to high-energy cosmic rays emanating from the sun, which affect the performance of the devices incorporating them. Thus studies of radiation damage help us design better reactor materials and more reliable transistors.

Solid-state physicists are interested in the nature of defects in various solids. Defects *can* be generated by quenching hot samples (Section 11.3), but the most satisfactory method of producing defects in a well-controlled manner involves radiation. This accounts for the popularity of this technique, and the use of reactors for this purpose. After the defects are created, their type and density can be studied by the methods discussed in Section 11.3.

Let us assume that the intensity of the incident collimated radiation beam is $I$, that is, $I$ is the number of particles, such as neutrons, per unit area per unit time. The intensity can also be written as

$$I = n\,v, \tag{11.35}$$

where $n$ is the number density and $v$ the velocity of the radiation particles. The incident beam is attenuated as it penetrates the solid, because the incident particles are scattered (and in some cases absorbed) by the atoms in the solid. The attenuation follows the usual exponential law, familiar in such situations [see Eq. (2.2) on the attenuation of x-rays],

$$I(x) = I_0\,e^{-x/l}, \tag{11.36}$$

where $x$ is the distance traveled by the beam in the solid and $l$ is a parameter of the solid. Since $I$ decreases very rapidly for $x > l$, the parameter $l$ is known as the *penetration depth* of the radiation.

The depth $l$ can be expressed in terms of the microscopic properties of the scattering atoms on the solid, and the interaction between these and the incident particles. One defines the *cross section* $\sigma$ of the scattering or target atom as the area "seen" by the incident particle. Thus an incident beam of unit area sees a cross section of $N\sigma$, where $N$ is the atomic concentration, and a fraction $N\sigma/1$ of the particles is scattered. Thus in a distance $dx$, the decrease in intensity is

$$-\,dI = (-\,N\sigma)I\,dx,$$

which, as a differential equation, can be integrated to yield (11.36), with

$$l = \frac{1}{N\sigma}. \tag{11.37}$$

Not surprisingly, the result is the same as the mean free path of an electron scattered by atoms in metals (Section 4.5). We may estimate the penetration depth for a neutron: The scattering is accomplished by the nuclei of the solid. Thus $\sigma \simeq \pi R^2$, where $R$ is the nuclear radius. Since $R$ is typically about $10^{-14}$ cm (somewhat smaller than the actual geometrical radius), $\sigma \simeq 10^{-28}\ \text{cm}^2$. (The area $10^{-28}\ \text{cm}^2$, referred to as a *barn*, is frequently used as a unit of area in nuclear physics.) Noting that $N \simeq 10^{29}\ \text{m}^{-3}$ in a solid, we find, upon substitution in (11.37), that $l \simeq 0.1$ m is the penetration depth of the neutron.

We are particularly interested here in the number and type of defects produced. The neutron interacts with the nuclei of the solid, and the loss of energy involved in the stopping of the particle is in part expended in displacing atoms from their crystalline sites, thereby creating Frenkel defects. There are several processes involved: First, the fast *fission neutrons*, of energy about 1.5 MeV from uranium, knock atoms out of their regular positions. Subsequently the slow (pile) neutrons continue to dislodge further atoms. Simultaneously, the atoms thus dislodged, known as *primary atoms*, may also have enough energy to displace other *secondary* atoms. The problem thus becomes quite complicated. The reader can find detailed explanations in the references at the end of the chapter.

Some estimates can be made, however. By regarding the collision as a hard-sphere type of interaction, one can show that the maximum energy that can be imparted to the target is

$$\Delta E = \frac{4M_i M}{(M_i + M)^2} E_i, \tag{11.38}$$

where $E_i$ is the energy of the incident particle, be it neutron or a primary atom, and $M_i$, $M$ are the masses of the incident and target particles, respectively. Thus if a neutron has $E_i = 1.5$ MeV, a primary Cu can acquire only about $10^5$ eV, that is, only about 6% of the incident energy is imparted to the atom. This primary atom is, however, very effective in dislodging further atoms because of the similarity in masses. In estimating the number of atoms to be dislodged further, we take the displacement energy as about 25 eV. This is much greater than the formation energy found by thermal means (Section 11.3), about 5 eV, because the atom in the thermal method has essentially an infinite time in which to be dislodged. In the radiation method, however, the atom must react almost instantaneously, or else the incident particle would pass it by. This requires higher energy. The number of atoms dislodged by the primary atoms is thus about $10^5/25 = 4000$ atoms. Given that the integrated intensity from the reactor is about $10^{23}$ fast neutron/m$^2$, then the number of dislodged atoms is about $4 \times 10^{26}$ per m$^3$, that is, about 1% of the total number of Cu atoms in the solid.

A particularly important type of defect, which is found in metals and other crystalline solids forming the cladding of nuclear reactors, is the *void*. A void is a cavity inside the solid; its size varies from a few angstroms to more than 1500 Å. The void is essentially empty, although a gas at very low concentration may be present in it.

Voids are created in solids which have been subjected to high doses of neutron radiation, for example, $10^{23}$ neutron/cm$^2$ at the moderately high temperatures present in fast reactor operations, for instance, 500°C. The creation of voids produces volume expansion in the substance, reaching as much as 15% or more at high radiation doses, and this leads after some years to deleterious effects on the substance. Consequently, the subject of voids has assumed great practical importance in the design of new reactors, and will even be more so in the yet-to-come fusion reactor, operating at very high temperatures.

A void is formed by the coalescence of a large number of vacancies. Initially these vacancies are created by irradiation at random points in the solid, but at moderately high temperatures these vacancies are quite mobile and cluster together to form voids. In order that the void may grow in size, a mechanism must operate to dispose of the interstitials which are generated simultaneously with the vacancies. If no suitable sink is found, the interstitials recombine with the vacancies, prohibiting further growth of voids. It is now certain that edge dislocations present in the solid act as sinks for the interstitials. This suggests the possibility of reducing the effect of voids by introducing impurities and other traps which reduce the mobility of vacancies and interstitials as well as the growth of voids. See R. Bullough and A. B. Lidiard, *Comments on Solid State Physics*, IV, 69 (1972); also A. Seeger, *ibid.*, IV, 79 (1972).

Neutron radiation damage has been discussed specifically because of its importance in reactor materials, but charged particles such as protons, $\alpha$-particles, electrons, etc., may also produce defects. These particles are rather ineffective in producing atomic defects, however. Thus the heavier charged particles, such as protons, lose most of their energy in exciting electrons, and, although such an *ionization* process is very important in insulators and semiconductors, it is not so in metals, in which the large number of free electrons quickly neutralizes the effect. In the case of electron radiation, the light charged particle, the electron, is further rendered ineffective in producing atomic defects because, since its mass is so small, it imparts very little energy to the much heavier atom. Just as in the case of a ball bouncing off a wall, the ball retains most of its kinetic energy.

## SUMMARY

### Imperfections

Real (as opposed to ideal) crystals usually contain several types of imperfections, such as substitutional and interstitial atoms, as well as vacancies or holes. Dislocations and surface defects are also usually present in crystals.

The number of vacancies is given by

$$N_v = N\, e^{-E_v/kT},$$

where $E_v$ is the formation energy of the vacancy.

### Diffusion

New atoms placed at a crystal surface diffuse through the crystal. The diffusion distance is found to be

$$\bar{x} = \sqrt{2\,DT}.$$

The diffusion coefficient increases exponentially with temperature, according to the formula

$$D = D_0\, e^{-Q/kT},$$

where $Q$ is the activation energy.

### Metallic alloys

Two elements may form a solid solution (alloy) if they satisfy the Hume–Rothery rules: The atoms must have comparable sizes and similar electronegativities. The crystals must have similar structures and similar electronegativities. The solute must have greater valence than the solvent.

A *phase diagram* is a graph which describes the melting characteristics of an alloy. It may be derived theoretically if the *free energy* of the alloy is given. This energy is defined as

$$F = E - TS.$$

The most stable phase or phase mixture is that having the minimum free energy.

### Dislocations

Dislocations greatly influence the mechanical properties and strength of metals. The reason why pure metals are usually soft and ductile is that they contain an appreciable number of dislocations which are free to move. Pure, dislocation-free metals are very strong.

## REFERENCES

### Imperfections

H. G. Van Bueren, 1960, *Imperfections in Crystals*, Amsterdam: North-Holland

A. L. Ruoff, 1973, *Materials Science*, Englewood Cliffs, N.J.: Prentice-Hall

W. Shockley, *et al.*, editors, 1952, *Imperfections in Nearly Perfect Crystals*, New York: John Wiley

J. I. Takamura, in W. Cohn, editor, 1970, *Physical Metallurgy*, Amsterdam: North-Holland

### Diffusion

P. G. Schewmon, 1963, *Diffusion in Solids*, New York: McGraw-Hill

L. A. Girifalco, 1964, *Atomic Migration in Crystals*, New York: Blaisdell

D. Lazarus, "Diffusion in Metals," in *Solid State Physics*, **10**, F. Seitz and D. Turnbull, editors, New York: Academic Press

### Alloys and thermodynamics

F. Rhines, 1956, *Phase Diagrams in Metallurgy*, New York: McGraw-Hill

R. A. Swalin, 1962, *Thermodynamics of Solids*, New York: John Wiley

W. Hume-Rothery and G. V. Raynor, 1954, *The Structure of Metals and Alloys*, third edition, London: The Institute of Metals

A. H. Cottrell, 1957, *Theoretical Structural Metallurgy*, London: St. Martin

A. G. Guy, 1951, *Elements of Physical Metallurgy*, Reading, Mass.: Addison-Wesley

W. G. Moffat, *et al.*, 1964, *The Structure and Properties of Materials*, Volume I, New York: John Wiley

W. Cohn, *op. cit.*

### Dislocations

W. C. Dash and A. G. Tweet, "Observing Dislocations in Crystals," *Scientific American*, **205,** 107 (1961)

A. H. Cottrell, 1964, *The Mechanical Properties of Matter*, New York: John Wiley

J. Weertman and J. R. Weertman, 1964, *Elementary Dislocation Theory*, New York: Macmillan

J. Friedel, 1964, *Dislocations*, Reading, Mass.: Addison-Wesley

A. H. Cottrell, 1953, *Dislocations and Plastic Flow in Crystals*, Oxford: Oxford University Press

W. T. Read, 1953, *Dislocations in Crystals*, New York: McGraw-Hill

N. F. Mott, 1956, *Atomic Structure and the Strength of Metals*, New York: Pergamon Press

### Ionic conductivity: the photographic process

F. C. Brown, 1967, *The Physics of Solids*, New York: W. A. Benjamin

N. F. Mott and R. W. Gurney, 1948, *Electronic Processes in Ionic Crystals*, second edition, Oxford: Oxford University Press; also in paperback, by Dover Press

### Radiation damage

A. C. Damask and G. J. Dienes, 1963, *Point Defects in Metals*, London: Gordon and Breach

*Radiation Damage in Solids*, Proc. of the International School of Physics, "Enrico Fermi," New York: Academic Press, 1962

## QUESTIONS

1. The text said that vacancy concentration is normally measured in quenched samples, at room temperature.
   a) Why is it necessary to quench the sample, rather than to cook it slowly?
   b) Is the quenched sample in thermal equilibrium?
   c) If the vacancies in a quenched sample are annealed out under adiabatic conditions, will the solid heat up or cool down? And by how much?
2. What is the justification for calling Eq. (11.16) the "lever formula?"
3. What is the meaning of the fact that the solidus and liquidus lines in the phase diagram converge at the endpoints?

## PROBLEMS

1. a) Calculate the atomic percentages of interstitials and vacancies at the melting point in Cu (1356°K). The formation energies for these defects in Cu are, respectively, 4.5 and 1.5 eV.
   b) Repeat the calculations at room temperature.
2. Verify that expression 11.7 satisfies both Fick's second law (11.6) and the initial conditions of the problem.
3. a) Carry out the integrations leading to the diffusion distance (11.8).
   b) Calculate the diffusion velocity, and explain physically why this velocity decreases in time, as it does.

4. The text estimated that an atom in a crystal diffuses a distance of about $1\mu$ in two years, if the lattice constant $d = 1$ Å and the jump frequency is 1 s. Estimate the distance the atom would travel in the same time interval if the atom were able to jump always in the same direction, e.g., to the right.
5. Other solutions to Fick's second law, besides the one reported in the text, are frequently quoted in the literature. These solutions correspond to boundary conditions different from those chosen here. Verify that the expression

$$c(x,t) = \frac{c_0}{2}\left[1 - \frac{2}{\sqrt{\pi}}\int_0^{x/2(Dt)^{1/2}} e^{-u^2}\,dn\right]$$

is also a solution of Fick's law corresponding to the following initial conditions: $c(x,0) = c_0$, for $x < 0$, and $= 0$ for $0 < x$. Plot $c(x,t)$ versus $x$ at various instants $(0 < t)$, and show that $c(0,t) = \frac{1}{2}$ at all times. [The term in the brackets involving the integral is known as the *error function*, and denoted by erf $(x/2(Dt)^{1/2})$.]
6. The diffusion activation energy of carbon in $\gamma$-iron (austenite) is $3.38 \times 10^4$ cal/mole, and $D_0 = 0.21$ cm$^2$/sec. Calculate the diffusion coefficient at 800°C and 1100°C.
7. The carburizing of steel is accomplished by placing iron in a carbon-rich atmosphere, and allowing sufficient time for the carbon atoms to diffuse through the solid. If you want to achieve a carbon concentration of 1% (in weight) at a depth of 3 mm after 10 hours of carburizing time at 1200°C, calculate the carbon concentration in weight per cent which must be maintained at the surface. Take the iron to be in the $\gamma$-phase, and use the data of Problem 6. [*Hint*: Use the solution given in Problem 5.]
8. The atomic size factor favors solid solubility for the following alloys. What is the effect of the relative valency factor in each case?

| *Solvent*: | Cu | Ge | Sn | Ag |
|---|---|---|---|---|
| | \| | \| | \| | \| |
| *Solute*: | Si | Si | Ag | Mg |

9. a) Construct the phase diagram for the Cu-Ni alloy, using the following data (Moffat, 1964).

| Weight % Ni : | 0 | 20 | 40 | 60 | 80 | 100 |
|---|---|---|---|---|---|---|
| Liquidus $T$ : | 1083 | 1195 | 1275 | 1345 | 1410 | 1453 |
| Solidus $T$ : | 1083 | 1135 | 1205 | 1290 | 1375 | 1453 |

   b) Starting with a liquid alloy of 60% Ni and cooling it gradually, state the composition of the solid that forms first.
   c) How much solid per kilogram can be extracted from the melt at 1300°C?
10. Establish the validity of Eq. (11.20) for the free energy.
11. Find the derivative of the mixing entropy $(\partial S/\partial c)$, and show that it is infinite at $c = 0$.
12. Referring to Fig. 11.13(a), show that the free energy for a phase mixture (where the concentrations of the phases are given by $c'$ and $c''$) is given by the straight line $F'\,F''$ in the average concentration range $c'' < c < c'$.
13. Prove the lever formula for a phase mixture whose free-energy diagram has the shape shown in Fig. 11.13(b).

14. Confirm that the free-energy diagrams of Figs. 11.15(a)–11.15(d) lead to the phase diagram 11.15(f). Indicate on this latter figure suitable values for the temperatures $T$, $T'$, $T''$, and $T'''$ indicated in the former figures.
15. The phase diagram for the Cu-Ag alloy is shown in Fig. 11.15(f).
    a) Confirm that the atomic % and weight % scales indicated are consistent with each other.
    b) Determine the atomic percentage of the $\alpha$-phase at the eutectic concentration just after solidification.
    c) Determine the percentage of the same phase at the temperature 850°C, and the Cu concentration in atomic %.
16. a) Starting with a Cu-Ag alloy in the liquid phase and 60% weight Cu, indicate the various phases which appear as the system is cooled progressively from the liquid to the solid phase.
    b) What is the weight fraction of the $\beta$ phase at 850°C?
17. Prove that the Fermi surface begins to touch the boundaries of the Brillouin zone in the fcc and bcc structures when the electron/atom ratios are 1.36 and 1.48, respectively. [Refer to Fig. 5.8.]
18. Show that the shear strain on any crystal plane vanishes if the solid is placed under hydrostatic pressure.
19. a) Show that in an fcc lattice the (111) planes have the highest atomic concentration.
    b) Show that the [100] direction in the (111) plane has the highest atomic concentration.

# CHAPTER 12 MATERIALS AND SOLID-STATE CHEMISTRY



*Invention breeds invention.*

Ralph Waldo Emerson

## 12.1 INTRODUCTION

This chapter deals with two topics: materials and solid-state chemistry. These topics are actually intimately related, because, to prepare and characterize new materials, scientists must understand their basic physical and chemical nature. The rapid advances in solid-state physics and chemistry in the past 20 years have been made possible by physicists and chemists working side by side, opening new vistas.

We shall first discuss amorphous semiconductors, and the effects of noncrystallinity on their band structure and mobility. We shall then describe liquid crystals—a class of substance exhibiting characteristics of both liquids and solids and their mechanical and optical properties. The mechanical and melting properties of polymers are also considered, and these properties are related to the long chainlike structure of the molecules that constitute them.

Many of the recent discoveries in physics have been taken over by chemists, and have now become standard tools for investigations in chemistry. Outstanding among these discoveries are nuclear magnetic resonance, electron spin resonance, and Mössbauer spectroscopies. We shall talk about all these, with particular reference to their applications in chemistry.

## 12.2 AMORPHOUS SEMICONDUCTORS

Many amorphous substances exhibit significant electrical conduction. We refer to these as *amorphous semiconductors*. This type of conduction is associated with electrons rather than ions in the solid, because the contribution of ions to the conductivity is usually very small. Although crystalline semiconductors have received most attention in the past (Chapters 6 and 7), scientists are expending great effort toward understanding amorphous ones. One reason for this increase in interest lies in the fact that some amorphous substances show certain unusual switching properties, which could be important in applications such as switching and memory devices. Also amorphous elements are usually cheaper to manufacture than crystalline ones, so the widespread use of amorphous elements in electronics could lead to a significant reduction in cost. Until recently, the electronic properties of amorphous semiconductors were poorly understood, because it is harder to treat electron states in a disordered solid than in a crystal, due to the absence of periodic symmetry. Although our understanding of crystalline structures rests on secure foundations today, there is still much to be learned in the field of amorphous semiconductors. This subject is a challenge not only to the physicist, but to the chemist and materials scientist, all of whose expertise is needed to unravel the many complexities of the topic. Undoubtedly many useful applications remain undiscovered.

For our discussion, let us divide amorphous semiconductors into four classes.

a) *Elemental amorphous semiconductors*, for example, Ge, Si, Sl, Te

b) *Covalent amorphous semiconductors* (binary), for example, $As_2Se_3$, GeTe

c) *Covalent amorphous semiconductors* (multicomponent), for example, chalcogenide, boride and arsenide glasses

d) *Ionic amorphous semiconductors*, for example, $Al_2O_3$, $V_2O_5$, and other transition-metal oxides.

In the first three classes, the atoms are held together by covalent bonds; in the last class, the binding is due primarily to ionic bonds. Since the ionic bonds involved are quite strong, of the order of 10 eV per bond, the electrons in class (d) are strongly bound to their ions, and are usually unable to participate in electrical conduction to any significant extent; we shall therefore omit these substances from further consideration.

Atomic order in a solid has an important bearing on the treatment of electronic states, as we have seen. So let us look into this question once more in connection with amorphous materials. Recall that the structure of a solid in the amorphous state is the same as that of a supercooled liquid; it is as though we are able to take a liquid and, at some instant "freeze" the position of every atom in the system. We recall from Section 1.8 that a liquid has a good short-range order: The positions of nearest neighbors are essentially the same as in the solid state. But a liquid, unlike a crystal, has no long-range order, so at a distance far from the atom in question, the other atoms appear to be randomly distributed.

The same situation prevails in an amorphous substance: Although long-range order is absent and far-away atoms seem to be randomly distributed, short-range order does exist. For instance, in amorphous Ge, each atom is surrounded by four nearest neighbors, forming the familiar tetrahedral bond, much as in the solid state. But if we look at the second-nearest neighbors, we discover that there are two different ways in which they can be arranged in such a way that the atoms at the apex of the tetrahedron are the centers of new tetrahedra. One of these arrangements leads to the fcc structure observed in crystalline Ge, the other to the wurtzite structure. In amorphous Ge both arrangements occur with essentially equal likelihood, and this leads to some disorder in the second-nearest neighbors. When this process is extended further and further away from the original atom, one discovers that the number of possible positions multiplies rapidly, resulting in complete disorder at long range. Our comments concerning Ge apply equally to Si, and also to other class (a) semiconductors, with appropriate modifications to accommodate the possibility of a different structure.

The type of disorder just discussed is a *positional* disorder. An additional type is encountered in the covalent semiconductors of classes (b) and (c). Thus in GeTe, for example, not only is there long-range disorder in the positions of the atoms, but even the chemical composition of the atom is uncertain, there being an equal probability of finding either a Ge or a Te atom at any position. This uncertainty is referred to as *compositional* disorder. Thus a binary amorphous semiconductor has both positional and compositional disorders, and thus is more disordered than an elemental semiconductor. There is even more compositional disorder when multicomponent substances in class (c) are considered. We should emphasize, however, that in spite of this, a good short-range order exists in all the substances discussed. For instance, even in the alloy $As_{20}Se_{50}Ge_{40}Te_{10}$, the atoms are so arranged that each Ge atom is surrounded by four nearest neighbors, forming a tetrahedral covalent bond. To explain the observed electronic properties of amorphous semiconductors, we need to use concepts both of short-range order and long-range disorder.

**Band structure**

We are interested now in electronic states in an amorphous semiconductor, since this knowledge is essential to the understanding of electrical and optical properties. Because of the extensive disorder present, the Bloch theorem (Section 5.3) does not hold here. And since this theorem is the basis of much of our treatment of electronic structure in crystals, many of the results derived in Section 5.3 do not apply directly to amorphous solids. In particular, the concept of the wave vector **k**, characterizing the electron function, is no longer meaningful. This also holds for the **k**-space and Brillouin zones. These concepts, which are direct consequences of the translational periodicity of a crystalline lattice, and which we found so useful in treating the electron states in crystals (Chapter 5), have to be discarded when we consider an amorphous solid.

Other concepts used in connection with crystals remain useful, however, even in disordered states. Figure 12.1(a) shows the density of states $g(E)$ for a crystalline semiconductor. The bottom of the conduction band (CB) is at $E_c$, and the top of the valence band (VB) at $E_v$. The range between these two energies, $E_v$ to $E_c$, is the energy gap, where no electron can exist in a perfectly pure crystal. The density of states vanishes completely in the entire range of the energy gap. Note that the edges of the CB and VB are infinitely sharp in the crystalline case.

Figure 12.1(b) shows the density-of-states function for the amorphous state of the same substance. The primary difference between Figs. 12.1(a) and 12.1(b) is that in 12.1(b) the density of states has extended into the gap from both the CB and the VB sides. Each of these bands now has a "tail" entirely within what was formerly a forbidden gap (the band tails are shaded in Fig. 12.1b). To understand this result, we may start with the crystalline state, begin to introduce some disorder, and then examine its effects on $g(E)$. Since we are allowed to introduce only long-range disorder, the effect of this on the energy levels is rather small (only a few percent), because an electron on a particular site interacts most strongly with nearest neighbors. In general, the effect of the disorder is therefore to shift the levels—up or down—by only a small amount throughout the band. There is one region, however, in which the effect of disorder is conspicuous: near the band edge. Here the effect of disorder is to displace some levels right into the energy gap, creating the band tail. Although the shift here may not be large, it is

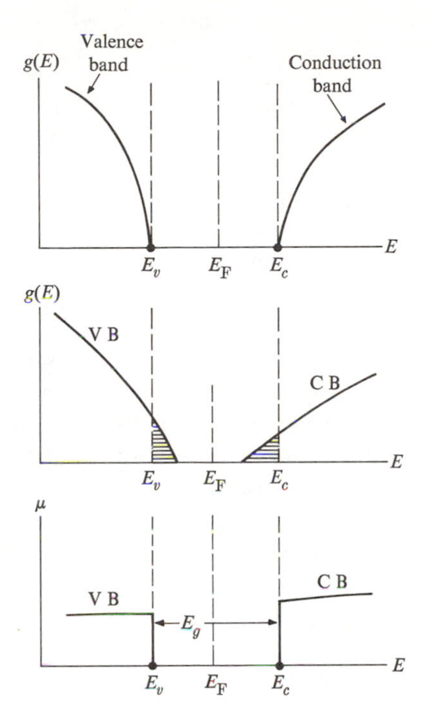


**Fig. 12.1** From top to bottom: density of states $g(E)$ versus energy $E$ for a crystalline semiconductor; $g(E)$ versus $E$ for an amorphous semiconductor; mobility $\mu$ versus $E$ for an amorphous semiconductor. Shaded regions in middle figure represent band tails introduced by the disorder.

significant, because the electron states in the tail have a different character from those in the remainder of the band. The band tailing occurs for both CB and VB, although the CB tail is likely to be larger because it is at a higher energy. (Explain!)

We must now make a clear distinction between *localized* and *delocalized* electron states. In a localized state, the electron is restricted to movement around only one particular atomic site, while in a delocalized state the electron is extended throughout the solid (existing partly at every atomic site). In the case of a crystal, all states are delocalized in accordance with the Bloch theorem (however, see Section 5.3). In the case of an amorphous solid, both types of states occur simultaneously. Those states in the main body of the band are delocalized just as are those in a crystal. On the other hand, the states in a band tail represent localized electrons. It is not too surprising that these latter states, falling in what was once an energy gap, are localized, as the reader will recall that the localized impurity states in a doped semiconductor did fall in the energy gap. In a certain sense we may well use the impurity model (Section 6.5) to treat the localized states in amorphous semiconductors. We may also speak of localized and delocalized holes in connection with the VB.

## Electronic conduction

The concept of delocalization is important in electronic conduction. A *delocalized electron* moves readily through the solid (Section 5.3). Since electrons are already distributed throughout the solid, they need only a little push—e.g., from an electric field—to set them adrift, carrying an electric current. This process is known as metallic conduction. By contrast, a *localized electron* is strongly bound to its site, and lies deep within its potential well, separated from the neighbors by high, thick potential barriers. It can move from one site to a neighboring one only if it is energetically excited above the potential barrier. But, since the barrier is usually about 1 eV, relatively few electrons are excited at room temperature. This process is known as *hopping*, and the thermal excitation process as *activation*.

Figure 12.1(c) illustrates this graphically by plotting the mobility $\mu$ of the electron as a function of the energy. Since the mobility of a localized electron is essentially zero, we see that for the CB, for example, the mobility drops sharply amd suddenly as the energy decreases from the main band to the band tail. A similar situation exists for the VB. So, although no sharp density-of-state gap exists, there is a sharp *mobility gap* (in the energy range in which $\mu = 0$), and this gap is approximately the same as the energy gap in the crystalline solid.

Let us now derive formulas for the conduction mobilities for delocalized and localized states. For the delocalized state, we may use the same argument we used in treating crystalline states (Section 6.7), and the result is the same as Eq. (6.31). That is,

$$\mu_0 = \frac{e\tau}{m^*}. \tag{12.1}$$

Note, however, that the collision time $\tau$ is now much shorter, due to the additional scattering caused by the disorder, which leads to a significant reduction in the mobility—by two orders of magnitude or so. The scattering of the electron due to the disorder is so strong that the mean free path is typically only a few times the interatomic distances, or about 10 Å.

A localized electron can drift through the solid by hopping between atomic sites only if it acquires the energy necessary to overcome the potential barrier. It acquires this excitation energy from thermal excitation of the solid. The problem is similar to the atomic diffusion case treated in Section 11.4 and the result is a hopping mobility of the form

$$\mu_H = Ae^{-W/kT}, \tag{12.2}$$

where $W$ is the activation energy. The mobility $\mu_H$ decreases rapidly with reduced temperature, and at low temperature is negligible. Even at ordinary temperatures $\mu_H$ is much smaller than $\mu_0$, typically three orders of magnitude less. Because it is so small, the hopping mobility will be neglected in the following discussion.

To calculate electrical conductivity—the quantity which is actually measured—one uses the relation (6.32), that is,

$$\sigma = ne\mu,$$

where $n$ is the concentration of carriers. In an intrinsic amorphous semiconductor, carriers are generated by exciting electrons from the VB to the CB across the gap, as in the crystalline case. Adding the contributions of both delocalized electrons and holes, one finds that

$$\sigma = \sigma_0 e^{-E_A/kT}, \tag{12.3}$$

where effective activation energy $E_A$, which is equal to $(Eg/2 + W)$, is typically about 0.5 eV. The conductivity increases rapidly with temperature because, as in the crystalline case, at higher temperature more free carriers are created. This prediction is confirmed in a general way by experiment on amorphous Ge, Si, and other substances, as shown in Fig. 12.2 for Si.

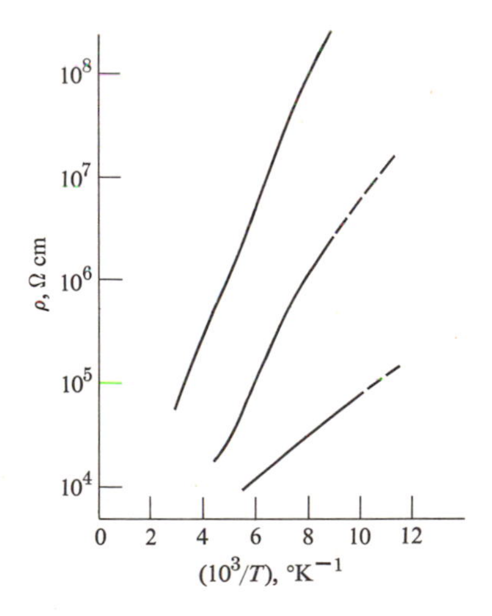


**Fig. 12.2** Resistivity $\rho$ versus $10^3/T$ for evaporated film of Si. The different curves correspond to various stages of growth and annealing. [Brodsky, *et al.*, *Phys. Rev.* **B1**, 2632 (1970)]

It is also noted experimentally that electrical conduction in amorphous semiconductors is insensitive to impurities, and that it is primarily $p$-type, i.e., the current is carried primarily by holes. The insensitivity to impurities, in marked contrast to the crystalline case, can be understood on the basis of the model of Fig. 12.1(b) by noting that if a donor—As, for example—is added, the extra electron can be accommodated in the band tail of the CB, where it contributes nothing to the current. The $p$-type character of the conduction can also be explained if one assumes that the band tail is larger for the CB than for the VB. In that case the Fermi level, which lies somewhere near the middle of the new energy gap, is closer to $E_v$ than $E_c$, resulting in more delocalized holes than delocalized electrons, leading to the $p$-type character indicated above.

To explain some of the properties of the chalcogenides, in which disorder becomes extensive, Cohen, Fritzsche, and Ovshinsky (CFO) proposed the model shown in Fig. 12.3: Here the two bands extend so far into the gap that they actually overlap each other. When such an overlap takes place, repopulation ensues, with electrons transferring from the higher region of the VB tail into the lower region of the CB tail. Since the states involved are localized, this results in the creation of large concentrations of positively and negatively charged centers, or traps. It should be apparent that electrical conduction in the CFO model obeys an equation of the same form as (12.3).

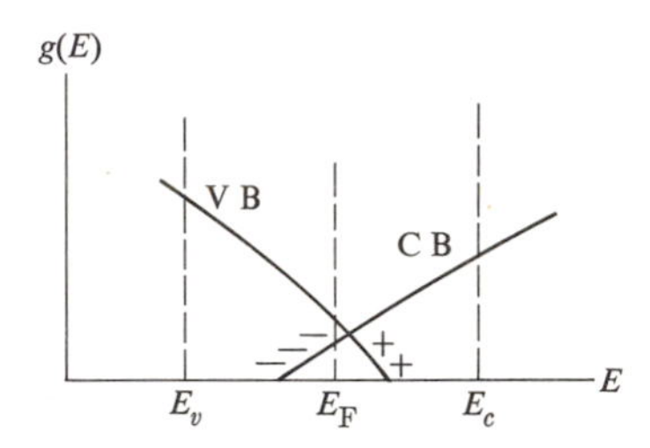


**Fig. 12.3** The CFO model. Positive and negative signs indicate ionization of impurities due to overlap of bands.

## Optical absorption

Optical absorption is a standard technique for investigating band structure, and it is therefore of interest to study absorption in amorphous semiconductors. As seen from Fig. 12.4, the absorption for Ge in the amorphous state is much the same as for the crystalline state, the main difference occurring near the fundamental absorption edge, where the cutoff frequency in the amorphous state is lower and not so sharply defined. This can be understood by noting that the absorption edge of the amorphous state is determined by exciting electrons from localized states in the VB to delocalized states in the CB. The diffused nature of the edge arises therefore from the diffused nature of the VB tail, and since this extends into the gap, it follows that the cutoff frequency is less than the crystalline absorption edge $(E_c - E_v)/h$. Note also that an absorption which involves exciting an electron from localized VB to localized CB states is not effective here, since absorption takes place only if the

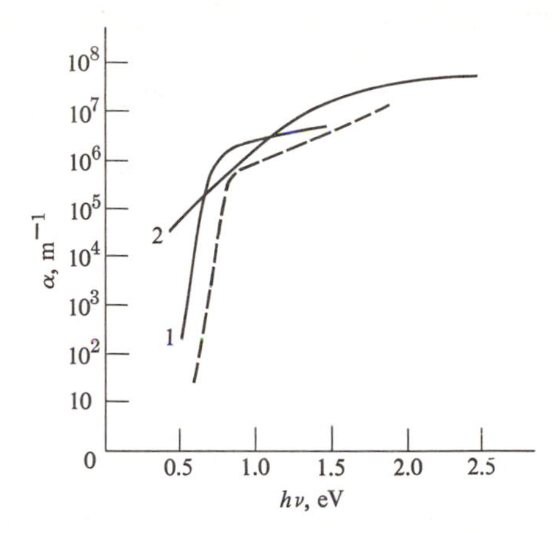


**Fig. 12.4** Absorption coefficient α versus photon energy in amorphous Ge. Curves 1 and 2 represent measurements by different workers. The dashed curve is for a single crystal. [From various sources; see Owen (1970)]

two states concerned are in the same spatial region, i.e., absorption is by independent atoms. This, however, is much weaker than absorption involving delocalized states, since these overlap over large spatial regions involving many atoms simultaneously.

## Switching

Some amorphous semiconductors exhibit characteristics which make them useful as switching and memory components in electronic circuits. Many different types of switching phenomena have been observed, the best known at present being that reported by Ovshinsky† in 1968, which is shown in Fig. 12.5.

If a voltage $V$ is applied across a special chalcogenide sample, the current increases along the line *Oab*. At the *threshold voltage* $V_T$ the sample undergoes a transition (*bc*) to a new state of extremely small resistance. In this On state, the $I$–$V$ characteristic is indicated by the line *ec*, where the current is essentially independent of the voltage. The voltage $V_H$ necessary to maintain the On state is known as the *holding voltage*. When the current is reduced below a certain point *e*, the sample switches back (*ea*) to the high-impedance, or Off state. The threshold voltage $V_T$ increases linearly with the thickness of the sample, and can readily be varied from 2.5 to 300 V, by increasing the thickness. On the other hand, the holding voltage is independent of the thickness, but can be varied between 0.5 and 1.5 V by varying the composition of the glass. The switching from the Off to the On state occurs very rapidly, in about 1 ns, while the On-to-Off

† S. R. Ovshinsky, *Phys. Rev. Letts.* **21**, 1450 (1968).

switching is much slower, taking about 1 $\mu$s. Furthermore, the device is symmetric and operates equally well with the reverse polarity. This type of device is now referred to as an Ovshinsky (or Ovonic) diode, after its discoverer.

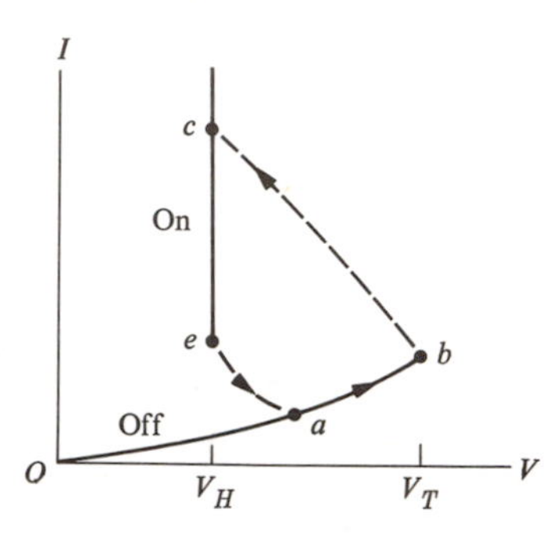


**Fig. 12.5** The current-voltage characteristics of an Ovonic diode. The dashed lines represent fast discontinuous changes.

Although there is not yet a complete understanding of the switching process and related phenomena, there is general agreement that the reason for the decrease in the resistance (in the On state) is that certain regions in the substance crystallize as a result of Joule heating, and that these form channels in which most of the current flows. Because the channels are crystalline, they have greater conductivity than the remainder of the sample, and consequently the sample as a whole has a smaller resistance. This accounts for the situation in a so-called read-mostly device, in which the material remains in the low-resistance state when the voltage is on. In a threshold switch, however, the crystallization process does not take place. For more details, refer to the February 1973 issue of *IEEE Transactions on Electronic Devices*, which is devoted to amorphous semiconductors.

### Xerography

One of the most familiar applications of amorphous semiconductors is the xerographic process. This involves depositing a thin film of amorphous selinium on a metallic substrate (usually Al), and the surface of the film is electrically charged all over by means of a corona discharge. The pattern of light to be copied is then allowed to fall on the film, causing the illuminated regions to be photoconductive, and the corresponding charge is allowed to leak away. The dark regions (dark conductivity about $10^{-16}\,\Omega\,\text{cm}^{-1}$) remain charged. A finely powdered, pigmented resin is then sprayed on the surface and clings to the charges. Finally the powdered pattern is transferred to a sheet of paper, and attached to it by heating.

## 12.3 LIQUID CRYSTALS

Another class of substances which has elicited much interest lately from chemists, physicists, and other materials scientists is liquid crystals. The interest is due in part to the promise these substances hold as potential electro-optical devices, but their unusual properties, which were hitherto not well understood, have also challenged the curiosity of these scientists.

In 1888, Reinitzer observed that, as he heated cholesteryl benzoate, the solid melted at 145°C into a liquid of a white, *turbid*, murky appearance. When this liquid is heated further, it undergoes another transition at 179°C, this time into a clear, *transparent* liquid. The substance was further investigated by Lehman, who found that the liquid actually exhibits optical anisotropy—i.e., birefringence—when in the turbid region, much as a crystal does. The unusual fact that the substance has the mechanical properties of a liquid—e.g., ability to flow and low viscosity—and the anisotropic optical properties of a crystal prompted Lehman to coin the descriptive name *liquid crystals*, which has been retained ever since. The liquid crystalline phase is often called the *mesophase*, and a substance having such character a *mesogen*.

The appearance of a liquid crystal, resembling that of a colloidal solution, led to early suggestions that such a substance is also a colloidal solution. We know that this is incorrect because a liquid crystal has well fixed lower and upper temperatures, an indication that we are dealing here with true phase transition, and hence a distinct phase of matter. X-ray studies also clearly establish that the substance has orientational order in the mesophase.

The molecules in liquid crystals are long and rodlike, a typical length range being 15–40 Å. The large anistropy of the molecule is essential for the appearance of the mesophase, as we shall see.

Liquid crystals are not rare substances. A large number of these is now known, essentially all being organic compounds, many of which contain aromatic molecules in their structure.

The mesogens discussed here are *thermotropic*, i.e., they are obtained by heating solids. There are also *lyotropic* liquid crystals, formed by dissolving certain crystals in suitable solutions. This latter group is relevant to the structure of biological membranes, which also appear to have a liquid crystalline structure.

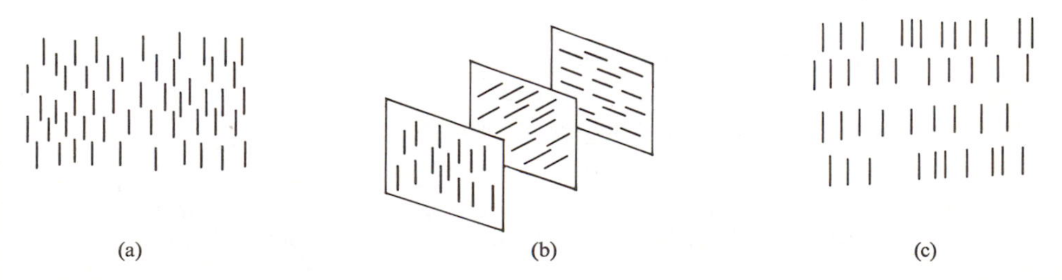


**Fig. 12.6** The (a) nematic, (b) cholesteric, and (c) smectic phases of liquid crystals.

## Classification

We have said that the molecules in a liquid crystal are long. In the mesophase these long molecules tend to align parallel to each other along a certain preferred direction. There are also additional structures present, on the basis of which Friedel divided liquid crystals into three different phases: *nematic*, *cholesteric*, and *smectic*.

i) The *nematic*† *phase* has the simplest structure. The molecules are parallel to each other, but otherwise their spatial distribution is random, as in a liquid (Fig. 12.6a). There is thus an orientational order, but the molecules are able to move around from one region to another as in a liquid—a fact responsible for the low viscosity. Each molecule is, of course, free to rotate around its axis, because of its rodlike shape. A liquid in the nematic phase also has a turbid appearance. An example of a nematic crystal is *p*-azoxyanisole, whose temperature range of existence is 116–136°C.

ii) In the *cholesteric phase*, the molecules are also aligned parallel to each other, but the direction of alignment twists progressively, resulting in a helical structure (Fig. 12.6b). Thus the substance consists of parallel sheets, or layers. In each sheet the molecules are aligned parallel to each other. The pitch of the helix is typically around 2000 Å, but this can be lengthened by the application of suitable external fields.

Because of the helical structure, a cholesteric substance exhibits optical activity, i.e., the plane of polarization of a light beam is rotated as it travels in the substance in a direction parallel to the axis of the helix. The amount of the optical activity is enormous in some cases, e.g., an activity of $6 \times 10^{4\circ}$/mm has been observed. That is, the plane of polarization is rotated through an angle of $6 \times 10^{4\circ}$ in a plate 1 mm thick, which can be compared with an activity of only 300°/mm in an ordinary organic compound.

Chemically, cholestrogens are usually ester cholesterols, a fact responsible for the name "cholesteric phase." An example is cholesteryl cinnamate, whose range of existence is 156–197°C. Mechanically, a cholesteric liquid has a somewhat higher viscosity than a nematic one.

iii) The structure of the *smectic*‡ *phase* is illustrated by Fig. 12.6(c). It consists of a series of layers, in which the molecules are all parallel to each other and normal to the layer plane. The layers interact only weakly, and can readily slip past each other, or be made to rotate relative to each other. It is these motions which are responsible for the liquid-like mechanical properties.

---

† The term *nematic* (meaning "threadlike" in Greek) alludes to the fact that these substances appear as long, thin filaments when they are viewed under a microscope.

‡ Smectic is from a Greek word implying association with soap, an allusion to the fact that first discovered substances of this kind were among soaps.

In addition to the orientational order, the molecules also exhibit a regularity in their distribution pattern within their own layer, i.e., a certain amount of spatial order exists in each layer. The type of spatial pattern as well as the amount of tilt of the direction of alignment relative to the layer plane (the direction of alignment is not always normal to the plane) have led to further finer subclassifications of the smectic phase. In the smectic $A$ phase, the axis is normal to the layers, while in the $C$ mesophase the axis is tilted. In the smectic $B$ phase the molecules are thought to have an hcp structure within the layers, and in the $D$ phase the structure is known to be cubic.

An example of a smectogen is ethyl *p*-azoxybenzoate, whose existence range is 114–120°C.

Of all mesophases, the smectic is closest to a solid structure. The only difference between a smectogen and a solid is the lamellar structure of the smectogen, which permits the slip and rotational motions. The smectogen is essentially a two-dimensional solid.

Some substances exhibit more than one type of mesophase, depending on the temperature. For instance, 4,4′-di-*n*-heptyloxyanoxybenzene is smectic in the range 74–95°C and nematic in the range 95–124°C. Above 124°C, the compound turns into a regular isotropic liquid. The fact that the smectic phase occurs at a lower temperature than the nematic is expected, inasmuch as the figure has a higher order.

**Orientational order and intermolecular forces**

We mentioned that the molecules in a mesophase are oriented with their long axes parallel to each other. At every point in space there is a *preferred* direction along which the molecules tend to align themselves. The system therefore has an *orientational order*. The preferred direction is specified by a unit vector $\mathbf{n}(\mathbf{r})$, which points along this direction everywhere in space; $\mathbf{n}(\mathbf{r})$ is known as the *director*. Since the preferred direction may vary from point to point, the director $\mathbf{n}(\mathbf{r})$ is a function of the position vector $\mathbf{r}$, a fact indicated explicitly in the notation.

Let us begin our discussion of order with the nematic phase, since it has the simplest structure. The molecules are not actually perfectly aligned along $\mathbf{n}(\mathbf{r})$, but fluctuate on both sides of this direction. The angular fluctuation may be appreciable—more than 10°—and the fluctuation time is usually very short, about $10^{-11}$s. To specify the degree of order, we define the orientational *order function* $S$ as

$$S = \overline{(3\cos^2\theta - 1)}/2, \tag{12.4}$$

where $\theta$ is the angle between the axis of a typical molecule and the director, and the bar signifies a time average over a whole period of molecular fluctuation. For a situation of perfect order, the molecule points along $\mathbf{n}(\mathbf{r})$ at all times—that is, $\theta = 0$—and consequently $S = 1$. For a complete absence of order, i.e., random orientation, all values of $\theta$ are equally likely, leading to $S = 0$. A partial order is therefore represented by a value of $S$ between zero and unity; the greater the order, the closer $S$ is to unity.

The definition of $S$ chosen in (12.4) is the most convenient, although other definitions are possible in principle, One may think that the simplest choice would be $S = \overline{\cos \theta}$, that is, the projection of the molecular length along the director, but this is inadmissible for the following reason: Despite its elongated shape, the molecule has a center of symmetry, and thus the two directions along the molecular axis are equivalent. In other words, the two orientations for the molecule, $\theta$ and $\pi - \theta$, are completely equivalent, and consequently the angular must be chosen so that its value is the same at $\theta$ and $\pi - \theta$. The function must therefore be even in $\cos \theta$ [recall $\cos(\pi - \theta) = -\cos \theta$], and the choice made in (12.4) is the simplest such function. [The angular function in (12.4) is the familiar Legendre function $P_2(\theta)$.]

[Also note that although the bar in (12.4) signifies a time average, it may alternatively be regarded as an average over the molecules in a particular neighborhood at a certain instant. This *ensemble* average leads to the same result as the time average, as asserted by Gibbs in his hypothesis in statistical mechanics.]

The order function $S$ depends on temperature. For $T < T_0$, where $T_0$ is the critical temperature—i.e., the temperature at which the mesogen melts into the isotropic liquid phase—$S$ decreases monotonously with $T$, and shows an abrupt jump from 0.4 to 0 at $T_0$ (Fig. 12.7). Also, when plotted as a function of the normalized temperature $T/T_0$, the curve has a universal validity; i.e., it is followed by all nematogens.

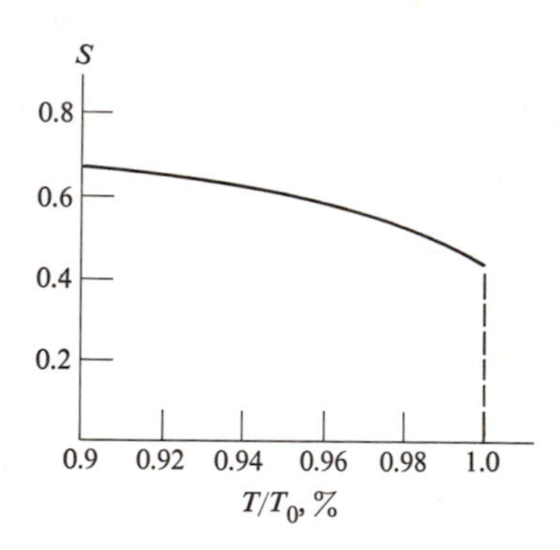


**Fig. 12.7** Orientational order $S$ versus $T/T_0$ for the nematic phase.

The order function may be measured by any of several techniques; the most direct method employs NMR spectoscropy.

We turn now to the forces responsible for the order. Since the order is spontaneous, it must be due to anisotropic intermolecular forces. A dipole–dipole electrical force of the form discussed in Section 9.2 would produce an orientational order, as first suggested by Born, but this cannot be entirely correct, since the molecules in many liquid crystals are nondipolar. But it can be shown by quantum considerations that two elongated molecules, $i$ and $j$, have an anisotropic potential

$$V_{ij} = - u(r_{ij})\,(3\cos^2\theta_{ij} - 1)/2, \tag{12.5}$$

where $r_{ij}$ is the intermolecular distance, $\theta_{ij}$ the angle between the axes, and

$$u(r_{ij}) = - b\,\Delta\,\alpha^2/r_{ij}^6; \tag{12.6}$$

$b$ is a positive constant. The quantity $\Delta\alpha = \alpha_{\parallel} - \alpha_{\perp}$ is the polarizability anistropy, i.e., the difference between the molecular polarizability along the molecular axis, $\alpha_{\parallel}$, and perpendicular to it, $\alpha_{\perp}$. The nature of the force here is the same as the polarization force discussed in connection with inert gas crystals (Section 1.10). The orientation force is stronger here because of the considerable asymmetry in molecular shape, but is still not very strong, the critical temperature being about 100°C.

By summing the intermolecular potential (12.5) over all molecules, and calculating the total free energy, one finds that this vanishes at

$$\bar{u} = 4.54\,kT, \tag{12.7}$$

where $\bar{u}$ is the average of $u(r_{ij})$ over the intermolecular distances. Equation (12.7) thus determines the critical temperature. Combining this with (12.5), one sees that $V_{ij} \sim T_0$, leading to the fact that $S(T) = S(T/T_0)$, and hence the universal character of the order–temperature curve, Fig. 12.7.

This discussion suggests that in principle any molecular substance with anisotropic molecules should exhibit a mesophase character. The fact that relatively few compounds do is explained by noting that the much stronger scalar intermolecular potential acting in addition to $V_{ij}$ of (12.5) usually causes the freezing of the liquid at a temperature higher than $T_0$, thus inhibiting the formation of the mesophase. To encourage the occurrence of the mesophase, one thus attempts either to increase the molecular anisotropy or to depress the freezing point. Many new liquid crystals have been synthesized on the basis of these rules.

Measurements on smectogens indicate that the order function is essentially independent of temperature. The explanation is that the temperature is so low that $S$ is close to its low-temperature limit.

### Elasticity

The orientation-inducing forces contribute to the elastic properties of a liquid crystal. The corresponding elastic energy is zero when the director $\mathbf{n}(\mathbf{r})$ is the same everywhere, but if the crystal is deformed the elastic energy increases in a manner depending on the type of deformation. The most general expression for the energy density, which must be even in $\mathbf{n}(\mathbf{r})$, is

$$E_e(\mathbf{r}) = \tfrac{1}{2}\{K_1[\nabla\cdot\mathbf{n}(\mathbf{r})]^2 + K_2[\mathbf{n}(\mathbf{r})\cdot\nabla\times\mathbf{n}(r)]^2 = + K_3[\mathbf{n}(\mathbf{r})\times(\nabla\times\mathbf{n})]^2\}, \tag{12.8}$$

where the first term on the right represents a pure divergence, and the second a pure twisting, and the last a pure bending of the field lines of the director (Fig. 12.8). The elastic constants $K_1$, $K_2$ and $K_3$ are small, $\simeq 10^{-6}$ dyne, and decrease rapidly as the temperature is raised.

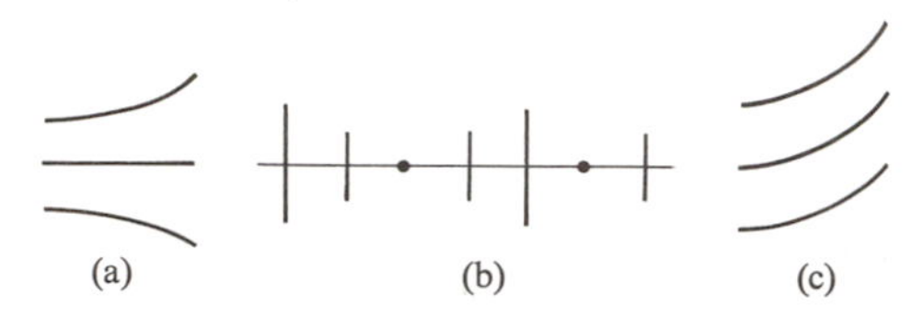


**Fig. 12.8** The (a) divergence, (b) twisting, and (c) bending deformations of the director.

## Magnetic effects

A magnetic field produces important effects in liquid crystals. A macroscopic free liquid crystal system is actually isotropic. The reason is that even though the system in any one small neighborhood is anisotropic, the director $\mathbf{n}(\mathbf{r})$ varies continuously from one region to another, so that the system as a whole is isotropic. (The situation is analogous to a ferromagnetic system, in which the random directions of the oriented domains result in an isotropic solid.) But when an external magnetic field is applied, the director tends to align with the field everywhere. The system is no longer isotropic, as can be detected, for example, by measuring the dielectric constants along and perpendicular to the field, $\epsilon_{\parallel}$ and $\epsilon_{\perp}$. Such measurement shows that $\epsilon_{\parallel} > \epsilon_{\perp}$. A complete orientation of the mesophase may be achieved by applying a field of a few kilogausses. This is a relatively small field, indicating once more that the internal forces involved are rather weak.

The reason for the alignment of $\mathbf{n}(\mathbf{r})$ with the magnetic field is that the magnetic susceptibility in the direction parallel to $\mathbf{n}(\mathbf{r})$, $\chi_{\parallel}$, is greater than in the perpendicular susceptibility, $\chi_{\perp}$. One can show that the density of the magnetic energy is

$$E_m = -(\Delta\chi)B^2(3\cos^2\phi - 1)/6, \tag{12.9}$$

where $\Delta\chi = \chi_{\parallel} - \chi_{\perp}$ is the susceptibility anisotropy and $\phi$ the angle between $\mathbf{n}(\mathbf{r})$ and the magnetic field $\mathbf{B}$. Note that $E_m$ is even in $\cos\theta$, as it should be, and that the angular factor is $P_2(\phi)$, except for a factor of $\frac{1}{2}$, resulting from the fact that the magnetization process takes place gradually as the field is raised from 0 to the final value of $B$. The energy is smallest at $\phi = 0$, where $\mathbf{n}(\mathbf{r})$ is parallel to $\mathbf{B}$, and greatest at $\phi = \pi/2$, where $\mathbf{n}(\mathbf{r})$ is normal to $\mathbf{B}$.

The fact that the director aligns with the field means that $\chi_{\perp} < \chi_{\parallel}$, which can be attributed to the fact that the molecules are more readily magnetized along their axes than in the normal direction. This can of course be related to the microscopic structure of the molecules. In particular one notes that in a chain of benzene-related rings, the magnetization is larger in the plane of the rings than in the normal direction.

Usually another effect must be considered in connection with magnetic measurements (which are often carried out by sandwiching the system between two glass plates): the surface. The effects of the surface on the orientation of the director are little understood at present, but it is known that if a pretreated glass surface has been rubbed many times, the system is uniformly oriented, with the director parallel to the surface and the rubbing direction.

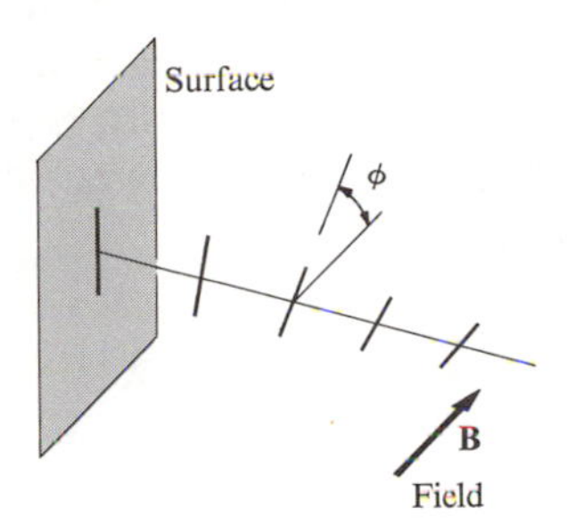


**Fig. 12.9** Twisting of the director due to a magnetic field in the region close to the surface.

Let us now study the combined effects of the surface and the magnetic field. Figure 12.9 shows that, as the distance from the surface $x$ increases, the director gradually aligns with the field. The alignment does not take place abruptly because of the elasticity of the medium. Since the field **B** is uniform, the director simply twists as $x$ increases. Thus

$$\nabla \cdot \mathbf{n}(\mathbf{r}) = \frac{d\phi}{dx} \qquad \text{and} \qquad E_e = \tfrac{1}{2}K_2(d\phi/dx)^2,$$

according to (12.8). The total energy is the sum of the elastic and magnetic energies

$$E = E_e + E_m = \tfrac{1}{2}\{K_2(d\phi/dx)^2 - \Delta\chi B\cos^2\phi\}, \tag{12.10}$$

where, in substituting for $E_m$ from Eq. (12.9), we have ignored the constant term, as it is irrelevant to the following discussion. The rate at which the director twists can now be found by minimizing the total energy (12.10), which leads to (see the problem section at the end of the chapter),

$$\tan(\phi/2) = e^{-x/\xi}, \tag{12.11}$$

where

$$\xi = (K_2/\Delta\chi)^{1/2}/B. \tag{12.12}$$

For small $x$—that is, very close to the surface—$\tan(\phi/2) \simeq 1$, and $\phi = \pi/2$, that is, **n** is normal to **B**. But as $\chi$ increases, $\tan(\phi/2)$ decreases and so does $\phi$—that is, **n** is approaching **B**, until at $x \gg \xi$, $\phi = 0$, and **n** is exactly along **B**. The length $\xi$, which represents the width of the transition region near the surface, is called the *coherence length*. This depends on $B$, and for $B = 5$ kG, $\xi$ is typically about $2\mu$.

Very interesting effects are produced when a magnetic field is applied to a cholesteric liquid crystal. Suppose that the field is normal to the helical axis, i.e., the field is parallel to the plane of the cholesteric sheets. In view of the above discussion, the field tends to align the molecules parallel to **B**, and thus "unwind the helix," but this is resisted by the internal molecular forces which have produced the helical structure in the first place. The result is a compromise; the effect of the field is to lengthen the pitch of the helix. A quantitative treatment is carried out by writing the total energy

$$E = \frac{1}{2}\left\{K_2\left(\frac{d\phi}{dx} - \frac{2\pi}{Z_0}\right)^2 - \Delta\chi B^2 \cos^2 \phi\right\}, \tag{12.13}$$

where $\phi$ is again measured from the direction of **B**. In writing the elastic energy, the first term on the right, we have subtracted the apparent strain associated with the free (natural) twist, $2\pi/Z_0$, leaving only the real strain, $d\phi/dx - 2\pi/Z_0$. The new pitch $Z$ is found by minimizing $E$ and solving the resulting equation. Although the procedure is straightforward, the solution of the differential equation is rather involved. The results are in good agreement with experiment (Fig. 12.10).

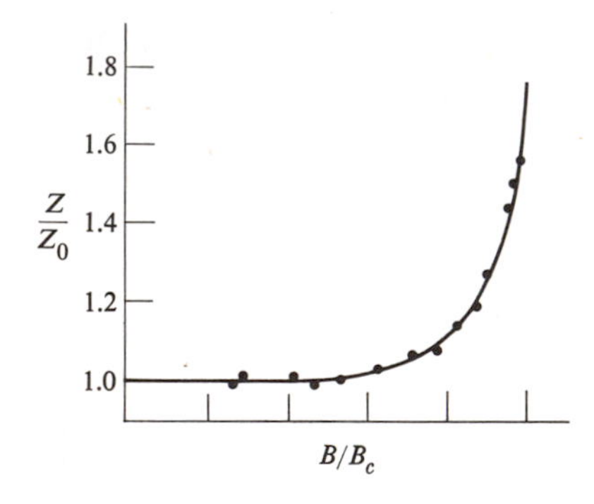


**Fig. 12.10** The pitch $Z/Z_0$ versus the magnetic field $B/B_c$ for the cholesteric mixture of cholestric acetate in 4, 4-dimethoxyazoxybenzine (III) at 119°C. [After R. B. Meyer, *Appl. Phys. Lett.* **14**, 208 (1969)]

The mathematical solution also shows another interesting result: The pitch of the helix becomes infinite at a *critical field* $B_c = \pi^2(K_2/\Delta\chi)^{1/2}/Z_0$. At this field the cholesteric structure disappears entirely, and the system enters a nematic phase. Such a field-induced transition from a cholesteric to a nematic phase has indeed been observed, and the observed field is in good agreement with theoretical predictions.

The effects of a magnetic field on a smectic phase are very slight, due to the fact that the internal forces are appreciably larger than those of the field. But even such a substance can be reoriented by the field, if this is applied to the isotropic liquid and the system is cooled through the critical temperature. The molecules in the isotropic phase, being free to rotate, align in clusters parallel to the field, and then serve as nuclei of growth for the smectic phase as the substance is cooled down.

**Optical properties**

The importance of the optical properties of liquid crystals has already been emphasized, when we stated that it was the anisotropy of the index of refraction which first led to the recognition of the mesophase as a distinct state of matter. We shall elaborate further on these properties here, beginning with the nematic and smectic phases. The cholesteric phase has its own peculiar optical characteristics, which will be considered subsequently.

In a completely oriented nematic or smectic phase, the index of refraction is anisotropic. Specifically, the system acts as a *uniaxial* medium, in which the index of refraction along the director $n'_{\parallel}$ is greater than the index of refraction in the normal direction $n'_{\perp}$ (the prime is used to distinguish the index of refraction from the director). The result $n'_{\perp} < n'_{\parallel}$ is attributed to the fact that the molecules are more easily polarized along their axes than in the perpendicular direction, as we have mentioned previously. The anisotropy in the refractive index leads to a large, positive birefringence, typically about 0.3, which is to be compared with the small value 0.01 in quartz.

*Dichroism* is also observed in liquid crystals, i.e., the absorption of a light wave depends on the directions of propagation and polarization of the wave. This property has been used in manufacturing polaroid plates from liquid crystal materials.

The turbid appearance of the nematic phase is due to the strong scattering of the light beam from the substance. This scattering is caused by the thermal fluctuations of the director around its equilibrium direction. The relaxation time for these fluctuations is about $10^{-7}$s, but the actual value depends on the temperature.

We have already remarked on the great optical activity in the cholesteric phase. Another interesting property in this phase is that the phase exhibits selective reflection, depending on the wavelength of the beam and the helical pitch. Regarding the substance as a periodic structure with a period equal to the pitch $Z_0$, and applying Bragg's law, one has

$$2\,Z_0 \sin\theta = \lambda.$$

For a typical value of $Z_0$, 2500 Å, the reflected beam falls in the visible range. It is this type of reflection which is responsible for the fascinating colors exhibited by thin films of cholestrogens. The reflected beam may also be modulated by a magnetic field which, as discussed above, lengthens the pitch and consequently shifts the beam toward the red side of the spectrum.

**Applications**

The properties of liquid crystals have been used in the development of many physical devices, particularly those of the electro-optical variety. These devices have not yet been put to use on a large scale, but it is hoped that they will soon.

Cholesteric substances are used for various purposes. Since the forces responsible for the helical structure are weak, even small perturbations of pressure, temperature, and electric or magnetic fields produce a sufficient change in the helical pitch to be readily detected by observing the light reflected from the substance. Thus cholestrogens are used to measure stresses and temperatures, as well as fields. They are also used as detectors of ultrasonic or electromagnetic radiation (because the energy absorbed by the substance raises its temperature) and in the manufacture of polaroid plates.

Nematic substances have been used in electro-optical display devices. When a thin layer is placed between two electrodes, the layer appears transparent at first because the substance is presumably oriented by the surface. If a voltage above a certain threshold value of, say, 5 V is applied across the electrodes, the compound suddenly turns murky white, i.e., scattering light. If one of the electrodes has a certain design on its surface, this design can be displayed optically, and modulated electrically. The physical process responsible for the murky appearance is probably the following: In the absence of voltage, the molecules are oriented with their axes parallel to the surface. When voltage is applied, ionic impurities in the substance are accelerated by the field and set into a drift motion between the electrodes. Since the drift velocity is normal to the axes of the molecules, the impurities collide frequently with the molecules, causing much turbulence, which is responsible for the light scattering (usually referred to in this context as *dynamic scattering*).

It also seems plausible that an ac voltage applied to a turbulent nematic may turn it into a transparent liquid, provided the frequency is high enough (the static voltage is presumed to be removed), because an ac field tends to orient the molecules parallel to it; and since these are free to rotate, they flip back and forth with the field. However, the ionic impurities, being massive, cannot follow the field at high frequency, and hence they remain stationary. This effect has indeed been observed at a frequency of 4000 Hz with a voltage of amplitude 50 V.

Another application of liquid crystals in display devices involves the operation of the liquid crystal in the so-called *twisted nematic mode*. The substance is sandwiched between two transparent electrodes, with two external polarizers placed adjacent to the electrodes, one on each side. The electrodes' surfaces are treated such that the axes molecules at the two electrodes are rotated at 90° relative to each other. The polarizers are also set such that their directions are 90° relative to each other. A plane-polarized light from the first polarizer has its polarization rotated as it passes the medium and thus passes through the second polarizer. If a voltage is applied, the molecular axes rotate, the polarization is not properly rotated in the medium, and little or no light passes the second polarizer. With a segmented display, the areas over which voltage is applied appear dark on a light background. (If the polarizers were set in parallel directions, these segments would appear bright on a dark background.)

The advantages of liquid crystal electro-optical devices are: (a) Low power consumption, since the device does not generate light but merely reflects it. (b) Clarity of image under normal lighting conditions (no dimming of the ambient light is necessary, as in the case of a conventional television screen). (c) Many liquid crystals are inexpensive and readily available.

## 12.4 POLYMERS

Polymers have molecules that are very long and chainlike, usually extending over several thousand angstroms. Because of their great length these molecules, which are usually organic, are referred to as *macromolecules*. Polymers include several classes of materials which we encounter frequently in our daily life, such as natural rubber, wood (which is primarily cellulose), hair, and skin. Synthetic polymers include foam rubber, plastics, many synthetic fibres (nylon, dacron, etc.), and adhesives, among other materials. Indeed the rapid advances in the technology of synthetic polymers are likely to produce a major impact on the materials we shall be using in the years to come. Some polymers are also important in the functions of biological organisms, but we shall postpone discussion of these *biopolymers* until Chapter 13.

Because of their molecular construction, polymers exhibit some common physical properties, and in this section we shall study these properties and show how they are related to the structure of the molecule.

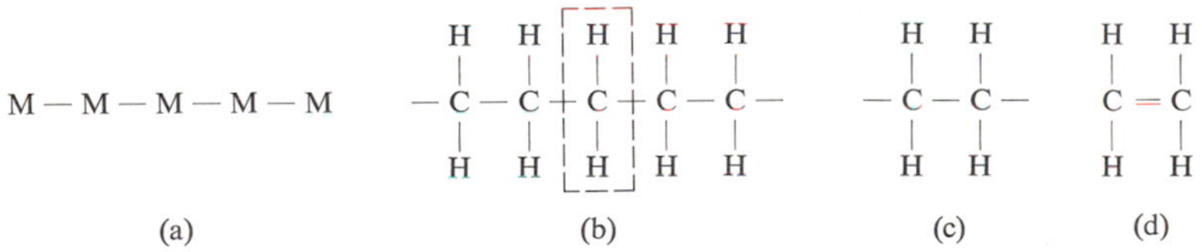

```
                              ┌ ─ ┐
                     H   H  | H |  H   H          H   H          H   H
                     |   |  | | |  |   |          |   |          |   |
M — M — M — M — M   —C — C ─┼ C ┼─ C — C—       —C — C—          C = C
                     |   |  | | |  |   |          |   |          |   |
                     H   H  | H |  H   H          H   H          H   H
                              └ ─ ┘
       (a)                    (b)                  (c)            (d)
```

**Fig. 12.11** (a) Arrangement of a polymer as a chain of monomers. (b) Structure of polyethylene; dashed line encloses the monomer. (c) Structure of ethylene group as it enters polyethylene. (d) Structure of free polyethylene molecule.

The structural arrangement of a single macromolecule is shown schematically in Fig. 12.11(a). It is composed of a repetition of building block M, called a *monomer*, i.e., the molecule is a chain of monomers. The binding forces holding the monomers together are usually covalent, or ionic, in nature, and consequently very strong. In addition, two neighboring macromolecules are held together by lateral weak van der Waals forces. A common polymer having a simple structure is polyethylene, which is shown in Fig. 12.11(b). The monomer here is the ethylene group, $C_2H_4$, whose structure is shown in Fig. 12.11(c). A free ethylene molecule in the gaseous state actually has the structure shown in

Fig. 12.11(d), but since the carbon atom has a proclivity for the classic tetrahedral bond, it requires only a little additional energy to break one of the carbon double bonds and "open up" the molecule, as indicated by Fig. 12.11(c). The group is now ready to join with other ethylene groups to form the macromolecule of Fig. 12.11(b). The number of monomers in a single macromolecule is called the degree of polymerization (DP), which is typically $10^4$, or even more.

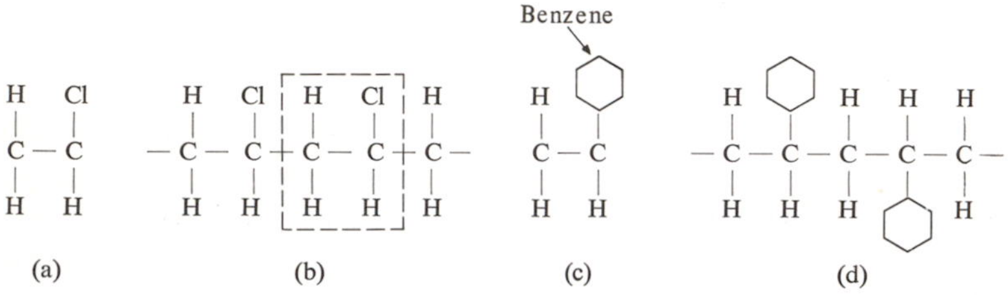

Benzene

(a) (b) (c) (d)

**Fig. 12.12** (a) The vinyl chloride group. (b) Polyvinyl chloride; dashed rectangle encloses the monomer. (c) The styrene group. (d) Polystyrene.

If one uses a vinyl chloride group ($C_2H_3Cl$), in which one of the hydrogen atoms in ethylene is replaced by chlorine, as shown in Fig. 12.12(a), the result is the polyvinyl chloride polymer illustrated in Fig. 12.12(b). It is also possible for one of the hydrogen atoms in the ethylene monomer to be replaced by a large and complex group. In a styrene monomer, for example, this side group is a benzene molecule, Fig. 12.12(c), and the resulting polystyrene macromolecule is shown in Fig. 12.12(d). The type of side group involved has an important bearing on the mechanical properties of the polymer. In the substances mentioned so far, the backbone of the molecule consisted of carbon atoms, but some of these may be replaced by other atoms, such as oxygen or sulfur; this also can influence the mechanical properties.

If the macromolecule has short chains attached to it, replacing some side groups, as shown in Fig. 12.13(a), we have a *branched* polymer. Note that the branch is attached to the main molecule by a strong covalent bond. A branch joining two long chains is called a *cross link*, and a polymer may contain a large

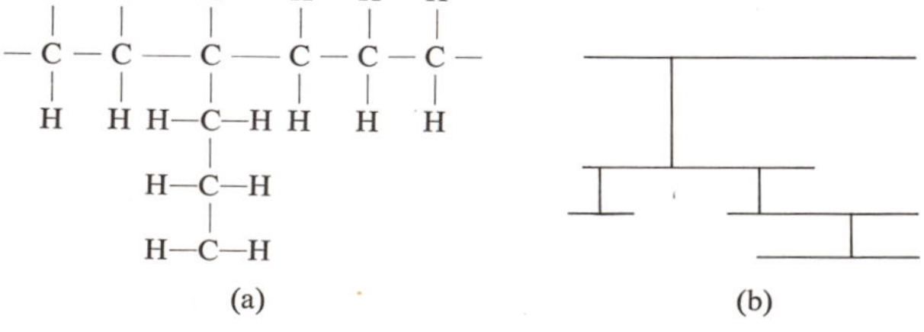

(a) (b)

**Fig. 12.13** (a) A short chain replaces a side group. (b) A branched polymer.

number of these. The effect of these links is to produce a polymer which is strong, but also one which tends to be brittle, as the links resist bending.

Polymeric substances are grouped according to the amount of cross-linkage present. A *thermosetting* polymer has a large number of cross links, forming what amounts to a complex network of long chains connected by short branches. Such material is strong at room temperature, and retains its shape; an example of this type is bakelite. At the same time, such material tends to be brittle, like glass, and cannot withstand any bending. When a thermosetting polymer is heated, many of the links are broken and the strength is reduced; this phenomenon is referred to as *degradation*. The other type of polymer is *thermoplastic*. Here the cross-linking is fairly limited. The material is weak and can readily be molded into any desired shape; when heated it becomes quite plastic.

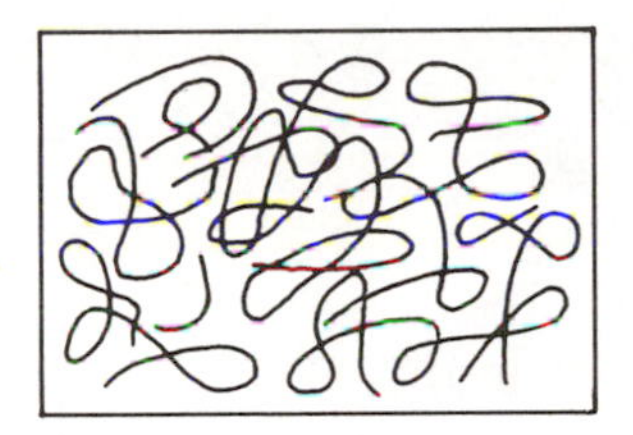

**Fig. 12.14** A polymer in the liquid state.

## Effects of temperature

One of the important characteristics of polymers is their sensitivity to temperature. At high enough temperature, a polymer exists in the liquid state, in which it usually has a thick, rubbery texture. Each molecule is folded around itself, and around others, many times over, resulting in a very complex molecular arrangement (Fig. 12.14), rather like the strands in a bowl of spaghetti. The molecules are constantly twisting and wriggling, due to thermal excitation, so that each molecule constantly changes its shape and position, but at any one instant the result is an amorphous distribution of molecular matter. When the temperature is lowered, changes take place in the system, and Fig. 12.15 illustrates this by plotting the

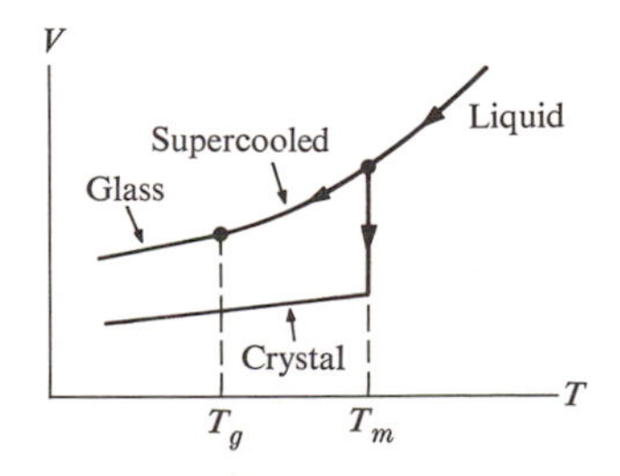


**Fig. 12.15** Various possible states in cooling of a polymer.

specific volume versus the temperature. The volume decreases gradually until the melting point $T_m$ is reached, whereupon, if the cooling is accomplished slowly, the polymer undergoes a discontinuous decrease in volume. The system is now in a crystalline state, and further reduction of the temperature causes a further decrease in the volume. The system is composed not of one single crystal, but of a large number of crystallites separated from each other by regions of supercooled liquid, as shown in Fig. 12.16.

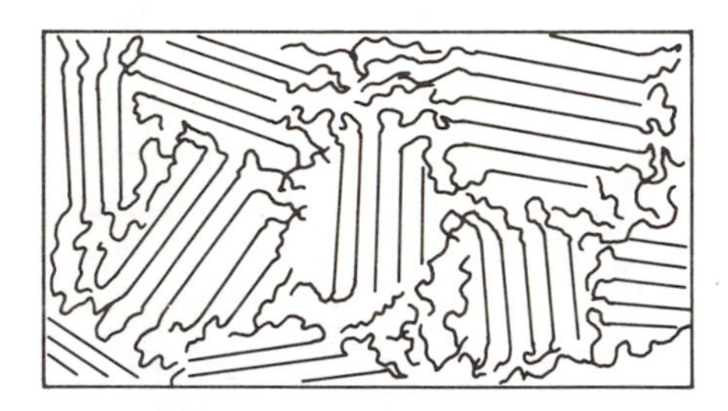

**Fig. 12.16** Fringed micell structure. [After P. J. Flory, (1953)]

This description of the polymeric crystalline state is referred to as the *fringed micell model.* Note that within each crystallite the macromolecules are aligned parallel to each other, somewhat as in a regular crystal. Note also that a single molecule may participate in several crystallites. However, under particularly favorable conditions, a truly single crystal can be grown from dilute polymer solution, such as the polyethylene shown in Fig. 12.17.

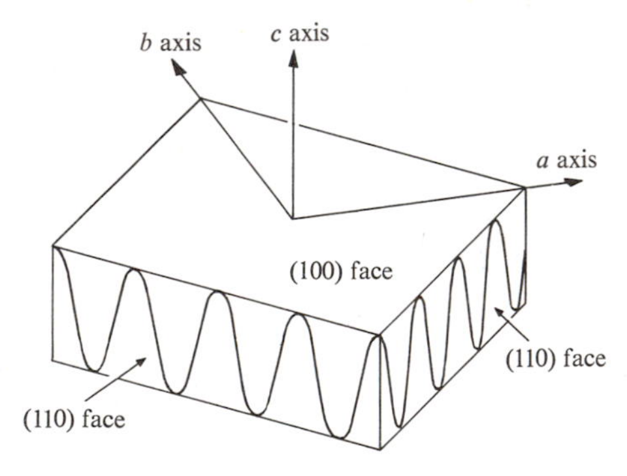


**Fig. 12.17** Folded-chain structure for polyethylene.

Under most circumstances a liquid polymer does not actually crystallize at the temperature $T_m$, but enters a supercooled liquid state, as shown in the upper curve of Fig. 12.15. Here the system behaves as a highly viscous liquid. The molecules are arranged randomly so that the structure is an amorphous one, but they continue to move and wriggle, though to a lesser extent than in the true liquid state. At some yet lower temperature $T_g$, the system undergoes another change to a new glassy, or vitreous, state. Here the system behaves as an amorphous solid which is strong and brittle, much as an ordinary glass is.

In the practical uses of polymers, the values of the temperatures $T_g$ and $T_m$ with respect to room temperature $T$ are of vital importance. If $T < T_g$, the substance is in the glassy state, and is strong and brittle. On the other hand, if $T_g < T < T_m$, the substance, as a highly viscous liquid, is plastic and ductile. Of course, in most applications, polymers are used in the glassy state, since only then do they have the required mechanical strength.

The values of $T_m$ and $T_g$ depend on the nature of the molecular bonds of the side-group molecules, and on the length and flexibility of the molecules. The stronger the bonds, the higher are these temperatures. However, since the bonding is due to weak forces, these temperatures are relatively low (100–200°C). The temperatures $T_m$ and $T_g$ may be raised if side molecules with polar bonds are introduced. By employing appropriate manufacturing techniques or varying chemical composition, in general one can arrange it so that $T_g$ and $T_m$ fall within a range suitable for the given application.

The reason that it is usually hard to achieve crystallization in polymers is primarily that the length of the molecules and the complexities of the side groups make it hard for the molecules to enter an ordered state. Thus polyethylene crystallizes quite readily because of the simplicity of its structure, but the chlorine atoms in polyvinyl chloride, being larger and more complex than hydrogen atoms, interfere with crystallization, and have the effect of depressing the melting point, or even preventing crystallization altogether. This applies even more forcefully to the effect of the benzene rings on the crystallization of polystyrene. The cross-linkage that may be present also inhibits the tendency of the molecule to go into the ordered state demanded by crystallization. Let us look at the liquid–crystal transition from a thermodynamic point of view. The change in free energy upon crystallization is (see Section 12.5)

$$\Delta F = \Delta E - T\,\Delta S, \tag{12.14}$$

where $\Delta E$ and $\Delta S$ are the changes in the internal energy and entropy of the system, respectively.

Now $\Delta E$ is negative because each molecule, upon crystallization, is at its equilibrium position, but its magnitude is small, since the forces involved are of the van der Waals type. By contrast, $\Delta S$ is large and negative, because the entropy of the liquid state far exceeds that of the crystalline state. To appreciate this, remember that a macromolecule can bend at every one of its many joints, and therefore has an enormous number of possible orientations. Since entropy increases with the number of possible orientations (see Section 11.5), there is a great amount of entropy associated with the liquid polymeric state. It follows therefore that the entropy term in (12.14) usually dominates the internal-energy term, that is, $\Delta F > 0$, and the system is prevented from crystallizing.

### Mechanical properties

Polymers exhibit a diverse range of physical properties, but it is the mechanical properties which are usually of prime interest. Mechanical properties depend on the state of the polymer. Here we shall concentrate mostly on the supercooled and glassy states. If a tensile stress is applied to a supercooled polymer, the substance flows plastically, as shown in Fig. 12.18, which depicts the strain as a function of time; the substance acts as a viscous liquid. Experiments show, however, that the response of the system also depends more precisely on the time scale of the applied stress, and that, if a rapidly alternating stress is applied, the supercooled polymer shows some elasticity. This property, combining both viscosity and elasticity, is referred to as *viscoelasticity*. A polymer in the glassy state also exhibits viscoelasticity, except that the viscoelastic strength is much larger than the strength in the supercooled state.

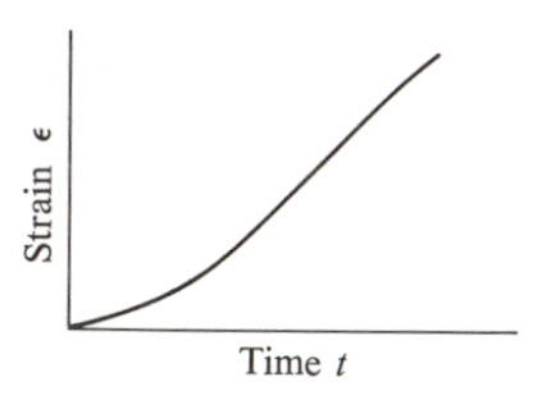


**Fig. 12.18** Strain $\epsilon$ versus time $t$ for a polymer, illustrating viscous property.

Another property is the great extensibility and flexibility of rubber. Under tensile stress, the sample may increase to several times its original length. A material capable of this behavior is known as an *elastomer*. Rubber is the best-known example.

All these properties can be explained in terms of molecular structure. We have seen that the molecules in a polymer are coiled in a very complicated manner around themselves and around each other, and furthermore, above $T_g$, they are wriggling about due to thermal excitation. When a tensile stress is applied, it acts by pulling at the ends of each molecule, causing it to *uncoil*. This is how the molecule, and consequently the sample, elongates. If the stress is maintained for a long time, the molecule, after the uncoiling process is completed, begins to slide past neighboring molecules. This sliding is an irreversible process. Once it has occurred, the polymer never returns to its original shape. Herein lies the physical basis for plastic flow and its attendant viscosity. Note, however, that before sliding takes place, the uncoiling process is reversible. When the stress is removed, each molecule coils back to its original shape. This is the basis for the elastic property mentioned above.

Let us now delve more deeply into the uncoiling mechanism. A straightforward, but incorrect, model suggests itself: that the folded molecules uncoil by sliding past each other as a result of the pulling action, much as a ball of string would straighten out if pulled at the ends. Since such sliding is irreversible, this model can account neither for the elastic property mentioned above nor for the substantial decrease in the Young's modulus observed as the temperature increases in the supercooled range ($T_g < T < T_m$).

The correct model is based on the fact that the uncoiling process is accomplished by rotations of the various segments in the backbone of the polymer molecule around the C—C bonds. The point is illustrated in Fig. 12.19, showing an ethane molecule connected by a single C—C bond. The right side of the molecule can rotate around the axis as shown, and may take up several positions, or *conformations*. These conformations are not necessarily all of the same energy, but if the energy differences involved are less than, or comparable to, $kT$, then all conformations are accessible, and the molecule flips back and forth between them as a result of the thermal excitations. The speed of the rotation increases rapidly with temperature, as in all similar processes. In a long molecule various segments of the molecule are incessantly rotating between available conformations, in a random fashion. When a stress is applied, the molecules accommodate this by rotating to those conformations which make the molecules longest without sliding taking place. Conversely, when the stress is removed, the molecule returns through segmental rotations to the shape with the greatest disorder, which is, more or less, the original shape.

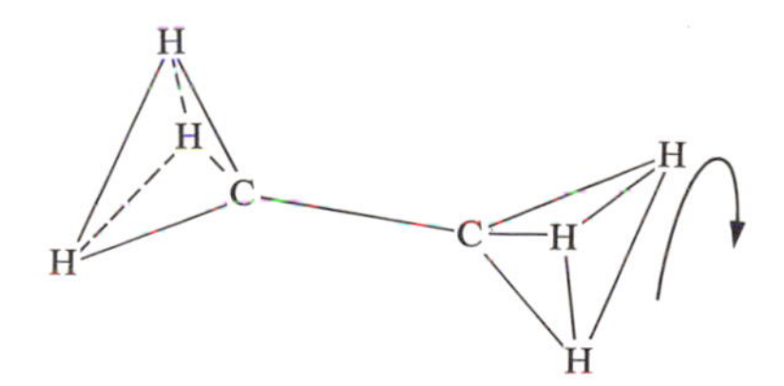


**Fig. 12.19** Possibility of rotation in ethane molecule.

The elasticity of rubber is explained in the same manner as the thermally induced uncoiling and coiling processes. The above model suggests that there is no energy change involved in stretching a piece of rubber. The resistance to stretching is caused not by the increase in internal energy, but by the decrease in entropy when the molecule uncoils. A useful analog is the case of an ideal gas, which resists compression because of the reduction in entropy associated with the decrease in volume, and not by an increase in energy, which remains constant under isothermal conditions. Just as the pressure of an ideal gas is proportional to its absolute temperature $T$, so the elastic constant (Young's modulus) is also proportional to $T$. The great elasticity of rubber is possible only when the side groups are simple enough not to interfere with the segmental rotations of the molecules.

### Electrical properties

Let us now take a look at the electrical, dielectric, and optical properties of polymers. Most pure polymers exhibit very small electrical conductivity; in fact, some of them are used for insulation purposes. The addition of impurities may significantly increase the electrical conductivity. Many *hydrophilic* (water-absorbing) polymers show good conduction when wet, and poor conduction when dry. This type of conductivity seems to be associated with the ionic conductivity of the protons. Generally speaking, *hydrophobic* (water-repelling) polymers are highly resistive.

The question of electronic conductivity in polymers is an interesting one, and some polymeric substances do, in fact, show appreciable conductivity of an electronic nature, but we shall postpone discussion of these to Chapter 13.

Dielectric properties are investigated by the use of a static or low-frequency electric field. Many polymers have high dielectric constants, and are sometimes used in the manufacture of capacitors. The polarization responsible for the dielectric property is primarily due to the polarization induced in the side groups, and is particularly large in polar side groups, such as chlorine and hydroxyl ions. The motion and orientation of these groups can be studied by measuring the frequency-dependent dielectric constant, and examining both the real and imaginary parts, as described in Section 8.9. The relaxation time is the inverse of the peak frequency of the imaginary part. These measurements indicate that one needs to introduce several relaxation times—not just a single one—which is expected, since some side groups are more mobile than others, depending on their local environments.

The optical properties of polymers are similar to those of other insulators. Since the frequency of the impressed field is large, only the electronic contribution to polarization is effective. Dipolar and atomic contributions cannot follow the field (Sections 8.6 and 8.8). Thus the index of refraction $n$ is determined primarily by the polarization of the clouds of electrons around the ionic centers and in the various bonds. In crystalline polymers the index of refraction is anisotropic, and the material exhibits optical birefrigence. Even amorphous substances may exhibit birefringence under some circumstances. For example, by stretching the substance, one can orient the planes of benzene rings of polystyrene in a certain direction. Since the $\pi$-electrons are more polarizable along the plane of the ring than perpendicular to it, the index of refraction is larger in the plane of the rings than in other directions, and the material becomes birefringent.

## 12.5 NUCLEAR MAGNETIC RESONANCE IN CHEMISTRY

Nuclear magnetic resonance (NMR) is one of the principal spectroscopic tools the modern chemist uses to study molecular structure. Other spectroscopic methods have been used—including optical, infrared, and even Raman spectroscopy—but the development of the NMR technique since the early 1950's has provided the chemist with one of the most accurate methods for determining molecular structure. The method can also be used in chemical analysis, and in studies of rates of chemical reaction.

We discussed the physical basis of the NMR technique in Section 9.13, in connection with the magnetic properties of matter. We shall review it here only briefly, with particular bias toward chemical applications. An atomic nucleus has a magnetic dipole moment $\mu$ which may be expressed as

$$\mu = g_n \mu_{Bn} I, \tag{12.15}$$

where $\mu_{Bn}$ is the nuclear magneton and $I$ the spin quantum number. The nuclear $g$-factor $g_n$ is a numerical constant which varies from one nucleus to another, and depends on the manner in which the moments of the nucleons, which make up the nucleus, are coupled to each other. The allowed values of the spin $I$ are 0, $\frac{1}{2}$, 1, etc. When $I = 0$, then $\mu_n = 0$, and the nucleus evinces no magnetic response and is of no further interest to us here. When $I > 0$, the nucleus exhibits magnetic response.

The nucleus of most interest in NMR is the proton, for which $I = \frac{1}{2}$. (Other nuclei commonly present in organic compounds, made up of carbon, hydrogen, and oxygen are $C^{12}$ and $O^{16}$, both of which are nonmagnetic.) This nucleus may be visualized, semiclassically, as a rotating spherical charge with the magnetic moment pointing along the axis of rotation. Those nuclei for which $I > \frac{1}{2}$ cannot be represented so simply, because in addition to their dipole moments they also have quadrupole and even higher moments, indicating a nonspherical distribution of nuclear charge. Since our interest lies primarily in the proton, we shall be concerned here only with the dipole moment.

When an external field $\mathscr{H}_0$ is applied to the sample,† the energy of the nucleus is split into $(2I + 1)$ sublevels, corresponding to this number of orientations of the nuclear moments relative to the field (note that the orientation direction is quantized, Section 8.2). For the proton, the multiplicity factor $2I + 1 = 2$, and hence the nuclear level splits into two sublevels, as shown in Fig. 12.20. (This is the nuclear analog of Zeeman splitting.)

The lower level corresponds to the proton moment pointing along the field, while the upper level corresponds to the moment pointing in the opposite direction. The energy difference between the two levels is $\Delta E = 2\mu_n \mathscr{H}_0$. As we said in Section 8.2, the system of nuclei is in resonance with an electromagnetic signal of frequency $\nu$ when the condition $h\nu = \Delta E$ is satisfied. That is

$$\nu = \frac{2\mu_n}{h} \mathscr{H}_0, \tag{12.16}$$

† We follow the common convention in NMR literature and use the cgs system in this section (and the next section also). Recall that 1 gauss or 1 oersted $= 10^{-4}$ Wb/m$^2$.

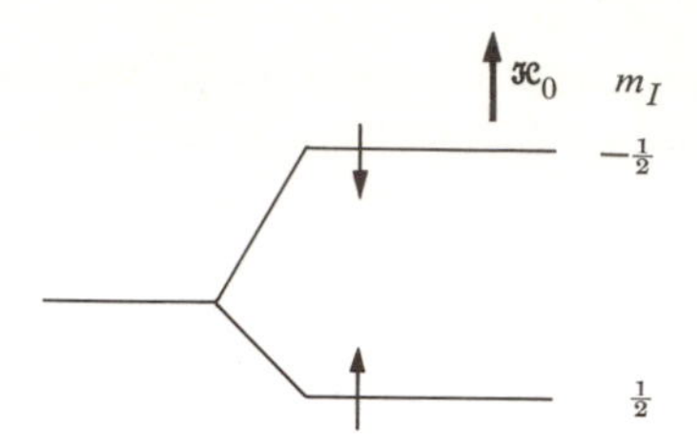


**Fig. 12.20** Two levels of a proton corresponding to two possible orientations in a magnetic field. Arrows at levels indicate orientations of the proton moment in these levels.

provided that the magnetic field of the signal is properly oriented relative to $\mathcal{H}_0$, the former being circularly polarized and normal to $\mathcal{H}_0$.† The resonance here reflects the fact that when (12.16) is satisfied, a proton in the lower level may absorb a photon from the signal in the upper level.

It is clear from (12.16) that by measuring the resonance frequency $\nu$ at a certain field, one may determine the nuclear moment $\mu_n$. Such information would be useful to the nuclear physicist interested in measuring nuclear moments, but it is of no use to the chemist whose interest lies in the environment outside the nucleus. The usefulness of NMR in chemistry, as in solid-state science, is based on the observation that the field felt by a nucleus inside the substance is not precisely equal to the external field $\mathcal{H}_0$. Rather this field is modified by a small field due to the environment in which the nucleus resides, and it is by measuring this additional field that we obtain information about the environment. The nucleus acts as our probe for investigating the internal structure through its monitoring of the environmental field.

Before discussing actual applications, let us say a little about experimental procedures: First, one holds the frequency fixed and varies the field, rather than the other way round, until resonance is achieved, because it is easier to vary the field than the frequency. Second, because the nuclear moment is so small compared with the electron moment (Section 9.13), the frequency $\nu$ lies in the radiofrequency (rf) range for the fields commonly used. This can be seen from (12.16), which may be written as

$$\nu = 2.13 g_n \mathcal{H}, \tag{12.17}$$

where $\nu$ is in MHz and $\mathcal{H}$ in kilo-oersteds. Thus, using $g_n \simeq 2.8$ for a free proton and $\mathcal{H}_0 = 10$ kOersted, one finds that $\nu \simeq 60$ MHz, which is in the rf range. The corresponding signal wavelength is about 1 meter. Spectroscopy in this range is easier than in the optical range because the circuit elements may be represented accurately by lumped parameters. In determining the internal magnetic field, one is not interested in the absolute value of this field because not only is

† If the signal is plane polarized, it may be resolved into appropriate circularly polarized waves, in the usual fashion, and only half the signal is effective.

it difficult to measure, but also it cannot be conveniently compared with theory, since the type of calculation involved is very complicated. One circumvents this by measuring only the relative field shift, by dissolving the substance in a standard liquid under standard conditions. One then compares the resonances of various protons in this substance, or with protons of a different substance dissolved in the same standard liquid at a different time. Several organic solvents have been used as reference liquids. These days, tetramethylsilane (TMS) is the one most favored, It is chemically inert, magnetically isotropic, and miscible with most organic solutes used.

Finally, the resolution in NMR spectroscopy is extremely good, about 1 part in $10^8$. To take full advantage of this fact, the external field $\mathscr{H}_0$ must be uniform throughout the sample, to the same degree of accuracy, in order that all protons see exactly the same external field.

The principal effect underlying the usefulness of the NMR technique in chemistry is the *chemical shift*. This refers to the fact that the field at the nucleus is not $\mathscr{H}_0$, but one which is modified by the chemical environment. As a result of the presence of $\mathscr{H}_0$, new electric currents are created in the electronic clouds surrounding the nucleus, and these produce a small field which opposes $\mathscr{H}_0$. That is, the induced currents act to magnetically shield the nucleus. Let us denote this shielding field by $\mathscr{H}_{sh}$. Then we may write, for the actual field seen by the nucleus,

$$\mathscr{H} = \mathscr{H}_0 - \mathscr{H}_{sh} = \mathscr{H}_0 - \sigma\mathscr{H}_0, \tag{12.18}$$

where we have indicated that the shielding field is proportional to $\mathscr{H}_0$, which is a reasonable supposition, since the induced currents are created by $\mathscr{H}_0$ itself. The proportionality parameter $\sigma$ is the shielding constant.

Let us illustrate this by an example. Figure 12.21(a) shows the low-resolution NMR spectrum for the protons in ethanol $C_2H_5OH$ (structure is shown in Fig. 12.21b). Three absorption lines are evident. Their intensities, as measured by the areas under the curve, are in the ratios 1:2:3. The lines are associated with the protons in the different radicals. There is one proton in the hydroxyl radical, in the methylene ($—CH_2—$) two equivalent protons, and in the methyl ($—CH_3$) three equivalent protons. This explains the above ratios, as the intensity for each radical is proportional to the number of equivalent protons therein. Since the observed frequency $\nu$ is fixed, the field $\mathscr{H}$ is the same for all lines, their differences lying in the different values of the shielding fields at the various resonance fields $\mathscr{H}_0$. The shielding fields are given by

$$\mathscr{H}_{sh} = \mathscr{H}_0 - \mathscr{H},$$

and hence the differences between the shielding fields at the various protons can be read directly from the figure. The shielding field increases from hydroxyl to methylene to methyl radicals. Although we cannot measure the absolute value of $\mathscr{H}_{sh}$, due to lack of knowledge of $\mathscr{H}$, the differences between the three different

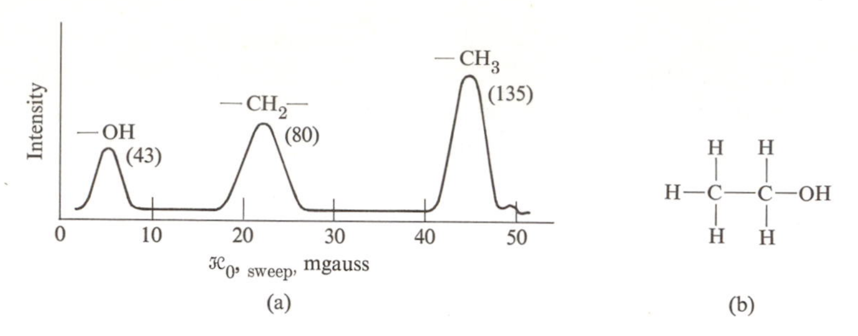


**Fig. 12.21** (a) Low-resolution NMR spectrum of protons in ethanol at 40 MHz and 9400 gauss: absorption intensity versus sweep field. Numbers in parentheses are experimental figures for areas under the corresponding peaks. [After Roberts (1959)] (b) The structure of ethanol.

$\mathscr{H}_{sh}$'s are indeed given by the differences between the peak fields in the figure. One now understands why the term "chemical shift" is used: The lines are shifted from each other by the shielding effect. It has also been demonstrated experimentally that the spacing between the lines increases in direct proportion to $\mathscr{H}_0$, when this field is varied, in accordance with the supposition made in (12.18).

In preparing tables of the chemical shift, one does not list $\sigma$, as it is far too small. Instead one lists a parameter $\delta$, which is defined as

$$\delta = \frac{(\mathscr{H}_{0,s} - \mathscr{H}_{0,r})}{\mathscr{H}_{0,r}} \times 10^6, \tag{12.19}$$

where $\mathscr{H}_{0,r}$ and $\mathscr{H}_{0,s}$ are, respectively, the resonance fields for a selected proton of the reference liquid and the proton of the substance under investigation which has been dissolved in the reference liquid. Using (12.18), one may write

$$\delta = (\sigma_s - \sigma_r)10^6,$$

showing that $\delta$ gives the relative change in the shielding field in parts per million. In fact, the so-called $\tau$-scale is commonly used, for convenience, where $\tau$ is defined as

$$\tau = 10 + \delta.$$

Table 12.1 lists the $\tau$-values for a few different groups of protons.

In principle, the procedure for using NMR in chemical analysis and determination of molecular structure is now clear. For use in chemical analysis, one can prepare a chart for the proton resonance fields for all available radicals (see the bibliography). In examining an unknown substance, you may compare your lines with those on the chart, and from this infer which protonic environments are present in the substance.

Here is an example of the use of charts in the determination of structure: Before the development of NMR techniques, the structure of diborane, $B_2H_6$,

**Table 12.1**
Observed Chemical Shifts of Protons in Some Aromatic Compounds (After Paudler, 1971)

| Compound | Group | Chemical shift, $\tau$ |
|---|---|---|
| Toluene | $—CH_3$ | 7.66 |
| Cumene | $—CH_3$ | 8.77 |
| Tetralin | $\alpha—CH_2—$ | 7.30 |
| | $\beta—CH_2—$ | 8.22 |
| Dibenzyl | $—CH_2—$ | 7.05 |
| Napthalene | $\alpha—CH=$ | 2.27 |
| | $\beta—CH=$ | 2.63 |

was unresolved between the two possibilities of the "bridge" structure and the ethane structure shown in Fig. 12.22. Since the observed spectrum indicates two different types of protons, the latter is ruled out, and the bridge structure is the correct one.

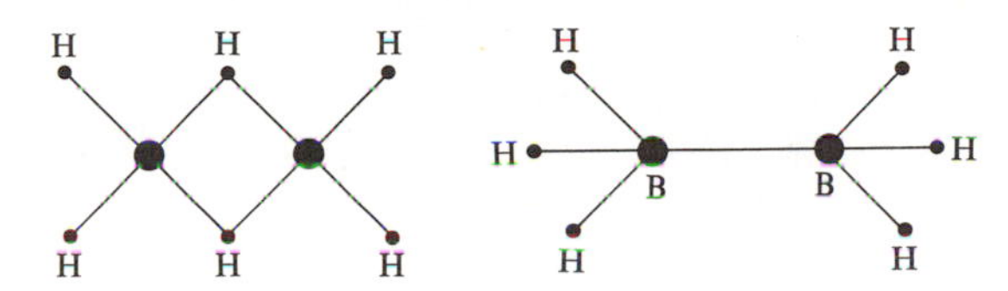


**Fig. 12.22** The two possible structures of diborane.

When you examine a resonance line more closely, using high-resolution techniques, you often find that it is composed of several finely spaced lines. The high-resolution spectrum for ethanol in Fig. 12.23 shows that the methylene and methyl lines are composed of four and three different lines, respectively. The total lines in the groups are still in the ratio 1:2:3, as before.

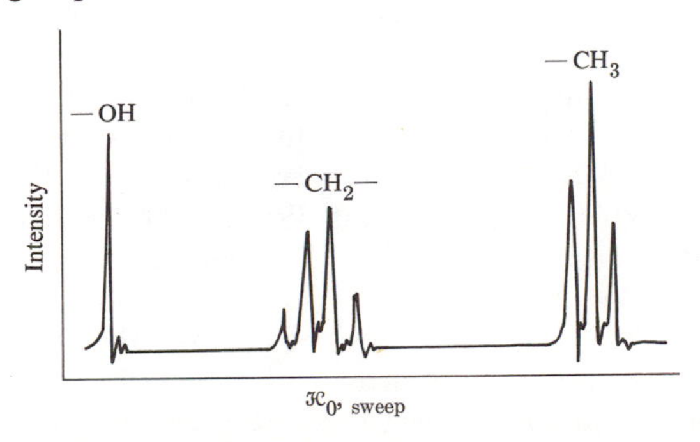


**Fig. 12.23** High-resolution NMR spectrum of ethanol.

The origin of line-splitting lies in the *spin–spin interaction* between the nuclei. Let us take the example of a proton in the methyl radical. Such a proton experiences a small magnetic field whose source is the dipole on the methylene radical (this in addition to the chemical shift discussed earlier), because, in effect, this radical acts as a tiny magnet. Now the field depends on the moment of the source dipole.

There are four ways in which the two moments can couple to each other, as shown in Fig. 12.24: Both moments are pointing upward, opposite to each other, or both downward. (Note that there are two different ways in which the protons may be oriented opposite each other, as shown in the figure.)

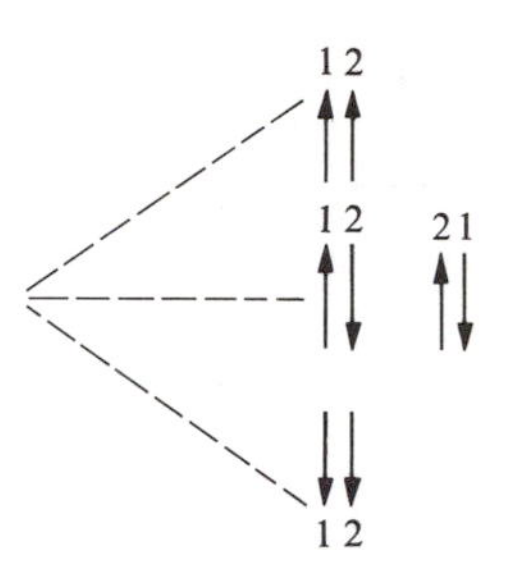


**Fig. 12.24** Four possible arrangements of the two proton moments in methylene group. Middle row indicates the two possibilities in which the moments cancel each other.

As time passes, the methylene radical occupies the various magnetic arrangements shown in the figure, with probability ratios 1:2:1 (why?). Each state has a different net dipole, and it is this which produces the field that acts on the resonating proton in the methyl group. It is clear, therefore, that the latter proton should split into three lines, in agreement with Fig. 12.23. The strongest line is due to the middle state of Fig. 12.24, and since this state has a zero moment, its field is zero and the line is actually undisplaced; the other two lines are placed symmetrically around it.

The number of high-resolution lines depends on the number of states available to the other radicals producing the field, and in turn the number of these states depends on how many equivalent protons are in the radical. The amount of splitting depends on the strength of the spin–spin interaction between the two radicals, and is denoted by $J$. This parameter $J$ depends strongly on the distance between the radicals, falling rapidly with increasing distance. (Note that the spacing of the multiplet $J$ is independent of the field $\mathscr{H}_0$, unlike the case of the chemical shift, which is proportional to $\mathscr{H}_0$.)

The same type of argument also shows that the line structure of the methylene line is a quartet, in agreement with Fig. 12.23.

A detailed investigation of the many features of the NMR spectrum—chemical shift, line splitting, intensities, etc.—can yield a wealth of information about a substance. Like any other powerful technique, the NMR method has grown immensely in recent years, and our brief coverage has highlighted only the basic aspects of the subject. You can find much more information in the references listed in the bibliography appended to this chapter. Applications of NMR in biology will be considered in Chapter 13.

## 12.6 ELECTRON SPIN RESONANCE IN CHEMISTRY

A perceptive reader, after the previous section on NMR, might ask whether a similar technique using electron spin resonance might be possible. Indeed it is, and the ESR technique is also widely used by chemists and materials scientists to investigate the microscopic properties of materials. This technique is also used increasingly in biological applications.

The physical basis of ESR, also called electron paramagnetic resonance (EPR), was discussed in Section 9.12. Here we shall review the subject only briefly, with the purpose of applying it to chemistry.

An electron in an atomic or molecular orbital has a magnetic moment $\boldsymbol{\mu}$, which may be expressed as

$$\boldsymbol{\mu} = - g\mu_B \mathbf{s}, \tag{12.20}$$

where $\mu_B$ is the Bohr magneton and $\mathbf{s}$ the spin quantum number vector of the electron.† The factor $g$ is 2 for a free electron, but in a substance the $g$-value may differ from this significantly because of the effects of the environment on the atomic orbital. The spin number $s$ may take the values 0, $\frac{1}{2}$, 1, etc., depending on the number of unpaired electrons and the manner in which they are coupled (Section 9.6).

When an external magnetic field $\mathscr{H}_0$ is applied to the sample, the electronic energy level splits into $(2s + 1)$ sublevels, corresponding to this number of orientations of the moment $\boldsymbol{\mu}$ relative to the direction of $\mathscr{H}_0$. This can be seen by noting that the additional energy arising from the interaction of the spin with the field, the Zeeman energy [see Eq. (9.36)], is

$$E_Z = - \boldsymbol{\mu} \cdot \mathscr{H}_0 = - g\mu_B m_s \mathscr{H}_0, \tag{12.21}$$

where we have used (12.20). The number $m_s$ is the projection of $\mathbf{s}$ along the $z$-axis and is called the *magnetic quantum number*. We recall from Section 5.6 that $m_s$ may take any of the values $s, s - 1, \ldots, - s$, which are $(2s + 1)$ in numbers; substitution of these into (12.21) leads to $(2s + 1)$ equally spaced energy levels.

Consider the simplest possible case: a single, unpaired electron for which $s = \frac{1}{2}$. In this case the original level splits into two sublevels, corresponding to $m_s = \frac{1}{2}$ and $m_s = - \frac{1}{2}$, as indicated in Fig. 12.22. The spacing between the levels is

$$\Delta E = 2g\mu_B \mathscr{H}_0. \tag{12.22}$$

† The vector $\mathbf{s}$ is defined as $\mathbf{S}/\hbar$ where $\mathbf{S}$ is the angular momentum vector see (Section A.4).

The ESR frequency $\nu$ is given by

$$\nu = \frac{\Delta E}{h} = \frac{2g\mu_B\mathscr{H}_0}{h}, \tag{12.23}$$

as in the NMR case. This resonance condition is due to the fact that an electron in the lower level can absorb a photon and make a transition to the upper level, flipping its spin in the process. Note that since $\mu_B$ is much larger than $\mu_n$, by a factor of about $10^3$, the frequency of the ESR is this much larger than the frequency of the NMR. This places ESR frequencies at about 1 GHz—in the microwave range. For example, if $\mathscr{H}_0 = 3.4$ kOersted, $g = 2$ are substituted into (12.23), one finds $\nu = 9.5$ GHz. In practice, the frequency is held fixed, and resonance is achieved by sweeping the field until condition (12.23) is satisfied. This is done for convenience, as we stated in connection with NMR.

The ESR of a free electron is not of interest in chemistry. What is of interest is to use ESR to study the internal structure of matter. One does this by comparing the spectrum for an electron inside the sample with that of a free electron. The two spectra differ in several respects. In the first place, the $g$-value for an electron in an atom or molecule is generally quite different from 2, the value for a free electron. The reason, as we recall from Section 9.6, is that $g$ depends on the way the spin and orbital angular momenta are coupled. But the orbital momentum is greatly affected by the environment (often quenched almost entirely, Section 9.6), and this fact is reflected in a different value for $g$. The $g$-value of a resonance line therefore gives information about the electron orbital in the molecule, and extensive tables for $g$ are available in the literature [see Bershon (1966)].

Another effect of the environment is to cause a splitting in the resonance line. Let us illustrate this effect for the simplest case: the hydrogen atom. The hydrogen electron, when placed in an external field $\mathscr{H}_0$, sees not only this field, but an additional small field due to the proton, because the proton acts as a tiny magnet which generates its own field that acts on the electron. This magnetic electron–nuclear coupling is referred to as *hyperfine interaction.* When we denote the hyperfine field by $\mathscr{H}_{hf}$, it follows that the total field seen by the electron is

$$\mathscr{H} = \mathscr{H}_0 + \mathscr{H}_{hf}.$$

Note, however, that $\mathscr{H}_{hf}$ depends on the orientation of the proton moment (the source). Since the proton has a spin number $I = \frac{1}{2}$, it has two different orientations, one parallel and the other opposite to $\mathscr{H}_0$. Therefore the electron sees two different fields

$$\mathscr{H} = \mathscr{H}_0 \pm \mathscr{H}_{hf}, \tag{12.24}$$

the upper corresponding to the proton moment parallel to $\mathscr{H}_0$. Substituting this into (12.21) for the Zeeman energy, one finds

$$E = E_Z + E_{hf} = -g\mu_B m_s(\mathscr{H}_0 \pm \mathscr{H}_{hf}). \tag{12.25}$$

Each Zeeman level is now doubly split by the hyperfine interaction. For the case of hydrogen, both the $m_s = \frac{1}{2}$ and $m_s = -\frac{1}{2}$ levels are doubly split, as shown in Fig. 12.25, with the splitting given by

$$\Delta E_{hf} = g\mu_B \mathscr{H}_{hf}. \tag{12.26}$$

In Fig. 12.25, the split levels are also labeled by the value of the proton magnetic spin number $m_I$. Note that since $\mathscr{H}_{hf}$ is usually much smaller than $\mathscr{H}_0$, hyperfine splitting is far smaller than Zeeman splitting.

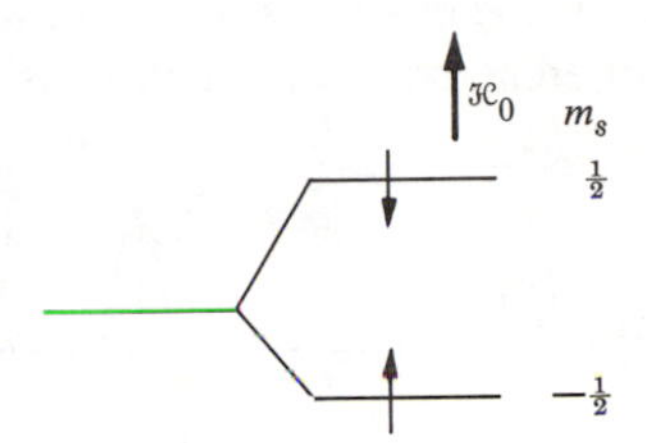


**Fig. 12.25** Splitting of an electron level in a magnetic field. Arrows at the levels indicate orientations of electron moment.

There are four levels in Fig. 12.26, and there are several possibilities for transitions between them; hence the possibility for several resonance frequencies. Note, however, that the transition $1 \rightarrow 2$ corresponds to the proton flipping its spin, the spin of the electron remaining unchanged. The process is thus one of nuclear resonance, which we examined in Section 12.5. This process, and the similar transition $3 \rightarrow 4$, will therefore be excluded from further discussion here.

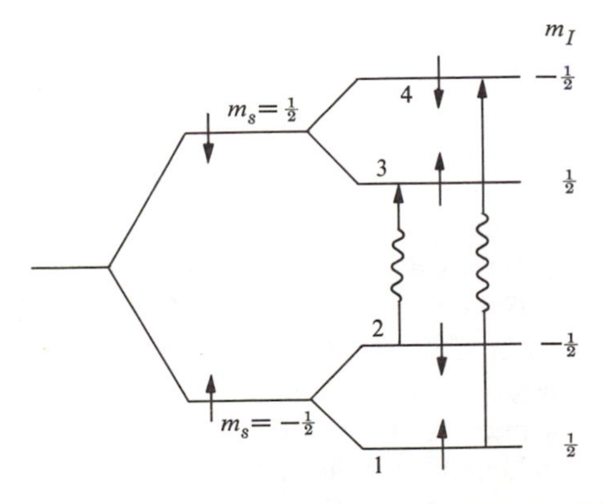


**Fig. 12.26** Zeeman and hyperfine splitting in hydrogen. (The hyperfine splitting is greatly exaggerated.) Arrows indicate orientations of electron and proton moments in the various levels. Wavy lines indicate allowed transitions.

We shall now show that the transitions $1 \rightarrow 3$ and $2 \rightarrow 4$ are forbidden by the selection rules. To see this, recall that the photon absorbed in the transition has a spin angular momentum of $\hbar$, and since the total angular momentum (of electron, proton, and photon) must be conserved, it follows that the allowed ESR processes must satisfy the relations

$$\Delta m_s = \pm 1, \qquad \Delta m_I = 0. \tag{12.27}$$

That is, $m_I$ must be conserved. The only allowed transitions are therefore the two that correspond to $1 \rightarrow 4$ and $2 \rightarrow 3$. If the external field were fixed, there would be two resonance frequencies, but since, in practice, the field is actually varied, one observes two different resonance fields, as shown in Fig. 12.27.

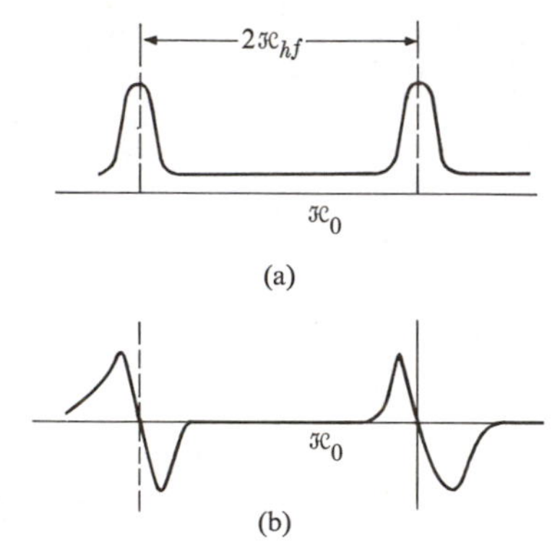


**Fig. 12.27** (a) Intensity of ESR absorption in hydrogen versus sweep field. (b) Intensity derivative.

We can see that the difference between these fields is twice the hyperfine field [note that the difference in energy between the two transitions is twice that of $\Delta E_{hf}$ of (12.26)]. That is,

$$\Delta \mathcal{H} = 2\mathcal{H}_{hf}, \tag{12.28}$$

and we have here a method for measuring $\mathcal{H}_{hf}$ as a measure of the strength of the hyperfine interaction. The quantity which is actually measured in ESR experiments is not the intensity itself, but its derivative; i.e., the slope of Fig. 12.27(a), which is shown in Fig. 12.27(b). The observed spectrum of hydrogen does indeed have this shape, with a line separation of 508 oersteds. This separation is very large compared with other observed separations, and is due to the fact that the hydrogenic electron, being in the 1s state, is piled rather heavily at the nucleus.

We have so far considered only the simplest possible case, and we now need to look into more complicated ones. If the nuclear spin $I > \frac{1}{2}$, each Zeeman level is split into more than two sublevels. Thus for $I = 1$, as in $^{14}N$, there are three hyperfine sublevels. Using the selection rules (12.27), we see that there are three resonance fields, equally spaced, with spacing equal to $2\mathscr{H}_{hf}$. Similarly, radicals containing $^{75}$As, $I = \frac{3}{2}$, exhibit a 4-line ESR spectrum.

A more interesting situation obtains when the electron interacts with more than one nucleus, as is often the case in molecules. Consider the case of the hydrogen molecule ion, $H_2^+$, in which the electron interacts with two protons. As a result, each Zeeman level splits into several levels; the number of levels is equal to the number of different states that the two protons can take.

There are four such possibilities, as indicated in Fig. 12.28(a), but the two possibilities shown in the middle are physically indistinguishable. Thus in $H_2^+$ each Zeeman level is split into three levels, the middle one being undisplaced, since it corresponds to $m_I = 0$. Using the selection rules (12.27), we see that there are three equally spaced lines, as shown in Fig. 12.28(b). Note, however, that the lines have intensities in the ratios 1:2:1. This can be explained by the fact that the middle line, due to $m_I = 0$, corresponds to the two possibilities in Fig. 12.28(a). (Note that the line multiplicity of Fig. 12.28(b) can be distinguished from the case of a single nucleus with $I = 1$ by the unequal intensities of the lines.)

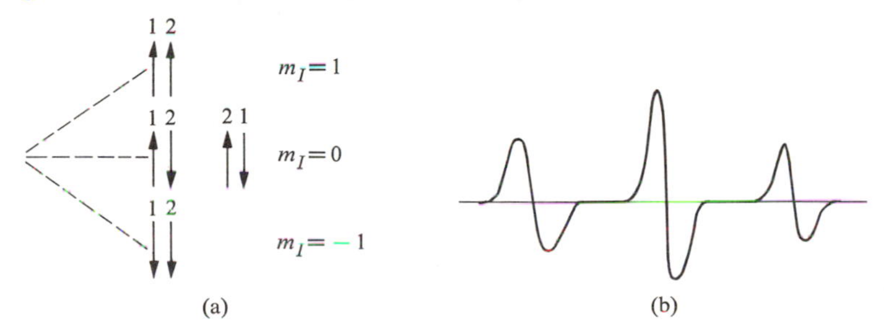


**Fig. 12.28** Hyperfine splitting of ESR line in hydrogen molecule ion $H_2^+$.

The situation is even more complicated when more than two nuclei are involved, as for example in the methyl radical $^{12}CH_3$, in which the electron on the C atom is acted on by the field of the three protons of hydrogen. You can show that there are four possibilities for the proton states, which occur in the ratios 1:3:3:1. The hyperfine spectrum for the methyl radical shown in Fig. 12.29 confirms this prediction. The line spacing here is 23 oersteds.

In the cases considered so far, all the magnetic nuclei in the molecule were equivalent. As an example of nonequivalent nuclei, consider the methyl radical $^{13}CH_3$. Note that $^{13}$C has a spin $I = \frac{1}{2}$. In addition to feeling the field of the three protons, the electron also feels the field due to the nucleus $^{13}$C. Since this nucleus has two different states, each of the above levels is doubly split by it. Because the odd electron in question is piled nearer to the carbon nucleus than to the proton, the hyperfine splitting due to the carbon nucleus is greater than that due to the proton, somewhat as shown in Fig. 12.30(a). The resulting spectrum consists of eight lines, as in Fig. 12.30(b). The lines, in fact, are close enough so that some of them overlap. The actual spectrum is shown in Fig. 12.30(c).

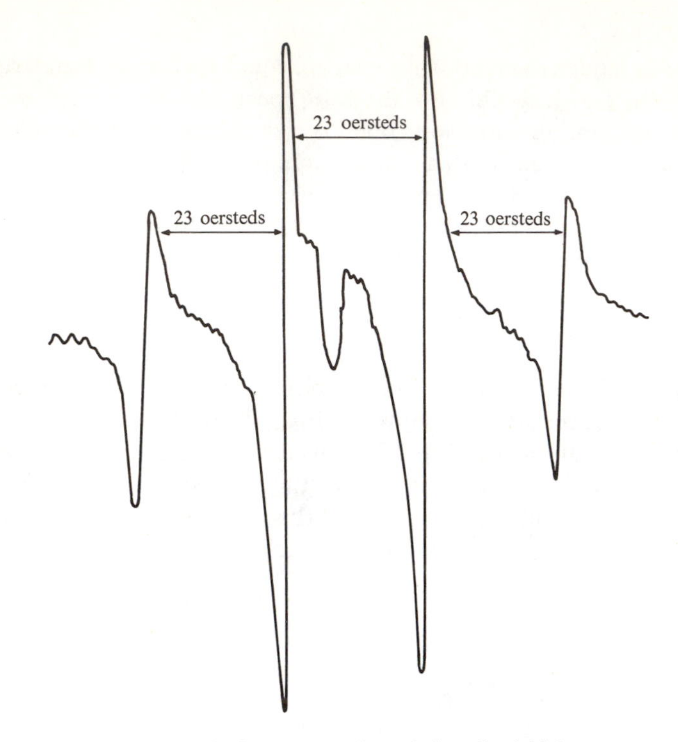


**Fig. 12.29** Spectrum of methyl radical $^{12}C_3$.

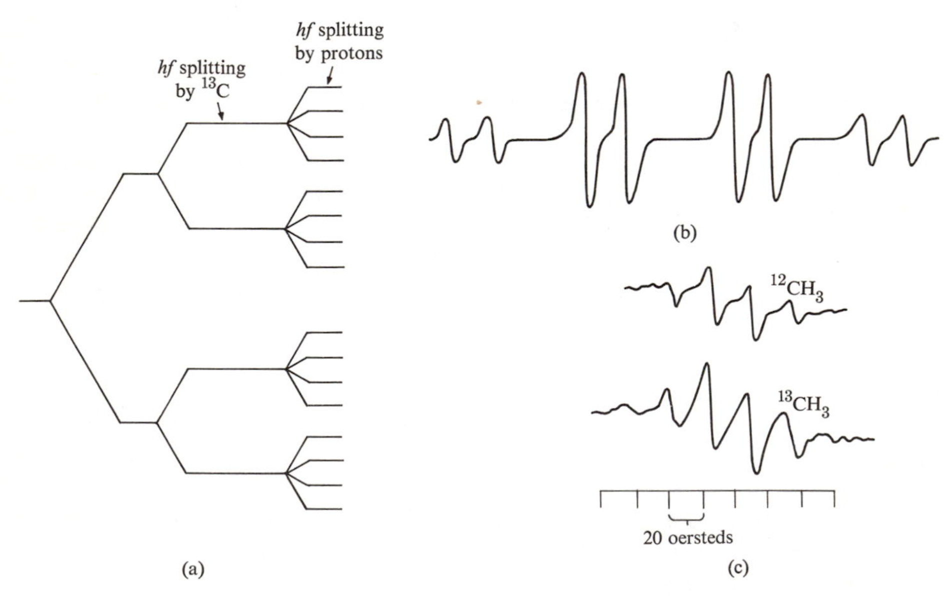


**Fig. 12.30** (a) Hyperfine splitting in methyl radical $^{13}CH_3$. (b) Hypothetical spectrum of this radical. (c) Observed spectrum of mixture of $^{12}CH_3$ and $^{13}CH_3$.

Now let us look at the type of electron–nuclear hyperfine interactions commonly encountered in molecules. There are two types: *dipolar interaction* and *contact interaction*. In dipolar interaction, the electron does not appreciably overlap the nucleus, and the interaction is a long-range magnetic dipole–dipole interaction (as for example in the methyl radical $^{12}CH_3$ considered above). By contrast, contact interaction refers to the case in which electrons pile up over the nucleus in question, as in splitting due to the $^{13}C$ nucleus in the methyl radical $^{13}CH_3$.

The two types of interaction have different characters, which can be differentiated experimentally. For instance, dipolar interaction is anisotropic, depending on the distribution of the nuclei relative to the external field $\mathscr{H}_0$, so that, as the substance is rotated, the lines move about to some extent. On the other hand, contact interaction is isotropic, since it depends only on the piling of electron charge at the nucleus. Usually the strength of contact interaction is a measure of the *s*-character of the electronic orbital. Recall from atomic physics (Section A.5) that only s-orbitals pile the electrons appreciably at the nucleus, while p, d, etc. orbitals show very little overlap with the nucleus.

The power of the ESR technique in studying molecular orbitals should now be evident. By examining the spectrum—the number of lines and their separations, intensities, character, etc.—one can glean a great deal of information. In fact, the ESR technique is the most accurate and detailed method now available for studying molecular orbitals in molecules and solids.

A new, but related, technique which is gaining recognition as a powerful, highly accurate spectroscopic method is the *double-resonance* technique. This involves both NMR and ESR processes used in tandem. For an example of one such type of resonance, called ENDOR (*electron–nuclear double resonance*) look back at Fig. 12.23. Suppose a strong microwave signal is used to causethe transition $1 \rightarrow 4$ in the system. After some time interval, the populations of the two levels are equalized,† and the ESR absorption becomes very weak. Suppose now that an rf signal, appropriate to the induced transition $4 \rightarrow 3$, is applied. This causes some of the electrons in level 4 to make transitions to level 3, an NMR process. As a result, the population in 4 is then less than in 1, and the ESR absorption rises sharply once more. The hyperfine splitting is thus obtained as a series of rf peaks, corresponding to differences in nuclear levels, and the resolution is often enormously improved.

## 12.7 CHEMICAL APPLICATIONS OF THE MÖSSBAUER EFFECT

The Mössbauer effect (ME) was discovered by R. Mössbauer while he was investigating $\gamma$-ray absorption in various nuclei. The discovery was announced in 1958, and in 1961 Mössbauer received the Nobel prize for this remarkable achievement, which has found wide applications not only in physics, but also in chemistry and biology.

---

† This is the phenomenon of saturation, referred to in Section 9.12.

Consider a nucleus in its excited state, whose energy is $E$ (Fig. 12.31). After a certain time, the nucleus makes a transition to the ground state, emitting a $\gamma$-ray photon in the process. (In the terminology of nuclear physics, the nucleus is radioactive.) The frequency of the photon is given by the Einstein relation $h\nu = E$. If this photon impinges on another identical nucleus in its ground state, the photon may be absorbed, resulting in the transfer of the nucleus to its excited state. This process, which is possible only because the energy of the photon is exactly equal to the energy of the excited state of the second nucleus, is a case of *resonant absorption*. It is analogous to the familiar resonance between two identical tuning forks. The energy of the $\gamma$-ray photon, typically of the order of $10^5$ eV, is much greater than the energy of the visible photon, about 5 eV, by virtue of the strong nuclear forces involved in the nuclear transition.

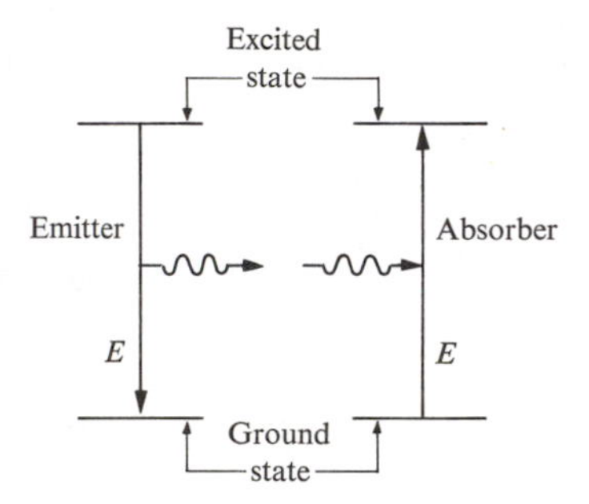


**Fig. 12.31** Resonance absorption.

As a matter of fact, the above resonant absorption does not take place, because when the emitting nucleus (emitter) ejects the photon, the nucleus recoils backward, absorbing a small fraction of the energy, so that, in effect, the photon's energy is slightly less than $E$. That is,

$$E_e = E - E_R, \tag{12.29}$$

where $E_e$ is the energy of the emitted photon and $E_R$ the recoil energy of the emitter. Similarly, the absorbing nucleus (the absorber) recoils forward as it absorbs the photon, acquiring some translational kinetic energy, and consequently, if the absorption is to take place, the photon's energy must be slightly greater than $E$. That is,

$$E_a = E + E_R, \tag{12.30}$$

where $E_a$ is the energy of the absorbed photon. Figure 12.32 shows the positions of $E_e$ and $E_a$ relative to the hypothetical recoil-free situation, and since $E_e < E_a$, the emitted photon does not appear to have enough energy to excite the second nucleus, which explains why resonant absorption is not usually observed in nuclear physics.

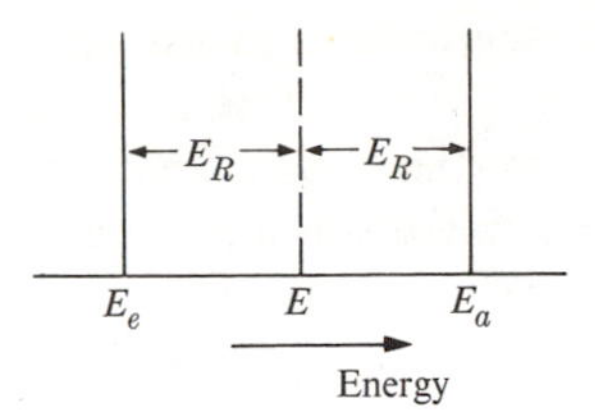


**Fig. 12.32** Energy shifts of emitter and absorber due to recoil motion.

The recoil energy $E_R$ can be calculated from the law of conservation of momentum. Applying this law to the emitter, we have

$$MV_R + h\nu/c = 0, \tag{12.31}$$

where $M$ and $V_R$ are the mass and recoil velocity of the emitter, respectively, and $h\nu/c$ is the momentum of the emitted photon. The recoil energy $E_R = \frac{1}{2}MV_R^2$, which, when we substitute for $V_R$ from (12.31), yields

$$E_R = \frac{1}{2}\frac{h^2\nu^2}{Mc^2}. \tag{12.32}$$

For a typical nucleus whose mass $M$ is 50 times the mass of the proton, one finds $E_R \simeq 0.01$ eV, which, though small, is significant because the energy levels of the nucleus are very sharp.

The situation described thus far represents the actual state of affairs up to the time Mössbauer made his observations. He found, to his surprise, however, that if the temperature of the system is lowered to the liquid helium range, a significant amount of $\gamma$-ray absorption actually does take place. The explanation, also supplied by Mössbauer, is that the system solidifies at such a low temperature. The nuclei are situated inside a solid, and furthermore, the atoms in the solid are essentially at rest. Since a nucleus or, equivalently, its atom, is strongly coupled to the remainder of the solid (Chapter 3), it follows that the emitting nucleus does not recoil individually, as in the gaseous state, but the solid recoils as a whole. Consequently the mass which should now be inserted in (12.32) is the mass of the entire solid. Since this mass is far greater than the mass of a single nucleus, the recoil energy is negligible. The same argument, of course, applies to the solid absorber, and we have here, in effect, a truly recoil-free situation, leading to resonant absorption, as described in the beginning of the section.

There is yet another aspect of the ME which makes it a highly useful tool: The absorption process can be modulated by rigidly moving either the emitter, the absorber, or both. Thus if the emitter moves toward the observer with a velocity $v$, the emitted photon undergoes a *Doppler shift*, according to the formula $\nu = \nu_0/(1 - v/c)$, where $\nu_0$ is the frequency of radiation from a stationary emitter. If the emitter and absorber are "tuned" to begin with, the motion of the emitter causes "detuning" and reduces the absorption. Conversely, if the emitter and absorber are detuned at the beginning, the motion of the emitter can be so arranged as to bring in the desired tuning.

It can be readily shown from the above Doppler formula that if $E_e$ and $E_a$ are the energies of the emitter and the absorber, respectively, then the velocity of the emitter required to establish the tuning is

$$\frac{v}{c} = \frac{E_a - E_e}{E_a}. \tag{12.33}$$

This affords the possibility of a high-accuracy velocity spectrometer, since $E_a$ and $E_e$ are usually known very accurately.

In solid-state physics and in chemistry, however, we are usually interested not in velocity measurements, but in energy levels and how they change when a given nucleus is placed in various solids. A typical usage of the ME in such situations is as follows: The emitter solid is doped by suitable radioactive nuclei under controlled standard conditions. The absorber is also doped by the same nuclei. The absorber solid may differ greatly from the emitter solid, and hence the way the energy levels of the nuclei are modified by the surrounding environment in the two solids may also differ. By studying the absorption of $\gamma$-rays and its dependence on the velocity of the emitter, one can study the environment in the absorber, in effect using the nuclei as microscopic probes. The most commonly used nucleus is $^{57}Fe$, but others of great chemical interest, such as $^{129}I$ and $^{119}Sn$, have also been used.

Let us now consider specific applications of the Mössbauer effect to chemistry. These applications rest on the following properties.

i) *The isomer shift.*† The shift of the nuclear levels, both ground and excited states, is brought about by the coulomb interaction between the active nucleus and the orbital electrons. Of all these electrons, only the s electrons have an appreciable effect, because only these overlap the nucleus and cause an appreciable coulomb interaction. It can be shown (Wertheim, 1964), that the net shift, including the energy displacements of both the ground and excited states, is

$$\Delta E = \frac{Ze^2}{10\epsilon_0}(R_{\text{ex}}^2 - R_{\text{gd}}^2)\,|\psi(0)|^2, \tag{12.34}$$

where $R_{\text{ex}}$ and $R_{\text{gd}}$ are the radii of the emitting nucleus in the excited and ground states. The quantity $\psi(0)$ is the wave function of the s electrons evaluated at the center of the nucleus. The presence of $|\psi(0)|^2$ in (12.34) is expected, since it represents the probability of the presence of the electron at the nucleus, i.e., the overlap of the electron with the nucleus.

---

† Two nuclei are *isomeric* if they contain the same number of protons. When a nucleus decays into another nucleus by the emission of a $\gamma$-ray, the two nuclei are isomeric, since the number of protons is the same, because no electrical charge was emitted.

The quantity observed in the ME is actually the difference in shifts between the absorber and emitter. Therefore

$$\Delta E_{\text{obs}} = \frac{Ze^2}{10\epsilon_0}(R_{\text{ex}}^2 - R_{\text{gd}}^2)\,[|\psi_a(0)|^2 - |\psi_e(0)|^2], \tag{12.35}$$

where the subscripts on the wave functions refer to the absorber and the emitter. Aside from the numerical factor, the shift consists of a product of two factors—one purely nuclear and the other purely atomic. Once the first factor is determined for a specific nucleus, Eq. (12.35) can be used to obtain the atomic factor under various conditions. It is evident once more that the ME does not determine the absolute value of $|\psi(0)|^2$ itself, but only the difference between its values in the emitter and absorber.

For example, consider iron-containing compounds, which we often encounter in chemistry and biochemistry, since many important biological molecules contain iron. In ionic salts, iron usually exists either as a divalent ($Fe^{2+}$) or trivalent ($Fe^{3+}$) ion. Measurements of chemical shift have shown that the shift is consistently larger in $Fe^{3+}$ than in $Fe^{2+}$. This is surprising, since both ions have the same number of outer s electrons ($3s^2$), and differ only in the number of d electrons—$Fe^{3+}(3d^5)$ and $Fe^{2+}(3d^6)$—which are not expected to produce any shift. However, the 3s electrons spend a fraction of their time outside the 3d shell, and during that time the nucleus is more screened (relative to the s electron) in $Fe^{2+}$ than in $Fe^{3+}$, because in $Fe^{3+}$ one more d electron has been ionized. One may say that the 3s electrons are more tightly pulled to the nucleus in $Fe^{3+}$ than in $Fe^{2+}$, and hence the larger shift. We see from this example that ME measurements yield information about not only s electrons, but other electrons as well.

As another example, the shifts of KI and $KIO_3$ are $-0.052$ and 0.16 cm/s, respectively. (The active nucleus is $^{129}I$ as absorber, and $^{129}Te$ as emitter.) The interpretation of these results is as follows: In the ionic compound KI, the iodine atom acquires an additional electron, resulting in an outer shell whose electronic structure is $5s^2p^6$. But in the iodate $KIO_3$, the iodine atom lies at the center of an octahedron whose corners are occupied by O atoms. There are six I–O mutually covalent orthogonal bonds, which we assume to be formed by the p electrons. Thus the p electrons are pulled toward the O atoms, causing a decrease in the screening on the s electron. That is, this causes a large shift, in agreement with experiment. The ME in this case sheds light on the nature of the chemical bond.

ii) *Quadrupole splitting.* Another source of interaction of a nucleus with its chemical environment relates to the coulomb interaction between the nucleus and its neighboring ions (the *ligands*). These ions produce an electric field at the nucleus. Since the nucleus has no electric dipole moment, the dipole interaction vanishes. However, a nucleus is not usually spherical in shape, but ellipsoidal. (This is so when the nuclear spin number $I > \frac{1}{2}$; see Section 12.5.) Because of this, the nucleus has an electrical *quadrupole moment*. This moment couples not to the ligand field itself, but to its gradient (evaluated at the nucleus), producing a shift in the energy level of the nucleus, which depends on the orientation of the nucleus relative to its environment. But since a nucleus has several allowed orientations (corresponding to allowed spin orientations), there are several possible shifts. That is, quadrupole coupling produces a *splitting* in the nuclear energy level. The character and magnitude of this splitting thus gives information about the environment.

The electric field gradient (EFG) is a tensor of 9 components: $V_{xx}, V_{yy}, V_{xy}$, etc., where $V_{xy} = \partial^2 V/\partial_x \partial_y$, etc., and $V$ is the coulomb potential of the ligands. By a suitable choice of axes, one can always reduce the number of components to three: $V_{xx}$, $V_{yy}$, $V_{zz}$, that is, the *principal elements*. Only two of these are independent because they must satisfy the Laplace equation $V_{xx} + V_{yy} + V_{zz} = 0$. The convention is to choose the two independent parameters as $V_{zz}$ (often denoted by $q$), and the asymmetry parameter $\eta = (V_{xx} - V_{yy})/V_{zz}$. The axis of highest symmetry is usually chosen to be the $z$-axis. If this axis has a 4-fold symmetry (octahedral coordination), the asymmetry parameter $\eta$ vanishes, and the gradient tensor then has cylindrical symmetry. Even a lower-symmetry 3-fold axis leads to a vanishing asymmetry parameter.

An example is the hydrated ferric chloride $FeCl_3 \cdot 6H_2O$, in which it has long been assumed that the iron ion is surrounded by an octahedral environment of water molecules (Fig. 12.33a). But the substance exhibits appreciable splitting, which suggests a symmetry which is lower than octahedral. Careful x-ray studies confirmed that the actual structure is another isomer, as shown in Fig. 12.33(b).

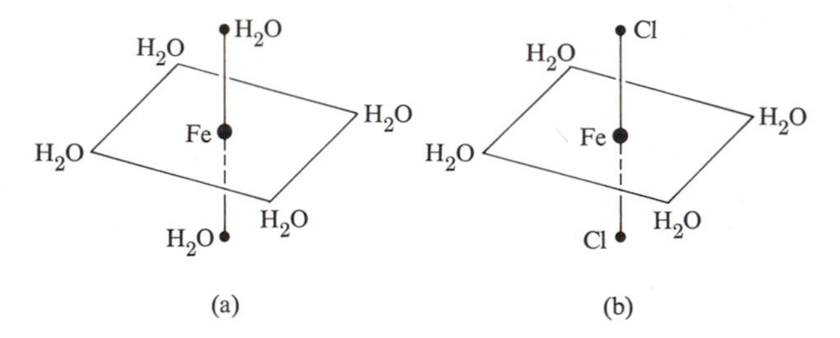


**Fig. 12.33** (a) Incorrect and (b) correct structures of $FeCl_3 \cdot 6H_2O$.

iii) *Magnetic hyperfine splitting.* If the nuclear state has a magnetic dipole moment ($I > 0$), the hyperfine interaction between the nucleus and the magnetic field of the orbital electrons splits the level into $(2I + 1)$ sublevels (Section 12.5). In general, both the ground and excited states of an ME-active nucleus split, and $\gamma$-radiation occurs between the magnetic sublevels of the excited state and those of the ground state. We can use the splitting of the line to determine the properties of the internal magnetic field, i.e., the hyperfine interaction. For example, in a ferromagnetic substance splitting should decrease as the temperature rises until it vanishes entirely at the Curie temperature. Thus the Curie point may be determined from ME measurements.

## REFERENCES

### Amorphous semiconductors

E. A. Owen, "Semiconducting Glasses," *Contemp. Phys.* **11,** 257 (1970)

D. Adler, "Amorphous Semiconductors," *Crit. Rev. Solid State Sci.* **2,** 317 (1971)

These articles, particularly the first one, contain references to hundreds of other relevant sources.

### Liquid crystals

I. G. Christyakov, 1967, *Sov. Phys.-Usp.* **9,** 551–573

J. L. Fergason, 1964, *Sci. Amer.*, **211,** 77–85

G. W. Gray, 1962, *Molecular Structure and the Properties of Liquid Crystals*, London: Academic Press

G. R. Luckhurst, 1972, *Phys. Bull.* **23,** 279–284

### Polymers

F. W. Billmeyer, 1962, *Textbook of Polymer Science*, New York: Interscience

F. Bueche, 1962, *The Physical Properties of Polymers*, New York: Interscience

A. V. Tobolsky, 1960, *Properties and Structures of Polymers*, New York: John Wiley

L. A. C. Treloar, 1949, *The Physics of Rubber Elasticity*, Oxford: Oxford University Press

T. Alfrey, Jr. and E. F. Gurnee, 1967, *Organic Polymers*, Englewood Cliffs, N.J.: Prentice-Hall

L. H. Van Vlack, 1963, *Elements of Materials Science*, Reading, Mass.: Addison-Wesley

M. Gordon, 1963, *High Polymers*, Reading, Mass.: Addison-Wesley

B. Wunderlich, 1969, *Crystalline High Polymers: Molecular Structure and Thermodynamics*, Americal Chemical Society

P. J. Flory, 1953, *Principles of Polymer Chemistry*, Ithaca, N.Y.: Cornell University Press

### NMR and ESR

P. B. Ayscough, 1967, *Electron Spin Resonance in Chemistry*, London: Methuen

A. Carrington and A. D. McLachlan, 1967, *Introduction to Nuclear Magnetic Resonance*, New York: Harper and Row

J. D. Robers, 1959, *Nuclear Magnetic Resonance*, New York: McGraw-Hill

L. M. Jackman, 1959, *Nuclear Magnetic Resonance Spectroscopy*, New York: Pergamon Press

W. W. Paudler, 1971, *Nuclear Magnetic Resonance*, Boston, Mass.: Allyn and Bacon

M. Bershon and J. C. Baird, 1966, *An introduction to Electron Paramagnetic Resonance*, New York: W. A. Benjamin

See also references under similar title in Chapter 10.

### Mössbauer effect

H. Fraunfelder, 1963, *The Mössbauer Effect*, New York: W. A. Benjamin

V. I. Gol'Dansky, 1964, *The Mössbauer Effect and its Applications in Chemistry*, New York: Consultants Bureau
L. May, editor, 1971, *An Introduction to Mössbauer Spectroscopy*, New York: Plenum
D. A. O'Conner, "The Mössbauer Effect," *Contemp. Phys.* **9**, 521, 1968
G. K. Wertheim, 1964, *Mössbauer Effect*, New York: Academic Press

## QUESTIONS

1. For the magnetic fields used, the magnetic energy is too small compared to the thermal energy, and hence the field does not orient single molecules; yet the field does orient the director. How do you resolve this apparent paradox?
2. Suppose that you prepare a mixture of two cholesteric liquid crystals which rotate the polarization in opposite senses. What is the phase of the product?
3. Could expression (12.8) be valid for a cholesteric liquid crystal? If not, find a plausible expression.
4. Show that the asymmetry parameter $\eta$ (in a Mössbauer effect) vanishes for a solid which has a 3-fold axis of symmetry.

## PROBLEMS

1. Read the articles by Adler (1971) and Owen (1970), and write a brief report.
2. Derive expression (12.3) for conductivity.
3. Prove that if the molecules in a nematic phase have random orientations, the order function $S$ vanishes.
4. Plot the intermolecular anisotropic potential in the nematic phase $V_{ij}$ versus the angle $\theta$ between the molecular axes of the two molecules involved, and point out the most favorable orientations.
5. Derive Eq. (12.9) for the orientational magnetic energy density.
6. Derive Eq. (12.11).
7. The molecular weight of a polyethylene molecule is 100,000. What is its length if the length of the C-C bond is 1.54 Å?
8. The monomer isoprene

$$\begin{array}{c} H_2C = \underset{\displaystyle CH_3}{\underset{|}{C}} - \underset{\displaystyle H}{\underset{|}{C}} = CH_2 \end{array}$$

   is the basic unit in natural rubber. Draw the complete molecular structure of rubber. What feature of this structure allows vulcanization to take place (the formation of sulfur cross links between adjacent chains)?
9. The difference in chemical shifts between two protons in a 60-MHz field is 700 Hz. What would be the difference in a 100-MHz field?
10. The proton resonance of a substance dissolved in TMS occurs at $-$ 500Hz relative to the standard. Calculate $\delta$ and $\tau$ for the proton.
11. The NMR spectrum of $^{19}F$ ($I = \frac{1}{2}$) in olefin, $C_3H_4F_2$, consists of two sets of peaks: A doublet of doublets with coupling constants at 45 and 10 Hz, respectively. The other set of peaks consists of a quadruplet with coupling constants of 45 and 8 Hz, respectively.

a) Determine the structure of this compound.
b) Predict the proton NMR spectrum for olefin.

12. The frequency-shift formula (12.33), derived in the text on the basis of the Doppler effect, may also be obtained from the laws of conservation of energy and momentum. Carry out this derivation.
13. Derive Eq. (12.34).

# CHAPTER 13 SOLID-STATE BIOPHYSICS



---

*What admits no doubt in my mind is that the Creator must have known a great deal of wave mechanics and solid state physics, and must have applied them.*

A. Szent-Györgyi, in *Introduction to a Submolecular Biology*

## 13.1 INTRODUCTION

Of all the scientific disciplines, molecular biology is undergoing the most rapid progress at the present time. Major breakthroughs are made almost every year, bringing us ever nearer to the understanding of life itself at its most fundamental level, the atomic–molecular level.

There are two reasons why solid-state physics is relevant in the study of molecular biology. These reasons prompted the inclusion of this chapter in the present work. First, the concepts of quantum mechanics are being increasingly applied to the study of biomolecules, and since many of these concepts have close parallels in solid-state physics, some of the theoretical techniques which have proved successful in solid-state physics can also be used in molecular biology. Second, accurate experimental techniques developed principally by solid-state physicists are being increasingly employed in the study of biomolecules and their structure. Thus x-ray diffraction is a standard technique of the molecular biologist, and other techniques—such as electron microscopy, ESR spectroscopy, etc.—are coming into further use every day. Modern biology is no longer a set of dry, empirical facts, but an exciting interplay of modern concepts of physics, chemistry, and engineering, all of which are finding their place in the unraveling of the problems of molecular biology. The structure of the collagen molecule, for example, was determined primarily by the great chemist, Linus Pauling, while of the three scientists (Watson, Crick, and Wilkins) responsible for the discovery of the DNA structure, two (Wilkins and Crick), are physicists by training.

This chapter presents a modest introduction to biology in a language that should be readily understood by the solid-state student. Though the subject matter may not closely resemble the typical solid-state coverage of the first twelve chapters of this book, it is based on concepts such as electron delocalization that will be well understood and appreciated by the reader. The material presented here covers almost the minimum background required by a student of physics who may contemplate entering the exciting field of molecular biology, or merely be interested in following current developments in the subject.

After this introduction, we present the quantum theory of delocalized electrons in biological molecules, particularly in benzene, in which this delocalization is especially important. We then define several "electronic indices", and indicate their relevance to the biological activity of the molecule. In the three remaining sections, the knowledge gained in the first part of the chapter is brought to bear on the study of nucleic acids, proteins, and miscellaneous topics, such as carcinogenesis.

If there is one unifying theme of this chapter, it is that of *electron delocalization*. Just as this profound concept is responsible for the most interesting phenomena in metals, semiconductors, and other solids, it is also of critical importance in biology. We quote from Pullman (1963, page 10): "The existence of delocalized $\pi$ electrons . . . is not only the essentially new property of conjugated molecules. It is also their most important property: The principal chemical, physico-chemical, and also, as will be seen later in detail, biochemical properties of such systems are determined by their $\pi$ electrons. The reason for this is that these electrons are much more *mobile* than the $\sigma$ electrons, and therefore participate more readily in chemical and biochemical processes."

## 13.2 BIOLOGICAL APPLICATIONS OF DELOCALIZATION IN MOLECULES

Biomolecules, unlike typical inorganic molecules, are usually very large, often containing several thousand atoms. In addition to carbon and hydrogen, the primary ingredients, these *macromolecules* often contain other atoms, such as nitrogen, oxygen, or phosphorus. In such a situation, the question of electron delocalization may be raised, and since this concept was an extremely important one in our understanding of the properties of metals and semiconductors, one may well ask whether delocalization also plays a significant role in biochemistry. We shall see that this is indeed the case, and the method closely parallels that previously employed in traditional solid-state physics.

We begin the discussion with the rather simple case of the benzene molecule (Fig. 13.1), a hexagonal ring with six C atoms at the corners, and an H atom at each of the C atoms. Some of the bonds are denoted as double bonds to satisfy the quadrivalent character of the C atom. Some of the electrons associated with the double bonds are not actually localized between specific atomic pairs, but revolve around the entire ring. These electrons, known as the *π-electrons*, are thus delocalized, and hence are of particular interest to us here.

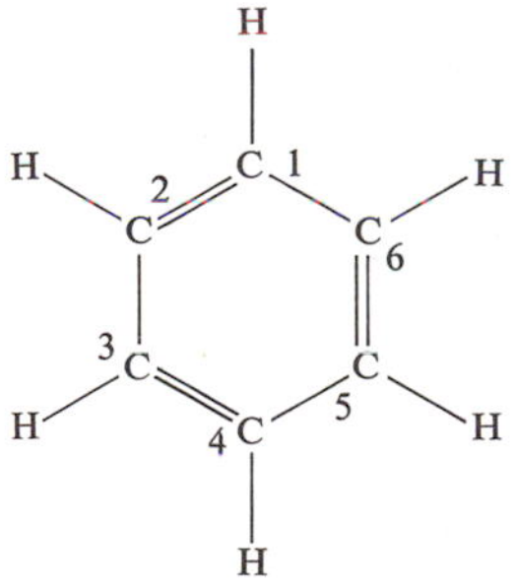

**Fig. 13.1** The benzene molecule.

Without getting too involved in details, we state that of the four electrons on each of the C atoms, three occupy hybridized sp orbitals, analogous to the hybridized orbitals discussed in Section A.8. These orbitals, known as $\sigma$-orbitals, are highly localized along the lines joining the atoms. The atomic $P_y$ orbitals, however (where $y$ is the direction normal to the plane of the ring), overlap with their neighboring atoms enough for the electrons occupying these orbitals to be able to jump from one atom to the next, and eventually to rotate around the ring, somewhat as in the case of crystalline solids. These are the $\pi$-electrons described above.

To proceed further, we have to solve the appropriate Schrödinger equation, and as this cannot be done exactly, an approximate procedure must be employed. It would appear from our brief description that the tight-binding model of Section 5.8 would be particularly appropriate here, since it introduces delocalization without a complete obliteration of localization. That is, the model occupies a middle ground between solid-state physics and traditional chemistry. This model, in fact, is widely used in biochemistry, in which it is known as the *Hückle method*, after the great chemist who first used it in this context.

As in Section 5.8, the molecular orbital (MO) is taken to be a linear combination of atomic orbitals. That is,

$$\psi = \sum_r c_r \phi_r, \tag{13.1}$$

where the $\phi_r$'s refer to the atomic $P_y$ orbitals of the various atoms in the ring, and the summation is over the six C atoms. The $c_r$'s are constants to be determined. The Schrödinger equation (SE) for a delocalized electron is

$$\left[ -\frac{\hbar^2}{2m} \nabla^2 + \sum_r V_r \right] \psi = E\psi, \tag{13.2}$$

where $V_r$ is the atomic potential of the $r$th atom, and hence $\sum V_r$ is the ring potential. To obtain the energy $E$ and the MO $\psi$, we follow a common approach in quantum mechanics: We multiply Eq. (13.2) from the left by $\phi_1^*$, $\phi_2^*$, etc., respectively, and integrate over space in each case. When one follows this procedure, one obtains a set of homogeneous algebraic equations in the $c_r$'s, and from the corresponding secular equation of this set one can solve for the energy $E$ (see Pullman, 1963, for details). One can write the energy as

$$E = \alpha + k\beta, \tag{13.3}$$

where $\alpha$ is the free atomic energy and $\beta$ the overlap integral between two neighboring C atoms. [$\alpha$ and $\beta$ are analogous to $E_v$ and $\gamma$, respectively, in Section 5.8. Note also that both $\alpha$ and $\beta$ are negative numbers for the same reason given there.] The parameter $k$ which thus specifies the energy is obtained from the secular equation.

For the case of benzene, the roots of this equation are $k = 2$, 1 (twice), $-1$ (twice), and $-2$. The first two roots lead to the energies $E_1 = \alpha + 2\beta$ and $E_2 = \alpha + \beta$, which, being of lower value than $\alpha$, lead to *bonding* orbitals, as in the case of the $H_2$ molecule (Section A.7). Two $\pi$-electrons (of opposite spins) occupy the first level, and four the second level, which exhausts all the six available electrons. The other energy levels lead to antibonding orbitals (why?), and are not occupied.

By inserting the various $k$ roots into the original equations, one can solve for the coefficients, the $c_r$'s, and hence determine how the electron, for each orbital, is distributed around the ring. Thus $|c_1|^2$ is the probability of the electron being at atom 1, $|c_2|^2$ the same for atom 2, and so forth. In the case of benzene, $|c_r|^2 = \frac{1}{6}$, as follows from the symmetry of the ring.

We shall illustrate the importance of delocalization by evaluating the energy of the $\pi$-electrons in the localized and delocalized models, respectively. In the localized case, the total energy is $6\alpha + 6\beta$, the second term arising from the fact that there are three double bonds, each occupied by two electrons. In the delocalized model, however, the total energy is $2(\alpha + 2\beta) + 4(\alpha + \beta) = 6\alpha + 8\beta$. Since $\beta$ is negative, the ring energy is reduced by the amount $2|\beta|$ due to delocalization. This is the factor responsible for the great stability of the benzene molecule (and other aromatic molecules). (The decrease in energy due to delocalization is known in chemistry as *resonance energy*.)

The tight-binding (TB) model can also be used in the treatment of *substituted* molecules, in which one or more of the C atoms is replaced by, for example, an N or O atom. A different $\alpha$ must be used for the new atom, of course, and also a new $\beta$ for bonds involving this atom, but otherwise the procedure remains unchanged.

The TB model also yields a great deal of useful information about the molecule, in addition to its binding energy. When these data (described below) are available, one knows much more about the behavior of the molecule than can be gleaned from the structural formula, which is simply a statement of the chemical composition.

In addition to the resonance energy, one may calculate other useful energy parameters. For instance, the *ionization potential*, which is the energy required to remove an electron from the molecule, is important because the smaller the ionization energy the greater the capacity of the molecule to lose one of its electrons—in other words, to act as an electron donor.

Another parameter is *electron affinity*, which is the energy needed to remove an electron from a singly charged negative ion. The larger the affinity the greater the capacity of the molecule to attract an electron and act as an *acceptor*. When a donor and acceptor happen to be close to each other, an electron is likely to transfer, forming a charge–transfer complex. Such a process occurs frequently in biochemistry.

Another important quantity is the *electronic charge* on the various atoms in the molecule. The electronic charge on the $r$th atom is (in units of $e$)

$$q_r = 2 \sum_{v} c_{r,v},$$

where the summation is over occupied MO's, and the factor 2 is due to the double occupancy of each orbital. Experimental information about $q_r$ for the various atoms can be obtained, for example, by NMR techniques (Section 12.5) because the greater the $q_r$ the larger the chemical shift.

Figure 13.2 shows the electronic charges in cytosine, as given by Pullman. NMR measurements (by Jardetsky) show that the proton on the $C_5$ atom has the greatest chemical shift. This is in agreement with the figure.

Also important are the *net charges* on the various atoms. The net charge on the $r$th atom is $Q_r = 1 - q_r$. When $0 < Q_r$, the atom acts as a center for interaction between the molecule and external negative ions, and vice versa. Good agreement between theory and experiment has been obtained in many cases.

The *bond order* $P_{rs}$ for a particular bond ($rs$) is a measure of how close this bond is to a pure double bond. It can be shown that $P_{rs} = 2\sum_v c_{r,v} c_{s,v}$, where $r$ and $s$ are the end atoms of the bond. For a pure double bond, $P_{rs} = 1$, but for most common molecules it is usually less than that. The various bond orders in the molecule can be investigated spectroscopically or calorimetrically, in a manner similar to the lattice vibrations discussed in Chapter 3, since the energies of single and double bonds differ widely.

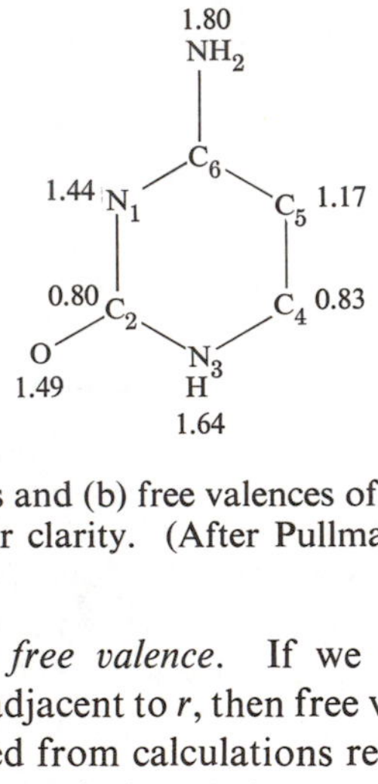

**Fig. 13.2** (a) Electronic charges and (b) free valences of various atoms in cytosine. Some of the H atoms are omitted for clarity. (After Pullman)

Finally, let us mention *free valence*. If we define $N_r = \sum_s P_{rs}$, where the summation is over all atoms adjacent to $r$, then free valence is defined as $F_r = \sqrt{3} - N_r$. [The term $\sqrt{3}$ is obtained from calculations related to the valence of carbon; see Pullman (1963).] When $F_r$ is large, the atom tends to act as a center for interaction of the molecule with external *free radicals*. Such entities have received increasing attention recently, and are thought to play a dominant role in biochemical processes.

Many workers, particularly the Pullmans, have applied the TB model to calculate the above parameters for various molecules, and were able to explain many biochemical phenomena. You will find a great many examples discussed in their book.

## 13.3 NUCLEIC ACIDS

The nucleic acids, DNA and RNA, are of great biological importance because they transmit the genetic code from parent to offspring. Both types of acids consist of very long molecules, called polynucleotides; the backbone of the molecules consists of sugar (ribose for RNA and deoxyribose for DNA) and phosphate groups. Attached to the molecule chain are side groups consisting of purine and pyrimidine rings, which are somewhat analogous to the benzene ring, but more complicated, and containing nitrogen atoms. As a matter of fact, the DNA molecule consists of not only one but *two* strands, which are entwined. They twist into a helical structure known as the *double helix* (Watson, 1968).

The two strands in DNA are bonded together via the hydrogen bond (Section 1.10) between the side rings on the chains. This hydrogen bond is simply a resonance energy due to the additional delocalization experienced by the $\pi$-electrons as the side rings fuse together.

Watson and Crick arrived at the double-helical structure of DNA from their interpretation of the x-ray diffraction pattern of the substance. The x-ray diffraction theory for helical structures can be developed in a manner analogous to that used for regular crystalline structures (Chapter 2). Although we shall not give details (see Dickerson, 1964). let us point out that helical structures have characteristic patterns which are distinguished by the absence of certain diffraction lines. The absence of these lines is taken as an indication of helical structures. From such observations, Watson and Crick determined that the pitch of the helix in DNA is 34 Å and its diameter 20 Å. The x-ray methods which play such a critical role in solid-state physics play much the same role in biology.†

**Radiation damage**

The study of radiation damage in biological materials is one of the most interesting fields in contemporary biology. Such studies not only afford a better understanding of biological materials, but also suggest means for protection against such damage. The human body is constantly bombarded by many types of radiation: from nuclear explosions, from television sets, from x-ray machines, and most of all from the sun itself.

When the DNA or other molecule is exposed to radiation, transformations take place, and a new set of product molecules emerges. The transformation is due, of course, to various rearrangements of atoms and ions, taking advantage of the energy absorbed from the incident radiation. This chemical reaction is not a simple one-step affair, but the result of several intermediate reactions which

---

† In recent years, the neutron diffraction technique (Section 2.11) has also come into increasing use in the study of the structure of biological molecules. The advantages of the neutron over the x-ray technique, as explained in Section 2.11, are: (a) The hydrogen atom, which is of great biological importance, is more readily detected by the neutron diffraction method. (b) Using the neutron diffraction method, one can distinguish between different isotopes of the same element, e.g., hydrogen versus deuterium. (c) Neutron radiation, having much smaller energy than x-radiation, is far less damaging to the biological sample.

take place in rapid succession. The initial and final substances are amenable to classical chemical techniques, but these techniques are not useful in the identification of the intermediate compounds; it is here that solid-state methods are especially useful.

Particularly common intermediates are free radicals—i.e., molecules containing single, unpaired electrons. Free radicals are highly reactive, and combine quickly with each other, producing stable molecules with paired electrons. But while a radical is in the free state, it possesses a net spin, and is consequently amenable to ESR analysis (Section 12.6).

Figure 13.3(a) shows the most abundant radical present in irradiated thymine-enriched DNA samples. It is produced from thymine by the rupture of a C=C bond and the addition of a hydrogen atom. The ESR spectrum of this radical, shown in Fig. 13.3(b) to consist of eight well-resolved lines, can be interpreted as follows (recall Section 12.6): A MO calculation shows that the unpaired electron resides primarily in the neighborhood of the two C atoms where the bond was ruptured. Thus the electron interacts most strongly with the protons in the methylene group ($CH_2$) and the methyl group ($CH_3$). Considering first the interaction with the $CH_2$, the ESR line should split into an equally spaced triplet. The interaction with the $CH_3$ then causes each of these sublines to split further into a quartet (the $CH_3$ rotates freely)—a total of twelve lines. It appears that some of these lines overlap, however, resulting in a diminution of the number to eight, as in Fig. 13.3(b).

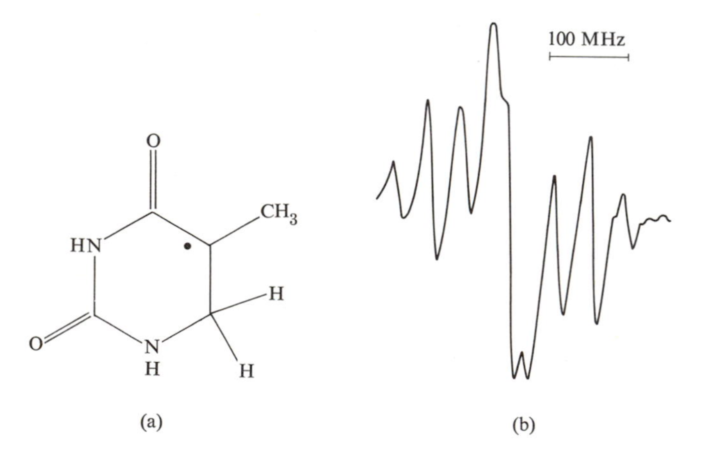


**Fig. 13.3** (a) Thymidine free radical (dot represents unpaired electron). (b) Derivative ESR spectrum of DNA, irradiated and recorded at 300°K.

## 13.4 PROTEINS

Proteins serve many vital functions in a living organism, and their functions vary over a wide range. Like polynucleotides, protein molecules are very long polymeric chains containing a large number—typically of the order of 1000—of repeat units, giving a typical molecular weight of $10^5$. However, unlike polynucleotides, in which the repeat units are nucleic acids, the repeat units in proteins are *amino acids*. There are 21 different amino acids. Depending on the acids and on the order in which they are arranged along the polymeric chain (the *polypeptide*), the various kinds of protein molecules are formed.

**Myoglobin and hemoglobin molecules**

Let us illustrate the application of solid-state methods to proteins by discussing the important *hemoglobin* molecule, which is the central character in the respiratory process. It transports oxygen from the lungs to the brain and muscles, and returns carbon dioxide to the lungs.

The active part of the hemoglobin molecule is the *heme* group, an iron-containing compound; $O_2$ and $CO_2$ are transported by attaching themselves to the Fe atom in this group. So the function of the hemoglobin molecule is greatly influenced by the electronic state of the Fe atom. Becase this atom is especially amenable to solid-state techniques, many measurements have been carried out on hemoglobin.

The hemoglobin molecule is very long, containing 574 amino acids and 4 heme groups. A related but simpler molecule is that of *myoglobin*, which has 150 amino acids and only one heme group. (Globin stores oxygen in the muscles.) This molecule is still far from simple, and it took Kendrew and Preutz years of painstaking effort to determine its structure. Their primary tool was x-ray diffraction, but they were also aided by ESR measurements.

Figure 13.4 shows the heme group in myoglobin, which consists of a conjugated planar molecule at whose center lies the Fe atom, surrounded by four nitrogen atoms. The heme plane is normal to the axis of the molecule. In addition to the four N atoms, the Fe atom is also attached to two other atoms (or groups), lying on opposite sides of the plane. One of these is the remainder of the molecule, the globin, and the other is normally either oxygen, carbon dioxide, or another oxygen compound. If the oxygen compound is replaced by another unit, the function of the molecule is altered. Sometimes this unit may be readily removed—e.g., as in fluorine—and in other cases, such as in cyanide, the removal is almost impossible and the function of the molecule is quenched irrevocably (poisoning).

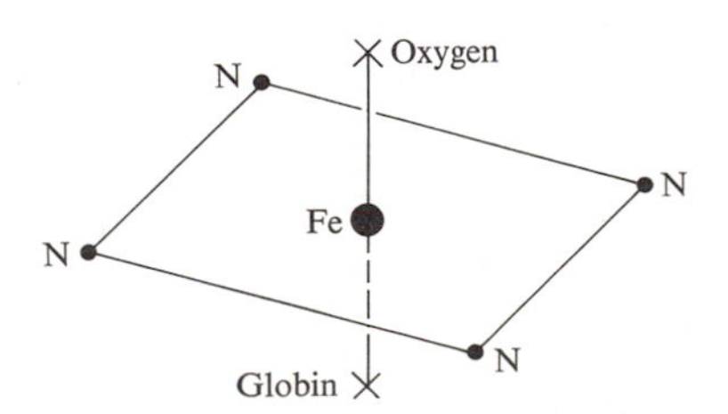


**Fig. 13.4** The heme group.

To consider the ESR aspect of the molecule, we must examine the spin state of the Fe atom. In fact, Fe exists as an ion, and there are two possible ionic states, the ferric ($Fe^{3+}$) and the ferrous ($Fe^{2+}$) states. Let us examine the simplest useful application, in which the ion is ferric. The ion contains five electrons (the neutral Fe atom has eight, but three of these have been transferred to adjacent atoms), which must be distributed among the 3d orbitals of $Fe^{3+}$. Since there are five such orbitals, the electrons may occupy these singly, or doubly if the electrons have opposite spins. To determine which possibility is the more stable, we refer to Hund's rule, which states that individual spins align themselves parallel to each other to the maximum extent allowed by the exclusion principle (Section 9.7). Thus the stable state in our case occurs where the electrons occupy the d orbital singly, with all the spins parallel to each other, giving a total spin of $s = 5 \times \frac{1}{2} = \frac{5}{2}$ for the $Fe^{3+}$ ion. (This is the so-called *high-spin state.*) This is encouraging because it means that the substance is magnetically active, and may exhibit ESR absorption.

When a magnetic field is applied to the ion, the spin angular momentum takes up various quantized orientations corresponding to the components $m_s = -s, -s+1, \ldots, s$ (Section 9.6), where $S_z = m_s \hbar$ is the angular-momentum component along the field. But there are two different fields that must be distinguished here: an *internal* field due to the interaction of $Fe^{3+}$ with its adjacent atoms (this field is normal to the heme plane), and the *external* field applied in the ESR experiment. The general discussion from this point on depends on the relative values of these fields (Ingram, 1969; Ayscough, 1967).

In our case, the internal field is far greater, and it is the one primarily responsible for the splitting of the magnetic level of the ion. The $Fe^{3+}$ level splits into three Zeeman sublevels, as shown in Fig. 13.5. Note that there are only three, not the expected six, Zeeman sublevels $[2s + 1 = 2(\frac{5}{2}) + 1 = 6]$ because the levels $m_s = +\frac{1}{2}$ and $-\frac{1}{2}$ are degenerate, as are $m_s = \pm\frac{3}{2}$ and $m_s = \pm\frac{5}{2}$. The reason for this degeneracy is the following: The splitting is caused primarily by the interaction with the atoms in the heme plane, and since the plane is symmetric relative to the up-down directions (normal to the plane), the orientations $m_s = +\frac{1}{2}$ and $-\frac{1}{2}$ along the axis of the molecule are physically equivalent, and so have the same energy. The same argument applies to the other sublevels. Note also that the splitting between the sublevels is large because the internal field is appreciable.

Now when an external field $\mathscr{H}$ is applied, each of the sublevels splits further into a doublet (Fig. 13.5), corresponding to the two possible values of $m_s$. In other words, the external field removes the degeneracy associated with the internal field. We need concern ourselves only with the doublet associated with $m_s = \frac{1}{2}$ and $-\frac{1}{2}$, because the photons involved in the usual ESR experiments don't have enough energy to make a transition between the widely split levels of the internal field. Also, at room temperature, only the lowest doublet is occupied. Thus in our experiment, we expect to obtain only *one* absorption line, corresponding to the transition $m_s = -\frac{1}{2}$ to $m_s = \frac{1}{2}$.

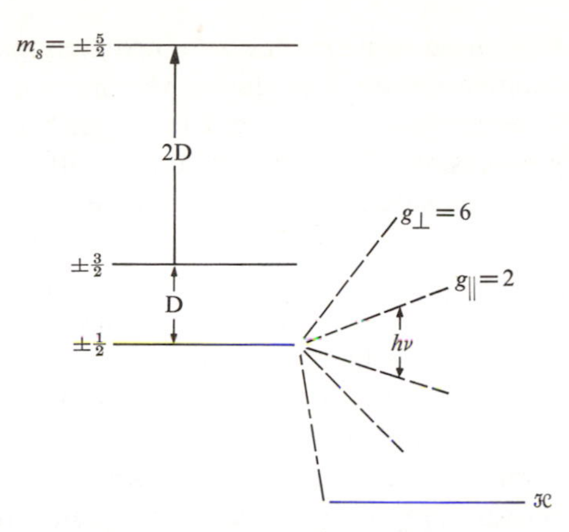


**Fig. 13.5** Splitting of magnetic level of ferric iron by internal field in heme group. Dashed lines represent splitting of lowest magnetic sublevel ($m_s = \pm\frac{1}{2}$) by the external field (not to scale).

But even this single line carries a useful piece of information: the orientation of the external field $\mathcal{H}$ relative to the heme plane. When $\mathcal{H}$ is normal to the plane, $\mathcal{H}$ is parallel to the internal field, and the Landé factor (Section 9.6) for $Fe^{3+}$ is $g_\| = 2$, as in the free-electron case. But when $\mathcal{H}$ is parallel to the plane, the effective value for this factor is much larger, namely $g_\perp = 6$. The reason for this large $g$-anisotropy is that the external field in the latter case is considerably modified by the internal field. So if one includes this complication by defining an effective $g$ (as if the internal field were absent), one then finds $g_\perp = 6$ (Ingram, 1969).

Quantitively, one has, for the different orientations,

$$\Delta E = h\nu = 2g_\| \mu_B \mathcal{H}_\| = 2g_\perp \mu_B \mathcal{H}_\perp, \tag{13.4}$$

where $\Delta E$ is the energy of the absorbed photon, $\nu$ is the standard frequency of the ESR spectrometer employed, and $\mathcal{H}_\|$ and $\mathcal{H}_\perp$ are the fields at which resonance is observed in the two different orientations. It is assumed, as is the customary practice, that resonance is achieved by holding the frequency fixed and sweeping the magnetic field.

The obvious conclusion from this discussion is that the orientation of the heme plane can be determined from ESR measurement. Thus if we rotate the myoglobin molecule in a myoglobin crystal until we obtain a $g$-value equal to 2, we can then be certain that the plane is normal to the field, or, equivalently, that the myoglobin molecule is parallel to the field. Results of x-ray diffraction then establish the orientation of the remainder of the molecule relative to this axis. The structure of the myoglobin molecule was actually determined in this manner.

Let us now move on to the hemoglobin molecule, which contains four heme groups. Their orientations relative to each other and to the remainder of the molecule are naturally of particular interest. Before ESR measurements were carried out, it was assumed that these heme planes were probably parallel to each other. However, ESR measurements show that this is not the case (Fig. 13.6).

The fact that there are four separate curves, rather than a single one, demonstrates that the planes are tilted relative to each other. It is possible, from this and other geometric considerations, to determine the angles between these planes. These ESR results were also used to investigate the structure of the hemoglobin molecule.

Our discussion has covered only the simplest aspect of the myoglobin and hemoglobin molecules. Nothing has been said, for instance, about the ferrous state, nor the effect of covalent bonding on the spin state. The effect of oxygen or other groups on the ESR spectrum is also important. Information on these and other related factors, can be obtained from ESR measurements; limitations of space require us here to simply refer you to the literature for further details (Ingram, 1969; Ayscough, 1966, and the references listed therein).

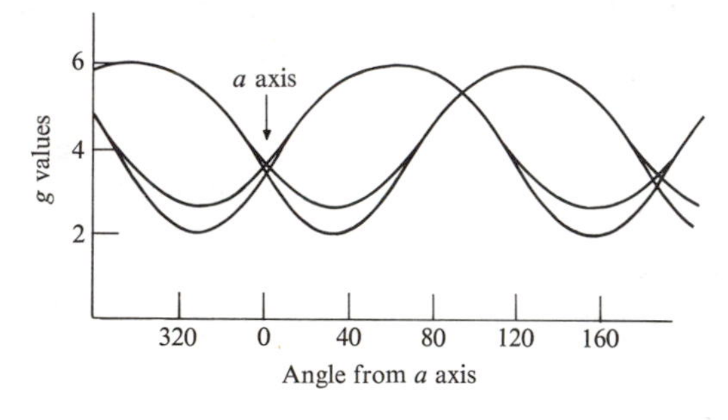


**Fig. 13.6** Anistropy of the $g$-values of the four heme groups in hemoglobin. (After Ingram)

Many experimenters have made Mössbauer measurements on hemoglobin which have yielded information on some of the questions raised above. Mössbauer measurements sometimes give information not available from ESR measurements, and the two techniques complement each other. By using Mössbauer measurements, for example, Long and Marshall were able to determine the splitting between the levels caused by the internal magnetic field (Fig. 13.5). For further discussion of biological applications of the Mössbauer technique, refer to the review by Maling and Weissblutt (Wyard, 1969), and the review by Johnson (1971), who provides many recent references.

Data obtained from NMR measurements made in recent years on hemoglobin and its derivatives confirm and/or complement data obtained by the other techniques discussed above. Resonating nuclei are the protons of the hydrogen atoms in the molecule. By comparing their chemical shifts in the heme group with those of other protons in the polypeptide chain, one can find at least partial explanations of the polypeptide–heme interaction, effects of oxygen bonding and its relation to structural changes during biochemical reactions, and many other phenomena. A recent brief review by Wüthrich and Schulman (1970), and its bibliography, could be a starting point for anyone interested in this promising area of biophysical research.

### Other electronic properties of proteins

We conclude this section on proteins by a discussion of their general electronic properties. We note here that electronic conjugation in polypeptide chains lies almost entirely in the peptide bonds themselves, the remainder of the molecule consisting of saturated amino acid fragments. This bond conjugation involves four $\pi$-electrons, two from the C=O bond and two from the lone pair of the N atom. The $\pi$ conjugation is less extensive in proteins than in nucleic acids, so the delocalization effects are somewhat less important in proteins, and consequently have received less attention. An exception is those few amino acids (four) which contain aromatic molecules in their side chains. In these four amino acids, delocalization is significant.

It was postulated many years ago by Szent-Györgyi that intrinsic electronic semiconduction takes place in some proteins, and that this conduction is responsible for the principal mode of energy transfer in these biosubstances. It has not yet been definitely established whether or not a significant amount of energy is in fact transported this way, but many experiments (by Eley and his associates) on dry proteins indicate that electronic conduction does take place, and that it has the familiar form

$$\sigma = \sigma_0 e^{-(E_g/2kT)}, \tag{13.5}$$

where $E_g$ is the energy gap [see Eq. (6.36)]. The energy gaps for myoglobin and hemoglobin are 2.97 and 2.75 eV, respectively. These values, showing that $E_g$ is close to 3 eV, indicate that electrons in these substances are not appreciably excited at room temperature, and that these materials are consequently good insulators. This is one of the major difficulties inherent in the Szent-Györgyi postulate regarding semiconduction as a mechanism for biochemical interaction.

The reader may well ask how semiconduction and its inherent delocalization is possible in proteins if the $\pi$-electrons are localized at the peptide bonds, as stated above. The answer is that the adjacent peptide linkages interact with each other via hydrogen bonds. By adopting this view, we see that the $\pi$-electrons extend to their neighboring groups, and eventually to the whole polypeptide, leading to the desired delocalization. Theoretical calculations along these lines (references in Pullman, 1963) indicate the presence of energy gaps which are in reasonable agreement with experiments. However, there are many complications involved in the comparison between theory and experiment, demanding extreme caution. Many of these difficulties are discussed by the Pullmans.

According to MO calculations, the aromatic amino acids are, in general, electron donors rather than acceptors. Their capacity in this regard, however, is rather poor, except for tryptophan, whose $k$-value is 0.53. This acid tends to form charge-transfer complexes (by losing its electron). Tryptophan's capacity as a donor may well be responsible for its role as a coenzyme in the metabolic process.

## 13.5 MISCELLANEOUS TOPICS

Let us further illustrate the application of solid-state physics to biological problems by a brief discussion of three important subjects. The choice of topics is somewhat arbitrary, and is intended only to indicate the potential of these methods in modern biology.

### Enzyme studies

Enzymes are protein substances which act as catalysts for biochemical reactions. In almost all cases the reaction is a multistage one, many of the intermediate compounds being free radicals. The ESR technique is especially useful in identifying these radicals, and hence in elucidating the microscopic nature of the reaction.

For example, peroxidase is an important enzyme which aids in the transfer of oxygen from hydrogen peroxide ($H_2O_2$) to other biochemical substances. If the biochemical substance is ascorbic acid (vitamin C), then the acid is oxydized, and the radical shown in Fig. 13.7(a) is expected. This is verified by ESR measurement (Fig. 13.7b), and, as anticipated, the level is doubly split by only one of the protons at the $\beta$ carbon.

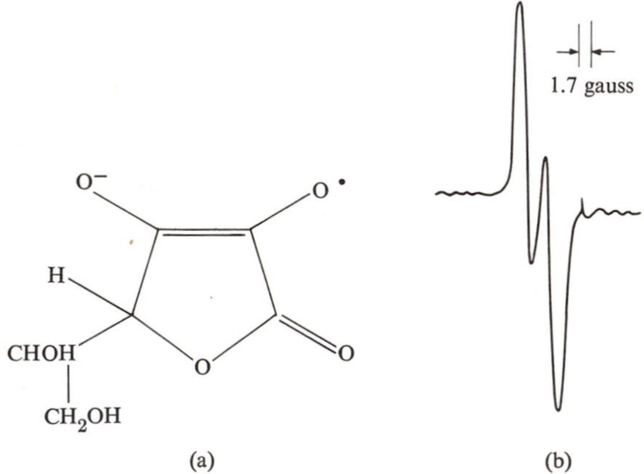


**Fig. 13.7** (a) Free radical of oxidized ascorbic acid. (b) ESR spectrum of radical in part (a). (After Piette, *et al.*)

Given the importance of enzyme functions and the power of the ESR technique, this method promises to be a most useful biological tool.

**Photosynthesis**

Plants synthesize sugar from carbon dioxide and water, with chlorophyl acting as a catalyst. Here again the process is multistage, and free radicals appear as intermediates. The process is also activated by light from the sun.

Green algae *Chlorella pyrenoidso* produce free-radical ESR signals upon illumination. The signal is absent when a mutant lacking chlorophyl is used, or after cessation of the illumination. The signal also increases with the concentration of chlorophyl.

**Carcinogenic activity**

Several proposals have been made to account for the carcinogenic (cancer-proding) activity of certain molecules in terms of their electronic structure. Although this complex problem has not yet been clarified to the point at which one of these proposals is definitely favored over the others, we shall describe briefly the most promising of these proposals, due to the Pullmans (1963). They postulated that carcinogenesis takes place through the interaction of highly reactive aromatic molecules with cellular material in the following manner: The carcinogenic aromatic molecule has certain highly reactive centers around its periphery. Once a cellular molecule comes in contact with such a center, the $\pi$-electrons of the carcinogenic molecule spread out throughout the system, thus binding the cellular material to the aromatic hydrocarbon.

To determine the centers of reactivity around the aromatic molecule, one introduces the concept of *localization energy*, which is the energy required to take one (or more) electrons out of the pool of $\pi$-electrons, and localize it at a particular C atom (or substituent) or a bond. If this energy is small, then such localization is readily achieved, and the particular atom or bond is suitable for strong reactivity with other reagents—nucleophilic, electrophilic, or free radicals. The localization energy can be calculated using the Hückle theory. The details can be found in Pullman (1963).

More specifically, the Pullmans developed the following criterion for a carcinogenic molecule. If it has two regions $K$ and $L$ (Fig. 13.8a), the localization energy for the $K$ region must be smaller than $3.31|\beta|$, and that of region $L$ greater than $5.66\,|\beta|$, $\beta$ being the overlap integral of Section 13.2. Consider, for instance, the anthracene molecule (Fig. 13.8b). The localization energy indices shown do not favor carcinogenesis according to the Pullman criterion, but as further rings are added, these indices change. Figure 13.8(c) shows that the particular molecule (1,2,5,6-dibenzanthracene) does satisfy the criterion of localization energies. Experiments confirm the carcinogenic activity of this molecule.

Another postulate for the carcinogenic mechanism involves transfer of an electron from the highest occupied level of the protein to an empty level in the associated hydrocarbon. In this case, one expects the donor and acceptor to exhibit ESR signals, because of the unpairing of the remaining electrons. Such signals have been observed in some carcinogenic reactions, indicating the formation of

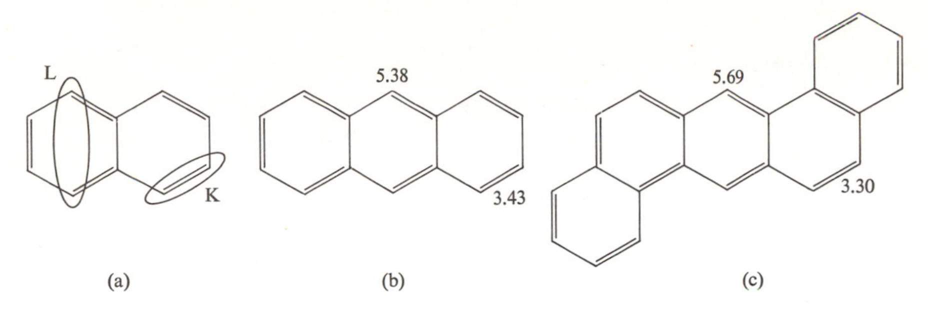


**Fig. 13.8** (a) The Pullmans' criterion for carcinogenesis. (b) Localization energies in anthracene. (c) Localization energies in 1,2,5,6-dibenzanthracene.

charge-transfer complexes, but no definite correlation has been established between *all* carcinogenic activities and the appearance of ESR signals. The situation is far from clear, and represents a major challenge to the modern biochemist.

## REFERENCES

I. Asimov, 1962, *The Genetic Code,* New York: New American Library

P. B. Ayscough, 1967, *Electron Spin Resonance in Chemistry,* London: Methuen

M. Bersohn and J. C. Baird, 1966, *Solid State Biophysics,* New York: McGraw-Hill

C. W. N. Cumper, 1966, *Wave Mechanics for Chemists,* New York: Academic Press

R. E. Dickerson in H. Neurath, editor, 1964, *The Protein,* New York: Academic Press

C. H. Haggis, editor, 1964, *Introduction to Molecular Biology*, New York: John Wiley

K. C. Holmes and D. M. Blow, 1966, *The Use of X-ray Diffraction in the Study of Protein and Nucleic Acid Structure,* New York: Interscience

D. J. E. Ingram, 1969, *Biological and Biochemical Applications of Electron Spin Resonance,* London: Adam Hilger

C. E. Johnson, 1971, "Mössbauer spectroscopy and biophysics," *Phys. Today* **24,** 2, 35

P. Karlson, 1968, *Introduction to Modern Biochemistry,* New York: Academic Press

J. Kendrew, 1966, *The Thread of Life: An Introduction to Molecular Biology,* Cambridge, Mass.: Harvard University Press

J. L. Kice and E. N. Marvell, 1966, *Modern Principles of Organic Chemistry,* New York: Macmillan

B. Pullman and A. Pullman, 1963, *Quantum Biochemistry*, New York: John Wiley

J. C. Phillips, 1969, *Covalent Bonding in Crystals, Molecules, and Polymers,* Chicago: University of Chicago Press

A. Szent-Györgyi, 1960, *Introduction to a Submolecular Biology,* New York: Academic Press

J. D. Watson, 1968, *The Double Helix,* New York: New American Library

M. Weissbluth, 1967, in *Structure and Bonding,* Vol. 2, New York: Springer-Verlag, edited by K. Jorgensen *et al.*

K. Wüthrich and R. G. Schulman, 1970, "Magnetic resonance in biology," *Phys. Today* **23,** 4, 43

S. J. Wyard, editor, 1969, *Solid State Biophysics,* New York: McGraw-Hill

# APPENDIX ELEMENTS OF QUANTUM MECHANICS

## A.1 BASIC CONCEPTS

Einstein assumed that electromagnetic radiation is composed of photons. The energy and momentum of each photon are given, respectively, by

$$E = h\nu = \hbar\omega \quad \text{(a)}, \qquad p = \frac{h\nu}{c} = \frac{h}{\lambda} = \hbar k \quad \text{(b)}, \tag{A.1}$$

where $\nu$ and $\lambda$ are the frequency and wavelength of the radiation, $\hbar = h/2\pi$, and $k = 2\pi/\lambda$, the wave vector of the wave. Equation (A.1a) is known as the *Einstein relation*.

DeBroglie assumed that Eqs. (A.1) apply also to particles. That is,

$$\omega = E/\hbar \quad \text{(a)}, \qquad \lambda = \frac{h}{p} \quad \text{or} \quad k = \frac{p}{\hbar} \quad \text{(b)}. \tag{A.2}$$

Heisenberg established the fact that the uncertainties in the position and momentum of a particle—that is, $\Delta x$ and $\Delta p$—satisfy the relation

$$\Delta x \, \Delta p \simeq \hbar. \tag{A.3}$$

A similar relation obtains also for time and energy,

$$\Delta t \, \Delta E \simeq \hbar, \tag{A.4}$$

where $\Delta t$ is usually identified with the lifetime of the particle.

## A.2 THE SCHRÖDINGER EQUATION

The deBroglie wave satisfies the Schrödinger equation (SE)

$$i\hbar \frac{\partial}{\partial t} \Psi(\mathbf{r},t) = \left[ -\frac{\hbar^2}{2m} \nabla^2 + V \right] \Psi(\mathbf{r}, t), \tag{A.5}$$

where $\Psi$ is the wave function of the particle and $V$ its potential energy. $|\Psi(\mathbf{r}, t)|^2 \, d^3r$ gives the probability of finding the particle in the volume element $d^3r$ at the instant $t$. The function must satisfy the normalization condition

$$\int |\Psi|^2 \, d^3r = 1, \tag{A.6}$$

the integration being over all space.

If $V$ is time-independent, the function $\Psi$ may be factorized as

$$\Psi = \psi(\mathbf{r}) \, e^{-i(E/\hbar)t},$$

and the space-dependent part $\psi(\mathbf{r})$, satisfies

$$\left[ -\frac{\hbar^2}{2m} \nabla^2 + V \right] \psi(\mathbf{r}) = E\psi(\mathbf{r}), \tag{A.7}$$

also known as the Schrödinger equation. Solving this equation subject to the appropriate boundary conditions yields the allowed energies and their corresponding wave functions.

## A.3 ONE-DIMENSIONAL EXAMPLES

Energies and wave functions of free particles are given by

$$E = \frac{\hbar^2 k^2}{2m} \quad \text{(a)}, \qquad \psi_k = Ae^{ikx} \quad \text{(b)}, \tag{A.8}$$

where $A$ is a constant; $k$ is the wave vector of the plane wave.

A particle in a box, of length $L$, has energies and wave functions as in (A.8), except that the vector $k$ is quantized as

$$k = n\frac{2\pi}{L}, \qquad n = 0, \pm 1, \pm 2, \text{etc.}, \tag{A.9}$$

which follows from the periodic boundary conditions (see Section 3.2). Thus

$$E_n = \frac{\hbar^2}{2m}\left(\frac{2\pi}{L}\right)^2 n^2 \quad \text{(a)}, \qquad \psi_n = \frac{1}{L^{1/2}} e^{i(2\pi/L)nx} \quad \text{(b)}. \tag{A.10}$$

Generalization to a three-dimensional box is straightforward.

The energies of a harmonic oscillator are given by

$$E_n = (n + \tfrac{1}{2})\hbar\omega, \qquad n = 0, 1, 2, \text{etc.}, \tag{A.11}$$

where $\omega$ is the classical natural frequency of the oscillator.

## A.4 THE ANGULAR MOMENTUM

The magnitude of the orbital angular momentum is quantized according to the formula

$$L = \sqrt{l(l+1)}\hbar, \tag{A.12}$$

where $l = 0, 1, 2$, etc. The states 0, 1, 2, etc., are referred to as s, p, d, etc., states.

The $z$-component of the angular momentum is also quantized according to

$$L_z = m_l \hbar, \tag{A.13}$$

where $m_l = -l, -l+1, \ldots, l-1$, or $l$.

The spin angular momentum is also quantized as in Eqs. (A.12) and (A.13), except that the only allowed value for $s$ is $s = \frac{1}{2}$.

## A.5 THE HYDROGEN ATOM; MULTIELECTRON ATOMS; PERIODIC TABLE OF THE ELEMENTS

When the coulomb potential is substituted in the Schrödinger equation, and this is solved, the allowed energies are found to be

$$E_n = -\frac{e^2}{2a_0}\frac{1}{n^2}, \tag{A.14}$$

where $n$, a positive integer, is known as the *principal quantum number*; $a_0 = 4\pi\hbar\epsilon_0/me^2$ is the first Bohr radius. The allowed orbital angular momentum number for each $n$ is $l = 0, 1, 2, \ldots, n - 1$, and for each $l$ the allowed values of $m_l$ are all the integers between $-l$ and $l$, inclusive. An electron can make the transition between two energy levels by the absorption or emission of a photon, provided that

$$\Delta E = \hbar\omega, \tag{A.15}$$

where $\Delta E$ is the difference in energy between the levels. This equation is known as the *Bohr frequency formula.*

The wave function for any state has the form

$$\psi_{nlm_l}(r, \theta, \phi) = R_{nl}(r)Y_{lm_l}(\theta, \phi), \tag{A.16}$$

where $r$, $\theta$, and $\phi$ are spherical polar coordinates. The radial function $R_{nl}$ is an oscillating function whose peaks determine the various atomic shells (Bohr orbits), and $Y_{lm_l}$, a so-called spherical harmonic, describes the rotation of the electron around the proton.

In multielectron atoms, the various electrons occupy the allowed states, beginning with the lowest energies, in accordance with the Pauli exclusion principle: A quantum state can accommodate at most two electrons of opposite spins. Each atomic shell—that is, a given value of $n$—can accommodate at most $2n^2$ electrons.

The outermost occupied shell, the *valence* shell, determines the chemical properties of the atom. If the valence shell is partially full, the atom is reactive. A completely full valence shell leads to an inert atom, e.g., helium.

Within each shell, the various subshells—i.e., various $l$'s—have different energies due to the manner in which the corresponding electrons are distributed relative to the nucleus. In particular, the $s$ subshell ($l = 0$) has the lowest energy because its electron has an appreciable probability of being very close to the nucleus.

Several important series of elements have significant magnetic properties related to their atomic characteristics. The *first transition series*, the row from Sc to Ni ($Z = 21$ to 28), has the outer 4s subshell occupied before the inner 3d subshell, due to the effect described above. (The periodic table of the elements is given inside the front cover.) The *second transition series*, extending from Y to Pd, is also similar due to the filling of the 5s before the 4d subshell. The rare-earth elements, or *lanthanides*, which extend from La to Lu ($Z = 57$ to 71), are also similar, in that the outer 6s subshell is filled before the 4f.

## A.6 PERTURBATION THEORY

Atoms are usually studied in the laboratory by applying external fields and observing their effects on the atomic properties. Both magnetic and electric fields alter the atomic spectrum (which may be observed by spectroscopic techniques), and from this one may gain information about the structure of the atom.

In the presence of the applied field, the potential becomes

$$V = V_0(\mathbf{r}) + V'(\mathbf{r}),$$

where $V_0(\mathbf{r})$ is the atomic potential and $V'(\mathbf{r})$ is the potential due to the field. In principle, one has to solve the SE again using the new potential, with the field included. Unfortunately, one can do this exactly in only a few special cases. However, if the field is weak, and hence the additional potential $V'(\mathbf{r})$ small, one may develop a satisfactory approximate procedure for calculating the energies and wave functions. The procedure is tantamount to a Taylor-series expansion in powers of the field. One can evaluate the energies and wave functions to the desired accuracy by including sufficiently high powers in the expansion.

The details of this method, known as *perturbation theory*, may be found in books on quantum mechanics. The results are

$$E_n \simeq E_n^{(0)} + \langle n|V'|n\rangle - \sum_m{}' \frac{|\langle m|V'|n\rangle|^2}{E_m^{(0)} - E_n^{(0)}} \tag{A.17}$$

and

$$\psi_n \simeq \psi_n^{(0)} - \sum_m{}' \frac{|\langle m|V'|n\rangle|^2}{E_m^{(0)} - E_n^{(0)}}. \tag{A.18}$$

Here $E_n^{(0)}$ and $\psi_n^{(0)}$ are the energy and wave function for an arbitrary level $n$ in the absence of the field—i.e., the unperturbed energy and wave function—while $E_n$ and $\psi_n$ are the corresponding quantities in the presence of the field. The pointed brackets have the following meanings:†

$$\langle n|V'|n\rangle \equiv \int \psi_n^{(0)*} V' \psi_n^{(0)} d^3r,$$

$$\langle m|V'|n\rangle \equiv \int \psi_m^{(0)*} V' \psi_n^{(0)} d^3r.$$

† The integral $\langle m|V'|n\rangle$ is referred to as the *matrix element* of the potential $V'$ between the states $\psi_n^{(0)}$ and $\psi_m^{(0)}$.

The summations in (A.17) and (A.18) are over all quantum states other than the $n$th one, which is the one under investigation. (The exclusion of the term $m = n$ from the sum is signified by the prime over the summation sign.) Both the energy and wave function are given to the second order in $V'$.

### The Zeeman effect

As an application of these results, let us consider the effect of a magnetic field on the spectrum of a hydrogen atom. To find the perturbation potential $V'(r)$, we note that, by virtue of its rotation, the electron has a magnetic moment $\boldsymbol{\mu} = -(e/2m)\,\mathbf{L}$, where $\mathbf{L}$ is the angular momentum. When an external field is applied, the dipole is coupled to it, and the potential energy is

$$V' = -\boldsymbol{\mu}\cdot\mathbf{B},$$

where $\mathbf{B}$ is the field (see Section 9.2). Assuming that the field is in the $z$-direction, we have

$$V' = -\mu_z B = +\left(\frac{e}{2m}\right) BL_z, \tag{A.19}$$

which is the perturbation potential we are seeking. This potential produces a shift in the energy given to the first order by

$$\langle n|V'|n\rangle = \frac{eB}{2m}\langle n|L_z|n\rangle, \tag{A.20}$$

according to (A.17). The shift is therefore proportional to the average value of the $z$-component of the angular momentum (recall the meaning of the angular bracket).

Let us apply this result to hydrogen. For the ground state, the 1s state, the angular momentum is zero. Thus $\langle 1s|L_z|1s\rangle = 0$, and there is no magnetic effect on that state, as shown in Fig. A.1. There is similarly no effect on the 2s state.

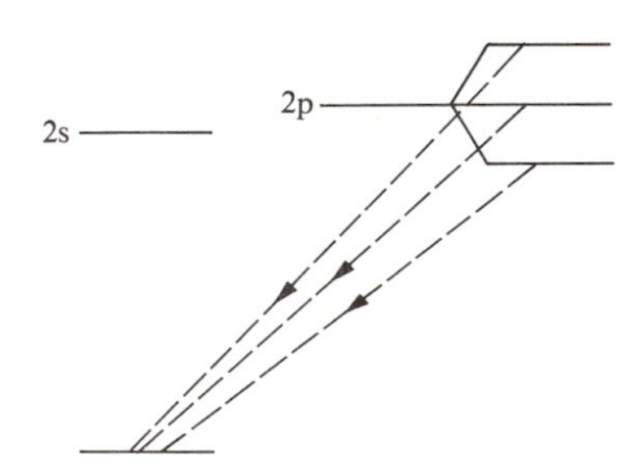


**Fig. A.1** The Zeeman effect. The s levels are unaffected by the magnetic field, while a p level splits into three sublevels.

The situation is different, however, for the 2p state. This corresponds to $l = 1$, and hence $\langle 2p\,|L_z|\,2p\rangle$ can take the values $-\hbar$, 0, or $\hbar$, corresponding to the three possible orientations of **L** relative to the $z$-axis (which is the direction of the field) (Section A.4). Thus the 2p level splits into three equidistant magnetic sublevels, with a spacing

$$\Delta E = \frac{eB}{2m}\hbar = \mu_{\mathrm{B}} B, \tag{A.21}$$

as shown in the figure. The quantity $\mu_{\mathrm{B}} = e\hbar/2m$, known as the *Bohr magneton*, has the value $9.27 \times 10^{-24}$ amp$\cdot$m$^2$.

In general, a subshell of angular momentum $l$ splits into $(2l + 1)$ equidistant levels, with a unit spacing given by (A.21). This splitting, engendered by the magnetic field, is known as the *Zeeman effect*. The effect is studied by observing the splitting of the various spectral lines as the field is turned on. For instance, the line due to the 2p → 1s transition is split into three lines because of the triple splitting of the 2p level. The Zeeman effect can thus be employed to determine the angular momentum of the various atomic states.

### Crystal-field splitting

When an atom is placed inside a crystal, the wave functions (or *atomic orbitals*) of the atom are altered, because the neighboring ions exert an electric field on the atomic electrons, which results in the distortion of the orbitals and splitting of the energy levels. This electric field is known as the *crystal field*. Its effect can be treated by perturbation theory, provided the field is not too large.

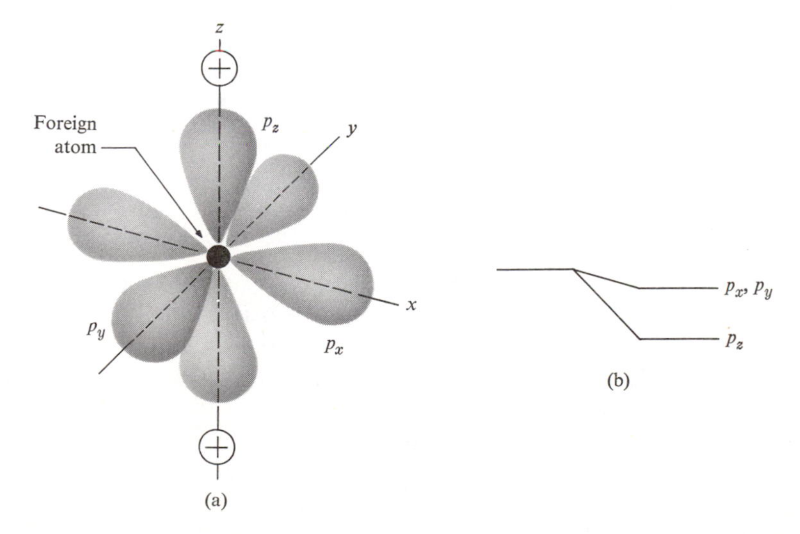


**Fig. A.2** Crystal-field splitting. (a) Charge distribution of the $p_x$, $p_y$, and $p_z$ orbitals. (b) Splitting of the orbitals' energies.

The crystal field depends on the number and geometrical arrangement of the neighboring ions. The most common coordination numbers are 2, 4, 6 (and 8), corresponding, respectively, to a linear, tetrahedral, octahedral (and square anti-prismatic) arrangement of the surrounding ions. By observing the splitting, one may determine the symmetry of the environment, which is equivalent to knowing the coordination number. We illustrate this by examining the effects on a $p$ orbital.

Suppose that the arrangement is linear, as shown in Fig. A.2(a), with two positive ions along the $z$-axis. The three $p$ orbitals are shown: $p_x$, $p_y$, and $p_z$. Note that the $p_z$ orbital deposits its electron primarily in the dumbbell-shaped distribution along the $z$-axis, where it is strongly attracted by the positive ions. Therefore the $p_z$ orbital is lowered in energy relative to the other two orbitals which lie along the $x$- and $y$-axes. Consequently the three orbitals, which were of equal energies, now acquire different energies, and the level is split, as shown in Fig. A.2(b). This *crystal-field splitting* is particularly significant in magnetic and optical properties of transition and rare-earth ions (Section 9.6), and also in electron paramagnetic resonance techniques (Section 9.12).

## A.7 THE HYDROGEN MOLECULE AND THE COVALENT BOND

When two hydrogen atoms are placed close together, they attract each other, and combine to form a hydrogen molecule, $H_2$, which is stable. The two atoms are held together by the two electrons present in the molecule, and we speak of the *hydrogen bond.* The orbitals of the electrons in this bond are distributed in a special fashion around the atoms. This double-electron bond, called a *covalent bond,* is present in other molecules as well.

Consider first the case of the hydrogen molecule ion, $H_2^+$. As an ionized $H_2$ molecule, it has only one electron, which moves in the field of the two protons (Fig. A.3a). We wish now to find the energies and wave functions for this molecule, particularly for the ground state. The potential energy is

$$V = \frac{e^2}{4\pi\epsilon_0 a} - \frac{e^2}{4\pi\epsilon_0 r_1} - \frac{e^2}{4\pi\epsilon_0 r_2}, \tag{A.22}$$

where the first term is due to the repulsion between the protons, and the last two are due to the attraction of the electron by the two protons. This potential is substituted into the SE, and the resulting differential equation is then solved. Although this problem can be solved analytically, the details are tedious and we prefer a simple approximate procedure.

When the electron is close to either proton, it behaves as a hydrogenic 1s atomic orbital. It is therefore reasonable to expect the molecular orbital for $H_2^+$ to be a linear combination of the two 1s orbitals centered at the two protons. There are two possibilities,

$$\psi_e = \psi_1 + \psi_2 \tag{A.23}$$

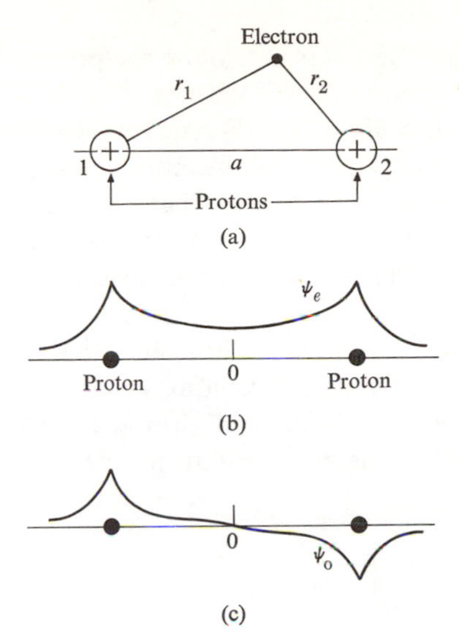


**Fig. A.3** (a) The hydrogen molecule ion. (b) The wave function $\psi_e$. (c) The wave function $\psi_o$.

and
$$\psi_o = \psi_1 - \psi_2, \tag{A.24}$$

where $\psi_1$ and $\psi_2$ represent the 1s states centered at the two protons, respectively, and the subscripts $e$ and $o$ signify even and odd combinations. Symmetry considerations preclude any other linear combinations, since the distribution of electron charge must be symmetric with respect to the two protons, and only these combinations satisfy this requirement (why?). The molecular orbitals $\psi_e$ and $\psi_o$ are sketched in Fig. A.3.

The charge distributions for these orbitals are given as $|\psi_e|^2$ and $|\psi_o|^2$ (Fig. A.4). It can be seen that $\psi_e$ deposits the electron primarily in the region between the

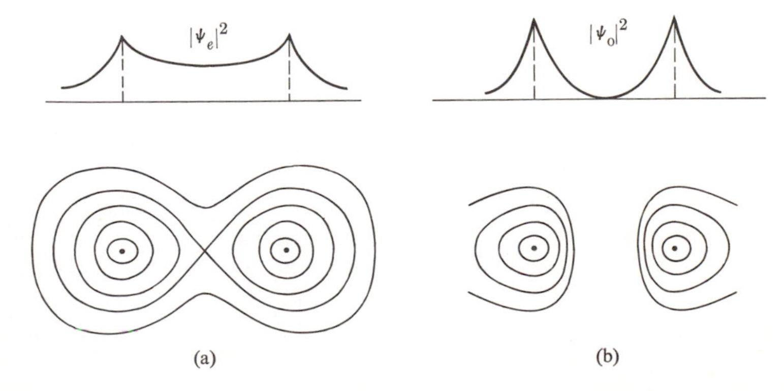


**Fig. A.4** (a) Charge distribution in profile and contour representations for the function $\psi_e$. (b) Charge distribution for $\psi_o$.

protons, while $\psi_o$ deposits the electron around the protons individually, and away from the intermediate region.

The two molecular orbitals have different energies, as illustrated in Fig. A.5, which shows the energies as a function of the internuclear distance. The even orbital, usually denoted $\sigma_g 1s$, has a lower energy than the odd orbital, $\sigma_u 1s$. Thus the electron favors the even orbital. Furthermore, the even orbital has a negative energy (the zero energy reference is that of a hydrogen atom—in its ground state—and a proton infinitely distant from each other). Thus is it a *bonding orbital* leading to a stable state. At the equilibrium situation, corresponding to the minimum energy, the internuclear separation is $a \simeq 2a_o \simeq 1.06$ Å, and the bonding energy is $-2.65$ eV. The odd orbital is *antibonding* (unstable), and has an energy of 10.2 eV at the equilibrium distance.

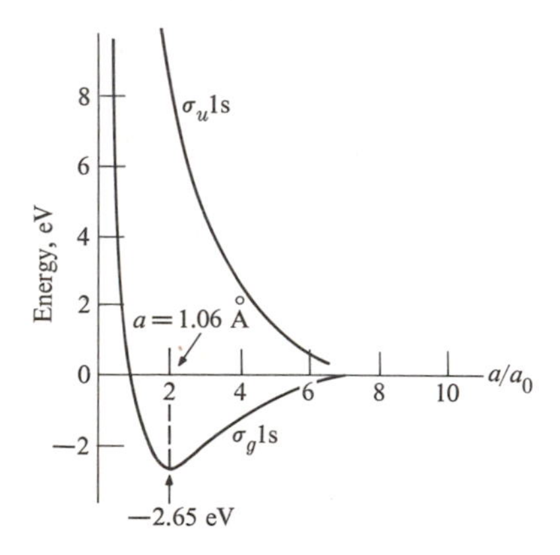


**Fig. A.5** Energies of ground and excited states for hydrogen molecule ion versus internuclear distance ($a_o = 0.53$ Å, the Bohr radius).

Recapitulating, we note that the $H_2^+$ molecule is a stable one. The repulsion between the protons is more than compensated for by the attraction between the electron and the protons. By adjusting its orbital properly, the electron is able to hold the protons together (like a glue!). This is what might be called a single-electron bond.

The above concepts can be readily adapted to the hydrogen molecule, which has two electrons. Both can occupy the bonding orbital $\sigma_g 1s$, provided that their spins are opposite to each other. Of course, the two electrons repel each other to some extent, and some adjustment for this must be made in the orbital. The energy of the $H_2$ molecule is shown in Fig. A.6 as a function of the internuclear distance.

The equilibrium separation is 0.74 Å, and the binding energy 4.48 eV (relative to two infinitely distant hydrogen atoms in their ground states). Since both elec-

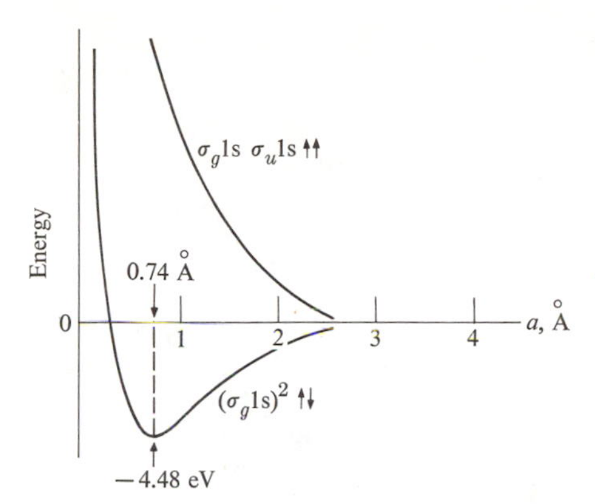


**Fig. A.6** Energies of ground and excited states for hydrogen molecule versus internuclear distance.

trons are in the $\sigma_g 1s$ state, the electrons are deposited between the nuclei, and hence are equally shared by the two protons. The concept of *electron sharing* in the covalent bonds is stressed repeatedly in the literature.

## A.8 DIRECTED BONDS

Carbon is an important chemical element. Both in molecules and solids, carbon forms tetrahedral bonds with its nearest neighbors. The carbon atom is positioned at the center of a tetrahedron, at whose four corners the neighboring atoms are located. The crystal structure in diamond, for instance, is such that each carbon atom is surrounded tetrahedrally by four other carbon atoms (Fig. A.7). Tetrahedral coordination occurs also in other elements of the fourth column in the periodic table, such as Si and Ge, as well as in many semiconducting compounds such as GaAs and InSb.

To explain the tetrahedral arrangement in diamond, we note that each C atom has four electrons in the second shell. Since there are four bonds joining the central atom to its neighbors, one may think of each bond as being covalent. Its two electrons are contributed, one by the central atom and the other by a neighboring atom. In this manner, each C atom surrounds itself by eight valence electrons, which is a stable structure in that the second shell of C is now completely full.

Although this reasoning is sound, it does not explain why the arrangement should be a regular tetrahedron, with the angle between the bonds 110°. To understand this, we must look more closely at the spatial distribution of the orbitals of the valence electrons. An isolated C atom has four valence electrons: two 2s electrons and two 2p electrons, the s electrons being slightly lower in energy. The s states are spherically symmetric, and the p states represent charge distributions

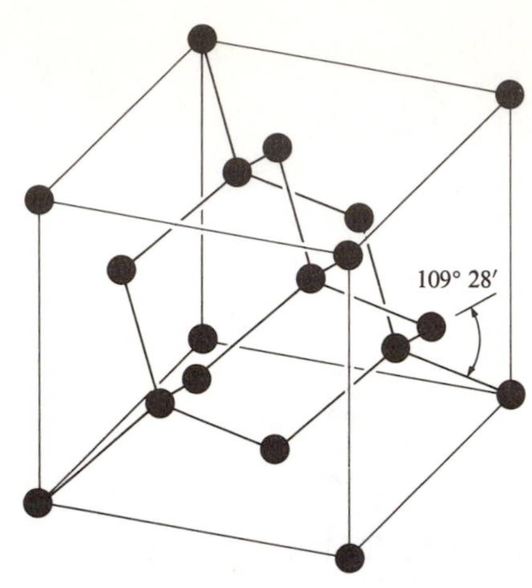


**Fig. A.7** The diamond structure and the tetrahedral bond.

lying along two of the three Cartesian axes. These states do not explain the observed spatial distribution of charge in diamond, in which the charges are distributed along the tetrahedral bonds. However, the situation can easily be remedied. We imagine that one of the 2s electrons is excited to one of the 2p states, resulting in a $1s2p^3$ configuration. This excitation is possible because the energy difference between the 2s and 2p orbitals is rather small. We now form the linear combinations

$$\begin{aligned}\psi_1 &= \tfrac{1}{2}(s + p_x + p_y + p_z)\\ \psi_2 &= \tfrac{1}{2}(s + p_x + p_y - p_z)\\ \psi_3 &= \tfrac{1}{2}(s + p_x - p_y - p_z)\\ \psi_4 &= \tfrac{1}{2}(s - p_x - p_y - p_z)\end{aligned} \qquad (A.25)$$

If one plots the densities $|\psi_1|^2$, $|\psi_2|^2$, etc., corresponding to these new orbitals, one finds that they are indeed distributed along the tetrahedral directions of Fig. A.7. This shows that these new orbitals give a better representation of the electrons' states than the old s, $p_x$, $p_y$, and $p_z$ orbitals.

By occupying the new orbitals, electrons of neighboring atoms can have a maximum degree of overlap, which is the primary rule for chemical stability. Even though some energy is required to excite a 2s electron to a 2p state, this is more than compensated for by the reduction in the energy of interaction with the adjacent atom. (We also see from this example that the lowest-energy electron configuration in a molecule may be different from the lowest-energy configuration in an isolated atom.)

The mixing of the s and p states in (A.25) is referred to as *hybridization*. The particular one operating in diamond is known as $sp^3$ hybridization. We see that, by forming different types of hybrids, one can arrive at many different kinds of directional bonds.

The $sp^3$ hybridization occurs also in Si and Ge. In Si, one 3s and three 3p states combine to form the four tetrahedral bonds, while in Ge the $sp^3$ hybridization involves one 4s and three 4p electrons.

## GENERAL REFERENCES

*Note:* * Advanced. ** Highly advanced. These labels indicate the quantum-mechanical and mathematical requirements for efficient comprehension of the work.

### Modern physics


R. M. Eisberg, 1961, *Fundamentals of Modern Physics*, New York: John Wiley

R. L. Sproull, 1963, *Modern Physics*, second edition, New York: John Wiley


### Thermodynamics and statistical physics


F. Reif, 1965, *Fundamentals of Statistical and Thermal Physics*, New York: McGraw-Hill


### Solid-state physics


W. R. Beard, 1965, *Electronics of Solids*, New York: McGraw-Hill

J. S. Blakemore, 1969, *Solid State Physics*, Philadelphia: W. B. Saunders

F. C. Brown, 1967, *The Physics of Solids*, New York: W. A. Benjamin

A. J. Dekker, 1957, *Solid State Physics*, Englewood Cliffs, N.J.: Prentice-Hall

H. J. Goldsmid, editor, 1968, *Problems in Solid State Physics*, New York: Academic Press

**W. A. Harrison, 1970, *Solid State Theory*, New York: McGraw-Hill

T. S. Hutchinson and D. C. Baird, 1968, *Engineering Solids*, second edition, New York: John Wiley

*C. Kittel, 1971, *Introduction to Solid State Physics*, fourth edition, New York: John Wiley

**C. Kittel, 1963, *Quantum Theory of Solids*, New York: John Wiley

**P. T. Landsberg, editor, 1969, *Solid State Theory*, New York: John Wiley

R. A. Levy, 1968, *Principles of Solid State Physics*, New York: Academic Press

J. P. McKelvey, 1966, *Solid State and Semiconductor Physics*, New York: Harper and Row

**J. D. Patterson, 1971, *Introduction to the Theory of Solid State Physics*, Reading, Mass.: Addison-Wesley

*F. Seitz, 1940, *Modern Theory of Solids*, New York: McGraw-Hill

*R. A. Smith, 1969, *Wave Mechanics of Crystalline Solids*, second edition, London: Chapman and Hall

**P. L. Taylor, 1970, *A Quantum Approach to the Solid State*, Englewood Cliffs, N.J.: Prentice-Hall

C. A. Wert and R. M. Thomson, 1970, *Physics of Solids*, second edition, New York: McGraw-Hill

**J. Ziman, 1972, *Principles of the Theory of Solids*, second edition, Cambridge: Cambridge University Press

**J. Ziman, 1960, *Electrons and Phonons*, Oxford: Oxford University Press

### Solid-state physics series

F. Seitz, D. Turnbull and H. Ehrenreich, editors, *Solid State Physics, Advances in Research and Applications*, various volumes, New York: Academic Press

(This series is referred to in the text as *Solid State Physics*.)

# INDEX

# INDEX